Dominated Operators

Mathematics and Its Applications

Volume 519

Dominated Operators

by

Anatoly G. Kusraev

Institute for Applied Mathematics and Informatics,
North Ossetian State University,
Vladikavkaz, Russia

and

Sobolev Institute of Mathematics,
Siberian Division,
Russian Academy of Sciences,
Novosibirsk, Russia

KLUWER ACADEMIC PUBLISHERS

DORDRECHT / BOSTON / LONDON

A C.I.P. Catalogue record for this book is available from the Library of Congress.

ISBN 0-7923-6485-6

Published by Kluwer Academic Publishers,
P.O. Box 17, 3300 AA Dordrecht, The Netherlands.

Sold and distributed in North, Central and South America
by Kluwer Academic Publishers,
101 Philip Drive, Norwell, MA 02061, U.S.A.

In all other countries, sold and distributed
by Kluwer Academic Publishers,
P.O. Box 322, 3300 AH Dordrecht, The Netherlands.

Printed on acid-free paper

This is a translation of the original Russian work of the same title,
Sobolev Institute Press © 1999, Novosibirsk–Vladikavkaz.
Translated by the author. Edited by S.S. Kutateladze.

Printed in the Netherlands

Contents

Foreword to the English Translation ix

Preface xi

Chapter 1. Boolean Algebras and Vector Lattices 1

§ 1.1. Boolean Algebras 2

§ 1.2. Representation of Boolean Algebras 10

§ 1.3. Vector Lattices 17

§ 1.4. Representation of Vector Lattices 25

§ 1.5. Normed Vector Lattices 33

§ 1.6. Comments 40

Chapter 2. Lattice-Normed Spaces 44

§ 2.1. Preliminaries 45

§ 2.2. Completion 54

§ 2.3. Examples 62

§ 2.4. Continuous Banach Bundles 70

§ 2.5. Measurable Banach Bundles 79

§ 2.6. Comments 84

Chapter 3. Positive Operators **89**

§ 3.1. Operators in Vector Lattices 90

§ 3.2. Fragments of a Positive Operator 99

§ 3.3. Orthomorphisms and Lattice Homomorphisms 108

§ 3.4. Maharam Operators 118

§ 3.5. Maharam's Extension of Positive Operators 126

§ 3.6. Comments 133

Chapter 4. Dominated Operators **141**

§ 4.1. The Space of Dominated Operators 142

§ 4.2. Decomposability of the Space of Dominated Operators 150

§ 4.3. Order Continuous Operators 156

§ 4.4. The Yosida–Hewitt-Type Theorems 163

§ 4.5. Extension of Dominated Operators 171

§ 4.6. Comments 179

Chapter 5. Disjointness Preserving Operators **187**

§ 5.1. Band Preserving Operators 188

§ 5.2. n-Disjoint Operators 195

§ 5.3. Weight-Shift-Weight Factorization 206

§ 5.4. Multiplicative Representation 214

§ 5.5. Decomposable Operators 221

§ 5.6. Comments 230

Chapter 6. Integral Operators **236**

§ 6.1. Vector Integration 237

§ 6.2. Integral Representation by Quasi-Radon Measures 248

§ 6.3. Functional Representation of Maharam's Extension 257

§ 6.4. Integral Operators 266

§ 6.5. Pseudointegral Operators 278

§ 6.6. Comments 286

Chapter 7. Operators in Spaces with Mixed Norm **291**

§ 7.1. Spaces with Mixed Norm 292

§ 7.2. Summing Operators 299

§ 7.3. Isometric Classification 307

§ 7.4. Kaplansky–Hilbert Modules 316

§ 7.5. AW^*-Algebras 325

§ 7.6. Comments 334

Chapter 8. Applications of Boolean-Valued Analysis **338**

§ 8.1. Real Numbers in Boolean-Valued Models 340

§ 8.2. Boolean-Valued Analysis of Vector Lattices 352

§ 8.3. Boolean-Valued Banach Spaces 362

§ 8.4. Involutive Banach Algebras 371

§ 8.5. Cyclically Compact Operators 379

§ 8.6. Comments 389

Appendix. Boolean-Valued Models **394**

References **413**

Symbol Index **435**

Subject Index **439**

Foreword to the English Translation

The topic of this book belongs to vector lattice theory. Dominated operators are remote relatives of the noble family of bounded linear operators and functionals residing in Banach space. The concept of a dominated operator was invented in the 1930s by *Leonid Vital′evich Kantorovich*, a renowned mathematician and a Nobel prize winner in economics. This concept utilizes the main sociological trick of functional analysis which rests on studying the population of some mathematical objects in order to reveal their individual features and acquired traits.

The theory of dominated operators has ripened in the recent decades mainly due to the contribution by Professor A. G. Kusraev and his students in Vladikavkaz and Novosibirsk.

A few years ago Professor A. G. Kusraev asked me to edit and introduce the English translation of his book. I undertook the task readily for two reasons: First, the topic of the book is attractive and close to my own area of research. Second, I feel proud of the achievements of Professor A. G. Kusraev, once a brilliant student of mine at Novosibirsk State University and now my inseparable fellow with whom I have been sharing many splendid days full of inspirational mathematics.

Unfortunately, the Russian version of this book is still in preparation in view of the unstable situation of the Northern Caucasus which hindered us in working on the translation. I hope that the reader will take this regrettable circumstance into account and forgive the inadvertent shortcomings of the inadequate communication between the author and the editor.

S. Kutateladze

Novosibirsk
Akademgorodok

Preface

The notion of a *dominated* or *majorized* operator rests on a simple idea that goes as far back as the Cauchy method of majorants. Loosely speaking, the idea can be expressed as follows. If an operator (equation) under study is dominated by another operator (equation), called a *dominant* or *majorant*, then the properties of the latter have a substantial influence on the properties of the former. Thus, operators or equations that have "nice" dominants must possess "nice" properties. In other words, an operator with a somehow qualified dominant must be qualified itself.

Mathematical tools, putting the idea of domination into a natural and complete form, were suggested by L. V. Kantorovich in 1935–36. He introduced the fundamental notion of a vector space normed by elements of a vector lattice and that of a linear operator between such spaces which is dominated by a positive linear or monotone sublinear operator. He also applied these notions to solving functional equations.

In the succeeding years many authors studied various particular cases of lattice-normed spaces and different classes of dominated operators. However, research was performed within and in the spirit of the theory of vector and normed lattices. So, it is not an exaggeration to say that dominated operators, as independent objects of investigation, were beyond the reach of specialists for half a century. As a consequence, the most important structural properties and some interesting applications of dominated operators have become available since recently.

By the early 1980s, certain qualitative changes took place in the theory of vector lattices. New methods were suggested, while the range of applications was considerably extended and enriched. Radically new ideas came from other branches of mathematics. All these facts provided necessary prerequisites for a deep study of dominated operators and led to formation of a novel theory of dominated operators. The aim of this book is to present the main results on dominated operators which were obtained in the last fifteen years, thus demonstrating a certain ripeness of the theory.

The book consists of eight chapters. Chapter 1 contains definitions and preliminary information about Boolean algebras and vector lattices. Chapter 1 is included mainly for fixing terminology and notation.

In Chapter 2 some structural properties are considered of a vector space endowed with a norm taking values in some vector lattice. Here, we address the questions of completing these spaces as well as representing them by means of continuous Banach bundles and measurable Banach bundles.

Chapter 3 is devoted to positive operators. The content of Chapter 3 is traditional except for some new results concerning fragments and order-interval-preserving extension of positive operators.

In Chapter 4 we study the general properties of dominated operators. A dominated operator has the least (or exact) dominant under rather weak assumptions. Assigning to each dominant operator its exact dominant we come to some vector norm with values in the vector lattice of regular operators. The central result of this chapter is decomposability of this lattice-normed space of dominants.

In Chapter 5 disjointness preserving and decomposable operators are considered. In particular, we give their analytic representations and decompositions into simpler parts.

Chapter 6 deals with integrality and pseudointegrality for dominated operators. It turns out that a dominated operator inherits these properties from every dominant. From this fact we deduce criteria for weak integrality and pseudointegrality of dominated operators. Several assertions about the general form of dominated operators are also stated.

Various classes of operators under study in functional analysis are often defined using the terms that mix norms and orders. This theme is developed in Chapter 7 in which we introduce some new classes of spaces and operators. The problem of isometric classification of spaces with mixed norm is also briefly discussed.

Chapter 8 is devoted to the so-called Boolean-valued analysis of vector lattices and dominated operators. Our starting point is the assertion claiming that each field of the reals in a Boolean-valued model gives rise to a universally complete vector lattice. Therefore, a huge part of the general theory of vector lattices admits a straightforward derivation by translating and interpreting the well known properties of the conventional reals. This chapter also exposes the Boolean-valued approach to more advanced sections of analysis such as lattice-normed spaces, involutive Banach algebras, etc. Elementary exposition of the apparatus of Boolean-valued model theory resides in the Appendix.

The diversity of topics and results we handle in the book determines the style of exposition. Information freely accessible to the reader is given without demonstration. All principal results are however furnished with complete proofs. Comments to all chapters contain some additional remarks and a guide to the literature. While writing the book, the author assumed the reader familiar with the standard courses in the theory of vector lattices and positive operators.

I seize the opportunity to express my gratitude to all those who helped me in preparation of the book. My pleasant debt is to acknowledge the financial support of the Sobolev Institute of Mathematics of the Siberian Division of the Russian Academy of Sciences, the North Ossetian Scientific Center, the North Ossetian State University, the Russian Foundation for Basic Research, the International Science Foundation, and the American Mathematical Society during the compilation of the monograph.

July, 1999 *A. G. Kusraev*

Chapter 1

Boolean Algebras and Vector Lattices

The present chapter collects some basic facts from the theories of Boolean algebras and vector lattices. It lays the foundation for the theory of lattice-normed spaces and dominated operators to which the book is devoted.

An elementary algebraic theory of Boolean algebras is briefly presented in Section 1.1. We expose the main constructions with Boolean algebras such as subalgebras, homomorphic images, ideals, factor algebras, Cartesian products, and partitions of unity (1.1.5). We state a very useful Exhaustion Principle for Boolean algebras and a pair of its corollaries (1.1.6). Some important examples of Boolean algebras are also provided: the algebras of regular open sets and clopen sets, the algebras of Borel sets modulo meager sets and measurable sets modulo negligible sets, etc. (1.1.7). Three representation theorems are proved in Section 1.2: the celebrated Stone Representation Theorem (1.2.3), the Loomis–Sikorski Theorem on representation of σ-complete Boolean algebras (1.2.6), and representation of multinormed Boolean algebras by measure spaces with direct sum property (1.2.10). Ogasawara's characterization of order completeness of Boolean algebras (1.2.4) and the Sikorski representation of a Boolean homomorphism (1.2.5) via the Stone space are given. The concept of lifting and the corresponding immersion of a measure space into the Stone space of the respective complete Boolean algebra are also presented (1.2.7, 1.2.8).

Section 1.3 begins with the notions of vector lattice and vector sublattice, ideal and factor lattice, order and relative uniform convergence, etc. (1.3.2, 1.3.4–1.3.6). The disjointness relation in a vector lattice provides the three basic Boolean algebras of bands, band projections, and fragments of an order-unity (1.3.3). These are isomorphic to one another for an order complete vector lattice with order-unity (1.3.7). The spectral function of an element of vector lattice and a list of its useful properties are given (1.3.8). The Freudenthal Spectral Theorem is stated as well as a characterization of general vector lattices in which the weak and strong forms of this theorem hold (1.3.9).

Three basic examples of universally complete vector lattices and some interconnections between them, as well as the representation of an arbitrary order complete vector lattice as an order-dense ideal in such spaces is the content of Section 1.4. The first example is the space of continuous functions on a quasiextremal compact space which assume possibly infinite values on a nowhere-dense set depending on a function (1.4.1, 1.4.2). This space is an order σ-complete vector lattice and even an order complete vector lattice provided that the underlying compact space is extremally disconnected (1.4.2). The second example is the space of all spectral functions with values in a σ-complete Boolean algebra (1.4.3). The Stone transform of a σ-complete Boolean algebra can be extended to a linear and lattice isomorphism of the corresponding space of spectral functions onto the space of extended valued continuous functions on the Stone space (1.4.4). Using the Freudenthal Spectral Theorem, we prove that an arbitrary order complete vector lattice is isomorphic to an order-dense ideal of the space of extended valued continuous functions on an extremal compact space (1.4.5, 1.4.6). The third example is the space of (equivalence classes of) measurable functions which is order complete provided that the underlying measure space has the direct sum property (1.4.7). In this case the space of essentially bounded measurable functions admits lifting (1.4.8); moreover, each lifting generates a linear and lattice isomorphism between the space of measurable functions and the space of extended valued continuous functions (1.4.9). Various conditions under which an abstract order complete vector space is isomorphic to an order-dense ideal in some space of measurable functions are also presented (1.4.10).

Section 1.5 starts with basic definitions and elementary facts of the theory of normed vector lattices (1.5.1, 1.5.2). We then discuss various useful characterizations of Banach lattices with order continuous norm (1.5.3), as well as monotonically complete Banach lattices and Banach lattices with order semicontinuous norm (1.5.4). Abstract M- and L^p-spaces are introduced (1.5.5) and the corresponding representation theorems are also stated (1.5.6). In particular, every Banach lattice has some AM-structure locally. This observation opens a way to some functional calculus in an arbitrary Banach lattice (1.5.7). Two more applications of a local AM-structure are stated: a disjointly complete Banach lattice has the projection property (1.5.10) and a Banach lattice is order complete if and only if it is disjointly complete (1.5.11). To prove these results, some order properties (1.5.8, 1.5.9) of an AM-space with unity are needed.

1.1. Boolean Algebras

In this section we sketch a minimum about Boolean algebras which we need in the sequel. A more explicit exposition may be found elsewhere; for instance, cf. [127, 283, 352, 384].

1.1.1. To fix terminology and notation, we recall some well-known notions.

An *ordered set* is a pair $(M, \leq)$, where $\leq$ is an order on M (see A.1.10). An ordered set is also called a *partially ordered set* or, briefly, a *poset*. It is in common parlance to apply all names of $(M, \leq)$ to the underlying set M of $(M, \leq)$. We indulge in doing the same elsewhere without further circumlocution.

An *upper bound* of a subset X of a poset M is an element $a \in M$ such that $x \leq a$ for all $x \in X$. A least element of the set of upper bounds of X is called a *least upper bound* or *supremum* of X and denoted by $\sup(X)$ or $\sup X$. In other words, $a = \sup(X)$ if and only if a is an upper bound of X and $a \leq b$ for every upper bound b of X. By reversal, i.e., by passing from the original order $\leq$ on a poset M to the *reverse* or *opposite order* $\leq^{-1}$ $(x \leq^{-1} \Leftrightarrow y \leq x)$, define a *lower bound* of a subset X of M and a *greatest lower bound*, $\inf(X)$ of X, also called an *infimum* of X and denoted $\inf X$. If a least upper or greatest lower bound of a set in M exists then it is unique and so deserves the definite article.

It can be easily checked that the following commutativity and associativity laws hold in every poset on duly stipulating existence of the suprema and infima in question:

(1) $\sup\limits_{\alpha \in \mathrm{A}} \sup\limits_{\beta \in \mathrm{B}} x_{\alpha,\beta} = \sup\limits_{\beta \in \mathrm{B}} \sup\limits_{\alpha \in \mathrm{A}} x_{\alpha,\beta}$;

(2) $\inf\limits_{\alpha \in \mathrm{A}} \inf\limits_{\beta \in \mathrm{B}} x_{\alpha,\beta} = \inf\limits_{\beta \in \mathrm{B}} \inf\limits_{\alpha \in \mathrm{A}} x_{\alpha,\beta}$;

(3) $\sup\big(\bigcup\limits_{\alpha \in \mathrm{A}} X_\alpha\big) = \sup\limits_{\alpha \in \mathrm{A}} \sup X_\alpha$;

(4) $\inf\big(\bigcup\limits_{\alpha \in \mathrm{A}} X_\alpha\big) = \inf\limits_{\alpha \in \mathrm{A}} \inf X_\alpha$.

1.1.2. A *lattice* is an ordered set L in which each pair $\{x, y\}$ has the *join* $x \vee y := \sup\{x, y\}$ and *meet* $x \wedge y := \inf\{x, y\}$. Given a subset X of a lattice L, we use the notation:

$$\bigvee X := \sup(X), \quad \bigwedge X := \inf(X),$$

$$\bigvee_{\alpha \in \mathrm{A}} x_\alpha := \bigvee\{x_\alpha : \alpha \in \mathrm{A}\}, \quad \bigwedge_{\alpha \in \mathrm{A}} x_\alpha := \bigwedge\{x_\alpha : \alpha \in \mathrm{A}\},$$

$$\bigvee_{k=1}^{n} x_k := x_1 \vee \dots \vee x_n := \sup\{x_1, \dots, x_n\},$$

$$\bigwedge_{k=1}^{n} x_k := x_1 \wedge \dots \wedge x_n := \inf\{x_1, \dots, x_n\}.$$

Here $(x_\alpha)_{\alpha \in \mathrm{A}}$ is a family in L, and $x_1, \dots, x_n$ stand for some members of L.

The binary operations *join* $(x, y) \mapsto x \vee y$ and *meet* $(x, y) \mapsto x \wedge y$ act in every lattice L and possess the following properties:

(1) *commutativity:*

$$x \vee y = y \vee x, \quad x \wedge y = y \wedge x;$$

(2) *associativity:*

$$x \vee (y \vee z) = (x \vee y) \vee z, \quad x \wedge (y \wedge z) = (x \wedge y) \wedge z.$$

By induction, from (2) we deduce that *each nonempty finite set in a lattice has the meet and join.* If each subset of a lattice L has the supremum and infimum then L is a *complete* lattice.

A lattice L is *distributive* provided that the following *distributive laws* are valid:

(3) $x \wedge (y \vee z) = (x \wedge y) \vee (x \wedge z)$;

(4) $x \vee (y \wedge z) = (x \vee y) \wedge (x \vee z)$.

If a lattice L has the least or greatest element then the former is called the *zero* of L and the latter, the *unity* of L. The zero and unity of L are solemnly denoted by $\mathbf{0}_L$ and $\mathbf{1}_L$. It is in common parlance to use the simpler symbols $\mathbf{0}$ and $\mathbf{1}$ and nicknames *zero* and *unity* provided that the context prompts the due details. Note also that $\mathbf{0}$ and $\mathbf{1}$ are neutral elements:

(5) $\mathbf{0} \vee x = x, \quad \mathbf{1} \wedge x = x.$

Specifying the general definitions, also note that $\bigvee \varnothing = \sup \varnothing := \mathbf{0}$ and $\bigwedge \varnothing = \inf \varnothing := \mathbf{1}$. A *complement* x^* of a member x of a lattice L with zero and unity is an element x^* of L such that

(6) $x \wedge x^* = \mathbf{0}, \quad x \vee x^* = \mathbf{1}.$

Elements x and y in L are *disjoint* if $x \wedge y = \mathbf{0}$. So, every element x is disjoint from any complement x^*. Note finally that if each element in L has at least one complement then we call L a *complemented lattice.* Recall by the way that a set U is *disjoint* whenever every two distinct members of U are disjoint. It is rather evident that an arbitrary lattice L may fail to have a complement to each element of L.

1.1.3. A *Boolean algebra* is a distributive complemented lattice with distinct zero and unity. Each element x of a Boolean algebra B has a unique complement denoted by x^*. This gives rise to the mapping $x \mapsto x^*$ $(x \in B)$ which is idempotent (i.e., $(\forall x \in B)\ (x^{**} := (x^*)^* = x)$) and presents a *dual isomorphism* or an *anti-isomorphism* of B onto itself (i.e., it is an order isomorphism between $(B, \leq)$ and $(B, \leq^{-1})$). The three operations $\vee$, $\wedge$, and $*$, living in every Boolean algebra B, are jointly referred to as *Boolean operations.*

A Boolean algebra B is *complete* (*σ-complete*), if each subset (countable subset) of B has a supremum and an infimum. It is in common parlance to speak of *σ-algebras* instead of σ-complete algebras.

Associated with a Boolean algebra B, the mappings $\bigvee, \bigwedge : \mathscr{P}(B) \to B$ are available that ascribe to a set in B its supremum and infimum, respectively. These mappings are sometimes referred to as *infinite operations*. The infinite operations obey many important rules among which we for instance mention the *De Morgan formulas*

$$\textbf{(1)}\ \Big(\bigvee_{\alpha\in\mathrm{A}} x_\alpha\Big)^* = \bigwedge_{\alpha\in\mathrm{A}} x_\alpha^*, \quad \Big(\bigwedge_{\alpha\in\mathrm{A}} x_\alpha\Big)^* = \bigvee_{\alpha\in\mathrm{A}} x_\alpha^*,$$

and the *infinite distributive laws*

$$\textbf{(2)}\ x \vee \bigwedge_{\alpha\in\mathrm{A}} x_\alpha = \bigwedge_{\alpha\in\mathrm{A}} x \vee x_\alpha,$$

$$\textbf{(3)}\ x \wedge \bigvee_{\alpha\in\mathrm{A}} x_\alpha = \bigvee_{\alpha\in\mathrm{A}} x \wedge x_\alpha,$$

with $x_\alpha \in B$ for all $\alpha \in \mathrm{A}$.

1.1.4. Let $\mathbf{2} := \mathbb{Z}_2 := \mathscr{P}(\{\varnothing\}) := \{\mathbf{0}, \mathbf{1}\}$ be the underlying set of the two-element Boolean algebra now viewed as a field with the following operations:

$$\mathbf{0}+\mathbf{0} := \mathbf{0}, \quad \mathbf{0}+\mathbf{1} = \mathbf{1}+\mathbf{0} := \mathbf{1}, \quad \mathbf{1}+\mathbf{1} := \mathbf{0},$$
$$\mathbf{0}\cdot\mathbf{1} = \mathbf{1}\cdot\mathbf{0} := \mathbf{0}, \quad \mathbf{0}\cdot\mathbf{0} := \mathbf{0}, \quad \mathbf{1}\cdot\mathbf{1} := \mathbf{1}.$$

Note that every member of $\mathbf{2}$ is idempotent.

Consider an arbitrary set B with the structure of an associative ring whose every element is idempotent: $(\forall b \in B)(b^2 = b)$. In this case B is called a *Boolean ring*. A Boolean ring is commutative and obeys the identity $b = -b$ for $b \in B$. Each Boolean ring is obviously a vector space and, at the same time, a commutative algebra over $\mathbf{2}$. Recall that the unity of an algebra differs from its zero by definition. So, we may and will identify the field $\mathbf{2}$ with the subring of a Boolean ring comprising the zero and unity of the latter. We usually reflect the practice in symbols by letting $\mathbf{0}$ stand for the zero and $\mathbf{1}$, for the unity of whatever ring. This agreement leads clearly to a rather popular notational collision: the addition and multiplication of $\mathbf{2}$ may be redefined on making $\mathbf{0}$ play the role of $\mathbf{1}$ and vice versa.

It is customary to endow a Boolean ring B with some order by the rule:

$$b_1 \leq b_2 \Leftrightarrow b_1 b_2 = b_1 \quad (b_1, b_2 \in B).$$

The poset $(B, \leq)$ obviously becomes a distributive lattice with the least element $\mathbf{0}$ and the greatest element $\mathbf{1}$. In this event, the lattice and ring operations are connected as follows:

$$x \vee y = x + y + xy, \quad x \wedge y = xy.$$

Moreover, to each element $b \in B$ there is a unique $b^* \in B$, the complement of b, such that

$$b^* \vee b = \mathbf{1}, \quad b^* \wedge b = \mathbf{0}.$$

Obviously, $b^* = \mathbf{1} + b$. Hence, each Boolean ring is a Boolean algebra under the above order.

In turn, we may transform a Boolean algebra B into a ring by putting

$$x + y := x \vartriangle y, \quad xy := x \wedge y \quad (x, y \in B),$$

where $x \vartriangle y := (x \wedge y^*) \vee (x^* \wedge y)$ is a *symmetric difference* of x and y. In this case $(B, +, \cdot, \mathbf{0}, \mathbf{1})$ becomes a unital Boolean ring whose natural order coincides with the initial order on B. Therefore, a Boolean algebra can be viewed as a unital algebra over $\mathbf{2}$ whose every element is idempotent.

1.1.5. Consider some methods of arranging new Boolean algebras.

(1) A nonempty subset B_0 of a Boolean algebra B is a *subalgebra* of B if B_0 is closed under the Boolean operations $\vee$, $\wedge$, and $*$; i.e., $\{x \vee y, x \wedge y, x^*\} \subset B_0$ for all $x, y \in B_0$. Under the order induced from B, every subalgebra B_0 is a Boolean algebra with the same zero and unity as those of B. In particular, $B_0 := \{\mathbf{0}_B, \mathbf{1}_B\}$ is a subalgebra of B. A subalgebra $B_0 \subset B$ is *regular* or *complete* (*σ-regular* or *σ-complete*) provided that for every set (countable set) A in B_0 the elements $\bigvee A$ and $\bigwedge A$, if exist in B, belong to B_0.

The intersection of every family of subalgebras is itself a subalgebra. The same holds true for regular (σ-regular) subalgebras, which makes the definition to follow sound. The least subalgebra of B containing a nonempty subset M of B is the *subalgebra generated by* M. The *regular* (*σ-regular*) *subalgebra generated by* M is introduced in much the same manner.

(2) An *ideal* of a Boolean algebra B is any nonempty set J in B obeying the conditions:

$$x \in J,\ y \in J \Rightarrow x \vee y \in J,$$
$$x \in J,\ y \leq x \Rightarrow y \in J.$$

The set $B_a := \{x \in B : x \leq a\}$, with $a \in B$, provides an example of an ideal of B. Such an ideal is called *principal*. If $\mathbf{0} \neq e \in B$ then the principal ideal B_e with the order induced from B is a Boolean algebra in its own right. The element e plays the role of unity in B_e. The lattice operations of B_e are inherited from B, and the complementation of B_e has the form $x \mapsto e - x := e \wedge x^*$ for all $x \in B$.

An ideal J is *proper* provided that $J \neq B$. A regular ideal of B is often called a *band* or *component* of B.

(3) Take Boolean algebras B and B' and a mapping $h : B \to B'$. Say that h *isotonic* or *monotone* if $(x \le y \Rightarrow h(x) \le h(y))$. (Note by the way an isotonic mapping from B to B' with the opposite order is *antitonic*.) Say that h is a *(Boolean) homomorphism*, if for all $x, y \in B$ the following equalities are fulfilled:

$$h(x \vee y) = h(x) \vee h(y),$$
$$h(x \wedge y) = h(x) \wedge h(y),$$
$$h(x^*) = h(x)^*.$$

Every homomorphism h is monotone and the image $h(B)$ of B is a subalgebra of B'. If h is bijective then we call h an *isomorphism*, and B and B', *isomorphic* Boolean algebras. An injective homomorphism is a *monomorphism*. A homomorphism h (of B to a complete B') is *complete* if h *preserves suprema and infima*; i.e., $h(\sup(U)) = \sup(h(U))$ and $h(\inf(V)) = \inf(h(V))$ for all $U \subset B$ and $V \subset B$ for which there are $\sup(U)$ and $\inf(V)$. Observe that $h : B \to B'$ is a *ring homomorphism* (see 1.1.4) if and only if h is a Boolean homomorphism from B into the Boolean algebra $B'_a := [\mathbf{0}, a]$ with $a := h(\mathbf{1}_B)$.

Given a set C and a bijection $h : B \to C$, we may furnish C with an order by putting $h(x) \le h(y)$ whenever $x \le y$. In this event, C turns into a Boolean algebra and h becomes an isomorphism between B and C.

(4) Let J be a proper ideal of a Boolean algebra B. Define the equivalence $\sim$ on B by the rule

$$x \sim y \Leftrightarrow x \mathbin{\triangle} y \in J \quad (x, y \in B).$$

Denote by φ the factor mapping of B onto the factor set $B/J := B/{\sim}$. Recall that φ is also called *canonical*. Given cosets (equivalence classes) u and v, i.e., members of B/J; agree to write $u \le v$ if and only if there are $x \in u$ and $y \in v$ satisfying $x \le y$. We have thus defined an order on B/J. In this event, B/J becomes a Boolean algebra which is called *factor algebra* or *quotient algebra* of B by J. The Boolean operations in B/J make φ a homomorphism. So, φ is referred to as the *factor homomorphism* of B onto B/J.

If $h : B \to B'$ is a homomorphism then $\ker(h) := \{x \in B : h(x) = \mathbf{0}\}$ is an ideal of B and there is a unique monomorphism $g : B/\ker(h) \to B'$ satisfying $g \circ \varphi = h$, where $\varphi : B \to B/\ker(h)$ is the factor homomorphism. Therefore, each homomorphic image of a Boolean algebra B is isomorphic to the factor algebra of B by a suitable ideal.

(5) Take a family of Boolean algebras $(B_\alpha)_{\alpha \in \mathrm{A}}$. Furnish the product $B := \prod_{\alpha \in \mathrm{A}} B_\alpha$ with the *coordinatewise order* or *product order* by putting $x \le y$ for

$x, y \in B$ whenever $x(\alpha) \leq y(\alpha)$ for all $\alpha \in \mathrm{A}$. In this event B becomes a Boolean algebra.

Each Boolean operation in B consists in implementing the respective operation in every coordinate Boolean algebra B_α, i.e., it is carried out *coordinatewise*. The zero $\mathbf{0}_B$ and unity $\mathbf{1}_B$ of B are as follows: $\mathbf{0}_B(\alpha) := \mathbf{0}_\alpha$ and $\mathbf{1}_B(\alpha) := \mathbf{1}_\alpha$ $(\alpha \in \mathrm{A})$, where $\mathbf{0}_\alpha$ and $\mathbf{1}_\alpha$ are the zero and unity in B_α. The Boolean algebra B is the *Cartesian product* or, simply, *product* of $(B_\alpha)_{\alpha \in \mathrm{A}}$.

(6) A disjoint subset of a Boolean algebra is an *antichain*. In other words, a subset A of B is an antichain provided that $a_1 \wedge a_2 = \mathbf{0}$ for all distinct $a_1, a_2 \in A$. If an antichain has the form $A := \{a_\xi : \xi \in \Xi\}$ then we presume that $a_\xi \wedge a_\eta = \mathbf{0}$ whenever $\xi \neq \eta$.

An antichain A in B is a *partition of an element* $b \in B$ and a *partition of unity*, when b is the unity of B, provided that $b = \bigvee A$.

Let $(b_\xi)_{\xi \in \Xi}$ be a partition of unity in B. According to 1.1.5 (2) $B_\xi := [\mathbf{0}, b_\xi]$ is a Boolean algebra with unity b_ξ.

The complete Boolean algebra B is isomorphic to $\prod_{\xi \in \Xi} B_\xi$. Some isomorphism is carried out by sending $b \in B$ to the mapping $\widetilde{b}$ by the rule $\widetilde{b}(\xi) := b \wedge b_\xi$ $(b \in B)$.

1.1.6. Recall a certain fundamental property of Boolean algebras. Let B be a Boolean algebra. A subset E of B *minorizes* a subset B_0 of B if to each $0 < b \in B_0$ there is an x in E such that $0 < x \leq b$. It is also in common parlance to call E *coinitial* or *minorant* set to B_0. Denote by u.b.(M) the set of all upper bounds of M.

Exhaustion Principle. *Let M be a nonempty subset of a Boolean algebra B. Assume given a subset E of B that minorizes the band B_0 of B generated by M. Then some antichain E_0 exists, $E_0 \subset E$, such that u.b.$(E_0) =$ u.b.(M) and to each $x \in E_0$ there is an element y in M satisfying $x \leq y$.*

$\triangleleft$ Consider the set $\mathfrak{A}$ of all antichains A with the following properties: (a) $A \subset E$; (b) to each $x \in A$ there is a $y \in M$ satisfying $x \leq y$. If $\mathbf{0} \neq y \in M$ then, by the minorant condition, $y \geq x$ for some $\mathbf{0} \neq x \in E$. Hence, $\{x\} \in \mathfrak{A}$ and $\mathfrak{A}$ is nonempty. The inclusion ordered set $\mathfrak{A}$ clearly obeys the hypotheses of the Kuratowski–Zorn Lemma. Therefore, there is a maximal element $E_0 \in \mathfrak{A}$. The property (b) from the definition of $\mathfrak{A}$ implies u.b.$(M) \subset$ u.b.(E_0). In particular, we are done if u.b.$(E_0) = \{\mathbf{1}\}$.

To prove the converse inclusion assume that $b_0 \notin$ u.b.(M) for some $b_0 \in$ u.b.(E_0), $b_0 \neq \mathbf{1}$. There exists an element $x \in M$, such that $x_0 := b_0^* \wedge x \neq \mathbf{0}$. By the minorant condition, $\mathbf{0} < y \leq x_0$ for some $y \in E$. The set $E_0 \cup \{y\}$ belongs to $\mathfrak{A}$ and has essentially more elements than E_0. This contradicts the fact that E_0 is maximal, and so u.b.$(E_0) \subset$ u.b.(M). $\triangleright$

(1) *To each nonempty set $M \subset B$ having the least upper bound, there is an antichain $A \subset B$ with the following properties: $\bigvee A = \bigvee M$ and, given $x \in A$, we may find y in M such that $x \le y$.*

$\triangleleft$ Choose the minorant $E := \bigcup_{y\in M}[\mathbf{0}, y]$ as a minorant for M and appeal to (1). $\triangleright$

(2) *A Boolean algebra is complete if and only if any antichain in it has the supremum.*

1.1.7. Examples.

(1) Recall that $\mathscr{P}(X)$ stands for the *powerset* of X comprising all subsets of X and also denoted by 2^X. Given a nonempty set X, observe that the powerset $\mathscr{P}(X)$ of X ordered by inclusion is a complete Boolean algebra. This algebra is often called the *boolean* of X. The Boolean operations on every boolean are the set-theoretic operations of union, intersection, and complementation.

(2) Let X be a topological space. Recall that a closed and open subset of X is called *clopen.* The collection of all clopen sets in X, ordered by inclusion, is a subalgebra of the boolean $\mathscr{P}(X)$. Denote this subalgebra by Clop(X). The Boolean operations in Clop(X) are inherited from $\mathscr{P}(X)$. Hence, they are set-theoretic. However, Clop(X) is not a regular subalgebra of $\mathscr{P}(X)$; i.e., the infinite operations in $\mathscr{P}(X)$ and Clop(X) may differ essentially.

(3) A closed subset F of a topological space X is called *regular* if $F = \mathrm{cl}(\mathrm{int}(F))$; i.e., if F coincides with the closure of the interior of F. By analogy, a *regular open set* G is defined by the formula $G = \mathrm{int}(\mathrm{cl}(G)$. Let $\mathrm{RC}\,(X)$ and $\mathrm{RO}\,(X)$ stand for the collections of all regular closed subsets and all regular open subsets, respectively, of X.

Furnished with the order by inclusion, $\mathrm{RC}\,(X)$ and $\mathrm{RO}\,(X)$ become complete Boolean algebras. The mapping $F \mapsto \mathrm{int}(F)$ $(F \in \mathrm{RC}\,(X))$ is an isomorphism between them. Despite $\mathrm{RC}\,(X)$ and $\mathrm{RO}\,(X)$ are included in the boolean $\mathscr{P}(X)$, they are not subalgebras of the latter. For instance, the Boolean operations on $\mathrm{RC}\,(X)$ take the form

$$E \vee F = E \cup F, \quad E \wedge F = \mathrm{cl}(\mathrm{int}(E \cap F)), \quad F^* = \mathrm{cl}(X - F).$$

(4) Denote by $\mathscr{B}or(X)$ the Borel σ-algebra of a topological space X (i.e., the σ-regular subalgebra of the boolean $\mathscr{P}(X)$ which is generated by the open sets of X). Consider the ideal $\mathscr{N}$ of $\mathscr{B}or(X)$ comprising the *meager* subsets of X (also called the first category sets in X). The factor algebra $\mathscr{B}or(X)/\mathscr{N}$ is a complete Boolean algebra called the *algebra of Borel sets modulo meager sets* or briefly the *Borel-by-meager algebra.*

We will arrive at an isomorphic algebra if instead of $\mathscr{B}or(X)$ we take the σ-algebra of sets with the Baire property. (A subset M of X has the *Baire property* if there is an open set G in X such that the symmetric difference $M \triangle G$ is a meager set.) If X is a *Baire* space, i.e., if X lacks nonempty open meager subsets, then the algebra in question is isomorphic to the algebra of regular closed sets $\mathrm{RC}(X)$.

(5) Assume given a nonempty set Ω, a σ-algebra $\mathscr{B} \subset \mathscr{P}(\Omega)$ of subsets of Ω, and a *measure* on $\mathscr{B}$ which is a positive countably additive function $\mu : \mathscr{B} \to \mathbb{R} \cup \{+\infty\}$, with $\mathbb{R}$ the *reals*. *Countable additivity*, as usual, means that

$$\mu\bigg(\bigcup_{n=1}^{\infty} A_n\bigg) = \sum_{n=1}^{\infty} \mu(A_n)$$

for every disjoint sequence (A_n) of $\mathscr{B}$.

A triple $(\Omega, \mathscr{B}, \mu)$ is said to be a *measure space* if the following conditions are met:

(a) if $A \subset \Omega$ and $A \cap K \in \mathscr{B}$ for each $K \in \mathscr{B}$ with $\mu(K) < +\infty$, then $A \in \mathscr{B}$;

(b) if $A \in \mathscr{B}$ and $\mu(A) = +\infty$ then there is $A_0 \in \mathscr{B}$ such that $A_0 \subset A$ and $0 < |A_0| < +\infty$;

(c) if $A \in \mathscr{B}$, $\mu(A) = 0$ and $A_0 \subset A$ then $A_0 \in \mathscr{B}$.

Let $\mathscr{N} := \{A \in \mathscr{B} : \mu(A) = 0\}$. Then $\mathscr{N}$ is a σ-complete ideal. The factor algebra $B(\Omega) := B(\Omega, \mathscr{B}, \mu) := \mathscr{B}/\mathscr{N}$ is also a σ-algebra called the *associated algebra* or the *algebra of measurable sets by measure zero sets.*

The measure space $(\Omega, \mathscr{B}, \mu)$ is said to possess the *direct sum property* if $\mathscr{B}$ contains a family $(A_\xi)_{\xi \in \Xi}$ of pairwise disjoint sets of finite measure such that the following holds: for every measurable subset $A \in \mathscr{B}$ of finite measure there exists a countable set of indices $\Theta \subset \Xi$ and a measure zero set $A_0 \in \mathscr{N}$ such that

$$A = A_0 \cup \bigg(\bigcup_{\xi \in \Theta} (A \cap A_\xi)\bigg).$$

If a measure space $(\Omega, \mathscr{B}, \mu)$ possesses the direct sum property then the associate Boolean algebra $B(\Omega, \mathscr{B}, \mu)$ is complete.

1.2. Representation of Boolean Algebras

The Stone Representation Theorem opens up a distinct possibility of representing a Boolean algebra as the Boolean algebra of clopen subsets of a compact space. The basic goal of this section is to prove this theorem and to describe some opportunities it affords.

1.2.1. Let B be an arbitrary Boolean algebra.

(1) A *character* of B is a Boolean homomorphism or, which is the same, a nonzero ring homomorphism $\chi : B \to \mathbf{2}$. Denote by $\mathrm{X}(B)$ the set of all characters of B and make $\mathrm{X}(B)$ into a topological space on furnishing it with the topology of pointwise convergence. To put it more explicitly, the topology on $\mathrm{X}(B)$ is induced by the product topology of $\mathbf{2}^B$, where we consider $\mathbf{2}$ with the unique compact topology on this set, the discrete topology of $\mathbf{2}$. Observe that, unless stated otherwise, we presume all topological spaces to be Hausdorff. Recall that a topological space X is *connected* whenever the only clopen subsets of X are $\varnothing$ and X. A topological space X is *totally disconnected* provided that each connected subspace of X is at most a singleton. The topological space $\mathbf{2}^B$, called sometimes a *Cantor discontinuum*, is compact and totally disconnected. A topological space with all these properties is a *Boolean space*. Evidently, $\mathrm{X}(B)$ is a closed subset of $\mathbf{2}^B$. Therefore, $\mathrm{X}(B)$ is itself is a Boolean space. Say that the Boolean space $\mathrm{X}(B)$ is the *character space* of a Boolean algebra B.

(2) Recall that a nonempty subset $\mathscr{F}$ of B is a *filter* on B provided that

$$x \in \mathscr{F},\ y \in \mathscr{F} \Rightarrow x \vee y \in \mathscr{F},$$
$$x \in \mathscr{F},\ x \leq y \Rightarrow y \in \mathscr{F}.$$

A filter other than B is *proper*. A maximal element of the inclusion-ordered set of all proper filters on B is an *ultrafilter* on B.

Let $U(B)$ stand for the set of all ultrafilters on B, and denote by $U(b)$ the set of ultrafilters containing b. We endow $U(B)$ with the topology with base $\{U(b) : b \in B\}$. This definition is sound since it is easy to check that $U(x \wedge y) = U(x) \cap U(y)$ $(x, y \in B)$; i.e., $U(B)$ is closed under finite intersections. The topological space $U(B)$ is often referred to as the *Stone space of* B and is denoted by $\mathscr{S}(B)$.

(3) Denote by $M(B)$ the set of all maximal (proper) ideals of a Boolean algebra B. An ideal here may be understood in accord with 1.1.5 (2) or in the conventional sense of ring theory. Clearly, a set J in B is an ideal of B if and only if $J^* := \{x^* : x \in J\}$ is a filter on B. Moreover, $J \in M(B) \leftrightarrow J^* \in U(B)$. Therefore, the mapping $J \mapsto J^*$ presents a bijection between $M(B)$ and $U(B)$. The set $M(B)$ is usually called the *maximal ideal space* of B and is always furnished with the inverse image topology translated from $U(B)$ which makes the mapping $J \mapsto J^*$ a homeomorphism.

1.2.2. Recall the prerequisites we need for applying the Gelfand transform in the case of a Boolean algebra.

(1) *A Boolean ring B is a field if and only if B is the pair of* $\mathbf{0}$ *and* $\mathbf{1}$. *Hence, there is a unique Boolean field to within isomorphism; namely,* $\mathbf{2}$.

$\lhd$ Indeed, a nonzero element $x \in B$ is invertible, and so the following implications are valid:

$$xx^{-1} = \mathbf{1} \Rightarrow xxx^{-1} = \mathbf{1} \Rightarrow xx^{-1} = x \Rightarrow x = \mathbf{1}. \ \rhd$$

Given $\chi \in \mathrm{X}(B)$, denote by χ^* the mapping $x \mapsto \chi(x)^*$ $(x \in B)$. Obviously, $\ker(\chi) := \{x \in B : \chi(x) = \mathbf{0}\}$ is an ideal, and $\ker(\chi)^*$ is a filter.

(2) *The mappings* $\chi \mapsto \ker(\chi)$ $(\chi \in \mathrm{X}(B))$ *and* $\chi \mapsto \ker(\chi)^*$ $(\chi \in \mathrm{X}(B))$ *are homeomorphisms of* $\mathrm{X}(B)$ *onto* $M(B)$ *and* $U(B)$, *respectively.*

$\lhd$ The mapping $\chi \mapsto \ker(\chi)$ is injective. If $J \in M(B)$ then B/J is a field and, by (1), B/J is isomorphic to $\mathbf{2}$. Fix such an isomorphism $\lambda : B/J \to \mathbf{2}$; and put $\chi := \lambda \circ \varphi$, where $\varphi : B \to B/J$ is the factor mapping. Obviously, $\ker(\chi) = J$ and so the mapping under discussion is bijective. The remaining claims are obvious. $\rhd$

(3) *For an* x *in* B *to equal zero it is necessary and sufficient that* $\chi(b) = \mathbf{0}$ *for all* $\chi \in \mathrm{X}(B)$.

$\lhd$ Assume that $x \neq \mathbf{0}$. Then the principal ideal $\{y \in B : y \leq x^*\}$ is proper, and so can be extended to a maximal ideal $J \in M(B)$. This claim, known as the *Krull Theorem*, is immediate from the Kuratowski–Zorn Lemma. By (2), $J = \ker(\chi)$ for a certain $\chi \in \mathrm{X}(B)$. Since $x \notin J$; therefore, $\chi(x) \neq \mathbf{0}$. $\rhd$

1.2.3. Stone Representation Theorem. *Each Boolean algebra* B *is isomorphic to the Boolean algebra of clopen sets of a Boolean space unique up to homeomorphism, the Stone space of* B.

$\lhd$ Denote by $C(\mathrm{X}(B), \mathbf{2})$ the algebra of continuous $\mathbf{2}$-valued functions on the character space $\mathrm{X}(B)$ of B which is a Boolean space. The *Gelfand transform* $\mathscr{G}_B$ assigns to an element $x \in B$ the $\mathbf{2}$-valued function

$$\widehat{x} : \chi \mapsto \chi(x) \quad (\chi \in \mathrm{X}(B)).$$

Obviously, $\mathscr{G}_B : B \to C(\mathrm{X}(B), \mathbf{2})$ is a injective homomorphism, i.e., a *monomorphism* (cf. 1.2.2 (3)). Take $f \in C(\mathrm{X}(B), \mathbf{2})$ and put $V_f := \{\chi \in \mathrm{X}(B) : f(\chi) = \mathbf{1}\}$. The set V_f is clopen. By the definition of the topology of $\mathrm{X}(B)$, there are $b_1, \dots, b_k \in B$ and $c_1, \dots, c_l \in B$ such that

$$V_f := \{\chi \in \mathrm{X}(B) : \chi(b_n) = \mathbf{1} \quad (n \leq k), \quad \chi(c_m) = \mathbf{0} \quad (m \leq l)\}.$$

Assign $b_0 := b_1 \wedge \dots \wedge b_k, c_0 := c_1 \vee \dots \vee c_l$ and $b := b_0 \wedge c_0^*$. The set V_f can be presented as follows:

$$\begin{aligned} V_f &= \{\chi \in \mathrm{X}(B) : \chi(b_0) = \mathbf{1}, \ \chi(c_0) = \mathbf{0}\} \\ &= \{\chi \in \mathrm{X}(B) : \chi(b) = \mathbf{1}\} = \{\chi \in \mathrm{X}(B) : \widehat{b}(\chi) = \mathbf{1}\}. \end{aligned}$$

Therefore, $f = \widehat{b}$, and so $\mathscr{G}_B$ is an isomorphism.

Assume now that Q_1 and Q_2 are Boolean spaces such that the mapping $h : C(Q_1, \mathbf{2}) \to C(Q_2, \mathbf{2})$ is an isomorphism of these algebras. If χ is a character of

$C(Q_2, \mathbf{2})$ then $\chi \circ h$ is a character of $C(Q_2, \mathbf{2})$. Hence, $\chi \mapsto \chi \circ h$ is a homeomorphism between the character spaces. On the other hand, the character space of $C(Q_k, \mathbf{2})$ is homeomorphic to Q_k. Therefore, the Boolean spaces Q_1 and Q_2 are homeomorphic. It suffices to note that the algebra $C(\mathrm{X}(B), \mathbf{2})$ is isomorphic to the algebra of clopen sets of the space $\mathrm{X}(B)$ and so, of the space $U(B)$ as well. $\rhd$

In view of this theorem, there is a mapping $B \mapsto \mathrm{Clop}(\mathscr{S}(B))$ which is occasionally called the Stone transform of B.

1.2.4. In the sequel we are mostly interested in complete and σ-complete Boolean algebras. The notion of a complete Boolean algebra is closely tied with that of an extremally disconnected compact space. Recall that a topological space Q is called *extremally* (*quasiextremally*) *disconnected* or simply *extremal* (*quasiextremal*) if the closure of an arbitrary open set (open F_σ-set) in it is open or, which is equivalent, the interior of an arbitrary closed set (closed G_δ-set) is closed. Clearly, an extremal (quasiextremal) space is totally disconnected.

Ogasawara Theorem. *A Boolean algebra is complete (σ-complete) if and only if its Stone space is extremal (quasiextremal).*

$\lhd$ We confine exposition to the case of complete Boolean algebra B. Assume further that h is an isomorphism of B on the algebra of clopen sets of the compact space $Q := U(B)$. Take an open set $G \subset Q$. Since Q is totally disconnected, $G = \bigcup \mathscr{U}$, where $\mathscr{U}$ stands for the union of all clopen subsets of G. Put $\mathscr{U}' := \{h^{-1}(U) : U \in \mathscr{U}\}$ and $b := \bigvee \mathscr{U}'$. The clopen set $h(b)$ is the closure of G. Indeed, $\mathrm{cl}(G) \subset h(b)$ and $h(b) \setminus \mathrm{cl}(G)$ is open. If the latter set is nonempty then $h(c) \subset h(b) \setminus \mathrm{cl}(G)$ for some $\mathbf{0} \neq c \in B$. This implies in turn that $h(c) \vee h(u) \leq h(b)$ for all $u \in \mathscr{U}'$, which contradicts the equality $b = \bigvee \mathscr{U}$. Therefore, $\mathrm{cl}(G) = h(b)$ is an open set.

Assume now that the compact space Q is extremal. Let $\mathscr{G}$ stand for some collection of clopen subsets of Q, and put $G := \bigcup \mathscr{G}$. The set G is open and the closure $\mathrm{cl}(G)$ of G must be open by the hypothesis about Q. Obviously, $\mathrm{cl}(G)$ is the least upper bound of $\mathscr{G}$ in the Boolean algebra of clopen sets $\mathrm{Clop}(Q)$. $\rhd$

1.2.5. Sikorski Theorem. *Assume that B and B' are Boolean algebras, and $h : B \to B'$ is a homomorphism between them. Denote by $\imath : B \to \mathrm{Clop}(\mathscr{S}(B))$ and $\imath' : B' \to \mathrm{Clop}(\mathscr{S}(B'))$ the Stone transforms of B and B'. There is a unique continuous mapping $\theta : \mathscr{S}(B') \to \mathscr{S}(B)$ such that*

$$h(x) = (\imath')^{-1}\theta^{-1}(\imath(x)) \quad (x \in B).$$

The mapping $h \mapsto \mathscr{S}(h) := \theta$ carries out a bijection between the sets of all homomorphisms from B to B' and the set of all continuous mappings from $\mathscr{S}(B')$ to $\mathscr{S}(B)$. If B'' is another Boolean algebra and $g : B' \to B''$ is a homomorphism, then $\mathscr{S}(g \circ h) = \mathscr{S}(h) \circ \mathscr{S}(g)$. Moreover, $\mathscr{S}(I_B) = I_{\mathscr{S}(B)}$.

◁ Denote $Q := \mathscr{S}(B)$ and $Q' := \mathscr{S}(B')$. If q' is an ultrafilter in B' then obviously $q := \{b \in B : \ h(b) \in q'\}$ is also an ultrafilter in B. Assigning $\theta(q') := q$ we arrive at the mapping $\theta : q' \in Q' \mapsto q \in Q$. Taking into consideration the equalities $\imath(b) = \{q \in Q : \ b \in q\}$ $(b \in B)$ and $\imath'(b') = \{q' \in Q' : \ b' \in q'\}$ $(b' \in B')$ we deduce

$$\begin{aligned} \imath' h(b) &= \{q' \in Q' : \ h(b) \in q'\} = \{q' \in Q' : \ b \in q\} \\ &= \{q' \in Q' : \ \theta(q') \in \imath(b)\} = \theta^{-1}(\imath(b)). \end{aligned}$$

In particular, $\theta(q') \in \imath(b)$ if and only if $q' \in \imath'(h(b))$; therefore θ is continuous. The remaining properties of θ are obvious. ▷

Two important particular cases of the situation under consideration are worthy of a special attention.

(1) *A Boolean algebra B_0 is isomorphic to a subalgebra of a Boolean algebra B if and only if the Stone space $\mathscr{S}(B_0)$ of B_0 is a continuous image of the Stone space $\mathscr{S}(B)$ of B.*

(2) *A Boolean algebra B' is a homomorphic image of a Boolean algebra B (or isomorphic with a factor algebra of B; see* 1.1.6 (4)*) if and only if the Stone space $\mathscr{S}(B')$ of B' is homeomorphic to a closed subset of the Stone space $\mathscr{S}(B)$ of B.*

1.2.6. Loomis–Sikorski Theorem. *Let Q be the Stone space of a Boolean σ-algebra B. Denote by $\mathrm{Clop}_\sigma(Q)$ the σ-algebra of subsets of Q which is generated by the set $\mathrm{Clop}(Q)$ of all clopen subsets of Q. Let Δ stand for the σ-ideal of $\mathrm{Clop}_\sigma(Q)$, comprising all meager sets. Then B is isomorphic with the factor algebra $\mathrm{Clop}_\sigma(Q)/\Delta$. If $\imath$ is an isomorphism of B onto $\mathrm{Clop}(Q)$ and φ is the factor mapping of $\mathrm{Clop}_\sigma(Q)$ onto the factor algebra $\mathrm{Clop}_\sigma(Q)/\Delta$ then the mapping $h := \varphi \circ \imath$ is an isomorphism of B onto $\mathrm{Clop}_\sigma(Q)/\Delta$.*

◁ Observe that h is a homomorphism as the composite of two homomorphisms. If $h(b) = 0$ then $\imath(b) \in \Delta$ and $\imath(b) = \varnothing$, since no nonempty clopen set is meager. Thus h is injective.

To prove that h is surjective put

$$\mathscr{F} := \{A \in \mathrm{Clop}_\sigma(Q) : \ (\exists b \in B)\, \varphi(A) = h(A)\}.$$

Since $\mathrm{Clop}(Q) \subset \mathscr{F} \subset \mathrm{Clop}_\sigma(Q)$, it suffices to observe that $\mathscr{F}$ is a σ-algebra. If $A \in \mathscr{F}$ then $\varphi(Q \setminus A) = h(b^*)$, so that $Q \setminus A \in \mathscr{F}$. Now, consider a sequence (A_n) of $\mathscr{F}$ and choose a sequence (b_n) of B such that $\varphi(A_n) = h(b_n)$. According to 1.1.7 (2) $\imath\left(\bigvee_{n=1}^\infty b_n\right) = A_0 \cup \bigcup_{n=1}^\infty \imath(b_n)$ with a nowhere-dense subset $A_0 \subset Q$.

Using this equality we easily deduce

$$\varphi\left(\bigcup_{n=1}^{\infty} A_n\right) = \varphi\left(A_0 \cup \bigcup_{n=1}^{\infty} A_n\right) = \varphi\left(A_0 \cup \bigcup_{n=1}^{\infty} \imath(b_n)\right)$$
$$= \varphi\left(\imath\left(\bigvee_{n=1}^{\infty} b_n\right)\right) = h\left(\bigvee_{n=1}^{\infty} b_n\right)$$

and the result follows. $\triangleright$

Birkhoff–Ulam Theorem. *Let Q be a compact space. For every Borel set $V \in \mathscr{B}or(Q)$ there exists a unique regular open set $h(V)$ such that the symmetric difference $V \triangle h(V)$ is meager. Let $\mathscr{N}$ stand for the σ-ideal of $\mathscr{B}or(Q)$, comprising all meager sets. Then h is an order σ-continuous homomorphism from $\mathscr{B}or(Q)$ onto* RO (Q). *Moreover, the kernel of h coincides with $\mathscr{N}$ and $\mathscr{B}or(Q)/\mathscr{N}$ is isomorphic to* RO (Q).

1.2.7. Consider a measure space $(\Omega, \mathscr{B}, \mu)$, and let $\varphi : \mathscr{B} \to B(\Omega)$ be the factor homomorphism. A Boolean homomorphism $\rho : B(\Omega) \to \mathscr{B}$ is called a *lifting* of the factor algebra $B(\Omega)$ if $\rho(A) \in A$ for each equivalence class $A \in B(\Omega)$. The latter means that $\varphi \circ \rho$ is the identity mapping on $B(\Omega)$; therefore, a lifting is a right-inverse of the homomorphism φ.

(1) *If a measure space $(\Omega, \mathscr{B}, \mu)$ possesses the direct sum property then the factor algebra $B(\Omega, \mathscr{B}, \mu)$ admits a lifting.*

$\triangleleft$ The proof can be found in [375]. $\triangleright$

(2) *Let ρ be a lifting of the factor algebra $B(\Omega)$. Then, for every family $(A_\xi)_{\xi\in\Xi}$ of elements of $B(\Omega)$, the union $A_s := \bigcup_{\xi\in\Xi} \rho(A_\xi)$ and the intersection $A_i := \bigcap_{\xi\in\Xi} \rho(A_\xi)$ are measurable and, moreover,*

$$\varphi(A_s) = \bigvee_{\xi\in\Xi} A_\xi, \quad \varphi(A_i) = \bigwedge_{\xi\in\Xi} A_\xi.$$

(3) For every point $\omega \in \Omega$, denote the ultrafilter $\{A \in B(\Omega) : \omega \in \rho(A)\}$ by $\tau(\omega)$. The mapping $\tau : \Omega \to Q$ thus constructed will be called the *canonical immersion* of Ω in Q corresponding to the lifting ρ.

1.2.8. Theorem. *Let ρ be a lifting of $B(\Omega, \mathscr{B}, \mu)$, let τ be the corresponding canonical immersion of Ω in the Stone space Q of the Boolean algebra $B(\Omega, \mathscr{B}, \mu)$, and let $\imath$ be the Stone transform from $B(\Omega)$ onto* Clop(Q). *The following are valid:*

(1) *$\rho(A) = \tau^{-1}(\imath(A))$ for each class $A \in B(\Omega)$;*

(2) *$\imath^{-1}(U) = \varphi(\tau^{-1}(A))$ for each clopen set $U \in$* Clop(Q);

(3) *the mapping $\tau : \Omega \to Q$ is Borel measurable and the image $\tau(\Omega)$ is dense in Q;*

(4) *the inverse image $\tau^{-1}(N)$ of every meager subset $N \subset Q$ is measurable in Ω and has zero measure.*

$\lhd$ Assertions (1) and (2) are straightforward. To prove (3) and (4), we consider an arbitrary open subset $V \subset Q$. Choose a family $(U_\xi)_{\xi\in\Xi}$ of clopen subsets with $V = \bigcup_{\xi\in\Xi} U_\xi$. Then (1) implies measurability of the inverse images $\tau^{-1}(U_\xi)$ for all $\xi \in \Xi$ and 1.2.7 (2) implies measurability of the set $\tau^{-1}(V) = \bigcup_{\xi\in\Xi} \tau^{-1}(U_\xi)$. The relation $\bigvee_{\xi\in\Xi} U_\xi = \mathrm{cl}(V)$ in the Boolean algebra $\mathrm{Clop}(Q)$, together with (2) and 1.2.7 (2), ensures the equality $\varphi(\tau^{-1}(V)) = \varphi(\tau^{-1}(\mathrm{cl}\, V)$. $\rhd$

1.2.9. A function $\mu : B \to \mathbb{R} \cup \{+\infty\}$ is called *additive*, *countably additive*, or *completely additive* if

$$\mu\Big(\bigvee_{\xi\in\Xi} x_\xi\Big) = \sum_{\xi\in\Xi} \mu(x_\xi)$$

for any finite, countable, or arbitrary antichain (x_ξ) of B, respectively. To eliminate the trivial additive function $\mu \equiv +\infty$ it is always assumed that $\mu(\mathbf{0}_B) = 0$. An additive function on Boolean algebra is often referred to as a *measure*.

We say that the function μ is *positive*, *strictly positive*, and *finite* if respectively $\mu(b) \geq 0$, $\mu(b) > 0$, and $\mu(b) < +\infty$ for all $0 \neq b \in B$. Finally, a positive μ is said to be *locally finite* if for any $b \in B$ with $0 < \mu(b)$ there exists $0 < b' \leq b$ such that $0 < \mu(b') < +\infty$. A positive countably additive locally finite measure on the Boolean algebra $\mathrm{Clop}_\sigma(Q)$ is called *normal* if it vanishes on the ideal of meager sets. An extremal space is said to be *hyperstonian* if there is a normal measure on $\mathrm{Clop}_\sigma(Q)$ strictly positive on $\mathrm{Clop}(Q)$.

Let $\mathscr{M}_+(B)$ denote the set of all finite completely additive positive measures on B. A complete Boolean algebra B is said to be *multinormed* if the set of all finite completely additive measures separates the points of B; in symbols, $(\forall 0 \neq b \in B)\,(\exists \mu \in \mathscr{M}_+(B))\, \mu(b) > 0$. In the case when there is a finite strictly positive completely additive measure μ on B, the pair (B, μ) is referred to as a *normed Boolean algebra.* A normed Boolean algebra (B, μ) can be endowed with the metric $\rho(x, y) := \mu(x \vartriangle y)$ and it is not difficult to observe that the metric space (B, ρ) is complete.

1.2.10. ***Theorem.*** *For a complete Boolean algebra B the following are equivalent:*

(1) *B is a multinormed Boolean algebra;*

(2) *B is isomorphic to the Cartesian product of a family of normed Boolean algebras;*

(3) *there exists a strictly positive locally finite completely additive measure on B;*

(4) *B is isomorphic to the associated algebra $B(\Omega, \mathscr{B}, \mu)$ for some measure space $(\Omega, \mathscr{B}, \mu)$ possessing the direct sum property;*

(5) *the Stone space $\mathscr{S}(B)$ is hyperstonian.*

$\lhd$ (1) $\Rightarrow$ (2): According to the Exhaustion Principle (see 1.1.6) we may choose a partition of unity $(b_\xi)_{\xi\in\Xi}$ in B and a family of positive completely additive measures $(\mu_\xi)_{\xi\in\Xi}$ such that $\mu_\xi(b_\xi) > 0$ for all $\xi \in \Xi$. If B_ξ is the principal ideal generated by b_ξ and the restriction of μ_ξ to B_ξ is denoted by the same symbol then (B_ξ, μ_ξ) is a normed Boolean algebra and B is isomorphic to the Cartesian product of the family $\big((B_\xi, \mu_\xi)\big)_{\xi\in\Xi}$.

(2) $\Rightarrow$ (3): The measure on B with the required properties can be defined by

$$\mu(b) := \sum_{\xi\in\Xi} \mu_\xi(b \wedge b_\xi) \quad (b \in B).$$

(3) $\Rightarrow$ (4): Observe first that if $\Omega := \mathscr{S}(B)$ then $\mathrm{Clop}_\sigma(\Omega)$ consists of the sets $U \triangle N$ where $U \in \mathrm{Clop}(\Omega)$ and $N \subset \Omega$ is meager. Now, let ν be a strictly positive locally finite completely additive measure on $\mathrm{Clop}(\Omega)$. If $\mathscr{B} := \mathrm{Clop}_\sigma(\Omega)$ and the measure μ on $\mathscr{B}$ is defined by letting $\mu(U \triangle N) := \nu(U)$ then $(\Omega, \mathscr{B}, \mu)$ is a measure space possessing the direct sum property and Boolean algebras $B(\Omega, \mathscr{B}, \mu)$ and $\mathrm{Clop}(\Omega)$ are isomorphic, see [162].

(4) $\Rightarrow$ (5): According to 1.2.8 a normal measure $\bar{\mu}$ on $\mathrm{Clop}_\sigma(Q)$ which is strictly positive on $\mathrm{Clop}(Q)$ can be obtain by putting $\bar{\mu}(A) := \mu(\tau^{-1}(A))$ $(a \in \mathrm{Clop}_\sigma(Q))$.

(5) $\Rightarrow$ (1): Assume that $Q := \mathscr{S}(B)$ is hyperstonian, and let μ be a normal measure on $\mathrm{Clop}_\sigma(Q)$ strictly positive on $\mathrm{Clop}(Q)$. Take an arbitrary $b \in B$. Since μ is locally finite there is a clopen set V with $\mu(V) > 0$ and $V \subset \imath(b)$ where $\imath : B \to \mathrm{Clop}(Q)$ is the Stone transform. Putting $\mu_b(x) := \mu(V \cap \imath(x))$ $(x \in B)$ we arrive at a finite positive completely additive measure μ_b on B with $\mu_b(b) = \mu(V) > 0$. $\rhd$

1.3. Vector Lattices

In this section we give some preliminaries to the theory of vector lattices; a more explicit exposition may be found elsewhere [15, 23, 145, 162, 163, 262, 336, 341, 388, 409].

1.3.1. Let $\mathbb{F}$ be a linearly ordered field. An *ordered vector space over* $\mathbb{F}$ is a pair $(E, \leq)$, where E is a vector space over $\mathbb{F}$ and $\leq$ is an order in E satisfying the following conditions:

(1) if $x \leq y$ and $u \leq v$ then $x + u \leq y + v$ whatever $x, y, u, v \in E$ might be;

(2) if $x \le y$ then $\lambda x \le \lambda y$ for all $x, y \in E$ and $0 \le \lambda \in \mathbb{F}$.

Informally speaking, we may "sum inequalities in E and multiply them by positive members of $\mathbb{F}$." This circumstance is worded as follows: $\le$ is an order compatible with vector space structure or, briefly, $\le$ is a *vector order.*

Equipping a vector space E over $\mathbb{F}$ with some vector order is equivalent to indicating a set $E_+ \subset E$, called the *positive cone* of E, with the following properties:

$$E_+ + E_+ \subset E_+, \quad \lambda E_+ \subset E_+ \ (0 \le \lambda \in \mathbb{F}), \quad E_+ \cap -E_+ = \{0\}.$$

Moreover, the order $\le$ and the cone E_+ are connected by the relation

$$x \le y \Leftrightarrow y - x \in E_+ \quad (x, y \in E).$$

The elements of E_+ are called *positive.*

An ordered vector space E is called *Archimedean* if for any pair of elements $x, y \in E$ the relation $(\forall n \in \mathbb{N})\ nx \le y$ implies $x \le 0$. In the sequel, all ordered vector spaces are assumed Archimedean.

1.3.2. A *vector lattice* is an ordered vector space that is also a lattice. Thereby in a vector lattice there exist a least upper bound $\sup\{x_1, \dots, x_n\} := x_1 \vee \dots \vee x_n$ and a greatest lower bound $\inf\{x_1, \dots, x_n\} := x_1 \wedge \dots \wedge x_n$ for every finite set $\{x_1, \dots, x_n\} \subset E$. In particular, every element x of a vector lattice has the *positive part* $x^+ := x \vee 0$, the *negative part* $x^- := (-x)^+ := -x \wedge 0$, and the *modulus* $|x| := x \vee (-x)$.

Let E be a vector lattice. For all $x, y, z \in E$ the following relations are valid:

(1) $x = x^+ - x^-$, $|x| = x^+ + x^- = x^+ \vee x^-$;

(2) $x \le y \Leftrightarrow x^+ \le y^+ \,\&\, y^- \le x^-$;

(3) $x \vee y = \frac{1}{2}(x + y + |x - y|)$, $x \wedge y = \frac{1}{2}(x + y - |x - y|)$;

(4) $|x| \vee |y| = \frac{1}{2}(|x + y| + |x - y|)$, $|x| \wedge |y| = \frac{1}{2}(|x + y| - |x - y|)$;

(5) $x + y = x \vee y + x \wedge y$, $|x - y| = x \vee y - x \wedge y$;

(6) $x + y \vee z = (x + y) \vee (x + z)$, $x + y \wedge z = (x + y) \wedge (x + z)$;

(7) $x, y, z \in E_+ \Rightarrow (x + y) \wedge z \le (x \wedge z) + (y \wedge z)$;

(8) $|x - y| = |x \vee z - y \vee z| + |x \wedge z - x \wedge z|$.

Let (x_α) and (y_α) be families in E for which $\sup(a_\alpha)$ and $\inf(y_\alpha)$ exist. Then for any $z \in E$ the *infinite distributive laws* are valid:

(9) $z \wedge \sup_\alpha(x_\alpha) = \sup_\alpha(z \wedge x_\alpha)$, $z \vee \inf_\alpha(y_\alpha) = \inf_\alpha(z \vee y_\alpha)$.

For the same (x_α), (y_α), and z the following useful relations are also true;

(10) $z + \sup_\alpha(x_\alpha) = \sup_\alpha(z + x_\alpha)$;

(11) $z + \inf_\alpha(y_\alpha) = \inf_\alpha(z + y_\alpha)$;

(12) $\sup_\alpha(x_\alpha) = -\inf_\alpha(-x_\alpha)$.

An *order interval* in E is a set of the form $[a,b] := \{x \in E : a \le x \le b\}$, where $a, b \in E$. The following constantly used property of vector lattices is frequently called the *Riesz Decomposition Property*.

(13) $[0, x + y] = [0, x] + [0, y]$ $(x, y \in E_+)$;

We indicate only two corollaries of (13).

(14) $(x_1 + \dots + x_n) \wedge y \le x_1 \wedge y + \dots + x_n \wedge y$ $(x_k, y \in E_+)$;

(15) $\bigwedge_{k=1}^{n} \sum_{l=1}^{m} x_{k,l} \le \sum_{j \in J} x_{1,j(1)} \wedge \dots \wedge x_{n,j(n)}$

where $x_{k,l} \in E_+$ and J is the set of all functions $j : \{1, \dots, n\} \to \{1, \dots, m\}$.

1.3.3. Two elements x and y are called *disjoint* if $|x| \wedge |y| = 0$. The disjointness of x and y is denoted by $x \perp y$.

(1) *The following properties of disjointness are easy from* 1.3.2:

$$x \perp y \Leftrightarrow |x + y| = |x - y| \Leftrightarrow |x| \vee |y| = |x| + |y|;$$
$$x^+ \perp x^-; \ (x - x \wedge y) \perp (y - x \wedge y);$$
$$x \perp y \Rightarrow |x + y| = |x| + |y|, \ (x + y)^+ = x^+ + y^+, \ (x + y)^- = x^- + y^-.$$

Let $u \in E_+$ and $e \wedge (u - e) = 0$ for some $0 \le e \in E$. Then e is said to be a *fragment*, or a *part*, or a *component* of u, or a *unit element with respect to* u.

(2) *The set* $\mathfrak{E}(u)$ *of all fragments of* u *with the order induced by* E *is a Boolean algebra. The lattice operations in* $\mathfrak{E}(u)$ *are taken from* E *and the Boolean complement has the form* $e^* := u - e$ $(e \in \mathfrak{E}(u))$.

The disjoint complement $M^\perp$ of a nonempty set $M \subset E$ is defined as

$$M^\perp := \{x \in E : (\forall y \in M)\, x \perp y\}.$$

A nonempty set K in E meeting the identity $K = K^{\perp\perp}$ is called a *band* (a *component* in the Russian literature) of E. Every band of the form $\{x\}^{\perp\perp}$ with $x \in E$ is called *principal*.

(3) *The inclusion-ordered set of all bands of* E *is denoted by* $\mathfrak{B}(E)$ *and presents a complete Boolean algebra. The Boolean operations of* $\mathfrak{B}(E)$ *take the shape:*

$$L \wedge K = L \cap K, \quad L \vee K = (L \cup K)^{\perp\perp}, \quad L^* = L^\perp \quad (L, K \in \mathfrak{B}(E)).$$

The Boolean algebra $\mathfrak{B}(E)$ is the *base* of E. Let K be a band of the vector lattice E. If there is an element $\sup\{u \in K : 0 \le u \le x\}$ in E then it is called the *band projection* of x onto K and is denoted by $[K]x$ (or $\pi_K x$). Given an arbitrary $x \in E$, we put $[K]x := [K]x^+ - [K]x^-$. The band projection of an element $x \in E$ onto K exists if and only if x is representable as $x = y + z$ with $y \in K$ and $z \in K^\perp$. Furthermore, $y = [K]x$ and $z = [K^\perp]x$. Assume that to each element $x \in E$ there is a band projection onto K, then the operator $x \mapsto [K]x$ $(x \in E)$ is a linear idempotent and $0 \le [K]x \le x$ for all $0 \le x \in E$, called a *band projection* or an *order projection*.

(4) *The set $\mathfrak{P}(E)$ of all band projections ordered by $\pi \le \rho \Leftrightarrow \pi \circ \rho = \pi$ is a Boolean algebra. The Boolean operations of $\mathfrak{P}(E)$ take the shape*

$$\pi \wedge \rho = \pi \circ \rho, \quad \pi \vee \rho = \pi + \rho - \pi \circ \rho, \quad \pi^* = I_E - \pi \quad (\pi, \rho \in (E)).$$

The band projection onto a principal band is called *principal.*

(5) *The principal projection $\pi_u := [u] := [u^{\perp\perp}]$, where $0 \le u \in E$, can be calculated by the following rule simpler than that indicated above:*

$$\pi_u x = \sup\{x \wedge (nu) : n \in \mathbb{N}\}.$$

A vector lattice E is said to have the *projection property* (*principal projection property*) if every band (every principal band) in $\mathfrak{B}(E)$ is a projection band. Clearly, every K_σ-space has the principal projection property.

1.3.4. The order relation in a vector lattice generates different types of convergence. Let $(\mathrm{A}, \le)$ be an upward-directed set. A net $(x_\alpha) := (x_\alpha)_{\alpha \in \mathrm{A}}$ in E is called *increasing* (*decreasing*) if $x_\alpha \le x_\beta$ $(x_\beta \le x_\alpha)$ for $\alpha \le \beta$ $(\alpha, \beta \in \mathrm{A})$.

We say that a net (x_α) in a vector lattice E *o-converges* to $x \in E$ if there exists a decreasing net $(e_\beta)_{\beta \in \mathrm{B}}$ in E such that $\inf\{e_\beta : \beta \in \mathrm{B}\} = 0$ and for each $\beta \in \mathrm{B}$ there is $\alpha(\beta) \in \mathrm{A}$ with $|x_\alpha - x| \le e_\beta$ $(\alpha \ge \alpha(\beta))$. In this event, we call x the *o-limit* of the net (x_α) and write $x = o\text{-}\lim x_\alpha$ or $x_\alpha \overset{(o)}{\to} x$.

If a net (e_β) in this definition is replaced by a sequence $(\lambda_n e)_{n \in \mathbb{N}}$, where $0 \le v \in E_+$ and $(\lambda_n)_{n \in \mathbb{N}}$ is a numerical sequence with $\lim_{n \to \infty} \lambda_n = 0$, then we say that a net $(x_\alpha)_{\alpha \in \mathrm{A}}$ *converges relatively uniformly* or more precisely *e-uniformly* to $x \in E$. The elements e and x are called the *regulator of convergence* and the *r-limit* of (x_α), respectively. The notations $x = r\text{-}\lim_{\alpha \in \mathrm{A}} x_\alpha$ and $x_\alpha \overset{(r)}{\to} x$ are also frequent.

A net $(x_\alpha)_{\alpha \in \mathrm{A}}$ is called o-fundamental (r-fundamental with regulator e) if the net $(x_\alpha - x_\beta)_{(\alpha,\beta) \in \mathrm{A} \times \mathrm{A}}$ o-converges (respectively, r-converges with regulator e) to zero. A vector lattice is said to be *relatively uniformly complete* if every r-fundamental sequence is r-convergent.

The presence of order convergence in a vector lattice allows us to determined the sum of an infinite family $(x_\xi)_{\xi\in\Xi}$. Indeed, given $\theta := \{\xi_1, \dots, \xi_n\} \in \mathscr{P}_{\text{fin}}(\Xi)$, put $y_\theta := x_{\xi_1} + \dots + x_{\xi_n}$. So, we arrive at the $(y_\theta)_{\theta\in\Theta}$, where $\Theta := \mathscr{P}_{\text{fin}}(\Xi)$ is naturally ordered by inclusion. Assuming that there is some x satisfying $x = o\text{-}\lim_{\theta\in\Theta} y_\theta$, we call the family (x_ξ) *summable in order* or *order summable* or *o-summable.* The element x is the *o-sum* of (x_ξ) and we write $x = o\text{-}\sum_{\xi\in\Xi} x_\xi$. Obviously, if $x_\xi \geq 0$ $(\xi \in \Xi)$ then for the o-sum of the family (x_ξ) to exist it is necessary and sufficient that the net $(y_\theta)_{\theta\in\Theta}$ has the supremum, in which case $o\text{-}\sum_{\xi\in\Xi} x_\xi = \sup_{\theta\in\Theta} y_\theta$. If (x_ξ) is a disjoint family then

$$o\text{-}\sum_{\xi\in\Xi} x_\xi = \sup_{\xi\in\Xi} x_\xi^+ - \sup_{\xi\in\Xi} x_\xi^- .$$

1.3.5. (1) A linear subspace J of a vector lattice is called an *order ideal* or o-ideal (or, finally, just an *ideal,* when it is clear from the context what is meant) if the inequality $|x| \leq |y|$ implies $x \in J$ for arbitrary $x \in E$ and $y \in J$. Every order ideal of a vector lattice is a vector lattice. If an ideal J possesses the additional property $J^{\perp\perp} = E$ (or, which is the same, $J^\perp = \{0\}$) then J is referred to as an *order-dense ideal* of E (the term "foundation" is current in the Russian literature).

(2) Let J be an ideal of a vector lattice E. Then the factor space $\widetilde{E} := E/J$ is also a vector lattice, provided that the order on $\widetilde{E}$ is determined by the positive cone $\varphi(E_+)$, with $\varphi : E \to \widetilde{E}$ standing for the canonical factor mapping.

The factor lattice E/J is Archimedean if and only if N is closed under relative uniform convergence. If E is an f-algebra and J is a ring and order ideal then E/N is an f-algebra and φ is algebra homomorphism. If E is a K_σ-space and J is sequentially order-closed then E/J is a K_σ-space and φ is sequentially order continuous.

(3) Denote by $\mathscr{I}(E)$ the set of all order ideals of E ordered by inclusion. Then $\mathscr{I}(E)$ is a complete lattice, with the lattice operations defined as $I \wedge J := I \cap J$ and $I \vee J := I + J$. Moreover, the lattice $\mathscr{I}(E)$ is distributive. The sublattice $\mathscr{I}_p(E)$ of principal ideals is also distributive.

1.3.6. (1) A *vector sublattice* is a vector subspace $E_0 \subset E$ such that $x \wedge y, x \vee y \in E_0$ for all $x, y \in E_0$. We say that a sublattice E_0 is *minorizing* if, for every $0 \neq x \in E_+$, there exists an element $x_0 \in E_0$ satisfying the inequalities $0 < x_0 \leq x$. We say that E_0 is a *majorizing* or *massive* sublattice if, for every $x \in E$, there exists $x_0 \in E_0$ such that $x \leq x_0$. Thus, E_0 is a minorizing or majorizing sublattice if and only if $E_+ \setminus \{0\} = E_+ + E_{0+} \setminus \{0\}$ and $E = E_+ + E_0$, respectively.

(2) A set in E is called (*order*) *bounded* (or *o-bounded*) if it is included in some order interval. The o-ideal generated by the element $0 \leq u \in E$ is the set $E(u) := \bigcup_{n=1}^\infty [-nu, nu]$; clearly, $E(u)$ is the smallest o-ideal in E containing u.

If $E(u) = E$ then we say that u is a *strong unity* or *strong order-unity* and E is a *vector lattice of bounded elements.* If $E(u)^{\perp\perp} = E$ then we say that u is an *order-unity* or *weak order-unity.* It is evident that an element $u \in E_+$ is an order-unity if $\{u\}^{\perp\perp} = E$; i.e., if E lacks nonzero elements disjoint from u.

(3) An element $x \geq 0$ of a vector lattice is called *discrete* if $[0, x] = [0, 1]x$; i.e., if $0 \leq y \leq x$ implies $y = \lambda x$ for some $0 \leq \lambda \leq 1$. A vector lattice E is called *discrete* or *atomic* if, for every $0 \neq y \in E_+$, there exists a discrete element $x \in E$ such that $0 < x \leq y$. If E lacks nonzero discrete elements then E is said to be *continuous* or *diffuse.*

1.3.7. A vector lattice is said to be (*conditionally*) *order complete* if each nonvoid order bounded set in it has least upper and greatest lower bounds. If, in a vector lattice, least upper and greatest lower bounds exist only for countable bounded sets, then it is called *countably order complete.* An order complete vector lattice and a countably order complete vector lattice are frequently referred to as a *Dedekind complete* vector lattice and a *Dedekind σ-complete* vector lattice or, in the Russian literature, *K-space*(= *Kantorovich space*) and a *K_σ-space*, respectively.

We say that a *K-space* (K_σ-space) is *universally complete* or *extended* if its every subset (countable subset) of pairwise disjoint elements is bounded.

(1) Theorem. *Let E be an arbitrary K-space. Then E has the projection property and the operation of projecting onto bands determines the isomorphism $K \mapsto [K]$ of the Boolean algebras $\mathfrak{B}(E)$ and $\mathfrak{P}(E)$. If there is an order-unity $\mathbf{1}$ in E then the mappings $\pi \mapsto \pi\mathbf{1}$ from $\mathfrak{P}(E)$ into $\mathfrak{E}(E)$ and $e \mapsto \{e\}^{\perp\perp}$ from $\mathfrak{E}(E)$ into $\mathfrak{B}(E)$ are isomorphisms of Boolean algebras, too.*

A K-space is *o-complete* in the sense that every o-fundamental net in it is o-convergent. Each K_σ-space and, hence, a K-space is Archimedean. Henceforth all vector lattices are presumed to be Archimedean.

Consider an order-bounded net $(e_\alpha)_{\alpha\in\mathrm{A}}$ in a K-space E , and let $e \in E$.

(2) *An order-bounded net $(e_\alpha)_{\alpha\in\mathrm{A}}$ o-converges to e if and only if the relation o-$\lim_{\alpha\in\mathrm{A}}[d]\big[(|e_\alpha - e| - d)^+\big] = 0$ holds in the Boolean algebra $\mathfrak{P}(E)$ for all positive $d \in E$.*

$\lhd$ It is easy to verify the necessity of the criterion in question. To prove its sufficiency, assign $e_0 := \inf_{\alpha\in\mathrm{A}} \sup_{\beta\geqslant\alpha} |e_\beta - e|$. If the net $(e_\alpha)_{\alpha\in\mathrm{A}}$ does not converge to e then $e_0 > 0$ and, thus, there are $\pi \in \mathfrak{P}(E)$, $d \in D$, and $n \in \mathbb{N}$ such that $0 < \pi d/n < e_0$. Therefore, for each index $\alpha \in \mathrm{A}$, we have

$$\sup_{\beta\geqslant\alpha}[d]\big[|e_\beta - e| > d/n\big] = [d]\Big[\sup_{\beta\geqslant\alpha}|e_\beta - e| > d/n\Big] \geqslant \pi,$$

which contradicts the convergence of $[d]\big[|e_\alpha - e| > d/n\big]$ to zero. $\rhd$

(3) *Suppose that E is a K-space with order-unity $\mathbf{1}$; while $(e_\alpha)_{\alpha\in\mathrm{A}}$ is a bounded net in E, and $e \in E$. Then $o\text{-}\lim_{\alpha\in\mathrm{A}} e_\alpha = e$ if and only if the relation $o\text{-}\lim_{\alpha\in\mathrm{A}} \left[(|e_\alpha - e| - \mathbf{1}/n)^+\right] = 0$ holds in the Boolean algebra $\mathfrak{P}(E)$ for all $n \in \mathbb{N}$.*

1.3.8. Let E be a K_σ-space with order-unity $\mathbf{1}$. We call the projection of the order-unity to the band $\{x\}^{\perp\perp}$ the *trace* of x and denoted it by e_x. Therefore, $e_x := \sup\{\mathbf{1} \wedge (n|x|) : n \in \mathbb{N}\}$. The trace e_x serves both as an order-unity of $\{x\}^{\perp\perp}$ and a unit element of E. Given a real λ, denote the trace of the positive part of $\lambda\mathbf{1} - x$ by e_λ^x; i.e., $e_\lambda^x := e_{(\lambda\mathbf{1}-x)^+}$. The function $\lambda \mapsto e_\lambda^x$ $(\lambda \in \mathbb{R})$ arising in this case is called the *spectral function* or *characteristic* of x.

Theorem. *Let E be an arbitrary K_σ-space with order-unity $\mathbf{1}$ and $\mathbb{P}$ be a dense subfield of $\mathbb{R}$. The spectral function $\lambda \mapsto e_\lambda^x$ $(\lambda \in \mathbb{R})$ of $x \in E$ has the following properties:*

(1) $(\forall\,\lambda,\mu \in \mathbb{R}) \quad (\lambda \le \mu \Rightarrow e_\lambda^x \le e_\mu^x)$;

(2) $e_{+\infty}^x := \bigvee_{\mu\in\mathbb{P}} e_\mu^x = \mathbf{1},\ e_{-\infty}^x := \bigwedge_{\mu\in\mathbb{P}} e_\mu^x = \mathbf{0}$;

(3) $e_\lambda^x = \bigvee\{e_\mu^x : \mu \in \mathbb{P}, \mu < \lambda\} \quad (\lambda \in \mathbb{R})$;

(4) $x \le y \Leftrightarrow (\forall\,\lambda \in \mathbb{P}) \quad (e_\lambda^y \le e_\lambda^x)$;

(5) $e_\lambda^{x+y} = \bigvee\{e_\mu^x \wedge e_\nu^y : \mu,\nu \in \mathbb{P},\ \mu+\nu = \lambda\}$;

(6) $e_\lambda^{x\cdot y} = \bigvee\{e_\mu^x \wedge e_\nu^y : 0 \le \mu,\ \nu \in \mathbb{P},\ \mu\nu = \lambda\} \quad (x \ge 0,\ y \ge 0)$;

(7) $e_\lambda^{-x} = \bigvee\{\mathbf{1} - e_{-\mu}^x : \mu \in \mathbb{P},\ \mu < \lambda\} = (\mathbf{1} - e_{-\lambda}^x) \cdot e_{(x+\lambda\mathbf{1})}$;

(8) $e_\lambda^{x\wedge y} = e_\lambda^x \vee e_\lambda^y;\quad e_\lambda^{x\vee y} = e_\lambda^x \wedge e_\lambda^y \quad (\lambda \in \mathbb{R})$;

(9) $x = \inf(A) \Leftrightarrow (\forall\,\lambda \in \mathbb{P})\,(e_\lambda^x = \bigvee\{e_\lambda^a : a \in A\})$;

(10) $e_\lambda^{|x|} = e_\lambda^x \wedge (\mathbf{1} - e_{-\lambda}^x) \wedge e_{x+\lambda\mathbf{1}} \quad (\lambda \in \mathbb{R})$;

(11) $e_\lambda^{\alpha x} = e_{\lambda/\alpha}^x\ (\alpha > 0),\ e_\lambda^{\alpha x} = e_{-\lambda/\alpha}^{-x}\ (\alpha < 0) \quad (\lambda \in \mathbb{R})$;

(12) $e_\lambda^{cx} = c \wedge e_\lambda^x + c^*\ (\lambda > 0),\ e_\lambda^{cx} = c \wedge e_\lambda^x\ (\lambda \le 0) \quad (c \in \mathfrak{C}(E))$.

A formula for $x = \sup(A)$ similar to (9) is generally not true. Nevertheless the following more complicated formula is valid:

$$x = \sup(A) \Leftrightarrow (\forall\,\lambda \in \mathbb{P}) \left(e_\lambda^x = \bigvee_{\nu<\lambda} \bigwedge \{e_\nu^a : a \in A\} \right).$$

If the function $\lambda \mapsto \inf\{e_\lambda^a : a \in A\}$ satisfies (3) then the spectral function for $x = \sup(A)$ is calculated by (9) in which sup and inf are interchanged.

1.3.9. (1) Freudenthal Spectral Theorem. *Let E be an arbitrary K_σ-space with an order-unity* **1**. *Every element $x \in E$ admits the representation*

$$x = \int_{-\infty}^{\infty} \lambda \, de_\lambda^x,$$

where the integral is understood to be the **1***-uniform limit of the integral sums*

$$x(\beta) := \sum_{n\in\mathbb{Z}} \tau_n (e_{t_{n+1}}^x - e_{t_n}^x), \quad t_n \le \tau_n \le t_{n+1},$$

as $\delta(\beta) := \sup_{n\in\mathbb{Z}}(t_{n+1} - t_n) \to 0$, where $\beta := (t_n)_{n\in\mathbb{Z}}$ is a partition of the real line.

In particular, the Freudenthal Spectral Theorem states that if E is a K_σ-space and $e \in E_+$, then every element $x \in E(e)$ can be e-uniformly approximated by linear combinations of fragments of e, i.e. by elements of the form $\sum_{k=1}^n \lambda_k e_k$, where $\lambda_1, \dots, \lambda_n \in \mathbb{R}$ and $e_1, \dots, e_n \in \mathfrak{C}(e)$. In the case when the latter holds in a vector lattice E we say that E possesses the *weak Freudenthal property*. It may happen that every element $x \in E(e)$ can be e-uniformly approximated by linear combinations $\sum_{k=1}^n \lambda_k \pi_k e$, where $\lambda_1, \dots, \lambda_n \in \mathbb{R}$ and $\pi_1, \dots, \pi_n \in \mathfrak{P}(E)$. Then a vector lattice E is said to possesses the *strong Freudenthal property*.

To characterize the spaces with strong Freudenthal property we need a definition. Two elements of a vector lattice E are *completely disjoint* if they lie in two disjoint projection bands of E. Clearly, a vector lattice with the principal projection property possesses the strong Freudenthal property. The converse is false.

(2) Theorem. *A vector lattice E possesses the strong Freudenthal property if and only if any two disjoint elements in E are completely disjoint. A vector lattice E possesses the weak Freudenthal property if and only if for any two elements $e, d \in E$ there exist disjoint elements $e_0 \in [0, e]$ and $d_0 \in [0, d]$ with $E(e + d) = E(e_0 + d_0)$.*

1.3.10. An *ordered algebra* over $\mathbb{F}$ is an ordered vector space E over $\mathbb{F}$ which is simultaneously an algebra over the same field and satisfies the following condition: if $x \ge 0$ and $y \ge 0$ then $xy \ge 0$ whatever $x, y \in E$ might be. To characterize the positive cone E_+ of an ordered algebra E, we must add to what was said in 1.3.1 the property $E_+ \cdot E_+ \subset E_+$. We say that E is a *lattice-ordered algebra* if E is a vector lattice and an ordered algebra simultaneously. A lattice-ordered algebra is an *f-algebra* if, for all $a, x, y \in E_+$, the condition $x \perp y = 0$ implies that $(ax) \perp y$ and $(xa) \perp y$. An f-algebra is called *faithful* or *exact* if, for arbitrary elements x and y, $xy = 0$ implies $x \perp y$. It is easy to show that an f-algebra is faithful if and only if it lacks nonzero nilpotent elements. The faithfulness of an f-algebra is equivalent to absence of any strictly positive element with zero square.

1.3.11. A *complex vector lattice* is defined to be the complexification $E \oplus iE$ (with i standing for the imaginary unity) of a real vector lattice E. Often it is additionally required that the modulus

$$|z| := \sup\{\mathrm{Re}(e^{i\theta} z) : 0 \leq \theta \leq \pi\}$$

exists for every element $z \in E \oplus iE$. In the case of a K-space or an arbitrary Banach lattice this requirement is automatically satisfied, since a complex K-space is the complexification of a real K-space. Speaking about order properties of a complex vector lattice $E \oplus iE$, we mean its real part E. The concepts of sublattice, ideal, band, projection, etc. are naturally translated to the case of a complex vector lattice by appropriate complexification.

1.4. Representation of Vector Lattices

In this section, we give some most important examples of function vector lattices. The question is studied of representing general vector lattices as those of measurable or continuous functions.

1.4.1. A special role in the theory of vector lattices is played by the spaces of continuous functions assuming possibly infinite values on a nowhere-dense set depending on a function. Before introducing these spaces, we need some preliminaries.

Given a function $f : Q \to \overline{\mathbb{R}}$ and $\lambda \in \overline{\mathbb{R}}$, put

$$\{f < \lambda\} := \{t \in Q : f(t) < \lambda\}, \quad \{f \leq \lambda\} := \{t \in Q : f(t) \leq \lambda\}.$$

(1) *Assume that Q is a topological space, Λ is a dense set in $\overline{\mathbb{R}}$, and $\lambda \mapsto U_\lambda$ $(\lambda \in \Lambda)$ is an increasing mapping from Λ to the inclusion ordered set $\mathscr{P}(Q)$. Then the following are equivalent:*

(a) *There is a unique continuous function $f : Q \to \overline{\mathbb{R}}$ satisfying*

$$\{f < \lambda\} \subset U_\lambda \subset \{f \leq \lambda\} \quad (\lambda \in \Lambda),$$

(b) *If $\lambda, \mu \in \Lambda$, and $\lambda < \mu$ then*

$$\mathrm{cl}(U_\lambda) \subset \mathrm{int}(U_\mu).$$

$\lhd$ The implication (a) $\Rightarrow$ (b) is evident.

Prove (b) $\Rightarrow$ (a). To this end, given $t \in Q$, put $f(t) := \inf\{\lambda \in \Lambda : t \in U_\lambda\}$. For the so-defined $f : Q \to \overline{\mathbb{R}}$, we easily see that $\{f < \lambda\} \subset U_\lambda \subset \{f \leq \lambda\}$. It is also clear that

$$\{f < \lambda\} = \bigcup\{U_\mu : \mu < \lambda \,\&\, \mu \in \Lambda\}, \quad \{f \leq \lambda\} = \bigcap\{U_\nu : \lambda < \nu \,\&\, \nu \in \Lambda\}.$$

Note that by now we have used only the fact that $\lambda \mapsto U_\lambda$ is an increasing mapping.

Consider the mappings

$$\lambda \mapsto V_\lambda := \operatorname{int}(U_\lambda), \quad \lambda \mapsto W_\lambda := \operatorname{cl}(U_\lambda) \quad (\lambda \in \Lambda).$$

These mappings are also increasing. So, the above implies that there are functions g and $h : Q \to \overline{\mathbb{R}}$ such that

$$\{g < \lambda\} \subset V_\lambda \subset \{g \leq \lambda\}, \quad \{h < \lambda\} \subset W_\lambda \subset \{h \leq \lambda\} \quad (\lambda \in \Lambda).$$

From the definition of W_λ it follows that $U_\mu \subset W_\lambda$ for $\mu < \lambda$. Since Λ is dense in $\mathbb{R}$, to all $t \in Q$ and $\nu > f(t)$ there are $\lambda,\ \mu \in \Lambda$ such that $f(t) < \mu < \lambda < \nu$, so that $t \in U_\mu \subset W_\lambda$ and $h(t) < \lambda < \nu$. Letting ν tend to $f(t)$, obtain $h(t) \leq f(t)$. The same inequality is immediate for $f(t) = +\infty$. By analogy, $V_\mu \subset U_\lambda$ for $\mu < \lambda$. Hence, $f(t) \leq g(t)$ for all $t \in Q$. Writing (b) as $W_\mu \subset V_\lambda$ $(\mu < \lambda)$, and arguing as above, conclude that $g(t) \leq h(t)$ for all $t \in Q$. Therefore, $f = g = h$. The fact that f is continuous follows from the equalities

$$\{f < \lambda\} = \{g < \lambda\} = \bigcup\{V_\mu : \mu < \lambda,\ \mu \in \Lambda\},$$
$$\{f \leq \lambda\} = \{h \leq \lambda\} = \bigcap\{W_\mu : \mu > \lambda,\ \mu \in \Lambda\},$$

since V_μ is open whereas W_μ is closed for all $\mu \in \Lambda$. $\rhd$

(2) *Let Q be a quasiextremal compact space. Assume that Q_0 is an open dense F_σ-set in Q and $f : Q_0 \to \mathbb{R}$ is a continuous function. Then there is a unique continuous function $\bar{f} : Q_0 \to \overline{\mathbb{R}}$ such that $f(t) = \bar{f}(t)$ $(t \in Q_0)$.*

$\lhd$ Indeed, if $U_\mu := \operatorname{cl}(\{f < \mu\})$ then the mapping $\mu \mapsto U_\mu$ $(\mu \in \mathbb{R})$ increases and meets the condition (b) of (1). Therefore, there is a unique function $\bar{f} : Q \to \overline{\mathbb{R}}$ satisfying $\{\bar{f} < \mu\} \subset U_\mu \subset \{\bar{f} \leq \mu\}$ $(\mu \in \mathbb{R})$. Obviously, in this case $\bar{f} \restriction Q_0 = f$, i.e. the restriction of $\bar{f}$ to Q_0 coincides with f. $\rhd$

1.4.2. Let Q be a quasiextremal compact space. Denote by $C_\infty(Q)$ the set of all continuous functions $x : Q \to \overline{\mathbb{R}}$ assuming the values $\pm\infty$ possibly on a nowhere-dense set. Order $C_\infty(Q)$ by assigning $x \leq y$ whenever $x(t) \leq y(t)$ for all $t \in Q$. Then, take $x, y \in C_\infty(Q)$ and put $Q_0 := \{|x| < +\infty\} \cap \{|y| < +\infty\}$. In this case Q_0 is open and dense in Q. According to 1.4.1 (2), there is a unique continuous function $z : Q \to \overline{\mathbb{R}}$ such that $z(t) = x(t) + y(t)$ for $t \in Q_0$. It is this function z that we declare the sum of x and y.

In an analogous way we define the product of a pair of elements. Identifying the number λ with the identically λ function on Q, we obtain the product of $x \in C_\infty(Q)$ and $\lambda \in \mathbb{R}$.

Clearly, the space $C_\infty(Q)$ with the operations and order introduced above is a vector lattice and a faithful f-algebra. The identically one function $\mathbf{1}$ is a ring and order-unity. The order ideal generated by $\mathbf{1}$ is the space $C(Q)$ of all continuous numeric functions on Q.

(1) *The space $C_\infty(Q)$ is a universally σ-complete K_σ-space.*

$\triangleleft$ Take a bounded increasing sequence (x_n) of elements of $C_\infty(Q)$. Put $V_\lambda := \bigcap_{n=1}^\infty \{x_n \leq \lambda\}$ and $U_\lambda := \operatorname{int} V_\lambda$. Then V_λ is a closed G_δ-set and, by assumption, U_λ is a clopen set. According to 1.4.1 (2) there is a unique function $x : Q \to \overline{\mathbb{R}}$ such that $\{x < \lambda\} \subset U_\lambda \subset \{x \leq \lambda\}$ for all $\lambda \in \mathbb{R}$. Now, it is not difficult to check that $x = \sup_n x_n$. Universally σ-completeness is obvious. $\triangleright$

(2) *The base of the vector lattice $C_\infty(Q)$ is isomorphic to the Boolean algebra of all regular open (closed) subsets of Q.*

$\triangleleft$ The same argument as used in (1) works. $\triangleright$

(3) *The space $C_\infty(Q)$ is an order complete vector lattice if and only if Q is extremal.*

$\triangleleft$ In the case when Q is extremal, the order completeness of $C_\infty(Q)$ may be proved as in (1). The converse follows from the Ogasawara Theorem 1.2.4, since Boolean algebras $\operatorname{Clop}(Q)$ and $\mathfrak{E}(\mathbf{1})$ are isomorphic. $\triangleright$

According to these arguments we may describe suprema and infima in $C_\infty(Q)$ as follows. If (x_α) is an order-bounded family in $C_\infty(Q)$ then $x = \sup_\alpha x_\alpha$ if and only if there exists a comeager subset $Q_0 \subset Q$ such that $x(t) = \sup_\alpha x_\alpha(t)$ for all $t \in Q_0$.

1.4.3. According to 1.3.8, to each element of a K_σ-space with order-unity there corresponds its spectral function; moreover, the operations transform in a rather definite way. This circumstance suggests that an arbitrary K_σ-space with unity can be realized as a space of "abstract spectral functions." We will expatiate upon this.

A *resolution of unity* or *resolution of the identity* in a Boolean algebra B is defined as a mapping $e : \mathbb{R} \to B$ satisfying the conditions

(1) $s \leq t \to e(s) \leq e(t) \quad (s, t \in \mathbb{R})$;

(2) $\bigvee_{t\in\mathbb{R}} e(t) = \mathbf{1}, \quad \bigwedge_{t\in\mathbb{R}} e(t) = \mathbf{0}$;

(3) $\bigvee_{s\in\mathbb{R}, s<t} e(s) = e(t) \quad (t \in \mathbb{R})$.

Let $\mathfrak{R}(B)$ be the set of all resolutions of unity in B. Introduce some order relation by the formula

$$e' \leq e'' \Leftrightarrow (\forall t \in \mathbb{R})(e''(t) \leq e'(t)) \quad (e', e'' \in \mathfrak{R}(B)).$$

Further, suppose that B is a σ-algebra and choose any dense countable subfield $\mathbb{P}$ of $\mathbb{R}$. By property (3), every resolution of unity is uniquely determined by its values on $\mathbb{P}$.

Given $e', e'' \in \mathfrak{R}(B)$, we may define the mapping

$$e : t \mapsto \bigvee\{e'(r) \wedge e''(s) : r, s \in \mathbb{P},\ r + s = t\} \quad (t \in \mathbb{P}),$$
$$e : t \mapsto \bigvee\{e(s) : s \in \mathbb{P},\ s < t\} \quad (t \in \mathbb{R})$$

which is obviously a resolution of unity in B. Putting $e' + e'' := e$, we obtain the structure of a commutative group in $\mathfrak{R}(B)$; moreover, $-e(t) = \bigvee\{\mathbf{1} - e(-s) : s \in \mathbb{P},\ s < t\}$ and the zero element $\bar{\mathbf{0}}$ have the form: $\bar{\mathbf{0}}(t) := \mathbf{1}$ if $t > 0$ and $\bar{\mathbf{0}}(t) := \mathbf{0}$ if $t \leq 0$. Set $\bar{\mathbf{1}}(t) := \mathbf{1}$ if $t > 1$ and $\bar{\mathbf{1}}(t) := \mathbf{0}$ if $t \leq 1$. Finally, define the product of an element $e \in \mathfrak{R}(B)$ and a real $\alpha \in \mathbb{R}$ by the rules

$$(\alpha e)(t) := e(t/\alpha) \quad (\alpha > 0,\ t \in \mathbb{R}),$$
$$(\alpha e)(t) := (-e)(-t/\alpha) \quad (\alpha < 0,\ t \in \mathbb{R}).$$

To each element $b \in B$ assign the resolution of unity $\bar{b}$ defined as $\bar{b}(t) := \mathbf{1}$ if $t > 1$, $\bar{b}(t) := b^* := \mathbf{1} - b$ if $0 \leq t < 1$, and $\bar{b}(t) := \mathbf{0}$ if $t \leq 0$.

Theorem. *Let B be a complete (σ-complete) Boolean algebra. The set $\mathfrak{R}(B)$ with the above-introduced operations and order represents a universally complete K-space (a universally σ-complete K_σ-space). The element $\bar{\mathbf{1}}$ serves as an order-unity and the mapping $b \mapsto \bar{b}$ ($b \in B$) is an isomorphism of Boolean algebras B and $\mathfrak{E}(\bar{\mathbf{1}})$.*

$\triangleleft$ The proof is elementary. Most effort is put into the routine calculation with resolutions of unity. $\triangleright$

1.4.4. Theorem. *Let E be a K_σ-space with order-unity $\mathbf{1}$ and $B := \mathfrak{E}(\mathbf{1})$. The mapping sending an element $x \in E$ to the spectral function $\lambda \mapsto e_\lambda^x$ ($\lambda \in \mathbb{R}$) is an isomorphism of E onto an order-dense ideal in $\mathfrak{R}(B)$. If E is universally σ-complete then E and $\mathfrak{R}(B)$ are isomorphic.*

$\triangleleft$ Denote by h the mapping from E to $\mathfrak{R}(B)$ we are interested in. By virtue of Theorem 1.3.8 h is linear and preserves order. Moreover, h is injective according to the Freudenthal Spectral Theorem 1.3.9. Prove that $h(E)$ is an order ideal in $\mathfrak{R}(B)$. Assume $0 \leq s \leq h(y)$ where $s \in \mathfrak{R}(B)$ and $y \in E$. Let $\lambda > \mu$, $b := s(\lambda) - s(\mu)$, and $y_0 := \mu b$. It can be easily checked that $\mu\bar{b} \leq \bar{b}s$, since

$$(\mu\bar{b})(t) = \begin{cases} \mathbf{1}, & \text{if } t > \mu, \\ b^*, & \text{if } 0 < t \leq \mu, \\ \mathbf{0}, & \text{if } t \leq 0; \end{cases} \qquad (\bar{b}s)(t) = \begin{cases} b \wedge s(t) + b^*, & \text{if } t > 0, \\ \mathbf{0}, & \text{if } t \leq 0. \end{cases}$$

Therefore, $h(y_0) = \mu h(b) = \mu\bar{b} \leq \bar{b}s \leq s \leq h(y)$ and we obtain $y_0 \leq y$. Finally, we deduce $\lambda(s(\lambda) - s(\mu)) = y_0 + (\lambda - \mu)b \leq y + (\lambda - \mu)\mathbf{1}$.

Now, consider a partition of the real axis of the form $\beta_N := (t_n)_{n\in\mathbb{Z}}$ where $t_n := n/N$ and $N \in \mathbb{N}$. The disjoint sum

$$\bar{x}(\beta_N) := \sum_{n\in\mathbb{Z}} t_{n+1}(s(t_{n+1}) - s(t_n))$$

exists in E, since $t_{n+1}(s(t_{n+1}) - s(t_n)) \le y + (1/N)\mathbf{1}$ as was proven above. Denote by A the sequence of all elements $\bar{x}(\beta_N)$. Every element of the form

$$\underline{x}(\beta_N) := \sum_{n\in\mathbb{Z}} t_n(s(t_{n+1}) - s(t_n))$$

is a lower bound of A. Therefore, there exists $x := \inf(A) := \inf\{\bar{x}(\beta)\}$. Note that

$$e_\lambda^{\bar{x}(\beta)} = \bigvee\{s(t_n) : t_n < \lambda\}.$$

Hence, by 1.3.8 (8), we infer

$$e_\lambda^x = \bigvee_{a\in A} e_\lambda^a = \bigvee_{t\in\mathbb{R},t<\lambda} s(t) = s(\lambda) \quad (\lambda \in \mathbb{R}).$$

Thereby $h(x) = s$ and $h(E)$ is an order ideal. Moreover, it can be easily observed that for a resolution of unity $0 \le s \in \mathfrak{R}(B)$ the following equivalences are true:

$$s \perp h(E) \Leftrightarrow (\forall b \in B)\, s \perp h(b) \Leftrightarrow (\forall b \in B)\, s \perp \bar{b}$$
$$\Leftrightarrow (\forall b \in B)\,(\forall t \ge 0)\, s(t) \vee \bar{b}(t) = \mathbf{1} \Leftrightarrow (\forall b \in B)\,(\forall 0 \le t < 1)\, s(t) \vee b^* = \mathbf{1}$$
$$\Leftrightarrow (\forall b \in B)\,(\forall 0 \le t < 1)\, s(t) \ge b \Leftrightarrow (\forall 0 \le t < 1) s(t) = \mathbf{1} \Leftrightarrow s = \mathbf{0}.$$

This proves that $h(E)$ is order dense in $\mathfrak{R}(B)$.

Thus, it remains to verify that h is surjective in the case when E is universally σ-complete. For an arbitrary resolution of unity s in the Boolean algebra $\mathbb{B}$ the disjoint sum

$$\bar{x}(\beta_N) := \sum_{n\in\mathbb{Z}} t_{n+1}(s(t_{n+1}) - s(t_n))$$

exists in the universally σ-complete K_σ-space E. The same argument as above applies, and we again arrive at $h(x) = s$ with $x := \inf\{\bar{x}(\beta)\}$. ▷

1.4.5. *Theorem.* *Suppose that E is a universally σ-complete K_σ-space with order-unity $\mathbf{1}$, Q is the Stone space of the Boolean algebra $\mathfrak{E}(\mathbf{1})$ and $\pi \mapsto \hat{\pi}$ is the canonical isomorphism from $\mathfrak{E}(\mathbf{1})$ onto $\mathrm{Clop}(Q)$. Then E is linearly and latticially* *isomorphic to $C_\infty(Q)$. Moreover, such an isomorphism may be carried out by sending $x \in E$ to $\widehat{x} : Q \to \overline{\mathbb{R}}$ by the rule*

$$\widehat{x}(q) := \inf\{\lambda \in \mathbb{R} : e_\lambda^x \in q\}.$$

$\lhd$ As was shown in 1.4.4, the K_σ-space E is isomorphic to the space of all B-valued spectral functions, with the function $\lambda \mapsto e_\lambda^x$ $(\lambda \in \mathbb{R})$ corresponding to $x \in E$. Assume that a clopen sunset U_λ of the Stone space Q corresponds to $e_\lambda^x \in B$. Then, by virtue of 1.4.1 (2), to every element $x \in E$ there corresponds a unique continuous function $\widehat{x} : Q \to \overline{\mathbb{R}}$ such that $\{\widehat{x} < \lambda\} \subset U_\lambda \subset \{\widehat{x} \le \lambda\}$. In this case, however,

$$\widehat{x}(q) = \inf\{\lambda \in \mathbb{R} : q \in U_\lambda\} = \inf\{\lambda \in \mathbb{R} : e_\lambda^x \in q\}.$$

The formulas $\bigwedge\{e_\lambda^x\} = \mathbf{0}$ and $\bigvee\{e_\lambda^x\} = \mathbf{1}$ (cf. 1.3.6 (2)) imply that the interior of the closed set $\bigcap\{U_\lambda : \lambda \in \mathbb{R}\}$ is empty while the open set $\bigcup\{U_\lambda : \lambda \in \mathbb{R}\}$ is dense in Q. Therefore, the function $\widehat{x}$ may assume the values $\pm\infty$ only on a nowhere-dense set and so $\widehat{x} \in C_\infty(Q)$.

We omit the elementary demonstration of the fact that $x \mapsto \widehat{x}$ is a linear and lattice isomorphism. $\rhd$

1.4.6. We now list a few corollaries to the above theorems.

(1) *Let E be an arbitrary K-space. Assume further that $\{e_\xi\}_{\xi\in\Xi}$ is a total disjoint positive family in E. Denote by Q the Stone space of the Boolean algebra of bands of $\mathfrak{B}(X)$. Then there is a unique linear and lattice isomorphism of E onto an order-dense ideal of the K-space $C_\infty(Q)$ such that e_ξ transforms into the characteristic function of a clopen subset Q_ξ of Q. This isomorphism sends $x \in E$ to the function $\widehat{x} : Q \to \overline{\mathbb{R}}$ acting by the rule*

$$\widehat{x}(q) := \inf\left\{\lambda \in \mathbb{R} : \{e_\lambda^\xi\}^{\perp\perp} \in q\right\} \quad (q \in Q_\xi),$$

where $\lambda \mapsto (e_\lambda^\xi)$ is the spectral function of the band projection of x to $\{e_\xi\}^{\perp\perp}$ with respect to the order-unity e_ξ.

(2) *A space E is a universally complete K-space (K-space of bounded elements) if and only if the image of E under the above isomorphism is the whole of $C_\infty(Q)$ (the subspace $C(Q)$ of all continuous functions on Q).*

(3) *Every universally complete K-space E can be endowed with a multiplication making E into a faithful commutative f-algebra. Moreover, if we require the order-unity fixed in E to be the multiplication unity, then this multiplication in E is unique.*

(4) *Let D be a subset of a K-space E with an upper bound $e \in E$. Then $\inf D = 0$ if and only if for every $o < \varepsilon \in \mathbb{R}$ there exists a partition of unity $(\pi_\xi)_{\xi\in\Xi}$ in $\mathfrak{P}(E)$ and a family $(d_\xi)_{\xi\in\Xi}$ in D such that $\pi_\xi d_\xi \le \varepsilon e$ for all $\xi \in \Xi$.*

$\lhd$ Without loss of generality we may assume that $\{e\}^{\perp\perp} = E = C_\infty(Q)$. If $\inf D = 0$ then, in view of 1.4.2, there is a comeager subset $Q_0 \subset Q$ such that $\inf\{d(t) : d \in D\} = 0$ for all $t \in Q_0$. Put $R_d := \operatorname{cl}\left(\{t \in Q : d(t) < \varepsilon e(t)\}\right)$. Then R_d is a clopen set and the least upper bound of the family $(R_d)_{d \in Q_0}$ in $\operatorname{Clop}(Q)$ is Q ($\bigcup_{d\in D} R_d$ contains Q_0 and is dense in Q). By the Exhaustion Principle there is a partition of unity $(Q_\xi)_{\xi\in\Xi}$ in $\operatorname{Clop}(Q)$ with the property: for each $\xi \in \Xi$ there is some $d_\xi \in D$ such that $Q_\xi \subset R_{d_\xi}$. Take the band projection corresponding to Q_ξ as π_ξ and we are done. The converse is evident. $\rhd$

1.4.7. (1) Let $(\Omega, \mathscr{A})$ be a *measurable space*, i.e., Ω is a nonempty set and $\mathscr{A}$ is a σ-algebra of its subsets. Denote by $\mathscr{M}(\Omega, \mathscr{A})$ the set of all real (complex) measurable functions on Ω and furnish $\mathscr{M}(\Omega, \mathscr{A})$ with the pointwise operations and order induced from $\mathbb{R}^\Omega$ (from $\mathbb{C}^\Omega$, with $\mathbb{C}$ the complex numbers). Choose some σ-complete ideal $\mathscr{N}$ of the algebra $\mathscr{A}$. Let N comprise the functions $f \in \mathscr{M}(\Omega, \mathscr{A})$ such that $\{t \in \Omega : f(t) \neq 0\} \in \mathscr{N}$. Assign $M(\Omega, \mathscr{A}, \mathscr{N}) := \mathscr{M}(\Omega, \mathscr{A})/N$. Then $\mathscr{M}(\Omega, \mathscr{A})$ and $M(\Omega, \mathscr{A}, \mathscr{N})$ are real (complex) K_σ-spaces and f-algebras at the same time.

(2) Consider an arbitrary measure space (Ω, Σ, μ), see 1.1.7 (5). Denote $L^0(\Omega, \mathscr{A}, \mu) := M(\Omega, \mathscr{A}, \mu) := M(\Omega, \mathscr{A}, \mu^{-1}(0))$. Again, $L^0(\Omega, \mathscr{A}, \mu)$ is a vector lattice. The coset $\mathbf{1}$ of the identically one function on Ω is an order and ring unity in $M(\Omega, \mathscr{A}, \mu)$. Denote by $L^\infty(\Omega, \mathscr{A}, \mu)$ the principal ideal in $L^0(\Omega, \mathscr{A}, \mu)$ generated by $\mathbf{1}$. Let $\mathscr{L}^\infty(\Omega) := \mathscr{L}^\infty(\Omega, \mathscr{A}, \mu)$ stand for the part of $\mathscr{M}(\Omega, \mathscr{A}$ comprising all essentially bounded functions.

If μ is a finite or σ-finite measure then $L^\infty(\Omega, \mathscr{A}, \mu)$ and $L^0(\Omega, \mathscr{A}, \mu)$ are order complete; moreover $L^0(\Omega, \mathscr{A}, \mu)$ is also universally complete. In general, the order completeness of $L^0(\Omega, \mathscr{A}, \mu)$ relates to the direct sum property for μ (see 1.1.7 (5) and 1.4.8 below).

(3) The space $L^0(\Omega, \mathscr{A}, \mu)$ is diffuse if and only if μ has no atoms. Recall that an *atom* of a measure μ is a set $A \in \mathscr{A}$ such that $\mu(A) > 0$ and $A' \in \mathscr{A}$, $A' \subset A$, implies either $\mu(A') = 0$ or $\mu(A') = \mu(A)$. The discreteness of $L^0(\Omega, \mathscr{A}, \mu)$ is equivalent to the fact that the measure μ is *purely atomic*, i.e., every set of nonzero measure contains an atom of μ. The equivalence class containing the identically unity function is an order and ring unity in $L^0(\Omega, \mathscr{A}, \mu)$.

1.4.8. A mapping $\rho : L^\infty(\Omega) \to \mathscr{L}^\infty(\Omega)$ is said to be a *lifting* of $L^\infty(\Omega)$ if for all $\alpha, \beta \in \mathbb{R}$ and $f, g \in L^\infty(\Omega)$ the following are true:

(a) $\rho(f) \in f$ and $\operatorname{dom}(\rho(f)) = \Omega$;

(b) if $f \leqslant g$, then $\rho(f) \leqslant \rho(g)$ everywhere on Ω;

(c) $\rho(\alpha f + \beta g) = \alpha\rho(f) + \beta\rho(g)$, $\rho(fg) = \rho(f)\rho(g)$, $\rho(f \vee g) = \rho(f) \vee \rho(g)$, $\rho(f \wedge g) = \rho(f) \wedge \rho(g)$;

(d) $\rho(0) = 0$ and $\rho(\mathbf{1}) = \mathbf{1}$ everywhere on Ω.

Theorem. *For a measure space $(\Omega, \mathscr{A}, \mu)$, the following are equivalent:*

(1) *$(\Omega, \mathscr{A}, \mu)$ possesses the direct sum property;*

(2) *$L^\infty(\Omega)$ admits a lifting;*

(3) *$L^0(\Omega, \mathscr{A}, \mu)$ is order complete and universally complete.*

The base of the K-space $L^0(\Omega, \mathscr{A}, \mu)$ is isomorphic to the Boolean algebra $B(\Omega, \mathscr{A}, \mu)$ of measurable sets modulo zero-measure sets.

1.4.9. Suppose that $(\Omega, \mathscr{A}, \mu)$ is a measure space with a direct sum property. Let ρ be a lifting of $L^\infty(\Omega, \mathscr{A}, \mu)$. and $\tau : \Omega \to Q$ be the corresponding canonical immersion of Ω into the Stone space Q of the Boolean algebra $B(\Omega, \mathscr{A}, \mu)$, see Denote by τ^* the mapping that sends each function $f \in C_\infty(Q)$ to the equivalence class of the measurable function $f \circ \tau$.

Theorem. *The mapping τ^* is a linear and order isomorphism from $C_\infty(Q)$ onto $L^0(\Omega, \mathscr{A}, \mu)$. The image of $C(Q)$ under the isomorphism τ^* coincides with $L^\infty(\Omega, \mathscr{A}, \mu)$. For $u \in L^0(\Omega, \mathscr{A}, \mu)$ the function $\widehat{u} := \tau^{*-1}(u)$ is called the Stone transform of u.*

1.4.10. Now, we will answer the question: which vector lattices are representable as order-dense ideals in $L^0(\Omega, \mathscr{A}, \mu)$? An order complete vector lattice E is said to be a *Kantorovich–Pinsker* space if its base $\mathfrak{B}(E)$ is a multinormed Boolean algebra (= there exists an essentially positive locally finite countably additive measure on $\mathfrak{B}(E)$, see 1.2.9), or equivalently, if E contains an order-dense ideal with a total set of order continuous functionals. It can be proved that if $(\Omega, \mathscr{A}, \mu)$ is a measure space possessing the direct sum property then $L^0(\Omega, \mathscr{A}, \mu)$ is a Kantorovich–Pinsker space.

Theorem. *Each Kantorovich–Pinsker space E is linearly and order isomorphic to an order-dense ideal of $L^0(\Omega)$ for a suitable measure space $(\Omega, \mathscr{A}, \mu)$ with the direct sum property. If an order-unity $\mathbf{1}$ is fixed in E, then among such isomorphisms there is a unique isomorphism taking $\mathbf{1}$ to the equivalence class of the identically one function on Ω. The space E is universally complete if and only if the image of each isomorphism from E onto an order-dense ideal of $L^0(\Omega)$ coincides with $L^0(\Omega)$.*

$\triangleleft$ The proof can be obtained by combining 1.2.10, 1.4.5, and 1.4.9. $\triangleright$

1.4.11. Let $(\Omega, \mathscr{A}, \mu)$ be a measure space with the direct sum property. An arbitrary order ideal E in $L^0(\Omega, \Sigma, \mu)$ is called an *ideal space* on (Ω, Σ, μ). We recall some basic properties of ideal function spaces. In this subsection $\widetilde{f} := f^\sim$ stands for the coset of a measurable function f.

(1) *Every ideal space is an order complete vector lattice.*

(2) *Let M be a subset in $L^0(\Omega, \mathscr{A}, \mu)$ unbounded from above. Then M contains a countable subset that is also unbounded from above.*

(3) *If μ is σ-finite then every nonempty order-bounded set M in ideal space contains a countable subset $(\widetilde{f_n})$ with $\sup(M) = \sup_n \widetilde{f_n}$. Moreover, $\widetilde{f} := \sup_n \widetilde{f_n}$ can be computed pointwise: $f(t) = \sup_n f_n(t)$ $(t \in \Omega)$.*

(4) *A sequence $(\widetilde{f_n}) \subset E$ o-converges to $\widetilde{f}$ if and only if it is order-bounded (in E) and $f_n(t) \to f(t)$ for almost all $t \in \Omega$.*

For $\widetilde{f} \in L^0(\Omega, \mathscr{A}, \mu)$ denote the *support* of $\widetilde{f}$ by $\operatorname{supp} \widetilde{f} := \operatorname{supp}(f) := \{t \in \Omega : f(t) \neq 0\}$. Let e_f stand for the coset of the characteristic function of $\operatorname{supp}(f)$. The *support of a nonempty subset* $M \subset L^0(\Omega, \mathscr{A}, \mu)$ is a measurable subset $\Omega_0 \subset \Omega$ such that $\sup\{e_f : f \in M\}$ is the coset of the characteristic function of Ω_0. Observe that $\operatorname{supp}(M)$ is determined up to a set of measure zero. We write $A \subset B \mod (\mu)$ if $\mu(B \setminus A) = 0$.

(5) *Let M be a nonempty subset of E, $\Omega_0 := \operatorname{supp}(M)$ and $\Omega_1 := \Omega \setminus \Omega_0$. Then*

$$M^{\perp} = \left\{\widetilde{f} \in E : \operatorname{supp}(f) \subset \Omega_1 \mod (\mu)\right\},$$

$$M^{\perp\perp} = \left\{\widetilde{f} \in E : \operatorname{supp}(f) \subset \Omega_0 \mod (\mu)\right\}.$$

(6) *Let K be a band in E, and let χ_K be the characteristic function of* $\operatorname{supp}(K)$. *Then the band projection $[K]$ has the form: $[K](\widetilde{f}) = (\chi_K f)^{\sim}$ $(\widetilde{f} \in E)$.*

(7) *An ideal space E is order dense in $L^0(\Omega, \mathscr{A}, \mu)$ if and only if* $\operatorname{supp}(E) = \Omega \mod (\mu)$. *If, in addition, μ is σ-finite then for every $0 \leq f \in L^0(\Omega, \mathscr{A}, \mu)$ there exists an increasing sequence (f_n) in E such that $f = o\text{-}\lim_n f_n$.*

1.5. Normed Vector Lattices

In this section we expose various classes of normed vector lattices determined by the interplay of order and norm. Some fundamental properties are given. A more detailed presentation can be found in [23, 162, 231, 242, 336, 341, 388].

1.5.1. Let E be a vector lattice. A norm $\|\cdot\|$ on E is called a *lattice norm* if $|x| \leq |y|$ implies $\|x\| \leq \|y\|$ for all $x, y \in E$. A lattice norm can be equivalently defined by following two relations: $0 \leq x \leq y \Rightarrow \|x\| \leq \|y\|$ $(x, y \in E)$ and $\|\,|x|\,\| = \|x\|$ $(x \in E)$. If $\|\cdot\|$ is a lattice norm on E, the pair $(E, \|\cdot\|)$ is called a *normed (vector) lattice.* A norm complete normed lattice is called a *Banach lattice.*

It is immediate from 1.3.2 (1, 8) and monotonicity of a lattice norm that the mappings $x \mapsto x^+$, $x \mapsto x^-$, $x \mapsto |x|$, and the lattice operations $(x, y) \mapsto x \vee y$

and $(x, y) \mapsto x \wedge y$ are uniformly continuous. In particular, the positive cone, the disjointness relation, and every band are closed. The closure of a vector sublattice (order ideal) of E is a vector sublattice (order ideal) of E. It is also clear that each of the lattice operations of a normed lattice E admits a unique continuous extension to the norm completion $\widetilde{E}$ so that $\widetilde{E}$ becomes a Banach lattice. We state also a useful criterion for norm completeness.

For an arbitrary normed vector lattice E the following are equivalent:

(1) *E is a Banach lattice;*

(2) *each increasing Cauchy sequence of positive elements in E has the least upper bound;*

(3) *every absolutely convergent series of positive elements in E is order convergent.*

If, in addition, E has the principal projection property then each of the following two conditions is equivalent to (1):

(4) *each laterally increasing Cauchy sequence of positive elements in E has the least upper bound;*

(5) *every absolutely convergent series of pairwise disjoint positive elements in E is order convergent.*

1.5.2. Consider an arbitrary vector lattice E. A linear functional f on E is called *order-bounded* if the image of every order interval in E under f is a bounded set in $\mathbb{R}$. Denote by $E^\sim$ the set of all order-bounded functionals on E. A functional f is positive if $f(x) \geq 0$ for all $x \in E_+$. The space $E^\sim$ becomes an ordered vector space if a vector order in it is defined by the cone $E^\sim_+$ of all positive functionals and is referred to as *order dual space.*

A linear functional $f \in E^\sim$ is called *order continuous* if $\lim_\alpha f(x_\alpha) = 0$ for every decreasing net (x_α) in E with $\inf x_\alpha = 0$. The set of all order continuous functionals is denoted by $E^\sim_n$. In the sequel E' stands for the space of all norm continuous functionals on a normed lattice E.

Each $x \in E$ defines by $\hat{x} : f \mapsto \langle x, f \rangle = f(x)$ $(f \in E^\sim)$ an order-bounded functional on $E^\sim$. The mapping assigning $\hat{x}$ to each $x \in E$ by $\langle f, \hat{x} \rangle = \langle x, f \rangle$ $(f \in E^\sim)$ is called the *canonical embedding in the second order dual* or the *evaluation mapping* This mapping preserves order and is injective provided that $E^\sim$ is point-separating.

We now list a few basic properties of order-bounded functionals and order duals.

(1) The order dual space is an order complete vector lattice.

(2) The set $E^\sim_n$ is a band in $E^\sim$.

(3) The evaluation mapping is an isomorphism from E onto a sublattice of $E^{\sim\sim}$ provided that $E^{\sim}$ is point-separating.

(4) Let E_0 be a massive subspace of an ordered vector space E. Each positive linear functional on E_0 has a positive linear extension to E.

(5) If E is a normed lattice then E' is an order ideal in $E^{\sim}$.

(6) If E is a normed lattice then E' is an order complete Banach lattice.

(7) If E is a Banach lattice then $E' = E^{\sim}$ and $E_n^{\sim}$ is a band in E'.

(8) Let E_0 be a vector sublattice of the normed vector lattice E. Each norm continuous positive linear functional on E_0 has a norm preserving positive linear extension to E.

1.5.3. A Banach lattice is said to have an *order continuous norm* (*σ-order continuous norm*) if $\lim_\alpha \|x_\alpha\| = 0$ for every downward directed net (sequence) (x_α) with $\inf_\alpha x_\alpha = 0$. In the Russian literature, order continuity is frequently called *Property* (A). We say that the norm in E is *laterally σ-continuous* if $\lim_n \|e_n\| = 0$ for every decreasing sequence (e_n) in E such that $(e_n - e_{n+1}) \perp e_{n+1}$ $(n \in \mathbb{N})$ and $\inf_n e_n = 0$.

Theorem. *For every Banach lattice E, the following are equivalent:*

(1) *The norm of E is order continuous;*

(2) *E is order complete and the norm of E is order continuous;*

(3) *E is order σ-complete and the norm of E is σ-order continuous;*

(4) *E is order complete and the norm of E is laterally σ-continuous;*

(5) *E is order σ-complete and each closed order ideal of E is a band;*

(6) *every closed order ideal of E is a projection band;*

(7) *every order interval of E is $\sigma(E, E')$-compact;*

(8) *every norm continuous linear functional on E is order continuous;*

(9) *the canonical embedding $E \to E''$ maps E onto an ideal of the Banach lattice E''.*

$\lhd$ The proof can be found in [242, 263, 336, 341]. $\rhd$

1.5.4. We say that a normed lattice E is *monotonically complete* or possesses *Property* (B) if for every increasing net (x_α) in E_+ with $\sup_\alpha \|x_\alpha\| < \infty$ there exists the supremum $x = \sup_\alpha x_\alpha$. Every monotonically complete normed lattice is order complete.

A vector lattice E is said to have the *order semicontinuous norm* or to possess *Property (C)* if $o\text{-}\lim_\alpha x_\alpha = x$ implies $\lim_\alpha \|x_\alpha\| = \|x\|$ for every increasing net (x_α) in E_+. The sequential variants of Properties (A), (B), and (C) are denoted by $(A)_\sigma$, $(B)_\sigma$, and $(C)_\sigma$, respectively. A Banach lattice possessing $(A)_\sigma$ and $(B)_\sigma$ is called a *KB-space*.

The significance of these properties for the theory of Banach lattices has various aspects. We emphasize only those related to the properties of canonical embedding into the second dual lattice.

(1) A Banach lattice E with $E_n^\sim$ point-separating is monotonically complete if and only if the canonical embedding $E \to (E_n^\sim)_n^\sim$ is an isometry.

(2) An order complete Banach lattice E with $E_n^\sim$ point-separating has order semicontinuous norm if and only if the canonical embedding E and $(E_n^\sim)_n^\sim$ are isomorphic under the canonical embedding.

(3) Let E be an order complete Banach lattice with $E_n^\sim$ point-separating. Then E is monotonically complete and has the order semicontinuous norm if and only if the image of E under the canonical embedding is the range of a positive contractive projection.

(4) A Banach lattice E is a KB-space if and only if the canonical embedding $\varkappa$ maps E onto a band in E''.

1.5.5. **(1)** A Banach lattice E is called an *abstract M-space* or *AM-space,* for short, if

$$\|x \vee y\| = \|x\| \vee \|y\| \quad (x, y \in E_+).$$

If the unit ball of an AM-space E contains a largest element e, then e is a strong order-unity and the unit ball of E coincides with the symmetric order interval $[-e, e]$. In this case E is said to be an *AM-space with unity.*

(2) Let E be an arbitrary vector lattice and $u \in E$. We may introduce the following seminorm in the principal ideal $E(u)$:

$$\|x\|_u := \inf\{\lambda \in \mathbb{R} : |x| \leq \lambda u\} \quad (x \in E(u)).$$

The seminorm $\|\cdot\|_u$ is a norm if and only if the lattice $E(u)$ is Archimedean. It can be easily seen that relative uniform convergence in E is the convergence in the norm of $(E(v), \|\cdot\|_v)$. Therefore, a vector lattice E is relatively uniformly complete if and only if the normed lattice $(E(u), \|\cdot\|_u)$ is complete for every $u \in E_+$.

Let E be a Banach space and $0 \neq u \in E$. Then $(E(u), \|\cdot\|_u)$ is an AM-space with unity u and the identity embedding $E(u) \to E$ is continuous.

This simple proposition asserts that every Banach lattice is locally arranged as an AM-space, which enables us to reduce the study of some properties of Banach lattices to that of AM-spaces. Such an approach is demonstrated below in 1.5.7, 1.5.11, and 1.5.12.

(3) A Banach lattice E is called an *AL^p-space* if

$$\|x + y\| = \left(\|x\|^p + \|y\|^p\right)^{\frac{1}{p}}$$

for all disjoint $x, y \in E_+$. Here $1 \leq p \leq \infty$ and, in case $p := \infty$, we let $(t^p + s^p)^{1/p} := \max\{s, t\}$ $(0 < s, t \in \mathbb{R})$. The terms AM-space and AL-space are conventionally used instead of AL^∞-space and AL^1-space, respectively. This definition of AM-space is equivalent to that given in (1) despite the fact that the required equality holds only for disjoint pairs of elements.

1.5.6. We recall three well-known facts from the theory of Banach lattices. Proofs can be found in [231, 242, 341].

(1) Theorem. *Let E be an AM-space. Then there exist a compact space Q and a family of triples $(t_\alpha, s_\alpha, \lambda_\alpha)_{\alpha \in \mathrm{A}}$ with $t_\alpha, s_\alpha \in Q$ and $0 \leq \lambda_\alpha < 1$ such that E is isometrically isomorphic to the closed sublattice*

$$F := \{x \in C(Q) : (\forall \alpha \in \mathrm{A})\ x(t_\alpha) = \lambda_\alpha x(s_\alpha)\}.$$

(2) Brothers Kreĭn–Kakutani Theorem. *Every AM-space with unity is linearly isometric and order isomorphic to the space of continuous functions $C(Q)$ on some compact space Q.*

(3) Theorem. *If $1 \leq p < +\infty$ then every AL^p-space is linearly isometric and order isomorphic to $L^p(\Omega, \mathscr{A}, \mu)$ for a suitable measure space $(\Omega, \mathscr{A}, \mu)$ with the direct sum property.*

1.5.7. The study of functions $f : \mathbb{R}^l \to \mathbb{R}$, for which $f(e_1, \dots, e_l)$ can naturally be defined for $e_1, \dots, e_l \in E$, is called the *functional calculus*. To assign some values to expressions of the form $f(e_1, \dots, e_l)$ we will use the local AM-structure of a Banach lattice. Denote by $\mathscr{H}(\mathbb{R}^l)$ the space of all real continuous functions f on $\mathbb{R}^l$ which are *positively homogeneous*, i.e. $f(\lambda t) = \lambda f(t)$ for all $t \in \mathbb{R}^n$ and $\lambda \geq 0$. Clearly, $\mathscr{H}(\mathbb{R}^n)$ is a vector lattice under pointwise operations.

Denote $S := \{(t_1, \dots, t_l) \in \mathbb{R}^l : |t_1| + \dots + |t_l| = 1\}$. Clearly, each function $f \in \mathscr{H}(\mathbb{R}^l)$ is completely defined by its values on S. Therefore, the restriction mapping $f \mapsto f|_S$ is a linear and lattice isomorphism from $\mathscr{H}(\mathbb{R}^l)$ onto $C(S)$. Thus, we may regard $\mathscr{H}(\mathbb{R}^l)$ as a Banach lattice with strong order-unity and norm $\|f\|_\infty := \sup\{|f(x)| : x \in S\}$.

We define $dx_j \in \mathscr{H}(\mathbb{R}^n)$ by $dx_j(t_1, \dots, t_n) = t_j$ $(j := 1, \dots, n)$.

Theorem. *Let E be a uniformly complete vector lattice, $e_1, \dots, e_l \in E$, and $e := e_1 + \dots + e_n$. Then there is a unique lattice homomorphism $h : \mathscr{H}(\mathbb{R}^l) \to E$ such that $h(dx_j) = e_j$ $(j := 1, \dots, l)$. Moreover, $h(\mathscr{H}(\mathbb{R}^l))$ is the e-uniform closure of the vector sublattice generated by $\{e_1, \dots, e_l\}$.*

◁ See [242, 341]. ▷

1.5.8. A *normed (Banach) ideal space* on (Ω, Σ, μ) is defined as an ideal space E on (Ω, Σ, μ) endowed with a lattice norm making E into a normed (Banach) space.

If a sequence x_n converges to x in the norm of a Banach ideal space E then $x_n \to x(\mu)$. If (x_n) is a Cauchy sequence in E then it converges in measure to some $x \in L^0(T, \Sigma, \mu)$.

Now, we consider some properties of the vector lattice $C(Q)$ of all continuous functions on a compact topological space Q. A vector lattice is called *disjointly complete* (*disjointly σ-complete* if every its order-bounded antichain (countable antichain) has supremum.

1.5.9. *Theorem.* *For a compact space Q, the following are equivalent:*

(1) *$C(Q)$ is order complete (σ-complete);*

(2) *$C(Q)$ is disjointly complete (σ-complete);*

(3) *Q is extremal (quasiextremal);*

(4) *$C(Q)$ possesses the projection property (principal projection property).*

$\triangleleft$ The implications (1) $\Rightarrow$ (2) and (1) $\Rightarrow$ (4) are obvious.

(2) $\Rightarrow$ (3): Take an open $G \subset Q$ such that $G = \bigcup_{n=1}^{\infty} F_n$ with F_n closed for all $n \in \mathbb{N}$. We will prove that $\operatorname{cl} G$ is open. Since every compact topological space is normal, we may find an open $G_1 \subset Q$ with $F_1 \subset G_1 \subset \operatorname{cl} G_1 \subset G$. By the same argument there is an open $G_2 \subset Q$ with $\operatorname{cl} G_1 \cup F_2 \subset G_2 \subset \operatorname{cl} G_2 \subset G$. By induction we may construct a sequence (G_n) of open subsets such that $\operatorname{cl} G_n \cup F_{n+1} \subset G_{n+1} \subset \operatorname{cl} G_{n+1} \subset G$. Clearly,

$$G = \bigcup_{n=1}^{\infty} G_n = \bigcup_{n=1}^{\infty} \operatorname{cl} G_n.$$

Without loss of generality, we may assume that all G_n differ from one another.

Now, we define a sequence of continuous functions (x_n) such that $0 \leq x_n \leq 1$ and

$$x_n(t) := \begin{cases} 1, & \text{if } t \in \operatorname{cl} G_{n-1}, \\ 0, & \text{if } t \in \operatorname{cl} G_{n-2} \cup (Q \setminus G_{n-1}), \end{cases}$$

where $G_{-1} := G_0 := \varnothing$. It is easily seen that each of the sequences (x_{3n}), (x_{3n-1}), and (x_{3n-2}) is pairwise disjoint. By assumption there exist $y_k := \sup_{n \in \mathbb{N}} x_{3n-k}$ $(k := 0, 1, 2)$. Put $y := y_0 \vee y_1 \vee y_2$. Obviously, $y_k(Q \setminus \operatorname{cl} G) = \{0\}$ $(k := 0, 1, 2)$, and therefore $y(Q \setminus \operatorname{cl} G) = \{0\}$. At the same time $y(G) = \{1\}$. Since y is continuous, we have $y(\operatorname{cl} G) = \{1\}$. Thus, $y = \chi_{\operatorname{cl} G}$ and $\operatorname{cl} G$ is clopen set.

Now, suppose that $C(Q)$ is disjointly complete. Then Q is quasiextremal as we have just proved. Moreover, the Boolean algebras $\operatorname{Clop}(Q)$ and $\mathfrak{C}(\mathbf{1})$ are isomorphic; therefore, Q is the Stone space of the Boolean algebra of clopen sets $\operatorname{Clop}(Q)$. Whence the required assertion follows from 1.1.6 (2) and the Ogasawara Theorem (1.2.4).

(3) $\Rightarrow$ (1): follows from 1.4.2 (1, 3).
(4) $\Rightarrow$ (3): See [341]. $\triangleright$

1.5.10. *Theorem.* *Let Q be a compact space and E be a vector sublattice in $C(Q)$ containing constants and separating the points of Q. If E is disjointly complete then Q is extremal and E possesses the projection property.*

$\triangleleft$ First, we prove that Q is extremal. In view of 1.5.9 it suffices to establish disjoint completeness of $C(Q)$. Take a pairwise disjoint family $(x_\xi)_{\xi\in\Xi}$, $0 \leq x_\xi \leq 1$, and put

$$z(t) := \begin{cases} x_\xi(t), & \text{if } x_\xi(t) > 0, \\ 0, & \text{if } t \in G := \operatorname{int}\{t \in Q : (\forall \xi \in \Xi) x_\xi(t) = 0\}. \end{cases}$$

Clearly, z is a continuous function defined on an open dense subset in Q. According to the Stone–Weierstrass Theorem, for each $\xi \in \Xi$ there exists a sequence $(y_{\xi,n} \in E_+)$, such that $0 \leq x_\xi - y_{\xi,n} \leq (1/n)\mathbf{1}$, $(n \in \mathbb{N})$. By assumption there exists $y_n := \sup_{\xi\in\Xi} y_{\xi,n}$. This sequence is uniformly convergent since $|y_{n+p} - y_n| \leq (1/n)\mathbf{1}$ $(n \in \mathbb{N})$. Put $x := \lim y_n$. Observe that $0 \leq z(t) - y_n(t) \leq (1/n)\mathbf{1}$ for all $t \in \operatorname{dom}(z)$; thus $x(t) = \lim y_n(t) = z(t)$ for all $t \in \operatorname{dom}(z)$. This means that x is a continuous extension of z onto Q; therefore $x = \sup_{\xi\in\Xi} x_\xi$.

Now, take an arbitrary element $e \in E_+$ and prove that $e\chi_{Q_0} \in E$ for any clopen set $Q_0 \subset Q$. Since E contains constants and separates points of Q, for each $q \in Q_0$ there exists $e_q \in E$ with $e_q(q) > e(q)$. Put $U_q := \{t \in Q : e_q(t) > e(t)\}$ and choose a finite subcover $U_{q_1}, \dots, U_{q_m}$ of Q_0. Then the following relations are consistent:

$$e_0 := e_{q_1} \vee \dots \vee e_{q_m} \in E;$$
$$e_0(t) = 0 \ (t \in Q \setminus Q_0); \quad e_0(t) \geq e(t) \ (t \in Q_0).$$

Now it is clear that $e \wedge e_0 = e\chi_{Q_0} \in E$. $\triangleright$

1.5.11. *Theorem.* *Every Archimedean disjointly complete vector lattice has the projection property. Every Archimedean disjointly σ-complete vector lattice has the principal projection property.*

$\triangleleft$ We consider the claim only for disjointly complete vector lattices. The disjointly σ-complete case is handled by the same arguments. Disjoint completeness of E is inherited by order ideals. In particular, all principal ideals $E(u)$ are disjointly complete. Clearly, the norm completion $\widetilde{E(u)}$ of $E(u)$ is an AM-space. Using 1.5.6 (2) we conclude that $\widetilde{E(u)}$ is isomorphic to $C(Q)$ for some compact space Q. The isomorphism brings $E(u)$ in a vector sublattice containing constants and separating points of Q. According to 1.5.10 E possesses the projection property. $\triangleright$

1.5.12. Theorem. *Let E be a Banach lattice or relatively uniformly complete vector lattice. Then the following are equivalent:*

(1) *E is order complete (order σ-complete);*

(2) *E is disjointly complete (disjointly σ-complete);*

(3) *E has the projection property (principal projection property).*

$\lhd$ The implication (1) $\Rightarrow$ (2) is obvious and (2) $\Rightarrow$ (3) follows from 1.5.11. To prove (3) $\Rightarrow$ (1) use the local AM-structure of a Banach lattice (relatively uniformly complete vector lattice) and apply (1) $\Leftrightarrow$ (4) from 1.5.9. $\rhd$

1.5.13. Let $\mathscr{A}$ be a Boolean algebra. Denote by $\mathrm{ba}(\mathscr{A})$ the set of all bounded finitely additive functions $\mu : \mathscr{A} \to \mathbb{R}$ with a vector structure induced from $\mathbb{R}^{\mathscr{A}}$. The set $\mathrm{ba}_+(\mathscr{A})$ of all positive bounded finitely additive functions is a cone and defines some structure of a vector lattice in $\mathrm{ba}(\mathscr{A})$. The lattice operations have the form:

$$\mu \vee \nu(a) = \sup\{\mu(b) + \mu(a \wedge b^*) : b \leq a\},$$
$$\mu \wedge \nu(a) = \inf\{\mu(b) + \mu(a \wedge b^*) : b \leq a\},$$
$$|\mu|(a) = \sup\{|\mu(b)| : b \leq a\},$$

where $\mu, \nu \in \mathrm{ba}(\mathscr{A})$. The *total variation* of μ is defined as $\|\mu\| := |\mu|(\mathbf{1})$ where $\mathbf{1}$ is the unity of $\mathscr{A}$. It is clear that the function $\mu \mapsto \|\mu\|$ is a lattice norm.

The vector lattice $\mathrm{ba}(\mathscr{A})$ *of bounded, finitely additive functions on a Boolean algebra is an AL-space. In particular,* $\mathrm{ba}(\mathscr{A})$ *is order complete and the norm of* $\mathrm{ba}(\mathscr{A})$ *is order continuous.*

1.6. Comments

1.6.1. (1) The theory of Boolean algebras originated from the classical work by G. Boole "An Investigation of the Laws of Thought, on Which Are Founded the Mathematical Theories of Logic and Probabilities" [49, 50]. The author himself formulated his intentions as follows: "The design of the following treatise is to investigate the fundamental laws of those operations of the mind by which reasoning is performed; to give expression to them in the language of a Calculus, and upon this foundation to establish the science of Logic and construct its method to make that method itself the basis of a general method for the application of the mathematical doctrine of probabilities; and, finally, to collect from the various elements of truth brought to view in the course of these inquiries some probable intimations concerning the nature and constitution of the human mind.... "

Pursuing this end, G. Boole carried out, in fact, the algebraization of the logical system lying behind the classical mathematical reasoning. In a result, he become the author of the algebraic system omnipresent under the name of Boolean algebra.

(2) Definition 1.1.2 looks somewhat strange at first sight. Indeed, it does not reveal the reasons for whatever distributive lattice to be called an algebra since the term "algebra" refers customarily to conventional objects (cf. Lie algebra, Banach algebra, C^*-algebra, etc.). The arising ambiguity is easily eliminated because a Boolean algebra is in fact an algebra over the two-element field. At the same time, it is perfectly natural to view Boolean algebras in different contexts at different angles. It is worth emphasizing that the particular Boolean algebras we deal with in functional analysis appear mostly as distributive complemented lattices.

(3) Recall that a *universal algebra* is an algebraic system without predicates. This concept makes available another definition of Boolean algebra. Namely, a Boolean algebra B is a universal algebra $(B, \vee, \wedge, *, \mathbf{0}, \mathbf{1})$ with two binary operations $\vee$ and $\wedge$, one unary operation $*$, and two distinguished elements $\mathbf{0}$ and $\mathbf{1}$ obeying the conditions: (a) $\vee$ and $\wedge$ are commutative and associative; (b) $\vee$ and $\wedge$ are both distributive relative to one another; (c) x and x^* complement one another; (d) $\mathbf{0}$ and $\mathbf{1}$ are neutral for $\vee$ and $\wedge$, respectively.

Conversely, given a universal algebra B of the above type, make B into a poset by letting $x \leq y$ whenever $x \wedge y = x$ for $x, y \in B$. In this event, note that $(B, \leq)$ is a distributive complemented lattice with join $\vee$, meet $\wedge$, complementation $*$, zero $\mathbf{0}$, and unity $\mathbf{1}$.

1.6.2. **(1)** Theorem 1.2.3 shows that every Boolean algebra is completely determined from its Stone space. In more detail, each property of a Boolean algebra B translates into the topological language, becoming a property of the Stone space $\mathscr{S}(B)$ of B. This way of studying Boolean algebras is the *Stone representation method.* A detailed presentation can be found in [144, 283, 352].

(2) Denote by Bool the category of Boolean algebras and Boolean homomorphisms, and let Comp stand for the category of compact spaces and continuous mappings. Then 1.2.3 and 1.2.5 may be paraphrased as follows:

Theorem. *The mapping $\mathscr{S}$ is a contravariant functor from the category* Bool *to the category* Comp.

(3) About measure spaces with the direct sum property see [78, 162, 239, 375]. The problem of whether lifting exists was formulated by A. Haar for the Lebesgue measure on the real axis and was solved by J. von Neumann in 1931. It was D. Maharam who proved that every σ-finite measure admits lifting [270]. The general case is handled in [375]. For the results in 1.2.7 and 1.2.8 as well as other aspects of lifting theory see in [78, 239, 375].

1.6.3. **(1)** In the history of functional analysis, the rise of the theory of ordered vector spaces is commonly attributed to the contribution of G. Birkhoff, L. V. Kantorovich, M. G. Kreĭn, H. Nakano, F. Riesz, H. Freudenthal, et al. At present, the theory of ordered vector spaces and its applications constitute a noble branch of mathematics representing, in fact, one of the main sections of contemporary functional analysis. The theory with a vast field of applications is thoroughly covered in many monographs (see [15, 19, 23, 100, 139, 145, 162, 163, 170, 184, 185, 197, 230, 231, 242, 262, 282, 336, 341, 388, 399, 409, 411]). Observe also the surveys [60–62] each with a rich reference list.

(2) The credit for finding the most important instance of ordered vector spaces, an order complete vector lattice or a K-space, is due to L. V. Kantorovich. This notion appeared in Kantorovich's first fundamental article on this topic [153] where he wrote, "In this note, I define a new type of space that I call a semiordered linear space. The introduction of such a space allows us to study linear operations of one abstract class (those with values in such a space) as linear functionals."

Here L. V. Kantorovich stated an important methodological principle, the *heuristic transfer principle* for K-spaces.

(3) The concept of universal completion for a K-space was introduced in another way by A. G. Pinsker (see [163]). He also proved existence of a universal completion unique to within isomorphism for an arbitrary K-space. The concepts of order-unity, fragment, and spectral function were introduced by H. Freudenthal. He also established Theorem 1.3.9 (1) (see [163, 388]).

(4) Weak and strong Freudenthal properties for general vector lattices were introduced and studied B. Lavrič [234]. Using the principal ideal lattice he also introduced the notion of zero-dimensional vector lattice with strong unity and proved that a vector lattice has the weak Freudenthal property if and only if every principal ideal in it is zero-dimensional. M. Pannenberg [315] extended Lavrič's notion of dimension zero to an arbitrary value of dimension by using the lattice $\mathscr{I}_p(E)$, see 1.3.5 (3). He also defined the topological stable rank $\operatorname{tsr}(E)$ for a Banach lattice E with quasi-interior positive elements and established that $\dim \mathscr{I}_p(E) = \operatorname{tsr}(E)$.

(5) Sometimes the use of Freudenthal's Spectral Theorem may be replaced by the following assertion: *In a Banach lattice E with the principal projection property every order interval is the norm closed convex hull of its own extreme points.* (The extreme points of the interval $[0, e]$ are exactly the fragments of e.) The assertion remains valid if E is a Banach lattice with quasi-interior positive elements having the topological stable rank zero, see M. Pannenberg [315].

1.6.4. (1) The fact that for a complete Boolean algebra B the set $\mathfrak{R}(B)$ of resolutions of unity is a universally complete K-space with base isomorphic to B (see 1.4.3 and 1.4.4) is due to L. V. Kantorovich [163]. Theorem 1.4.4 was obtained by A. G. Pinsker (see [163]). The representation of an arbitrary K-space as an order-dense ideal in $C_\infty(Q)$ was established independently by B. Z. Vulikh and

T. Ogasawara (see [163, 388]).

(2) The theory of vector lattices and K-spaces arose historically before the theory of general ideal spaces which began developing in the 1950s with research by J. Diedonnè, G. G. Lorentz, I. Galperin, H. W. Ellis, A. C. Zaanen, W. A. J. Luxemburg, et al. The synthesis of these theories occurred in the 1960s in the works of W. A. J. Luxemburg, A. C. Zaanen, and G. Ya. Lozanovskiĭ (see [61, 409]). After the papers of Diedonnè, one of the most common terms for an ideal space is a *Köthe space*. The term "ideal space" was coined within the school of M. A. Krasnosel′skiĭ on integral operators and equations (P. P. Zabreĭko, P. E. Sobolevskiĭ, Ya. B. Rutitskiĭ et al.) due to the fact that an ideal space is actually an o-ideal in L^0 (see [188, 189]).

1.6.5. (1) The material presented in 1.5.1–1.5.8 is traditional for the theory of normed lattices and can be found in [23, 162, 231, 242, 263, 336, 341, 409].

(2) Theorems 1.5.10, 1.5.11 and the equivalence (1) $\Leftrightarrow$ (2) in 1.5.12 belong to A. I. Veksler and V. A. Geĭler [381].

(2) In this book we never touch the theory of positive operators in a Banach space with cone which originated with M. G. Kreĭn's articles written in the 1940s and developed later by M. A. Krasnosel′skiĭ in Russia and by many research groups in other countries. The history, state of the art, and various applications are reflected in [35, 66, 139, 170, 184, 185, 187, 188, 285, 389, 390, 399].

Chapter 2

Lattice-Normed Spaces

In this chapter we consider structural properties of a vector space with some norm taking values in a vector lattice. Such a vector space is called a *lattice-normed space*; an LNS for short. The most important peculiarities of LNSs are connected with the decomposability property (2.1.1 (4)). The latter allows us, in particular, to indicate a complete Boolean algebra of linear projections in a lattice-normed space, which is isomorphic and closely related to the Boolean algebra of band projections of the norm lattice (2.1.3, 2.1.4). Moreover, a decomposable LNS admits a compatible module structure over a certain ring of orthomorphisms (2.1.8). These facts are closely related with the disjointness relation induced by the vector norm (2.1.2). If an LNS is simultaneously a vector lattice then there is another disjointness relation connected with its lattice structure. Some simple interrelation between them is reflected in the notions of norm-indecomposable and norm-n-decomposable elements (2.1.9). It turns out that every norm-n-decomposable elements is very often the sum of n norm-indecomposable elements (2.1.10). Partitions of unity in a Boolean algebra lead to the operation of mixing elements in a lattice-normed space (2.2.1). If there exists a mixing of every (vector) norm-bounded family in a lattice-normed space then the space is called disjointly complete (2.1.5). For instance, such are Banach–Kantorovich spaces (BKSs), i.e. decomposable and order complete lattice-normed spaces (2.2.1). Moreover, a decomposable lattice-normed space is order complete if and only if it is disjointly (laterally) complete and complete with respect to relative uniform convergence (2.2.2). In addition, every LNS has a lattice norm completion that can be obtained by consecutively applying the operations of mixing and closure with respect to relative uniform convergence (2.2.8, 2.2.9).

LNSs consisting of continuous and measurable vector-functions (Section 2.3) are most frequent in analysis. A more general example of an LNS is presented by a space of sections of a continuous Banach bundle (CBB) (2.4.7). It turns out that each BKS is linearly isometric to a space of almost global sections of some continuous Banach bundle (2.4.10). Thus, an LNS admits some functional representation.

This representation is uniquely determined in the class of ample continuous Banach bundle (2.4.6). Every continuous Banach bundle over an extremal compact space has a unique ample hull to within isometry (2.4.11). The notion of ample continuous Banach bundle allows us, in particular, to introduce the CBB of operator spaces (2.4.13).

The notion of measurable Banach bundle (MBB) is a measurable analog of a continuity structure. Measurable sections are defined as the limits of almost everywhere convergent sequences (on the subsets of finite measure) of elements of some set of sections given axiomatically and called a measurability structure (2.5.1). Under some assumptions the space of equivalence classes of measurable sections (2.5.2) of a measurable Banach bundle appears to be a *bo*-complete LNS (2.5.3). The notion of lifting in a factor space of measurable sections of an MBB is introduced (2.5.5). Measurable Banach bundles with lifting are called liftable. There is a constructive connection between liftable measurable Banach bundles and complete continuous Banach bundles. A method for constructing a liftable MBB given an arbitrary complete CBB over the corresponding extremal compact space is described in 2.5.8 and 2.5.9. Moreover, this method is universal: every liftable MBB is so obtainable (2.5.10). These facts enable us to represent Banach–Kantorovich spaces as the spaces of measurable functions associated with measurable Banach bundles (2.5.11–2.5.13).

2.1. Preliminaries

In this section, we introduce basic notions and consider simple properties of vector norms.

2.1.1. Consider a vector space X and a real vector lattice E. (All vector lattices under consideration are assumed Archimedean.) A mapping $|\cdot| : X \to E_+$ is a *vector* (E*-valued*) *norm* if it satisfies the following axioms:

(1) $|x| = 0 \Leftrightarrow x = 0 \quad (x \in X)$;

(2) $|\lambda x| = |\lambda||x| \quad (\lambda \in \mathbb{R},\ x \in X)$;

(3) $|x + y| \le |x| + |y| \quad (x, y \in X)$.

A vector norm is called a *decomposable* or *Kantorovich norm* if

(4) for all $e_1, e_2 \in E_+$ and $x \in X$, from $|x| = e_1 + e_2$ it follows that there exist $x_1, x_2 \in X$ such that $x = x_1 + x_2$ and $|x_k| = e_k$ $(k := 1, 2)$.

In the case when condition (4) is valid only for disjoint $e_1, e_2 \in E_+$, the norm is said to be *disjointly-decomposable* or, in short, *d-decomposable.*

A triple $(X, |\cdot|, E)$ (in brief (X, E), $(X, |\cdot|)$, or X with the default parameters omitted) is a *lattice-normed space* (*over* E) if $|\cdot|$ is an E-valued norm in the vector space X. The space E is called the *norm lattice* of X. If the norm $|\cdot|$ is decomposable (d-decomposable) then the space $(X, |\cdot|)$ itself is called decomposable

(d-decomposable). In the sequel we use the abbreviation LNS for "lattice-normed space."

2.1.2. If $|x| \wedge |y| = 0$ then we call the elements $x, y \in X$ *disjoint* and write $x \perp y$. As in the case of a vector lattice, a set of the form $M^\perp := \{x \in X : (\forall y \in M)\ x \perp y\}$, with $\varnothing \neq M \subset X$, is called a *band* or a *component*. The symbol $\mathscr{B}(X)$ denotes the set of all bands ordered by inclusion. We say that $K \in \mathscr{B}(X)$ is a *projection band* if $K \oplus K^\perp = X$. The projection $h(\pi)$ onto the band K along the band $K^\perp$ is called a *band projection*. A lattice-normed space X is said to have the *projection property* whenever every band in X is a projection band. For uniformity, we often write $\mathfrak{B}(X)$ instead of $\mathscr{B}(X)$ and take liberty of using the terminology from the theory of vector lattices. Such a treatment is admissible as long as X is considered as a LNS by itself. But, if X is simultaneously a vector lattice, we should be more accurate to avoid confusions, see 2.1.4 and 2.1.9 below. Given $L \subset E$ and $M \subset X$, we let by definition $h(L) := \{x \in X : |x| \in L\}$ and $|M| := \{|x| : x \in M\}$. It is clear that $|h(L)| \subset L \cap |X|$.

(1) *Suppose that every band of the vector lattice $E_0 := |X|^{\perp\perp}$ contains the norm of some nonzero element. Then $\mathfrak{B}(X)$ is a complete Boolean algebra and the mapping $L \mapsto h(L)$ is an isomorphism of the Boolean algebras $\mathfrak{B}(|X|^{\perp\perp})$ and $\mathfrak{B}(X)$.*

$\lhd$ It is clear that the mapping h preserves the intersection of any nonempty family of bands. Therefore, h preserves infima, since in the algebras under consideration they coincide with intersections. Moreover, $h(\{0\}) = \{0\}$ and $h(|X|^{\perp\perp}) = X$. Thus, it is sufficient to establish that $h(L^\perp) = h(L)^\perp$ for $L \in \mathfrak{B}(|X|^{\perp\perp})$. The inclusion $h(L^\perp) \subset h(L)^\perp$ is obvious. If $0 \neq x \in h(L)^\perp$ then $|x|$ is disjoint from all the elements of the form $|y|$ in L. At the same time, $x \notin h(L^\perp)$ implies that $0 < e \leq |x|$ for some $e \in L_+$. Therefore, in the band $\{e\}^{\perp\perp}$ there are no elements of the form $|y| \neq 0$, which contradicts our assumption. $\rhd$

(2) *If elements $x, y \in X$ are disjoint, then $|x + y| = |x| + |y|$.*

$\lhd$ Indeed, from the relations $|x| \wedge |y| = 0$ and $|x| \leq |x+y| + |y|$ we infer that

$$|x| \leq (|x+y| + |y|) \wedge |x| \leq |x+y| \wedge |x| \leq |x+y|.$$

Similarly, $|y| \leq |x+y|$; therefore,

$$|x| + |y| = |x| \vee |y| \leq |x+y|. \ \rhd$$

(3) *For every pair of disjoint elements $e_1, e_2 \in E$ the decomposition $x = x_1 + x_2$ with $|x_1| = e_1$ and $|x_2| = e_2$ is unique.*

◁ Assume $|x_1| = |y_1| = e_1$, $|x_2| = |y_2| = e_2$, and $x = x_1 + x_2 = y_1 + y_2$. Then $x_1 - y_1 \perp y_2 - x_2$, since $|x_1 - y_1| \le |x_1| + |y_1| = 2e_1$ and $|x_2 - y_2| \le 2e_2$. By (2) $0 = |(x_1 - y_1) + (x_2 - y_2)| = |x_1 - y_1| + |x_2 - y_2|$, whence $x_1 = y_1$ and $x_2 = y_2$. ▷

(4) *Assume the hypotheses of Proposition (1) to be satisfied. Suppose that X is d-decomposable and there exists a band projection π onto $L \in \mathfrak{B}(E_0)$. Then the projection $h(\pi)$ onto the band $K := h(L)$ along the band $K^\perp$ exists and, moreover, $\pi|x| = |h(\pi)x|$ for all $x \in X$.*

◁ In view of the d-decomposability condition, for every $x \in X$, there are $x_1, x_2 \in X$ such that $x = x_1 + x_2$, $|x_1| = \pi|x|$, and $|x_2| = \pi^\perp|x|$. This means that X is the direct sum of the bands K and $K^\perp$. Let $h(\pi)$ be the projection onto K along $K^\perp$. By the definition of the isomorphism h, we have $h(\pi)x \in K = h(\pi E_0)$, i.e., $|h(\pi)x| \in \pi E_0$. Thus, $\pi^\perp|h(\pi)x| = 0$, i.e., $\pi|h(\pi)x| = |h(\pi)x|$. Since $h(\pi)x$ and $h(\pi^\perp)x$ are disjoint, in view of (2) we have

$$\pi|x| = \pi\big(|h(\pi)x| + |h(\pi^\perp)x|\big) = \pi|h(\pi)x|.$$

Consequently, $\pi|x| = \pi|h(\pi)x| = |h(\pi)x|$. ▷

2.1.3. In the sequel, by a *Boolean algebra of projections* in a vector space X we mean a set $\mathscr{B}$ of commuting idempotent linear operators that act in X. Moreover, the Boolean operations have the following form:

$$\pi \wedge \rho := \pi \circ \rho = \rho \circ \pi, \quad \pi \vee \rho = \pi + \rho - \pi \circ \rho, \quad \pi^* = I_X - \pi \quad (\pi, \rho \in \mathscr{B}),$$

and the zero and identity operators in X serve as the zero and unity of the Boolean algebra $\mathscr{B}$.

Suppose that $E_0 := |X|^{\perp\perp}$ is a vector lattice with the projection property and the space X is d-decomposable. Then X have the projection property. Moreover, there exists a complete Boolean algebra $\mathscr{B}$ of band projections in X and an isomorphism h from $\mathfrak{P}(E_0)$ onto $\mathscr{B}$ such that

$$b|x| = |h(b)x| \quad (b \in \mathfrak{P}(E_0),\ x \in X).$$

◁ A nonzero band $L \in \mathfrak{B}(E_0)$ cannot be disjoint from the set $|X|$. Hence, for some $x \in X$, we have $|x| \notin L^\perp$. If π is the projection onto L then the element $\pi|x|$ differs from zero. Due to d-decomposability of X, we have $|x_0| = \pi|x| \in L$ for some $x_0 \in X$. Thus, we may apply 2.1.2 (1, 4). To every band $K \in \mathfrak{B}(X)$ there corresponds the projection π_K along $K^\perp$. Assign $\mathscr{B} := \{\pi_K : K \in \mathfrak{B}(X)\}$. It is clear that $\mathscr{B}$ is a complete Boolean algebra of projections. With an order projection $\rho \in \mathfrak{P}(E_0)$, we associate the projection π_K with $K := h(\rho E_0)$. Denote the mapping thus obtained by the same letter h. Then h is an isomorphism of the Boolean algebras $\mathfrak{P}(E_0)$ and $\mathscr{B}$. The required property of the isomorphism h follows from (3). ▷

We identify the Boolean algebras $\mathfrak{P}(E_0)$ and $\mathscr{P}(X) := \mathscr{B}$ and write $\pi|x| = |\pi x|$ $(x \in X,\ \pi \in \mathfrak{P}(E_0))$.

2.1.4. Suppose that X is a vector lattice. The norm $|\cdot|$ is *monotone* if $|x| \leq |y|$ implies $|x| \leq |y|$ for $x, y \in X$. We say that $|\cdot|$ is *order semicontinuous* (*o*-semicontinuous for short) if $\sup|x_\alpha| = |\sup x_\alpha|$ for each increasing net $(x_\alpha) \subset X$ with a least upper bound $x \in X$ and *order continuous* (*o*-continuous for short) if $\inf|x_\alpha| = 0$ for any decreasing net $(x_\alpha) \subset X$ with $\inf_\alpha x_\alpha = 0$.

If X is a vector lattice with a monotone norm then $\mathscr{P}(X)$ consists of positive projections. If, in addition, X is order complete and the norm is order semicontinuous then $\mathscr{P}(X)$ is a complete subalgebra of the Boolean algebra $\mathfrak{P}(X)$.

$\lhd$ For a vector lattice with a monotone norm we have obviously that $h(\pi)$ is an order ideal in X for any $\pi \in \mathfrak{P}(E_0)$. If $0 \leq x \in h(\pi)$ and $0 \leq y \in h(\pi^\perp)$ then $|x \wedge y| \in \pi(E) \cap \pi^\perp(E) = \{0\}$ by monotonicity. Hence, $x \wedge y = 0$ and x and y are disjoint not only in the sense of Subsection 2.1.2 but also in the sense of the order relation in X. Therefore the projection operator corresponding to the decomposition $X = h(\pi) \oplus h(\pi^\perp)$ is positive.

Now, assume that X is order complete and $|\cdot|$ is order semicontinuous. It suffices to prove that $h(\pi)$ is order-closed in X. Let $(x_\alpha) \subset h(\pi)_+$ be a bounded increasing net. Since the net $|x_\alpha|$ is bounded in E, by order semicontinuity of the norm we have $\sup|x_\alpha| = |\sup x_\alpha| \in \pi(E)$. Therefore $\sup x_\alpha \in h(\pi)$ and $h(\pi)$ is a band. $\rhd$

2.1.5. We say that a net $(x_\alpha)_{\alpha \in \mathrm{A}}$ *bo-converges* to an element $x \in X$ and write $x = bo\text{-}\lim x_\alpha$ if there exists a decreasing net $(e_\gamma)_{\gamma \in \Gamma}$ in E such that $\inf_{\gamma \in \Gamma} e_\gamma = 0$ and, for every $\gamma \in \Gamma$, there is an index $\alpha(\gamma) \in \mathrm{A}$ such that $|x - x_\alpha| \leq e_\gamma$ for all $\alpha \geq \alpha(\gamma)$. Given an element $e \in E_+$, let the following condition be satisfied: for every number $\varepsilon > 0$, there is an index $\alpha(\varepsilon) \in \mathrm{A}$ such that $|x - x_\alpha| \leq \varepsilon e$ for all $\alpha \geq \alpha(\varepsilon)$. Then we say that (x_α) *br-converges* to x and write $x = br\text{-}\lim x_\alpha$. A net (x_α) is said to be *bo-fundamental (br-fundamental)* if the net $(x_\alpha - x_\beta)_{(\alpha,\beta) \in \mathrm{A} \times \mathrm{A}}$ *bo*-converges (*br*-converges) to zero. A lattice-normed space is called *bo-complete (br-complete)* if every *bo*-fundamental (*br*-fundamental) net in it *bo*-converges (*br*-converges) to an element of the space.

Take a family $(x_\xi)_{\xi \in \Xi}$ and associate with it the net $(y_\alpha)_{\alpha \in \mathrm{A}}$, where $\mathrm{A} := \mathscr{P}_{\mathrm{fin}}(\Xi)$ is the set of all finite subsets of Ξ and $y_\alpha := \sum_{\xi \in \alpha} x_\xi$. If $x := bo\text{-}\lim y_\alpha$ exists then the family (x_ξ) is said to be *bo-summable* and x is its *sum*. It is conventional to write $x = bo\text{-}\sum_{\xi \in \Xi} x_\xi$ in this case. Sometimes, if X is not a vector lattice, we write o and r instead of bo and br in the terms like o-complete.

A set $M \subset X$ is called *norm-bounded* if there exists an $e \in E_+$ such that $|x| \leq e$ for all $x \in M$. A space X is called disjointly complete or *d-complete* if every norm-bounded set in X of pairwise disjoint elements is *bo*-summable.

2.1.6. A subspace $X_0 \subset X$ is called a *bo-ideal* if, for $x \in X$ and $x_0 \in X_0$, from $|x| \leq |x_0|$ it follows that $x \in X_0$. In this subsection we assume that X is decomposable.

(1) *A subspace $X_0 \subset X$ is a bo-ideal if and only if $X_0 = h(L)$ for some o-ideal $L \subset E$.*

$\lhd$ Sufficiency is obvious. In order to prove necessity, we take an arbitrary o-ideal $X_0 \subset X$. Let L be the bo-ideal generated by the set $|X_0|$. It is clear that $X_0 \subset h(L)$. If $x \in h(L)$ then $|x| \le |u_1| + \dots + |u_n|$ for suitable $u_1, \dots, u_n \in X_0$. Due to decomposability of X, we have the representation $x = x_1 + \dots + x_n$, where $|x_k| \le |u_k|$ $(k := 1, \dots, n)$. By the definition of bo-ideal, $x_k \in X_0$; hence, $x \in X_0$. Thus, $X_0 = h(L)$. $\rhd$

(2) *The set $|h(L)|$ minorizes $|X|^{\perp\perp} \cap L$ for every o-ideal $L \subset E$. In particular,* 2.1.2 (1) *holds.*

$\lhd$ Indeed, if $0 < e \in |X|^{\perp\perp} \cap L$ then there exists an element $0 < e_0 \le e$ such that $e_0 \le |x_1| + \dots + |x_n|$ for some $x_1, \dots, x_n \in X$. Due to decomposability of X, the representation $e_0 = |u_1| + \dots + |u_n|$ holds, where $u_1, \dots, u_n \in X$. For at least one of the numbers k, we have $|u_k| \ne 0$. Since $e \in L$ and $|u_k| \le e$, it follows that $u_k \in h(L)$. $\rhd$

(3) *An order ideal $K \subset X$ is a band if and only if it is bo-closed.*

$\lhd$ Necessity is obvious. Suppose that K is a closed bo-ideal. According to (1), we have $K = h(L)$, where L is a bo-ideal in E. Demonstrate that $K = h(L^{\perp\perp})$. If $x \in h(L^{\perp\perp})$ then $|x| = \sup \mathscr{E}$, where $\mathscr{E} := \{e \in L : 0 \le e \le |x|\}$. By decomposability of X, we may choose an x_e so that $|x - x_e| = |x| - e$ and $|x_e| = e$. The set $\mathscr{E}$ is directed upward and the net $(x_e)_{e \in \mathscr{E}}$ o-converges to x. Consequently, $x \in K$ and $K \supset h(L^{\perp\perp})$. The reverse inclusion is obvious. $\rhd$

2.1.7. Under the assumption that (X, E) is a decomposable LNS, we establish three auxiliary facts.

(1) *Suppose that an element $x \in X$ and an increasing sequence $(a_n) \subset E_+$ are given and, moreover, $(a_n) \le |x|$ $(n \in \mathbb{N})$. Then there exists a sequence $(x_n) \subset X$ such that, for all $n \in \mathbb{N}$ and $m > n$, we have*

$$|x_n| = a_n, \quad |x - x_n| = |x| - a_n, \quad |x_m - x_n| = a_m - a_n.$$

$\lhd$ Denote $b_n := |x| - a_n$ $(n \in \mathbb{N})$, $b_0 := |x|$. By decomposability of X, we may choose sequences $(u_n) \subset X$ and $(v_n) \subset X$ so that

$$\begin{array}{lll} x = u_1 + v_1, & |u_1| = a_1, & |v_1| = b_1, \\ u_{n+1} + v_{n+1} = v_n, & |v_{n+1}| = b_{n+1}, & |u_{n+1}| = b_n - b_{n+1}. \end{array}$$

Assign $x_n = \sum_{k=1}^{n} u_k$. Then $x = x_n + v_n$ and the following relations hold:

$$|x_n| \le \sum_{k=1}^{n} |u_k| = \sum_{k=1}^{n} (b_{k-1} - b_k) = b_0 - b_n = a_n.$$

At the same time, $|x| \le |x_n| + |v_n| \le a_n + b_n = |x|$; consequently, $|x_n| = a_n$. Thus, for $m > n$, we have

$$|x_m - x_n| = \left|\sum_{k=n+1}^{m} u_k\right| \le \sum_{k=n+1}^{m} |u_k| = a_m - a_n \le |x_m - x_n|. \ \triangleright$$

(2) *If (X, E) is d-complete then it has the projection property. In particular, every d-complete Archimedean vector lattice has the projection property.*

$\triangleleft$ Given $y \in X$, denote by $E(y)$ the o-ideal in E that is generated by the element $|y|$. Assign $X(y) := \{x \in X : |x| \in E(y)\}$. It is easy to see that $(X(y), E(y))$ is a decomposable LNS. If K is a band in X and $K(y) := K \cap E(y)$ then projections of y onto K and $K(y)$ exist simultaneously and they are equal. At the same time $E(y)$ is disjointly complete. Indeed, for an arbitrary family (e_ξ) that is bounded in $E(y)$ and consists of disjoint elements, there is a family $(x_\xi) \subset X$ such that $|x_\xi| = e_\xi$. Therefore, for $x = bo\text{-}\sum x_\xi$, we have $|x| = o\text{-}\sum |x_\xi| = \sup e_\xi \in E(y)$. Thus, in view of 2.1.2 (3), it remains to apply 1.5.11. $\triangleright$

(3) *If (X, E) is d-complete and r-complete then $E_0 := |X|^{\perp\perp}$ is a K-space and $|X| = E_{0+}$.*

$\triangleleft$ According to (2) X has the projection property. Let $\hat{E}$ be the Dedekind completion of E. Since $E(y)$ is d-complete, we deduce that, for every $e \in \hat{E}(y)_+$, there is an increasing sequence $e_n \in E(y)_+$ that r-converges to e. Using (1), choose a sequence $(x_n) \subset X$ such that $|x_n| = e_n$ and $|x_m - x_n| \le e_m - e_n$ $(m > n)$. It is clear that the sequence (x_n) is r-fundamental and, due to r-completeness of X, $x = r\text{-}\lim x_n \in X$ exists. Moreover, $|x| = r\text{-}\lim e_n = e$. Consequently, $E(y) = \hat{E}(y)$. Next, let e be an arbitrary positive element in the Dedekind completion of the lattice $|X|^{\perp\perp}$. Then, for every maximal disjoint family $(e_\xi) \subset |X|$, we may choose a partition (π_n^ξ) of the identity projection in $\hat{E}$ so that $\pi_n^\xi e \le n e_\xi$, i.e., $\pi_n^\xi e \in |X|$. Therefore, there is a family $(x_n^\xi) \subset X$ and an $x \in X$ such that $|x_n^\xi| = \pi_n^\xi e \le e$ and $|x| = e$. Consequently, $E_0 = |X|^{\perp\perp}$ is a K-space and $|X| = E_{0+}$. $\triangleright$

2.1.8. Let A be a sublattice and subring in $\operatorname{Orth}(E)$ containing $\mathfrak{P}(E)$. We say that a lattice-normed space X over E *admits a compatible module structure over A* if X can be endowed with a structure of a faithful unitary A-module such that:

(a) the natural representation of A in X defines an isomorphism of the Boolean algebras $\mathfrak{P}(E)$ and $\mathscr{P}(X)$ from 2.1.3;

(b) $|ax| = |a||x|$ $(a \in A,\ x \in X)$.

In the case when X is a vector lattice we assume, in addition to (a) and (b), the following:

(c) $\mathscr{B}(X)$ is a complete subalgebra of the Boolean algebra of bands $\mathfrak{B}(X)$.

Let X be a d-decomposable lattice-normed space over a vector lattice E with $E = |X|^{\perp\perp}$ and let A be a sublattice and subring in $\mathrm{Orth}(E)$ *containing $\mathfrak{P}(E)$. Each of the following assertions implies that X admits a compatible module structure over A:*

(1) *A is the algebra of finite rank elements;*

(2) *E is order σ-complete, $A := \mathscr{L}(E)$, and X is br-complete;*

(3) *E is order σ-complete, $A = \mathrm{Orth}(E)$, and X is sequentially bo-complete;*

(4) *E is order complete, $A = \mathrm{Orth}(E)$, and X is a vector lattice with an order semicontinuous monotone norm.*

$\lhd$ Let $a \in A$ be a finite rank element, i.e. $a = \sum \lambda_k \pi_k$ where $\lambda_1, \dots, \lambda_n \in \mathbb{R}$ and $\pi_1, \dots, \pi_n$ is a finite partition of unity in $\mathfrak{P}(E)$. Then we put $ax := \sum \lambda_k \pi_k x$. Taking into account 2.1.2 (2), 2.1.3, and the identification of the Boolean algebras $\mathfrak{P}(E_0)$ and $\mathscr{P}(X)$,

$$|ax| = \left|\sum \lambda_k \pi_k x\right| = \sum |\lambda_k| \pi_k |x| = a|x|.$$

Now an arbitrary $a \in A$ is the order limit of an increasing sequence of finite rank elements $(a_n) \subset A$. The sequence $(a_n x) \subset X$ is *bo*-fundamental, since $|a_n x - a_m x| = |a_n - a_m||x| \to 0$. Therefore, we may assign $ax := bo\text{-}\lim a_n x$. Moreover,

$$|ax| = |bo\text{-}\lim a_n x| = o\text{-}\lim |a_n||x| = a|x|.$$

The remaining part of the proof is straightforward. $\rhd$

2.1.9. Let X be a lattice-normed vector lattice over E. In this case we have two different disjointness relations in X, one induced by vector norm as defined in 2.1.2 and the other determined by the order relation in X, see 1.3.3. They will be referred to, if need be, as *order disjointness* and *norm disjointness*, respectively. An element $x \in X$ is said to be *norm-n-decomposable* if $\bigwedge_{k=0}^{n} |x_k| = 0$ whenever $|x| = \sum_{k=0}^{n} x_k$ and $x_0, x_1, \dots, x_n$ are pairwise order disjoint elements in X_+. A norm-1-decomposable element is called *norm-indecomposable.*

If E is a vector lattice with the principal projection property then we may state the following equivalent definition: x is norm-n-decomposable if and only if for every collection of pairwise order disjoint elements $x_0, x_1, \dots, x_n \in X_+$ with $x = x_0 + x_1 + \dots + x_n$ there exists a partition of unity $\pi_0 \dots \pi_n$ in $\mathfrak{P}(F)$ such that $\pi_k x_k = 0$ for all $k = 0, 1, \dots, n$. Indeed, $e_0 \wedge \dots \wedge e_n = 0$ is equivalent to $\pi_k e_k = 0$ $(\forall k = 0, 1, \dots, n)$ for a suitable partition of unity $\pi_0 \dots \pi_n$ in $\mathfrak{P}(F)$.

In the case when E admits compatible module structure over $\mathrm{Orth}(E)$ we may give one more equivalent definition by replacing the partition of unity with the collection of positive orthomorphisms $\sigma_0, \sigma_1, \dots, \sigma_n$ such that $\sigma_0+\sigma_1+\dots+\sigma_n = I_F$.

(1) *The sum of n norm-indecomposable elements is norm-n-decomposable element.*

$\lhd$ Let $x = u_1 + \dots + u_n$ where $u_1, \dots, u_n$ are norm-indecomposable elements in X. Without loss of generality, we may assume that $u_k \geq 0$ $(k := 1, \dots, n)$. Take pairwise order disjoint elements $x_0, x_1, \dots, x_n \in X$ with $x_0 + x_1 + \dots + x_n = u_1 + \dots + u_n$. Using the Riesz Decomposition Property 1.3.2 (13), choose $u_{k,l} \in X_+$ such that

$$u_k = \sum_{l=1}^{n} u_{k,l} \ (k := 1, \dots, n), \quad x_l = \sum_{k=0}^{n} u_{k,l} \ (l := 0, 1, \dots, n).$$

From 1.3.2 (15) we deduce

$$\begin{aligned} 0 \leq \bigwedge_{l=0}^{n} |x_l| &= \bigwedge_{l=0}^{n} \Big| \sum_{k=1}^{n} u_{k,l} \Big| \leq \bigwedge_{l=0}^{n} \sum_{k=1}^{n} |u_{k,l}| \\ &\leq \sum_{j \in J} |u_{j(0),0}| \wedge |u_{j(1),1}| \wedge \dots \wedge |u_{j(n),n}| \end{aligned}$$

where J is the set of all functions $j : \{0, 1, \dots, m\} \to \{1, \dots, m\}$. At least two indices in $\{j(0), \dots, j(n)\}$ are equal, say $j(r) = j(s) = m$, $0 \leq r, s \leq n$, $r \neq s$. Therefore,

$$|u_{j(0),0}| \wedge |u_{j(1),1}| \wedge \dots \wedge |u_{j(n),n}| \leq |u_{m,r}| \wedge |u_{m,s}| = 0.$$

The latter follows from the relations $0 \leq u_{m,r} \leq x_r$, $0 \leq u_{m,s} \leq x_s$, $u_m = u_{m,0} + \dots + u_{m,n}$, since x_r and x_s are order disjoint and u_m is norm-indecomposable. $\rhd$

(2) *Let E be a vector lattice with the principal projection property and X be a d-decomposable lattice-normed vector lattice over E. Let $x \in X$ be a positive norm-n-decomposable element and $x = x_0 + x_1 + \dots + x_{n-1}$ for some collection of pairwise disjoint $x_0 \dots x_{n-1}$. If π is a band projection in E with $\pi(E) \subset \{|x_0| \wedge |x_1| \wedge \dots \wedge |x_{n-1}|\}^{\perp\perp}$ then $y := \pi(x_0)$ is norm-indecomposable and $z := \pi(x - x_0)$ norm-$(n-1)$-decomposable; moreover, y and z are disjoint fragments of x.*

$\lhd$ Let $0 \leq y_1 \leq y$, $0 \leq y_2 \leq y$, $y_1 \wedge y_2 = 0$, and $y_1 + y_2 = y$. Prove that $|y_1| \wedge |y_2| = 0$. Let π_k be the band projection onto $\{|x_k|\}^{\perp\perp}$ and put $\pi :=$

$\pi_0\pi_1 \dots \pi_{n-1}$. Then $n+1$ elements $\pi x_1, \dots, \pi x_{n-1}, y_1, y_2$ are pairwise disjoint and their supremum is equal to

$$\pi x_1 + \pi x_2 + \cdots + \pi x_{n-1} + y_1 + y_2 = \pi(x - x_0) + \pi(x_0) = \pi x.$$

Since any fragment of a norm-n-decomposable element is also norm-n-decomposable we obtain $|\pi x_1| \wedge \cdots \wedge |\pi x_{n-1}| \wedge |y_1| \wedge |y_2| = 0$, whence $\pi e = 0$, $e := \pi(|x_1| \wedge \cdots \wedge |x_{n-1}| \wedge |y_1| \wedge |y_2|)$. At the same time $\pi^\perp e = 0$, since $\pi^\perp e \leq \pi^\perp |y_1| \leq \pi^\perp |\pi x_0| = 0$, so that $e = 0$. It follows from this that $\pi(|y_1| \wedge |y_2|) = 0$, because π is the band projection onto $(|x_0| \wedge \cdots \wedge |x_{n-1}|)^{\perp\perp}$. Thus, $|y_1| \wedge |y_2| = \pi^\perp(|y_1| \wedge |y_2|) \leq \pi^\perp |\pi x_0| = 0$. Clearly y is a fragment of x.

Suppose now $z = z_1 + \cdots + z_n$ and $z_k \perp x_l$ $(k \neq l)$. Then πx is the sum of $n+1$ pairwise disjoint collection $\pi x_0, z_1, , \dots, z_n$ and $|\pi x_0| \wedge |z_1| \wedge \cdots \wedge |z_n| = 0$, since πx is norm-n-decomposable. It follows from this that $\pi e = 0$, $\pi(|x_0| \wedge |z_1| \wedge \cdots \wedge |z_n|) = 0$. At the same time $\pi^\perp e = 0$, so that $e = 0$. Consequently, $\pi(|z_1| \wedge \cdots \wedge |z_n|) = 0$ and $|z_1| \wedge \cdots \wedge |z_n| = 0$. $\triangleright$

2.1.10. *Theorem.* *Let E be a vector lattice with the principal projection property and X be a d-complete lattice-normed vector lattice over E. Then every norm-n-decomposable element $x \in X$ there exist n norm-indecomposable elements $x_1, \dots, x_n$ such that $x = x_1 + \cdots + x_n$.*

$\triangleleft$ The proof is by induction on n. The case $n = 1$ is a tautology. Suppose the claim is true for $n - 1$. By the Kuratowski–Zorn Lemma there is a maximal set $\mathscr{P}$ of pairwise disjoint band projections in E such that for each $\pi \in \mathscr{P}$ there exist n pairwise disjoint elements $x_0, x_1, \dots, x_{n-1} \in X$ such that $\sum_{i=0}^{n-1} x_i = x$ and $\pi(E) \subset (|x_0| \wedge |x_1| \wedge \cdots \wedge |x_{n-1}|)^{\perp\perp}$.

Using 2.1.9 (2), we may construct a functions $\pi \mapsto x_\pi$ defined on $\mathscr{P}$ such that x_π is positive and norm-indecomposable, $y_\pi := \pi x - x_\pi$ is positive norm-$(n-1)$-decomposable, $x_\pi = \pi x_\pi$, and x_π and y_π are disjoint fragments of x. Since the family $(x_\pi)_{\pi \in \mathscr{P}}$ is disjoint and X is d-complete we may define $y := \bigvee \{x_\pi : \pi \in \mathscr{P}\}$ and $z := x - y$. Observe that y and z are disjoint fragments of x.

By definition $\pi y = x_\pi$ for every $\pi \in \mathscr{P}$, so that, setting $\rho := \sup \mathscr{P}$, we obtain $\rho y = y$ and $\rho^\perp y = 0$. If $y = y_1 + y_2$ with $y_1 \wedge y_2 = 0$ then $\pi y_1 + \pi y_2 = x_\pi$ for $\pi \in \mathscr{P}$ and $\pi(|y_1| \wedge |y_2|) = 0$, since x_π is norm-indecomposable; thus $\rho(|y_1| \wedge |y_2|) = 0$. But $\rho^\perp(|y_1| \wedge |y_2|) \leq \rho |y| = 0$, whence $|y_1| \wedge |y_2| = 0$. Assume now that $z = z_0 + z_1 + \cdots + z_{n-1}$ for a disjoint collection $z_0, z_1, \dots, z_{n-1} \in X_+$ and put $e := |z_0| \wedge \cdots \wedge |z_{n-1}|)$. Then πe for every $\pi \in \mathscr{P}$, because $\pi z = y_\pi$ is $n-1$-decomposable. At the same time $x = (z_0 + y) + z_1 + \cdots + z_{n-1}$ and, taking into consideration the definition of $\mathscr{P}$, we obtain $\pi^\perp(|z_0| \wedge \cdots \wedge |z_{n-1}|) \leq \pi^\perp(|z_0 + y| \wedge |z_1| \wedge \cdots \wedge |z_{n-1}|) = 0$. We conclude that $\bigwedge_{i=0}^{n-1} |z_i| = 0$. Thus, we have proved that y is norm-indecomposable, z is norm-$n-1$-decomposable and $x = y + z$ and we are done in the case of positive x. For an arbitrary norm-n-indecomposable x the modulus $|x|$ is also norm-n-indecomposable and by above proved fact $|x| = y_1 + \dots + y_n$ with y_k norm-indecomposable. In view of the Riesz Decomposition Property $x^+ = u_1 + \dots + u_n$, $x^- = v_1 + \dots + v_n$, $u_k + v_k = y_k$ $(u_k, v_k \in X;\ k := 1, \dots, n)$. The element $x_k := u_k - v_k$ is norm-indecomposable, since $y_k = |x_k|$ and $x = x_1 + \dots + x_n$. $\triangleright$

2.2. Completion

Here we briefly consider the questions of completing lattice-normed spaces.

2.2.1. Every decomposable o-complete lattice-normed space is called a *Banach–Kantorovich space* (a *BKS* for short). If a Banach–Kantorovich space is in addition a vector lattice and the norm is monotone then it is called a *Banach–Kantorovich lattice.* Let (X, E) be a Banach–Kantorovich space and, moreover, $E = |X|^{\perp\perp}$. According to 2.1.2 (2), the Boolean algebras $\mathfrak{P}(E)$ and $\mathscr{B}(X)$ can be identified and $\pi|x| = |\pi x|$ $\big(\pi \in \mathfrak{P}(E),\ x \in X\big)$.

For every bounded family $(x_\xi)_{\xi\in\Xi}$ in X and every partition of unity $(\pi_\xi)_{\xi\in\Xi}$ in $\mathfrak{P}(X)$, the sum $x := o\text{-}\sum_{\xi\in\Xi} \pi_\xi x_\xi$ exists. Moreover, x is a unique element in X satisfying the relations $\pi_\xi x = \pi_\xi x_\xi$ $(\xi \in \Xi)$.

$\triangleleft$ If $e := \sup|x_\xi|$ then, for $\alpha, \beta \in \mathscr{P}_{\mathrm{fin}}(\Xi)$, we have

$$|y_\alpha - y_\beta| = \Big|\sum_{\xi\in\alpha\triangle\beta} \pi_\xi x_\xi\Big| \le \Big(\sum_{\xi\in\alpha\triangle\beta} \pi_\xi\Big) e \le e,$$

where $y_\gamma := \sum_{\xi\in\gamma} \pi_\xi x_\xi$ and $\alpha\triangle\beta$ is the symmetric difference of the sets α and β. Hence it is clear that the net (y_α) is bo-fundamental; hence, $x := bo\text{-}\lim y_\alpha$ exists. $\triangleright$

This proposition, in particular, ensures d-completeness of all BKSs. Moreover, the definitions imply readily that every BKS is br-complete as well. Thus, an arbitrary BKS (X, E) is d-complete and br-complete; therefore, by 2.1.7 (3) we have $E = |X|^{\perp\perp}$ and $E_+ = |X|$.

2.2.2. Let $\mathbb{B}$ be a complete Boolean algebra and let A be an arbitrary nonempty set. Assign

$$\mathbb{B}(A) := \Big\{\nu : A \to \mathbb{B} : (\forall \alpha, \beta \in A)\ \big(\alpha \ne \beta \Rightarrow \nu(\alpha) \wedge \nu(\beta) = 0\big) \wedge \bigvee_{\alpha\in A} \nu(\alpha) = 1\Big\}.$$

Thus, $\mathbb{B}(A)$ is the set of all partitions of unity in $\mathbb{B}$ that are indexed by elements of the set A. If A is an ordered set then we may order the set $\mathbb{B}(A)$ as well:

$$\nu \le \mu \Leftrightarrow (\forall \alpha, \beta \in A)\ \big(\nu(\alpha) \wedge \mu(\beta) \ne 0 \Rightarrow \alpha \le \beta\big) \quad \big(\nu, \mu \in \mathbb{B}(A)\big).$$

It is easy to show that this relation is actually a partial order in $\mathbb{B}(A)$. If A is directed upward (downward) then so does $\mathbb{B}(A)$. Let Q be the Stone space of the algebra $\mathbb{B}$. Identifying an element $\nu(\alpha)$ with a clopen subset of Q, we construct the mapping $\bar{\nu} : Q_\nu \to A$, $Q_\nu := \cup\{\nu(\alpha) : \alpha \in \mathrm{A}\}$, by letting $\bar{\nu}(q) = \alpha$ whenever $q \in \nu(\alpha)$. Thus, $\bar{\nu}$ is a step-function that takes the value α on $\nu(\alpha)$. Moreover, $\nu \le \mu \Rightarrow (\forall q \in Q_\nu \cap Q_\mu)\,(\bar{\nu}(q) \le \bar{\mu}(q))$. We shall use the set $\mathbb{B}(A)$ in the proof of the following important completeness criterion.

2.2.3. *Theorem.* *A decomposable lattice-normed space is order complete if and only if it is disjointly complete and complete with respect to relative uniform convergence.*

$\lhd$ Necessity was noted in 2.2.1. We will prove sufficiency. Suppose that a decomposable lattice-normed space X is d-complete and r-complete. According to 2.1.7 (3), we may assume without loss of generality that $E := |X|^{\perp\perp}$ is a K-space. Take an o-fundamental net $(x_\alpha)_{\alpha\in\mathrm{A}} \subset X$. Let $\mathbb{B} := \mathfrak{P}(X)$ be the base of X. Given $\nu \in \mathbb{B}(A)$, assign $x_\nu := \sum_{\alpha\in\mathrm{A}} \nu(\alpha)x_\alpha$. The element x_ν is well defined due to the d-completeness of X. A new net $(x_\nu)_{\nu\in\mathbb{B}(A)}$ thus appeared. Show that the net is r-fundamental. Choose a decreasing net $(e_\alpha)_{\alpha\in\mathrm{A}} \subset E$ so that $\inf e_\alpha = 0$ and $|x_\alpha - x_\beta| \le e_\gamma \le e$ whenever $\alpha, \beta \ge \gamma$ $(\alpha, \beta, \gamma \in A)$, where $e \in E$. Take an arbitrary number $\varepsilon > 0$. There exist a partition of unity $(\rho_\xi)_{\xi\in\Xi}$ in the algebra $\mathbb{B}$ and a mapping $\varphi : \Xi \to A$ such that $\rho_\xi e_{\varphi(\xi)} \le \varepsilon e$ $(\xi \in \Xi)$. This is easily deduced from the properties of o-convergence in a K-space. Assign $\pi_\alpha := \sup\{\rho_\xi : \varphi(\xi) = \alpha,\ \xi \in \Xi\}$ and $\pi_\alpha = 0$ in case $\alpha \notin \operatorname{im}\varphi$. Then $(\pi_\alpha)_{\alpha\in\mathrm{A}}$ is a partition of unity in $\mathbb{B}$ and $\pi_\alpha e_\alpha \le \varepsilon e$ $(\alpha \in \mathrm{A})$. Denote this partition of unity by $\nu(\varepsilon)$. Show that $|x_{\nu'} - x_\nu| \le \varepsilon e$ whenever $\nu, \nu' \ge \nu(\varepsilon)$. Let $\nu = (\rho_\alpha)_{\alpha\in\mathrm{A}}$ and $\nu' = (\tau_\alpha)_{\alpha\in\mathrm{A}}$. If $\pi_\alpha\rho_\beta\tau_\gamma \ne 0$ then $\beta, \gamma \ge \alpha$; consequently,

$$\begin{aligned}\pi_\alpha\rho_\beta\tau_\gamma|x_\nu - x_{\nu'}| &= \pi_\alpha|\rho_\beta\tau_\gamma x_\gamma - \rho_\beta\tau_\gamma x_\beta| \\ &= \pi_\alpha\rho_\beta\tau_\gamma|x_\gamma - x_\beta| \le \pi_\alpha e_\alpha \le \varepsilon e.\end{aligned}$$

Summing up over α, β, and γ, we find $|x_\nu - x_{\nu'}| \le \varepsilon e$ for $\nu, \nu' \ge \nu(\varepsilon)$. Thus, the net (x_ν) is r-fundamental and, due to the r-completeness, $x := r\text{-}\lim x_\nu$ exists. It is clear that $|x - x_\nu| \le \varepsilon e$ whenever $\nu \ge \nu(\varepsilon)$.

For any fixed index $\gamma \in A$, construct a special partition of unity $\nu := (\rho_\alpha)_{\alpha\in\mathrm{A}}$ by letting $\rho_\alpha = \pi_\alpha$ in case $\alpha \ne \gamma$ and $\alpha \ne \beta$, $\rho_\gamma = 0$, and $\rho_\beta = \pi_\gamma \vee \pi_\beta$, where $\beta \in A$ and $\beta \ge \gamma$. Then $\nu \ge \nu(\varepsilon)$ and

$$\pi_\gamma|x - x_\beta| = \pi_\gamma\rho_\beta|x - x_\nu| = \pi_\gamma|x - x_\nu| \le \varepsilon e.$$

Thus, $\pi_\gamma|x - x_\beta| \le \varepsilon e$ for all $\beta, \gamma \in A$, $\beta \ge \gamma$. Put $c_\gamma := \sup\{|x - x_\beta| : \beta \ge \gamma\}$ and $c := \inf\{c_\gamma : \gamma \in A\}$. Then $\pi_\gamma c \le \pi_\gamma c_\gamma \le \varepsilon e$. Since $\gamma \in A$ and $\varepsilon > 0$ are arbitrary, we have $c = 0$; hence, $o\text{-}\lim|x - x_\alpha| = 0$. $\rhd$

2.2.4. As before, let mE be a universal completion of a vector lattice E; moreover, let a unity $\mathbf{1}$ be fixed together with the respective multiplication structure in mE. Take a lattice-normed space X with $|X|^{\perp\perp} = E$. The *operator-dual space* X^* is defined as follows. An operator $x^* : X \to mE$ belongs to X^* if and only if there exists an element $0 \leq c \in mE$ such that

$$\langle x, x^* \rangle := x^*(x) \leq c\,|x| \quad (x \in X).$$

The least element $0 \leq c \in mE$ satisfying the indicated relation exists. This element is denoted by $|x^*|$. It is easy to see that the mapping $x^* \mapsto |x^*|$ is an mE-valued norm in X^* and the following inequality holds:

$$\langle x, x^* \rangle \leq |x|\,|x^*| \quad (x \in X).$$

(1) *If $x^* \in X^*$ then $|\pi x^*| = \pi |x^*|$ for every projection $\pi \in \mathfrak{P}(mE)$.*

$\lhd$ From the inequality $\langle x, x^* \rangle \leq c|x|$ it follows readily that $\langle x, \pi x^* \rangle \leq \pi c|x|$; therefore, $|\pi x^*| \leq \pi |x^*|$. On the other hand, $\langle x, x^* \rangle \leq (|\pi x^*| + |\pi^\perp x^*|)\,|x|$ for all $x \in X$; hence, $|x^*| \leq |\pi x^*| + |\pi^\perp x^*| \leq |\pi x^*| + \pi^\perp |x^*|$. Hence we deduce $\pi |x^*| \leq |\pi x^*|$. $\rhd$

(2) *The space X^* is d-complete and r-complete.*

$\lhd$ Take a partition of unity (π_ξ) in $\mathfrak{P}(mE)$, a family $(x^*_\xi) \subset X^*$, and an element $x \in X$. In a universally complete K-space there exist $\langle x, x^* \rangle := \sum \pi_\xi \langle x, x^*_\xi \rangle$ and $e := \sum \pi_\xi |x^*_\xi|$; moreover, $\langle x, x^* \rangle \leq e|x|$. Hence it is clear that $x^* \in X^*$. Since $\pi_\xi x^* = \pi_\xi x^*_\xi$, we have $x^* = \sum \pi_\xi x^*_\xi$ in view of (1). Thus, d-completeness is proven. Completeness with respect to relative uniform convergence is established below in a more general situation (see 4.2.1). $\rhd$

(3) *The space X^* is a universally complete BKS.*

$\lhd$ Decomposability of X^* follows from (1) due to r-completeness, and it remains to refer to Theorem 2.2.3. $\rhd$

2.2.5. According to 2.2.4 (3), the second operator-dual space, X^{**}, is a universally complete BKS. The canonical embedding $\varkappa : X \to X^{**}$ is defined, as usual, by the formula

$$\langle x^*, \varkappa x \rangle := \varkappa x(x^*) = \langle x, x^* \rangle \quad (x^* \in X^*).$$

The canonical embedding $\varkappa$ is a linear isometry.

$\lhd$ We only need to show that $\varkappa$ preserves the norm, i.e., $|\varkappa x| = |x|$ $(x \in X)$. Observe first that $|\varkappa x| \leq |x|$, since $\langle x^*, \varkappa x \rangle \leq |x|\,|x^*|$ $(x^* \in X^*)$. Next, since the operator $|\cdot|$ is sublinear, therefore, by the Hahn–Banach–Kantorovich Theorem, there exists an operator $x^* : X \to mE$ such that $|x| = \langle x, x^* \rangle$ and $\langle y, x^* \rangle \leq |y|$ $(y \in X)$. Hence it is clear that $|x^*| \leq \mathbf{1}$; therefore, $|x| = \langle x^*, \varkappa x \rangle \leq |\varkappa x|\,|x^*| \leq |\varkappa x|$. $\rhd$

2.2.6. By a *universal completion* (an *order completion* or, in short, a *bo-completion*) of a lattice-normed space (X, E) we mean a universally complete BKS (Y, mE) (respectively, a BKS (Y, oE)) together with a linear isometry $\imath : X \to Y$ such that each universally complete o-complete subspace of (Y, mE) (respectively, any decomposable o-complete subspace of (Y, oE)) containing $\imath X$ coincides with Y. Here oE is a Dedekind completion of the vector lattice E, and mE is, as before, a universal completion of oE; moreover, we assume $E \subset oE \subset mE$, see 1.1.8. Given a set U in a lattice-normed space Y, we assign

$$r(U) := \{y = br\text{-}\lim_{n\to\infty} y_n : \quad (y_n)_{n\in\mathrm{N}} \subset U\},$$
$$o(U) := \{y = bo\text{-}\lim y_\alpha : \quad (y_\alpha)_{\alpha\in\mathrm{A}} \subset U\},$$
$$d(U) := \Big\{y = bo\text{-}\sum_{\xi\in\Xi} \pi_\xi y_\xi : (y_\xi)_{\xi\in\Xi} \subset U\Big\},$$

where A is an arbitrary directed set, (π_ξ) is an arbitrary partition of unity in $\mathfrak{P}(Y)$, and the limits and the sums exist in Y. Let $r_0(U)$ be the part of $r(U)$ containing the limits of sequences in U converging with regulator $\mathbf{1}$, and let $d_0(U)$ be the part of $d(U)$ containing the finite sums. Observe the following simple relations: $d_0(d_0(U)) = d_0(U)$, $d(d_0(U)) = d(U)$. Now denote $mX := rd(\varkappa X)$, where $\varkappa$ is the canonical embedding $X \to X^{**}$ of 2.2.5 and the operations d and r are calculated in the universally complete BKS (X^{**}, mE).

2.2.7. *If Y is a decomposable BKS then, for any $U \subset Y$, the following hold:*

(1) $dd(U) = d(U)$;

(2) $dr_0d(U) = r_0d(U)$;

(3) $rd(U) = r_0d(U)$;

(4) $rrd(U) = rd(U)$.

$\lhd$ In each of the desired equalities, the inclusion $\supset$ is obvious; therefore, the reasoning below concerns the reverse inclusion.

(1): Take a family $(y_\xi)_{\xi\in\Xi}$ in $d(U)$ and a partition of unity $(\pi_\xi)_{\xi\in\Xi}$ in the algebra $\mathfrak{P}(Y)$. Represent an element $y_\xi \in d(U)$ as $y_\xi = \sum_{u\in U} \pi_{u,\xi} u$, where $(\pi_{u,\xi})_{u\in U}$ is a partition of unity in $\mathfrak{P}(Y)$ for each $\xi \in \Xi$. Assign $y := \sum_{\xi\in\Xi} \pi_\xi y_\xi$ and $\pi_{(\xi,u)} := \pi_\xi \circ \pi_{u,\xi}$ $(\xi \in \Xi,\ u \in U)$. If $(\xi, u) \neq (\eta, v)$ then

$$\pi_{(\xi,u)} \circ \pi_{(\eta,v)} = \pi_\xi \circ \pi_\eta \circ \pi_{u,\xi} \circ \pi_{v,\eta} = 0.$$

At the same time,

$$\bigvee_{(\xi,u)\in\Xi\times U} \pi_{(\xi,u)} = \bigvee_{\xi\in\Xi} \pi_\xi \circ \Big(\bigvee_{n\in U} \pi_{u,\xi}\Big) = \mathbf{1};$$

consequently, $(\pi_\lambda)_{\lambda\in\Xi\times U}$ is a partition of unity in $\mathfrak{P}(Y)$. Now observe that

$$\pi_{(\xi,u)}y = \pi_{u,\xi}(\pi_\xi y) = \pi_{u,\xi}(\pi_\xi y_\xi) = \pi_\xi(\pi_{u,\xi}u) = \pi_{(\xi,u)}u.$$

Thus, $y = \sum_{(\xi,u)} \pi_{(\xi,u)}u$ and $y \in d(U)$.

(2): Now let the family (y_ξ) be included in $r_0 d(U)$. Fix a number $\varepsilon > 0$ and, for each $\xi \in \Xi$, choose a $u_\xi \in d(U)$ so that $|y_\xi - u_\xi| \le \varepsilon\mathbf{1}$. If $u = \sum \pi_\xi u_\xi$ and y is as above, then $|y - u| \le \varepsilon\mathbf{1}$. However, due to (1), $u \in d(U)$; therefore, $y \in r_0 d(U)$.

(3): Suppose that a sequence $(y_n)_{n\in\mathbb{N}}$ in $d(U)$ converges to an element $y \in Y$ with regulator $e \in mE$. Choose a partition of unity $(\pi_\xi)_{\xi\in\Xi}$ in $\mathfrak{P}(mE)$ so that $\pi_\xi e \le \lambda_\xi \mathbf{1}$ for suitable $\lambda_\xi > 0$. Take an arbitrary $\varepsilon > 0$. For each $\xi \in \Xi$, there exists an index $n(\xi) \in \mathbb{N}$ such that $\pi_\xi|y - y_{n(\xi)}| \le \varepsilon\mathbf{1}$. Assign $u = \sum_{\xi\in\Xi} \pi_\xi y_{n(\xi)}$ and observe that $u \in d(U)$ due to (1). It remains to take account of the inequality $|y - u| \le \varepsilon\mathbf{1}$, and we arrive at the conclusion $y \in r_0 d(U)$.

(4): First, note that the operation r_0 is a topological closure and, therefore, $r_0 r_0(A) = r_0(A)$ for each $A \subset Y$. However, the latter can be easily proven directly. Next, applying (2) and (3), we may write

$$\begin{aligned} rrd(U) &= rr_0 d(U) = rdr_0 d(U) = r_0 dr_0 d(U) = r_0 r_0 d(U) \\ &= r_0 d(U) = rd(U). \ \triangleright \end{aligned}$$

2.2.8. *Theorem.* *Each lattice-normed space has a universal completion that is unique to within linear isometry. The space (mX, mE) serves as a universal completion for (X, E) and is referred to as the universal completion of (X, E) in the sequel.*

$\triangleleft$ The d-completeness of the space mX follows from 2.2.7 (2, 3), and its r-completeness is clear from 2.2.7 (4). By Theorem 2.2.3, mX is a BKS. Moreover, mX is universally complete, since the operation d in the definition of mX is calculated in the universally complete space (X^{**}, mE), see 2.2.6. The embedding $\imath := \varkappa : X \to mX$ is a linear isometry (2.2.5). If Y is a decomposable BKS and $\imath X \subset Y \subset mX$, then the values of the operations d and r are always in Y; therefore,

$$mX = rd(\imath X) \subset rd(Y) \subset Y \subset mX,$$

i.e., $Y = mX$. Thus, mX is a universal completion for the space X.

Assume that Y' is one more universal completion for X and let $\imath' : X \to Y'$ be the corresponding isometric embedding. The operator $h := \imath' \circ \imath^{-1} : \imath X \to Y'$ is a linear isometry. We may extend h onto $d(\imath X)$ and next onto $rd(\imath X)$ preserving linearity and isometry. Moreover, $h(mX)$ is a universally complete BKS and $\imath' X \subset h(mX) \subset Y'$; consequently, $h(mX) = Y'$. $\triangleright$

2.2.9. In this subsection X is a d-decomposable lattice-normed space over a vector lattice E possessing the principal projection property.

(1) *Let $U \subset X$ and $d_0(U) = U$. Then for every $x \in X$ there exists a net $(u_\alpha)_{\alpha\in\mathrm{A}}$ in U such that the net $\big(|x - u_\alpha|\big)_{\alpha\in\mathrm{A}}$ decreases and $\{|x - u_\alpha| : \alpha \in \mathrm{A}\} = \{|x - u| : u \in U\}$. In particular, o-$\lim_{\alpha\in\mathrm{A}}|x - u_\alpha| = \inf_{u\in U}|x - u|$.*

$\lhd$ Take $x \in X$ and introduce some equivalence relation and preorder in U by the following formulas:

$$u \sim v \;\Leftrightarrow\; |x - u| = |x - v|,$$
$$u \preccurlyeq v \;\Leftrightarrow\; |x - u| \geq |x - v|.$$

Since E is a lattice with the principal projection property we may choose a projection $\pi \in \mathfrak{P}(X)$ such that $\pi|x - u| + \pi^\perp|x - v| = |x - u| \wedge |x - v|$. The element $w := \pi u + \pi^\perp v$ lies in U since $d_0(U) = U$. Moreover,

$$|x - w| = |x - u| \wedge |x - v|,$$

whence $u \preccurlyeq w$ and $v \preccurlyeq w$. Thus, the preordered set $(U, \preccurlyeq)$ is directed upward. Hence it is clear that the factor set $\mathrm{A} := U/{\sim}$, endowed with the factor order, is an upward-directed ordered set. Now, the desired net $(u_\alpha)_{\alpha\in\mathrm{A}}$ is defined by choosing $u_\alpha \in \alpha$ $(\alpha \in \mathrm{A})$. $\rhd$

We say that a set $U \subset X$ *approximates an element* $x \in X$ if $\inf_{u\in U}|x - u| = 0$ and that U *approximates a set* $V \subset X$ if U approximates every element of V. In case if U approximates each $x \in X$ we say that U is an *approximating set.*

(2) *If V approximates U and W approximates V, then W approximates U.*

$\lhd$ For an arbitrary $u \in U$, denote $e := \inf_{w\in W}|w - u|$ and suppose that $e \neq 0$. Since $\inf_{v\in V}|v - u| = 0$, there exist $v \in V$ and a band projection ρ such that $\rho|v - u| < \rho e/2$. By the same arguments, $\inf_{w\in W}|w - v| = 0$ and there are $w \in W$ and a band projection π with $\pi|w - v| < \pi\rho e/2$. Thus we arrive at the contradiction

$$\pi\rho e \leqslant \pi\rho|w - u| \leqslant \pi\rho|w - v| + \pi\rho|v - u| < \pi\rho e/2 + \pi\rho e/2 = \pi\rho e. \ \rhd$$

(3) *Let $U \subset$ and $x \in X$. Then U approximates x if and only if x is the bo-limit of some net of elements of $d_0(U)$.*

$\lhd$ Indeed, if U approximates x then $\inf\big\{|x - u| : u \in d_0(U)\big\} = 0$. Thus, in view of (1) there is a net $(u_\alpha)_{\alpha\in\mathrm{A}}$ in $d_0(U)$ such that o-$\lim|x - u_\alpha| = 0$. Conversely, if x is a bo-limit of a net containing in $d_0(U)$ then $d_0(U)$ approximates x. It remains to apply (2), since U approximates $d_0(U)$. $\rhd$

2.2.10. Assume that E is an order complete vector lattice, B is a complete Boolean algebra, and $h : \mathfrak{P}(E) \to B$ is a ring homomorphism. Say that a net $(\pi_\alpha)_{\alpha\in\mathrm{A}}$ in $\mathfrak{P}(E)$ *h-converges to zero* and write $h\text{-}\lim_{\alpha\in\mathrm{A}} \pi_\alpha = \mathbf{0}$ provided that $o\text{-}\lim_{\alpha\in\mathrm{A}} \pi_\alpha = \mathbf{0}$ and $o\text{-}\lim_{\alpha\in\mathrm{A}} h(\pi_\alpha) = \mathbf{0}$ in the Boolean algebras $\mathfrak{P}(E)$ and B respectively. We say that a net $(e_\alpha)_{\alpha\in\mathrm{A}}$ in E *h-converges to* $e \in E$ and write $h\text{-}\lim_{\alpha\in\mathrm{A}} e_\alpha = e$ if the net $(e_\alpha)_{\alpha\in\mathrm{A}}$ is bounded beginning with some member and $h\text{-}\lim_{\alpha\in\mathrm{A}}[d]\big[(|e_\alpha - e| - d)^+\big] = \mathbf{0}$ for all positive $d \in E$. In this case, we call the element e the *h-limit* of the net $(e_\alpha)_{\alpha\in\mathrm{A}}$. Finally, we say that a net $(u_\alpha)_{\alpha\in\mathrm{A}}$ in X *h-converges to* $u \in X$ and write $h\text{-}\lim_{\alpha\in\mathrm{A}} u_\alpha = u$ if $h\text{-}\lim_{\alpha\in\mathrm{A}} \lvert u_\alpha - u \rvert = 0$. In this case, we call the element u the *h-limit* of the net $(u_\alpha)_{\alpha\in\mathrm{A}}$. For an arbitrary family $(e_\xi)_{\xi\in\Xi}$ in E_+, the notation $h\text{-}\inf_{\xi\in\Xi} e_\xi = 0$ means that $\inf_{\xi\in\Xi} h\big([d]\big[(e_\xi - d)^+\big]\big) = \mathbf{0}$ for all positive $d \in E$.

(1) *If $h\text{-}\lim_{\alpha\in\mathrm{A}} e_\alpha = e$ or $h\text{-}\inf_{\xi\in\Xi} e_\xi = 0$, then $o\text{-}\lim_{\alpha\in\mathrm{A}} e_\alpha = e$ or $\inf_{\xi\in\Xi} e_\xi = 0$, respectively. The converse is also true provided that the homomorphism h is o-continuous.*

Let U be a subset of an LNS X. We say that U *h-approximates an element* $x \in X$ if $h\text{-}\inf_{u\in U} \lvert x - u \rvert = 0$. We say that U *h-approximates a set* $V \subset X$ if U h-approximates every element of V. A subset of an LNS X is called *h-approximating* if it h-approximates X. Every h-approximating set is approximating and, in case the homomorphism h is o-continuous, the notions of approximating and h-approximating set coincide.

(2) *Suppose that V h-approximates U and W h-approximates V. Then W h-approximates U.*

$\lhd$ Consider an arbitrary element $u \in U$, fix a positive element d of the norm lattice, and assign $b := \inf_{w\in W} h\big([d][(\lvert x - z \rvert - d)^+]\big)$. By 2.2.9 (2), it is sufficient to establish the equality $b = 0$. For simplicity, we assume that $h([d]) = \mathbf{1}$. Suppose to the contrary that $b \neq 0$. Then, in view of $\inf_{v\in V} h([(\lvert v - u \rvert - d/2)^+]) = \mathbf{0}$, there is an element $v \in V$ such that $b_0 := b \wedge h([\lvert v - u \rvert - d/2)^+]) < b$. Similarly, in view of the equality $\inf_{w\in W} h([(\lvert w - v \rvert - d/2)^+]) = \mathbf{0}$, there is an element $w \in W$ such that $(b \backslash b_0) \wedge h([\lvert w - x \rvert - d/2)^+]) < (b \backslash b_0)$. It is easy to verify that w satisfies the inequality $b \wedge h([(\lvert w - u \rvert - d)^+]) < b$, which contradicts the definition of b. $\rhd$

(3) *A set $U \subset X$ h-approximates an element $x \in X$ if and only if x is the h-limit of some net in $d_0(U)$.*

$\lhd$ If U h-approximates x then, in view of 2.2.9 (1), there exists a net $(v_\alpha)_{\alpha\in\mathrm{A}}$ in $d_0(U)$ such that the net $\big(\lvert x - v_\alpha \rvert\big)_{\alpha\in\mathrm{A}}$ decreases and

$$\{\lvert x - v_\alpha \rvert : \alpha \in \mathrm{A}\} = \{\lvert x - v \rvert : v \in d_0(U)\}.$$

It remains to observe that $h\text{-}\lim_{\alpha\in\mathrm{A}} \lvert x - v_\alpha \rvert = 0$.

Conversely, if x is the h-limit of a net in $d_0(U)$, then $d_0(U)$ h-approximates x. It remains to observe that U h-approximates $d_0(U)$ and to use (2). $\rhd$

2.2.11. (1) *Theorem.* *For each lattice-normed space X, there exists a bo-completion which is unique to within linear isometry and is referred to as the bo-completion of X in the sequel.*

$\triangleleft$ Recall that $E \subset oE \subset mE$. Assign

$$\hat{X} = \{x \in mX : |x| \in oE\}.$$

Then $\hat{X}$ is an order completion of the space X. $\triangleright$

We always assume that any lattice-normed space X is contained in its *bo*-completion $\hat{X}$.

(2) *For the bo-completion $\hat{X}$ of the space X, we have $\hat{X} = rdX$. If X is decomposable and $E_0 := |X|^{\perp\perp}$ is a vector lattice with the principal projection property, then also $\hat{X} = oX$.*

$\triangleleft$ The first part of the assertion ensues from 2.2.7 and 2.2.8. Take $\hat{x} \in \hat{X}$ and choose a net $(x_\alpha) \subset X$ that *bo*-converges to $\hat{x}$. Apply Proposition 2.2.10 (1) with $U := X$ and $X := \hat{X}$. There exists a net $(x_\alpha)_{\alpha \in \mathrm{A}}$ in X such that the net $(|\hat{x} - x_\alpha|)_{\alpha \in \mathrm{A}}$ decreases and $e := \inf_{x \in X} |\hat{x} - x| = \inf_{\alpha \in \mathrm{A}} |\hat{x} - x_\alpha|$, where the infimum is calculated in oE. In view of the equality $\hat{X} = rdX$, for an arbitrary number $\varepsilon > 0$, we may find a family $(x_\xi) \subset X$ and a partition of unity $(\pi_\xi) \subset \mathfrak{P}(X)$ so that $|\hat{x} - \sum \pi_\xi x_\xi| \le \varepsilon |x|$. So, we may write

$$e = o\text{-}\sum \pi_\xi e \le o\text{-}\sum \pi_\xi |\hat{x} - x_\xi| = \Big|\hat{x} - bo\text{-}\sum \pi_\xi x_\xi\Big| \le \varepsilon |\hat{x}|.$$

Therefore, $e = 0$ and $\hat{x} = bo\text{-}\lim x_\alpha$. $\triangleright$

2.2.12. *Let X be a vector lattice with a monotone norm. Then its norm completion $\hat{X}$ is a Banach–Kantorovich lattice and X is a sublattice in $\hat{X}$.*

$\triangleleft$ Denote $Y_+ := rd(X_+)$. Then Y_+ is a cone. Indeed, Y_+ is a wedge, since the operations r and d are linear. At the same time, if $\pm y \in Y_+$ then, by definition, for every $0 < \varepsilon \in \mathbb{R}$ there are a partition of unity $(\pi_\xi) \subset \mathfrak{P}(E)$ and families (x'_ξ) and (x''_ξ) in X_+ such that

$$\Big|y - bo\text{-}\sum \pi_\xi x'_\xi\Big| \le \varepsilon |y|, \quad \Big|y + bo\text{-}\sum \pi_\xi x''_\xi\Big| \le \varepsilon |y|.$$

From these inequalities we deduce

$$|2y| \le \Big|y - bo\text{-}\sum \pi_\xi x'_\xi\Big| + \Big|y + bo\text{-}\sum \pi_\xi x''_\xi\Big| \le 2\varepsilon |y|.$$

Therefore, $|y| = 0$ and $y = 0$. Further details are obvious. $\triangleright$

2.2.13. Let X be an arbitrary, not necessarily d-decomposable, lattice-normed space over an arbitrary vector lattice E. Suppose that a d-decomposable lattice-normed space $\overline{X}$ over E includes X as a subspace with the induced norm. We say that $\overline{X}$ is a *d-decomposable hull* of X, if $\overline{X} = d_0(X)$, i.e., if every member of $\overline{X}$ is represented as a finite sum $\pi_1 x_1 + \dots + \pi_n x_n$, where $x_1, \dots, x_n \in X$ and pairwise disjoint $\pi_1, \dots, \pi_n \in \mathfrak{P}(E)$.

(1) *Suppose that a vector lattice E possesses the principal projection property. Then every (not necessarily d-decomposable) lattice-normed space over E has a d-decomposable hull that is unique to within an isometry.*

$\lhd$ Let mX be a universal completion of a lattice-normed space X. Define $\overline{X}$ to be the set of all finite sums $\sum_{k=1}^n \pi_k x_k$, where $x_1, \dots, x_n \in X$ and $(\pi_k)_{k=1}^n$ is a partition of unity in the Boolean algebra $\mathfrak{P}(E)$. It is easy to see that $\overline{X}$ is an LNS over E and is a d-decomposable hull of X. Uniqueness of this d-decomposable hull is obvious. $\rhd$

(2) We describe one more useful construction. Let E and F be K-spaces and let X be an LNS over E. Suppose that a mapping $S : E_+ \to F_+$ is subadditive $\big(S(e_1 + e_2) \leqslant Se_1 + Se_2\big)$, positively homogeneous $\big(S(\lambda e) = \lambda Se\ (\lambda \in \mathbb{R}_+)\big)$, and increasing $\big(0 \leqslant e_1 \leqslant e_2 \Rightarrow Se_1 \leqslant Se_2\big)$.

Consider the vector subspace $X_0 := \{x \in X : S(|x|) = 0\}$ and denote by $S_X x$ the coset in X/X_0 containing an $x \in X$. It is easy to check that the space X/X_0 is a lattice-normed space with respect to the norm $|S_X x| := S(|x|)$. Observe that the LNS X/X_0 need not be d-decomposable. We call a d-decomposable hull of the LNS X/X_0 the *norm transformation of X by means of S* and denote it by SX. The linear operator $S_X : X \to SX$ is called the *operator of norm transformation of X by means of S.*

2.3. Examples

In this section we consider important examples of spaces of continuous, weakly continuous, measurable, and weakly measurable vector-functions that can be naturally normed by means of a vector lattice.

2.3.1. We begin with the simplest extreme cases, namely, vector lattices and normed spaces. If $X = E$ then the modulus of an element can be taken as its vector norm: $|x| := |x| = x \vee (-x)\quad (x \in E)$. Decomposability of this norm is easy from the Riesz Decomposition Property holding in every vector lattice.

If $E = \mathbb{R}$ then X is a normed space. We shall use the conventional notation for the norm, $\|\cdot\|$, and omit references to the order structure of the norm lattice.

2.3.2. Let Q be a topological space and let Y be a normed space. Let $X := C_b(Q, Y)$ be the space of bounded continuous vector-functions from Q into

Y. Assign $E := C_b(Q, \mathbb{R})$. Given $f \in X$, define its vector norm $|f|$ by the relation $|f| : t \mapsto \|f(t)\|$ $(t \in Q)$. Then $|\cdot|$ is a decomposable norm. Indeed, assume that $|f| = h_1 + h_2$ for some $h_1, h_2 \in E$. Define a vector-function $f_1 : Q \to Y$ so that $f_1(t) = h_1(t)f(t)/\|f(t)\|$ whenever $f(t) \neq 0$ and $f_1(t) = 0$ whenever $f(t) = 0$. Then $f_1 \in X$ and $f_2 := f - f_1 \in X$; moreover, $|f_k| = h_k$ $(k := 1, 2)$. The space X is br-complete if and only if Y is a Banach space.

2.3.3. Suppose that E is an order-dense ideal of the universally complete vector lattice $C_\infty(Q)$, where Q is an extremal compact space. Let $C_\infty(Q, X)$ be the set of cosets of continuous vector-functions u that act from comeager subsets $\mathrm{dom}(u) \subset Q$ into some normed space X. Recall that a set is called *comeager* if its complement is meager. Vector-functions u and v are equivalent if $u(t) = v(t)$ whenever $t \in \mathrm{dom}(u) \cap \mathrm{dom}(v)$.

The set $C_\infty(Q, X)$ is endowed, in a natural way, with the structure of a module over $C_\infty(Q)$. Moreover, the continuous extension of the pointwise norm defines a decomposable norm on $C_\infty(Q, X)$ with values in $C_\infty(Q)$. Indeed, given any $z \in C_\infty(Q, X)$, there exists a unique function $x_z \in C_\infty(Q)$ such that $\|u(t)\| = x_z(t)$ $\big(t \in \mathrm{dom}(u)\big)$, for every representative u of the coset z. Assign $|z| := x_z$ and

$$E(X) := \{z \in C_\infty(Q, X) : |z| \in E\}.$$

If X is a Banach space then $E(X)$ presents a Banach–Kantorovich space with $C_\infty(Q, X)$ its universal completion.

◁ The assertion can be easily deduced from 2.2.3. It also follows immediately from a more general fact which will be proved in the next section (see 2.4.6 and 2.4.7). ▷

Denote by $C_\#(Q, X)$ the part of $C_\infty(Q, Y)$ consisting of cosets z with $|z| \in C(Q)$. Thus, $C_\#(Q, X) := E(X)$ where $E := C(Q)$. Observe that $C_\#(Q, X)$ is also a Banach–Kantorovich space, whereas the space $C(Q, X)$ of everywhere defined continuous functions from Q to X, being a lattice-normed space over $C(Q)$ (see 2.3.2) is not d-complete in general (see 2.4.9). Thus, $C(Q, X)$ does not coincide with the space $C_\#(Q, X)$, unless Q is finite or X is finite-dimensional.

2.3.4. For the same X and E define an embedding of the *algebraic tensor product* $E \otimes X$ into the space $E(X)$. Associate with an element $z = \sum_{k=1}^n e_k \otimes x_k \in E \otimes X$ the coset $\bar{z}$ of the vector-function

$$q \mapsto \bar{z}(q) := \sum_{k=1}^n e_k(q) x_k \quad \Big(q \in Q(z) := \Big\{s \in Q : \sum_{k=1}^n |e_k(s)| < +\infty\Big\}\Big).$$

We also endow the space $E \otimes X$ with the norm

$$|z|_0 := \inf\Big\{\sum_{k=1}^n |e_k| \|x_k\|\Big\} \quad (z \in E \otimes X),$$

where the infimum is taken over all representations $z = \sum_{k=1}^{n} e_k \otimes x_k$.

(1) *The mapping* $\iota : z \mapsto \bar{z}$ *from* $E \otimes X$ *into* $E(X)$ *is a linear isometry.*

$\triangleleft$ Clearly, ι is a linear embedding. If $z = \sum_{k=1}^{n} e_k \otimes x_k$ then

$$\lvert z \rvert \leq \left\| \sum_{k=1}^{n} e_k(\cdot) x_k \right\| \leq \sum_{k=1}^{n} |e_k(\cdot)| \|x_k\|;$$

hence, $\lvert z \rvert \leq \lvert z \rvert_0$. For each $s \in Q(z)$ we may choose a function $\phi_s \in E$ such that $\phi_s(s) = 1$. Put $x_s := \sum_{k=1}^{n} e_k(s) x_k \in X$ and $e'_k := e_k - \phi_s e_k(s)$. Then $z = \phi_s x_s + \sum_{k=1}^{n} e'_k x_k$ and for an arbitrary $s \in Q(z)$ we have

$$\begin{aligned} \lvert z \rvert_0(s) &\leq \Big(|\phi_s| \|x_s\| + \sum_{k=1}^{n} |e'_k| \|x_k\| \Big)(s) \\ &= \Big(|\phi_s(\cdot)| \|x_s\| + \sum_{k=1}^{n} |e_k(\cdot) - \phi_s(\cdot) e_k(s)| \|x_k\| \Big)(s) \\ &= |\phi_s(s)| \|x_s\| = \left\| \sum_{k=1}^{n} e_k(s) x_k \right\| = \|z(s)\| = \lvert z \rvert. \ \triangleright \end{aligned}$$

Thus, $E \otimes X$ can be identified with a subspace of the LNS $E(X)$. It turns out that $E \otimes X$ is, in a sense, dense in $E(X)$.

(2) *For every* $z \in E(X)$ *and every* $\varepsilon > 0$, *there exist a family* $\big(Q(\xi)\big)_{\xi \in \Xi}$ *of pairwise disjoint clopen sets* $Q(\xi) \subset Q$ *whose union is dense in* Q *and a family* $(z_\xi)_{\xi \in \Xi}$ *in* $E \otimes X$ *such that* $\big\lvert z|_{Q(\xi)} - z_\xi|_{Q(\xi)} \big\rvert \leq \varepsilon \lvert z \rvert$ $(\xi \in \Xi)$.

$\triangleleft$ It is sufficient to consider the case in which $\{\lvert z \rvert\}^{\perp\perp} = E$ and the identical unity is contained in E. For every point $q \in \mathrm{dom}(z) \subset Q$, choose a clopen neighborhood $U(q)$ so that

$$\|z(q) - z(q')\| \leq \varepsilon \|z(q')\| \quad (q' \in U(q)).$$

In the Boolean algebra of clopen subsets of Q there exists a partition of unity $\big(Q(\xi)\big)_{\xi \in \Xi}$ which possesses the following property: for each $\xi \in \Xi$, there is a point $q_\xi \in Q$ such that $Q(\xi) \subset U(q_\xi)$. Now, if we assign $z_\xi := \chi_{Q(\xi)} z(q_\xi)$ then

$$\|z(q') - z_\xi(q')\| = \|z(q') - z(q_\xi)\| \leq \varepsilon \|z(q')\| \quad \big(q' \in Q(\xi)\big).$$

Thus,

$$\chi_{Q(\xi)} \lvert z - z_\xi \rvert \leq \varepsilon \lvert z \rvert,$$

which is equivalent to the desired inequality. $\triangleright$

2.3.5. Now we turn to defining a space of weakly continuous vector-functions which is similar to $E(X)$. We suppose that $Z \subset X'$ is a *norming* subspace (for X), i.e.,

$$\|x\|_X = \sup\{|\langle x, z\rangle| : z \in Z,\ \|z\| \leq 1\} \quad (x \in X).$$

Here, as usual, X' is the dual space and $\langle \cdot\,, \cdot\rangle$ is the canonical bilinear form of the duality $X \leftrightarrow X'$.

Denote by $\mathscr{M}$ the set of $\sigma(X, Z)$-continuous vector-functions $u : \operatorname{dom}(u) \to X$ such that $\operatorname{dom}(u)$ is a comeager set in Q. Consider the factor set $C_\infty(Q, X|Z) := \mathscr{M}/\!\sim$, where $u \sim v$ means that $u(t) = v(t)$ $\big(t \in \operatorname{dom}(u) \cap \operatorname{dom}(v)\big)$. The set $C_\infty(Q, X|Z)$ can be made into a vector space in the natural way: if $\widetilde{u}$ stands for the cosets of $u \in \mathscr{M}$ then a sum $\lambda\widetilde{u} + \mu\widetilde{v}$ is interpreted as the coset of the pointwise sum $\lambda u(t) + \mu v(t)$, $t \in \operatorname{dom}(u) \cap \operatorname{dom}(v)$. For $u \in \mathscr{M}$ and $z \in Z$ denote by $\langle u, z\rangle$ the unique continuous extension of the function $t \mapsto \langle u(t), z\rangle$ $(t \in \operatorname{dom}(u))$ onto the entire Q. If $u \sim v$ then clearly $\langle u, z\rangle = \langle v, z\rangle$; thus, for $w \in C_\infty(Q, X|Z)$ and an arbitrary $u \in w$ we may put $\langle w, z\rangle := \langle u, z\rangle$. The set $R(u) := \big\{\langle u, z\rangle : z \in Z,$ $\|z\| \leq 1\big\}$ is order-bounded in $C_\infty(Q)$, it is pointwise bounded on the comeager set $\operatorname{dom}(u)$. Therefore, for an arbitrary $u \in w$ we may assign

$$\lvert w \rvert := \lvert u \rvert := \sup\big\{\langle u, z\rangle : z \in Z,\ \|z\| \leq 1\big\},$$

where supremum is taken in $C_\infty(Q)$. Observe that the function $\|u(\cdot)\| : t \mapsto \|u(t)\|$ $(t \in \operatorname{dom}(u))$ is pointwise the least upper bound of the same family $R(u)$. Therefore, the functions $\lvert u \rvert$ and $\|u(\cdot)\|$ coincide on a comeager subset of Q. Nevertheless, these function may differ on $\operatorname{dom}(u)$.

It is easy to see that $\lvert\cdot\rvert$ is a decomposable norm with values in $C_\infty(Q)$. Moreover, $C_\infty(Q, X|Z)$ is naturally endowed with the structure of a faithful module over the ring $C_\infty(Q)$. Put

$$E_w(X, Z) := \big\{u \in C_\infty(Q, X|Z) : \lvert u \rvert \in E\big\}.$$

Indicate an important particular case $E_w(X') := E_w(X', X)$ that appears when $X := X'$ and $Z := X \subset X''$.

If X is a Banach space then for every order-dense ideal $E \subset C_\infty(Q)$ the set $E_w(X, Z)$, endowed with the operations and E-valued norm $\lvert\cdot\rvert$ induced from $C_\infty(Q, X|Z)$, is a Banach–Kantorovich space over E and $C_\infty(Q, X|Z)$ is its universal completion. In particular, $E_w(X')$ is a Banach–Kantorovich space over E.

2.3.6. Let X, Q, and E be the same as in 2.3.5. Let Y be one more normed space, with $Z \subset Y'$ a norming space for Y and $\mathscr{L}(X, Y)$ standing for the space of all bounded linear operators from X to Y. Denote by $\mathfrak{M}_Q(X, Y)$ the set of all operator-functions $K : \operatorname{dom}(K) \to \mathscr{L}(X, Y)$ that satisfy the following conditions:

(a) $Q_0 := \operatorname{dom}(K)$ is a comeager subset of Q; (b) the function $s \mapsto \langle K(s)x, z\rangle$ $(s \in Q_0)$ is continuous for all $x \in X$ and $z \in Z$. It follows from the definition that, given $x \in X$ and $z \in Z$, there exists a unique element $u \in C_\infty(Q)$ for which $u(t) = \langle K(t)x, z\rangle$ $(t \in Q_0)$. Assign $\langle Kx, z\rangle := u$. Just as in 2.3.5 there exists a $\varphi \in C_\infty(Q)$ such that $\|K(s)\| \le \varphi(s)$ $(s \in Q_0)$. Thus, it is clear that

$$|K| := \sup\{\langle Kx, z\rangle : \|x\| \le 1,\ \|z\| \le 1\} \le \varphi$$

exists in the $C_\infty(Q)$. Introduce an equivalence in $\mathfrak{M}_Q(X, Y)$ by letting $K \sim L$ whenever $K(s) = L(s)$ for all $s \in \operatorname{dom}(K) \cap \operatorname{dom}(L)$. If $\widetilde{K}$ is the coset of an operator-function K then we assign $|\widetilde{K}| := |K|$. This definition is sound since $|K| = |L|$ for equivalent K and L. Now introduce the space

$$E_w\big(\mathscr{L}(X, Y), Z\big) := \{K \in \mathfrak{M}_Q(X, Y)/\!\sim\ : |K| \in E\}.$$

It is easy to verify that $E_w\big(\mathscr{L}(X, Y), Z\big)$ is a decomposable LNS. It coincides with $E_w(X, Z)$ whenever $X := \mathscr{L}(X, Y)$ and $Z := X \otimes Z$. An important particular case is presented by the space

$$E_w\big(\mathscr{L}(X, Y')\big) := E_w\big(\mathscr{L}(X, Y'), Y\big) := E_w\big((X \hat{\otimes} Y)'\big),$$

where $X \hat{\otimes} Y$ is the *projective tensor product* of the spaces X and Y.

If Y is a Banach space then for every order-dense ideal $E \subset C_\infty(Q)$ the set $E_w\big(\mathscr{L}(X, Y), Z\big)$ is a Banach–Kantorovich space over E and $C_\infty(Q, \mathscr{L}(X, Y)|Z)$ is its universal completion. In particular, $E_w\big(\mathscr{L}(X, Y')\big)$ is a Banach–Kantorovich space over E.

2.3.7. Let (Ω, Σ, μ) be a measure space with the direct sum property, let E be an order-dense ideal in $L^0(\Omega, \Sigma, \mu)$, and let X be a normed space. Let $L^0(\mu, X) := L^0(\Omega, \Sigma, \mu, X)$ be the space of cosets of Bochner μ-measurable vector-functions acting from Ω into X. As usual, vector-functions are equivalent if they assume equal values at almost all points of the set Ω. If $\widetilde{u} \in L^0(\mu, X)$ is the coset of a measurable vector-function of $u : \Omega \to X$ then $t \mapsto \|u(t)\|$ $(t \in \Omega)$ is a scalar measurable function whose coset is denoted by the symbol $|\widetilde{z}| \in L^0(\mu)$. Assign by definition

$$E(X) := \{u \in L^0(\mu, X) : |u| \in E\}.$$

Then $\big(E(X), E\big)$ is a lattice-normed space with decomposable norm. Obviously, $L^p(\mu, X)$ $(1 \le p \le \infty)$ coincides with $E(X)$, where $E = L^p(\mu)$.

If X is a Banach space then $E(X)$ is a Banach–Kantorovich space and $L^0(\mu, X)$ is its universal completion.

$\triangleleft$ The assertion can be easily deduced from 2.2.2. It also follows immediately from a more general fact which will be proved in Section 2.5 (see 2.5.4 (3)). $\triangleright$

2.3.8. Suppose, as above, that (Ω, Σ, μ) is a measure space with the direct sum property. Let ρ be a lifting of $L^\infty(\Omega, \Sigma, \mu)$ and $\tau : \Omega \to Q$ be the corresponding canonical immersion of Ω into the Stone space Q of the Boolean algebra $B(\Omega, \Sigma, \mu)$, see 1.4.9. Denote by $\tau^* := \tau_X^*$ the mapping that sends each function $f \in C_\infty(Q, X)$ to the equivalence class of the measurable vector-function $f \circ \tau$.

(1) *The mapping τ^* is a linear isometry (in the sense of vector norms) from $C_\infty(Q, X)$ onto $L^0(\Omega, \Sigma, \mu, X)$. The image of $C_\#(Q, X)$ under τ^* coincides with the subspace $L^\infty(\Omega, \Sigma, \mu, X)$.*

$\lhd$ A more general fact is proved in 2.5.9. $\rhd$

The *algebraic tensor product* $E \otimes X$ is identified with the set of vector-functions of the form

$$f(t) = \sum_{i=1}^{n} e_i(t)x_i \quad (e_i \in E,\ x_i \in X).$$

It is clear that $E \otimes X$ is a linear subset in $E(X)$.

(2) *Let E be an ideal space and let X be a Banach space. Then $E \otimes X$ is dense in $E(X)$ with respect to the vector norm.*

$\lhd$ It is readily apparent from (1) and 2.3.4. $\rhd$

2.3.9. Now we consider the lattice-normed space $E[F]$ of measurable functions of two variables defined on the product of some measure spaces (Ω, Σ, μ) and (Ω', Σ', μ'). To avoid technical inconveniences assume that (Ω', Σ', μ') is σ-finite; however, the definition and main properties remain valid in the general case.

Let E be an ideal space on (Ω, Σ, μ) and let F be a Banach ideal space on (Ω', Σ', μ'). Denote by $E[F]$ the space of all measurable functions K on $\Omega' \times \Omega$ satisfying the following: (a) the coset of the function $s \mapsto K(s, t)$ belongs to F for almost all $t \in \Omega$; (b) the function $t \mapsto \|K(\cdot, t)\|_F$ is measurable and its coset $|K|$ belongs to E.

(1) *Suppose that a norm on F is o-semicontinuous. Then $E[F]$ is a Banach–Kantorovich space over E and an ideal space on $\Omega' \times \Omega$.*

$\lhd$ There are two subtle reasonings in the proof. The first is to demonstrate that *o*-semicontinuity of the norm in F provides measurability for the function $|K|$ (see [162, XI.1.4]). It follows from this that $E[F]$ is a linear space and an ideal space on $\Omega' \times \Omega$. Thus, $E[F]$ is also a decomposable *d*-complete lattice-normed space over E. The second is to prove *bo*-completeness of $E[F]$. In virtue of Theorem 2.2.3 we need only to check that this space is *br*-complete. The latter is equivalent to *br*-completeness of all spaces $E_e[F]$ where $e \in E_+$ and $E_e := E(e)$. Define a scalar norm $\|\!|\cdot|\!\|$ on $E_e[F]$ by $\|\!|K|\!\| := \|\,|K|\,\|_e$. In 7.1.2 we will prove that $E_e[F]$ is *br*-complete if and only if $(E_e[F], \|\!|\cdot|\!\|)$ is norm complete. It can be also easily seen that E_e is an AM-space and $(E_e[F], \|\!|\cdot|\!\|)$ is a normed lattice. In view of 1.5.1 (5), it remains to prove that for every disjoint sequence (K_n) of positive functions in $E_e[F]$ there exists $K := \sup_n K_n$ provided that $\sum_{n=1}^{\infty} |\!|\!| K_n |\!|\!| < \infty$. The sought least upper bound K exists in a universally complete vector space $L^0(\Omega' \times \Omega, \nu \times \mu)$. Show that $K \in E_e[F]$. Observe that the series $\sum_{n=1}^{\infty} |K_n|$ is absolutely summable and its sum exists, since $(E_e, \|\cdot\|_e)$ is norm complete. From this we deduce that for almost all $t \in \Omega$ the series $\sum_{n=1}^{\infty} K_n(\cdot, t)$ is absolutely summable in F. If $K_t(\cdot)$ stands for the sum of this series then $K_t(s) = K(s,t)$ for almost all $s \in \Omega'$ in view of 1.5.8. Thus $K(\cdot, t) \in F$ for almost all $t \in \Omega$. Moreover, $|K| \leq \sum_{n=1}^{\infty} |K_n|$ and $|K| \in E_e$. $\triangleright$

(2) *Let F be a Banach ideal space on (Ω', Σ', μ') with an order semi-continuous norm. For every measurable vector-function $f : \Omega \to F$, there is a measurable function $K(s,t)$ on $(\Omega' \times \Omega)$ such that, for almost all $t \in \Omega$ the equality $K(s,t) = (f(t))(s)$ holds for almost all $s \in \Omega'$. The correspondence $f \mapsto K$ implements an isometric embedding of $E(F)$ onto a closed subspace and a sublattice of $E[F]$.*

$\triangleleft$ Without loss of generality we may assume that (Ω, Σ, μ) is a finite measure space. Approximate f by a sequence (f_n) of finite-valued functions in the sense of almost everywhere norm convergence in F:

$$\|f_n(t) - f(t)\|_F \to 0 \quad \text{for almost all} \quad t \in \Omega.$$

Every function f_n generates a measurable function $K_n(\cdot, \cdot)$ by formula $K_n(s,t) = (f_n(t))(s)$; moreover, $|K_n - K_m|(t) = \|f_n(t) - f_m(t)\|_F \to 0$ almost everywhere. Since $E[F]$ is norm o-complete there is $\widetilde{K} \in E[F]$ with $|\widetilde{K}_n - \widetilde{K}| \to 0$ where $\widetilde{L}$ stands for the coset of measurable function L. Now the required identity follows from the relations:

$$\|f(t) - K(\cdot, t)\|_F \leq \|f(t) - f_n(t)\|_F + |K_n - K|(t) \to 0. \ \triangleright$$

(3) *Let the measure μ be not purely atomic. Then the lattice-normed spaces $E(F)$ and $E[F]$ coincide (under embedding (2)) if and only if F is a Banach ideal space with order continuous norm.*

2.3.10. Now we introduce the measurable version of the space $E_w(X)$. Take the same E and X as in 2.3.7, and a norming subspace $Z \subset X'$, see 2.3.5. A vector-function $u : \Omega \to X$ is called *$\sigma(X, Z)$-measurable* or simply *Z-measurable* if, for each $z \in Z$, the function $t \mapsto \langle u(t), z\rangle$ $(t \in \Omega)$ is measurable. Denote the coset of the last function by the symbol $\langle u, z\rangle$, so that $\langle u, z\rangle \in L^0(\mu)$. Let $\mathscr{M}(\Omega, X|Z)$ be the set of Z-measurable vector-functions $u : \Omega \to X$. We say that Z-measurable vector-functions u, v are Z-equivalent and write $u \simeq v$ if measurable functions $\langle u, z\rangle$ and $\langle v, z\rangle$ are equal almost everywhere for each $z \in Z$. Consider the factor set

$L^0(\mu, X|Z) := L^0(\Omega, \Sigma, \mu, X|Z) := \mathscr{M}(\Omega, X|Z)/\simeq$ and define vector space structure in it by setting $\alpha\widetilde{u} + \beta\widetilde{v} := (\alpha u + \beta v)^{\sim}$. For a coset $\widetilde{u} \in L^0(\mu, X|Z)$ with $u \in \mathscr{M}(\Omega, X|Z)$ put $\langle \widetilde{u}, z\rangle := \langle u, z\rangle$.

Observe that the set $R(\widetilde{u}) := \{\langle \widetilde{u}, z\rangle : z \in Z, \|z\| \le 1\}$ is order-bounded in $L^0(\mu)$. Otherwise, according to 1.4.11 (2), one can choose a sequence (f_n) in $R(\widetilde{u})$ which is unbounded from above. But this is a contradiction, since the function $f(t) := \sup_n f_n(t)$ is measurable and $|f(t)| \le \|u(t)\|_X < \infty$ $(t \in \Omega)$. Given $\widetilde{u} \in \mathscr{M}/\simeq$, assign

$$|u| := \sup\{\langle u, z\rangle : z \in Z, \|z\| \le 1\},$$

where the supremum is taken in the $L^0(\Omega, \Sigma, \mu)$. It is easy to verify that $L^0(\mu, X|Z)$ is a decomposable lattice-normed space over $L^0(\mu)$. Define now the set

$$E_w(X, Z) := \{u \in L^0(\mu, X|Z) : |u| \in E\}.$$

In the same manner as in 2.3.7, we indicate an important particular case, when X is a dual Banach space ($X := X'$) and Z is its predual ($Z := X \subset X''$). The notation $E_w(X') := E_w(X', X)$ is conventional.

(1) *For every order ideal $E \subset L^0(\mu)$ the space $E_w(X')$ endowed with the operations and E-valued norm $|\cdot|$ induced from $L^0(\Omega, \Sigma, \mu, X'|X)$ is a Banach–Kantorovich space over $L^0(\Omega, \Sigma, \mu)$ and $L^0(\Omega, \Sigma, \mu, X'|X)$ is its universal completion.*

It is evident that every measurable vector-function is weakly measurable and $u \sim v$ implies $u \simeq v$ for every pair of measurable vector-functions u and v. Therefore, there is a mapping (called the canonical embedding) assigning to each $\widetilde{u} \in L^0(\mu, X)$ the coset $\{v \in \mathscr{M}(\mu, X|Z) : v \simeq u\} \in L^0(\mu, X|Z)$.

(2) *The canonical embedding $L^0(\mu, X) \to L^0(\mu, X|Z)$ is a linear isometry.*

2.3.11. Take the same E and X as in 2.3.10. Let Y be one more normed space and $Z \subset Y'$ be a norming subspace. An operator-function $K : \Omega \to \mathscr{L}(X, Y)$ is called *Z-weakly measurable* if, for all $x \in X$ and $z \in Z$, the function $\langle z, Kx\rangle : t \mapsto \langle z, K(t)x\rangle$ $(t \in \Omega)$ is measurable. Denote by $\mathfrak{M}_\mu(X, Y)$ the set of all Z-weakly measurable operator-functions $K : \Omega \to \mathscr{L}(X, Y)$.

Introduce some equivalence relation $\sim$ in $\mathfrak{M}_\mu(X, Y)$ by letting $K \sim L$ if and only if, for each pair $(x, z) \in X \times Z$, the measurable functions $\langle z, Kx\rangle$ and $\langle z, Lx\rangle$ coincide almost everywhere. For a $K \in \mathfrak{M}_\mu(X, Y)$, as well as for the corresponding coset $\widetilde{K}$, assign

$$|K| = |\widetilde{K}| := \sup\{\langle z, K_x\rangle : \|x\| \le 1, \|z\| \le 1\},$$

where the supremum is taken in $L^0(\mu)$. This definition is sound, since the same reasoning as in 2.3.10 shows that $|K|$ is the least upper bound of an order-bounded set in $L^0(\mu)$. Now put

$$E_w(\mathscr{L}(X,Y),Z) := \{K \in \mathfrak{M}_\mu(X,Y)/\sim \,:\, |K| \in E\}.$$

This notation agrees with 2.3.10, since it is easily seen that $E_w(\mathscr{L}(X,Y),Z) = E_w(X,Z)$ whenever $X := \mathscr{L}(X,Y)$ and $Z := X \otimes Z$. We also note that $E_w(X') = E_w(\mathscr{L}(X,\mathbb{R}))$. Again, we distinguish an important particular case

$$E_w(\mathscr{L}(X,Y')) := E_w(\mathscr{L}(X,Y'),Y) := E_w((X\hat{\otimes}Y)'),$$

where $X\hat{\otimes}Y$ is the projective tensor product of the spaces X and Y.

The following assertions hold:

(1) $E_w(\mathscr{L}(X,Y),Z)$ *is a decomposable lattice-normed space;*

(2) $E_w(\mathscr{L}(X,Y'))$ *is a Banach–Kantorovich space.*

2.3.12. An operator-function $K : \Omega \to \mathscr{L}(X,Y)$ is called *simply measurable* if, for each $x \in X$, the vector-function $Kx : t \mapsto K(t)x$ $(t \in \Omega)$ is measurable. It is clear that any simply measurable operator-function is weakly measurable. Let $\mathfrak{M}_\mu^s(X,Y)$ be the part of $\mathfrak{M}_\mu(X,Y)$ consisting of simply measurable operator-functions with values in $\mathscr{L}(X,Y)$. (As usual, we assume that $Y \subset Z'$ and $\mathscr{L}(X,Y) \subset \mathscr{L}(X,Z')$.) Assign

$$E_{ws}(\mathscr{L}(X,Y)) := \{K \in \mathfrak{M}_\mu^s(X,Y)/\approx \,:\, |K| \in E\},$$

where $|K|$ is the same as in 2.3.11 and the equivalence $\approx$ is defined as follows: $L \approx K$ if and only if, for each $x \in X$, the measurable vector-functions Kx and Lx coincide almost everywhere. If $K \in \mathfrak{M}_\mu^s(X,Y)$ then we also have

$$|K| = \sup\{\|Kx\| : \|x\| \le 1\},$$

where $\|Kx\|$ is the measurable function $t \mapsto \|K(t)x\|$ $(t \in \Omega)$. Observe that $E(Y) = E_{ws}(\mathscr{L}(\mathbb{R},Y))$ and $E_w(X') = E_{ws}(\mathscr{L}(X,\mathbb{R}))$.

2.4. Continuous Banach Bundles

The main subject of this section is the following assertion: Every Banach–Kantorovich space is linearly isomorphic to the space of almost global sections of some continuous Banach bundle.

2.4.1. First of all, we give a general definition of a Banach bundle. Let Q be a topological space. A *bundle* over Q is defined to be a continuous surjection $\sigma : \mathscr{X} \to Q$ of a topological space $\mathscr{X}$ onto Q. A nonempty set $\mathscr{X}_q = \sigma^{-1}(q)$ is called the *stalk* at a point $q \in Q$. A mapping s from a nonempty set $\operatorname{dom}(s) \subset Q$ into $\mathscr{X}$ is called a *section* over $\operatorname{dom}(s)$ if $s(q) \in \mathscr{X}_q$ for each $q \in \operatorname{dom}(s)$. A continuous section s is called *local, almost global,* or *global,* whenever its domain of definition $\operatorname{dom}(s)$ is respectively an open proper subset, a comeager subset, or the whole of Q. A bundle $\sigma : \mathscr{X} \to Q$ is often identified with the mapping $q \mapsto \mathscr{X}_q$ $(q \in Q)$, and the notation $\mathscr{X}(q)$ is used instead of $\mathscr{X}_q$. Moreover, the bundle is sometimes denoted by $\mathscr{X}$, with the default parameters σ and Q omitted.

A set of sections S is called *stalkwise dense* in $\mathscr{X}$ if the set $\{s(q) : s \in S\}$ is dense in $\mathscr{X}(q)$ for every $q \in Q$. The *product* of two bundles $\sigma : \mathscr{X} \to Q$ and $\sigma' : \mathscr{X}' \to Q$ is the bundle $\tau : \mathscr{X} \times_Q \mathscr{X}' \to Q$ defined by the following formulas:

$$\mathscr{X} \times_Q \mathscr{X}' := \{(x, x') \in \mathscr{X} \times \mathscr{X}' : \sigma(x) = \sigma'(x')\},$$
$$\tau : (x, x') \mapsto \sigma(x) = \sigma'(x') \quad ((x, x') \in \mathscr{X} \times_Q \mathscr{X}').$$

2.4.2. A *continuous Banach bundle* over Q is defined to be a bundle $\sigma : \mathscr{X} \to Q$ that satisfies the following four conditions:

(1) every stalk $\mathscr{X}(q)$ is a Banach space with the norm $\|\cdot\|_q$, and the topology defined by the norm coincides with that induced;

(2) addition, $\mathscr{X} \times_Q \mathscr{X} \ni (x, y) \mapsto x + y \in \mathscr{X}$, scalar multiplication, $\mathbb{R} \times \mathscr{X} \ni (\alpha, x) \mapsto \alpha x \in \mathscr{X}$, and the zero section, $Q \ni q \mapsto 0_q \in \mathscr{X}(q)$, are continuous mappings;

(3) the sets of the shape

$$U(s, \varepsilon) := \{x \in \mathscr{X} : \sigma(x) \in \operatorname{dom}(s),\ \|x - s(\sigma(x))\|_{\sigma(x)} < \varepsilon\},$$

where $\varepsilon > 0$ and s is a local section, form a base for the topology on $\mathscr{X}$;

(4) for each point $x \in \mathscr{X}$ and every number $\varepsilon > 0$, there exists a local section s such that $\sigma(x) \in \operatorname{dom}(s)$ and $\|x - s(\sigma(x))\|_{\sigma(x)} < \varepsilon$.

In the case of a paracompact space Q, the above definition can be simplified. Namely, instead of (3) and (4) we may require the following:

(3′) every neighborhood of zero $0_q \in \mathscr{X}_q$ contains a neighborhood of the form $U(s, \varepsilon)$, where $\varepsilon > 0$ and s is the restriction of the zero section $q \mapsto 0_q$ to some open set;

(4′) the mapping σ is open, and $\|\cdot\| : \mathscr{X} \to \mathbb{R}$, $\|x\| := \|x\|_{\sigma(x)}$, is upper semicontinuous.

In this case a continuous Banach bundle has a very useful property:

(5) *For each $x \in \mathscr{X}$, there exists a global continuous section s such that $s(\sigma(x)) = x$.*

2.4.3. It is sometimes convenient to define a topology for a Banach bundle with the help of a certain set of sections. Take a surjective mapping $\sigma : \mathscr{X} \to Q$ and assume that $\mathscr{X}_q := \sigma^{-1}(q)$ is a Banach space for each $q \in Q$. The latter amounts to considering a mapping $q \mapsto \mathscr{X}_q$ from Q into the class of Banach spaces.

Introduce the structure of a vector space in the set of all global sections $S(Q, \mathscr{X})$ by letting $(\alpha u + \beta v)(q) = \alpha u(q) + \beta v(q)$ $(q \in Q)$ for $\alpha, \beta \in \mathbb{R}$ and $u, v \in S(Q, \mathscr{X})$. For each section $s \in S(Q, \mathscr{X})$ we consider its pointwise norm $|\!|\!|s|\!|\!| : q \mapsto \|s(q)\|_q$ $(q \in Q)$.

Now let $\mathscr{C}$ be a set of global sections that satisfies the following conditions:

(1) $\mathscr{C}$ is a vector subspace of $S(Q, \mathscr{X})$;

(2) for each $s \in \mathscr{C}$, the function $|\!|\!|s|\!|\!|$ is continuous;

(3) for each $q \in Q$, the set $\{s(q) : s \in \mathscr{C}\}$ is dense in the stalk $\mathscr{X}_q$.

With the help of $\mathscr{C}$, called a *continuity structure* in $\mathscr{X}$, we may introduce a topology in $\mathscr{X}$ by taking the family $\{U(s, \varepsilon)\}$ as a basis for the topology, where $\varepsilon > 0$ and s is the restriction of some section in $\mathscr{C}$ to an arbitrary open set. Then $\sigma : \mathscr{X} \to Q$ is a continuous Banach bundle. Moreover, a section $u \in S(Q, \mathscr{X})$ is continuous if and only if the function $q \mapsto \|u - s\|_q$ $(q \in Q)$ is continuous for every section $s \in \mathscr{C}$.

Let $C(Q, \mathscr{X})$ denote the space of all global continuous sections of $\mathscr{X}$.

2.4.4. Consider an important example of a Banach bundle. Assume that $(X, (\|\cdot\|_q)_{q \in Q})$ is a multinormed space, where Q is still a topological space. For each $q \in Q$, assign $X_q := \{x \in X : \|x\|_q = 0\}$ and denote by $\mathscr{X}_q$ the completion of the normed factor space X/X_q. With each element $x \in X$ associate a global section $\imath(x) \in S(Q, \mathscr{X})$ by the formula $\imath(x)(q) = \varphi_q(x)$, where $\varphi_q : X_q \to X/X_q \subset \mathscr{X}_q$ is the factor homomorphism. The set $\mathscr{C} := \imath(X)$ satisfies conditions 2.4.3 (1–3) if and only if, for each $x \in X$, the function $q \mapsto \|x\|_q$ $(q \in Q)$ is continuous. If the last condition is satisfied then, due to 2.4.3, $\mathscr{X}$ becomes a continuous Banach bundle. In this case, the constructed Banach bundle $\mathscr{X}$ is said to be *associated with the multinormed space* $(X, (\|\cdot\|_q)_{q \in Q})$, and the mapping $\imath : X \to C(Q, \mathscr{X})$ is called the *canonical embedding.*

2.4.5. We shall need the following auxiliary fact:

Let $(Q_\xi)_{\xi \in \Xi}$ be a family of pairwise disjoint open sets whose union is dense in Q. If P_ξ is a comeager subset of Q_ξ for each $\xi \in \Xi$, then the union $P := \bigcup_{\xi \in \Xi} P_\xi$ is again a comeager set in Q.

$\lhd$ For each $\xi \in \Xi$, choose a sequence $(G_{n,\xi})_{n \in \mathbb{N}}$ of open dense subsets of Q_ξ such that $\bigcap_{n \in \mathbb{N}} G_{n,\xi} \subset P_\xi$ $(\xi \in \Xi)$. Since Q_ξ are pairwise disjoint, we have

$$\bigcup_{\xi \in \Xi} \bigcap_{n \in \mathbb{N}} G_{n,\xi} = \bigcap_{n \in \mathbb{N}} \bigcup_{\xi \in \Xi} G_{n,\xi}.$$

Hence it is clear that if $G_n := \bigcup_{\xi\in\Xi} G_{n,\xi}$ then $P \supset \bigcap_{n\in\mathbb{N}} G_n$. Thus, P is a comeager set, since G_n is open and dense in Q for each $n \in \mathbb{N}$. $\triangleright$

2.4.6. Now suppose that Q is an extremal compact space. Take a continuous Banach bundle $\mathscr{X}$ over Q. If u is an almost global section of the bundle $\mathscr{X}$ then the function $q \mapsto \|u(q)\|_q$ is defined and continuous on a comeager set $\mathrm{dom}(u)$. Consequently, there exists a unique function $|u| \in C_\infty(Q)$ such that $|u|(q) = \|u(q)\|_q$ $(q \in \mathrm{dom}(u))$. Introduce an equivalence relation in the set of almost global sections $\mathfrak{M}(Q, \mathscr{X})$ by letting $u \sim v$ if $u(q) = v(q)$ whenever $q \in \mathrm{dom}(u) \cap \mathrm{dom}(v)$. For equivalent u and v, we have $|u| = |v|$; therefore, we may define $|\tilde{u}| := |u|$, where $\tilde{u}$ is the coset of the almost global section u. Denote the factor set $\mathfrak{M}(Q, \mathscr{X})/{\sim}$ by $C_\infty(Q, \mathscr{X})$. The set $C_\infty(Q, \mathscr{X})$ can be naturally made a lattice-normed space. For instance, by the element $\tilde{u} + \tilde{v}$ we mean the coset of the almost global section $q \mapsto u(q) + v(q)$ $(q \in \mathrm{dom}(u) \cap \mathrm{dom}(v))$. If E is an order ideal in $C_\infty(Q)$ then we assign

$$E(\mathscr{X}) := \{u \in C_\infty(Q, \mathscr{X}) : |u| \in E\}.$$

In each coset $\tilde{u}$, there exists a unique section $\bar{u} \in \tilde{u}$ such that $\mathrm{dom}(v) \subset \mathrm{dom}(\bar{u})$ for all $v \in \tilde{u}$. The section $\bar{u}$ is called *extended.* The space $C_\infty(Q, \mathscr{X})$ can thus be represented as the space of all extended almost global sections of the bundle $\mathscr{X}$.

The constant mapping $\mathscr{X} = Q \times \{X\}$ that associates with each point of a topological space Q the same Banach space X is a simple example of a Banach bundle. If the totality $\mathscr{C}$ of constant functions $c : Q \to X$ is taken as a continuity structure $\mathscr{C}$, then the continuous sections of the CBB $(\mathscr{X}, \mathscr{C})$ are exactly continuous functions $u : Q \to X$. Moreover, in this case we have $C_\infty(Q, \mathscr{X}) = C_\infty(Q, X)$ and $E(\mathscr{X}) = E(X)$. It will be shown in 2.4.8 that the space $C(Q, X)$ is a Banach–Kantorovich space if and only if the trivial bundle $Q \times \{X\}$ is ample.

2.4.7. Theorem. *Let $\mathscr{X}$ be a continuous Banach bundle over an extremal compact space Q and let E be an order-dense ideal in $C_\infty(Q)$. Then $E(\mathscr{X})$ is a Banach–Kantorovich space and $C_\infty(Q, \mathscr{X})$ is its universal completion.*

$\triangleleft$ It is sufficient to establish that $C_\infty(Q, \mathscr{X})$ is a universally complete Banach–Kantorovich space. Take a pairwise disjoint family $(u_\xi)_{\xi\in\Xi}$ in $C_\infty(Q, \mathscr{X})$. Let Q_ξ be the least clopen set in Q on the complement of which u_ξ equals zero. Then $Q_\xi \cap Q_\eta = \varnothing$ for $\xi \neq \eta$. Without loss of generality we may assume that $(Q_\xi)_{\xi\in\Xi}$ is a partition of unity in the Boolean algebra of clopen subsets of Q. Assign $G_\xi := Q_\xi \cap \mathrm{dom}(u_\xi)$ and $G := \bigcup_{\xi\in\Xi} G_\xi$. Define a section u_0 by the formulas

$$\mathrm{dom}(u) := G, \quad u_0|_{G_\xi} = u_\xi|_{G_\xi} \quad (\xi \in \Xi).$$

The set G is comeager due to 2.4.5, and the section u_0 is continuous by definition. Consequently, u_0 determines a unique element $u \in C_\infty(Q, \mathscr{X})$ for which $\pi_\xi u_\xi = \pi_\xi u$

($\xi \in \Xi$), where π_ξ is the projection induced by the operator of multiplication by the characteristic function of the set G_ξ. If θ is a finite set in Ξ and $u_\theta = \sum_{\xi\in\theta} u_\xi$, then $|u - u_\theta| \le \pi_\theta e$, where $\pi_\theta = \sup\{\pi_\xi : \xi \in \theta\}$; therefore, $u = bo\text{-}\lim u_\theta$.

Now let a sequence $(u_n)_{n\in\mathbb{N}}$ in $C_\infty(Q, \mathscr{X})$ be r-fundamental with regulator $e \in C_\infty(Q)$. On the comeager set $G := \operatorname{dom}(e) \cap \bigcap_{n=1}^\infty \operatorname{dom}(u_n)$, we may define a section u by the formula $u(q) := \lim u_n(q)$. Existence of the limits follows from the estimate $\|u_n(q) - u_m(q)\| \le \lambda_k e(q)\ \ (n, m \ge k)$, where $\lambda_n \to 0$. After passing to the limit as $m \to \infty$, it becomes clear from the last inequality that u_n converges to u locally uniformly. Consequently, the section u is continuous. It remains to observe that $|u_n - u| \le \lambda_k e\ \ (n \ge k)$, i.e., $br\text{-}\lim_{n\to\infty} u_n = u$.

Thus, $C_\infty(Q, \mathscr{X})$ is d-complete and r-complete and, therefore, the required assertion ensues from 2.2.3. ▷

2.4.8. A continuous Banach bundle $\mathscr{X}$ over an extremal compact space Q is called *ample* (or *complete*) if every bounded almost global continuous section of it can be extended to a global continuous section. Introduce the notation

$$C_\#(Q, \mathscr{X}) := \{u \in C_\infty(Q, \mathscr{X}) : |u| \in C(Q)\},$$

i.e., $C_\#(Q, \mathscr{X}) = E(\mathscr{X})$ for $E := C(Q)$. Let $C(Q, \mathscr{X})$ denote the set of all global continuous sections of $\mathscr{X}$.

The following assertions are equivalent:

(1) *the bundle $\mathscr{X}$ is ample;*

(2) $C_\#(Q, \mathscr{X}) = C(Q, \mathscr{X})$;

(3) *$C(Q, \mathscr{X})$ is a Banach–Kantorovich space.*

◁ The implications (1) ⇒ (2) ⇒ (3) are obvious. Suppose that (3) is satisfied and show that $\mathscr{X}$ is an ample bundle. Take a section $u \in C_\#(Q, \mathscr{X})$. There exist a partition of unity $(Q_\xi)_{\xi\in\Xi}$ in the Boolean algebra of clopen subsets of Q and a family of global continuous section $(u_\xi)_{\xi\in\Xi}$ such that $\|u(q) - u_\xi(q)\|_q \le 1/n$ for all $q \in Q_\xi \cap \operatorname{dom}(u)$ and $\xi \in \Xi$. This assertion can be easily deduced from 2.4.2 (5), since $\operatorname{dom}(u)$ is dense in Q. Let π_ξ be the projection in $C_\infty(Q, \mathscr{X})$ that corresponds to the set Q_ξ. Due to bo-completeness of $C(Q, \mathscr{X})$, there exists a section $v_n \in C(Q, \mathscr{X})$ with $\pi_\xi v_n = \pi_\xi u_\xi$ ($\xi \in \Xi$). It follows from this that $|u - v_n| \le (1/n)\mathbf{1}$. Thus, the sequence (v_n) in $C(Q, \mathscr{X})$ converges relatively uniformly to u. Again we may apply bo-completeness of $C(Q, \mathscr{X})$ and conclude that $u \in C(Q, \mathscr{X})$, as required. ▷

Let us show that assertions (1)–(3) are also equivalent to the following:

(4) *every bounded continuous section of the bundle $\mathscr{X}$ that is defined on a dense subset of Q can be extended to a global continuous section.*

It is not difficult to show that the trivial bundle $Q \times \{X\}$ is complete if and only if Q is a finite space or X is a finite-dimensional space.

2.4.9. Let $\mathscr{X}$ and $\mathscr{Y}$ be continuous Banach bundles over the same compact space Q. A mapping $h : q \mapsto h_q := h(q) \in \mathscr{L}(\mathscr{X}_q, \mathscr{Y}_q)$ $(q \in Q)$ is called a *homomorphism* from $\mathscr{X}$ into $\mathscr{Y}$ if h_q is a bounded linear operator from $\mathscr{X}_q$ into $\mathscr{Y}_q$ for all $q \in Q$, and, for each continuous section $u \in C(Q, \mathscr{X})$, the mapping $hu := h \otimes u : q \mapsto h_q\big(u(q)\big)$ $(q \in Q)$ is continuous.

(1) *If h is a homomorphism from $\mathscr{X}$ into $\mathscr{Y}$ then the function $|\!|\!|h|\!|\!| : q \mapsto \|h(q)\|$ is bounded.*

$\lhd$ For every $q \in Q$ define a linear operator $T_q : C(Q, \mathscr{X}) \to \mathscr{Y}(q)$ by setting $T_q u = h(q)u(q)$. Endow the space $C(Q, \mathscr{X})$ with the uniform norm $\|u\|_\infty := \sup\{\|u(q)\|_{\mathscr{X}(q)};\ q \in Q\}$ and observe that $\|T_q\| \leq \|h(q)\|$. Moreover,

$$\sup_{q \in Q} \|T_q u\| = \sup_{q \in Q} \|h \otimes u\| < \infty$$

for every $u \in C(Q, \mathscr{X})$. We may apply the Uniform Boundedness Principle and derive $\sup_{q \in Q} \|T_q\| < \infty$. It remains to note that

$$\|h(q)\| = \sup\{\|h(q)x\| : x \in \mathscr{X}(q),\ \|x\| \leqslant 1\}$$
$$= \sup\{\|h(q)u(q)\| :\ u \in C(Q, \mathscr{X}),\ \|u\|_\infty \leqslant 1\} = \|T_q\|. \ \rhd$$

We say that a homomorphism h is an *isometry* if h_q is an isometry for each $q \in Q$. The set of all homomorphism from $\mathscr{X}$ into $\mathscr{Y}$ denote by $\mathrm{Hom}_Q(\mathscr{X}, \mathscr{Y})$. If h is an isometry of $\mathscr{X}$ onto $\mathscr{Y}$ then the mapping $u \mapsto hu \quad \big(u \in C(Q, \mathscr{X})\big)$ is a linear isometry of the lattice-normed spaces $C(Q, \mathscr{X})$ and $C(Q, \mathscr{Y})$. The converse assertion is also valid; however, a stronger assertion holds in the case of ample bundles.

(2) *Suppose that $\mathscr{X}$ and $\mathscr{Y}$ are ample Banach bundles over an extremal compact space Q. Then $\mathscr{X}$ and $\mathscr{Y}$ are isometric if and only if the LNSs $C_\infty(Q, \mathscr{X})$ and $C_\infty(Q, \mathscr{Y})$ are linearly isometric.*

$\lhd$ Let J be a linear isometry of $C_\infty(Q, \mathscr{X})$ onto $C_\infty(Q, \mathscr{Y})$. Then $u \in C(Q, \mathscr{X})$ if and only if $Ju \in C(Q, \mathscr{Y})$, since $C = C_\#$ for any ample bundle (see 2.4.8 (2)). Given $u \in C(Q, \mathscr{X})$ and $q \in Q$, assign $h_q\big(u(q)\big) := (Ju)(q)$. It is easy to see that

$$\|h_q\big(u(q)\big)\|_q = \|(Ju)(q)\|_q = |Ju|(q) = |u|(q) = \|u(q)\|_q,$$

i.e., h_q is a linear isometry of $\mathscr{X}_q$ into $\mathscr{Y}_q$. Moreover, it is obvious that $hu \in C(Q, \mathscr{Y})$ whenever $u \in C(Q, \mathscr{X})$. For each $y \in \mathscr{Y}_q$, we may choose a global continuous section v so that $v(q) = y$. If $u := J^{-1}(v)$ then $h_q\big(u(q)\big) = y$; hence, the operator h_q is surjective. Consequently, h is an isometry of $\mathscr{X}$ onto $\mathscr{Y}$. $\rhd$

2.4.10. Theorem. *Every Banach–Kantorovich space X over an order-dense ideal $E \subset C_\infty(Q)$ is linearly isometric to $E(\mathscr{X})$ for some ample continuous Banach bundle $\mathscr{X}$ over Q. Moreover, such a bundle $\mathscr{X}$ is unique to within linear isometry.*

$\triangleleft$ Let $\widetilde{X}$ be the universal completion of the space X (see 2.2.8). Assign $X_0 := \{x \in \widetilde{X} : |x| \in C(Q)\}$ and $\|x\|_q := |x|(q)$ $(q \in Q,\ x \in X_0)$. It is clear that $(\|\cdot\|_q)_{q\in Q}$ is a multinorm in X_0. Let $\mathscr{X}$ be the continuous Banach bundle associated with the multinorm and let $\imath : X_0 \to C(Q, \mathscr{X})$ be the canonical embedding (see 2.4.4). From the definition of the topology in $\mathscr{X}$ (see 2.4.3) it is clear that an arbitrary section $u \in C(Q, \mathscr{X})$ goes through the ε-tube defined by some section of the form $\imath(x)$, $x \in X_0$, on a suitable clopen neighborhood of each point of Q. Consequently, for every $\varepsilon > 0$, we may choose a partition of unity $(Q_\xi)_{\xi\in\Xi}$ in the Boolean algebra of clopen subsets of Q and a family $(x_\xi)_{\xi\in\Xi}$ of elements in X_0 such that $\|u(q)-\imath(x_\xi)(q)\|_q \le \varepsilon$ $(q \in Q_\xi, \xi \in \Xi)$; i.e., $|\pi_\xi u-\pi_\xi \imath(x_\xi)| \le \varepsilon\mathbf{1}$ $(\xi \in \Xi)$, where π_ξ is the projection induced by the operator of multiplication by the characteristic function of the set Q_ξ. In the space X_0 there exists an element $x_\varepsilon = \sum \pi_\xi x_\xi$, for which $|\imath(x_\varepsilon) - u| \le \varepsilon\mathbf{1}$. Due to br-completeness of X_0, the limit $x := br\text{-}\lim x_\varepsilon$ also exists as $\varepsilon \to 0$; moreover, $\imath(x) = u$ and $|x| = |u|$. Thus, $\imath(X_0) = C(Q, \mathscr{X})$, and, hence, $C(Q, \mathscr{X})$ is a Banach–Kantorovich space. According to 2.4.8, the bundle $\mathscr{X}$ is ample. The uniqueness assertion follows from 2.4.9. $\triangleright$

2.4.11. A continuous Banach bundle $\mathscr{X}$ is called a *subbundle* of a continuous Banach bundle $\overline{\mathscr{X}}$ if $\mathscr{X}(q)$ is a Banach subspace of $\overline{\mathscr{X}}(q)$ for every $q \in Q$ and, moreover, $C(Q, \mathscr{X}) = C(Q, \overline{\mathscr{X}}) \cap S(Q, \mathscr{X})$. A subbundle $\mathscr{X}$ is called *dense* in $\overline{\mathscr{X}}$ if every section $u \in C(Q, \overline{\mathscr{X}})$ takes the values $u(q) \in \mathscr{X}(q)$ on a dense subset of Q. An *ample hull* or *completion* of a continuous Banach bundle $\mathscr{X}$ is any ample continuous Banach bundle $\overline{\mathscr{X}}$ containing $\mathscr{X}$ as a dense subbundle.

(1) Theorem. *Every continuous Banach bundle over an extremal compact space has an ample hull unique to within isometry.*

$\triangleleft$ Consider a continuous Banach bundle $\mathscr{X}$ over Q and endow the space $C_\#(Q, \mathscr{X})$ with the multinorm $(\|\cdot\|_q)_{q\in Q}$, by writing $\|u\|_q := |u|(q)$ for all $u \in C_\#(Q, \mathscr{X})$ and $q \in Q$. According to 2.4.4, the multinormed space thus obtained generates a continuous Banach bundle over Q that we denote by $\overline{\mathscr{X}}$. Let $\imath$: $C_\#(Q, \mathscr{X}) \to C(Q, \overline{\mathscr{X}})$ be the corresponding canonical embedding. At each point $q \in Q$ it is possible to define an isometric monomorphism $h(q) : \mathscr{X}(q) \to \overline{\mathscr{X}}(q)$, satisfying the equality $h(q)u(q) = \imath(u)(q)$ for all $u \in C(Q, \mathscr{X})$.

By Definition 2.4.9 it follows that h is a homomorphism from $\mathscr{X}$ into $\overline{\mathscr{X}}$ and, thus, performs an isometry from $\mathscr{X}$ onto a Banach subbundle of $\overline{\mathscr{X}}$. According to above definitions, this subbundle is dense in $\overline{\mathscr{X}}$. Uniqueness for the completion of a continuous Banach bundle to within isometry follows from 2.4.9 (2). $\triangleright$

(2) Theorem. *Let $\mathscr{X}$ and $\mathscr{Y}$ be CBBs over Q, let $\overline{\mathscr{X}}$ and $\overline{\mathscr{Y}}$ be respective ample hulls, and let D be a comeager subset of Q. Then every homomorphism $h \in \mathrm{Hom}_D(\mathscr{X}, \mathscr{Y})$ has a unique extension $\overline{h} \in \mathrm{Hom}_D(\overline{\mathscr{X}}, \overline{\mathscr{Y}})$. Moreover, the pointwise norms $|h|$ and $|\overline{h}|$ coincide on a comeager subset of Q.*

$\lhd$ Take $h \in \mathrm{Hom}_D(\mathscr{X}, \mathscr{Y})$. For arbitrary $q_0 \in D$ and $\bar{x} \in \overline{\mathscr{X}}(q_0)$ define $\overline{h}(q_0)\bar{x} \in \overline{\mathscr{Y}}(q_0)$. To this end choose a section $\bar{u} \in C(Q, \overline{\mathscr{X}})$ such that $\bar{u}(q_0) = \bar{x}$ and denote by u the restriction of $\bar{u}$ to comeager subset $\{q \in Q : \bar{u}(q) \in \mathscr{X}(q)\}$. Then $h \otimes u$ is a continuous almost global section of $\mathscr{Y}$. Let $\overline{h \otimes u} \in C_\infty(Q, \overline{\mathscr{Y}})$ be a maximal extension of $h \otimes u$ in $\overline{\mathscr{Y}}$. According to 2.4.9 (1) $\|h\|$ is a locally bounded function. Therefore, $\|H \otimes u\|$ is bounded on a neighborhood of q_0 and, consequently, $q_0 \in \mathrm{dom}(\overline{h \otimes u})$. Now, we put $\overline{h}(q_0)\overline{x} := \overline{h \otimes u}(q_0)$. Clearly, $\overline{h \otimes u}(q_0)$ does not depend on the choice of the section $\bar{u} \in C(Q, \overline{\mathscr{X}})$ satisfying $\bar{u}(q) = \bar{x}$ and, hence, the desired homomorphism $\overline{h}$ is correctly defined. The uniqueness of $\overline{h}$ follows from the equality $\overline{H}(q)\big(\mathrm{ext}_{\overline{\mathscr{X}}}(u)(q)\big) = \mathrm{ext}_{\overline{\mathscr{Y}}}(H \otimes u)(q)$ for all $u \in C_\#(Q, \mathscr{X})$ and $q \in Q$.

Since $C_\infty(Q, \mathscr{X})$ is a universal completion of $C(Q, \mathscr{X})$, for every section $u \in C_\infty(Q, \mathscr{X})$, $|u| \leqslant 1$, one can find a net $(u_\alpha)_{\alpha \in \mathrm{A}}$ in $C(Q, \mathscr{X})$ *bo*-converging to u and satisfying the inequality $|u_\alpha| \leqslant 1$. Then $|h \otimes u_\alpha - h \otimes u| \leqslant \|\overline{h}\| |u_\alpha - u|$ holds on $D \cap \mathrm{dom}(u)$, so that $|h \otimes u_\alpha|$ *bo*-converges to $|h \otimes u|$. Thus, $|h \otimes u| \leqslant \sup_{\alpha \in \mathrm{A}} |h \otimes u_\alpha|$ and, for $q \in D$, we deduce

$$
\begin{aligned}
\|h\|(q) &= \sup\{|h \otimes u|(q) : u \in C(Q, \mathscr{X}),\ |u| \leqslant 1\}, \\
&\geq \sup\{|h \otimes u|(q) : u \in C_\infty(Q, \mathscr{X}),\ |u| \leqslant 1\} = \|\overline{h}\|(q).
\end{aligned}
$$

The reverse inequality is obvious. $\rhd$

(3) *Let $\mathscr{X}$ be a continuous Banach bundle, with $\overline{\mathscr{X}}$ an ample hull of $\mathscr{X}$. Then the spaces $C_\infty(Q, \mathscr{X})$ and $C_\infty(Q, \overline{\mathscr{X}})$ coincide.*

$\lhd$ This follows from the definition of dense subbundle, since it is not difficult to check that a Banach bundle $\mathscr{X}$ is a dense subbundle of a continuous Banach bundle $\overline{\mathscr{X}}$ if and only if every section $u \in C_\infty(Q, \overline{\mathscr{X}})$ takes the values $u(q) \in \mathscr{X}(q)$ on a comeager subset of Q. $\rhd$

2.4.12. Now we introduce the continuous Banach bundle $B(\mathscr{X}, \mathscr{Y})$ whose continuous sections are homomorphisms from a continuous Banach bundle $\mathscr{X}$ into a continuous Banach bundle $\mathscr{Y}$. Consider a nonempty extremal compact space Q, and continuous Banach bundles $\mathscr{X}$ and $\mathscr{Y}$ over Q.

(1) *Let D be a dense subset of Q, and let a mapping $h : q \in D \mapsto h(q) \in \mathscr{L}\big(\mathscr{X}(q), \mathscr{Y}(q)\big)$ be such that $hu \in C(D, \mathscr{Y})$ for all $u \in C(Q, \mathscr{X})$. Suppose that the bundle $\mathscr{X}$ is ample. Then the pointwise norm $\|h\|$ is continuous.*

In view of 2.4.2 (5), the equality

$$
\|h\|(q) = \sup\{|hu|(q) : u \in C(Q, \mathscr{X}),\ |u| \leqslant 1\}
$$

is valid for all $q \in D$ which implies that the function $|||h|||$ is lower semicontinuous. In order to prove upper semicontinuity for $|||h|||$ at an arbitrary point $q \in D$ we assume that $|||h|||(q) < \lambda$ and establish the inequality $|||h|||(p) < \lambda$ for all elements p in a neighborhood of q within D. Assume the contrary. Then, taking a number λ_0 so that $|||h|||(q) < \lambda_0 < \lambda$ and defining $V = \{p \in D : |||h|||(p) > \lambda_0\}$ we conclude that $q \in \operatorname{cl} V$. For every element $p \in V$ let a section $u_p \in C(Q, \mathscr{X})$ satisfy $|u_p| \leqslant 1$ and $|hu_p|(p) > \lambda_0$. In view of continuity of the function $|hu_p|$ at each point $p \in V$ there is a clopen neighborhood $V_p \subset Q$ such that $|hu_p| > \lambda_0$ on $V_p \cap D$. It is easy to verify that the following equality holds: $\operatorname{cl} \bigcup_{p\in V} V_p = \operatorname{cl} V$. The Exhaustion Principle implies existence of a family $(U_p)_{p\in V}$ of pairwise disjoint clopen subsets of Q, satisfying the conditions $U_p \subset V_p$ $(p \in V)$ and $\operatorname{cl} \bigcup_{p\in V} U_p = \operatorname{cl} V$. Since $\mathscr{X}$ is ample, the continuous bounded section $\bigcup_{p\in V} u_p|_{U_p} \cup 0|_{Q\setminus \operatorname{cl} V}$ of $\mathscr{X}$ over the dense subset $\bigcup_{p\in V} U_p \cup (Q\setminus \operatorname{cl} V)$ can be extended to a global section $u \in C(Q, \mathscr{X})$. Obviously, $|u| \leqslant 1$. Since the sections u and u_p coincide on U_p, it follows that $|hu| \geqslant \lambda_0$ on $\operatorname{cl} V$. The last assertion contradicts the inequality $|||h|||(q) < \lambda_0$. $\rhd$

(2) *Every homomorphism $h \in \operatorname{Hom}_Q(\mathscr{X}, \mathscr{Y})$ has a continuous pointwise norm $|||h|||$ provided that the bundle $\mathscr{X}$ is ample.*

The following assertion is a version of the Stone–Weierstrass Theorem for a CBB over an extremal compact space. The section coinciding with u on a clopen set V and vanishing on its complement is denoted by $[V]u$.

(3) Theorem. *Let a vector subspace $\mathscr{U} \subset C(Q, \mathscr{X})$ be stalkwise dense in $\mathscr{X}$ and contain $[V]u$ for all elements $u \in \mathscr{U}$ and clopen subsets $K \subset Q$. Then $\mathscr{U}$ is uniformly dense in $C(Q, \mathscr{X})$.*

$\lhd$ Suppose that a subspace $\mathscr{U} \subset C(Q, \mathscr{X})$ meets the hypotheses of the theorem. We will fix a section $v \in C(Q, \mathscr{X})$ and, given an arbitrary $\varepsilon > 0$, construct an element $u \in \mathscr{U}$ satisfying the inequality $\|u - v\|_\infty \leq \varepsilon$. Given a point $q \in Q$ choose a section $u_q \in \mathscr{U}$ such that $\|u_q(q) - v(q)\| < \varepsilon$. In addition, denote by V_q a clopen neighborhood of q on which $|u_q - v| < \varepsilon$. Let us refine a finite subcover $V_{q_1}, \dots, V_{q_n}$ from the open cover $(V_q)_{q\in Q}$ of the compact Q. According to the Exhaustion Principle (see 1.1.6) there is a partition $W_1, \dots, W_n$ of Q into clopen subsets such that $W_k \subset V_{q_k}$ for all $k = 1, \dots, n$. It is clear that the section $u = \sum_{k=1}^n [W_k] u_{q_k}$ is the desired one. $\rhd$

2.4.13. Theorem. *Suppose that $\mathscr{X}$ is an ample bundle. Then there exists a (unique) continuous Banach bundle $B(\mathscr{X}, \mathscr{Y})$ over Q such that the stalk $B(\mathscr{X}, \mathscr{Y})(q)$ over each point $q \in Q$ is a Banach subspace of $B(\mathscr{X}(q), \mathscr{Y}(q))$ and $C(Q, B(\mathscr{X}, \mathscr{Y})) = \operatorname{Hom}_Q(\mathscr{X}, \mathscr{Y})$.*

$\lhd$ Consider the discrete Banach bundle over Q with the stalk over each point $q \in Q$ equal to the closure of the subspace $\{h(q) : h \in \operatorname{Hom}_Q(\mathscr{X}, \mathscr{Y})\}$ in the bundle $B(\mathscr{X}(q), \mathscr{Y}(q))$. In view of 2.4.12 (2), the space $\operatorname{Hom}_Q(\mathscr{X}, \mathscr{Y})$ is a continuity structure in this bundle and thus makes it into the desired continuous Banach bundle $B(\mathscr{X}, \mathscr{Y})$. The equality

$$C(Q, B(\mathscr{X}, \mathscr{Y})) = \mathrm{Hom}_Q(\mathscr{X}, \mathscr{Y})$$

ensues from 2.4.12 (3). ▷

2.4.14. Theorem. *Suppose that $\mathscr{X}$ is an ample bundle and D is a comeager subset of Q and $\mathscr{U} \subset C(D, \mathscr{X})$ is stalkwise dense in $\mathscr{X}$ on D. Let a mapping $h : q \in D \mapsto h(q) \in B(\mathscr{X}(q), \mathscr{Y}(q))$ be such that $h \otimes u \in C(D, \mathscr{Y})$ for all elements $u \in \mathscr{U}$. Then $h \in C(D, B(\mathscr{X}, \mathscr{Y}))$.*

◁ Let $\overline{\mathscr{Y}}$ be an ample hull of $\mathscr{Y}$. Using 2.4.11 (2), prove that there exists a section

$$\overline{h} \in C_\infty(Q, B(\mathscr{X}, \overline{\mathscr{Y}}))$$

such that the operator $\overline{h}(q)$ is an extension of $h(q)$ for all $q \in D \cap \mathrm{dom}(\overline{h})$. Moreover, h and $\overline{h}$ coincide on $D \cap \mathrm{dom}(\overline{h})$. Assume that there exists an element $q_0 \in D \backslash \mathrm{dom}(\overline{h})$. For every $q \in D := \mathrm{dom}(|h|)$ choose a section $u_q \in C(Q, \mathscr{X})$ and a clopen set $U_q \subset Q$ such that $q \in U_q$ and the inequalities $|u_q| \leqslant 1$ and $|h \otimes u_q| \geqslant |h| - \varepsilon$ hold on U_q. According to the Exhaustion Principle we may find a family $(V_q)_{q \in D}$ of pairwise disjoint clopen sets in Q such that $V_q \subset U_q$ $(q \in D)$ and the union of all V_q is dense in Q. Let $\bar{u} \in C(Q, \mathscr{X})$ be a section coinciding with u_q on V_q. Then $|\overline{h} \otimes \bar{u}| \geqslant |\overline{h}| - 1$. Since $\mathscr{U}$ is stalkwise dense in $\mathscr{X}$ on D there is an element $u \in \mathscr{U}$ with $\|u(q_0) - \bar{u}(q_0)\| < 1/2$; thus, $|u - \bar{u}| \leqslant 1/2$ in a neighborhood of q and we deduce

$$\|h(q_0)u(q_0)\| = |h \otimes u|(q_0) = |\overline{h} \otimes u|(q_0)$$
$$\geqslant (|\overline{h} \otimes u| - |\overline{h}| \cdot |u - \bar{u}|)(q_0) \geqslant (\frac{1}{2}|\overline{h}| - 1)(q_0) = \infty.$$

The contradiction shows that $D \subset \mathrm{dom}\,\overline{h}$. ▷

2.5. Measurable Banach Bundles

In this section, we introduce the notion and establish some elementary properties of measurable sections obtained by means of a measurability structure and prove that every Banach–Kantorovich space over an ideal space is linearly isometric to the space of measurable sections of a measurable Banach bundle.

2.5.1. Throughout the section, (Ω, Σ, μ) is a nonzero measure space. Let $\mathscr{X}$ be a Banach bundle over Ω. Denote by $S_\sim(\Omega, \mathscr{X})$ the set of all sections of $\mathscr{X}$ defined almost everywhere on Q. We call a set of sections $\mathscr{C} \subset S_\sim(\Omega, \mathscr{X})$ a *measurability structure* on $\mathscr{X}$, if it satisfies the following three conditions:

(a) $\lambda_1 c_1 + \lambda_2 c_2 \in \mathscr{C}$ for all $\lambda_1, \lambda_2 \in \mathbb{R}$ and $c_1, c_2 \in \mathscr{C}$;

(b) the pointwise norm $\|c\| : \Omega \to \mathbb{R}$ of every element $c \in \mathscr{C}$ is measurable;

(c) the set $\mathscr{C}$ is stalkwise dense in $\mathscr{X}$.

If $\mathscr{C}$ is a measurability structure in $\mathscr{X}$ then we call the pair $(\mathscr{X}, \mathscr{C})$ a *measurable Banach bundle* over Ω. We shall usually write simply $\mathscr{X}$ instead of $(\mathscr{X}, \mathscr{C})$ and denote the measurability structure $\mathscr{C}$ by $\mathscr{C}_{\mathscr{X}}$.

Let $(\mathscr{X}, \mathscr{C})$ be a measurable Banach bundle over Ω. We say that $s \in S_\sim(\Omega, \mathscr{X})$ is a *$\mathscr{C}$-step-section* (or simply a *step-section*, if it is clear which measurability structure is meant), if $s = \sum_{k=1}^n [A_k] c_k$ for some $n \in \mathbb{N}$, $A_1, \dots, A_n \in \Sigma$ and $c_1, \dots, c_n \in \mathscr{C}$. A section $u \in S_\sim(\Omega, \mathscr{X})$ is called *$\mathscr{C}$-measurable* (or simply *measurable*) if, for every $K \in \Sigma$ with $\mu(K) < +\infty$, there is a sequence $(s_n)_{n \in \mathbb{N}}$ of $\mathscr{C}$-step-sections such that $s_n(\omega) \to u(\omega)$ for almost all $\omega \in K$. The set of all $\mathscr{C}$-measurable sections of $\mathscr{X}$ is denoted by $\mathscr{M}(\Omega, \Sigma, \mu, \mathscr{X} | \mathscr{C})$ or $\mathscr{M}(\Omega, \mathscr{X})$ for brevity.

2.5.2. Suppose that $\mathscr{X}$ is a measurable Banach bundle over Ω. We consider the equivalence relation $\sim$ in the set $\mathscr{M}(\Omega, \mathscr{X})$ which is the coincidence almost everywhere: $u \sim v$ means that $u(\omega) = v(\omega)$ for almost all $\omega \in \Omega$. The coset containing an element $u \in \mathscr{M}(\Omega, \mathscr{X})$ is denoted by $u^\sim$. The factor set $\mathscr{M}(\Omega, \mathscr{X})/{\sim}$ is made into a vector space in the natural way: we write $\lambda u^\sim + \mu v^\sim = (\lambda u + \mu v)^\sim$ for $\lambda, \mu \in \mathbb{R}$ and $u, v \in \mathscr{M}(\Omega, \mathscr{X})$.

In addition, for every element $u^\sim \in \mathscr{M}(\Omega, \mathscr{X})/{\sim}$ we may define its (vector) *norm* $|u^\sim| := \|u\|^\sim \in M(\Omega)$. It is clear that the pair $\big(\mathscr{M}(\Omega, \mathscr{X})/{\sim}, |\cdot|\big)$ is an LNS over $L^0(\Omega, \Sigma, \mu)$; we denote it by $L^0(\Omega, \Sigma, \mu, \mathscr{X})$, or shortly $L^0(\Omega, \mathscr{X})$. Note that the space $L^0(\Omega, \mathscr{X})$ can be endowed with the natural structure of a module over the ring $L^0(\Omega, \Sigma, \mu)$; as follows: $e^\sim u^\sim := (eu)^\sim$ for all $e \in \mathscr{M}(\Omega)$ and $u \in \mathscr{M}(\Omega, \mathscr{X})$.

2.5.3. Theorem. *If a measure space (Ω, Σ, μ) possesses the direct sum property and if $\mathscr{X}$ is a measurable Banach bundle over Ω, then $L^0(\Omega, \mathscr{X})$ is a Banach–Kantorovich space over $L^0(\Omega, \Sigma, \mu)$.*

$\triangleleft$ Decomposability of the LNS $L^0(\Omega, \mathscr{X})$ is obvious. In view of 2.2.3 for proving *bo*-completeness of $L^0(\Omega, \mathscr{X})$, it suffices to establish its *d*- and *br*-completeness.

Let $(u_\xi^\sim)_{\xi \in \Xi}$ be a family of elements of $L^0(\Omega, \mathscr{X})$, with pairwise disjoint supports $\mathbf{A}_\xi \in B(\Omega)$ (see 1.4.11). Denote by $\bar{\Xi}$ the enrichment $\Xi \cup \{\infty\}$ of the set Ξ with a new element ∞ and define $u_\infty := 0$ and $\mathbf{A}_\infty := \big(\sup_{\xi \in \Xi} \mathbf{A}_\xi\big)^*$. Fix a lifting ρ of $L^\infty(\Omega)$ (see 1.4.9) and denote $A_\xi := \rho(\mathbf{A}_\xi)$ for each $\xi \in \bar{\Xi}$. In view of 1.2.7 (2), the union $\bigcup_{\xi \in \bar{\Xi}} A_\xi$ is measurable and differs from Ω by a set of measure zero. It is easy to verify that the section $\bigcup_{\xi \in \bar{\Xi}} u_\xi\big|_{A_\xi}$ is measurable and the corresponding class is the desired sum $bo\text{-}\sum_{\xi \in \Xi} u_\xi^\sim \in L^0(\Omega, \mathscr{X})$.

Now, suppose that a sequence $(u_n^\sim)_{n\in\mathbb{N}}$ in $L^0(\Omega, \mathscr{X})$ is *br*-fundamental. Then, for almost all $\omega \in \Omega$ the sequence $\big(u_n(\omega)\big)_{n\in\mathbb{N}}$ is fundamental. Due to completeness of the stalks of $\mathscr{X}$, there exists a section $u \in S_\sim(\Omega, \mathscr{X})$ to which the sequence $(u_n)_{n\in\mathbb{N}}$ converges almost everywhere. It is clear that the section u is measurable and the corresponding class is the desired *br*-limit of the sequence $(u_n^\sim)_{n\in\mathbb{N}}$. $\triangleright$

2.5.4. Some important corollaries of the above theorem are now available.

(1) ***Theorem.*** *If (Ω, Σ, μ) is a measure space possessing the direct sum property and E is an ideal of $L^0(\Omega)$ then the set $E(\mathscr{X}) := \{\mathbf{u} \in L^0(\Omega, \mathscr{X}) : |\mathbf{u}| \in E\}$ endowed with the operations induced from $L^0(\Omega, \mathscr{X})$ is a Banach–Kantorovich over E.*

(2) The symbol $\mathscr{L}^\infty(\Omega, \mathscr{X})$ stands for the set $\{u \in \mathscr{M}(\Omega, \mathscr{X}) : \|u\| \in \mathscr{L}^\infty(\Omega)\}$ and its elements are called *(essentially) bounded measurable sections* of $\mathscr{X}$. The totality of equivalence classes constituted by essentially bounded sections is denoted by $L^\infty(\Omega, \mathscr{X})$. Obviously, the space $L^\infty(\Omega, \mathscr{X})$ coincides with $E(\mathscr{X})$, where $E = L^\infty(\Omega, \Sigma, \mu)$. In particular, $L^\infty(\Omega, \mathscr{X})$ is a Banach–Kantorovich space over $L^\infty(\Omega, \Sigma, \mu)$,

(3) Let X be a Banach space. Consider the trivial Banach bundle $\mathscr{X} = \Omega \times \{X\}$ and let the totality of constant functions $c : \Omega \to X$ be taken as the measurability structure of $\mathscr{X}$. Then the set $\mathscr{M}(\Omega, \mathscr{X})$ consists exactly of all Bochner measurable X-valued functions defined almost everywhere in Ω: $\mathscr{M}(\Omega, X) = \mathscr{M}(\Omega, \mathscr{X})$. Consequently, $L^0(\Omega, X) = L^0(\Omega, \mathscr{X})$. In particular, $L^0(\Omega, X)$, $L^\infty(\Omega, X)$, and $E(X)$, where E is an order ideal in $L^0(\Omega, \Sigma, \mu)$, are Banach–Kantorovich spaces over $L^0(\Omega, \Sigma, \mu)$.

2.5.5. Let $\mathscr{X}$ be a measurable Banach bundle over Ω. Consider a lifting $\rho : L^\infty(\Omega) \to \mathscr{L}^\infty(\Omega)$ (see 1.4.8). We call a mapping $\rho_{\mathscr{X}} : L^\infty(\Omega, \mathscr{X}) \to \mathscr{L}^\infty(\Omega, \mathscr{X})$ a *lifting* of $L^\infty(\Omega, \mathscr{X})$ *associated with* ρ if, for all $u, v \in L^\infty(\Omega, \mathscr{X})$ and $e \in L^\infty(\Omega)$, the following relations hold:

(1) $\rho_{\mathscr{X}}(u) \in u^\sim$ and $\operatorname{dom}(\rho_{\mathscr{X}}(u)) = \Omega$;

(2) $\|\rho_{\mathscr{X}}(u)\| = \rho(|u|)$;

(3) $\rho_{\mathscr{X}}(u + v) = \rho_{\mathscr{X}}(u) + \rho_{\mathscr{X}}(v)$;

(4) $\rho_{\mathscr{X}}(eu) = \rho(e)\rho_{\mathscr{X}}(u)$;

(5) the set $\{\rho_{\mathscr{X}}(u) : u \in L^\infty(\Omega, \mathscr{X})\}$ is stalkwise dense in $\mathscr{X}$.

In case there exist a lifting of $L^\infty(\Omega)$ and a lifting of associated with it, we say that $\mathscr{X}$ is a *liftable measurable Banach bundle.*

2.5.6. Let ρ be a lifting of $L^\infty(\Omega)$ and let $\mathscr{X}$ and $\mathscr{Y}$ be liftable measurable Banach bundles over Ω. We call the bundles $\mathscr{X}$ and $\mathscr{Y}$ *ρ-isometric*, if their liftings

$\rho_{\mathscr{X}}$ and $\rho_{\mathscr{Y}}$ are associated with ρ and there exists an isometry h from $\mathscr{X}$ onto $\mathscr{Y}$ such that $\rho_{\mathscr{Y}}(hu) = h\rho_{\mathscr{X}}(u)$ for all $u \in L^\infty(\Omega, \mathscr{X})$.

Two points $\omega_1, \omega_2 \in \Omega$ are said to be *ρ-indistinguishable* if $\rho(f)(\omega_1) = \rho(f)(\omega_2)$ for every class $f \in L^\infty(\Omega)$. We say that a measurable Banach bundle $\mathscr{X}$ is *invariant with respect to ρ*, or *ρ-invariant*, if $\mathscr{X}(\omega_1) = \mathscr{X}(\omega_2)$ and $\rho(u)(\omega_1) = \rho(u)(\omega_2)$ for all $u \in L^\infty(\Omega, \mathscr{X})$ for arbitrary ρ-indistinguishable points $\omega_1, \omega_2 \in \Omega$.

2.5.7. The following assertion enables us to assume without loss of generality that every measurable Banach bundle under consideration is invariant with respect to the corresponding lifting.

For every measurable Banach bundle over Ω having a lifting associated with ρ, there is a ρ-stationary liftable measurable Banach bundle ρ-isometric to it.

$\lhd$ Suppose that a measurable Banach bundle $\mathscr{X}$ meets the hypothesis of the assertion. Fix an arbitrary pair of $E\rho$-indistinguishable points $\omega_1, \omega_2 \in \Omega$. It suffices to construct an isometry $\imath$ from the stalk $\mathscr{X}(\omega_1)$ onto $\mathscr{X}(\omega_2)$ such that $\imath\big(\rho(u)(\omega_1)\big) = \rho(u)(\omega_2)$ for all $u \in L^\infty(\Omega, \mathscr{X})$. Indeed, in this case, the stalks over indistinguishable points can be identified by means of such isometries.

For each $\omega \in \Omega$, denote by $\mathscr{X}_0(\omega)$ the subspace $\{\rho(u)(\omega) : u \in L^\infty(\Omega, \mathscr{X})\}$ of the stalk $\mathscr{X}(\omega)$. Let classes $u, v \in L^\infty(\Omega, \mathscr{X})$ be such that $\rho(u)(\omega_1) = \rho(v)(\omega_1)$. Then $\rho(u)(\omega_2) = \rho(v)(\omega_2)$, because from the properties of lifting it follows that

$$\|\rho(u)(\omega_2) - \rho(v)(\omega_2)\| = \rho(|u-v|)(\omega_2) = \rho(|u-v|)(\omega_1) = \|\rho(u)(\omega_1) - \rho(v)(\omega_1)\| = 0.$$

This enables us to consider a bijection $\imath_0 : \mathscr{X}_0(\omega_1) \to \mathscr{X}_0(\omega_2)$, defined by the rule $\imath_0\big(\rho(u)(\omega_1)\big) = \rho(u)(\omega_2)$ for every class $u \in L^\infty(\Omega, \mathscr{X})$. Due to the fact that the subspaces $\mathscr{X}_0(\omega)$ are dense in the corresponding stalks $\mathscr{X}(\omega)$, the isometry $\imath_0$ can be extended to the desired isometry $\imath : \mathscr{X}(\omega_1) \to \mathscr{X}(\omega_2)$. $\rhd$

2.5.8. Suppose that Q is the Stone space of the Boolean algebra $B(\Omega)$ and $\tau : \Omega \to Q$ is the canonical immersion of Ω into Q corresponding to the lifting ρ of $L^\infty(\Omega)$. Let $\mathscr{Y}$ be an ample continuous Banach bundle over Q and $\mathscr{X} = \mathscr{Y} \circ \tau$.

If $\mathscr{C}$ is a continuous structure in $\mathscr{Y}$, then the set $\mathscr{C} \circ \tau$ is a measurability structure in $\mathscr{X}$, since $\|c \circ \tau\| = \|c\| \circ \tau$ and $\|c\| \circ \tau$ is a measurable function. The bundle $\mathscr{Y} \circ \tau$ is always regarded as a measurable Banach bundle with respect to the measurability structure $\mathscr{C} \circ \tau$.

For every $v \in C_\infty(Q, \mathscr{Y})$ the composite $v \circ \tau$ is a measurable section of the bundle $\mathscr{X}$.

$\lhd$ This is evident. $\rhd$

For $v \in C_\infty(Q, \mathscr{Y})$ we denote by $(v \circ \tau)^\sim$ the coset containing an element $v \circ \tau$.

2.5.9. Theorem. *The mapping $v \mapsto (v \circ \tau)^\sim$ is an isometry from the BKS, $C_\infty(Q, \mathscr{Y})$ onto $L^0(\Omega, \mathscr{X})$ and is associated with the isomorphism $\big(e \mapsto (e \circ \tau)^\sim\big)$: $C_\infty(Q) \to L^0(\Omega)$. The image of $C(Q, \mathscr{Y})$ under this isometry is $L^\infty(\Omega, \mathscr{X})$.*

$\lhd$ We need only to prove that an almost everywhere defined section u of the bundle $\mathscr{X}$ is measurable if and only if $u \sim v \circ \tau$ for some element $v \in C_\infty(Q, \mathscr{Y})$. In other words, it should be verified that $\mathscr{U} = \mathscr{L}^0(\Omega, \mathscr{X})$, where

$$\mathscr{U} := \{u \in S_\sim(\Omega, \mathscr{X}) : \big(\exists v \in C_\infty(Q, \mathscr{Y})\big)\, u \sim v \circ \tau\}.$$

We confine ourself to the case of a σ-finite measure space. In view of 2.5.8 $\mathscr{U} \subset \mathscr{L}^0(\Omega, \mathscr{X})$. Let us prove the reverse inclusion.

Let $v \in C_\infty(Q, \mathscr{Y})$, $A \in \Sigma$ and $\hat{A} := \hat{A}^\sim$ be the clopen set in Q corresponding to A under the Stone transform. Then

$$[A](v \circ \tau) \sim [\rho(A^\sim)](v \circ \tau) = \big[\tau^{-1}(\hat{A})\big](v \circ \tau) = \big([\hat{A}]v\big) \circ \tau,$$

which implies that the set $\mathscr{U}$ contains the fragments of all its elements. Thus, $\mathscr{U}$ contains also the set of all $\mathscr{C}$-step-sections. Suppose that $u \in \mathscr{L}^0(\Omega, \mathscr{X})$. By the definition of measurability in 2.5.1, there exists a sequence of step-sections $u_n \sim v_n \circ \tau$ that converges to a section $u \in S_\sim(\Omega, \mathscr{X})$ almost everywhere. Then we have $o\text{-}\lim_{n\to\infty} \big(|\!|\!| u_n - u |\!|\!|\big)^\sim = o\text{-}\lim | u_n^\sim - u^\sim | = 0$ in the K-space $L^0(\Omega, \Sigma, \mu)$; thus, $|u_n^\sim - u_m^\sim| \to 0$. Since the correspondence $f \mapsto (f \circ \tau)^\sim$ is an order isomorphism of vector lattices $L^0(\Omega, \Sigma, \mu)$ and $C_\infty(Q)$, it follows that $o\text{-}\lim_{n\to\infty} | v_n - v_m | = 0$ in $C_\infty(Q)$. Due to o-completeness of the LNS $C_\infty(Q, \mathscr{Y})$, there exists the o-limit $v \in C_\infty(Q, \mathscr{Y})$ of the sequence (v_n). Obviously, $u \sim v \circ \tau$. $\rhd$

The inverse isometry from $L^0(\Omega, \mathscr{X})$ onto $C_\infty(Q, \mathscr{Y})$ is defined by the rule $u \mapsto \hat{u}$, where $\hat{u}$ is a unique section in $C_\infty(Q, \mathscr{Y})$, representing u as $(\hat{u} \circ \tau)^\sim$. We refer to the section $\hat{u}$ as the *Stone transform* of u.

The mapping $u \mapsto \hat{u} \circ \tau$ is a lifting of $L^\infty(\Omega, \mathscr{X})$ associated with ρ. Endowed with this lifting, the measurable Banach bundle $\mathscr{X}$ is ρ-stable.

Theorem 2.5.9 describes a method of constructing a liftable measurable Banach bundle given a complete continuous Banach bundle over the corresponding Stone space. The following result shows that every liftable measurable Banach bundle can be obtained exactly in such a way.

2.5.10. Theorem. *Let $\mathscr{X}$ be a ρ-invariant measurable Banach bundle over Ω that has a lifting associated with ρ. Then there exists an ample continuous Banach bundle $\hat{\mathscr{X}}$ over Q unique to within an isometry and such that $\mathscr{X} = \hat{\mathscr{X}} \circ \tau$ and $\rho(u) = \hat{u} \circ \tau$ for all $u \in L^\infty(\Omega, \mathscr{X})$.*

$\lhd$ Since $\mathscr{X}$ is ρ-stationary, we may define a Banach bundle $\mathscr{Y}$ over $\tau(\Omega)$ by the formula $\mathscr{Y}\big(\tau(\omega)\big) := \mathscr{X}(\omega)$ and endow it with the continuity structure $\big\{\rho(u) \circ \tau^{-1} : u \in L^\infty(\Omega, \mathscr{X})\big\}$.

Denote by $\mathscr{U}$ the vector space $C^b(D, \mathscr{Y})$ with the multinorm $\big(\|\cdot\|_q\big)_{q\in Q}$, $\|u\|_q = |u|(q)$, where $|u| : Q \to \mathbb{R}$ is a unique continuous extension of the bounded function $q \mapsto \|u(q)\|$ $(q \in D)$. Let $\hat{\mathscr{X}}$ be the CBB over Q generated by the multinormed space $\mathscr{U}$ and let $\imath : \mathscr{U} \to C(Q, \hat{\mathscr{X}})$ be corresponding canonical embedding (see 2.4.4). Since $\mathscr{U}$ is norm complete, stalkwise dense in $\hat{\mathscr{X}}$ and invariant with respect to band projections, we may prove that $\imath(\mathscr{U}) = C(Q, \hat{\mathscr{X}})$. For each point $q \in D$ we may define a linear isometry $h(q)$ between the stalks $\mathscr{X}(q)$ and $\hat{\mathscr{X}}(q)$ satisfying the equality $h(q)u(q) = \imath(u)(q)$ for all $u \in \mathscr{U}$. It is easy to see that h is an isometry from $\mathscr{X}$ onto $\hat{\mathscr{X}}|_D$. Consequently, $\hat{\mathscr{X}}$ is the desired continuous Banach bundle. Most of the necessary properties of the bundle $\hat{\mathscr{X}}$ are easily verified. Prove that the bundle is ample.

It follows from the construction of the bundle $\hat{\mathscr{X}}$ that for each class $u \in L^\infty(\Omega, \mathscr{X})$, the bounded section $\rho(u) \circ \tau \in C\big(\tau(\Omega), \mathscr{Y}\big)$ can be extended to an element of $C(Q, \hat{\mathscr{X}})$ that will be denoted by $\hat{u}$. It can be easily verified that the mapping $u \mapsto \hat{u}$ is an isometry from the LNS $L^\infty(\Omega, \mathscr{X})$ onto $C(Q, \hat{\mathscr{X}})$ associated with the isomorphism $(e \mapsto \hat{e})$ from $L^\infty(\Omega)$ onto $C(Q)$, see 1.4.9. It follows that the image of $L^\infty(\Omega, \mathscr{X})$ under this isometry is o-dense in $C(Q, \hat{\mathscr{X}})$. In view of the o-completeness of $L^\infty(\Omega, \mathscr{X})$, this image coincides with $C(Q, \hat{\mathscr{X}})$ and, hence, the LNS $C(Q, \hat{\mathscr{X}})$ is o-complete as well. Uniqueness of the bundle $\hat{\mathscr{X}}$ follows from 2.4.10. ▷

We call the ample continuous Banach bundle $\hat{\mathscr{X}}$ in the statement of the last theorem the *Stone transform* of the measurable Banach bundle $\mathscr{X}$.

Three facts presented below are immediate consequences of applying Theorem 2.5.9 and the representation result 2.4.9.

2.5.11. Theorem. *Every Banach–Kantorovich space X over an order-dense ideal $E \subset L^0(\Omega)$ is linearly isometric to $E(\mathscr{X})$ for some liftable measurable Banach bundle $\mathscr{X}$ over Ω. Moreover, such a bundle $\mathscr{X}$ is unique to within a linear isometry.*

2.5.12. Theorem. *For every Banach–Kantorovich X over E and every isomorphism $\imath$ from the K-space E onto an order-dense ideal $F \subset L^0(\Omega)$ there exist a liftable measurable Banach bundle $\mathscr{X}$ over Ω (unique to within isometry) and some isometry from X onto $F(\mathscr{X})$ associated with $\imath$.*

2.5.13. Theorem. *For every Banach–Kantorovich space X over a Kantorovich–Pinsker space, there exist a measurability structure Ω, possessing the direct sum property, an order-dense ideal $F \subset L^0(\Omega)$, and a liftable measurable Banach bundle $\mathscr{X}$ over Ω such that the LNSs X and $F(\mathscr{X})$ are isometric.*

2.6. Comments

2.6.1. (1) The concept of lattice-normed space was introduced for the first time by L. V. Kantorovich in 1936 [154]. These are vector spaces normed by elements of a vector lattice. Somewhat earlier, G. Kurepa [192] considered "espaces pseudodistanciés," i.e. a space with a metric that takes values in an ordered vector space. First applications of vector norms and metrics were related to the method of successive approximations in numerical analysis, see [154, 161, 163, 182, 340].

(2) It is worth stressing that [157] is the very paper in which the unusual decomposability axiom (see 2.1.1 (4)) for an abstract norm appeared for the first. Paradoxically, this axiom (4) was often omitted as inessential in the further research by other authors. The profound importance of (4) was rediscovered in connection with Boolean-valued analysis (see [196, 197]).

(3) The connection between decomposability and existence of a Boolean algebra of projections in an LNS was discovered by A. G. Kusraev [196, 197]. Spaces with a fixed Boolean algebra of linear projections and a coordinated order (the so-called coordinated spaces) were studied by J. L. B. Cooper [69, 70]. Assertions 2.1.7 (1, 3) were obtained in [181].

(4) The notion of discrete element is important in the structure theory of vector lattices, see [163, 263, 388]. Discrete functionals and discrete (real-valued) measures are well studied and have a simple structure [139, 326]. As was shown by J. A. Crenshaw [71], under certain mild conditions a discrete element in the lattice of order-bounded operators is completely defined by a discrete functional on the domain vector lattice and a discrete element of the target vector lattice. Thus discrete operators comprise a poor class.

At the same time there is a series of interesting results in which the concept of module discreteness plays the central role (see for instance [42, 197, 215, 220, 344, 393]). This motivates the study of concepts analogous to module discreteness, module atomicity, and module indecomposability in lattice-normed spaces. The notion of norm-n-indecomposable element from 2.1.9 and Theorem 2.1.10 are due to V. A. Radnaev [322, 323].

2.6.2. (1) The completeness criterion 2.2.3 was stated by A. G. Kusraev in [196] under the condition that the norm lattice E is order complete. In [197], this was proven in a more general situation of spaces with decomposable vector multinorms. The assumption of order completeness for E was removed in [181]. For an Archimedean vector lattice (the case of $X = E$) the indicated fact was established by A. I. Veksler and V. A. Geĭler [381].

(2) The concept of universal completion (maximal extension) for an arbitrary K-space was introduced and studied by A. G. Pinsker; see [163]. He established, in particular, that any K-space has a unique, to within an isomorphism, maximal extension. Assertion 2.2.8, which is a generalization of the Pinsker Theorem for LNSs, was essentially obtained in [197]. As regards Theorem 2.2.11 (1) on order completion of an LNS, see [197, 218]. The assertion $\hat{X} = oX$ in 2.2.11 (2) is due to A. E. Gutman. The properties of approximating sets and their applications to order-bounded operators were studied in [120].

(3) The operator-dual space X^* in 2.2.4, introduced here for constructing the maximal extension mX, is of independent interest; it was studied in [197]. In particular, an LNS X is operator-reflexive in the sense that $X^{**} = \varkappa(X)$ if and only if the set $\{x \in X : |x| \leq e\}$ is weakly cyclically compact for each element $e \in E_+$ (see [197]). More about the conception of cyclical compactness will be said in Section 8.5.

2.6.3. **(1)** Information about measurable functions with values in a Banach space and, particularly, in the space of bounded linear operators, is presented in [74, 78, 87]. A general idea of constructing spaces in Section 1.2 can be expressed as follows: If X is a Banach (or locally convex) space and E is a function space then we may associate with them a class Z of vector-functions (measurable or continuous) by requiring that $f \in Z$ if and only if $l \circ f \in E$ for each $l \in X'$, etc. (see [54, 92, 238, 319]). This idea was elaborated while developing the theory of vector integration [51, 74, 78, 87, 91, 375].

(2) From 2.3.4 it is clear that the algebraic tensor product $E \otimes X$ is *bo*-dense in the space $E(X)$. Denseness of $E \otimes X$ in $E(X)$ with respect to the scalar norm is connected with order continuity of a Banach lattice E: see Phuong-Cac [319], A. V. Bukhvalov [55], and V. L. Levin [237, 238]. The vector norm in $E \otimes X$, that is introduced in 1.2.10, is analogous to Levin's cross-norm, see [237, 238]. Assertion 2.3.4 (1) is proven as in [238, Theorem 4.2].

(3) The tensor product of Banach lattice E and an arbitrary Banach space with cross-norm induced by the space of regular operators was introduced and learned by V. L. Levin [238]. V. T. Khudalov [169, 170] studied a similar cross-norm for a Banach space with regular cone. An important method for constructing tensor products of vector and Banach lattices was proposed by D. Fremlin [98, 99]. G. Shotaev [350, 351] constructed tensor products of lattice-normed spaces and of Banach spaces with mixed norm.

(4) In the study of measurable vector-functions the following question is of interest: If $X = F$ is a Banach ideal space of functions of a variable s then every vector-function $\vec{f} : T \to X = F$ generates a function of two variables by the formula $\phi(s,t) = (\vec{f}(t))(s)$. However, $\phi(\cdot,\cdot)$ may fail to be measurable as a function of two variables even in the simplest cases (for instance, for the measure space $[0,1]$ with the Lebesgue measure and $X = L^2(0,1)$). Indeed, W. Sierpinski constructed an example of a subset of a square which is Lebesgue nonmeasurable and has at most two common points with every straight line. The characteristic function ϕ_0 of this set is nonmeasurable as a function of two variables whereas the corresponding vector-function f is the zero function since $\phi_0(s,t) = 0$ almost everywhere for all t. By adding ϕ_0 to nonzero measurable vector-functions, we obtain similar examples with nonzero vector-functions.

(5) The fact that Property (C) in F implies measurability for the function $|K|(t) := \|K(\cdot,t)\|_F$ (see 2.3.9 (1)) was established by W. A. J. Luxemburg. It can be also deduced from one result of Yu. I. Gribanov [115]; see [162, XI.1.3, XI.1.4]. Proposition 2.3.9 (3) is essentially due to H. W. Ellis [90]. The case of an atomic measure is of no avail since then all functions are measurable.

(6) A much more difficult question is the representation $f(t)(\cdot) = K(\cdot,t)$ for weakly measurable functions. One of the possible solutions to this problem rests on the Bukhvalov integrality criterion 6.5.4 (see [59, 228]).

If F is a Banach ideal space with order continuous norm then $E(F') = E_s(F')$.

2.6.4. (1) The invention of Banach bundles is customarily connected with the name of J. von Neumann who proposed in 1937 some ideas about varying Banach spaces. The corresponding formal descriptions appeared about 1950 in the papers of R. Godement, I. Kaplansky, and I. M. Gelfand and M. A. Naĭmark. Presently, the theory of continuous Banach bundles is a rather wide area of research. An adequate description for the current state of this theory can be found in the surveys and references of [95]. The monograph [104] contains a detailed discussion of most of the necessary notions related to CBBs. Continuous Banach bundles are often used for representing various functional-analytical objects, see [95, 104, 131, 339].

(2) In the article [218] by A. G. Kusraev and V. Z. Strizhevskiĭ, the space $E(\mathscr{X})$ was introduced of almost global sections of a continuous Banach bundle $\mathscr{X}$, and Theorem 2.4.7 was established; it was also shown that each BKS is linearly isometric to $E(\mathscr{X})$ for a suitable $\mathfrak{X}$; however, uniqueness of the bundle $\mathscr{X}$ was not established (cf. the existence result in Theorem 2.4.10). A. E. Gutman found a class of uniqueness for Theorem 1.4.7, the class of ample continuous Banach bundles, and proved Theorem 1.4.10 [123]. For a detailed presentation of these and other interesting results see [118–124]. The material of Section 2.4 shows that ample CBBs have other advantages. For instance, only after assuming the completeness of a CBB $\mathscr{X}$ it becomes possible to introduce the CBB $\mathscr{L}(\mathscr{X},\mathscr{Y})$ and the dual CBB $\mathscr{X}'$ (2.4.13); this result is also due to A. E. Gutman [118].

2.6.5. (1) The idea of a measurability structure has been proposed by N. Dinculeanu as early as in 1966, but has not been much studied since then. A different approach to defining measurability of sections has been prevalent so far in the papers on Banach bundles. Namely, let $\mathscr{X}$ be a continuous Banach bundle over a locally compact space Ω with a fixed Radon measure. A section $u \in S(\Omega,\mathscr{X})$ is called measurable if, for every compact $K \subset \Omega$, there exists a sequence of continuous sections $u_n \in S(\Omega,\mathscr{X})$ converging to u almost everywhere on K.

(2) In Section 2.5 we follow A. E. Gutman [119]. The way of introducing measurable sections in Section 2.5 is similar to Daniel's construction and is formally more general than the traditional topological approach. As is seen from 1.5.10, liftable measurable Banach bundles in the class of all MBBs occupy in a sense the same place as complete continuous Banach bundles in the class of all CBBs. The connection between liftable measurable Banach bundles and complete CBBs, described in 2.5.8–2.5.10, enables us to transfer the facts of the theory of complete CBBs to the case of a measurable Banach bundle; see [119].

Chapter 3

Positive Operators

In this chapter we briefly present some basic results on positive operators in vector lattices. Two natural classes of operators are closely related to positive operators: the set of differences of positive operators called the space of regular operators, and the space of order-bounded operators (3.1.1). The fundamental Riesz–Kantorovich Theorem (3.1.2) claims that, under natural conditions, the space of regular operators is an order complete vector lattice and that a regular operator can be described as an order-bounded operator, i.e., an operator carrying order-bounded sets into order-bounded sets. Explicit expressions for finite and infinite lattice operations as well as for the moduli, positive and negative parts of order-bounded operators comprise the so-called order calculus. One of the approaches leans upon the concept of a generating set of projections (3.1.5). All these facts require order completeness of the target vector lattice. Order complete vector lattices appear by necessity also in the problem of dominated extension of linear operators: such an extension is possible in a mass setting if and only if the target vector lattice is order complete (the Hahn–Banach–Kantorovich Theorem (3.1.7) and the Bonnice–Silvermann–To Theorem (3.1.8)).

Simplest representatives of the class of order-bounded operators are orthomorphisms (3.3.2) and lattice homomorphisms (3.3.1). Hahn–Banach–Kantorovich-type theorems are also true for lattice homomorphisms (3.3.10, 3.3.11). This fact is closely connected with the extreme structure of convex sets of operators (3.3.7–3.3.9).

Among order bounded operators are important classes of order continuous and order σ-continuous operators (3.2.1). Their antipodes, called singular and σ-singular operators, are distinguish by the property of vanishing on an order-dense or σ-order-dense ideal, respectively (3.2.1). Order continuous and order σ-continuous operators comprise bands (3.2.3 (2)). Moreover, these bands are the disjoint complements of the sets of singular and σ-singular bands under the appropriate conditions (3.2.3 (1)). It is a long tradition stemming from measure theory to look for some explicit description of different fragments of an operator or measure. Several results in this direction are presented in 3.2.5–3.2.9 and 3.3.13.

Next we consider the class of the so-called Maharam operators. A Maharam operator is a "full-valued" or order interval preserving order continuous operator. One of the main peculiarities of a Maharam operator is that the Boolean algebra of its fragments is isomorphic to the base of the domain vector lattice (3.4.5). Variants of the Hahn Decomposition Property and the Radon–Nikodým Theorem are valid for Maharam operators (3.4.6 (1), 3.4.9). An important example of a Maharam operator is the transformation assigning to each order-bounded operator its restriction to a fixed massive sublattice of the domain vector lattice (3.5.2 (6)). This fact enables us to prove the existence of simultaneous extension from an arbitrary massive sublattice (3.4.11).

Maharam operators may be useful in studying different classes of operators. Given a positive operator, we may always extend it to a Maharam operator by a construction very much resembling the Daniell construction of the Lebesgue integral (3.5.2, 5.5.3). The domain of the so-extended operator is then approximated in a way by the domain vector lattice of the initial operator (3.5.6–3.5.8). The band generated by the initial operator is linear and order isomorphic to that of the resulting Maharam operators (3.5.5). These results are used to obtain some description for the Boolean algebra of fragments of a positive operator by up-and-down procedures (3.5.9, 3.5.10).

3.1. Operators in Vector Lattices

In this section we introduce spaces of order-bounded operators and present several fundamental results which are systematically used in the sequel.

3.1.1. Let E and F be vector lattices. A linear operator $T : E \to F$ is said to be *positive* provided that $T(E_+) \subset F_+$; *regular* provided that it can be represented as a difference of two positive operators; last, *order-bounded* or *o-bounded* provided that T carries every order-bounded subset in E onto an order-bounded subset in F. The set of all regular, order-bounded, and positive operators from E into F is denoted by $L^r(E, F)$, $L^\sim(E, F)$, and $L_+(E, F) := L^\sim(E, F)_+$, respectively. The classes $L^r(E, F)$ and $L^\sim(E, F)$ are vector subspaces in the vector space $L(E, F)$ of all linear operators from E to F, provided with the order relation: $S \geq T \Leftrightarrow S - T \geq 0$.

Every positive operator is obviously order-bounded and the difference of order-bounded operators is order-bounded, too. Thus, every regular operator is order-bounded. The converse may fail but is also true in the case of order complete F. The latter follows immediately from the famous Riesz–Kantorovich Theorem to be proved in the next subsection. First we give a simple auxiliary fact.

Let E be a vector lattice, let X be an arbitrary real vector space, and let U be an additive and positively homogeneous mapping from E_+ to X:

$$U(x+y) = Ux + Uy, \quad U(\lambda x) = \lambda Ux \quad (0 \leq \lambda \in \mathbb{R};\ x, y \in E_+).$$

Then U admits a unique linear extension T to the whole vector lattice E. If in addition F is a vector lattice and $U(E_+) \subset F_+$ then T is positive.

$\triangleleft$ The uniqueness of extension is obvious from the formula $x = x^+ - x^-$; therefore, it remains to prove existence. For every $x \in E$, we put $Tx = Ux^+ - Ux^-$ and observe that T is the sought extension. Indeed, for $z = x - y$, $x, y \in E_+$ we have $z^+ - z^- = x - y$ or $z^+ + y = x + z^-$. Therefore, $Uz^+ + Uy = Ux + Uz^-$ by hypotheses, whence $Tz = Uz^+ - Uz^- = Ux - Uy = Tx - Ty$. Now, we may conclude that T is additive on E. Additivity implies that $T(-x) = -Tx$ for all $x \in E$, which involves homogeneity of T by virtue of its positive homogeneity. $\triangleright$

3.1.2. *Riesz–Kantorovich Theorem.* *If E is a vector lattice and F is some K-space then the set of order-bounded operators $L^\sim(E, F)$ ordered by the cone of positive operators $L^\sim(E, F)_+$ is a K-space. Moreover, for every $S, T \in L^\sim(E, F)$ and $x \in E_+$, the following formulas hold:*

(1) $(S \vee T)x = \sup\{Sx_1 + Tx_2 : x_1, x_2 \geqslant 0, x = x_1 + x_2\}$;

(2) $(S \wedge T)x = \inf\{Sx_1 + Tx_2 : x_1, x_2 \geqslant 0, x = x_1 + x_2\}$;

(3) $S^+x = \sup\{Sy : 0 \leq y \leq x\}$;

(4) $S^-x = -\inf\{Sy : 0 \leq y \leq x\}$;

(5) $|S|x = \sup\{|Sy| : |y| \leq x\}$;

(6) $|S|x = \sup\left\{\sum_{i=1}^n |Sx_i| : x_1, \dots, x_n \geqslant 0, x = \sum_{i=1}^n x_i, n \in \mathbb{N}\right\}$;

(7) $|Sx| \leq |S|(|x|)$ $(x \in E)$.

$\triangleleft$ To establish that some order vector space L is a vector lattice, it suffices to check that, for every $x \in L$, there exists $|x| = x \vee (-x)$. Then it remains to use the formulas 1.3.2 (3).

Denote by Ux the right-hand side of the formula (5) and prove that the operator $U : E_+ \to F$ is additive and positively homogeneous as in 2.1.1. Observe that for a fixed $x \in E_+$ the set $\{y \in E : |y| \leq x\}$ is bounded and, since $S \in L^\sim(E, F)$, its image $\{|Sy| : |y| \leq x\}$ is also bounded in F. Thus, the supremum in (5) exists and the operator U is correctly defined. Positive homogeneity of U is obvious and it remains to prove its additivity. Formula (5) can be slightly simplified. By virtue of associativity of bounds, we have

$$Ux = \sup\{|Sy| : |y| \leq x\} = \sup\{(Sy) \vee (S(-y)) : |y| \leq x\} = \sup\{Sy : |y| \leq x\}.$$

Take now $x_1, x_2 \in E_+$ and prove that $U(x_1 + x_2) = Ux_1 + Ux_2$. If y_1 and y_2 are such that $|y_1| \le x_1$ and $|y_2| \le x_2$ then $|y_1 + y_2| \le x_1 + x_2$ and

$$Sy_1 + Sy_2 = S(y_1 + y_2) \le U(x_1 + x_2),$$

whence

$$Ux_1 + Ux_2 \le U(x_1 + x_2).$$

Prove the reverse inequality. Let $|y| \le x_1 + x_2$. Put $y_1 := x_1 \wedge y^+ - x_1 \wedge y^-$, $y_2 := y - y_1$. It is easy to check that $|y_1| \le x_1$, $|y_2| \le x_2$, whence

$$|Sy| = |S(y_1 + y_2)| \le |Sy_1| + |Sy_2| \le Ux_1 + Ux_2$$

which proves additivity. In view of 3.1.1, the operator U admits a unique linear extension to E which we denote by $|S|$. Check that $|S| = S \vee (-S)$. From a simplified version of the formula for U it is obvious that $[S] \ge \pm S$ and that $V \ge \pm S$ yields $V \ge [S]$ by which the claim is verified.

Thus, we established that $L^\sim(E, F)$ is a vector lattice and the modulus of an operator in the vector lattice can be calculated by formula (5). Formulas (1)–(4) and (7) easily follow from (5) as was mentioned above. Prove that the vector lattice $L^\sim(E, F)$ is order complete. To this end it suffices to prove existence of a least upper bound of an arbitrary bounded above set. Without loss of generality, we may assume that the set under study is directed upward. Thus, let M be an upward-directed set of operators and let $V \in L^\sim(E, F)$ be an upper bound of it. Given $x \in E_+$, we put

$$Tx := \sup\{Sx : S \in M\}.$$

The supremum exists since $Sx \le Vx$ for all $S \in M$ and all $x \in E_+$. Again we may apply the proposition from 3.1.1 to the operator U. Let T be an extension of U to E. Then T belongs to $L^\sim(E, F)$, since $S \le T \le V$ for every $S \in M$. It is now seen that $T = \sup M$. Thus, we verified that $L^\sim(E, F)$ is order complete.

It remains to prove the formula (6). According to (7), the right-hand side in (6), say f, is less or equal to the left-hand one. To prove the reverse inequality fix $x \in E_+$ and take y, $|y| \le x$. Then we have

$$Sy = Sy^+ - Sy^- \le |Sy^+| + |Sy^-| + |S(x - |y|)| \le f$$

and the claim easily follows. $\triangleright$

3.1.3. Let E and F be vector lattices with F order complete. Assume that given a positive operator $S : G \to F$ on an order ideal $G \subset E$. If the set $S([0, e] \cap G)$ is order-bounded in F for every $e \in E_+$ then we may define

$$\mathscr{E}_G(S)e := \sup\{Sg : g \in G,\ 0 \le g \le e\} := \sup\{S(g \wedge e) : g \in G\} \quad (e \in E_+).$$

An operator $\mathscr{E}_G(S) : E_+ \to F$ is additive and positively homogeneous, so it can be extended to E by differences, see 3.1.1. The resulting operator, called the *minimal extension* of S, is denoted by the same symbol $\mathscr{E}(S) := \mathscr{E}_G(S)$. Specify some properties of minimal extensions.

(1) *The minimal extension $\mathscr{E}(S)$ agrees with S on G and vanishes on $G^{\perp}$.*

$\lhd$ It follows directly from the definition of minimal extension. $\rhd$

Let $L^{\mathrm{ext}}(G, F)$ denote the set of regular operator $S \in L^{\sim}(G, F)$ such that S^+ and S^- admit minimal extension. It is evident that $S \in L^{\mathrm{ext}}(G, F)$ if and only if there is a positive operator $\widetilde{S} \in L^{\sim}(E, F)$, a dominant for S, such that $|S|(g) \leq \widetilde{S}(g)$ $(g \in G^+)$. Since the operator $\mathscr{E}_G$ defined on $L^{\mathrm{ext}}(G, F)_+$ is additive and positively homogeneous, it can be extended by differences as in 3.1.1 and we obtain a positive operator $\mathscr{E}_G : L^{\mathrm{ext}}(G, F) \to L^{\sim}(E, F)$ which is called the *minimal extension operator*. Let $\mathscr{R}_G : L^{\sim}(E, F) \to L^{\sim}(G, F)$ be the *restriction operator* $T \mapsto T|_G$. The restriction operator is also positive.

(2) *The minimal extension operator $\mathscr{E}_G$ preserves order continuity and sequential order continuity.*

$\lhd$ Ensure that $\mathscr{E}_G$ preserves sequential order continuity. Take $0 < S \in L^{\mathrm{ext}}_{n\sigma}(G, F)$ and denote $\widetilde{S} := \mathscr{E}_G S$. For any increasing sequence $(e_n)_{n \in \mathbb{N}}$ in E_+ with $\sup_n e_n = e$ we have

$$\begin{aligned}\widetilde{S}(e) &= \sup\{S(g \wedge e) : g \in G\} = \sup\{\sup_n S(g \wedge e_n) : g \in G\} \\ &= \sup_n \sup\{S(g \wedge e_n : g \in G\} = \sup_n \widetilde{S}(e_n).\end{aligned}$$

Thus, $\widetilde{S}(e_n) \nearrow \widetilde{S}(e)$. Now, the relation $e_n \searrow 0$ implies that $\widetilde{S}(e_1 - e_n) \nearrow \widetilde{S}(e_1)$, whence $\widetilde{S}(e_n) \searrow 0$. $\rhd$

The *null ideal* of an operator $T \in L^{\sim}(E, F)$ is the set $\mathscr{N}(U) = \{x \in E : U(|x|) = 0\}$. The disjoint complement $\mathscr{N}(U)^{\perp}$ is called the *carrier* or the *band of essential positivity* of the operator T and is denoted by $\mathscr{C}_T$. In case when $\mathscr{N}(U)^{\perp} = E$ we say that U is *essentially positive*.

(3) *The operator $\pi_G := \mathscr{E}_G \circ \mathscr{R}_G$ is a band projection in $L^{\sim}(E, F)$ and*

$$\pi_G^{\perp} L^{\sim}(E, F) = \{T \in L^{\sim}(E, F) : G \subset \mathscr{N}(T)\}.$$

$\lhd$ It is sufficient to prove that $S := \pi_G T$ is a fragment of T for every positive $T \in L^{\sim}(E, F)$. Let $V := S \wedge (T - S)$. If $x \in E_+$ then $Sx_\alpha \nearrow Sx$ for some increasing net $(x_\alpha) \subset G$ according to the definition of π_G. Since $V \leq S$, we have $V(x - x_\alpha) \leq S(x - x_\alpha) \searrow 0$. In virtue of (1) V vanishes on G; therefore, $Vx = o\text{-}\lim Vx_\alpha = 0$. $\rhd$

(4) *The subspace $L^{\text{ext}}(G,F)$ is a band in $L^{\sim}(G,F)$ linearly and lattically isomorphic to $\pi_G L^{\sim}(E,F)$ under the minimal extension operator.*

3.1.4. Thus, we have constructed an isotonic mapping $G \mapsto \pi_G$ from the lattice $\mathscr{I}(E)$ to the Boolean algebra $\mathfrak{P}(L^{\sim}(E,F))$. Now, consider some properties of this mapping. In case $G = e^{\perp\perp}$ we put $\pi_{[e]} := \pi_G$.

(1) *Let π be a band projection in E and $G = \pi(E)$. Then $\pi_G T = T \circ \pi$ for every $T \in L^{\sim}(E,F)$. In particular, $\pi_{[e]}T = T \circ [e]$, where $[e]$ stands for the band projection onto $\{e\}^{\perp\perp}$.*

$\lhd$ According to the definition of minimal extension $\pi_G T x = \sup\{T(\pi(g \wedge x)) : g \in G\} = \sup\{T(g \wedge \pi(x)) : g \in G\} = T(\pi x)$. $\rhd$

(2) *Suppose that C and D are order ideals in E and $G = C \cap D$. Then $\pi_G = \pi_C \wedge \pi_D$. In particular, π_G and π_D are disjoint if and only if $C \cap D = \{0\}$.*

$\lhd$ Since $C \cap D = \{c \wedge d : c \in C,\ d \in D\}$, for $0 \le x \in E$ and $0 \le T \in L^{\sim}(E,F)$ we deduce

$$\begin{aligned}\pi_G T(x) &= \sup\{T(x \wedge c \wedge d) : c \in C,\ d \in D\}\\ &= \sup_{c \in C} \sup\{T(x \wedge c \wedge d) : d \in D\} = \pi_C(\pi_D T)(x).\end{aligned}$$

Thus, $\pi_G = \pi_C \circ \pi_D = \pi_C \wedge \pi_D$. $\rhd$

(3) *Suppose that C and D are order ideals in E and $G = C + D$. Then $\pi_G = \pi_C \vee \pi_D$.*

$\lhd$ Monotonicity of $G \mapsto \pi_G$ implies $\pi_G \ge \pi_C \vee \pi_D$. If $u \in [0,x] \cap (C + D)$ then $u = c + d$ for some $c \in C_+$ and $d \in D_+$ and for $x_1 := c$, $x_2 := x - c$ we have $x = x_1 + x_2$, $c \in [0, x_1] \cap C$, and $d \in [0, x_2] \cap D$. Therefore,

$$Tu = Tc + Td \le \pi_C T(x_1) + \pi_D T(x_2) \le (\pi_D T \vee \pi_D T)x \le (\pi_D \vee \pi_D)Tx$$

and $\pi_G T(x) = \sup Tu \le (\pi_D \vee \pi_D)Tx$. $\rhd$

(4) *Let (G_α) be an increasing family of order ideals in E and $G := \bigcup G_\alpha$. Then G is also an order ideal and $\pi_G = \sup_\alpha \pi_{G_\alpha}$. Moreover, the representation holds:*

$$\pi_G^{\perp} L^{\sim}(E,F) = \{T \in L^{\sim}(E,F) : (\exists \alpha)\, G_\alpha \subset \mathscr{N}(T)\}.$$

$\lhd$ This follows immediately from the definitions on using associativity of least upper bounds:

$$(\pi_G T)x = \sup_{g \in G} T(g \wedge x) = \sup_\alpha \sup_{g \in G_\alpha} T(g \wedge x) = (\sup_\alpha \pi_{G_\alpha} T)x. \ \rhd$$

If $G := E(e)$ is the ideal generated by $e \in E_+$ then we write π_e instead of π_G.

(5) The following representation for π_e is valid:

$$\pi_e Tx = \sup_n T(ne \wedge x) \quad (x \in E^+,\, T \in L^+(E,F)),$$
$$\pi_e Tx = \pi_e Tx^+ - \pi_e Tx^- \quad (x \in E,\, T \in L^+(E,F)),$$
$$\pi_e T = \pi_e T^+ - \pi_e T^- \quad (T \in L^\sim(E,F)).$$

3.1.5. Very often it is useful to have formulas for the calculation of $(T \vee S)x$ and $(T \wedge S)x$ using the fragments of T and S. Different approaches to this problem can be unified by means of the following notion. A set of projections $\mathscr{P} \subset \mathfrak{P}(L^\sim(E,F))$ is said to be *generating* if for all $T \in L^+(E,F)$ and $x \in E$ we have

$$Tx^+ = \sup\{pTx : p \in \mathscr{P}\}.$$

As an easy example we cite the following. To each band projection $\pi \in \mathfrak{P}(E)$ assign the band projection $\hat{\pi}T \mapsto T \circ \pi$ acting in $L^\sim(E,F)$ and denoted by $\mathscr{P}^\circ$ the set of all such band projections.

(1) *Put* $\mathscr{P}^\pi := \{\pi_e : e \in E_+\}$. *Then* $\mathscr{P}^\pi$ *is a generating set of projections in* $L^\sim(E,F)$.

$\triangleleft$ Indeed, if $e := x^+$ then $\pi_e Tx^+ = Tx^+$ and $\pi_e Tx^- = 0$ by 3.1.3 (1); therefore $\pi_e Tx = \pi_e Tx^+ - \pi_e Tx^- = Tx^+$. $\triangleright$

(2) *If a vector lattice* E *has the strong Freudenthal property then* $\mathscr{P}^\circ$ *is a generating set of projections in* $L^\sim(E,F)$.

$\triangleleft$ According to 1.3.9 (2) the strong Freudenthal property implies that for every $x \in E$ there is a band projection π with $x^+ = \pi x$. Therefore, $T(x^+) = \hat{\pi}Tx$ for all $T \in L^\sim(E,F)$. $\triangleright$

(3) Theorem. *Let* E *and* F *be vector lattices with* F *order complete. A set* $\mathscr{P}$ *of band projections in* $L^\sim(E,F)$ *is generating if and only if for any* $T, S \in L^\sim(E,F)$ *and* $x \in E^+$ *the following formulas hold:*

$$(T \wedge S)x = \inf\{pTx + p^\perp Sx :\, p \in \mathscr{P}\},$$
$$(T \vee S)x = \sup\{pTx + p^\perp Sx :\, p \in \mathscr{P}\}.$$

$\triangleleft$ We confine exposition to sufficiency. Let $\mathscr{P}$ be a generating set of projections. In virtue of 1.3.2 (10) and 1.3.2 (1) it is sufficient to prove $\inf\{pTx + p^\perp Sx\} = 0$ provided that $T \wedge S = 0$. First observe that

$$Vx^+ - \sup\{pVx :\, p \in \mathscr{P}\} = Vx^+$$
$$+ \inf\{-pVx^+ + pVx^- :\, p \in \mathscr{P}\} = \inf\{p^\perp Vx^+ + pVx^- :\, p \in \mathscr{P}\}.$$

Thus, the identities $\inf\{pVx^+ + p^\perp Vx^- : p \in \mathscr{P}\} = 0$ and $Vx^+ = \sup\{pVx : p \in \mathscr{P}\}$ are equivalent for all $V \in L^+(E, F)$ and $x \in E$. Note next that

$$pTx = p(T(2y) + T(x - 2y)) \le 2Ty + pT(x - 2y)^-,$$
$$p^\perp Sx = p^\perp(S(2x - 2y) + S(2y - x)) \le 2S(x - y) + p^\perp S(2y - x)^+$$

for $p \in \mathscr{P}$ and $0 \le y \le x$, $x \in E^+$. Adding these inequalities and passing to the infimum we deduce

$$\inf_p\{pTx + p^\perp Sx\} \le 2\inf_y\{Ty + S(x - y)\}$$
$$+ \inf_p\{p(T + S)(2y - x)^- + p^\perp(T + S)(2y - x)^+\} = 0$$

and the result follows. $\triangleright$

3.1.6. Observe the following useful corollary to 3.1.2 and 3.1.5 (2).

(1) *If E has the strong Freudenthal property then it is possible to calculate the supremum in 3.1.2 (1), 3.1.2 (6), and the infimum in 3.1.2 (2) over partitions of $x \in E_+$ into the disjoint sum: $x_1 \perp x_2$ in 3.1.2 (1, 2) and $x_k \perp x_l$ $(k \ne l)$ in 3.1.2 (6).*

$\triangleleft$ The assertions concerning 3.1.2 (1, 2) follow immediately from 3.1.5 (2, 3). Hence, we may write

$$|T|x := (T \vee (-T))x = \sup\{Tx_1 - Tx_2 : x = x_1 + x_2,\ x_1 \perp x_2,\ x_1, x_2 \in E_+\}$$
$$\le \sup\{|Tx_1| + |Tx_2| : x = x_1 + x_2,\ x_1 \perp x_2,\ x_1, x_2 \in E_+\}.$$

This proves the inequality $\le$ in 3.1.2 (5), while the reverse inequality is obvious. $\triangleright$

(2) *Let E and F be the same as in Theorem 3.1.2. Then operators T and S in $L^+(E, F)$ are disjoint if and only if for every $e \in E_+$ and $0 < \varepsilon \in \mathbb{R}$ there exist a partition of unity $\pi_\alpha \subset \mathfrak{P}(F)$ and a family $(e_\alpha) \subset E$, $0 \le e_\alpha \le e$, such that for all α the inequalities hold:*

$$\pi_\alpha Te_\alpha \le \varepsilon Te, \quad \pi_\alpha S(e - e_\alpha) \le \varepsilon Se.$$

If, in addition, E has the strong Freudenthal property then we may assume that $(e_\alpha) \subset \mathfrak{E}(e)$, i.e., e_α is a fragment of e for every α.

$\triangleleft$ Suppose that

$$(S \wedge T)e := \inf\{Te + S(e - x) : 0 \le x \le e\} = 0$$

for some $e \in E_+$. Put $u := Se \wedge Te + \pi(Se+Te)$, where π is the band projection onto $\{Se \wedge Te\}^\perp$. As is seen, $Se + Te \in \{u\}^{\perp\perp}$, $[Se]u \le Se$ and $[Te]u \le Te$ (as before, $[a]$ stands for the band projection onto $\{a\}^{\perp\perp}$). According to 1.4.6 (4) and (1) there are a partition of unity $\pi_\alpha \subset \mathfrak{P}(F)$ and a family $(e_\alpha) \subset [0, e] \subset E$ such that $\pi_\alpha(Te_\alpha + S(e - e_\alpha)) \le \varepsilon u$ for all α. Thus, $\pi_\alpha Te_\alpha \le \varepsilon[Te]u$ and $\pi_\alpha S(e - e_\alpha) \le \varepsilon[Se]u$. The converse is obvious. ▷

(3) *Let E be a vector lattice and let $\mathscr{P}$ be a generating set of band projections in $L^\sim(E, F)$. Then operators T and S in $L^+(E, F)$ are disjoint if and only if for every $e \in E_+$ and $0 < \varepsilon \in \mathbb{R}$ there exist a partition of unity $\pi_\alpha \subset \mathfrak{P}(F)$ and a family $(p_\alpha) \subset \mathscr{P}$, such that for all α the inequalities hold:*

$$\pi_\alpha p_\alpha Te \le \varepsilon Te, \quad \pi_\alpha p_\alpha^\perp Se \le \varepsilon Se.$$

3.1.7. We will conclude this section with the problem of extension of positive operators and dominated extension of linear operators.

Kantorovich Theorem. *Let X be a preordered vector space, let X_0 be a massive subspace in X, and let F be a K-space. Then each positive operator $T_0 : X_0 \to F$ has a positive extension $T : X \to F$.*

◁ First, let $X = X_0 \oplus X_1$, where X_1 is a one-dimensional subspace, $X_1 := \{\alpha x_1 : \alpha \in \mathbb{R}\}$. Since X_0 is massive and T_0 is positive, the set $U := \{T_0 x^0 : x^0 \in X_0,\ x^0 \ge x_1\}$ is nonempty and bounded below. Assign

$$Tx := \{T_0 x_0 + \alpha f : x = x_0 + \alpha x_1,\ x_0 \in X_0,\ \alpha \in \mathbb{R}\}$$

where $f := \inf U$. It is clear that T is a linear operator from X to F and $T|_{X_0} = T_0$. So, only the positivity of T needs checking. If $x = x_0 + \alpha x_1$ and $x \ge 0$, then the case of α equal to 0 is trivial. If $\alpha > 0$ then $x_1 \ge -x_0/\alpha$. This implies that $-T_0 x_0/\alpha \le f$, i.e., $Tx \in F_+$. In a similar way for $\alpha < 0$ observe that $x_1 \le -x_0/\alpha$. Thus, $f \le -T_0 x_0/\alpha$ and, finally, $Tx = T_0 x_0 + \alpha f \in F_+$.

Now let $\mathscr{S}$ be the collection of linear operators $S : \operatorname{dom}(S) \to F$ such that S extends T_0 and $S(\operatorname{dom}(S)_+) \subset F_+$. Clearly, $\mathscr{S}$ is inductive in order by inclusion and so, by the Kuratowski–Zorn Lemma, $\mathscr{S}$ has a maximal element T. If $x_1 \in X \setminus \operatorname{dom}(T)$, apply the above proved result with $X := \operatorname{dom}(T) \oplus X_1$, $X_0 := \operatorname{dom}(T)$, $T_0 := T$, and $X_1 := \mathbb{R}x_1$ to obtain an extension of T. But this contradicts the maximality of T; thus, T is a sought operator. ▷

3.1.8. Let X be an arbitrary real vector space. An operator $p : X \to E$ is called *sublinear* if $p(x + y) \le p(x) + p(y)$ and $p(\lambda x) = \lambda p(x)$ for all $x, y \in X$ and $0 \le \lambda \in \mathbb{R}$. The collection of all linear operators from X into E dominated by p is called the *support set* of p and denoted by ∂p; symbolically,

$$\partial p := \{T \in L(X, E) : (\forall x \in X)\, Tx \le p(x)\},$$

where $L(X, E)$ is the space of all linear operators from X to E. A member of ∂p is a *supporting operator* of p. The *epigraph* $\operatorname{epi}(p)$ of p is defined by $\operatorname{epi}(p) := \{(x, e) \in X \times E : e \geq p(x)\}$. The epigraph of any sublinear operator p is a cone and defines in $X \times E$ a preorder relation, so that $(x, e) \leq (x', e')$ if and only if $e' - e \geq p(x - x')$.

(1) *Let $U \in L(E, F)$ and $V \in L(X, F)$. Then the operator $\mathscr{B} : (x, e) \mapsto Ve - Ux$ is positive on the space $X \times E$ endowed with the above-mentioned preorder if and only if $V \geq 0$ and $U \in \partial(V \circ P)$.*

$\triangleleft$ Suppose $\mathscr{B} \geq 0$. Since $(e, 0) \geq 0$ for every $e \in E_+$, we conclude $Ve = \mathscr{B}(0, e) \geq 0$. Thus, for $(x, e) \geq 0$ we have $Ve - Ux \geq 0$ and $Ux \leq Ve \leq Vp(x)$. The converse is similar. $\triangleright$

Assume that X_0 is a subspace of X and $T_0 : X_0 \to E$ is a linear operator such that $T_0 x \leq p(x)$ for all $x \in X_0$. If for every such X, X_0, T_0 and p, there exists an operator $T \in \partial p$ that is an extension of T_0 from X_0 to the whole of X, then we say that E *admits dominated extension of linear operators* or have the *dominated extension property*. A complete characterization of ordered vector spaces admitting dominated extension of linear operators resides in the following two theorems.

(2) Hahn–Banach–Kantorovich Theorem. *Every order complete vector lattice has the dominated extension property.*

$\triangleleft$ The claim follows from the Kantorovich Theorem on using (1) with $V = I_F$. $\triangleright$

(3) Bonnice–Silvermann–To Theorem. *Every ordered vector space admitting dominated extension of linear operators is order complete.*

3.1.9. In the following corollaries E is a K-space and p is a sublinear operator acting from a vector space X into E.

(1) *Each sublinear operator is the upper envelope of its support set, i.e. the next representation holds*

$$p(x) = \sup\{Tx : T \in \partial p\} \quad (x \in X).$$

Moreover, for an arbitrary point $x_0 \in X$ there exists a linear operator T from X into E supporting p at x_0, i.e. such that $Tx_0 = p(x_0)$ and $T \in \partial p$.

$\triangleleft$ Put $X_0 = \{\lambda x_0 : \lambda \in \mathbb{R}\}$ and define the linear operator $T_0 : X_0 \to E$ by $T(\lambda x_0) := \lambda p(x_0)$. For $\lambda \geq 0$ we have $T(\lambda x_0) = \lambda p(x_0) = p(\lambda x_0)$. If $\lambda < 0$ then

$$T_0(\lambda x_0) = \lambda p(x_0) = -|\lambda| p(x_0) \leq p(-|\lambda| x_0) = p(\lambda x_0).$$

Thus T_0 supports the restriction $p \restriction X_0$. By the Hahn–Banach–Kantorovich Theorem there exists an extension T of the operator T_0 to the whole space X dominated by p on X. This ensures that $T \in \partial p$ and $Tx_0 = T_0 x_0 = p(x_0)$. $\triangleright$

(2) *Let Y be another vector space and let T be a linear operator from Y into X. Then*

$$\partial(p \circ T) = \partial p \circ T.$$

$\lhd$ Consider an arbitrary element S from $\partial(p \circ T)$. Clearly, $-p(T(-y)) \le Sy \le p(Ty)$. Therefore, $Ty = 0$ implies $Sy = 0$. This means that $\ker(T) \subset \ker(S)$, where $\ker(R) := R^{-1}(0)$ is the kernel of the operator R. Consequently, the equation $\mathscr{X} \circ T = S$ is solvable for an unknown linear operator $\mathscr{X} : T(Y) \to E$. By assumption, every solution U_0 to this equation satisfies the inequality $U_0 x_0 \le p(x_0)$ for all $x_0 \in X_0 := T(Y)$. Therefore, according to the Hahn–Banach–Kantorovich Theorem, there exists an extension $U \in L(X, E)$ of the operator U_0 supporting the sublinear operator p. Thus $U \in \partial p$ and $U \circ T = S$, i.e. $S \in \partial p \circ T$. The reverse inclusion is obvious. $\rhd$

It should be stressed that if T is an identical embedding of a subspace X_0 into the space X, then the proposition exactly expresses the dominated extension property. In this connection Proposition 1.4.14 (2) is often referred to as the *Hahn–Banach formula.*

(3) *Let X_0 be a massive sublattice of a vector lattice X and let $S_0 : X_0 \to E$ be a positive operator with $S_0 x_0 \le T x_0$ $(x_0 \in X_0)$. Then there exists a positive extension $S : X \to E$ of S_0 such that $S \le T$.*

$\lhd$ Put $p(x) := T(x^+)$ $(x \in X)$. Then $\partial p = [0, T]$ and $S_0 x_0 \le S_0(x_0^+) \le T(x_0^+) = p(x_0^+)$ for all $x_0 \in X_0$. By the Hahn–Banach–Kantorovich Theorem there is an extension S of S_0 with $S \in \partial p$. $\rhd$

3.2. Fragments of a Positive Operator

The aim of the section is to obtain some basic formulas for calculating specific fragments of a positive operator. Simple fragments associated with order ideals in the domain of the operator prove to be very useful for this purpose.

3.2.1. An operator $T : E \to F$ is called *order continuous* (*order σ-continuous*) if Tx_α order converges to Tx for each net $(x_\alpha)_{\alpha \in \mathrm{A}}$ (each sequence $(x_\alpha)_{\alpha \in \mathbb{N}}$) in E with order limit x. The set of all order continuous regular operators (order σ-continuous regular operators) with the vector and order structure induced from $L^\sim(E, F)$ is denoted by $L_n^\sim(E, F)$ $(L_{n\sigma}^\sim(E, F))$. If $F = \mathbb{R}$ then we shall write $E_n^\sim$ rather than $L_n^\sim(E, \mathbb{R})$, see 1.5.2.

A positive operator $T \in L^\sim(E, F)$ is order continuous (order σ-continuous) if and only if $Tx_\alpha \xrightarrow{o} 0$ for every decreasing net (sequence) (x_α) in E with $\inf_\alpha x_\alpha = 0$.

$\lhd$ The proof can be found in [388] (Lemmas VIII.3.1 and VIII.4.3). $\rhd$

An ideal G in E is said to be a σ-order-dense ideal if for any $e \in E_+$ there is an increasing sequence $(g_n) \subset G_+$ such that $e = \sup g_n$. An operator $T \in L^\sim(E, F)$ is called *singular* if it vanishes on some order-dense ideal $G \subset E$. If T vanishes on a σ-order-dense ideal, then U is called *σ-singular.* The sets of singular and σ-singular operators are denoted by $L_s^\sim(E,F)$ and $L_{s\sigma}^\sim(E,F)$, respectively. It is easily seen from 3.1.2 (5, 7) that $L_s^\sim(E,F)$ and $L_{s\sigma}^\sim(E,F)$ are order ideals in $L^\sim(E,F)$.

We say that a vector lattice E *is rich in σ-order-dense ideals* if for any $x \in E_+$ there exists an element $e \in E_+$ such that $x \in \{e\}^{\perp\perp}$ and $\{e\}^\perp + \{e\}^{\perp\perp}$ is a σ-order-dense ideal in E. Each of the following types of vector lattices is rich in σ-order-dense ideals:

(a) vector lattice with a weak order-unity;

(b) vector lattice with the countable sup property;

(c) vector lattice with the principal projection property.

(d) vector lattice E in which for any $x \in E$ there exists a countable antichain (e_n), $e_k \perp e_j$ $(j \neq k)$, such that $\{e_n : n \in \mathbb{N}\}^\perp = E$ and $x \in \{e_1\}^{\perp\perp}$.

3.2.2. Theorem. *Let E and F be vector lattices with E rich in σ-order-dense ideals and F order complete. An operator $T \in L^\sim(E,F)$ is σ-order continuous if and only if it is disjoint to all σ-singular operators:*

$$L_\sigma^\sim(E,F) = L_{s\sigma}^\sim(E,F)^\perp.$$

$\triangleleft$ The inclusion $L_{n\sigma}^\sim(E,F) \subset L_{s\sigma}^\sim(E,F)^\perp$ is true without any additional assumption on E. Indeed, if $0 \le U \in L_{n\sigma}^\sim(E,F)$ and $0 \le V \in L_{s\sigma}^\sim(E,F)$, then $T := U \wedge V \ge 0$ is σ-continuous and vanishes on a σ-order-dense ideal $G \subset E$; thus, $T = 0$. The sets $L_{n\sigma}^\sim(E,F)$ and $L_{s\sigma}^\sim(E,F)$ are thus seen to be disjoint, since they are order ideals.

Now, suppose that an operator $0 < U \in L_{n\sigma}^\sim(E,F)^\perp$ is not σ-order continuous. Take a sequence $x_n \searrow 0$, with $f := \inf_n Ux_n > 0$. By assumption there exists an element $e \in E_+$ such that $g_1 \in \{e\}^{\perp\perp}$ and $G = \{e\}^\perp + \{e\}^{\perp\perp}$ is a σ-order-dense ideal. It is no restriction to assume that $x_1 = e$, since we may put $e := e \vee x_1$ if necessary.

We will find a nonzero fragment of U vanishing on some σ-order-dense ideal. Since $U(e) \ge f > 0$, we may choose a number $0 < \varepsilon < 1$ and a nonzero band projection ρ in F such that $\rho f > \varepsilon \rho U(e) > 0$. Let $e_n := (x_n - \varepsilon e)^+$, $\pi_n := \pi_{e_n}$, and $V := \inf_n \rho(\pi_n U)$, where π_{e_n} is the band projection defined in 3.1.4 (5). Clearly $e_n \searrow 0$, the sequence (π_n) is decreasing, and V is a fragment of U. Moreover, this fragment is nonzero, since

$$V(e_1) = \inf_n \sup_m \rho U(me_n \wedge e_1) \ge \inf_n \rho U(e_n) \ge \rho(f - \varepsilon U(e)) > 0,$$

Prove that V vanishes on the σ-order-dense ideal G generated by the increasing sequence $g_n = (\varepsilon e - x_n)^+$ and the order ideal $\{e\}^\perp$, in symbols:

$$G = \{e\}^\perp + \bigcup \{E(g_n)\}.$$

For $x := g_n$ we have

$$Vx \leq \pi_n U(g_n) = \sup_m U(me_n \wedge g_n) = 0,$$

since e_n and g_n are disjoint. If $0 \leq x \in \{e\}^\perp = \{x\}_1^\perp$, then $Vx \leq \pi_1 U(x) = 0$. Thus $V|_G = 0$.

It remains to check that G is a σ-order-dense ideal. For an arbitrary $x \in E_+$ we may find increasing sequences $(e''_n) \subset \{e\}^\perp$ and $(e'_n) \subset \{e\}^{\perp\perp}$ such that $x = \sup_n(e''_n + e'_n)$. Moreover, $e = \sup_k g_k/\varepsilon$ and $e'_n = \sup_m me \wedge e'_n$ $(k, n, m \in \mathbb{N})$, whence

$$x = \sup_{m,n,k} (e''_n + (mg_k/\varepsilon) \wedge e'_n) = \sup_n (e''_n + (ng_n/\varepsilon) \wedge e'_n). \ \triangleright$$

3.2.3. Let E and F be vector lattices with F order complete.

(1) *Theorem.* *An operator $T \in L^\sim(E, F)$ is order continuous if and only if it is disjoint from all singular operators:*

$$L_n^\sim(E, F) = L_s^\sim(E, F)^\perp.$$

$\triangleleft$ The proof can be obtained as a simple modification of the reasoning of 3.2.2, on replacing the sequences x_n and e_n by nets x_α and $e_\alpha = (x_\alpha - \varepsilon x_{\alpha_0})$, $\alpha \geq \alpha_0$, and taking into consideration that the ideal $\{x_{\alpha_0}\}^{\perp\perp} + \{x_{\alpha_0}\}^\perp$ is always order-dense. $\triangleright$

(2) *The spaces $L_n^\sim(E, F)$ and $L_{n\sigma}^\sim(E, F)$ are bands in $L^\sim(E, F)$.*

$\triangleleft$ The statement concerning $L_n^\sim(E, F)$ is immediate from (1). To prove the second part we denote by $L_{[e]}^\sim(E, F)$ the set of all regular operators U such that the restriction of U to $\{e\}^{\perp\perp}$ lies in $L_{n\sigma}^\sim(\{e\}^{\perp\perp}, F)$. According to 3.1.3 (4) and 3.2.2 $L_{[e]}^\sim(E, F)$ is a band. Now, it remains to observe that

$$L_{n\sigma}^\sim(E, F) = \bigcap_{e \in E_+} L_{[e]}^\sim(E, F). \ \triangleright$$

(3) *Let $\hat{E}$ be an order completion of E. Then every positive order continuous operator $T : E \to F$ has a unique positive extension $\hat{T} : \hat{E} \to F$ which is also order continuous. The correspondence $T \mapsto \hat{T}$ extends by differences to an isomorphism from $L_n^\sim(E, F)$ onto $L_n^\sim(\hat{E}, F)$.*

$\triangleleft$ Let $\hat{T}$ be an arbitrary positive extension of T to $\hat{E}$. According to (1) it suffices to prove that $\hat{T} \perp L_s^\sim(\hat{E}, F)$. Note that if $0 \leq \hat{S} \in L_s^\sim(\hat{E}, F)$ then $S := \hat{S}|_E$ is also singular, since $G \cap E$ is an order-dense ideal in E for every order-dense ideal $G \subset \hat{E}$. Thus, $T \perp S$. Using 3.1.2 (2), it is easy to observe that $\hat{S} \perp \hat{T}$. $\triangleright$

3.2.4. Let $\mathscr{F}$ be a filter in the lattice $\mathscr{I}(E)$ of order ideals in E. Then the set of $\mathscr{F}$-singular operators

$$L_{\mathscr{F}}(E,F) = \{T \in L^{\sim}(E,F) : \mathscr{N}_T \in \mathscr{F}\}$$

is an order ideal in $L^{\sim}(E,F)$.

(1) *The operator*

$$\pi_{\mathscr{F}} = \inf\{\pi_G : G \in \mathscr{F}\}$$

is the band projection onto $L_{\mathscr{F}}(E,F)^{\perp}$.

◁ By definition $\pi_{\mathscr{F}}$ is a band projection in $L^{\sim}(E,F)$. Denote by π the projection onto $L_{\mathscr{F}}(E,F)^{\perp}$. If $T \in L_{\mathscr{F}}(E,F)$ and $G := \mathscr{N}_T$, then $\pi_{\mathscr{F}}T \leq \pi_G T = 0$. Thus $\pi_{\mathscr{F}} \leq \pi$.

Conversely, suppose that $T \in L_{\mathscr{F}}(E,F)^{\perp}$ and $G \in \mathscr{F}$. Then $T - \pi_G T$ is contained in $L_{\mathscr{F}}(E,F)^{\perp}$. At the same time $T - \pi_G T \in L_{\mathscr{F}}(E,F)$, since $T - \pi_G T$ vanishes on G. Thus,

$$T - \pi_G T \in L_{\mathscr{F}}(E,F) \cap L_{\mathscr{F}}(E,F)^{\perp} = 0.$$

Thus, we conclude that $T - \pi_G T = 0$, whence $\pi_{\mathscr{F}} \geq \pi$. ▷

Denote by $\mathscr{I}_d(E)$ and $\mathscr{I}_{\sigma d}(E)$ the sets of all order-dense ideals and σ-order-dense ideals in E. Assume that T_n and $T_{n\sigma}$ stand for o-continuous and σ-order continuous parts of T. If we take in (1) $\mathscr{F} := \mathscr{I}_d(E)$ or $\mathscr{F} := \mathscr{I}_{\sigma d}(E)$ then, by virtue of 3.2.2 and 3.2.3, we arrive at the following:

(2) *For an order continuous part of a positive operator* T *we have*

$$T_n = \inf\{\pi_G T : G \in \mathscr{I}_d(E)\}.$$

(3) *If, in addition, a vector lattice* E *is rich in* σ*-order-dense ideals then the formula is also valid:*

$$T_{n\sigma} = \inf\{\pi_G T : G \in \mathscr{I}_{\sigma d}(E)\}.$$

◁ According to 3.2.4 (1) the operators π_n and $\pi_{n\sigma}$, defined by

$$\pi_n = \inf\{\pi_G : G \in \mathscr{I}_d(E)\}, \quad \pi_{n\sigma} = \inf\{\pi_G : G \in \mathscr{I}_{\sigma d}(E)\},$$

are the band projections onto $L_s^{\sim}(E,F)^{\perp}$ and $L_{s\sigma}^{\sim}(E,F)^{\perp}$ respectively. But we know that $L_n^{\sim}(E,F) = L_s^{\sim}(E,F)^{\perp}$ and if E is rich in σ-order-dense ideals then we have $L_\sigma^{\sim}(E,F) = L_{s\sigma}^{\sim}(E,F)^{\perp}$. Therefore $\pi_n T = T_n$ and $\pi_{n\sigma} T = T_{n\sigma}$. ▷

3.2.5. Using projections of type π_G, we may localize assumptions in Theorem 3.2.2 and deduce further formulas for calculating σ-order continuous parts. Let $L^{\sim}_{s\sigma(G)}(E,F)$ be a subset of $L^{\sim}(E,F)$ consisting of the operators vanishing on σ-order-dense ideals in G, i.e.,

$$L_{s\sigma(G)}(E,F) = \{T : \mathscr{N}_T \cap G \in \mathscr{I}_{\sigma d}(G)\}.$$

(1) *Let $G \in \mathscr{I}(E)$ be an ideal rich in σ-order-dense ideals. For $T \in L^{\sim}(E,F)$ the operator $\pi_G T$ is σ-order continuous if and only if it is disjoint from all operators σ-singular on G:*

$$\pi_G L^{\sim}_{n\sigma}(E,F) = \pi_G L^{\sim}_{s\sigma(G)}(E,F)^{\perp}.$$

$\lhd$ If $G \in \mathscr{I}(E)$ is rich in σ-order-dense ideals then $L^{\sim}_{n\sigma}(G,F) = L^{\sim}_{s\sigma}(G,F)^{\perp}$ by Theorem 3.2.2. Therefore, $L^{\text{ext}}_{n\sigma}(G,F) = L^{\text{ext}}_{s\sigma}(G,F)^{\perp}$, where the disjoint complement is taken in $L^{\text{ext}}(G,F)$. Indeed, the inclusion $\subset$ is obvious. If $0 \le T \in L^{\text{ext}}(G,F)$, $T \perp L^{\text{ext}}_{s\sigma}(G,F)$, and an operator $0 \le S \le T$ is such that $S \in L^{\sim}_{s\sigma}(G,F)$, then actually $S \in L^{\text{ext}}_{s\sigma}(G,F)$ and consequently $S = 0$. This proves the converse inclusion. Using 3.1.3 (2, 4), we may now deduce

$$\pi_G L^{\sim}_{n\sigma}(E,F) = \mathscr{E}_G\big(L^{\text{ext}}_{n\sigma}(G,F)\big) = \mathscr{E}_G\big(L^{\text{ext}}_{s\sigma}(G,F)^{\perp}\big) = \pi_G L_{s\sigma(G)}(E,F)^{\perp}. \ \rhd$$

For every $0 \le T \in L^{\sim}(E,F)$ the following representations hold:

(2) $\pi_G T_{n\sigma} = \inf\{\pi_D T : D \in \mathscr{I}_{\sigma d}(G)\}$;

(3) $T_{n\sigma} x = \inf\{\pi_D T x : D \in \mathscr{I}_{\sigma d}(G)\} \quad (x \in G_+)$;

(4) $T_{n\sigma} = \sup_{e \in E_+} \inf\{\pi_D T : D \in \mathscr{I}_{\sigma d}(\{e\}^{\perp\perp})\}$.

$\lhd$ By 3.2.4 (1) the operator $\pi = \inf\{\pi_D : D \cap G \in \mathscr{I}_{\sigma d}(G)\}$ is the band projection onto $L_{s\sigma(G)}(E,F)^{\perp}$. Therefore,

$$\pi_G \pi = \inf\{\pi_G \pi_D : D \cap G \in \mathscr{I}_{\sigma d}(G)\} = \inf\{\pi_C : C \in \mathscr{I}_{\sigma d}(G)\}$$

is the band projection onto $\pi_G L_{s\sigma(G)}(E,F)^{\perp}$. This proves (2). The formula (3) follows from 3.2.1 (a) and (4) is true, since $\{e\}^{\perp\perp}$ is rich in σ-order-dense ideals for every $e \in E_+$. $\rhd$

3.2.6. **(1)** *For any operator $0 \le T \in L^{\sim}(E,F)$ the following representations hold:*

$$T_n x = \inf\{\sup_{\alpha} T x_\alpha : x_\alpha \nearrow x\} \quad (x \in E_+),$$

$$T_{n\sigma} x = \inf\{\sup_{n} T x_n : x_n \nearrow x\} \quad (x \in E_+).$$

◁ Denote by $h(T,x)$ the right-hand side of the second formula. Choose a sequence $x_n \nearrow x$ and consider the set $\mathscr{G}(x_n)$ of all order ideals in $\{x\}^{\perp\perp}$ containing (x_n). By definition $\sup_n Tx_n \le \pi_G Tx$ for each $G \in \mathscr{G}(x_n)$, whence $h(T,x) \le \inf\{\pi_G Tx : G \in \mathscr{G}(x_n)\}$. At the same time each σ-order-dense ideal G contains a sequence $x'_n \nearrow x$ and, by definition, $G \subset \mathscr{G}(x'_n)$. This implies, in virtue of 3.2.4 (3), that $h(T,x) \le T_{n\sigma}x$, thereby $0 \le h(T - T_{n\sigma}, x) \le (T - T_{n\sigma})_{n\sigma}x = 0$. Moreover, $h(T_{n\sigma}, x) = T_{n\sigma}x$, since $T_{n\sigma}$ is sequentially order continuous. Using the additivity of $h(\cdot, x)$, we obtain

$$h(T,x) = h(T_{n\sigma}, x) + h(T - T_{n\sigma}, x) = T_{n\sigma}x. \ \triangleright$$

(2) *If E is a vector lattice with the strong Freudenthal property then for any $0 \le T \in L^\sim(E,F)$ we have*

$$T_n x = \inf\{\sup_\alpha T\pi_\alpha x : \pi_\alpha \nearrow [x]\} \quad (x \in E_+),$$
$$T_{n\sigma} x = \inf\{\sup_n T\pi_n x : \pi_n \nearrow [x]\} \quad (x \in E_+),$$

where π_α, π_n are band projections in E and $[x]$ is the band projection onto $\{x\}^{\perp\perp}$.

◁ The proof is in a similar spirit. ▷

3.2.7. Theorem. *Let $\mathscr{S}$ be an upward-directed set in $L^\sim(E,F)_+$ and σ denote the band projection onto $\mathscr{S}^{\perp\perp}$. Then for every $0 \le T \in L^\sim(E,F)$ and $e \in E_+$ the following representations hold:*

$$(\sigma^\perp T)e = \inf_{\substack{0<\varepsilon\in\mathbb{R}\\ S\in\mathscr{S}}} \sup\{\pi Tx : \pi \in \mathfrak{P}(F),\ 0 \le x \le e,\ \pi Sx \le \varepsilon Se\},$$
$$(\sigma T)e = \sup_{\substack{0<\varepsilon\in\mathbb{R}\\ S\in\mathscr{S}}} \inf\{\pi Tx + \pi^\perp Te : \pi \in \mathfrak{P}(F),\ 0 \le x \le e,\ \pi S(e-x) \le \varepsilon Se\}.$$

◁ It is sufficient to prove only the second identity. Denote by $g(T)$ the right-hand side of the sought relation. Take an arbitrary $S \in \mathscr{S}$ and $U \in L^\sim(E,F)$ with $U \le (nS) \wedge T$. The last inequality implies in particular that U belongs to the order ideal generated by $\mathscr{S}$. For a fixed $0 < \varepsilon \in \mathbb{R}$ choose any $x \in [0,e]$ and $\pi \in \mathfrak{P}(F)$ so that $\pi S(e-x) \le \varepsilon Se$. Then

$$Ue \le \pi U(e-x) + \pi Tx + \pi^\perp Te \le \varepsilon nSe + (\pi Tx + \pi^\perp Te).$$

Thus,

$$\begin{aligned} Ue \le \varepsilon nSe + \inf\{\pi Tx + \pi^\perp Te : \pi \in \mathfrak{P}(F),\ x \in [0,e]\, \pi S(e-x) \le \varepsilon Se \\ \le \varepsilon nSe + g(T). \end{aligned}$$

Since $\varepsilon > 0$ is arbitrary, we conclude $Ue \leq g(T)$.

By hypotheses $\sigma T \in \mathscr{S}^{\perp\perp}$; therefore, $\sigma T = \sup U_\alpha$ for an increasing net (U_α) in the order ideal generated by $\mathscr{S}$. Hence, $\sigma Te = \sup\{U_\alpha e\} \leq g(T)$ or $\sigma Te = \sup U_\alpha e \leq g(T)$.

Prove the reverse inequality. To this end observe first that

$$\begin{aligned}
&(\sigma T)e + g(\sigma^\perp T)\\
&= \sup_{\substack{0<\varepsilon\in\mathbb{R}\\ S\in\mathscr{S}}} \inf_{\substack{\pi\in\mathfrak{P}(F), x\in[0,e]\\ \pi S(e-x)\leq\varepsilon Se}} \{(\sigma T)e + \pi\sigma^\perp Tx + \pi^\perp\sigma^\perp Te\}\\
&\geq \sup_{\substack{0<\varepsilon\in\mathbb{R}\\ S\in\mathscr{S}}} \inf_{\substack{\pi\in\mathfrak{P}(F), x\in[0,e]\\ \pi S(e-x)\leq\varepsilon Se}} \{\pi((\sigma T)x + \sigma^\perp Tx) + \pi^\perp Te\} = g(T).
\end{aligned}$$

In addition, we may use the already-proven formula $g(T) \geq \sigma Te$ for $T := \sigma T$; therefore,

$$(\sigma T)e + g(\sigma^\perp T) \geq g(T) \geq g(\sigma T) + g(\sigma^\perp T) \geq (\sigma T)e + g(\sigma^\perp T),$$

$(\sigma T)e + g(\sigma^\perp T) = g(T)$. Thus, it remains to prove that $g(\sigma^\perp T) = 0$.

Let $U := \sigma^\perp T$ and $S \in \mathscr{S}$. Since $U \perp S$, we may choose, according to 3.1.6 (2), a partition of unity $\pi_\alpha \subset \mathfrak{P}(F)$ and a family $(e_\alpha) \subset E$, $0 \leq e_\alpha \leq e$, such that

$$\begin{aligned}
&\varepsilon Ue \geq \pi_\alpha Ue_\alpha = \pi_\alpha(\pi_\alpha Ue_\alpha + \pi_\alpha^\perp Ue)\\
&\geq \pi_\alpha \inf\{\pi Ux + \pi^\perp Ue : \pi \in \mathfrak{P}(F),\ 0 \leq x \leq e,\ \pi S(e-x) \leq \varepsilon Se\}.
\end{aligned}$$

Hence, $\varepsilon Ue \geq \pi_\alpha g(U)$ for all α and $\varepsilon Ue \geq g(U)$. Since the inequality is valid for an arbitrary $\varepsilon > 0$, we deduce $g(U) = 0$. $\triangleright$

3.2.8. We present several simple corollaries to Theorem 3.2.7. Take positive operators T and S in $L^\sim(E,F)$. Denote by T_S the band projection of T onto $\{S\}^{\perp\perp}$.

(1) If E is an arbitrary vector lattice then the representations are valid:

$$\begin{aligned}
(T - T_S)e &= \inf_{0<\varepsilon\in\mathbb{R}} \sup\{\pi Tx : \pi \in \mathfrak{P}(F),\ 0 \leq x \leq e,\ \pi Sx \leq \varepsilon Se\},\\
(T_S)e &= \sup_{0<\varepsilon\in\mathbb{R}} \inf\{\pi Tx + \pi^\perp Te : \pi \in \mathfrak{P}(F),\ 0 \leq x \leq e,\ \pi S(e-x) \leq \varepsilon Se\}.
\end{aligned}$$

$\triangleleft$ We need only to put $\mathscr{S} := \{S\}^{\perp\perp}$ in 3.2.7. $\triangleright$

(2) The formulas from (1) admit the following refinement:

$$(T - T_S)e = [Se]^{\perp}Te + \inf_{0<\varepsilon\in\mathbb{R}} \sup\{\pi Tx :$$
$$\pi \in \mathfrak{P}(F),\ \pi \le [Se],\ 0 \le x \le e,\ \pi Sx \le \varepsilon Se\},$$

$$(T_S)e = [Se]Te + \sup_{0<\varepsilon\in\mathbb{R}} \inf\{\pi Tx$$
$$+([Se]^{\perp} - \pi)Te : \pi \in \mathfrak{P}(F),\ \pi \le [Se]^{\perp},\ 0 \le x \le e,\ \pi S(e-x) \le \varepsilon Se\}.$$

$\lhd$ These formulas are equivalent to $[Se]^{\perp}T_S e = 0$ and $[Se](T - T_S)e = 0$ respectively. Using (1), we may calculate

$$\begin{aligned}[Se]^{\perp}T_S e &= \sup_{0<\varepsilon\in\mathbb{R}} \inf_{x,\pi}\{[Se]^{\perp}\pi Tx + [Se]^{\perp}\pi^{\perp}Te\}\\ &\le \sup_{0<\varepsilon\in\mathbb{R}} \inf_{x}\{[Se]^{\perp}Tx + [Se]^{\perp}[Se]Te\}\\ &\le \sup_{0<\varepsilon\in\mathbb{R}} \inf_{0\le x\le e}\{[Se]^{\perp}Tx\} = 0. \ \rhd\end{aligned}$$

(3) Let $0 \le S := \varphi \otimes u$ be a rank-one operator, i.e., $Sx = \varphi(x)u$ for some $u \in F_+$ and $\varphi \in L^{\sim}(E,\mathbb{R})_+$. Then

$$T_S e = \sup_{0<\varepsilon\in\mathbb{R}} \inf_{0\le x\le e}\{[u]Tx : \varphi(e-x) \le \varepsilon\varphi(e)\}.$$

$\lhd$ If $0 \ne \pi \le [u]$ then $\pi Sx \le \varepsilon Se$ means that $\varphi(x)\pi u \le \varepsilon\varphi(e)u$ and is equivalent to $\varphi(x) \le \varepsilon\varphi(e)$. At the same time the inequality $\pi Sx \le \varepsilon Se$ does not depend on π if $\varphi(e) \ne 0$, since in this case $[Se] = [u]$. Thus the supremum in (1) is attained at $\pi = [u]$ and we come to the relation

$$(T - T_S)e = \inf_{0<\varepsilon\in\mathbb{R}} \sup_{0\le x\le e}\{[u]Tx : \varphi(x) \le \varepsilon\varphi(e)\} + [u]^{\perp}Te\}$$

which is equivalent to the sought representation. $\rhd$

(4) Now let $J(E,F) = (L^{\sim}(E,\mathbb{R})\otimes F)^{\perp\perp}$ be the band of *almost integral operators*. Then the band projection σ onto $J(E,F)$ has the following representation:

$$(\sigma T)e = \sup_{\substack{0<\varepsilon\in\mathbb{R}\\ 0\le\varphi\in E^{\sim}}} \inf\{Tx : \varphi(e) \le \varepsilon\varphi(e-x),\ 0 \le e \le x\}.$$

(5) If $\varphi \in E^{\sim}$ is an essentially positive o-continuous functional then

$$(T - \sigma T)x = \inf_{\varepsilon \in \mathbb{R}_+} \sup_e \{Te : \varphi(e) \leq \varepsilon\varphi(x), 0 \leq e \leq x\}.$$

3.2.9. (1) *Theorem.* *Let E be an arbitrary vector lattice and $\mathscr{P}$ be a generating set of projections. Then the representation is valid:*

$$(T - T_S)e = \inf_{0<\varepsilon\in\mathbb{R}} \sup\{\pi pTe : \pi \in \mathfrak{P}(F),\, p \in \mathscr{P},\, \pi pSe \leq \varepsilon Se\},$$
$$(T_S)e = \sup_{0<\varepsilon\in\mathbb{R}} \inf\{(\pi p)^{\perp}Te : \pi \in \mathfrak{P}(F),\, p \in \mathscr{P},\, \pi pSe \leq \varepsilon Se\}.$$

◁ The proof is quite analogous to that of 3.2.8 (1). The only difference is that one should use 3.1.5 (3) instead of 3.1.2 (1, 2). ▷

(2) Assume that E is a vector lattice with the strong Freudenthal property and $0 \leq S, T \in L^{\sim}(E, F)$ are arbitrary. Then for each $e \in E_+$ we have

$$(T - T_S)e = \inf_{0<\varepsilon\in\mathbb{R}} \sup\{\pi T\sigma e : \pi \in \mathfrak{P}(F),\, \sigma \in \mathfrak{P}(E),\, \pi S\sigma e \leq \varepsilon Se\},$$
$$(T_S)e = \sup_{0<\varepsilon\in\mathbb{R}} \inf\{(T - \pi T\sigma)e : \pi \in \mathfrak{P}(F),\, \sigma \in \mathfrak{P}(E),\, \pi S\sigma e \leq \varepsilon Se\}.$$

◁ The proof is straightforward from 3.1.4 (1), 3.1.5 (2), and (1). ▷

(3) If $E = F$ and $S = I_E$ is the identity operator in E then from (1) we may easily deduce:

$$(T - T_S)e = \sup\{\pi^{\perp}T\pi e : \pi \in \mathfrak{P}(E)\}.$$

In particular, T belongs to the band $\{I_E\}^{\perp\perp}$ if and only if it is band preserving, i.e. T is an orthomorphism.

3.2.10. *Theorem.* *Consider the following assertions:*

(1) $T \in \{S\}^{\perp\perp}$;

(2) *for every $e \in E_+$ and $0 < \varepsilon \in \mathbb{R}$ there exits $0 < \delta \in F$, $[Se] \leq [\delta]$, $\delta \leq Se$, such that $\pi Sx \leq \delta$ implies $\pi Tx \leq \varepsilon Te$ for all $\pi \in \mathfrak{P}(F)$ and $x \in [0, e]$;*

(3) *for every $e \in E_+$ and $0 < \varepsilon \in \mathbb{R}$ there exits $0 < \delta \in F$, $[Se] \leq [\delta]$, $\delta \leq Se$, such that $\pi S\sigma e \leq \delta$ implies $\pi T\sigma e \leq \varepsilon Te$ for all $\pi \in \mathfrak{P}(F)$ and $\sigma \in \mathfrak{P}(E)$;*

(4) *for every $e \in E_+$ and $0 < \varepsilon \in \mathbb{R}$ there exits $0 < \delta \in F$, $[Se] \leq [\delta]$, $\delta \leq Se$, such that $\pi pSe \leq \delta$ implies $\pi pTe \leq \varepsilon Te$ for all $\pi \in \mathfrak{P}(F)$ and $p \in \mathscr{P}$.*

For an arbitrary vector lattice E we have (1) $\Leftrightarrow$ (2). In case when E has the strong Freudenthal property (1) $\Leftrightarrow$ (3). If $\mathscr{P}$ is a generating set of projections in $L^{\sim}(E, F)$ then (1) $\Leftrightarrow$ (4).

$\lhd$ We shall prove only one of the required equivalences, say (1) $\Leftrightarrow$ (2). Other equivalences are proven in a quite similar way. Fix $e \in E_+$ and define the decreasing sequence $(r_n(e))$ by putting

$$r_n(e) := \sup\left\{\pi Tx : \pi \in \mathfrak{P}(F),\, 0 \le x \le e,\, \pi Sx \le \frac{1}{n} Se\right\}.$$

According to 3.2.8 (1) $T \in \{S\}^{\perp\perp}$ if and only if $\inf_n r_n(e) = 0$ for all $e \in E_+$. But the last relation is equivalent to the following: for every $e \in E_+$ and $0 < \varepsilon \in \mathbb{R}$ there exits a partition of unity $(\pi_n)_{n\in\mathbb{N}}$ in $\mathfrak{P}(F)$ such that

$$\pi Sx \le \frac{1}{n} Se \Rightarrow \pi_n \pi Tx \le \varepsilon Te$$

for whatever $\pi \in \mathfrak{P}(F)$ and $x \in [0, e]$. Put

$$\delta := \sum_{n=1}^{\infty} \frac{1}{n} \pi_n Se.$$

Since the inequalities $\pi_n \pi Sx \le \frac{1}{n} Se$ $(n \in \mathbb{N})$ and $\pi Sx \le \delta$ are equivalent, the last assertion can be rewritten as

$$\begin{gathered}(\forall e \in E_+)(\forall \varepsilon \in \mathbb{R}, \varepsilon > 0)(\exists\, \delta \in F, \delta > 0)(\forall x \in [0, e])(\forall \pi \in \mathfrak{P}(F))\\ (\pi Sx \le \delta \Rightarrow \pi Tx \le \varepsilon Te).\end{gathered}$$

This completes the proof. $\rhd$

3.3. Orthomorphisms and Lattice Homomorphisms

Several useful results on the structure of orthomorphisms and lattice homomorphisms in vector lattices as well as a Hahn–Banach–Kantorovich-type theorem for lattice homomorphisms comprise the subject of the current section.

3.3.1. Take vector lattices E and F. A linear operator $T : E \to F$ is called a *lattice homomorphism* if one (and hence all) of the following identities holds:

$$\begin{aligned} T(x \vee y) &= Tx \vee Ty \quad (x, y \in E),\\ T(x \wedge y) &= Tx \wedge Ty \quad (x, y \in E),\\ T(|x|) &= |Tx| \quad (x \in E).\end{aligned}$$

It is clear that a lattice homomorphism preserves suprema and infima of nonempty finite sets and also preserves the positive and negative parts of every element. An injective lattice homomorphism is called a *lattice* (rarely, an *order*) *monomorphism*, an *isomorphic embedding* and even *lattice isomorphism* of E into F. If a lattice homomorphism $T : E \to F$ is a bijection then we say that E and F are *lattice* (or *order*) *isomorphic* or that T provides the *order isomorphism* between E and F. The set of all lattice homomorphisms from E into F is denoted by $\mathrm{Hom}(E, F)$.

Let $L_a^\sim(E, F)$ be the band in $L^\sim(E, F)$ generated by $\mathrm{Hom}(E, F)$ and $L_d^\sim(E, F)$ be its disjoint complement:

$$L_a^\sim(E, F) = \mathrm{Hom}(E, F)^{\perp\perp}, \quad L_d^\sim(E, F) = \mathrm{Hom}(E, F)^{\perp}.$$

An operator $T \in L^\sim(E, F)$ is said to be *atomic* (*diffuse*) if $T \in L_a^\sim(E, F)$ (respectively $T \in L_d^\sim(E, F)$). A linear operator $T : E \to F$ is said to be disjointness preserving if $Tx \perp Ty$ whenever $x \perp y$.

Meyer Theorem. *Let T be an order-bounded disjointness preserving operator from a vector lattice E to a vector lattice F. Then T has the positive part T^+, negative part T^-, and modulus $|T|$ of T; moreover, $T^+x = (Tx)^+$ and $T^-x = (Tx)^-$ for $0 \le x \in E$, and $|Tx| = |T|(|x|)$ for $x \in E$.*

3.3.2. Consider a vector lattice E and some its sublattice $D \subset E$. A linear operator T from D into E is said to be *band preserving* or *nonextending* if one (and hence all) of the following identities holds:

$$\begin{aligned} Te \in \{e\}^{\perp\perp} \quad & (e \in D), \\ e \perp f \Rightarrow Te \perp f \quad & (e \in D,\ f \in E), \\ T(K \cap D) \subset K \quad & (K \in \mathfrak{B}(E)). \end{aligned}$$

A band preserving operator need not be order-bounded (see Section 5.1). The set of all order-bounded band preserving operators from D to some vector sublattice $D' \subset E$ is denoted by $\mathrm{Orth}(D, D')$. An order-bounded band preserving operator $\alpha : D \to E$ defined on an order-dense ideal $D \subset E$ is named an *extended orthomorphism* in E. Since an extended orthomorphism is disjointness preserving, it is also regular according to 3.3.1. Let $\mathrm{Orth}(D, E)$ denote the set of all extended orthomorphisms defined on a fixed order-dense ideal D.

(1) *Every extended orthomorphism in a vector lattice is order continuous. Any two extended orthomorphisms commute.*

The space of all extended orthomorphisms $\mathrm{Orth}^\infty(E)$ is defined as follows. Denote by $\mathfrak{M}$ the collection of all pairs (D, π), where D is an order-dense ideal in E and $\pi \in \mathrm{Orth}(D, E)$. Elements (D, π) and (D', π') in $\mathfrak{M}$ are announced to be equivalent (in writing $(D, \pi) \sim (D', \pi')$) provided that the orthomorphisms π and π' coincide on the intersection $D \cap D'$. The factor set of $\mathfrak{M}/\sim$ modulo this equivalence relation is denoted by $\mathrm{Orth}^\infty(E)$. The set $\mathrm{Orth}^\infty(E)$ under the pointwise addition, scalar multiplication, and lattice operations becomes a vector lattice. Moreover, it becomes an ordered algebra under composition. An element $\alpha \in \mathrm{Orth}^\infty(E)$, defined on the whole space E, is called an *orthomorphism*. The collection of all orthomorphisms in E is denoted by $\mathrm{Orth}(E)$. Let $\mathscr{Z}(E)$ be the o-ideal generated by the identity operator I_E in $L^\sim(E)$. The space $\mathscr{Z}(E)$ is often called the *ideal center* of the vector lattice E. Identify every orthomorphism $\pi \in \mathrm{Orth}(E)$ with the corresponding equivalence class in $\mathrm{Orth}^\infty(E)$.

(2) *The space* $\mathrm{Orth}^\infty(E)$ *is a laterally complete faithful f-algebra with unity* I_E. *Moreover,* $\mathrm{Orth}(E)$ *is an f-subalgebra of* $\mathrm{Orth}^\infty(E)$ *and* $\mathscr{Z}(E)$ *is an f-subalgebra of* $\mathrm{Orth}(E)$.

(3) *Every Archimedean f-algebra E with unity* $\mathbf{1}$ *is algebraically and latticially isomorphic to the f-algebra of orthomorphisms in E. Moreover, the ideal in E generated by* $\mathbf{1}$ *is mapped onto* $\mathscr{Z}(E)$.

(4) *If E is an order complete vector lattice then* $\mathrm{Orth}^\infty(E)$ *is a universally complete vector lattice and* $\mathrm{Orth}(E)$ *and* $\mathscr{Z}(E)$ *are order-dense ideals.*

(5) *Let E and F be order-dense ideals in a universally complete vector lattice G with a fixed order and ring unity. Then for every orthomorphism* $\pi \in \mathrm{Orth}(E,F)$ *there exists a unique* $g \in G$ *such that* $\pi x = g \cdot x$ $(x \in E)$.

3.3.3. Kutateladze Theorem. *Let E be a vector lattice and let F be a K-space. A positive operator $T : E \to F$ is a lattice homomorphism if and only if, for every operator $S : E \to F$ satisfying $0 \leqslant S \leqslant T$, there is an orthomorphism $\rho \in \mathrm{Orth}(F)$ such that $0 \leqslant \rho \leqslant I_F$ and $S = \rho \circ T$. In other words, T is a lattice homomorphism if and only if $[0, T] = [0, I_F] \circ T$.*

$\lhd$ Let T be a lattice homomorphism and $0 \le S \le T$. Then $\ker(T) = \mathscr{N}_T \subset \mathscr{N}_S \subset \ker(S)$ and there exists a linear operator ρ_0: $F_0 \to F$, $F_0 := T(E)$, such that $S = \rho_0 \circ T$. It is easy to observe that F_0 is sublattice and $0 \le \rho_0 \le I_F|_{F_0}$. In view of 3.1.9 (3) ρ_0 admits extension to a linear operator $\rho : F \to F$ with $0 \le \rho \le I_F$ and we are done.

Conversely, assume that $[0, T] = [0, I_F] \circ T$ and take a pair of disjoint elements $x, y \in E_+$. Since $\pi_x T \le T$, it follows by hypothesis that there exists an orthomorphism $\rho \in \mathrm{Orth}(F)$ with $\pi_x T = \rho \circ T$. In virtue of 3.1.3 (1) $\pi_x T$ vanishes on $E(x)^\perp$ and agrees with T on $E(x)$. Therefore, ρ coincides with the identity mapping on the ideal $F(Tx)$ and vanishes at Ty. Thus,

$$Tx \wedge Ty = \rho(Tx \wedge Ty) = \rho Tx \wedge \rho Ty = Tx \wedge 0 = 0.$$

This shows that T is a lattice homomorphism. $\rhd$

3.3.4. Now we list some simple consequences of Theorem 3.3.3 on assuming that E and F are vector lattices, with F order complete.

(1) *Let $T : E \to F$ be a lattice homomorphism and $F = T(E)^{\perp\perp}$. Then there exists an isomorphism φ from the Boolean algebra $\mathfrak{G}(T)$ of fragments of T onto $\mathfrak{P}(F)$ such that $\varphi(S) \circ T = S$ for all $S \in \mathfrak{G}(T)$.*

$\lhd$ Assign to each projection $\pi \in \mathfrak{P}(F)$ the operator $\psi(\pi) := \pi \circ T$. Then ψ is a Boolean homomorphism from $\mathfrak{P}(F)$ into $\mathfrak{C}(T)$. If $\psi(\pi) = 0$ for some $\pi \in \mathfrak{P}(F)$ then π vanishes on the image of T and $\pi = 0$, since $F = T(E)^{\perp\perp}$. This means that ψ indeed is a monomorphism and it remains to establish surjectivity of ψ.

Let $S \in \mathfrak{G}(T)$. According to 3.3.3 $S = \alpha T$ for some $\alpha \in \mathrm{Orth}(F)$. Since $\alpha T \wedge (I - \alpha)T = 0$, we have

$$0 = \inf\{\alpha T e_1 + (I - \alpha)T e_2 : e_1 + e_2 = e,\ e_1, e_2 \in E_+\} \geq (\alpha \wedge (I - \alpha))Te$$

for every $e \in E_+$. Therefore, $\alpha \wedge (I - \alpha)$ vanishes on $T(E)$, whence $\alpha \wedge (I - \alpha) = 0$ and $\alpha \in \mathfrak{P}(F)$. The surjectivity of ψ is thus proven. It remains to observe that $\varphi := \psi^{-1}$ is the isomorphism sought since $\varphi(S) \circ T = \psi \circ \varphi(S) = S$. $\rhd$

Let $\Phi : E \to G$ be a lattice homomorphism. Assume that $G = \Phi(E)^{\perp\perp}$ is a K-space. In the universal completion mG, we fix the multiplicative structure that is uniquely determined by a choice of unity. Let F be an order ideal in mG, and $L^{\Phi}(E, F)$ be the set of all regular operators $S : E \to F$ such that $S \in \{\Phi\}^{\perp\perp}$, where S and Φ are regarded as operators from E into mG. Assign $F' := \{f \in mF : f \cdot \Phi(E) \subset F\}$.

(2) *The set F' is an order ideal in mG which is linearly and order isomorphic to the K-space $\mathscr{L}^{\Phi}(E, F)$. Such an isomorphism is performed by associating with an element $f \in F'$ the operator S_f by the formula*

$$S_f(e) = f \cdot \Phi(e) \quad (e \in E).$$

(4) Operators S_1 and S_2 of the band $\{T\}^{\perp\perp}$ are disjoint if and only if their images $\mathrm{im}(S_1)$ and $\mathrm{im}(S_2)$ are disjoint.

(5) Observe that the order ideal $\mathscr{I}(T)$ and the band $\{T\}^{\perp\perp}$ generated by T admit a simple description. Namely, if $\mathrm{Orth}(T, F) := \mathrm{Orth}(G, F)$, where G is the order ideal in F generated by $T(E)$ then for every lattice homomorphism T we have

$$\mathscr{I}(T) = \mathscr{Z}(F) \circ T, \quad \{T\}^{\perp\perp} = \mathrm{Orth}(T, F) \circ T.$$

3.3.5. We now discuss Hahn–Banach-type extension theorems for lattice homomorphisms. First we establish some auxiliary facts. Let X be an ordered vector space and let F be a K-space. A positive operator $T : X \to F$ is said to be *discrete* if $[0, T] = [0, I_F] \circ T$. Here $[0, T] := \{S \in L^{\sim}(X, F) : 0 \leq S \leq T\}$ and $[0, I_F] := \{\alpha \in L^{\sim}(F) : 0 \leq S \leq \alpha\}$ are order intervals in $L^{\sim}(X, F)$ and $L^{\sim}(F)$, respectively. Thus, in virtue of 3.3.3, if E is a vector lattice then a discrete operator is just a lattice homomorphism. It is noteworthy that discrete operators exist not only on vector lattices. There is a natural interconnection between discrete operators and extreme points of support sets. Recall that the operator $T \in \partial p$ is called *extreme* if the conditions $T_1, T_2 \in \partial p$ and $T_1 + T_2 = 2T$ imply that $T = T_1 = T_2$. The set of extreme operators in the support set ∂p is denoted by $\mathrm{Ch}(p)$. It is seen that this definition agrees with the conventional concept of an extreme point of a convex set. A positive cone X_+ of an ordered vector space X is said to be *reproducing* if $X = X_+ - X_+$.

Let X be an ordered vector space and T be an F-valued discrete operator on X. Then either $F = \{0\}$ or X_+ is a reproducing cone.

$\lhd$ Let $\mathfrak{X} := X_+ - X_+$ and $x_0 \in X \setminus \mathfrak{X}$. Now let f be a functional on X such that $\ker(f) \supset \mathfrak{X}$ and $f(x_0) = 1$. Denote by $f \otimes e$ the operator $x \mapsto f(x)e$ $(x \in X)$. Clearly, $T + f \otimes Tx_0 \in [0, T]$ whence $Tx_0 + Tx_0 = \alpha Tx_0$ for some $\alpha \in [0, I_E]$. This implies $Tx_0 = 0$. Thus we conclude that either $X = \mathfrak{X}$ or $T = 0$. In the latter case if $F \neq \{0\}$ and $e \in F \setminus \{0\}$ then $T + f \otimes e \in [0, T]$, and again $\mathfrak{X} = X$. $\rhd$

3.3.6. Kantorovich Theorem for a Discrete Operator. *Let X be an ordered vector space and F be a K-space. Further, let X_0 be a massive subspace of X and let $T_0 : X_0 \to F$ be a discrete operator on X_0. Then there exists a discrete extension T of the operator T_0 to the space X.*

$\lhd$ First let $X = X_0 + \mathbb{R}x_1$ for some $x_1 \in X \setminus X_0$. Define $T : X \to F$ by $Tx_1 := \inf\{T_0x_0 : x_0 \in X_0,\, x_1 \leq x_0\}$ and $T_0 := T|_{X_0}$. It is clear that T is a positive operator (since X_0 is massive). If $T' \in [0, T]$, then by assumption there is an orthomorphism $\alpha \in [0, I_F]$ with $T'|_{X_0} = \alpha T|_{X_0}$, i.e. $T'x_0 = \alpha Tx_0$ for all $x_0 \in X_0$. Since the equalities

$$
\begin{aligned}
T'x_1 &= \inf\{T'x_0 : x_0 \in X_0,\, x_1 \leq x_0\} = \alpha Tx_1;\\
(T - T')x_1 &= \inf\{(T - T')x_0 : x_0 \in X_0,\, x_1 \leq x_0\} = (I_E - \alpha) \circ Tx_1
\end{aligned}
$$

hold, it follows that $T' = \alpha \circ T$.

Now let $X = \bigcup_t X_t$, where (X_t) is an upward-directed (by inclusion) family of subspaces containing X_0. Assume the positive operators $T_t : X_t \to E$ to be discrete and T_s to be a restriction of T_t to X_s whenever $X_s \subset X_t$. Consider the extension T of the operator T_0 to X defined by the relation $Tx := T_tx$ $(x \in X_t)$. Introduce two sublinear operators $P, P_t : X \to F$ by the formulas

$$
\begin{aligned}
P(x) &:= \inf\{Tx' : x' \in X,\, 0 \leq x',\, x \leq x'\},\\
P_t(x) &:= \inf\{Tx' : x' \in X_t,\, 0 \leq x',\, x \leq x'\}.
\end{aligned}
$$

According to (1), P and P_t are correctly defined and, in addition,

$$\partial P = [0, T], \quad \partial P_t = [0, T_t].$$

Since T_t is discrete, we have $(T_t x)^+ = P_t(x)$ for x and t with $x \in X_t$. From this we deduce $(Tx)^+ = (T_t x)^+ = P_t(x) \geq P(x) \geq (Tx)^+$. Consequently, $(Tx)^+ = P(x)$ for all $x \in X$. Calculating support sets, we obtain $[0, I_E] \circ T = \partial(x \mapsto (x)^+) \circ T = \partial(x \mapsto (Tx)^+) = \partial P = [0, T]$, i.e. T is a discrete operator. By construction, $T|_{X_0} = T_0$. ▷

3.3.7. *Let p be a sublinear operator from a vector space to a vector lattice E and let T be a positive operator from E to a K-space F. For an operator S in ∂P the following are equivalent:*

(1) *the operator $T \circ S$ belongs to $\mathrm{Ch}(T \circ P)$;*

(2) *for all operators $0 \leq T_1, T_2 \in L^{\sim}(E, F)$ and $S_1, S_2 \in L(X, F)$ satisfying the conditions*

$$T_1 + T_2 = T, \quad T \circ S = S_1 + S_2, \quad S_1 \in \partial(T_1 \circ P), \quad S_2 \in \partial(T_2 \circ P),$$

the equalities $T_1 \circ S = S_1$ and $T_2 \circ S = S_2$ hold;

(3) *for the operator $\mathscr{A} : (x, e) \mapsto e - Sx$, defined on the ordered space $X \times E$ with positive cone $\mathrm{epi}(P)$, the equality holds*

$$[0, T] \circ \mathscr{A} = [0, T \circ \mathscr{A}].$$

◁ (1) ⇒ (2): If the operators T_1, T_2, S_1, S_2 satisfy the premises of (2) then the relations hold

$$\begin{aligned} 2T \circ S = (T_1 \circ S + S_2) + (S_1 + T_2 \circ S); \\ T_1 \circ S + S_2 \in \partial(T \circ P), \\ S_1 + T_2 \circ S \in \partial(T \circ P). \end{aligned}$$

Thus by (1) we have $T \circ S = T_1 \circ S + S_2 = S_1 + T_2 \circ S$, whence $T_1 \circ S = S_1$ and $T_2 \circ S = S_2$.

(2) ⇒ (3): First of all observe that the operator $\mathscr{B} : (x, y) \mapsto Uy - Vx$, where $U \in L(E, F)$ and $V \in L(X, F)$, is positive on the space $X \times E$ endowed with the above-mentioned order, if and only if $U \in L^+(E, F)$ and $V \in \partial(U \circ P)$, see 3.1.7 (1).

It follows that the operator $\mathscr{A}$ is positive since $S \in \partial P$. Therefore $[0, T] \circ \mathscr{A} \subset [0, T \circ \mathscr{A}]$. If (2) holds for S, and the operator $\mathscr{B}$ is positive and less than the operator $T \circ \mathscr{A}$, then $U \geq 0$ and $T - U \geq 0$. Moreover, $V \in \partial(U \circ P)$ and we have $T \circ S - V \in \partial((T - U) \circ P)$. Applying (2) for the operators $T_1 := U$, $T_2 := T - U$, $S_1 := V$, and $S_2 := T \circ S - V$, we obtain $V = S_1 = T_1 \circ S = U \circ S$. Thus, $\mathscr{B} = U \circ \mathscr{A}$, where $U \in [0, T]$.

(3) $\Rightarrow$ (1): Let S satisfy (3) and $T \circ S = \alpha_1 S_1 + \alpha_2 S_2$, where $S_1, S_2 \in \partial(T \circ P)$ and $\alpha_1, \alpha_2 \geq 0$, $\alpha_1 + \alpha_2 = 1$. Consider an operator $\mathscr{B}(x, y) := \alpha_1 Ty - \alpha_1 S_1 x$. Then $\mathscr{B} \in [0, T \circ \mathscr{A}]$ according to the relations

$$(T \circ \mathscr{A} - \mathscr{B})(\cdot, 0) = -\alpha_2 S_2, \quad (T \circ \mathscr{A} - \mathscr{B})(0, \cdot) = \alpha_1 T.$$

Thus the equation $T_1 \circ \mathscr{A} = \mathscr{B}$ holds for some $T_1 \in [0, T]$. In other words, $T_1 = \alpha_1 T$ and $T_1 \circ S = \alpha_1 S_1$. This means that $S_1 = T \circ S$. $\triangleright$

3.3.8. Kutateladze–Milman Theorem. *Let $P : Y \to E$ be a sublinear operator acting from a vector space Y into a K-space E. For $T \in L(X, Y)$ the following holds:*

$$\operatorname{Ch}(P \circ T) \subset \operatorname{Ch}(P) \circ T.$$

$\triangleleft$ Let $U \in \operatorname{Ch}(P \circ T)$. Clearly, $U = V \circ T$ for some $V \in \partial P$ in view of 3.1.9 (2). Let V_0 be the restriction of V onto $\operatorname{im}(T)$. Obviously, V_0 lies in $\operatorname{Ch}(P \circ \iota)$, where ι is the identical embedding of $\operatorname{im}(T)$ into Y. Therefore, by 3.3.7, the operator $\mathscr{V}_0 : \operatorname{im}(T) \times E \to E$, acting as

$$\mathscr{V}_0 : (y, e) \mapsto e - V_0 e \quad (y \in \operatorname{im}(T),\ e \in E),$$

is discrete in the ordered space $Y \times E$ with positive cone $\operatorname{epi}(P)$. The subspace $\operatorname{im}(T) \times E$ is obviously massive in $Y \times E$. Thus, according to 3.3.6, there is a discrete extension $\mathscr{V}$ of the operator $\mathscr{V}_0$. Undoubtedly, the operator $Sy := \mathscr{V}(y, 0)$ belongs to the support set ∂P being an extreme point there, and moreover, agrees with V on the image $\operatorname{im}(T)$. In other words, $U = V \circ T = S \circ T$ and $S \in \operatorname{Ch}(P)$. $\triangleright$

3.3.9. Let X be an arbitrary vector space. A sublinear operator $p : X \to F$ is called a *seminorm* if $p(x) = p(-x)$ for all $x \in X$. Given a seminorm p, define the set of linear operators $\mathscr{L}(p) \subset L(X, F)$ and the mapping $|\cdot| : \mathscr{L}(p) \to \operatorname{Orth}(F)$ as

$$\mathscr{L}(p) := \{T \in L(X, F) : (\exists \gamma \in \operatorname{Orth}(F))\, T \in \partial(\gamma \circ p)\},$$
$$|T| := \inf\{\gamma \in \operatorname{Orth}(F)_+ : T \in \partial(\gamma \circ p)\} \quad (T \in \mathscr{L}(p)).$$

(1) *The triple $(\mathscr{L}(p), |\cdot|, \operatorname{Orth}(F))$ is a Banach–Kantorovich space provided that $p(x) = 0$ implies $x = 0$.*

$\triangleleft$ Obviously, the operator $|\cdot|$ takes positive values and satisfy 2.1.1 (2, 3). If $|T| = 0$ then for every $0 < \varepsilon \in \mathbb{R}$ there exist a partition of unity $(\pi_\xi)_{\xi \in \Xi}$ in $\mathfrak{P}(F)$ and a family $(\rho_\xi)_{\xi \in \Xi}$ in $\operatorname{Orth}(F)$ such that $\pi_\xi \rho_\xi \leq \varepsilon I_F$ and $T \in \partial(\rho_\xi p)$ for all $\xi \in \Xi$. Thus, $\pi_\xi |Tx| \leq \pi_\xi \rho_\xi p(x) \leq \varepsilon p(x)$, whence $|Tx| \leq \varepsilon p(x)$ and $T = 0$.

Take a band projection π and an orthomorphism γ in F and observe that $\pi\circ T\in\partial(\gamma p)$ if and only if $\pi\circ T\in\partial(\pi\circ\gamma\circ p)$. Moreover, since $T\in\partial(\alpha\circ p)$ for some $\alpha\in\mathrm{Orth}(F)$, for every $\gamma\in\mathrm{Orth}(F)$ with $\pi\circ T\in\partial(\gamma\circ p)$ there exists $\gamma':=\pi\gamma+\pi^{\perp}\alpha$ such that $T\in\partial(\gamma'\circ p)$ and $\pi\gamma=\pi\gamma'$. Taking into consideration these properties we may deduce:

$$\begin{aligned}|\pi\circ T| &= \inf\{0\leq\gamma\in\mathrm{Orth}(F):\ \pi\circ T\in\partial(\gamma\circ p)\}\\ &= \inf\{\pi\circ\gamma\in\mathrm{Orth}(F)_+,\ \pi\circ T\in\partial(\gamma\circ p)\}\\ &= \inf\{\pi\circ\gamma\in\mathrm{Orth}(F)_+,\ T\in\partial(\gamma\circ p)\}=\pi\circ|T|.\end{aligned}$$

Thus, identifying π with the band projection $\overline{\pi}$ in $\mathrm{Orth}(F)$ given by $\rho\mapsto\pi\circ\rho$, we conclude that $\pi|T|=\pi|T|$ and in particular $|\cdot|$ is d-decomposable. Suppose the sequences $(T_n)\subset\mathscr{L}(p)$, $(\lambda_n)\subset\mathbb{R}$, and an orthomorphism $\rho\in\mathrm{Orth}(F)$ satisfy the relation $\lim_n\lambda_n=0$ and $|T_n-T_m|\leq\lambda_k\rho$ for all $m,n\geq k$. Then $|T_nx-T_mx|\leq\lambda_k\rho p(x)$ and the sequence (T_nx) is r-fundamental in F. Therefore, we may define the linear operator $Tx:=r\text{-}\lim_n T_nx$ $(x\in X)$. Since $|T-T_n|\leq\lambda_k\rho$ for $n\geq k$, we conclude that $T\in\mathscr{L}(p)$ and $r\text{-}\lim|T-T_n|=0$. Consequently, $\mathscr{L}(p)$ is br-complete and also decomposable according to 2.1.9 (2). Disjoint completeness of $\mathscr{L}(p)$ can be easily checked. It remains to apply 2.2.3. $\rhd$

A sublinear operator $p:E\to F$ is said to be a *submorphism* if $p(x\vee y)=p(x)\vee p(y)$ for all $x,y\in E$. If $p(x)=p(|x|)$ $x\in E$ and the mapping $x\mapsto p(x^+)$ is submorphism then p is called an M-seminorm. We say that p is positive if $p(x)\geq 0$ for every $x\in X$. Denote by $\partial_h p$ the set of all lattice homomorphisms in ∂p. In the next two propositions p is an arbitrary positive submorphism.

(2) *Let T be a lattice homomorphism in ∂p. Then there exists a $\rho\in\mathrm{Orth}(F)$ such that $0\leq\rho\leq I_F$ and $T\in\mathrm{Ch}(\rho\circ p)$.*

$\lhd$ Take a lattice homomorphism S in ∂p. Assign $\lambda(S):=\inf\{\gamma\in\mathrm{Orth}(F)_+:\ S\in\partial(\gamma\circ p)\}$. According to (1) $\lambda(\alpha S)=\alpha\lambda(S)$ for every $0\leq\alpha\in\mathrm{Orth}(F)$. Let $\mathscr{S}:=(-S,1):(x,f)\mapsto f-Sx$. Suppose that an order relation in $X\times F$ is given by the cone $\mathrm{epi}(\gamma p)$. The operator $\mathscr{B}:=(-B,\beta):(x,f)\mapsto\beta f-Bx$ belongs to $[0,\mathscr{S}]$. Then, by definition $Bx\leq\beta\circ\gamma p(x)$ and $(S-B)x\leq(1-\beta)\gamma p(x)$ for all $x\in X$. Putting $x\leq 0$, we obtain $B\in[0,B]$ and, in view of 3.3.3, $B=\rho S$ for some $\rho\in[0,I_F]$. Thus, we arrive at $\rho S\in\partial(\gamma\beta p)$ and $(1-\rho)S\in\partial((1-\beta)\gamma p)$. From this, using the homogeneity of $\lambda(\cdot)$, we deduce $\rho\gamma=\lambda(\rho S)\leq\gamma\beta$ and $(1-\rho)\gamma=\lambda((1-\rho)S)\leq(1-\beta)\gamma$, whence $\rho\gamma=\beta\gamma$ and $\rho=\beta$. The equality $[0,\mathscr{S}]=[0,I_F]\circ\mathscr{S}$ is thereby proved, since $\mathscr{B}=(-B,\beta)=(-\rho S,\rho)=\rho(-S,1)=\rho\mathscr{S}$. According to 3.3.7 $S\in\mathrm{Ch}(\gamma\circ p)$. $\rhd$

(3) *Every extreme point of ∂p is a lattice homomorphism.*

$\lhd$ Given an operator $0\leq B\leq S$ define $\mathscr{B}_0:X\times\{0\}\to F$ by $\mathscr{B}_0(x,0):=-Bx$. Observe that $Bx\leq B(x^+)\leq S(x^+)\leq p(x^+)=p(x)^+=p(x)$, i.e., $B\in\partial p$ and the same is true for $S - B$. From this it follows that $0 \leq \mathscr{B}_0 \leq \mathscr{S}|_{X \times \{0\}}$, where $\mathscr{S} : (x, f) \mapsto f - Sx$ and $X \times F$ is ordered by epi p. According to 3.1.9 (3) there exists an extension $\mathscr{B} = (-B, \beta)$ of $\mathscr{B}_0$ to $X \times F$ such that $0 \leq \mathscr{B} \leq \mathscr{S}$. In view of 3.3.7 $\mathscr{B} = \rho \circ \mathscr{S}$ for some $\rho \in [0, I_F]$, whence $\beta = \rho$ and $B = \rho S$. It remains to apply Theorem 3.3.3. $\triangleright$

3.3.10. *Theorem.* *Let X and E be vector lattices, and let F be a K-space. Let $p : E \to F$ be a positive submorphism and $T : X \to E$ be a lattice homomorphism. Then the Hahn–Banach formula for lattice homomorphisms is valid:*

$$\partial_h(p \circ T) = (\partial_h p) \circ T.$$

$\triangleleft$ It suffice to prove that $\partial_h(p \circ T) \subset (\partial_h p) \circ T$. To this end, take $S \in \partial_h(p \circ T)$ and observe that $p \circ T$ is a submorphism. From 3.3.9 it follows that there exist an orthomorphism $\rho \in [0, I_F]$ and an operator $U \in \mathrm{Ch}(p \circ T)$ such that $S = \rho \circ U$. By the Kutateladze–Milman Theorem we see that $U = V \circ T$ for some $V \in \mathrm{Ch}(p)$. Thus, $S = \rho \circ V \circ T$. By virtue of 3.3.9 $\rho \circ V$ is a lattice homomorphism. Since $p \geq 0$, $0 \leq \rho \leq I_F$, and $V \in \partial p$, we have $\rho \circ V \in \partial p$, which completes the proof. $\triangleright$

3.3.11. We give two Hahn–Banach-type theorems for lattice homomorphisms.

(1) *Theorem.* *Let p be a positive submorphism from E to F, and let T_0 be a lattice homomorphism from a sublattice $E_0 \subset E$ to F such that $T_0 x \leq p(x)$ $(x \in E_0)$. Then T_0 extends to a lattice homomorphism $T : E \to F$ with $Tx \leq p(x)$ $(x \in E)$.*

$\triangleleft$ The claim follows immediately if in 3.3.10 we substitute the identical embedding of E_0 into E for T. $\triangleright$

(2) *Theorem.* *Let E_0 be a massive sublattice in E and let T_0 be a lattice homomorphism from E_0 to F. Then T_0 extends to a lattice homomorphism $T : E \to F$.*

$\triangleleft$ This fact follows directly from 3.3.3 and 3.3.8. It is possible to produce another proof combining 3.1.8 (3) and 3.3.10 and observing that the sublinear operator p from 3.1.8 (3) is a submorphism if T_0 is a lattice homomorphism. $\triangleright$

3.3.12. For any positive operator $T \in L^{\sim}(E, F)$ define the mapping $p_T : E \to F_+$ by

$$p_T(x) := \inf \{Tx_1 \vee \dots \vee Tx_n : |x| \leq x_1 \vee \dots \vee x_n;\ x_1, \dots, x_n \in E_+,\ n \in \mathbb{N}\}.$$

(1) *The mapping p_T is an M-seminorm.*

(2) *For every $T \in L^{\sim}(E, F)$ the representation holds:*

$$p_T(x) = \sup\{Sx : S \in \mathrm{Hom}(E, F),\ 0 \leq S \leq T\} \quad (x \in E_+).$$

◁ Given $x_0 \in E_+$, denote by $E_0 := \mathbb{R}x_0$ the vector sublattice generated by x_0, and put $T_0(\lambda x_0) := \lambda p_T(x_0)$, $\lambda \in \mathbb{R}$. By 3.3.11 (1) there exists $T \in \partial p_T \cap \mathrm{Hom}(E,F)$ such that $Tx_0 = T_0x_0 = p_T(x_0)$. ▷

(2) *If E possesses the principal projection property then*

$$p_T(x) = \inf\{Tx_1 \vee \dots \vee Tx_n : |x| = x_1 + \dots + x_n;\\ x_1, \dots, x_n \in E_+,\ x_k \perp x_l\ (k \neq l),\ n \in \mathbb{N}\}.$$

3.3.13. (1) *An order-bounded operator $T : E \to F$ is diffuse if and only if*

$$0 = \inf\{|T|x_1 \vee \dots \vee |T|x_n : |x| \leq x_1 \vee \dots \vee x_n;\\ x_1, \dots, x_n \in E_+,\ n \in \mathbb{N}\} \quad (x \in E).$$

◁ It is easy to see that $T \in L^{\sim}_d(E,F)$ if and only if no nontrivial homomorphism is dominated by $|T|$. This is equivalent to $p_T = 0$ by virtue of 3.3.12 (2). ▷

(2) *Let $T \in L^{\sim}(E,F)$ and assume that S is the band projection of T onto $L^{\sim}_a(E,F)$. Then the following representation holds:*

$$Sx = \sup\left\{\sum_{k=1}^{n} p_T(x_k) : x = \sum_{k=1}^{n} x_k;\ x_1, \dots, x_n \in E_+,\ n \in \mathbb{N}\right\}$$

for all $x \in E_+$.

◁ Denote by Ux the right-hand side of the desired formula. In view of 3.3.4 (5) each positive operator from $L^{\sim}_a(E,F)$ is the supremum of a disjoint set in $\mathrm{Hom}(E,F)$, since there is a maximal disjoint family in $\mathrm{Hom}(E,F)$. Therefore,

$$S = \sup\{V \in \mathrm{Hom}(E,F) : 0 \leq V \leq T\}.$$

Since each V on the right-hand side of this formula satisfies the inequality $Vx \leq \sum_{k=1}^{n} p_T(x_k)$ whenever $x_k \in E_+$ and $x = \sum_{k=1}^{n} x_k$, we find that $Sx \leq Ux$. At the same time 3.3.12 (2) implies $p_T(x) \leq Sx$, $x \in E_+$, so that

$$S = \sup\{V \in L^{\sim}(E,F)_+ : p_T(u) \leq Su\ (u \in E_+)\}.$$

Each V on the right-hand side of the formula satisfies $\sum_{k=1}^{n} p_T(x_k) \leq Vx$, whence $Ux \leq Sx$. Thus, $S = U$, which completes the proof. ▷

3.4. Maharam Operators

Under discussion is some class of order continuous positive operators that behave in many instances like functionals. A Radon–Nikodým-type theorem for such operators is established.

3.4.1. Let E be a vector lattice, F be some K-spaces, and let T be a positive operator from E into F. Say that T possesses the *Maharam property* if, for every $x \in E_+$ and $0 \le f \le Tx \in F_+$ there exists $0 \le e \le x$ such that $f = Te$. A positive order continuous operator with the Maharam property is called a *Maharam operator*. Observe that $T \in L(E,F)_+$ possesses the Maharam property if only if the equality $T([0,x]) = [0, Tx]$ holds for any $x \in E_+$. Thus, a Maharam operator is exactly an order continuous order intervals preserving positive operator.

Recall that the *carrier* $\mathscr{C}_T$ of T, is defined by $\mathscr{C}_T := \{x \in E : T(|x|) = 0\}^{\perp}$. Let in addition $F_T := \{T(|x|) : x \in E\}^{\perp\perp}$, and let $\mathscr{D}_m(T)$ be the greatest order-dense ideal in the universal completion $m(E)$ of the space E to which T is extendible by o-continuity. Thus, $z \in \mathscr{D}_m(T)$ if and only if $z \in m(E)$ and the set $\{T(x) : 0 \le x \le |z|\}$ is bounded in F. Say that a linear operator $S : E \to F$ is *absolutely continuous with respect to* T and write $S \ll T$ if $S(x) \in \{T(x)\}^{\perp\perp}$ for all $x \in E$.

For $\alpha \in \operatorname{Orth}(m(F))$ we set $\mathscr{D}(\alpha) := \{e \in F : \alpha e \in F\}$. Observe that the orthomorphism algebra $\operatorname{Orth}(m(F))$ is a universally complete K-space. Moreover, the correspondence $\alpha \mapsto (\alpha, \mathscr{D}(\alpha))$ implements a bijection from $\operatorname{Orth}(m(F))$ onto $\operatorname{Orth}^{\infty}(F)$. Thus, $\operatorname{Orth}^{\infty}(F)$ is an f-algebra and a universally complete K-space.

3.4.2. Examples.

(1) Every positive functional possesses the Maharam property.

(2) Each positive orthomorphism acting in a K-space is a Maharam operator.

(3) Let F be an arbitrary K-space, and let $\mathfrak{A}$ be an arbitrary set. Denote by $l_1(\mathfrak{A}, F)$ the set of all o-summable families of elements in F indexed by $\mathfrak{A}$:

$$l_1(\mathfrak{A}, F) := \left\{ (e_\alpha)_{\alpha \in \mathfrak{A}} \in F^{\mathfrak{A}} : o\text{-}\sum_{\alpha \in \mathfrak{A}} |e_\alpha| \in F \right\}.$$

Define an operators T from $l_1(\mathfrak{A}, F)$ into F by the formula

$$Tz := \sum_{\alpha \in \mathfrak{A}} e_\alpha \quad (z := (e_\alpha)_{\alpha \in \mathfrak{A}} \in l_1(\mathfrak{A}, F)).$$

Then $l_1(\mathfrak{A}, F)$ with its natural vector structure and order is a K-space and T is a Maharam operator.

(4) Let (Q, Σ, μ) be a probability space, and let F be a Banach lattice. Consider the space $E := L^1(Q, \Sigma, \mu, F)$ of Bochner integrable F-valued functions, and let $T : E \to F$ denote the Bochner integral

$$Tf := \int\limits_Q f \, d\mu \quad (f \in E).$$

If the Banach lattice F is order complete and has order continuous norm then E is a K-space under the natural order $f \geq 0 \Leftrightarrow f(t) \geq 0$ for almost all $t \in Q$) and T is a Maharam operator.

(5) Take again a probability space (Q, Σ, μ) and let Σ_0 and μ_0 be a σ-subalgebra of Σ and the restriction of μ to Σ_0. The conditional expectation operator $\mathscr{E}(\cdot, \Sigma_0)$ is a Maharam operator from $L^1(Q, \Sigma, \mu)$ onto $L^1(Q, \Sigma_0, \mu_0)$. The restriction of $\mathscr{E}(\cdot, \Sigma_0)$ to $L^p(Q, \Sigma, \mu)$ is also a Maharam operator from $L^p(Q, \Sigma, \mu)$ to $L^p(Q, \Sigma_0, \mu_0)$. These facts follow immediately from the simple properties of conditional expectation.

(6) Let E_0 be a massive vector sublattice of a vector lattice E. The latter means that $E_0 + E_+ = E$. Given an operator $T \in L^\sim(E, F)$ denote by $\varrho(T)$ the restriction of T to E_0. If F is a K-space then the restriction operator $\varrho : L^\sim(E, F) \to L^\sim(E_0, F)$ is a surjective and essentially positive Maharam operator (see 3.1.2 (3)).

3.4.3. Theorem. *Let E and F be some K-spaces, and let T be a Maharam operator from E into F. Then there exists a lattice isomorphism h from the universally complete K-space* $\mathrm{Orth}^\infty(F_T)$ *onto the order complete K-subspace in* $\mathrm{Orth}^\infty(E_T)$ *such that the following conditions hold:*

(1) *$h(\mathfrak{Pr}(F_T))$ is an order complete subalgebra of the Boolean algebra $\mathfrak{Pr}(E_T)$;*

(2) *$h(\mathscr{Z}(F_T))$ is a sublattice and subring in $\mathscr{Z}(E_T)$;*

(3) *for every positive o-continuous operator $S : E \to F$ absolutely continuous with respect to T, the equality $\pi \circ S(x) = S \circ h(\pi)(x)$ holds for all $\pi \in \mathrm{Orth}^\infty(F_T)_+$ and $x \in \mathscr{D}(h(\pi))$; in particular, S is a Maharam operator.*

$\triangleleft$ Without loss of generality, we may assume that $E = E_T$ and $F = F_T$. Put

$$|e| := T(|e|) \quad (e \in E).$$

It is easily seen that $|\cdot|$ is a vector norm on E. Axioms 2.1.1 (1–3) are obvious. If $|e| = f_1 + f_2$ for some $e \in E_+$ and $f_1, f_2 \in F_+$, $f_1 \perp f_2$, then by the Maharam property there exist $e_1, e_2 \in [0, |e|]$ such that $Te_k = f_k$ $(k := 1, 2)$. It is easy to observe that $T(e_1 \wedge e_2) = 0$ and $T(|e| - e_1 + e_2) = 0$. Therefore, $e_1 \perp e_2$ and $|e| = e_1 + e_2$, since T is essentially positive. So, we have proved that $|\!\cdot\!|$ is d-decomposable. It is seen from the definitions that the norm $|\!\cdot\!|$ is monotone and semicontinuous. Now, from 2.1.4 we deduce the existence of a Boolean isomorphism $h : \mathfrak{P}(F) \to \mathfrak{P}(E)$ satisfying (1).

The isomorphism h is uniquely extended to the isomorphism from the space $\mathrm{Orth}^\infty(F)$ onto the order complete subspace in $\mathrm{Orth}^\infty(E)$ constituted by those elements in $\mathrm{Orth}^\infty(E)$ whose spectral functions take their values in the Boolean algebra $\mathscr{B} = h(\mathfrak{Pr}(F))$. Denote this isomorphism by the same symbol h. If $\alpha := \sum_{l=1}^n \alpha_l, \pi_l$, where $\lambda_1, \dots, \lambda_n \in \mathbb{R}_+$ and $\{\pi_1, \dots, \pi_n\}$ is a partition of unity in the algebra $\mathfrak{Pr}(F)$, then, obviously, $\pi_l \circ \alpha \circ S = \pi_l \circ S(\lambda_l h(\pi_l)) = \pi_l \circ S \circ h(\alpha)$ for all l. Summing over l yields $\alpha \circ S = S \circ h(\alpha)$. Finally, if $\alpha \in \mathrm{Orth}^\infty(F)_+$ then $\alpha = \sup(\alpha_\xi)$ for some upward-directed family (α_ξ) in $\mathscr{L}(F)$. Whereas the elements of $\mathscr{L}(F)$ are the r-limits of orthomorphisms of the form $\sum_{l=1}^n \lambda_l \pi_l$. Thus, to complete the proof of (1) and (2), it remains to appeal to o-continuity of the operator S.

To prove (3) consider a positive operator $S : E \to F$ absolutely continuous with respect to T. By the definition of the isomorphism (see 2.1.2 (4)) h, for $\pi \in \mathfrak{B}(F)$ and $x \in E$ we have

$$S \circ h(\pi)x \in \{T \circ h(\pi)x\}^{\perp\perp} \subset \pi(F).$$

Consequently, $\pi^\perp \circ S \circ h(\pi) = 0$ or $S \circ h(\pi) = \pi \circ S \circ h(\pi)$. Replacing π by $\pi^\perp$ in the previous argument, we obtain $\pi \circ S \circ h(\pi^\perp) = 0$. Therefore, $\pi \circ S = \pi \circ S \circ h(\pi)$. But then $\pi \circ S = \pi \circ S \circ h(\pi)$. We thus arrive at the sought relation $\pi \circ S = S \circ h(\pi)$. ▷

3.4.4. *Let E and F be some K-spaces and let $T : E \to F$ be a regular operator such that $|T|$ is a Maharam operator. If $(Tx)^+ > 0$ for some $x \in E_+$ then there exists a projection $\pi \in \mathfrak{Pr}(E)$ such that $T(\pi x) > 0$ and the operator $T \circ \pi$ is positive.*

◁ Suppose $Tx \not\geq 0$ and look at the set Π of all projections $\pi \in \mathfrak{Pr}(E)$ meeting the inequality $0 \geq T \circ \pi x$. It is easy to see that $\Pi \neq \varnothing$. Hence, by order continuity of the operator T, every chain in Π is bounded above. Consequently, by the Kuratowski–Zorn Lemma, there exist a maximal element π_0 of the set Π. If the projection $0 < \pi_1 \leq \pi_0^\perp$ is such that $T \circ \pi_1 x \leq 0$ then

$$T \circ (\pi_1 + \pi_0)x \leq T \circ \pi_1 x + T \circ \pi_0 x \leq 0,$$

and we arrive at a contradiction: $\pi_0 < \pi_0 + \pi_1 \in \Pi$. Hence, $T \circ \pi_1 x \not\leq 0$ for every $0 \neq \pi_1 \in [0, \pi_0^\perp]$. We now demonstrate that every such projection satisfies the inequality $T \circ \pi_1 x \geq 0$. To this end, we suppose that $\pi_1 \neq 0$, $\pi_1 \perp \pi_0$, and

$(T \circ \pi_1 x)^- > 0$. Let ρ be the band projection onto the band generated by the element $(T \circ \pi_1 x)^-$. Then $0 > \rho \circ T \circ \pi_1 x$ and, in view of Theorem 3.4.3, we have $T \circ h(\rho)\pi_1 x < 0$. This in particular implies that $h(\rho) \circ \pi_1 > 0$; and since $h(\rho) \circ \pi_1 \perp \pi_0$, by the above-mentioned property of the projection π_0 we obtain $T \circ h(\rho) \circ \pi_1 x \nleq 0$. This contradiction shows that $T \circ \pi_1 x > 0$ for all $\pi_1 \neq 0$, $\pi_1 \perp \pi_0$. Finally let $[x]$ be the band projection onto the band generated by the element x. Then $\pi := \pi_0^\perp \circ [x]$ is a sought projection. Indeed, $T \circ \pi x = T \circ \pi_0^\perp x$ and $Tx = T \circ \pi_0^\perp x - (-T \circ \pi_0 x)$. It follows that $T \circ \pi_0^\perp x \geq (Tx)^+ > 0$. On the other hand, if $0 \leq y \in \{x\}^{\perp\perp}$ and $(e_\lambda^y)_{\lambda \in \mathbb{R}}$ is the spectral function (or characteristic) of the element y relative to x then $e_\lambda^y = 0$ for $\lambda < 0$ and, for $\lambda \geq 0$, we have $T \circ \pi(e_\lambda^y) = T \circ \pi_0^\perp(e_\lambda^y) = T \circ \pi_0^\perp \circ [e_\lambda^y]x \geq 0$. Appealing to the Freudenthal Spectral Theorem, we finally obtain:

$$T \circ \pi(y) = T \circ \pi \left(\int_0^\infty \lambda \, de_\lambda^y \right) \int_0^\infty \lambda \, d(T \circ \pi(e_\lambda^y)\,) \geq 0. \ \triangleright$$

3.4.5. Theorem. *Let E and F be some K-spaces and let $T : E \to F$ be a Maharam operator. Then there exists an isomorphism φ from the Boolean algebra $\mathfrak{E}\,(T)$ of fragments of T onto $\mathfrak{P}\,(E_T)$ such that $T \circ \varphi(S) = S$ for all $S \in \mathfrak{E}\,(T)$.*

$\triangleleft$ Again we may assume that $E = E_T$. Let T_0 stand for the unique o-continuous extension of the operator T to $\mathscr{D}_m\,(T)$. Then, for each operator $S \in \mathfrak{E}\,(T_0)$, its restriction to E is an isomorphism of the Boolean algebras $\mathfrak{E}\,(T_0)$ and $\mathfrak{E}\,(T)$. Similarly, the Boolean algebras $\mathfrak{Pr}\,(E)$ and $\mathfrak{Pr}\,(\mathscr{D}_m\,(T)\,)$ are isomorphic. Thus, without loss of generality, we may assume $E = \mathscr{D}_m\,(T)$. Assign to each projection $\pi \in \mathfrak{Pr}\,(E)$ the operator $\psi\,(\pi) := T \circ \pi$. Then ψ is an increasing mapping from $\mathfrak{Pr}\,(T)$ into $\{T\}^{\perp\perp}$ and $\psi(0) = 0$ and $\psi(I_E) = T$. Clearly, if the projections π and ρ are disjoint then the carriers of the operators $\psi(\pi)$ and $\psi(\rho)$ are disjoint too; therefore, $\psi(\pi) \perp \psi(\rho)$. Moreover, for $\pi \in \mathfrak{Pr}\,(E)$, the equalities

$$\psi(I_E - \pi) = T \circ\, (I_E - \pi) = T - T \circ \pi = T - \psi(\pi)$$

are valid. Consequently, $\psi(\pi^\perp) = \psi(\pi)^\perp$. Thus, $\psi(\pi) \in \mathfrak{E}\,(T)$ for all $\pi \in \mathfrak{Pr}\,(E)$.

Consider now two arbitrary projections π_1 and $\pi_2 \in \mathfrak{Pr}\,(E)$. Since the projections $\rho_l := \pi_l - \pi_1 \circ \pi_2$ $(l := 1, 2)$ are disjoint, the operators $\psi(\rho_1)$ and $\psi(\rho_2)$ are disjoint too. On the other hand,

$$\psi(\pi_1) \wedge \psi(\pi_2) - \psi(\pi_1 \wedge \pi_2) = T \circ \pi_1 \wedge T \circ \pi_2 - T \circ \pi_1 \circ \pi_2$$

$$= (T \circ \pi_1 - T \circ \pi_1 \circ \pi_2) \wedge (T \circ \pi_2 - T \circ \pi_1 \circ \pi_2) = \psi(\rho_1) \wedge \psi(\rho_2) = 0.$$

Hence, $\psi(\pi_1 \wedge \pi_2) = \psi(\pi_1) \wedge \psi(\pi_2)$. Thus, ψ is a homomorphism from the Boolean algebra $\mathfrak{Pr}(E)$ into the Boolean algebra $\mathfrak{E}(T)$. Essential positivity of the operator T implies that if $\psi(\pi) = 0$ for some $\pi \in \mathfrak{Pr}(E)$ then $\pi = 0$. This means that ψ indeed is a monomorphism and it remains to establish surjectivity for it.

Let $S \in \mathfrak{E}(T)$ and look at the set

$$\Pi := \{\pi \in \mathrm{Orth}(E)_+ : T \circ \pi \leq S\}.$$

Using the Kuratowski–Zorn Lemma, we now demonstrate that Π contains a maximal element. Indeed, Π is nonempty and, for a linearly ordered set $(\pi_\xi)_{\xi\in\Xi}$ in Π the set $(T \circ \pi_\xi)_{\xi\in\Xi}$ is bounded since it is included $[0, S]$. But then the assumption $E = \mathscr{D}_m(T)$ implies that $(\pi_\xi)_{\xi\in\Xi}$ is a bounded set. If $\pi_0 := \sup\{\pi_\xi : \xi \in \Xi\}$ then $\pi_0 = o\text{-}\lim \pi_\xi$ and, by order continuity of the operator T, we have:

$$T \circ \pi_0 = T \circ (o\text{-}\lim \pi_\xi) = \text{o-}\lim T \circ \pi_\xi \leq S,$$

i.e. $\pi_0 \in \Pi$. Thus, there is a maximal element $\pi \in \Pi$ in the set Π. Show that $T \circ \pi = S$. To this end, suppose the contrary and let the operator $S_1 := S - T \circ \pi$ take strictly positive value at some $0 < x_0 \in E$. Then, for a suitable $0 < \varepsilon < 1$ and $0 \neq \rho \in \mathfrak{Pr}(F)$, we have $\rho\,(S_1 x_0 - \varepsilon\rho \circ T x_0) > 0$. The operator $\rho \circ |S_1 - \varepsilon T|$ is absolutely continuous with respect to T and, by Theorem 3.4.3, is a Maharam operator. According to 3.4.4, there exists a projection $\pi_\varepsilon \in \mathfrak{Pr}(E)$ such that $(S_1 - \varepsilon T) \circ \pi_\varepsilon x_0 > 0$ and $(S_1 - \varepsilon T) \circ \pi_\varepsilon \geq 0$. The former relations implies that $\pi_\varepsilon > 0$ and, from the latter, we have $T(\pi + \varepsilon\pi_\varepsilon) \leq S$. Thus, $\pi < \pi + \varepsilon\pi_\varepsilon \in \Pi$, which contradicts the maximality of π in Π. This substantiates the relation $S = T \circ \pi$. Further, by assumption, $S \wedge (T - S) = 0$; hence, $0 = (T \circ \pi) \wedge (T \circ (I_X - \pi)) \geq T\,(\pi \wedge (I_X - \pi)) \geq 0$. In view of essential positivity of T, the last leads to the equality $\pi \wedge (I_E - \pi) = \varnothing$ which is equivalent to the containment $\pi \in \mathfrak{Pr}(E)$. The surjectivity of ψ is thus proven. It remains to observe that $\varphi := \psi^{-1}$ is the isomorphism sought since $T \circ \varphi(S) = \psi \circ \varphi\,(S) = S$. $\rhd$

3.4.6. Observe the following corollary to Theorems 3.4.4 and 3.4.5.

(1) *Let E, F, and T be the same as in Theorem 3.4.3. Then operators S_1 and S_2 of the band $\{T\}^{\perp\perp}$ are disjoint if and only if their carriers $\mathscr{C}_{S_1}$ and $\mathscr{C}_{S_2}$ are disjoint.*

$\lhd$ Disjointness of the carriers $\mathscr{C}_{S_1}$ and $\mathscr{C}_{S_2}$ obviously implies that of the operators S_1 and S_2 (this fact does not depend on the Maharam property and is valid for arbitrary regular operators). To prove the converse, we first observe that if T_1 and T_2 are positive o-continuous operators and $T_1 \in \{T_2\}^{\perp\perp}$ then $\mathscr{C}_{T_1} \subset \mathscr{C}_{T_2}$. Indeed, assuming the contrary, we may find a projection π such that $0 < T_1 \circ \pi \leq T_1$ and $\mathscr{C}_{T_1\circ\pi} \perp \mathscr{C}_{T_2}$, which contradicts the relation $T_1 \in \{T_2\}^{\perp\perp}$ by virtue of the previous remark.

Let now S_1 and S_2 be disjoint. Then the band projections T_1 and T_2 of T onto $\{S_1\}^{\perp\perp}$ and $\{S_2\}^{\perp\perp}$ are disjoint too. On the other hand, $\mathscr{C}_{S_l} = \mathscr{C}_{T_l}$ by the previous remarks. By Theorem 3.4.5, $\mathscr{C}_{T_1} \perp \mathscr{C}_{T_2}$ has to be valid; therefore, $\mathscr{C}_{S_1} \perp \mathscr{C}_{S_2}$. ▷

(2) *Hahn Decomposition Theorem.* *Let E and F be some K-spaces and let $T : E \to F$ be an order-bounded operator with $|T|$ a Maharam operator. Then there exists a band projection $\pi \in \mathfrak{P}(E)$ such that $T^+ = T \circ \pi$ and $T^- = T \circ \pi^\perp$.*

◁ Apply Theorem 3.4.5 to the operator $|T|$ and put $\pi := \varphi(T^+)$. ▷

3.4.7. In the sequel, one more fact is needed on representations of order continuous operators. Let E and F be some K-spaces; and let $m(E)$ be, as usual, the universal completion of the space E with a fixed ring and order-unity $\mathbf{1}$. Suppose that, on some order-dense ideal $\mathscr{D}(\Phi) \subset m(E)$, an essentially positive Maharam operator Φ is defined which acts on F and $\mathscr{D}(\Phi) = \mathscr{D}_m(\Phi)$. Let $E_0 := E \cap \mathscr{D}(\Phi)$, let Φ_0 be the restriction of the operator Φ to the order-dense ideal E_0, and regard Φ_0 as an order-unity in the band $\{\Phi_0\}^{\perp\perp} \subset L^\sim(E_0, F)$.

Denote by the symbol $\mathscr{L}_\Phi(E, F)$ the set of all o-continuous regular operators from E into F whose restriction to E_0 belongs to the band $\{\Phi_0\}^{\perp\perp}$, i.e.

$$\mathscr{L}_\Phi(E, F) := \{S \in L_n^\sim(E, F) : S \restriction E_0 \in \{\Phi_0\}^{\perp\perp}\}.$$

As is seen, an operator S belongs to $\mathscr{L}_\Phi(E, F)$ if and only if S results from the extension of some $S_0 \in \{\Phi_0\}^{\perp\perp}$ by o-continuity. This in particular implies that $\mathscr{L}_\Phi(E, F)$ is a band in $L_n^\sim(E, F)$.

Consider the set $E' \subset m(E)$ defined by the relation

$$E' := \{x' \in m(E) : x' \cdot E \subset \mathscr{D}(\Phi)\}.$$

3.4.8. Theorem. *The set E' is an order-dense ideal in $m(E)$ linearly and latticially isomorphic to the space $\mathscr{L}_\Phi(E, F)$. The isomorphism may be implemented by assigning the operator $S_{x'} \in \mathscr{L}_\Phi(E, F)$ to an element $x' \in E'$ by the formula*

$$S_{x'}(x) = \Phi(x \cdot x') \quad (x \in E).$$

◁ The fact that E' is an order ideal in $m(E)$ follows immediately from the definitions. On the other hand, the bases of the spaces $\mathscr{L}_\Phi(E, F)$ and $m(E)$ are isomorphic by 3.4.5. Therefore, E' becomes an order-dense ideal in $m(E)$ if we establish the sought isomorphism of the spaces $m(E)$ and $\mathscr{L}_\Phi(E, F)$.

Obviously, if $x' \in E'$ then $S_{x'}$ is an order continuous regular operator from E into F. Observe that Φ_0 is a Maharam operator. Consequently, if $e \in \mathfrak{E}(\mathbf{1})$, i.e., e is a unit element of $\mathbf{1}$, then, by Theorem 3.4.5, the operator S_e is a fragment of Φ_0

and therefore $S_e \in \{\Phi_0\}^{\perp\perp}$. Let $(e_\lambda^{x'})_{\lambda\in\mathbb{R}}$ be the spectral function of x'. Then, by the Freudenthal Spectral Theorem,

$$x' = \int_{-\infty}^{\infty} \lambda \, de_\lambda^{x'},$$

where the integral on the right-hand side is the r-limit of integral sums of the form $\sum_{-\infty}^{\infty} l_n (e_{\lambda_{n+1}} - e_{\lambda_n})$, $l_n \in (\lambda_n, \lambda_{n+1}), \lambda_n \to +\infty$, and $\lambda_{-n} \to -\infty$ for $n \to +\infty$. It follows that the operator $S_{x'}$ has the representation

$$S_{x'}(x) = \int_{-\infty}^{\infty} \lambda \, d(\Phi(x \cdot e_\lambda^{x'})),$$

i.e., the operator $S_{x'}$ is made from operators of the form S_e, $e := e_\lambda^{x'}$ by summation and passage to the o-limit. Since every band is closed under these operations, we have $S_{x'} \in \{\Phi_0\}^{\perp\perp}$. Thus, $S_0 \in \{\Phi_0\}^{\perp\perp}$ and $S_{x'} \in \mathscr{L}_\Phi(E, F)$. Also, it is clear that the correspondence $x' \mapsto S_{x'}$ is an injective linear operator from E' into $\mathscr{L}_\Phi(E, F)$ and $x' \geq 0$ if and only if $S_{x'} \geq 0$.

It remains to show that, for every $S \in \mathscr{L}_\Phi(E, F)$, there exists an $x' \in E'$ such that $S = S_{x'}$. Indeed, let T be the restriction of S to E_0 and consider the spectral function $(e_\lambda^T)_{\lambda\in\mathbb{R}}$ of the operator T (with respect to the order-unity Φ_0). By virtue of 3.4.5, the family $(h(e_\lambda^T))_{\lambda\in\mathbb{R}}$ is a resolution of unity in $\mathfrak{C}(\mathbf{1})$; consequently, for some $x' \in m(F)$, we have $e_\lambda^{x'} = h(e_\lambda^T)$ for all $\lambda \in \mathbb{R}$. Moreover,

$$e_\lambda^T(x) = \Phi(x \cdot e_\lambda^{x'})$$

for all $\lambda \in \mathbb{R}$ and $x \in E_0$. Appealing now to the Freudenthal Spectral Theorem and using some elementary properties of o-summable families, given $x \in E_+$ we obtain the relations

$$Tx = \left(\int_{-\infty}^{\infty} \lambda \, d(e_\lambda^T) \right) x = \int_{-\infty}^{\infty} \lambda \, d(e_\lambda^T(x))$$

$$= \int_{-\infty}^{\infty} \lambda d\Phi(x \cdot e_\lambda^{x'}) = \Phi\left(x \cdot \int_{-\infty}^{\infty} \lambda de_\lambda^{x'} \right) = \Phi(x \cdot x').$$

Suppose now that $x \in E_+$, $(x_\alpha) \subset E'$ and $\sup(x_\alpha) = x$. Then $\Phi(x_\alpha \cdot x') \leq S(x)$ and since $\mathscr{D}(\Phi) = \mathscr{D}_m(\Phi)$, the family $(x_\alpha \cdot x')$ is bounded in $\mathscr{D}(\Phi)$. Hence, $x \cdot x' \in \mathscr{D}(\Phi)$ and $Sx = \Phi(x \cdot x')$. Thus, $x' \in E'$ and the sought representation holds. $\rhd$

3.4.9. Luxemburg–Schep Theorem. *Let E and F be order complete vector lattices; and let S and T be positive order continuous operators from E to F, with T possessing the Maharam property. Then the following are equivalent:*

(1) $S \in \{T\}^{\perp\perp}$;

(2) $S \ll T$;

(3) *there exists an orthomorphism $0 \leq \rho \in \mathrm{Orth}^\infty(E)$ such that $Sx = T(\rho x)$ for all $x \in \mathscr{D}(\rho)$;*

(4) *there exists a sequence of orthomorphisms $(\rho_n) \subset \mathrm{Orth}(E)$ such that $Sx = \sup_n T(\rho_n x)$ for all $x \in E$.*

◁ The implication (1) ⇒ (2) is trivial and the equivalences (1) ⇔ (3) and (1) ⇔ (4) can be easily deduced from 3.4.8. Prove (2) ⇒ (1). Suppose that $S_0 \in \{T\}^\perp$ and $0 \leq S_0 \leq S$. Then the operator $S_0 + T$ is absolutely continuous with respect to T and, by virtue of Theorem 3.4.3, possesses the Maharam property. According to 3.4.6 (1) $\mathscr{C}_{S_0} \perp \mathscr{C}_T$. Now, if $0 \leq x \in \mathscr{C}_{S_0}$ then $Tx = 0$ and $S_0 x \in \{Tx\}^{\perp\perp} = \{0\}$. Thus, $S_0 = 0$ and this completes the proof. ▷

3.4.10. Theorem. *Let E and F be K-spaces, and let $T(E)^{\perp\perp} = F$. Suppose that every increasing net (e_α) in E_+ is order-bounded provided that (Te_α) is order-bounded in F. Then there exists an order continuous lattice homomorphism $S : F \to E$ such that $T \circ S = I_F$.*

◁ Observe that $T(E) = F$ and $\mathscr{C}_T = E$ by assumption. First suppose that F contains an order-unity $\mathbf{1}$. Then $\mathbf{1} = Te$ for some $0 \leq e \in E$. For every $f \in F(\mathbf{1})$ there is a unique orthomorphism $\pi_f \in \mathscr{Z}(F)$ with $\pi_f \mathbf{1} = f$. Moreover, $f \mapsto \pi_f$ is an isomorphism of the vector lattices $F(\mathbf{1})$ and $\mathscr{Z}(F)$. Let h be the lattice homomorphism of Theorem 3.4.3. Put $S(f) := h(\pi_f)e$ $(f \in F(\mathbf{1}))$. Then S is an order continuous lattice homomorphism from $F(\mathbf{1})$ to E and $(T \circ S)f = T(h(\pi_f)e) = \pi_f Te = \pi_f \mathbf{1} = f$. Observe further that a minimal extension of S to the whole of F, if such an extension exists, would be the sought operator. Take $f \in F$ and put $f_n := f \wedge (n\mathbf{1})$ $(n \in \mathbb{N})$. Then $f_n \in F(\mathbf{1})$ and for $e_n := S(f_n)$ we have $T(e_n) = f_n \leq f$. Thus (e_n) is bounded in E and we may set $Sf := \sup_n Se_n$.

In the case when F lacks order-unity we consider a family $(f_\xi)_{\xi \in \Xi}$ of positive pairwise disjoint elements with $F = \{f_\xi : \xi \in \Xi\}^{\perp\perp}$. Then $f_\xi = T(e_\xi)$ for some $e_\xi \in E$ and the family $(e_\xi)_{\xi \in \Xi}$ is pairwise disjoint, since T is essentially positive. Apply the above-proven fact to the operator $\pi_\xi \circ T$, where π_ξ is the band projection onto $\{f_\xi\}^{\perp\perp}$. Thus, there is an operator $S_\xi : \pi_\xi F \to h(\pi_\xi)E$ such that $T \circ S_\xi$ is identity mapping on $\pi_\xi F$. Put

$$Sf := o\text{-}\sum_{\xi \in \Xi} S_\xi([f_\xi]f) \quad (f \in F_+).$$

To ensure that the operator S is soundly defined, take a finite set $\theta := \{\xi_1, \dots, \xi_n\}$ in Ξ and put

$$g_\theta := \sum_{k=1}^{n} S_{\xi_k}([f_{\xi_k}]f), \quad d_\theta := \sum_{k=1}^{n} [f_{\xi_k}]f.$$

Then (d_θ) and (g_θ) are increasing nets and $Tg_\theta = d_\theta$. By hypothesis (g_θ) is bounded, since (d_θ) is bounded above by f. Now it is an easy exercise to check that S is the desired operator. $\rhd$

3.4.11. Theorem. *Let E_0 be a massive vector sublattice of a vector lattice E and let F be a K-space. Then there exists an order continuous lattice homomorphism ε from $L^\sim(E_0, F)$ to $L^\sim(E, F)$ such that $\varrho \circ \varepsilon$ is the identical mapping on $L^\sim(E_0, F)$.*

$\lhd$ We know already that ϱ is an essentially positive Maharam operator from $L^\sim(E, F)$ onto $L^\sim(E_0, F)$, see 3.4.2 (6). Take an increasing net (T_α) in $L^\sim(E, F)$ and denote $S_\alpha := \varrho(T_\alpha)$. Suppose that the net (S_α) has an upper bound, say $S \in L^\sim(E_0, F)$. For an arbitrary $x \in E$ choose $x_0 \in E_0$ with $x \le x_0$. Then $T_\alpha x \le T_\alpha x_0 \le S_\alpha x_0 \le S_0 x_0$. Thus, (T_α) is bounded above and we may apply Theorem 3.4.10. $\rhd$

3.5. Maharam's Extension of Positive Operators

The main problem discussed in this section is the extension of an arbitrary positive operator to an order interval preserving order continuous operator, i.e., the *Maharam extension.* The structural properties of such extension allows us to deduce some results on approximation of the Boolean algebra of fragments of a positive operator by elementary fragments.

3.5.1. Consider vector lattices X and F, with F order complete, and an operator $\Phi \in L_+(X, F)$. Suppose that Φ is essentially positive, i.e., $\mathscr{N}(\Phi) = \{0\}$; and put $|x| := \Phi(|x|)$ $(x \in X)$. Then $(X, |\cdot|)$ is a lattice-normed space. Denote by $L^1(\Phi)$ the order norm completion (bo-completion) of X. According to Theorems 2.2.8 and 2.2.11 (1, 2) $L^1(\Phi)$ is a Banach–Kantorovich lattice. This completion enjoys the following properties (see 2.2.7–2.2.12):

(1) there is an isometric isomorphism ι from X onto the vector sublattice $\iota(X) \subset L^1(\Phi)$;

(2) there is a lattice and ring isomorphism h from $\mathscr{L}(E)$ into $\mathscr{L}(L^1(\Phi))$ such that $\pi\Phi(x) = |h(\pi)\iota(x)|$ $(x \in X_+,\ \pi \in \mathscr{L}(E)_+)$;

(3) for every $z \in L^1(\Phi)$ and $0 < \varepsilon \in \mathbb{R}$ there are a partition (π_ξ) of the projection $[|z|]$ and a family $(x_\xi) \subset X$ such that

$$|z - z_\varepsilon| \le \varepsilon |z|, \quad z_\varepsilon := \sum \pi_\xi \iota(x_\xi).$$

Let $\overline{X}$ be the order ideal in $L^1(\Phi)$ generated by $\iota(X)$.

3.5.2. Theorem. *For an arbitrary essentially positive operator $\Phi \in L^{\sim}(X,F)$ the space $L^1(\Phi)$ is an order complete Banach–Kantorovich lattice with an additive order continuous norm; moreover, $\iota(X)^{\perp\perp} = L^1(\Phi)$ and $\overline{X}$ is an order-dense ideal in $L^1(\Phi)$.*

$\lhd$ It is evident that the above defined norm in X is additive on X_+. Take $x', x'' \in d\iota(X)_+$ and choose a partition of unity $(\pi_\xi) \subset \mathfrak{F}$ and two families $(x'_\xi) \subset X_+$ and $(x''_\xi) \subset X_+$ such that $\pi_\xi x' = \pi_\xi \iota(x'_\xi)$ and $\pi_\xi x'' = \pi_\xi \iota(x''_\xi)$. Taking into account simple properties of a decomposable norm, we may deduce

$$\pi_\xi |x' + x''| = \pi_\xi |\iota(x'_\xi) + \iota(x''_\xi)| = \pi_\xi \big(|\iota(x'_\xi)| + |\iota(x''_\xi)|\big) = \pi_\xi \big(|x'| + |x''|\big).$$

Thus, the norm is additive on $d\iota(X)_+$. Additivity on a larger cone $L^1(\Phi)_+$ is easily proven by passing to the limit in view of the representation $L^1(\Phi) = rd\iota(X)$ from 2.2.11 (2).

Suppose now that a net $(z_\alpha) \subset L^1(\Phi)$ decreases to zero and put $e := \inf |z_\alpha|$. If $\alpha \leq \beta$ then

$$|z_\beta - z_\alpha| = |z_\beta| - |z_\alpha| = (|z_\beta| - e) - (|z_\alpha| - e) \to 0.$$

Thus, the net is *bo*-fundamental and there exists $z := bo\text{-}\lim z_\alpha$. Since $0 \leq z \leq z_\alpha$, we conclude that $z = 0$. At the same time $e := o\text{-}\lim |z_\alpha| = 0$ and order continuity of the norm is proven. Finally, assume that a net $(z_\alpha) \subset L^1(\Phi)$ is increasing and bounded above. Then the net $(|z_\alpha|) \subset F$ is also increasing and bounded above. In view of additivity of the norm the net (z_α) is again *bo*-fundamental and there exists $z := bo\text{-}\lim z_\alpha$. It is easily seen that $z = \sup z_\alpha$.

Now we check the identity $\iota(X)^{\perp\perp} = L^1(\Phi)$. According to 2.2.9 the subspace $\iota(X)$ is dense in $L^1(\Phi)$ in the sense of 3.5.1 (3). If $y \in L^1(\Phi)$ and $y \perp \iota(X)$ then $\pi_\xi y \perp \pi_\xi \iota(x_\xi)$; therefore, $y \perp z_\varepsilon$. Since bands are *br*-closed, we complete the proof by passage to the limit as ε tends to zero. $\rhd$

3.5.3. Recall that the space $L^1(\Phi)$ can be endowed with the structure of an A-module for $A := \mathrm{Orth}(F)$. Moreover, in view of 2.1.4 and 2.1.9, the natural representation of A in $L^1(\Phi)$ defines an isomorphism of A onto a sublattice and subalgebra in $\mathrm{Orth}\, L^1(\Phi)$.

Since the norm $|\cdot|$ is additive on the cone $L^1(\Phi)_+$, we may define the essentially positive operator $\tilde{\Phi} : L^1(\Phi) \to F$ by putting

$$\tilde{\Phi} : z \to |z^+| - |z^-| \quad (z \in L^1(\Phi)).$$

Clearly, order continuity of the norm implies that $\tilde{\Phi}$ is an order continuous operator. Moreover, $\tilde{\Phi}$ is order interval preserving, and hence, a Maharam operator. Indeed, if $0 \le e \le \tilde{\Phi}(z)$, $z \in L^1(\Phi)_+$, then there is an orthomorphism $0 \le \alpha \le I_F$ such that $e = \alpha\tilde{\Phi}(z)$. By virtue of A_+-homogeneity of the norm $e = \alpha|z| = |\alpha z| = \tilde{\Phi}(\alpha z)$. At the same time $\alpha z \in [0, z]$ and $\tilde{\Phi}([0, z]) = [0, \tilde{\Phi}(z)]$.

If $A_0 := \mathscr{L}(F)$ then $\overline{X}$ is an A_0-submodule in $L^1(\Phi)$. Moreover, the restriction $\overline{\Phi} := \tilde{\Phi}|_{\overline{X}}$ of $\tilde{\Phi}$ to $\overline{X}$ is also a Maharam operator.

3.5.4. *Theorem.* *For every operator $S \in \{\Phi\}^{\perp\perp}$ there exists a unique operator $\overline{S} \in \{\overline{\Phi}\}^{\perp\perp}$ such that $S = \overline{S} \circ \iota$. The assignment $S \mapsto \overline{S}$ implements an isomorphism of K-spaces $\{\Phi\}^{\perp\perp}$ and $\{\overline{\Phi}\}^{\perp\perp}$.*

$\lhd$ First of all observe that every positive (and thus every order-bounded) operator $S : X \to F$ admits a unique extension S_0 to the vector sublattice $d\iota(X)$ which commutes with all projections in $\mathfrak{P}(F)$. To demonstrate we need only to put

$$S_0 z := \sum \pi_\xi S(x_\xi) \quad \Big(z := \sum \pi_\xi \iota(x_\xi) \in d\iota(X)\Big).$$

The definition is correct because $|z| \le \iota x$ for some $x \in X$ and $\pi_\xi S(x_\xi) \le S(x)$ for all ξ. Let S lie in the order ideal generated by Φ, i.e., $|S| \le C\Phi$. Then $|S_0 z| \le C|z|$. Therefore, S_0 is norm r-continuous and admits a unique extension $\overline{S}$ to the space $\overline{X} = rd\iota(X)$ by br-continuity. It is easy to check that $\overline{S}$ commutes with all projections in $\mathfrak{P}(F)$ and lies in the order ideal generated by $\overline{\Phi}$.

Take now an increasing net (S_α) of positive operators in the order ideal generated by Φ such that $S := \sup S_\alpha \in \{\Phi\}^{\perp\perp}$. If $z \in \overline{X}$ then $|z| \le \iota(x)$ for some $x \in X$; therefore we may estimate $|\overline{S}_\alpha z| \le \overline{S}_\alpha(|z|) \le Sx$. Thus it is correct to define some positive operator by putting

$$\overline{S}z := \sup \overline{S}_\alpha z \quad (z \in \overline{X}_+).$$

Obviously, $\overline{S}$ is a positive operator and $\overline{S} = \sup \overline{S}_\alpha \in \{\overline{S\Phi}\}^{\perp\perp}$. It is easily seen that the mapping $S \mapsto \overline{S}$ is linear, positive, and one-to-one. It remains to show that an operator $S \in \{\Phi\}^{\perp\perp}$ has only one extension lying in $\{\overline{\Phi}\}^{\perp\perp}$. Observe that every operator $T \in \{\overline{\Phi}\}^{\perp\perp}$ commutes with all projections in $\mathfrak{P}(F)$, since $\overline{\Phi}$ is a Maharam operator, see 3.4.3 (3). Therefore, if T vanishes on $\iota(X)$ then

$$T\Big(\sum \pi_\xi \iota(x_\xi)\Big) = \sum \pi_\xi T(\iota(x_\xi)) = 0.$$

Since T is br-continuous and $d\iota(X)$ is br-dense in $\overline{X}$, we conclude that T is the zero operator. $\rhd$

3.5.5. (1) *Theorem.* *The space $\{\Phi\}^{\perp\perp}$ is isomorphic to the ideal $X' := \{u \in m\overline{X} : u\cdot\iota(X) \subset L^1(\Phi)\}$. The isomorphism is implemented by assigning the operator $S_u \in L^\sim(X, F)$ to an element $u \in X'$ by the formula*

$$S_u(x) = \overline{\Phi}(u \cdot \iota x) \quad (x \in X).$$

$\lhd$ The proof is immediate from 3.4.8 and 3.5.4. $\rhd$

(2) *The space* $\mathrm{Hom}_A(\overline{X}, F)$ *of all order-bounded order continuous A-linear operators (A-module homomorphisms) from* $\overline{X}$ *to* F *coincides with the band* $\{\overline{\Phi}\}^{\perp\perp}$.

$\lhd$ If $S \in \{\overline{\Phi}\}^{\perp\perp}$ then, according to 3.4.3, S is A-linear and $S \in \mathrm{Hom}_A(\overline{X}, F)$. Conversely, let $0 < S \in \mathrm{Hom}_A(\overline{X}, F)$ and $S \perp \overline{\Phi}$. Then $T := S + \overline{\Phi}$ is a Maharam operator and $S, \overline{\Phi} \in \{T\}^{\perp\perp}$. From 3.4.6 (1) we conclude that $C_S \perp C_{\overline{\Phi}}$. Since $\overline{\Phi}$ is essentially positive, $C_{\overline{\Phi}} = \overline{X}$ and $C_S = \{0\}$. Thus $S = 0$ and this completes the proof. $\rhd$

3.5.6. Introduce two sets $\mathscr{M}_0$ and $\mathscr{M}$ in $\overline{X}$. The former consists of the elements representable as $o\text{-}\sum \rho_\xi \iota(x_\xi)$, where $(\rho_\xi) \subset \mathfrak{P}(F)$ is an arbitrary partition of unity and $(x_\xi) \subset X$ is an order-bounded set. The latter comprises finite sums $\sum_{k=1}^{n} \rho_k \cdot \iota(x_k)$, where $\rho_k \in \mathfrak{P}(F)$, $x_k \in X$ $(k = 1, \dots, n)$. Clearly, $\mathscr{M}$ and $\mathscr{M}_0$ are vector sublattices in $\overline{X}$ and $\iota(X) \subset \mathscr{M} \subset \mathscr{M}_0$.

Given a set M in $\overline{X}$, denote by $M^{\downarrow}$ the set of all elements $z \in \overline{X}$, of the form $z = \inf z_\alpha$, where $(z_\alpha) \subset M$ is a downward directed net. The set $M^{\uparrow}$ is defined similarly on using upward-directed nets. If we take sequences instead of nets in these definitions then the corresponding sets are denoted by $M^{\downharpoonleft}$ and $M^{\upharpoonleft}$. More precisely, $M^{\downharpoonleft}$ is the set of all $y = \inf z_n$ in $\overline{X}$, where $(z_n) \subset M$ is a decreasing sequence. Finally, we set $M^{\downarrow\uparrow} := (M^{\downarrow})^{\uparrow}$ and $M^{\downharpoonleft\upharpoonleft} := (M^{\downharpoonleft})^{\upharpoonleft}$.

An element $y \in \overline{X}$ *belongs to* $\mathscr{M}^{\downarrow\uparrow}$ *if and only if for arbitrary* $m \in \overline{X}$ *with* $m > |y|$ *and* $n \in \mathbb{N}$ *there exists* $c \in \mathscr{M}^{\downarrow\uparrow}$ *such that* $c \le y$ *and* $\overline{\Phi}(y - c) \le \frac{1}{n}\overline{\Phi}(m)$.

$\lhd$ If $y \in \mathscr{M}^{\downarrow\uparrow}$, then $y = \sup \varphi_y$, where $\varphi_y = \{c \in \mathscr{M}^{\downarrow} : c \le y\}$. Since $\mathscr{M}$ is a sublattice in $\overline{X}$, the set $\mathscr{M}^{\downarrow}$ is also a sublattice; therefore, φ_y is directed upward. In view of order continuity of $\overline{\Phi}$ on $\overline{X}$ we conclude

$$\overline{\Phi} = \sup \overline{\Phi}(\varphi_y) := \sup\{\overline{\Phi}(c) : c \in \varphi_y\}.$$

The set $\iota(X)$ is a massive sublattice in $\overline{X}$ and we may choose $m \in \mathscr{M}$ with $|y| \le m$. Evidently $\overline{\Phi}(y) = \sup \Phi(\varphi_y \cap [-m, m]) = o\text{-}\lim \overline{\Phi}(\varphi_y \cap [-m, m])$. Now, by the properties of o-convergence in K-space, for each $n \in \mathbb{N}$ there exist a partition of unity $(\rho_\xi) \subset \mathfrak{P}(F)$ and a family $(c_\xi) \subset \varphi_y \cap [-m, m]$ such that

$$\rho_\xi(\overline{\Phi}(y) - \overline{\Phi}(c_\xi)) \le \frac{1}{n}\rho_\xi \overline{\Phi}(m)$$

for every ξ.

Denote $c := \inf\{\rho_\xi c_\xi + \rho_\xi^{\perp} m\}$ and observe that $c \in \mathscr{M}^{\downarrow}$ and $c \le y$. Since $\overline{\Phi}$ is A-linear and order continuous, we have $\rho_\xi \overline{\Phi}(c) = \rho_\xi \overline{\Phi}(c_\xi)$ for all ξ and $\overline{\Phi}(y - c) \le \frac{1}{n}\overline{\Phi}(m)$. Take a pair of elements y and m in $\overline{X}$ with $|y| \le m$. If for every $n \in \mathbb{N}$ there exists $c_n \in \mathscr{M}^{\downarrow}$ such that $c \le y$ and $\overline{\Phi}(y - c_n) < \frac{1}{n}\overline{\Phi}(m)$, then $y \in \mathscr{M}^{\downarrow\uparrow}$. Indeed, if $c = \sup c_n$, then $c \le y$ and $\overline{\Phi}(c) = \sup \overline{\Phi}(c_n) = \overline{\Phi}(y)$; thus, $y = c \in \mathscr{M}^{\downarrow\uparrow}$, since $\overline{\Phi}$ is essentially positive. $\rhd$

3.5.7. Theorem. *The following assertions hold:*

(1) $\overline{X} = \mathscr{M}^{\downarrow\uparrow} = \mathscr{M}^{\uparrow\downarrow}$;

(2) $\overline{X} = \mathscr{M}_0^{\downarrow\uparrow} = \mathscr{M}_0^{\uparrow\downarrow}$.

If F satisfy the countable chain condition then

(3) $\overline{X} = \mathscr{M}^{\downarrow\uparrow} = \mathscr{M}^{\uparrow\downarrow}$.

$\lhd$ We confine exposition to (1). Show that $\mathscr{M}^{\downarrow\uparrow}$ is a conditionally order complete sublattice in $\overline{X}$. It is clear that $\mathscr{M}^{\downarrow\uparrow}$ is conditionally complete upper semilattice, since $(\mathscr{M}^{\downarrow\uparrow})^{\uparrow} = \mathscr{M}^{\downarrow\uparrow}$. Prove the identity $\mathscr{M}^{\downarrow\uparrow\downarrow} = \mathscr{M}^{\downarrow\uparrow}$. Take $z \in \mathscr{M}^{\downarrow\uparrow\downarrow}$. There exists $m \in \mathscr{M}$ such that $|z| \le m$ and for every $0 < \varepsilon \in \mathbb{R}$ and $n \in \mathbb{N}$ one can choose a partition of unity $(\rho_\xi) \subset \mathfrak{P}(F)$ and a family $(y_\xi) \subset \mathscr{M}^{\downarrow\uparrow}$ satisfying the relations:

$$|y_\xi| \le m, \quad y_\xi \ge z, \quad \rho_\xi \overline{\Phi}(y_\xi - z) \le 2^{-n}\varepsilon\rho_\xi\overline{\Phi}(m).$$

Clearly, for $y_n := \sup\{\rho_\xi y_\xi - \rho_\xi^{\perp} m\}$, we have $y_n \in \mathscr{M}^{\downarrow\uparrow}$, $|y_n| \le m$, $y_n \ge z$, and $\overline{\Phi}(y_n - z) \le 2^{-n}\varepsilon\overline{\Phi}(m)$. For each $n \in \mathbb{N}$, according to the above-proved description of $\mathscr{M}^{\downarrow\uparrow}$, choose $c_n \in \mathscr{M}^{\downarrow}$ with $c_n \le y_n$, $|c_n| \le m$, and $\overline{\Phi}(y_n - c_n) \le 2^{-n}\varepsilon\overline{\Phi}(m)$. Denote $c = \inf c_n$ and observe that $c \in \mathscr{M}^{\downarrow}$, $|c| \le m$, and $c = \inf c_n \le \inf y_n = z$. Prove the inequality $\overline{\Phi}(z - c) \le 2\varepsilon\overline{\Phi}(m)$ which implies that $z \in \mathscr{M}^{\downarrow\uparrow}$ and $\mathscr{M}^{\downarrow\uparrow\downarrow} = \mathscr{M}^{\downarrow\uparrow}$.

Consider a sequence $c'_n = \inf\{c_k : k = 1, \dots, n\}$. Clearly, $\overline{\Phi}(z - c) = \overline{\Phi}(|z - c|) = o\text{-}\lim_n \overline{\Phi}(|z - c'_n|)$. At the same time $|z - c'_n| \le \sum_{k=1}^n |z - c_k|$, whence

$$\overline{\Phi}(|z - c'_n|) \le \sum_{k=1}^{n} \overline{\Phi}(|z - c_k|) \le \sum_{k=1}^{n} \big(\overline{\Phi}(|z - y_k|) + \overline{\Phi}(|y_k - c_k|)\big) \le 2\varepsilon\overline{\Phi}(m).$$

Thus, $\overline{\Phi}(z - c) \le 2\varepsilon\overline{\Phi}(m)$.

Now, using additivity of operations $(\cdot)^{\downarrow}$ and $(\cdot)^{\uparrow}$, observe that $\mathscr{M}^{\downarrow\uparrow}$ is a vector space:

$$\mathscr{M}^{\downarrow\uparrow} + \mathscr{M}^{\downarrow\uparrow} = \mathscr{M}^{\downarrow\uparrow},$$
$$\mathscr{M}^{\downarrow\uparrow} - \mathscr{M}^{\downarrow\uparrow} = \mathscr{M}^{\downarrow\uparrow} + \mathscr{M}^{\uparrow\downarrow} = \mathscr{M}^{\downarrow\uparrow\downarrow} = \mathscr{M}^{\downarrow\uparrow}.$$

Finally we conclude that $\mathscr{M}^{\downarrow\uparrow}$ is a K-space embedded in $\overline{X}$ as an order-closed sublattice.

Observe that the restriction operator $\mathscr{R} : \overline{T} \mapsto \overline{T}|_{\mathscr{M}^{\downarrow\uparrow}}$ is an isomorphism of $\mathrm{Hom}_A(\overline{X}, F)$ onto $\mathrm{Hom}_A(\mathscr{M}^{\downarrow\uparrow}, F)$. Indeed, if an operator $\overline{T}$ vanishes on $\mathscr{M}^{\downarrow\uparrow}$ then T vanishes on X and $\overline{T} \circ \imath = 0$; thus $\overline{T} = 0$ by virtue of 3.5.4. This proves that $\mathscr{R}$ is injective. For an arbitrary $\hat{S} \in \mathrm{Hom}_A(\mathscr{M}^{\downarrow\uparrow}, F)$ put $S := \hat{S} \circ \imath$ and let $\overline{S}$ be a unique order continuous operator from $\overline{X}$ to F for which $\overline{S} \circ \imath = S$ (see

Theorem 3.5.4). Then $\overline{S}$ and $\hat{S}$ agree on $\imath(X)$, and so they coincide on $\mathscr{M}^{\downarrow\uparrow}$. Thus, the vector lattices $\mathrm{Hom}_A(\overline{X}, F)$, $\mathrm{Hom}_A(\mathscr{M}^{\downarrow\uparrow}, F)$, and $\{\overline{\Phi}\}^{\perp\perp}$ are isomorphic. It follows from 3.4.5 that the bases of the K-spaces $\mathscr{M}^{\downarrow\uparrow}$ and $\overline{X}$ are also isomorphic, whence $\mathscr{M}^{\downarrow\uparrow} = \overline{X}$. ▷

3.5.8. We observe the following simple consequences:

(1) $L^1(\overline{\Phi}) = \mathscr{M}^{\downarrow\uparrow}$.

◁ The proof is immediate from 3.5.2 and 3.5.7 (1). ▷

(2) *The space $\overline{X}$ is the order completion of the vector sublattice*

$$W := \mathscr{M}^{\downarrow} - \mathscr{M}^{\downarrow} := \{y_1 - y_2 : y_1, y_2 \in \mathscr{M}^{\downarrow}\}.$$

In particular, the restriction of each $T \in \{\overline{\Phi}\}^{\perp\perp}$ to W is an order continuous A-linear operator.

◁ Obviously, $\mathscr{M}^{\downarrow}$ is a cone and hence $\mathscr{M}^{\downarrow} - \mathscr{M}^{\downarrow}$ is a vector subspace. Moreover, W is a sublattice since $(y_1 - y_2) \vee 0 = y_1 \vee y_2 - y_2 \in W$.

Since $\overline{X} = \mathscr{M}^{\downarrow\uparrow}$, for an arbitrary $0 < y \in \overline{X}$ there is $0 < w \in W$ such that $w \leq y$, i.e. W is a massive sublattice. It remains to observe that $\overline{X} = W^{\uparrow} = W^{\downarrow}$. ▷

3.5.9. According to 3.5.4 and 3.5.5 (2), the vector lattices $\overline{X}$, $L^1(\Phi)$, $\{\Phi\}^{\perp\perp}$, and $\{\overline{\Phi}\}^{\perp\perp}$ have isomorphic bases. In the rest of the section we will give a detailed description for bases for $\overline{X}$ and $\{\Phi\}^{\perp\perp}$. As usual, we denote by $[\iota x]$ the band projection in $\overline{X}$ onto $\{\iota(x)\}^{\perp\perp}$.

(1) *For $x \in X_+$ the projection π_x has the representation: $\pi_x(\Phi) = \overline{\Phi} \circ [\iota x] \circ \iota$.*

◁ Indeed, using 3.1.4 (5) and order continuity of $\overline{\Phi}$, we deduce

$$\begin{aligned}\pi_x(\Phi)y &= \sup\{\Phi(y \wedge nx) : n \in \mathbb{N}\} \\ &= \sup\{\overline{\Phi}(\iota(y) \wedge n\iota(x)) : n \in \mathbb{N}\} \\ &= \overline{\Phi}(\sup\{\iota(y) \wedge n\iota(x)) \\ &= \overline{\Phi} \circ [\iota x](\iota(y))). \ \triangleright\end{aligned}$$

Denote by $\mathscr{S}(\overline{X})$ and $\mathscr{S}(\Phi)$ the sets of all projections in $\overline{X}$ and the set of all fragments of Φ representable in the form

$$\bigvee_{k=1}^{n} \rho_k[\iota x_k] \quad \text{and} \quad \bigvee_{k=1}^{n} \rho_k \pi_{x_k}(\Phi),$$

where $(x_k) \subset X_+$, $(\rho_k) \subset \mathfrak{P}(F)$, $n \in \mathbb{N}$. Given a band K in X, denote by $\langle K \rangle$ the band projection in $\overline{X}$ onto $(\iota K)^{\perp\perp}$, i.e. $\langle K \rangle := [\iota K]$. Put $\langle x \rangle := [\iota(\{x\}^{\perp\perp})]$ and $\pi_{\langle x \rangle} := \pi_{\{x\}^{\perp\perp}}$ $(x \in X)$.

(2) *For every band K in X the projection π_K has the representation:* $\pi_K(\Phi) = (\overline{\Phi} \circ \langle K \rangle) \circ \iota$. *In particular,* $\pi_{\langle x \rangle}(\Phi) = (\overline{\Phi} \circ \langle x \rangle) \circ \iota$.

$\lhd$ The proof is similar to (1). $\rhd$

Let $\mathscr{C}(\overline{X})$ denote the set of projections $\bigvee_{k=1}^n \rho_k \cdot \langle x_k \rangle$, and let $\mathscr{C}(\Phi)$ be the set of fragments of the form $\bigvee_{k=1}^n \rho_k \cdot \pi_{\langle x_k \rangle}$ where $n \in \mathbb{N}$, $(\rho_k) \subset \mathfrak{P}(F)$, and $(x_k) \in X$. In the case when X is a vector lattice with the principal projection property we may consider one more set $\mathscr{A}(\Phi)$ consisting of the fragments of Φ representable as $\bigvee_{k=1}^n \rho_k \circ \Phi \circ [x_k]$, where $n \in \mathbb{N}$, $(\rho_k) \subset \mathfrak{P}(F)$, and $[x_k]$ is the band projection in X onto $\{x_k\}^{\perp\perp}$.

3.5.10. *The following are valid:*

(1) $\mathfrak{P}(\overline{X}) = \mathscr{S}(\overline{X})^{\downarrow\uparrow}$;

(2) $\mathfrak{E}(\Phi) = \mathscr{S}(\Phi)^{\downarrow\uparrow}$.

$\lhd$ By definition $[\iota y] \in \mathscr{S}(\overline{X})$ for each $y \in \mathscr{M}$. If $0 \le y \in \mathscr{M}^{\downarrow}$, we may choose $x \in X_+$ and a family $(m_\xi)_{\xi \in \Xi} \subset \mathscr{M}$ such that $\iota x \ge m_\xi \ge y$ $(\xi \in \Xi)$ and $y = \inf m_\xi$. It is not difficult to verify that

$$[y] = \bigvee_{n \in \mathbb{N}} \bigwedge_{\xi \in \Xi} \left[\left(m_\xi - \frac{1}{n} \iota x \right)^+ \right].$$

Since the element $y_{n,\xi} = (m_\xi - \frac{1}{n}\iota x)^+$ is contained in $\mathscr{M}$, it follows $[y_{n,\xi}] \in \mathscr{S}$ and $[y] \in \mathscr{S}^{\downarrow\uparrow}$. An arbitrary projection $\pi \in \mathfrak{P}(\overline{X})$ has the representation $\pi = \sup\{[y] : y \in \overline{X}_+, \pi y = y\}$. Thus, taking 3.5.8 into consideration we arrive at the desired inclusion $\pi \in \left((\mathscr{S}^{\downarrow\uparrow})^{\uparrow}\right)^{\uparrow} = \mathscr{S}^{\downarrow\uparrow}$. $\rhd$

3.5.11. *The following are valid:*

(1) $\mathfrak{P}(\overline{X}) = \mathscr{C}(\overline{X})^{\uparrow\downarrow\uparrow}$;

(2) $\mathfrak{E}(\Phi) = \mathscr{C}(\Phi)^{\uparrow\downarrow\uparrow}$.

If X has the principal projection property then

(3) $\mathfrak{E}(\Phi) = \mathscr{A}^{\uparrow\downarrow\uparrow}$.

$\lhd$ It is sufficient to show that $[\iota x] \in \mathscr{C}^{\uparrow\downarrow}$ for every $x \in X_+$. Then $\mathscr{S}(\overline{X}) \subset \mathscr{C}(\overline{X})^{\uparrow\downarrow}$, so that

$$\mathfrak{P}(\overline{X}) = \mathscr{S}(\overline{X})^{\downarrow\uparrow} \subset \left(\mathscr{C}(\overline{X})^{\uparrow\downarrow} \right)^{\downarrow\uparrow} = \mathscr{C}(\overline{X})^{\uparrow\downarrow\uparrow} \subset \mathfrak{P}(\overline{X}).$$

Thus, what we need is only to justify the representation:

$$[\iota x] = \bigwedge_{t \in X_+} \bigvee_{n \in \mathbb{N}} \langle (nx - t)^+ \rangle.$$

Denote $\sigma_t := \bigwedge_n \langle (nx - t)^+ \rangle$ and $\sigma = \bigwedge_t \sigma_t$. It is not difficult to observe that $\sigma_t \geq [\iota x]$ for all $t \in X_+$. For an arbitrary projection $\rho \in \mathfrak{P}(\overline{X})$ with $\rho \wedge [\iota x] = 0$ put $\rho_t := \rho \wedge [\iota t]$ $(t \in X_+)$. Then $\rho_t \leq [\iota(t - nx)^+] \leq \langle (t - nx)^+ \rangle$ for every $n \in \mathbb{N}$. Since $\langle (t - nx)^+ \rangle \wedge \langle (nx - t)^+ \rangle = 0$ it follows $\rho_t \wedge \langle (nx - t)^+ \rangle = 0$ and $\rho_t \wedge \sigma_t = \bigwedge_n (\rho_t \wedge \langle (nx - t)^+ \rangle = 0$. From this we obtain the following identities:

$$\rho_t \wedge \sigma = 0, \quad \rho \wedge \sigma = \sup \rho_t \wedge \sigma = 0.$$

Now, putting $\rho = [\iota x]^\perp$, we arrive at the desired inequalities $[\iota x] \leq \sigma \leq [\iota x]$.

(2): It follows from (1) and 3.5.9 (2).

(3): It is sufficient to observe that $(\overline{\Phi} \circ \langle x \rangle) \circ \iota = \Phi \circ \pi_x$. ▷

3.6. Comments

3.6.1. (1) In [157], L. V. Kantorovich laid grounds for the theory of regular operators in K-spaces. Also, the Riesz–Kantorovich Theorem (3.1.2) appeared in this article for the first time. F. Riesz [331] formulated an analogous assertion for the space of continuous linear functionals over the lattice $C[a, b]$ in his famous talk at the International Mathematical Congress in Bologna in 1928 and thereby became enlisted in the cohort of the founders of the theory of ordered vector spaces.

(2) The minimal extension operator and its properties are well known (see, for instance, [23]). The set of $e \in mE$ for which $S([0, |e|] \cap G)$ is order-bounded in F is a maximal order-dense ideal in mE to which the operator S admits the minimal extension. Let $p : G \to F$ be an increasing ($0 \leq e_1 \leq e_2$ implies $p(e_1) \leq p(e_2)$) sublinear operator with $p(g) = p(g^+)$ $(g \in G)$. Denote by E the set of all $e \in mG$ for which $p([0, |e|] \cap G)$ is order-bounded in F. The operator $p_m : E \to F$ defined by $p_m(e) := \sup\{p(g) : g \in G, 0 \leq g \leq e^+\}$ is also sublinear and

$$\partial p_m = \{\pi_G S : S \in L^\sim(E, F),\, S|_G \in \partial p\}.$$

This construction is, in particular, a basis for Yu. A. Abramovich's maximal extension of a vector lattice [1–3].

(3) The operators of the form $\pi_G S$, where $G \in \mathscr{I}(E)$, have long been in use (see, for instance, [388]) but seem to be considered explicitly in A. R. Schep's article [337], see also [23]. The idea to employ fragments of the form $\pi_G S$ and π_e instead of the fragments $S \circ \pi$ belongs to E. V. Kolesnikov [14, 175, 176].

(4) Yu. A. Abramovich [3] developed a version of the calculus of 3.1.2 in which suprema and infima can be taken over partitions of the argument into disjoint parts, see 3.1.6 (2). For the modulus of a regular operator, this fact was independently established by W. A. J. Luxemburg and A. C. Zaanen [263], see also [23, 228, 409].

(5) The concept of a generating set of projections as well as Theorem 3.1.5 (3) belongs to S. S. Kutateladze [227]. The main idea of [227] is as follows: The fragments of a positive operator U are the extreme points of the order interval $[0, U]$. The latter set coincides with the supporting set ∂p of the sublinear operator $p(x) := Ux^+$. Thereby, studying the fragments of a positive operator reduces to describing the extremal structure of a supporting set. Such a description for a general sublinear operator was obtained for the first time in the article [221] by S. S. Kutateladze (for a detailed exposition, see [209]).

(6) The problem of dominated extension of linear operators originates with the Hahn–Banach Theorem; see [63] for its history. Theorem 3.1.8 (2) was discovered by L. V. Kantorovich in 1935 [153]. The equivalence between the extension and the least upper bound properties (Theorem 3.1.8 (3)) was first established by W. Bonnice and R. Silvermann [48] and T.-O. To [374]; an elegant proof with decisive simplifications is due to A. D. Ioffe [138]; see also [209]. Theorem 3.1.17 is due to L. V. Kantorovich [159].

(7) Theorem 3.1.8 (2) can be considered as an exemplar application of the heuristic transfer principle for K-spaces (see 1.6.3 (2)). It claims that the *Kantorovich principle* is valid in relation to the classical Dominated Extension Theorem; i.e., we may replace the reals in the standard Hahn–Banach Theorem by elements of an arbitrary K-space and a linear functional by a linear operator with values in this K-space.

3.6.2. (1) Theorem 3.2.2 was proved by E. V. Kolesnikov [178]; of course, it is well known at least when E is a K_σ-space. Theorems 3.2.3 (2) and 3.2.3 (3) are due to Ogasawara [302] and A. I. Veksler [376]. In [178] E. V. Kolesnikov suggested some simple localization method that enabled him to obtain the formulas of 3.2.5 for calculating the order continuous and σ-order continuous parts. The formulas 3.2.6 (1) took their final form (see [337] and [20]) gradually in the works of various authors (W. A. J. Luxemburg, A. C. Zaanen, C. D. Aliprantis, A. R. Schep, and P. van Eldik); a piece of this history can be learned from [23, 409]. Formulas 3.2.6 (2) are just variants of 3.2.6 (1) in the case when the vector lattice E has the strong Freudenthal property.

(2) The projection formulas of 3.2.7 and 3.2.8 (1) were established by E. V. Kolesnikov [177]. Using nonstandard methods of analysis, S. S. Kutateladze independently found 3.2.8 (1) in [227] at E. V. Kolesnikov's request. The particular cases given in 3.2.8. (3, 4) were earlier obtained by C. D. Aliprantis and O. Burkinshaw [22]. Kutateladze's method of generating sets (see 3.6.1 (5)) works also for projection formulas: Theorem 3.2.9 (1) have been proven in [227]. Of course, this result enables us to find various projection formulas by taking concrete generating sets of projections. A particular case of Kutateladze's formulas, presented in

3.2.9 (2), was earlier obtained by A. G. Kusraev and V. Z. Strizhevskiĭ in [218]. Equivalences (1) $\Leftrightarrow$ (3) and (1) $\Leftrightarrow$ (4) in Theorem 3.2.10 were established in [218] and [227], respectively. A variant of the equivalence (1) $\Leftrightarrow$ (3) in the case when $F_n^\sim$ separates the points of F was earlier obtained in [20]; see also [23].

(3) The *shadow* of an operator $S : E \to F$ is the mapping $\operatorname{shdw}(S) : \mathfrak{P}(E) \to \mathfrak{P}(E)$ defined by $\operatorname{shdw}(\pi) = [S\pi(E)]$. Thus, $\operatorname{shdw}(\pi)$ is the band projection onto $(S\pi(X))^{\perp\perp}$, see 5.2.2. Denote

$$\langle \operatorname{shdw}(S)\rangle := \{T \in L^\sim(E,F) : (\forall \sigma \in \mathfrak{P}(E))\ \operatorname{shdw}(T)\sigma \le \operatorname{shdw}(S)\sigma\}.$$

It is easy to observe that $\langle \operatorname{shdw}(S)\rangle$ is a band containing $\{S\}^{\perp\perp}$. Let $[\operatorname{shdw}(S)]$ and $[\operatorname{shdw}(S)]_n$ be the band projections onto $\langle \operatorname{shdw}(S)\rangle$ and $\langle \operatorname{shdw}(S)\rangle \cap L_n^\sim(E,F)$, respectively. Denote by Δ and Δ_n the sets of all finite and arbitrary partitions of unity in $\mathfrak{P}(E)$ E. V. Kolesnikov proved that the following projection formulas hold:

$$[\operatorname{shdw}(S)]T = \inf\left\{\sum_{k=1}^{n} \operatorname{shdw}(\sigma_k)T\sigma_k : (\sigma_1,\dots,\sigma_n) \in \Delta,\ n \in \mathbb{N}\right\},$$

$$[\operatorname{shdw}(S)]_n T = \inf\left\{o\text{-}\sum_{\alpha\in \mathrm{A}} \operatorname{shdw}(\sigma_\alpha)T\sigma_\alpha : (\sigma_\alpha)_{\alpha\in \mathrm{A}} \in \Delta_n\right\}.$$

3.6.3. (1) The basic properties of lattice homomorphisms are presented in [23]. Theorem of 3.3.1 is due to M. Meyer [86, 282]. Theorem 3.3.3 was announced in [219] and proved in [220] by S. S. Kutateladze. He obtained it as a simple consequence of his powerful *canonical sublinear operator method*; for details see [209]. Another proof was found by W. A. J. Luxemburg and A. R. Schep [261]. Propositions 3.3.4 (1–5) are easy consequences of Kutateladze's Theorem. V. A. Radnaev [324] demonstrated that the converse of 3.3.4 (4) is also true: *a positive operator $T \in L^\sim(E,F)$ is a lattice homomorphism if and only if for every $0 \le T_1, T_2 \le T$ the relation $T_1 \perp T_2$ implies $T_1(E) \perp T_2(E)$.* Lattice homomorphisms in vector lattices are closely related to disjointness preserving operators to which Chapter 5 is devoted; see Section 5.6 for the relevant comments.

(2) The theory of orthomorphisms stems from H. Nakano [287]. Orthomorphisms were studied by many authors under various names: *dilatators* (H. Nakano [287]), *essentially positive operators* (G. Birkhoff [46]), *polar preserving endomorphisms* (P. F. Conrad and J. E. Diem [68]), *multiplication operators* (R. C. Buck [53] and A. W. Wickstead [396]), and *stabilisateurs* (M. Meyer [281]). The main stages of this development are reflected in the books by A. Bigard, K. Keimel, and S. Wolfenstein [44], C. D. Aliprantis and O. Burkinshaw [23], A. C. Zaanen [409] etc.; see also the survey by A. V. Bukhvalov [60]. The results of this book are covered by [23, 409]. Available is an extensive bibliography on the theory of orthomorphisms; we indicated a portion of it connected with the subjects we discuss in the sequel: [4, 12, 13, 39, 45, 86, 122, 123, 135, 259, 261, 279, 310, 311, 321, 408].

(3) In 3.3.5–3.3.8 a portion of Kutateladze's theory of extreme operators is exposed. The main results (Theorems 3.3.6, 3.3.7 and 3.3.8) were established in [221, 223]. In the same papers the following operator variant of the classical Kreĭn–Milman Theorem was suggested:

Take a K-space F and let $0 \leq T \in L^{\sim}(E, F)$. Call an operator $S \in \partial p$ a T-*extreme point* of ∂p if $T \circ S \in \mathrm{Ch}(T \circ p)$. Say that S is an *o-extreme point* of ∂p or p if S is a T-*extreme point* of ∂p for every positive o-continuous operators $T : E \to F$ and every K-space F. Denote the set of all o-extreme points by $\mathscr{E}_0(P)$.

Kreĭn–Milman Theorem for o-Extreme Points. *Every sublinear operator $p : X \to E$ is the upper envelope of the set of its o-extreme points. Symbolically,*

$$p(x) = \sup\{Tx : T \in \mathscr{E}_0(p)\} \quad (x \in X).$$

Moreover, the least upper bound on the right-hand side is attained for each $x \in X$.

These and other facts from [221, 223, 226] give sound grounds for studying the extreme structure of convex sets of linear operators. The set $\mathscr{E}(T)$ of all positive extensions of a positive operator T defined on a massive subspace G of a vector lattice E coincides with the supporting set ∂p whenever $p : E \to F$ is defined by $p(e) := \inf\{Tg : e \leq g,\, g \in G\}$ $(e \in E)$. Now, it follows from the Kreĭn–Milman Theorem that the convex set $\mathscr{E}(T)$ has extreme points. This simple corollary was independently proved by Z. Lipecki [244] (see also [23]). About other extension results by Z. Lipecki see [243, 245, 248, 249].

(4) The Hahn–Banach formula for lattice homomorphisms (Theorem 3.3.10) was established by V. A. Radnaev [324]. Radnaev's approach is based upon Kutateladze's machinery for calculating supporting sets and their extreme boundaries. The auxiliary facts of 3.3.9 belong to V. A. Radnaev [324] (3.3.9 (1)) and G. J. H. M. Buskes and A. C. M. van Rooij [64] (3.3.9 (2, 3)). The important corollary 3.3.11 (1) was obtained by G. J. H. M. Buskes and A. C. M. van Rooij in [64]. Theorem 3.3.11 (2), conventionally referred to as the Lipecki–Luxemburg–Schep Theorem because of [260] and [244] (see, for instance, [23, 41, 63, 64]), was announced in [221] and proved in [223] by S. S. Kutateladze. (Actually S. S. Kutateladze proved Theorem 3.3.6 which contains 3.3.11 (2) as a particular case in view of 3.3.3.) Various approaches (mysteriously, except those by S. S. Kutateladze) to Hahn–Banach-type theorems for lattice homomorphisms are discussed in C. B. Bernau [41]; see also a nice survey by G. Buskes [63] in which the history, interconnections, and part of numerous generalizations of the Hahn–Banach Theorem are collected.

(5) In [210] it was indicated that developing a calculus of extreme points is desirable and the following problem was formulated:

For which sublinear operators $p : X \to E$ and $q : E \to F$ does the relation $\mathrm{Ch}(q \circ p) \subset \mathrm{Ch}(q) \circ \mathrm{Ch}(p)$ (or $\mathrm{Ch}(q \circ p) = \mathrm{Ch}(q) \circ \mathrm{Ch}(p)$) hold? The same question relates to o-extreme points: For which p and q does the relation $\mathscr{E}_0(q \circ p) \subset \mathscr{E}_0(q) \circ \mathscr{E}_0(p)$, or $\mathscr{E}_0(q \circ p) = \mathscr{E}_0(q) \circ \mathscr{E}_0(p)$ hold?

Of course, the problem is partially motivated by Kutateladze–Milman's Theorem 3.3.8. V. A. Radnaev observed in [324] that $\mathrm{Ch}(\pi p) = \pi\,\mathrm{Ch}(p)$ for a sublinear operator $p : X \to E$ and a positive orthomorphism $\pi \in \mathrm{Orth}(E)$. Radnaev's formula of 3.3.10 can be also considered as a partial answer to the above-mentioned questions, since according to 3.3.9 (2, 3) it can be rewritten equivalently as

$$\bigcup_{0 \le \pi \le I_E} \pi\,\mathrm{Ch}(p \circ T) = \Bigg(\bigcup_{0 \le \pi \le I_E} \pi\,\mathrm{Ch}(p) \Bigg) \circ T.$$

(6) The results of 3.3.12 and 3.3.13 are from the article by C. B. Huijsmans and B. de Pagter [136]. This article also contains some interesting applications of 3.3.13 (1, 2). In the case when E is a vector lattice with the strong Freudenthal property and F is order complete E. V. Kolesnikov gave another projection formula: If $0 \le T \in L^{\sim}(E, F)$ and T_d is the band projection of T onto $L^{\sim}_d(E, F)$ then

$$T_d = \inf \big\{ \sup\{\pi T \sigma : \pi T \sigma e \le \varepsilon T e\} : 0 < \varepsilon \in \mathbb{R},\ 0 \le e \in E \big\}.$$

(The infimum and supremum are taken in $L^{\sim}(E, F)$.) To obtain a pointwise formula we must make the set under the supremum on the right-hand side upward-directed by adding finite disjoint suprema:

$$T_d(x) = \inf_{0<\varepsilon\in\mathbb{R}} \sup \Bigg\{ \sum_{k,l \le n} \pi_k T \sigma_{k,l} x : \pi_k T \sigma_{k,l} \le \varepsilon T e,\ \pi_k \in \mathfrak{P}(F),\ \sigma_{k,l} \in \mathfrak{P}(E),$$
$$\pi_k \circ \pi_l = 0\,(k \ne l),\ \sigma_{k,l} \circ \sigma_{k,m} = 0\,(l \ne m)\,(k, l, m \le n \in \mathbb{N}) \Bigg\}.$$

The corresponding test for a diffuse operator is stated as follows: A positive operator $T : E \to F$ is *diffuse* if and only if

$$0 = \inf \{T\sigma_1 e \vee \dots \vee T\sigma_n e : e = \sigma_1 e + \dots + \sigma_n e,\ \sigma_1, \dots, \sigma_n \in \mathfrak{P}(E),\ n \in \mathbb{N}\}$$

for every $e \in E_+$.

(7) Let E and F be the same as in (6). If $S : E \to F$ is a lattice homomorphism then the projection $[S]T$ of a positive operator $T : E \to F$ onto the band $\{S\}^{\perp\perp}$ is calculated by the formula (E. V. Kolesnikov [179]):

$$[S]T = \inf\Big\{\sum_{k=1}^{n} \operatorname{shdw}(\sigma_k)T\sigma_k : \sigma_k \in \mathfrak{P}(E),$$

$$\sum_{k=1}^{n} \sigma_k = I_E,\ \sigma_k \circ \sigma_l = 0\,(k \neq l),\ n \in \mathbb{N}\Big\}.$$

3.6.4. (1) In the series of papers [266–269], D. Maharam developed an original approach to the study of positive operators in the spaces of measurable functions. A brief description for the method and main results are collected in her survey [271]. W. A. J. Luxemburg and A. R. Schep [261] extended a portion of Maharam's theory connected with the Radon–Nikodým Theorem to the case of positive operators in vector lattices. The terms "Maharam property" and "Maharam operator" (3.4.1) were introduced in [261] and [195, 197]. In [266–269] operators with Maharam property are referred to as *full-valued*.

(2) Theorems in 3.4.3, 3.4.6 (2), and 3.4.9 were obtained by W. A. J. Luxemburg and A. R. Schep in [261]. Theorems 3.4.8, 3.4.10, and 3.4.11 belong to A. G. Kusraev [193, 195, 197]. In the scalar case ($F = \mathbb{R}$) Theorem 3.4.8, for the functionals in the order ideal generated by Φ, was proved by B. Z. Vulikh (see [163, 388]); the result was announced by G. Ya. Lozanovskiĭ in [254] for $\mathscr{L}_\Phi(E, \mathbb{R})$, see also [391]. Another proof in the scalar case was given by M. M. Rice [327]. Theorem 3.4.10 for $E := C(X)$ and $F := C(Y)$, where X and Y are compact spaces with Y totally disconnected, was obtained by H. P. Lotz [252].

(3) Denote by $\Pi(E)$ the set of all partitions of unity in $\mathfrak{P}(E)$. Let E and F be order complete vector lattices and let $S : E \to F$ be a Maharam operator. Then the projection $[S]T$ of a positive operator $T : E \to F$ onto the band $\{S\}^{\perp\perp}$ can be calculated by the formula (see E. V. Kolesnikov [179]):

$$[S]T = \inf\Big\{ o\text{-}\sum_{\alpha \in \mathrm{A}} \operatorname{shdw}(\sigma_\alpha)T\sigma_\alpha : (\sigma_\alpha)_{\alpha \in \mathrm{A}} \in \Pi(E)\Big\}.$$

(4) The facts, exposed in Section 3.4, are sufficient to reveal some analogy between Maharam operators and positive o-continuous linear functionals and to hint the conjecture: *Each fact on functionals of the indicated form ought to have a parallel variant for a Maharam operator.* The theory of Boolean-valued models discloses full profundity of such an analogy and allows us to transform the above heuristic argument into an exact research method. We give without proof only one result in this direction (see A. G. Kusraev [195, 197]). As in 8.1.1, we assume that B is a complete Boolean algebra and $\mathscr{R}$ stands for the reals in the Boolean-valued universe $\mathbf{V}^{(B)}$.

Theorem. *Let X be an arbitrary K-space and let E be a universally complete K-space $\mathscr{R}{\downarrow}$. Assume that $\Phi : X \to E$ is a Maharam operator such that $X = X_\Phi = \mathscr{D}_m(\Phi)$ and $E = E_\Phi$. Then there exist elements $\mathscr{X}$ and $\varphi \in \mathbf{V}^{(B)}$ such that the following hold:*

(i) *$[\![\mathscr{X}$ is a K-space and $\varphi : \mathscr{X} \to \mathscr{R}$ is a positive o-continuous functional and $\mathscr{X} = \mathscr{X}_\varphi = \mathscr{D}_m(\varphi)]\!] = \mathbf{1}$;*

(ii) *if $X' := \mathscr{X}{\downarrow}$ and $\Phi' = \varphi{\downarrow}$ then X' is a K-space and $\Phi' : X' \to E$ is a Maharam operator;*

(iii) *there exists a linear and lattice isomorphism h from X onto X' such that $\Phi = \Phi' \circ h$;*

(iv) *for a linear operator Ψ, the containment $\Psi \in \{\Phi\}^{\perp\perp}$ is true if and only if there exist $\psi \in \mathbf{V}^{(B)}$ for which $\psi \in \{\varphi\}^{\perp\perp}$ holds inside $\mathbf{V}^{(B)}$ and $\Psi = (\psi{\downarrow}) \circ h$.*

About various applications of the above results on Maharam operators and some extension of this theory to sublinear and convex operators see [197, 209].

(5) The operator ε from 3.4.11 is conventionally called a *simultaneous extension operator*. Thus, Theorem 3.4.11 states existence of an order continuous simultaneous extension from a massive sublattice to the ambient vector lattice. This fact is interesting to compare with the following result obtained by T. Ando [25] (see also [231, 241]):

Theorem. *Let E be a Banach lattice of dimension ≥ 3. Then the following are equivalent:*

(i) *For every sublattice E_0 of E there is a positive simultaneous extension operator of norm one from E_0' to E';*

(ii) *E is an AL^p-space for some $1 \leq p < \infty$, or an AM-space.*

(6) Let G be a universally complete vector lattice with fixed order-unity $\mathbf{1}$ and let M denote the order ideal in G generated by $\mathbf{1}$. Let E and F be arbitrary sublattices of G. For $f \in E^\sim$ and $e \in E_+$ define $f_e \in M^\sim$ by $f_e(x) := f(ex)$ $(x \in M)$. Order-bounded functionals $f \in E^\sim$ and $g \in F^\sim$ are said to be *disjoint* if $|f|_e$ and $|g|_c$ are disjoint in $M^\sim$ for every $e \in E_+$ and $c \in F_+$; in this event we write $f \sqcup g$. The following result was established in [193] by using the simultaneous extension operator from 3.4.11.

Theorem. *Assume that $\mathbf{1}'$ and $\mathbf{1}''$ are fixed order-unities in the universally complete spaces $m(E^\sim)$ and $m(M^\sim)$. Then there exists a unique pair $(\mathscr{R}_E, \mathscr{V}_E)$ such that*

(i) *$\mathscr{V}_E$ is an o-closed sublattice in $m(M^\sim)$ and $\mathscr{R}_E$ is a linear and lattice isomorphism of $m(E^\sim)$ onto $\mathscr{V}_E$;*

(ii) *$\mathscr{R}_E(\mathbf{1}') = \pi(\mathbf{1}'')$, where $\pi = [\mathscr{V}_E^{\perp\perp}]$;*

(iii) *for every $f \in E^\sim$ and $g \in M^\sim$ the relations $f \sqcup g$ and $\mathscr{R}_E(f) \perp g$ are equivalent.*

In the particular case when E is an order complete vector lattice (i.e. E is an order ideal in G) the result was obtained by G. Ya. Lozanovskiĭ [255, 391], see also [60, 61].

3.6.5. (1) In 3.5 we follow G. P. Akilov, E. V. Kolesnikov, and A. G. Kusraev [14]. The construction of 3.5.1–3.5.8 stems from D. Maharam's theory of positive operators [267–269]. About further development of this construction, see Section 6.3.

(2) In 3.5.10 (1, 2) and 3.5.11 (1–3) every fragment of a positive operator is obtained from its simpler fragments by up and down procedures. Similar assertions are often referred to as *up-down theorems*. The first up-down theorem (3.5.11 (3)) was established by B. de Pagter [309], see also [20, 23]. However, it involved two essential constraints: F should admit a total set of o-continuous functionals, and E must be order complete (or at least possess the principal projection property). The first constraint was eliminated in [218] and the second, in [14]. Of course, in the latter case the set of simple fragments is essentially different (see 3.5.9).

(3) A general up-down theorem was obtained by S. S. Kutateladze [227]. Namely, he proved that if $\mathscr{P}$ is a generating set of projections in $L^\sim(E, F)$ (where E and F are vector lattices with F order complete) then $\mathfrak{E}(\Phi) = \mathscr{P}^\vee(\Phi)^{\uparrow\downarrow\uparrow}$, where $\mathscr{P}^\vee(\Phi)$ comprises of fragments representable as $\sum_{k=1}^\infty \pi_k \circ (\rho_k \Phi)$ with pairwise disjoint $\pi_k \in \mathfrak{P}(F)$ and arbitrary $\rho_k \in \mathscr{P}$. All formulas from 3.5.11 can be deduced from Kutateladze's Up-Down Theorem by specifying generating sets.

Chapter 4

Dominated Operators

Dominated operators are frequent in analysis. The most familiar examples are the Bochner integral, the conditional expectation operator, strongly or weakly integral operators, and weighted shift operators.

In this chapter we study general properties of dominated operators. A dominated operator has the least (or exact) dominant under rather weak assumptions. For instance, it is sufficient to require that the domain of an operator be a decomposable space and the norm lattice of the image be order complete (4.1.2). There exist explicit formulas for calculating the exact dominant (4.1.5 and 4.1.8) which are rather useful for studying interrelation between an operator and its exact dominant. Using these formulas, it is possible for instance to show that a dominated operator is order continuous (= completely additive) if and only if so is its exact dominant (4.3.2, 4.3.6). From this fact we easily deduce that a dominated operator is order continuous if and only if it is completely additive (4.3.7). Moreover, an order continuous dominated operator from a decomposable LNS into an order complete LNS admits a unique extension onto the order completion of the domain with the exact dominant preserved (4.3.3). Therefore, in some cases it is possible in particular to improve formulas for calculating exact dominants (4.3.4 and 4.3.5).

The central result of the present chapter is decomposability of the dominant norm. If X is a decomposable lattice-normed space and Y is a Banach–Kantorovich space then the exact dominant determines a decomposable norm in the space $M(X,Y)$ of dominated operators; hence, $M(X,Y)$ is a Banach–Kantorovich space (4.2.6). The proof of this fact in particular uses the results on the Boolean algebra of fragments of a positive operator (3.5.10). Decomposability of the dominant norm implies for instance that a dominated operator admits unique decomposition into the disjoint sum of an order continuous operator and an order singular operator.

The decomposition in which singularity is understood in the sense of disjointness from all order continuous operators is referred to as the *weak form of the Yosida–Hewitt decomposition* and is valid for dominated operators under the above general hypotheses (4.4.2). Singularity means often that an operator vanishes on a "huge" subset of its domain. In this case decomposition into the o-continuous and singular parts is called the *strong form of the Yosida–Hewitt decomposition* and is also valid if X and Y are normed by order complete vector lattices with the Egorov property and the countable sup property, respectively (4.4.10, 4.4.11). Explicit formulas can be written down for calculating the order continuous summand (4.4.3, 4.4.4). An internal description is also given for almost integral operators, i.e., operators whose exact dominants belong to the band generated by finite-rank order continuous operators (4.3.10).

Extension of a positive (dominated) operator by order continuity is among the traditional problems. It is natural to attempt to prove extension results by means of the Daniell construction of the Lebesgue integral. In the case when the target space is weakly σ-distributive a direct adaptation of the Daniell extension method can be successfully implemented resulting in a sequentially order continuous extension to the Baire completion of the domain lattice (4.5.3–4.5.9). The extension to the Borel completion is also accessible but requires more burdensome (σ, ∞)-distributivity assumption (4.5.10).

4.1. The Space of Dominated Operators

In this section we introduce the notion of a dominated operator. Under not very restrictive conditions, every dominated operator has the exact dominant. Explicit formulas for calculating the exact dominant are also derived.

4.1.1. Consider lattice-normed spaces (X, E) and (Y, F), a linear operator $T : X \to Y$, and a positive operator $S : E \to F$. If the condition

$$|Tx| \le S(|x|) \quad (x \in X)$$

is satisfied then we say that S *dominates* or *majorizes* T or that S is a *dominant* or *majorant* for T. In this situation, T is called a *dominated* or *majorizable* operator. Let $\operatorname{maj}(T)$ be the set of all dominants of the operator T. It is clear that $\operatorname{maj}(T)$ is a convex set in the K-space $L^{\sim}(E, F)$. If there is a least element in $\operatorname{maj}(T)$ with respect to the order induced from $L^{\sim}(E, F)$ then it is called the *least* or *exact dominant* of T and denoted by $|T|$. Consequently, $|T|$ is a positive operator from E into F; moreover, $|T| = \inf \operatorname{maj}(T) \in \operatorname{maj}(T)$ and the inequality

$$|Tx| \le |T|(|x|) \quad (x \in X)$$

holds. The set of all dominated operators from X into Y is denoted by $M(X, Y)$. Thus,

$$T \in M(X, Y) \iff \operatorname{maj}(T) \ne \varnothing.$$

4.1.2. *Suppose that X is decomposable and F is order complete. Then every dominated operator $T : X \to Y$ has the exact dominant $|T|$.*

$\lhd$ Observe first that $\operatorname{maj}(T)$ is a lower semilattice in $L^{\sim}(T, F)$, i.e., if $S_1, S_2 \in \operatorname{maj}(T)$ then $S_1 \wedge S_2 \in \operatorname{maj}(T)$. Indeed, if $|x| = e_1 + e_2$, where $x \in X$ and $e_1, e_2 \in E_+$, then the representation $x = x_1 + x_2$, $|x_k| = e_k$ $(k := 1, 2)$ holds. Consequently, the following inequalities are valid:

$$|Tx| \leq |Tx_1| + |Tx_2| \leq S_1 e_1 + S_2 e_2.$$

Passage to the infimum over e_1 and e_2, $e_1 + e_2 = |x|$, yields $|Tx| \leq (S_1 \wedge S_2)(|x|)$. Thus, $S_1 \wedge S_2 \in \operatorname{maj}(T)$ and the set $\operatorname{maj}(T)$ is downward directed; therefore, the infimum $S := \inf \operatorname{maj}(T)$ can be calculated pointwise on the cone E_+. It follows that

$$|Tx| \leq \inf \{S'(|x|) : S' \in \operatorname{maj}(T)\} = S(|x|) \quad (x \in X).$$

Hence, $S \in \operatorname{maj}(T)$, i.e., $S = |T|$. $\rhd$

4.1.3. We now list the main types of dominated operators most frequent in the literature.

(1) If $E = F = \mathbb{R}$ then X and Y are normed spaces and $M(X, Y) = \mathscr{L}(X, Y)$ is the space of all bounded linear operators from X into Y. The exact dominant of an operator $T \in M(X, Y)$ is determined by the positive number that is the norm of T:

$$\|T\| = \sup\{\|Tx\| : \ \|x\| \leq 1\}.$$

(2) Suppose that $X = E$, $Y = F$, and the modulus of an element serves as the lattice norm both in E and F. Then $M(X, Y)$ coincides with the space $L^r(E, F)$ of all regular operators from E into F. The exact dominant of an operator $T \in M(X, Y)$ is its modulus, $|T|$. If F is a K-space then we have

$$|T|x = \sup\{Tu : \ |u| \leq x\} \quad (x \in E_+).$$

(3) Suppose that $E = \mathbb{R}$ and $Y = F$. Then an operator $T : X \to F$ is dominated if and only if the set $\{Tx : \ x \in X, \ \|x\| \leq 1\}$ is order-bounded in F. If F is a K-space then the exact dominant has the form

$$|T| = \sup\{Tx : \ \|x\| \leq 1\}.$$

(Actually, the exact dominant is presented by the mapping $t \mapsto t|T|$ $(t \in \mathbb{R})$.) The element $|T|$ is called the *abstract norm* of T. So, in case F is a K-space, the set $M(X, F)$ coincides with the set $L_A(X, F)$ of operators with abstract norm.

(4) Suppose now that $X = E$ and $F = \mathbb{R}$. Then the fact that an operator $T : E \to Y$ is dominated means existence of a positive functional e^* on E such that

$$\|Te\| \leq \langle |e|, e^* \rangle \quad (e \in E).$$

The exact dominant can be calculated as follows:

$$|T|e = \sup\left\{ \sum_{k=1}^{n} \|Te_k\| : \; e_1, \dots, e_n \in E_+, \; \sum_{k=1}^{n} e_k = e, \; n \in \mathbb{N} \right\} \quad (e \in E_+).$$

(5) Suppose that $E = F$ and $\Lambda \subset \mathrm{Orth}(E)$ is a sublattice and subring in $\mathrm{Orth}(E)$. We call an operator $T : X \to Y$ *bounded in the sense of vector norm* if there exists an orthomorphism $0 \le \alpha \in \lambda\Lambda$ such that $|Tx| \le \alpha|x|$ for all $x \in X$. Denote the set of all such operators by $\mathscr{L}_\Lambda(X, Y)$.

As in 2.1.2, we call elements $x \in X$ and $y \in Y$ disjoint and write $x \perp y$ whenever $|x| \wedge |y| = 0$. An operator T is said to be *band preserving* if $x \perp y$ implies $Tx \perp y$ for all $x \in X$ and $y \in Y$. By an *orthomorphism* we mean any operator that preserves bands and takes norm o-bounded sets into norm o-bounded sets. Let $\mathrm{Orth}(X, Y)$ be the set of all orthomorphisms from X into Y. It is possible to show (see below 4.1.9 and 4.1.11) that if E is a K-space and X and Y are decomposable then $\mathrm{Orth}(X, Y) \subset M(X, Y)$. Moreover, $\mathrm{Orth}(X, Y) = \mathscr{L}_{\mathrm{Orth}}(X, Y)$ and the following formula is valid:

$$|T|e = \sup\{|Tx| : |x| \le e\} \quad (e \in E_+).$$

Observe that, in each of the cases considered, the exact dominant can be found by a formula of special type. In the sequel, we give general formulas for calculating exact dominants.

4.1.4. Consider lattice-normed spaces X and Y with norm lattices E and F, respectively. We assume that X is decomposable and F is order complete. Denote by E_{0+} the conic hull of the set $|X| := \{|x| : \; x \in X\}$, i.e., the set of elements of the form $\sum_{k=1}^{n} |x_k|$, where $x_1, \dots, x_n \in X$ and $n \in \mathbb{N}$. The following assertion holds:

The cone E_{0+} is an order ideal in E_+. If E possesses the principal projection property then $E_{0+} = |X|$.

$\lhd$ If $e \in E_+$ and $x_1, \dots, x_n \in X$ are such that $e \le |x_1| + \dots + |x_n|$ then $e = e_1 + \dots + e_n$ for some $0 \le e_k \le |x_k|$ $(k := 1, \dots, n)$. By decomposability of X, there are $y_1, \dots, y_n \in X$ such that $|y_k| = e_k$ $(k := 1, \dots, n)$. Thus, $e \in E_{0+}$; therefore, E_{0+} is an order ideal in E.

Next, if $e = e_1 + e_2$, where $e_k := |x_k|$ and $x_k \in X$ $(k := 1, 2)$, then $e \le 2e_1 \vee e_2$ and in view of what was proven, we only need to show that $e_1 \vee e_2 \in |X|$. At the same time, for the projection π onto the band $\{e_1 - e_1 \wedge e_2\}^{\perp\perp}$, we have

$$e_1 \vee e_2 = \pi e_1 + \pi^\perp e_2 = |\pi x_1| + |\pi^\perp x_2| = |\pi x_1 + \pi^\perp x_2|,$$

i.e., $e_1 \vee e_2 \in |X|$. $\rhd$

4.1.5. Theorem. *The exact dominant of an arbitrary operator $T \in M(X,Y)$ can be calculated by the following formulas:*

$$|T|e = \sup\left\{\sum_{k=1}^{n}|Tx_k| : \ x_1,\dots,x_n \in X, \ \sum_{k=1}^{n}|x_k| = e, \ n \in \mathbb{N}\right\} \quad (e \in E_{0+}),$$
$$|T|e = \sup\{|T|e_0 : \ e_0 \in E_{0+}, \ e_0 \le e\} \quad (e \in E_+),$$
$$|T|e = |T|e^+ - |T|e^- \quad (e \in E).$$

$\lhd$ Denote by Se the right-hand side of the first of the desired formulas. Since, for all $x_1,\dots,x_n \in X$, we have

$$\sum_{k=1}^{n}|Tx_k| \le |T|\Bigg(\sum_{k=1}^{n}|x_k|\Bigg) = |T|e,$$

a mapping $S : E_{0+} \to F$ is well defined. If $u_1,\dots,u_n \in X$ and $v_1,\dots,v_m \in X$ are such that $\sum_{k=1}^{n}|u_k| = e$ and $\sum_{k=1}^{m}|v_k| = f$, then $e, f \in E_{0+}$ and, by the definition of the operator S, we have

$$\sum_{k=1}^{n}|Tu_k| + \sum_{k=1}^{m}|Tv_k| \le S(e+f).$$

Passing to the supremum over all finite families $\{u_1,\dots,u_n\}$ and $\{v_1,\dots,v_n\}$ of the form under consideration, we obtain $Se + Sf \le S(e+f)$.

Suppose that $e + f = \sum_{k=1}^{n}|x_k|$. By decomposability of the norm, in X there exist finite families $e_1,\dots,e_n \in E_{0+}$, $f_1,\dots,f_n \in E_{0+}$, $u_1,\dots,u_n \in X$, and $v_1,\dots,v_n \in X$, for which the following system of conditions is consistent:

$$f = f_1 + \dots + f_n, \ \ e = e_1 + \dots + e_n,$$
$$x_k = u_k + v_k, \ \ |u_k| = e_k, \ \ |v_k| = f_k \quad (k = 1,\dots,n).$$

Whence we infer

$$\sum_{k=1}^{n}|Tx_k| \le \sum_{k=1}^{n}|Tu_k| + \sum_{k=1}^{n}|Tv_k| \le Se + Sf.$$

Again, passage to the supremum over families $x_1,\dots,x_n$ yields the inequality $S(e+f) \le S(e) + S(f)$. So, S is an additive operator. Moreover, it is obvious that $S(\lambda e) = \lambda Se$ for $\lambda \ge 0$. Now, extend S from E_{0+} onto E_+ by letting $Se := \sup\{Se_0 : \ e_0 \le e, \ e_0 \in E_{0+}\}$. Since $Se_0 \le |T|e_0 \le |T|e$ for $e_0 \le e$, the supremum in the definition of Se exists; moreover, $Se \le |T|e$ $(e \in E_+)$. It is easy to see that S is an additive and positively homogeneous operator from E_+ into F. Finally, letting $Se = Se^+ - Se^-$ for $e \in E$, we obtain some positive operator $S : E \to F$; moreover, $S \le |T|$. On the other hand, for $x \in X$, we have $|Tx| \le S|x|$, i.e., S is a dominant for T and, hence, $S \ge |T|$. Finally, we obtain $S = |T|$. $\rhd$

4.1.6. Observe a useful corollary to Theorem 4.1.5.

A linear operator $T : X \to Y$ is dominated if and only if, for each $e \in E_+$, the set

$$\mathscr{U}(e) := \left\{ \sum_{k=1}^{n} |Tx_k| : x_1, \dots, x_n \in X, \sum_{k=1}^{n} |x_k| \le e, \ n \in \mathbb{N} \right\}$$

is order-bounded. In this case, $|T|e = \sup \mathscr{U}(e)$ $(e \in E_+)$.

$\lhd$ Necessity is obvious. Suppose that $\mathscr{U}(e)$ is an o-bounded set for each $e \in E_+$. Assign $Ue = \sup \mathscr{U}(e)$. Define an operator $S : E_+ \to F$ in the same way as in the proof of the previous theorem. By o-boundedness of $\mathscr{U}(e)$, these definitions are sound; moreover, $Se \le Ue$ $(e \in E_+)$. Suppose that, for some $x_1, \dots, x_n \in X$, $y_1, \dots, y_m \in X$, and $e \in E_+$, we have

$$|x_1| + \dots + |x_n| \le e = |y_1| + \dots + |y_m|.$$

The element $f := e - |x_1| - \dots - |x_n|$ can be represented as $f = f_1 + \dots + f_m$, where $0 \le f_k \le |y_k|$ $(k = 1, \dots, m)$. By decomposability of X, there exist $z_1, \dots, z_m \in X$ such that $f_k = |z_k|$ $(k = 1, \dots, m)$. So, for $u := |Tx_1| + \dots + |Tx_n|$ in $\mathscr{U}(e)$, we have $u \le u + |Tz_1| + \dots + |Tz_m| \le Se$. Since the choice of $u \in \mathscr{U}(e)$ is arbitrary, $Ue \le Se$ holds for $e \in E_{0+}$. We also observe that $Ue = \sup\{Se_0 : e_0 \le e, \ e_0 \in E_{0+}\}$. Thus, the operators S and U coincide on E_+. It remains to take account of the inequality $|Tx| \le U(|x|)$ $(x \in X)$ as well as linearity and positivity of the operator S, and apply 4.1.5. $\rhd$

In the first formula of 4.1.5, we may calculate the supremum over disjoint families $x_1, \dots, x_n \in X$. In order to obtain such an improvement, we need some auxiliary fact in which E is assumed to have the strong Freudenthal property.

4.1.7. *Suppose that a mapping $S : E_+ \to F_+$ satisfies the following:* **(a)** *S is positively homogeneous, i.e., $S(\lambda e) = \lambda Se$ $(e \in E_+, \lambda \in \mathbb{R}_+)$;* **(b)** *$S$ is subadditive, i.e., $S(e+f) \le Se + Sf$ $(e, f \in E_+)$;* **(c)** *S is orthogonally additive, i.e., $S(e+f) = Se + Sf$ $(e, f \in E_+, e \wedge f = 0)$;* **(d)** *$S$ has a linear dominant, i.e., $Se \le Ue$ $(e \in E_+)$ for some linear operator $U : E \to F$. Then S is additive.*

$\lhd$ Take arbitrary $e, f \in E_+$ and assign $c := e + f$. Denote by $\mathscr{F}(c)$ the set of all elements in E_+ of the form $\sum_{k=1}^{n} \lambda_k \pi_k c$, where $\lambda_1, \dots, \lambda_n \in \mathbb{R}_+$ and $\pi_1, \dots, \pi_n$ are pairwise disjoint projections in E with the sum equal to unity. If $e', e'' \in \mathscr{F}(c)$ are such that $e' = \sum_{k=1}^{n} \lambda'_k \pi_k c$ and $e'' = \sum_{k=1}^{n} \lambda''_k \pi_k c$ and, moreover, $\lambda'_k \le \lambda''_k$ $(k := 1, \dots, n)$, then, in view of (a) and (c), we have

$$Se' = \sum_{k=1}^{n} \lambda'_k S \pi_k c \le \sum_{k=1}^{n} \lambda''_k S \pi_k c = Se''.$$

This means that S is isotonic on $\mathscr{F}(c)$. For every $0 < \varepsilon \in \mathbb{R}$, we may choose $e', e'' \in \mathscr{F}(c)$ so that $e' \leq e \leq e''$, $e - e' \leq \varepsilon c$, and $e'' - e \leq \varepsilon c$. Applying (a), (b), and (d), we may write

$$Se - Se' \leq S(e - e') \leq U(e - e') \leq \varepsilon Uc,$$

since the operator U is positive. By the same reason, $Se'' - Se \leq \varepsilon Uc$. Then, in view of the fact that the operator S is isotonic on $\mathscr{F}(c)$, we have $Se' - Se \leq Se'' - Se \leq \varepsilon Uc$; consequently, $|Se - Se'| \leq \varepsilon Uc$. It is clear that the same inequality holds for e''.

Now, choose pairwise disjoint projections $\pi_1, \dots, \pi_n$, $\pi_1 + \dots + \pi_n = I_E$, and nonnegative numbers $\lambda_1, \dots, \lambda_n$, $\mu_1, \dots, \mu_n$ so that, for the elements $e' := \sum_{k=1}^n \lambda_k \pi_k c$ and $f' = \sum_{k=1}^n \mu_k \pi_k c$, we have $e' \leq e$, $f' \leq f$, $e - e' \leq \varepsilon c$, and $f - f' \leq \varepsilon c$. Then, in view of what was proven,

$$|Se - Se'| \leq \varepsilon Uc,$$
$$|Sf - Sf'| \leq \varepsilon Uc,$$
$$|S(f + e) - S(f' + e')| \leq 2\varepsilon Uc.$$

On the other hand, applying (a) and (c), we have

$$S(e' + f') = S\Big(\sum_{k=1}^n (\lambda_k + \mu_k)\pi_k c\Big) = \sum_{k=1}^n (\lambda_k + \mu_k) S(\pi_k c)$$
$$= S\Big(\sum_{k=1}^n \lambda_k \pi_k c\Big) + S\Big(\sum_{k=1}^n \mu_k \pi_k c\Big) = Se' + Sf'.$$

Taking account of the properties of the elements e' and f', we may write

$$|S(e + f) - Se - Sf| \leq |Se - Se'| + |Sf - Sf'| + |S(e + f) - S(e' + f')| \leq 4\varepsilon Uc.$$

Arbitrariness of the choice of ε yields $S(e + f) = Se + Sf$. ▷

4.1.8. Theorem. *Suppose that E is a vector lattice with the strong Freudenthal property, X is decomposable, and F is order complete. Then, for every operator $T \in M(X, Y)$, the following formula is valid:*

$$|T|e = \sup\Big\{ \sum_{k=1}^n |T\pi_k x| : x \in X, \ |x| = e, \ \pi_1, \dots, \pi_n \in \mathfrak{P}(X),$$
$$\pi_k \circ \pi_l = 0 \ (l \neq k), \ \sum_{k=1}^n \pi_k = I_X, \ n \in \mathbb{N}\Big\} \quad (e \in E_{0+}).$$

$\triangleleft$ Denote by Se the right-hand side of the desired equality. Demonstrate that the operator $S : E_{0+} \to F$ satisfies the conditions (a)–(d) of 4.1.7. According to 4.1.4, $E_0 := E_{0+} - E_{0+}$ is an o-ideal in E and $E_{0+} = |X|$; therefore, Se is defined for each $e \in E_{0+}$. Positive homogeneity of S is beyond any doubt. It is also obvious that $Se \le |T|e$ $(e \in E_{0+})$. Let $e, f \in E_{0+}$ and $x \in X$ be such that $|x| = e + f$. By decomposability of X, we have $x = y + z$, $|y| = e$, and $|z| = f$ for some $y, z \in X$. Take pairwise disjoint projections $\pi_1, \dots, \pi_n \in \mathfrak{P}(X)$, $\pi_1 + \dots + \pi_n = I_X$. It is clear that

$$\sum_{k=1}^{n} |T\pi_k x| \le \sum_{k=1}^{n} |T\pi_k y| + \sum_{k=1}^{n} |T\pi_k z| \le Se + Sf.$$

Passage to the supremum over $x, \pi_1, \dots, \pi_n$ under consideration yields $S(e + f) \le Se + Sf$. The reverse inequality is valid for e and f disjoint. Indeed, let π be a projection with $\pi e = e$ and $\pi^\perp f = f$ and let $\rho := \pi^\perp$. If $\rho_1, \dots, \rho_m$ are pairwise disjoint projections in $\mathfrak{P}(X)$ with the sum equal to unity, then the projections $\pi\pi_k, \rho\rho_l$ $(k := 1, \dots, n;\ l := 1, \dots, m)$ are pairwise disjoint too and their sum is equal to I_X. Moreover, $y = \pi x$ and $z = \rho x$. Thus, the following estimates hold:

$$\sum_{k=1}^{n} |T\pi_k y| + \sum_{l=1}^{m} |T\rho_l z| = \sum_{k=1}^{n} |T\pi\pi_k x| + \sum_{l=1}^{m} |T\rho_l \rho x| \le S(e + f).$$

Passing to the supremum over $y, z, \pi_1, \dots, \pi_n, \rho_1, \dots, \rho_m$, we obtain the desired inequality. According to 4.1.7, the operator S is additive. Extend S onto the entire E as in 4.1.5. Then S is a positive operator, $S \le |T|$, and $|Tx| \le S(|x|)$ $(x \in X)$. Whence it is clear that $S = |T|$. $\triangleright$

4.1.9. *Suppose that (X, E) and (Y, F) are the same as in 4.1.8. An operator $T : X \to Y$ is dominated if and only if, for each $e \in E_+$, the set*

$$\mathscr{U}(e) := \Big\{ \sum_{k=1}^{n} |T \circ \pi_k x| : x \in X,\ |x| \le e,\ \pi_1, \dots, \pi_n \in \mathfrak{P}(X),$$
$$\pi_k \circ \pi_l = 0\ (l \ne k),\ \sum_{k=1}^{n} \pi_k = I_X,\ n \in \mathbb{N} \Big\}$$

is order-bounded. In this case, the exact dominant can be calculated as in 4.1.8.

$\triangleleft$ Proposition 4.1.7 remains valid if we weaken condition 4.1.7 (d) and require that S have a sublinear dominant, i.e., the operator $U : E \to F$ is sublinear and isotonic. If the sets under consideration are order-bounded then the operator $Ue := \sup \mathscr{U}(e^+)$ $(e \in E)$ is sublinear and isotonic on the cone E_+. The operator $S : E_+ \to F_+$ can be defined in the same way as in 4.1.8 if we replace $|T|$ by U in the reasoning presented there. Whence we infer that S is additive. Since $|Tx| \le S(|x|)$ $(x \in X)$, we have $T \in M(X, Y)$ and from 4.1.8 it follows that $S = |T|$. $\triangleright$

4.1.10. We call a linear operator $T : X \to Y$ *subdominated* if there exists an isotonic sublinear operator $P : E \to F$, called a *subdominant* of T, such that $|Tx| \le P(|x|)$ $(x \in X)$. A subdominant P is called *normal* if $Pe = P(e^+)$ $(e \in E)$.

If X is a decomposable space and F is an order complete lattice then, for each subdominated operator T, the least normal subdominant $[T]$ exists and can be calculated by the formula

$$[T]e := \sup\{|Tx| : |x| \le e^+\} \quad (e \in E).$$

Under the same assumptions, an operator T is subdominated if, for each $e \in E_+$, the set $\{|Tx| : |x| \le e\}$ is bounded in F. Subdominated operators are important in their own right; however, we are interested here in the question, when the least subdominant $[T]$ coincides with the exact dominant $|T|$ and, consequently, the latter can be calculated by the simpler formula indicated above.

An operator $T : X \to Y$ is said to *preserve disjointness* if $x_1 \perp x_2$ implies $Tx_1 \perp Tx_2$ for all $x_1, x_2 \in X$.

4.1.11. Theorem. *Let X be decomposable, let F be order complete, and consider an operator $T : X \to Y$. Suppose that one of the following conditions is satisfied:*

(1) *$Y = F$;*

(2) *T is disjointness preserving and E possesses the principal projection property.*

In this case, if T is a subdominated operator then it is dominated and $|T|e = [T]e$ $(e \in E_+)$.

$\lhd$ (1): Taking account of the formula of 4.1.10 for calculating $[T]$ as well as associativity of suprema, we may write

$$[T]e = \sup\{Tx : |x| \le e\} \quad (e \in E_+).$$

If $|x_k| \le e_k$ $(k := 1, 2)$, $x = x_1 + x_2$, and $e = e_1 + e_2$, then $|x| \le e$ and $[T]e \ge Tx = Tx_1 + Tx_2$. Whence, after passing to suprema, we obtain $[T]e \ge [T]e_1 + [Te_2]$. Thus, $[T]$ is additive on the cone E_+ and its extension onto E is a dominant for T.

(2): The operator T is dominated in view of 4.1.9; therefore, it is sufficient to show that $|T|e \le [T]e$ $(e \in E_+)$. To ensure take pairwise disjoint projections $\pi_1, \dots, \pi_n \in \mathfrak{P}(X)$. By hypothesis, $T(\pi_k x) \perp T(\pi_l x)$ $(k \ne l)$; therefore,

$$\sum_{k=1}^{n} |T(\pi_k x)| = \left| T\Big(\sum_{k=1}^{n} \pi_k\Big) x \right|.$$

Using Theorem 4.1.8, we may write

$$|T|e = \sup\{|Tx| : \ x \in X, \ |x| = e\} \le [T]e \quad (e \in E_{0+}). \ \rhd$$

4.2. Decomposability of the Space of Dominated Operators

In this section we establish that the set of dominated operators becomes a BKS under the dominant norm. The main difficulty rests in proving decomposability of the dominant norm.

4.2.1. As we have seen in 4.1.2, in the case of a decomposable lattice-normed space X and an order complete vector lattice F the dominant norm $|T| \in L^{\sim}(E,F)$ is correctly defined for each dominated operator $T : X \to Y$. So, the set $M(X,Y)$ with the mapping $|\cdot| : M(X,Y) \to L^{\sim}$ is a lattice-normed space too. The axioms of a norm are obvious; the triangle inequality ensues from the inclusion $\mathrm{maj}(T_1+T_2) \supset \mathrm{maj}(T_1)+\mathrm{maj}(T_2)$. Decomposability of the dominant norm is nontrivial; it is proven below in 4.2.2–4.2.6.

If X is decomposable and Y is bo-complete, then the lattice-normed space $M(X,Y)$ is bo-complete too.

$\lhd$ The proof follows the standard scheme. Let (T_α) be a *bo*-fundamental net in $M(X,Y)$, i.e., for $\alpha, \beta \ge \gamma$, we have $|T_\alpha - T_\beta| \le S_\gamma$, where the net S_γ decreases and converges to zero in $L^{\sim}(E,F)$. Then, in view of the norm inequality $|T_\alpha x - T_\beta x| \le S_\gamma(|x|)$, the net $(T_\alpha x)$ is also *bo*-fundamental for every $x \in X$. Since Y is norm *bo*-complete, there exists a linear operator $T : X \to Y$ acting by the formula $Tx = bo\text{-}\lim T_\alpha x$. Passage to the limit over α in the above norm inequality gives $|Tx - T_\beta x| \le S_\gamma(|x|)$. Thus, $|Tx| \le |Tx - T_\beta x| + |T_\beta x| \le S_\gamma(|x|) + |T_\beta|(|x|)$, and the operator T is dominated. Next, for $x_1, \dots, x_n \in X$ such that $|x_1| + \dots + |x_n| \le e$, we have

$$\sum_{k=1}^{n} |T_\alpha x_k - T_\beta x_k| \le \sum_{k=1}^{n} S_\gamma(|x_k|) \le S_\gamma(e).$$

Passing to the order limit over α and taking the supremum of the result over all finite families $\{x_1, \dots, x_n\}$ of the above-considered form yields $|T - T_\beta| \le S_\gamma$ for all $\beta \ge \gamma$, i.e., $bo\text{-}\lim T_\alpha = T$. $\rhd$

4.2.2. Theorem. *Assume that X is decomposable and Y is bo-complete. Let A be an arbitrary order ideal in E. For every $T \in M(X,Y)$ there exists a unique operator $\pi_A T \in M(X,Y)$ such that $|\pi_A T| = \pi_A |T|$ and $|T - \pi_A T| = |T| - \pi_A |T|$.*

$\lhd$ Denote $\Phi := |T|$. We call a net $(x_\alpha)_{0 \le \alpha \in \mathrm{A}} \subset X$ *shearing* for an $x \in X$ (*with respect to* A) if $o\text{-}\lim \pi_A \Phi(|x - x_\alpha|) = 0$ and $o\text{-}\lim \pi_A^{\perp} \Phi(|x_\alpha|) = 0$. For such a net, the limit $bo\text{-}\lim Tx_\alpha$ exists. Indeed, for all $\beta \geq \alpha$ we have

$$|Tx_\beta - Tx_\alpha| \leq \Phi(|x_\beta - x_\alpha|) \leq \pi_A\Phi(|x - x_\beta|)$$
$$+\pi_A\Phi(|x - x_\alpha|) + \pi_A^\perp\Phi(|x_\alpha|) + \pi_A^\perp\Phi(|x_\beta|).$$

Each of the four summands on the right-hand side of the inequality o-converges to zero. Hence, the net (Tx_α) is bo-fundamental and the desired bo-limit exists by bo-completeness of Y. Similarly, we may check that if (y_α) is another shearing net for x then $o\text{-}\lim|Tx_\alpha - Ty_\alpha| = 0$. A shearing net for an arbitrary point $x \in X$ can be constructed as follows: Assign $u_\alpha := |x| \wedge \alpha$ and $v_\alpha := |x| - u_\alpha = (|x| - \alpha)^+$, where $0 \leq \alpha \in \mathrm{A}$. By decomposability of X, there exists a net $(x_\alpha) \subset X$ such that $|x_\alpha| = u_\alpha$ and $|x - x_\alpha| = v_\alpha$ for all α. It is clear that

$$\pi_A\Phi(|x - x_\alpha|) = \pi_A\Phi(v_\alpha) \xrightarrow{(o)} \pi_A\pi_A^\perp\Phi(|x|) = 0,$$
$$\pi_A^\perp\Phi(|x_\alpha|) = \pi_A^\perp\Phi(u_\alpha) \xrightarrow{(o)} \pi_A^\perp\pi_A\Phi(|x|) = 0.$$

So, an operator $\pi_A T : X \to Y$ is well defined by the formula $\pi_A Tx := bo\text{-}\lim Tx_\alpha$, where (x_α) is a shearing net for x. Next, observe that if (y_α) and (z_α) are shearing nets for y and z, respectively, then $(y_\alpha + z_\alpha)$ is a shearing net for $y+z$. Whence, taking into account the definition of the operator $\pi_A T$, we easily infer $\pi_A T(y+z) = \pi_A Ty + \pi_A Tz$. The fact that $\pi_A T$ is dominated is clear from the relation

$$|\pi_A Tx| = o\text{-}\lim|Tx_\alpha| \leq o\text{-}\lim \Phi(|x_\alpha|) = \pi_A\Phi(|x|).$$

So, $\pi_A T \in M(X, Y)$ and $|\pi_A T| \leq \pi_A\Phi$. For the operator $T - \pi_A T$, we may write analogous relations:

$$|(T - \pi_A T)x| = o\text{-}\lim|Tx - Tx_\alpha| \leq o\text{-}\lim \Phi(|x - x_\alpha|) = \pi_A^\perp\Phi(|x|).$$

Thus, $|T - \pi_A T| \leq \pi_A^\perp\Phi$. Taking account of the inequalities obtained, we derive

$$|T| \leq |\pi_A T| + |T - \pi_A T| \leq \pi_A\Phi + \pi_A^\perp\Phi = \Phi = |T|;$$

consequently, $|\pi_A T| = \pi_A\Phi$ and $|T - \pi_A T| = \pi_A^\perp\Phi$.

To prove uniqueness, assume that there is an operator T' with the same properties as $\pi_A T$. Then

$$\pi_A|T' - \pi_A T| \leq \pi_A(|T - \pi_A T| + |T - T'|) = \pi_A(2\pi_A^\perp|T|) = 0;$$
$$\pi_A^\perp|T' - \pi_A T| \leq \pi_A^\perp(|\pi_A T| + |T'|) = \pi_A^\perp(2\pi_A|T|) = 0,$$

whence we easily infer $|T' - \pi_A T| = 0$ or $T' = \pi_A T$. $\triangleright$

4.2.3. Several comments are in order in connection with 4.2.2.

(1) *The mapping* $\pi_A : T \to \pi_A T$ *is a linear projection in* $M(X,Y)$.

$\lhd$ We know that π_A is a band projection in $L^{\sim}$ (see 3.1.3 (3)). Therefore, by replacing T with $\pi_A T$ in $|T - \pi_A T| = |T| - \pi_A|T|$ we infer $|\pi_A T - \pi_A^2 T| = |\pi_A T| - \pi_A^2|T| = 0$. $\rhd$

(2) *We put* $\pi_e T := \pi_A T$ *if* A *is the principal ideal generated by* $e \in E_+$. *In this case the projection* π_e *in* $M(X,Y)$ *can be defined using shearing sequences instead of nets.*

$\lhd$ Indeed, we may construct a shearing sequence for an arbitrary point $x \in X$ in a similar way. Assign $u_n := |x| \wedge (ne)$ and $v_n := |x| - u_n = (|x| - ne)^+$. By decomposability of X, there exists a sequence $(x_n) \subset X$ such that $|x_n| = u_n$ and $|x - x_n| = v_n$ $(n \in \mathbb{N})$. Exactly in the same manner as in 4.2.2 we prove that (x_n) is a shearing sequence for x and all shearing sequences have the same order limit. So, the operator $\pi_e T \in M(X,Y)$ is well defined by the formula $\pi_e Tx := bo\text{-}\lim Tx_n$, where (x_n) is a shearing sequence for x. $\rhd$

(3) It is also useful to extract the next assertion from the proof presented in 4.2.2: For each $x \in X$, the following hold:

$$\begin{aligned}(\pi_e T)x &= bo\text{-}\lim\{Tx_n : \ |x - x_n| = |x| - |x_n|, \ |x_n| = |x| \wedge (ne)\},\\ \pi_A Tx &= bo\text{-}\lim\{Tx_\alpha : \ |x_\alpha| = \alpha, \ |x - x_\alpha| = |x| - \alpha, \ 0 \leq \alpha \in \mathrm{A}\}.\end{aligned}$$

(4) *Suppose that* A *and* B *are order ideals in* E *with* $A \cap B = \{0\}$ *and* $C = A + B$. *Then* $\pi_C T := \pi_A T + \pi_B T$.

$\lhd$ Let $(x_c)_{0 \leq c \in C}$ be a shearing net for x with respect to C satisfying the conditions $|x_c| = c$, $|x - x_c| = |x| - c$, $0 \leq c \in C$. Since $c = a + b$, where $a \in A$, $b \in B$, and $a \perp b$, in view of the decomposability assumption, there is a representation $x_c = x_a + x_b$ with $|x_a| = a$, $|x - x_a| = |x| - a$, $|x_b| = b$, $|x - x_b| = |x| - b$. Thus $(x_a)_{0 \leq a \in A}$ and $(x_b)_{0 \leq b \in B}$ are shearing nets for x with respect to A and B. According to (3) we conclude that $\pi_C Tx := bo\text{-}\lim Tx_c = bo\text{-}\lim Tx_a + bo\text{-}\lim Tx_b = \pi_A Tx + \pi_B Tx$. $\rhd$

4.2.4. (1) *Consider a net of dominated operators* (T_α) *such that for some* $T_0 \in M(X,Y)$ *we have* $|T_\alpha| \wedge |T_0 - T_\alpha| = 0$ *for all* α *and there exists* $S := o\text{-}\lim |T_\alpha|$. *Then* $S \in \mathfrak{E}(|T_0|)$ *and the equality* $T := bo\text{-}\lim T_\alpha$ *correctly define a dominated operator* $T : X \to Y$ *with* $|T| = S$.

$\lhd$ By assumption $T_\alpha \perp (T_0 - T_\alpha)$; therefore, $|T_0 - T_\alpha| = |T_0| - |T_\alpha|$ in view of 2.1.2 (2). Thus, $(|T_\alpha|) \subset \mathfrak{E}(|T_0|)$ and $S \in \mathfrak{E}(|T_0|)$. Denote $\Phi := |T_0|$ for simplicity. Since $|T_\alpha - T_\beta| \leq 2\Phi$, we obtain the following chain:

$$\begin{aligned}|T_\alpha - T_\beta| &= (2\Phi - 2S) \wedge |T_\alpha - T_\beta| + (2S) \wedge |T_\alpha - T_\beta|\\ &\leq (2\Phi - 2S) \wedge (|T_\alpha| + |T_\beta|) + (2S) \wedge (|T_0 - T_\alpha| + |T_0 - T_\beta|)\\ &= (2\Phi - 2S) \wedge (|T_\alpha| + |T_\beta|) + (2S) \wedge (2|T_0| - |T_\alpha| - |T_\beta|).\end{aligned}$$

Whence it is clear that the net T_α is fundamental. According to 4.2.1, there exists an operator $T := bo\text{-}\lim T_\alpha$. This T is a linear operator. Moreover, the following hold:

$$|Tx| = o\text{-}\lim|T_\alpha x| \le o\text{-}\lim|T_\alpha|(|x|) \le o\text{-}\lim|T_\alpha|(|x|) = S(|x|).$$

Consequently, $|T| \le S$. Similarly, $|T_0 - T| \le \Phi - S$; therefore, as in 4.2.2, we infer $|T| = S$ and $|T_0 - T| = \Phi - S$. $\triangleright$

(2) *Let (A_α) be an increasing family of order ideals in E and $A := \bigcup A_\alpha$. Then for an arbitrary dominated operator $T : X \to Y$ the equality holds $\pi_A T = bo\text{-}\lim \pi_{A_\alpha} T$.*

$\triangleleft$ This is immediate from (1) and 3.1.4 (4). $\triangleright$

4.2.5. *The dominant norm $|\cdot| : M(X,Y) \to L^\sim(E,F)$ is disjointly-decomposable when Y is a Banach–Kantorovich space.*

$\triangleleft$ Take pairwise disjoint projections $\sigma_1, \dots, \sigma_n \in \mathfrak{P}(F)$ and positive elements $e_1, \dots, e_n \in E$. Assign

$$\sigma := \bigvee_{k=1}^{n} \sigma_k \pi_k, \quad \sigma T := \sigma_1 \circ (\pi_1 T) + \dots + \sigma_n \circ (\pi_n T),$$

where $\pi_k := \pi_{e_k}$ $(k := 1, \dots, n)$. Then $\sigma^\perp = \sigma_1 \pi_1^\perp + \dots + \sigma_n \pi_n^\perp$ and according to what was proven in Theorem 4.2.2, we have

$$|\sigma T| = \sigma_1 |\pi_1 T| + \dots + \sigma_n |\pi_n T| = \sigma_1 \pi_1 |T| + \dots + \sigma_1 \pi_1 |T| = \sigma |T|,$$
$$|\sigma^\perp T| = \sigma_1 |\pi_1^\perp T| + \dots + \sigma_n |\pi_n^\perp T| = \sigma_1 \pi_1^\perp |T| + \dots + \sigma_n \pi_n^\perp |T| = \sigma^\perp |T|.$$

Next, take $\sigma \in \mathscr{S}(|T|)^\downarrow$ (see 3.5.9). There exists a decreasing net (σ_α) of projections in $\mathscr{S}(|T|)$ such that $\sigma = \inf \sigma_\alpha$. For each α, an operator $\sigma_\alpha T \in M(X,Y)$ is defined, moreover, $|\sigma_\alpha T| = \sigma_\alpha |T|$ and $|\sigma_\alpha^\perp T| = \sigma_\alpha^\perp |T|$. According to 4.2.4 (1) there exists a dominated operator $\sigma T := bo\text{-}\lim \sigma_\alpha T$; in addition, $|\sigma T| = \sigma |T|$, $|\sigma^\perp T| = \sigma^\perp |T|$. By the same reasoning, we may establish the latter equalities for the case in which $\sigma \in \mathscr{S}(T)^{\downarrow\uparrow}$. Thus, for arbitrary fragments $\Phi_1 := \sigma|T|$ and $\Phi_2 := \sigma^\perp |T|$ of the dominant norm, we have $T = T_1 + T_2$ and $|T_k| = \Phi_k$ $(k = 1, 2)$ whenever $T_1 = \sigma T$ and $T_2 = \sigma^\perp T$. The application of 3.5.10 (2) completes the proof. $\triangleright$

4.2.6. Theorem. *Let X be a decomposable lattice-normed space and let Y be a BKS. Then the space of dominated operators $M(X,Y)$ is a BKS.*

$\triangleleft$ The proof is contained in 4.2.2–4.2.5, since any *bo*-complete *d*-decomposable lattice-normed space is decomposable. The latter is an easy consequence of 1.1.3 and the Freudenthal Spectral Theorem. $\triangleright$

4.2.7. So, under the hypotheses of Theorem 4.2.6, to each dominated operator $T : X \to Y$ and each representation $|T| = S_1 + S_2$, where $0 \leq S_1, S_2 : E \to F$, there exist dominated operators $T_1, T_2 : X \to Y$ such that $T = T_1 + T_2$ and $|T_k| = S_k$ $(k := 1, 2)$. If the operators S_1 and S_2 are disjoint then there exists a unique pair of operators T_1 and T_2 satisfying the condition under consideration. This fact follows from the general properties of a decomposable norm, see 2.1.2 (3). Thus, to every fragment S of the operator T there corresponds a unique operator $U \in M(X, Y)$ such that $|U| = S$. The operator U is called a *fragment* of T. The set $\mathfrak{E}(T)$ of all fragments of T has the following form:

$$\mathfrak{E}(T) = \{U \in M(X, Y) : \ |U| \wedge |T - U| = 0\}.$$

The above implies that the sets $\mathfrak{E}(T)$ and $\mathfrak{E}(|T|)$ are bijective; hence, the structure of a complete Boolean algebra can be transferred to $\mathfrak{E}(T)$. In the same way as in 3.5.9, we may introduce the set $\mathscr{S}(T)$; moreover, the equality $\mathfrak{E}(T) = \mathscr{S}(T)^{\downarrow\uparrow}$ holds.

4.2.8. In the remaining part of this section we briefly consider the decomposability phenomena for vector-valued measures. Let $\mathscr{A}$ be an arbitrary Boolean algebra. A mapping μ defined on $\mathscr{A}$ and acting into an arbitrary vector space Z is called a (*finitely additive vector*) *measure* if $\mu(a_1 \vee a_2) = \mu(a_1) + \mu(a_2)$ for every pair of disjoint members $a_1, a_2 \in \mathscr{A}$. Let $Z = F$ be a vector lattice. A measure $\mu : \mathscr{A} \to F$ is called *bounded* whenever $\mu(\mathscr{A})$ is an order-bounded set in F. If, in addition, $\mu(a) \geq 0$ for all $a \in \mathscr{A}$ then μ is said to be *positive*. We denote by $\mathrm{ba}(\mathscr{A}, F)$ $\big(\mathrm{ba}_+(\mathscr{A}, F)\big)$ the space of all bounded (positive) vector measures. It is obvious that the space $\mathrm{ba}(\mathscr{A}, F)$, ordered by the cone of positive measures $\mathrm{ba}_+(\mathscr{A}, F)$, is an ordered vector space.

Suppose that Y is a lattice-normed space over a vector lattice F and take a vector measure $\mu : \mathscr{A} \to Y$. If $|\mu(a)| \leq \nu(a)$ $a \in \mathscr{A}$ for some positive vector measure $\nu : \mathscr{A} \to F$, then μ is termed a *dominated vector measure* and the measure ν is a *dominant* of μ. Denote by $\mathrm{da}(\mathscr{A}, Y)$ the vector space of all dominated vector measures defined on $\mathscr{A}$ and acting into Y. If there is a least element in the set $\mathrm{maj}\,(\mu)$ of all dominants of μ (with respect to the order induced from $\mathrm{ba}(\mathscr{A}, F)$) then it is called the *least* or *exact dominant* of μ and is denoted by $|\mu|$.

Let $S(\mathscr{A})$ denote the set of all simple functions defined on the Stone space $Q := \mathscr{S}(\mathscr{A})$. More precisely, $S(\mathscr{A})$ consists of scalar-valued functions $x : Q \to \mathbb{R}$ of the form $x = \sum_{k=1}^{n} \lambda_k \chi_{\hat{a_k}}$ where $a_1, \dots, a_n$ are pairwise disjoint members of $\mathscr{A}$ and $\hat{a_k}$ is the clopen subset of Q corresponding to a_k under the Stone transform. It is evident that $S(\mathscr{A})$ is a vector subspace and sublattice in $\mathbb{R}^Q$. Now we may easily check that the formula $I_\mu(x) := \sum_{k=1}^{n} \lambda_k \mu(a_k)$ defines correctly a linear map I_μ from the space of simple functions $S(\mathscr{A})$ into Y. Let $C(\mathscr{A})$ denote the closure of $S(\mathscr{A})$ in $\mathbb{R}^Q$ with respect to uniform convergence. Then $C(\mathscr{A})$ is also a vector lattice and I_μ has a unique extension to $C(\mathscr{A})$ which is denoted by the same symbol.

The mapping $\mu \mapsto I_\mu$ is a one-to-one linear correspondence between the vector spaces $M(S(\mathscr{A}), Y)$ (or $M(C(\mathscr{A}), Y)$) and $\operatorname{da}(\mathscr{A}, Y)$. Moreover, for every positive measure $\nu : \mathscr{A} \to F$ we have

$$\nu \in \operatorname{maj}(\mu) \Longleftrightarrow I_\nu \in \operatorname{maj}(I_\mu).$$

4.2.9. This correspondence between measures and operators can be used in revealing some important properties of the space of dominated operators. Now we assume that F is order complete. According to 4.1.2 $M(S(\mathscr{A}), Y)$ is a lattice-normed space, since $S(\mathscr{A})$ is a vector lattice. Moreover, $M(S(\mathscr{A}), Y)$ is a Banach–Kantorovich space if so is Y, see 4.2.6. Taking into consideration the correspondence $\mu \leftrightarrow I_\mu$ we may infer various consequences concerning the space of dominated measures $\operatorname{da}(\mathscr{A}, Y)$.

(1) *Each dominated vector measure $\mu \in \operatorname{da}(\mathscr{A}, Y)$ has the least dominant $|\mu|$ which is calculated by the formula:*

$$|\mu|(a) = \sup\left\{\sum_{k=1}^{n} |\mu(a_k)| : a_1, \dots, a_n \in \mathscr{A},\ a_k \wedge a_l = 0\,(k \neq l),\ \bigvee_{k=1}^{n} a_k = a\right\}.$$

◁ According to 4.2.8 $\nu = \inf \operatorname{maj}(\mu)$ if and only if $I_\nu = \inf \operatorname{maj}(I_\mu)$. It follows that $|I_\mu| = I_{|\mu|}$ and it remains to apply 4.1.8. ▷

(2) *A vector measure $\nu : \mathscr{A} \to F$ is dominated if and only if ν is bounded; thus $\operatorname{ba}(\mathscr{A}, F) = \operatorname{da}(\mathscr{A}, F)$. Moreover, $\operatorname{ba}(\mathscr{A}, F)$ is an order complete vector lattice and, for all $\nu, \nu_1, \nu_2 \in \operatorname{ba}(\mathscr{A}, F)$ and every $a \in \mathscr{A}$, the following hold:*

$$\begin{aligned}
(\nu_1 \vee \nu_2)(a) &= \sup\{\nu_1(a_1) + \nu_2(a_2) : a_1 \vee a_2 = a,\ a_1 \wedge a_2 = 0\};\\
(\nu_1 \wedge \nu_2)(a) &= \inf\{\nu_1(a_1) + \nu_2(a_2) : a_1 \vee a_2 = a,\ a_1 \wedge a_2 = 0\};\\
\nu^+(a) &= \sup\{\nu(a') : a' \leq a\};\\
\nu^-(a) &= -\inf\{\nu(a') : a' \leq a\};\\
|\nu|(a) &= \sup\{|\nu(a')| : a' \leq a\}.
\end{aligned}$$

◁ It is easily seen that the correspondence $\nu \leftrightarrow I_\nu$ implements an order isomorphism between $\operatorname{ba}(\mathscr{A}, F)$ and $M(S(\mathscr{A}), F) = L^\sim(S(\mathscr{A}), F)$. Application of 4.1.11 (1) and 3.1.2 completes the proof. ▷

(3) *The space $\operatorname{da}(\mathscr{A}, Y)$ is a lattice-normed space over $\operatorname{ba}(\mathscr{A}, F)$. In addition, $\operatorname{da}(\mathscr{A}, Y)$ is a Banach–Kantorovich space if so is Y.*

◁ Since the correspondence $\mu \leftrightarrow I_\mu$ is a linear isometry between $\operatorname{da}(\mathscr{A}, Y)$ and $M(S(\mathscr{A}), Y)$, appealing to 4.2.6, end the proof. ▷

It follows from (3) that the space of dominated vector measures is decomposable. In particular, the Lebesgue-type decomposition theorem is valid for dominated measures.

(4) *Let $\mu : \mathscr{A} \to Y$ be a dominated measure and let $\nu : \mathscr{A} \to F$ be a positive measure. Then there exist a unique pair of dominated measures μ_1 and μ_2 in $\mathrm{da}(\mathscr{A}, Y)$ such that*

$$|\mu_1| \in \lambda^{\perp\perp}, \quad |\mu_2| \perp \lambda,$$
$$\mu = \mu_1 + \mu_2, \quad |\mu| = |\mu_1| + |\mu_2|.$$

$\lhd$ Let π be the projection in $\mathrm{ba}(\mathscr{A}, F)$ onto the band $\{\lambda\}^{\perp\perp}$. According to (3) $\mathrm{da}(\mathscr{A}, Y)$ is decomposable. Thus, in view of 2.1.2 (4), we may put $\mu_1 := \widetilde{\pi}\mu$ and $\mu_1 := \widetilde{\pi}^{\perp}\mu$ where $\widetilde{\pi} := h(\pi)$ $\rhd$

4.3. Order Continuous Operators

Here, we establish a simple but rather useful fact: a dominated operator is order continuous if and only if it has an order continuous dominant. Next, we consider a result on extension of a dominated operator. We also give new formulas for calculating an exact dominant in the case of an order continuous operator, and a criterion for order continuity.

4.3.1. An operator $T : X \to Y$ is *norm order continuous* or *bo-continuous* if for every net $(x_\alpha) \subset X$, from the equality $bo\text{-}\lim_\alpha x_\alpha = 0$ it follows that $bo\text{-}\lim_\alpha Tx_\alpha = 0$. The set of all dominated bo-continuous operators is denoted by $M_n(X, Y)$. So, the containment $T \in M_n(X, Y)$ means that $T \in M(X, Y)$ and from $|x_\alpha| \xrightarrow{(o)} 0$ it follows that $|Tx_\alpha| \xrightarrow{(o)} 0$. By considering only bo-convergent sequences, we introduce the notion of a *sequentially bo-continuous* or *σbo-continuous* operator; the set of all these operators is denoted by $M_{\sigma n}(X, Y)$. The results below have sequential versions for operators in $M_{\sigma n}$; however, we restrict exposition to considering o-continuous operators.

4.3.2. Theorem. *Let X be decomposable and let F be order complete. Then a dominated operator $T : X \to Y$ is bo-continuous if and only if its exact dominant is o-continuous; i.e.,*

$$T \in M_n(X, Y) \iff |T| \in L_n(E, F).$$

$\lhd$ Suppose that $T \in M_n(X, Y)$. Take an $e \in E_{0+}$ and an increasing net $(e_\alpha) \subset E_{0+}$, $e := \sup e_\alpha$. Assign

$$f := \sup_\alpha \sup \left\{ \sum_{k=1}^{n} |Tx_k| : \ x_1, \dots, x_n \in X, \ \sum_{k=1}^{n} |x_k| = e_\alpha, \ n \in \mathbb{N} \right\}.$$

According to 4.1.5, we have $f = \sup_\alpha|T|e_\alpha \le |T|e$. Demonstrate that $|T|e \le f$. To this end, consider a finite family $\{x_1, \dots, x_n\} \subset X$ with the property $|x_1| + \dots + |x_n| = e$. Given α, we associate with each $k := 1, \dots, n$ a representation $x_k := u_{k,\alpha} + v_{k,\alpha}$ such that $|x_k| = |u_{k,\alpha}| + |v_{k,\alpha}|$, $\sum_{k=1}^n |u_{k,\alpha}| = e_\alpha$, and $\sum_{k=1}^n |v_{k,\alpha}| = e - e_\alpha$. Since $e - e_\alpha$ decreases and o-converges to zero, we have $|x_k - u_{k,\alpha}| = |v_{k,\alpha}| \xrightarrow{(o)} 0$ for each $k := 1, \dots, n$. Thus,

$$\sum_{k=1}^n |Tx_k| = o\text{-}\lim_\alpha \left(\sum_{k=1}^n |Tu_{k,\alpha}| \right).$$

At the same time, given β, we have

$$\sum_{k=1}^n |Tu_{k,\beta}| \le \sup \left\{ \sum_{k=1}^n |Tu_k| : u_1, \dots, u_n \in X,\ \sum_{k=1}^n |u_k| = e_\beta,\ n \in \mathbb{N} \right\}$$
$$= |T|e_\beta \le \sup_\alpha |T|e_\alpha = f.$$

Passage to the o-limit over β in the last relation yields $\sum_{k=1}^n |Tx_k| \le f$. Finally, taking the supremum over $\{x_1, \dots, x_n\}$, we obtain $|T|e \le f$. Whence it follows that the operator $|T|$ is o-continuous on the ideal $E_0 := E_{0+} - E_{0+}$ (see 4.1.4). In view of 4.1.5, $|T|$ coincides on the entire E with its minimal extension from the ideal E_0. However, it is easy to show that the minimal extension of an o-continuous operator is o-continuous, see 3.1.3 (2). The converse is obvious. ▷

4.3.3. Theorem. *Suppose that (X, E) is a decomposable LNS and $(\tilde{X}, \tilde{E})$ is its bo-completion. Let (Y, F) be a bo-complete LNS. For an arbitrary norm order continuous dominated operator $T : X \to Y$, there exists a unique norm order continuous extension $\tilde{T}$ onto $\tilde{X}$ with the exact dominant preserved. The latter means that $\tilde{T} \in M(\tilde{X}, Y)$ and $|\tilde{T}|$ is an extension of the operator $|T| : E \to F$ onto $\tilde{E}$ by o-continuity.*

◁ Suppose that $T \in M_n(X, Y)$. By Theorem 4.3.2, the operator $|T| \in L_n(E, F)$ is order continuous; hence, it admits a unique o-continuous extension $0 \le \Phi \in L_n(\tilde{E}, F)$, see 3.2.3 (3). We will show existence of an extension $\tilde{T}$ such that $|\tilde{T}| = \Phi$.

First, define the operator $\tilde{T}$ at the elements of the form πx, where $x \in X$ and $\pi \in \mathfrak{P}(\tilde{X})$. Assign $A := \{e \in E_+ :\ \pi e = e,\ e \le |x|\}$. For $e \in A$, there exists an element $x_e \in X$ such that $|x_e| = e$ and $|x - x_e| = |x| - e$. The net $(x_e)_{e \in A}$ bo-converges to πx, since $\pi x_e = x_e$ and

$$|\pi x - x_e| = \pi|x| - e \xrightarrow{(o)} 0.$$

By bo-continuity of the operator T, the net $(Tx_e)_{e\in A}$ is bo-fundamental. Taking bo-completeness of Y into account, we may define $\tilde{T}(\pi x) := bo\text{-}\lim_e Tx_e$. Passage to the limit in the inequality $|Tx_e| \le |T|(|x_e|)$ yields $|\tilde{T}(\pi x)| \le \Phi(|\pi x|)$. Now, let $E_0 \subset \tilde{E}$ and $X_0 \subset \tilde{X}$ consist respectively of the elements of the form $\sum_{k=1}^n \pi_k e_k$ and $\sum_{k=1}^n \pi_k x_k$, where $e_k \in E$, $x_k \in X$, and $\pi_k \in \mathfrak{P}(\tilde{X})$. Then (X_0, E_0) is an LNS with decomposable norm, moreover, E_0 is a sublattice with the projection property. We can define $\tilde{T}$ on X_0 by the equality

$$\tilde{T}\left(\sum_{k=1}^n \pi_k x_k\right) = \sum_{k=1}^n \tilde{T}(\pi_k x_k).$$

We obtain an operator $\tilde{T} \in M(X_0, Y)$ with dominant Φ.

Next, we use Proposition 2.2.11 (2), according to which X_0 is o-dense in $\tilde{X}$. Any element $x \in \tilde{X}$ has the form $x = bo\text{-}\lim x_\alpha$ for some net $(x_\alpha)_{\alpha\in\mathrm{A}}$ in X_0. Repeating the reasoning presented above, we may define an operator $\tilde{T} \in M(\tilde{X}, Y)$ by letting $\tilde{T}x = bo\text{-}\lim \tilde{T}x_\alpha$. In this case, $|\tilde{T}x| \le \Phi(|x|)$ $(x \in \tilde{X})$. In particular, $|\tilde{T}| \le \Phi$ and the operator $\tilde{T}$ is bo-continuous. In view of 4.3.2, $|\tilde{T}|$ is a bo-continuous operator too. It is clear that the restriction of $|\tilde{T}|$ onto E is greater or equal to $|T|$. So, the o-continuous operators $|\tilde{T}|$ and Φ coincide on E; consequently, $\Phi = |\tilde{T}|$. ▷

4.3.4. Theorem. *Suppose that X is decomposable, F is order complete, and an operator $T \in M(X, Y)$ is norm order continuous. Then the exact dominant $|T|$ is calculated by the following formula:*

$$|T|e = \sup\left\{\sum_{k=1}^n |Tx_k| : \ x_k \perp x_l \ (k \ne l), \ |x_k| \le e \quad (k := 1, \dots, n), \right.$$
$$\left. x_1, \dots, x_n \in X, \ n \in \mathbb{N}\right\} \quad (e \in E_+).$$

◁ Denote by Se the right-hand side of the desired equality. It is obvious that the operator $S : E_+ \to F_+$ is positively homogeneous and isotonic. Arguments similar to 4.1.5 and 4.1.8 show that S is subadditive and orthogonal additive. It is also clear that $Se \le |T|e$ $(e \in E_+)$. Next, according to 4.3.2, the operator $|T|$ is order continuous. If a net (e_α) increases and o-converges to e then, due to the fact that S is isotonic and subadditive, we obtain

$$0 \le Se - Se_\alpha \le S(e - e_\alpha) \le |T| \, (e - e_\alpha) \to 0.$$

Whence it follows order continuity of S. Let $\tilde{E}$ be the o-completion of E. For $\tilde{e} \in \tilde{E}_+$, assign $\tilde{S}\tilde{e} := \sup\{Se : e \in E_+, \, e \le \tilde{e}\}$. By o-continuity of S, it is also true that

$$\tilde{S}\tilde{e} = \inf\{Se : \ e \in E_+, \ e \ge \tilde{e}\} \quad (\tilde{e} \in \tilde{E}_+).$$

From these formulas it follows that the operator $\tilde{S} : \tilde{E_+} \to F$ satisfies all conditions of 4.1.7, hence, is additive. Then the operator S is additive too and, therefore, $Se = |T|e$ $(e \in E_+)$. $\triangleright$

4.3.5. We point out three more formulas for calculating the exact dominant of a *bo*-continuous operator.

(1) *Under the hypotheses of Theorem* 4.3.4, *the following formula is valid:*

$$|\tilde{T}|e = \sup\Big\{\sum_{k=1}^{n} |Tx_k| : x_k \perp x_l \ (k \neq l),\ x_k \in X,$$
$$|x_k| \le e \ (k := 1, \dots, n),\ n \in \mathbb{N}\Big\} \quad (e \in \tilde{E_+}).$$

$\triangleleft$ The claim follows from 4.3.3 and 4.3.4. $\triangleright$

(2) *Let X be a Banach space and let $\tilde{E}$ be the o-completion of the vector lattice E. If $T \in M_n(\tilde{E}(X), Y)$ then*

$$|T|\tilde{e} = \sup\Big\{\sum_{k=1}^{n} |T(x_k \otimes e_k)| : \ e_k \perp e_l \ (k \neq l),\ e_k \in E_+,\ e_k \le \tilde{e},$$
$$x_k \in X,\ \|x_k\| \le 1 \ (k := 1, \dots, n),\ n \in \mathbb{N}\Big\} \quad (\tilde{e} \in \tilde{E_+}).$$

$\triangleleft$ Let X_0 be the set of elements in $\tilde{E}(X)$ which have the form $u := \sum_{k=1}^{n} x_k \otimes e_k$, where $x_1, \dots, x_n \in X$ are arbitrary and $e_1, \dots, e_n \in E$ are pairwise disjoint. In the representation of an element u, we may assume $\|x_k\| \le 1$, since otherwise we may change $x_k := x_k/\|x_k\|$ and $e_k := \|x_k\|e_k$. It is easy to see that (X_0, E) is a decomposable LNS. Consequently, for $\tilde{e} \in E_+$, the desired formula follows from 4.3.4. For $\tilde{e} \in \tilde{E_+}$, we should use the formula $|T|\tilde{e} = \sup\{|T|e : \ e \in E_+,\ e \le \tilde{e}\}$ which is valid due to *o*-continuity of the operator $|T|$. $\triangleright$

(3) *If the hypotheses of (2) are met and $E = \tilde{E}$, then*

$$|T|e = \sup\Big\{\sum_{k=1}^{n} |T(x_k \otimes e_k)| : \ x_k \in X,\ \|x_k\| \le 1,$$
$$e_k \in E_+ \ (k := 1, \dots, n),\ e_k \perp e_l \ (k \neq l),\ \sum_{k=1}^{n} e_k = e,\ n \in \mathbb{N}\Big\} \quad (e \in E_+).$$

$\triangleleft$ In this situation, we may apply Theorem 4.1.8 to the restriction T_0 of the operator T onto X_0. Thus, we obtain the desired formula for $|T_0|$. Taking account of the fact that $E(X)$ is the *bo*-completion of X_0 (see 2.3.4), we may apply Theorem 4.3.3, according to which $|T| = |T_0|$. $\triangleright$

4.3.6. An operator $T : X \to Y$ is called *completely additive* if, for every o-summable family $(x_\alpha)_{\alpha\in\mathrm{A}}$ of pairwise disjoint elements ($x_\alpha \perp x_\beta$, $\alpha \neq \beta$), the family (Tx_α) is summable too and, moreover,

$$T\Bigl(bo\text{-}\sum_{\alpha\in\mathrm{A}} x_\alpha\Bigr) = bo\text{-}\sum_{\alpha\in\mathrm{A}} Tx_\alpha.$$

Suppose that the hypotheses of Theorem 4.1.8 are met. A dominated operator $T \in M(X,Y)$ is completely additive if and only if so is its exact dominant $|T|$.

$\lhd$ Assume that an operator T is completely additive. Take $x \in X$ and a finite partition of unity $\pi_1, \dots, \pi_n \in \mathfrak{P}(X)$. Assign $e := |x| = \sum_{\alpha\in\mathrm{A}} e_\alpha$, where $e_\alpha \in E_+$ and $e_\alpha \perp e_\beta$ ($\alpha \neq \beta$), and let ρ_α be the projection onto the band $\{e_\alpha\}^{\perp\perp}$. Taking into account additivity of the operator T, we may write

$$\sum_{k=1}^{n} |T(\pi_k x)| = \sum_{k=1}^{n} \Bigl|T\Bigl(\sum_{\alpha\in\mathrm{A}} \rho_\alpha \pi_k x\Bigr)\Bigr| = \sum_{k=1}^{n} \Bigl|\sum_{\alpha\in\mathrm{A}} T(\rho_\alpha \pi_k x)\Bigr|$$

$$\le \sum_{\alpha\in\mathrm{A}} \sum_{k=1}^{n} |T(\pi_k \rho_\alpha x)| \le \sum_{\alpha\in\mathrm{A}} |T|(\rho_\alpha |x|).$$

Whence, according to Theorem 4.1.8, we infer

$$|T|e \le \sum_{\alpha\in\mathrm{A}} |T|e_\alpha.$$

Since the reverse inequality is obvious, we obtain complete additivity of the operator $|T|$ on the cone E_{0+}. If otherwise $e, e_\alpha \in E_+$ then, for an arbitrary $e' \in E_{0+}$, $e' \le e$, in view of what was proven, we have

$$|T|e' = \sum_{\alpha\in\mathrm{A}} |T|(\rho_\alpha e') \le \sum_{\alpha\in\mathrm{A}} |T|e_\alpha.$$

Hence, after passing to the supremum over e', we again obtain $|T|e \le \sum |T|e_\alpha$. Finally, if $e \in E$ then

$$|T|e = |T|e^+ - |T|e^- = \sum_{\alpha\in\mathrm{A}} |T|\rho_\alpha e^+ - \sum_{\alpha\in\mathrm{A}} |T|\rho_\alpha e^-$$

$$= \sum_{\alpha\in\mathrm{A}} |T|(\rho_\alpha e^+ - \rho_\alpha e^-) = \sum_{\alpha\in\mathrm{A}} |T|(\rho_\alpha e).$$

So, if T is completely additive then so is $|T|$. The converse is obvious. $\rhd$

4.3.7. Theorem. *Let E be a vector lattice with the principal projection property, let X be decomposable, and let F be order complete. For a dominated operator $T \in M(X, Y)$, the following are equivalent:*

(1) *T is bo-continuous;*

(2) *T is completely additive.*

$\triangleleft$ The implication (1) $\Rightarrow$ (2) is obvious. We will prove (2) $\Rightarrow$ (1). It is sufficient to show that $|T|$ is an order continuous operator. According to 4.3.6, the operator $|T|$ is completely additive. Take a decreasing net $(e_\alpha)_{\alpha \in \mathrm{A}}$ in E which o-converges to zero. Fix an index $\alpha_0 \in A$ and assign $e_0 := e_{\alpha_0}$. For an arbitrary number $\varepsilon > 0$, there exists a partition of unity $(\rho_\alpha)_{\alpha \geq \alpha_0}$ in $\mathfrak{P}(E)$ such that $\rho_\alpha e_\alpha \leq \varepsilon e_0$ $(\alpha \geq \alpha_0)$. Assign $f := \inf_{\alpha \in \mathrm{A}} |T| e_\alpha$ and $\pi_\theta := \sum_{\alpha \in \theta} \rho_\alpha$, where θ is a finite subset of $\{\alpha \in \mathrm{A} : \ \alpha \geq \alpha_0\}$. Choose $\beta \in A$ so that $\beta \geq \alpha$ for all $\alpha \in \theta$. Then $\pi_\alpha e_\beta \leq \pi_\alpha e_\alpha \leq \varepsilon e_0$ $(\alpha \in \theta)$; hence,

$$\begin{aligned} f \leq |T| e_\beta &= |T|(\pi_\theta e_\beta) + |T|(I_E - \pi_\theta) e_\beta \\ &\leq \varepsilon |T| e_0 + |T|(I_E - \pi_\theta) e_0. \end{aligned}$$

From complete additivity of the operator $|T|$ it follows that

$$o\text{-}\lim_{\theta} |T|(I_E - \pi_\theta) e_0 = 0;$$

therefore, $f \leq \varepsilon |T| e_0$. Since $\varepsilon > 0$ is arbitrary, $f = 0$. Consequently, the operator $|T|$ is order continuous. $\triangleright$

4.3.8. Let E be a sublattice of a vector lattice G. Say that a net $(x_\alpha) \subset X$ is *G-convergent* to $x \in X$ if there exists a decreasing net $(e_\gamma) \subset E$ such that $\inf_\gamma e_\gamma = 0$ in G and, for every γ, there is an index $\alpha(\gamma)$ such that $|x - x_\alpha| \leq e_\gamma$ for all $\alpha \geq \alpha(\gamma)$. An operator $T : X \to Y$ is called *G-continuous* (*sequentially G-continuous*) if o-$\lim_\alpha Tx_\alpha = 0$ (o-$\lim_n Tx_n = 0$) for every net $(x_\alpha) \subset X$ (sequence $(x_n) \subset X$) G-converging to zero. The set of all dominated (regular) G-continuous operators is denoted by $M_G(X, Y)$ (by $L_G^{\sim}(E, F)$). In the sequential case we use in this notation σG instead of G.

Theorem. *Let X be decomposable and let F be order complete. Then a dominated operator $T : X \to Y$ is G-continuous (sequentially G-continuous) if and only if so is its exact dominant; i.e.,*

$$\begin{aligned} T \in M_G(X, Y) &\Leftrightarrow |T| \in L_G(E, F), \\ T \in M_{\sigma G}(X, Y) &\Leftrightarrow |T| \in L_{\sigma G}(E, F). \end{aligned}$$

$\triangleleft$ The proof proceeds in much the same way as in 4.3.2. $\triangleright$

4.3.9. Consider one more band of $M(X,Y)$. Let $J(E,F)$ be the band of $L^{\sim}(E,F)$ generated by the set of finite-rank o-continuous operators: $J(E,F) := (E_n^{\sim} \otimes F)^{\perp\perp}$. Define

$$M_J(X,Y) := \{T \in M(X,Y) : |T| \in J(E,F)\}.$$

Almost integral operators we call the elements of $M_J(X,Y)$. In order to obtain some internal description for the band M_J, we need the notion of $*$-convergence by P. S. Aleksandrov and P. S. Urysohn. We recall that a sequence $(e_n)_{n\in\mathbb{N}}$ in E is called *$*$-convergent* to $e \in E$ (with respect to o-convergence), $e_n \xrightarrow{(*)} e$ in writing, if every subsequence $(e_{n_k})_{k\in\mathbb{N}}$ of it contains a subsequence $(e_{n_{k_l}})_{l\in\mathbb{N}}$ o-convergent to x. We call an operator $T : X \to Y$ *$*$-o-continuous* if it is bo-continuous and takes every bounded sequence $*$-convergent to zero ($x_n \in X$, $|x_n| \xrightarrow{(*)} 0$) into a sequence o-convergent to zero ($Tx_n \xrightarrow{(o)} 0$). Denote by $M_{*n}(X,Y)$ the set of all $*$-o-continuous dominated operators from X into Y and assign $L_{*n}(E,F) := M_{*n}(E,F)$.

It is easy to see that $M_{*n}(X,Y)$ is a band. Indeed, from the definitions it follows immediately that $M_{*n}(X,Y)$ is an order ideal. Next, if $x_n \xrightarrow{(*)} 0$ and $T_\alpha \xrightarrow{(o)} T$, where $x_n \in X$, $|x_n| \leq e \in E$, and $T_\alpha \in M_{*n}(X,Y)$, then $Tx_n \xrightarrow{(o)} 0$, since $|Tx_n| \leq |T - T_\alpha|(e) + |T_\alpha x_n|$. A K-space F is said to be *regular* if for every sequence (A_n) of subsets $A_n \subset F$ with $o\text{-}\liminf A_n = 0$ there exist finite subsets A'_n such that $o\text{-}\liminf A'_n = 0$.

4.3.10. Theorem. *Let X and Y be BKSs and let the norm lattice F be a regular K-space. A dominated operator from X into Y is almost integral if and only if it is $*$-o-continuous:*

$$M_J(X,Y) = M_{*n}(X,Y).$$

$\lhd$ Every functional $e' \in E_n^{\sim}$ is $*$-o-continuous, therefore, $E_n^{\sim} \otimes F \subset L_{*n}(E,F)$. Since $L_{*n}(E,F)$ is a band, we have $J(E,F) \subset L_{*n}(E,F)$; hence, $M_J(X,Y) \subset M_{*n}(X,Y)$.

Conversely, suppose that the operator $T \in M(X,Y)$ is $*$-o-continuous. By decomposability of the least dominant (see 4.2.6), there exists an operator $U \in M(X,Y)$ such that $S := |U| \perp J(E,F)$, $|T-U| \in J(E,F)$, and $|T| = |U| + |T-U|$. If we prove that $S = 0$ then $U = 0$ and $T \in M_J(X,Y)$. First of all, observe that the operator U is $*$-o-continuous, since T and $T - U$ do. By Theorem 4.3.2, the operator S is o-continuous; consequently, without loss of generality, we may assume that there is a weak order-unity $\mathbf{1} \in F$ and an essentially positive functional $f \in E_n^{\sim}$. Let $e := |x|$ for some $x \in X$. The relation $S \perp f \otimes \mathbf{1}$ implies

$$0 = (S \wedge f \otimes \mathbf{1})e = \inf_{0 \leq e_0 \leq e} \{S(e - e_0) + f(e_0) \cdot \mathbf{1}\}.$$

Assign $A_n := \{S(e - e_0) + f(e_0) \cdot \mathbf{1} : \ 0 \le e_0 \le e, \ f(e_0) \le 1/n\}$. It is clear that $A_{n+1} \subset A_n$ and $\inf(A_n) = 0$ for each $n \in \mathbb{N}$. According to our assumption about F, there exist finite sets $A'_n \subset A_n$ such that $o\text{-}\lim_{n\to\infty} \inf(A'_n) = 0$. There exists a strictly increasing sequence $(k(n))_{n\in\mathbb{N}}$ of naturals and a sequence $(e_m)_{m\in\mathbb{N}}$ in E such that $0 \le e_m \le e$ and $A'_n = \{S(e - e_m) + f(e_m) \cdot \mathbf{1} : \ k(n) \le m < k(n+1)\}$. Since the functional $f \in E_n^{\sim}$ is essentially positive and $f(e_m) \to 0$, we have $e_m \xrightarrow{(*)} 0$. Consider a sequence $(x_m)_{m\in\mathbb{N}}$ in X, for which $|x_m| = e_m$ and $|x - x_m| = e - e_m$ (see 2.1.7 (1)). It is clear that (x_m) $*$-converges to zero; thus, $Ux_m \xrightarrow{(o)} 0$. So, the following equalities hold:

$$o\text{-}\lim_{m\to\infty} |Ux_m| = 0,$$

$$o\text{-}\lim_{n\to\infty} \inf\{S(e - e_m) : \ k(n) \le m < k(n+1)\} = 0.$$

For an arbitrary $\varepsilon > 0$, there exist partitions of unity $(\pi_m)_{m\in\mathbb{N}}$ and $(\rho_n)_{n\in\mathbb{N}}$ in the Boolean algebra $\mathfrak{P}(F)$ such that

$$\pi_m |Ux_k| \le \varepsilon \mathbf{1} \quad (k, m \in \mathbb{N}, \ k \ge m),$$

$$\rho_n \inf\{S(e - e_k) : \ k(p) \le k < k(p+1)\} \le \varepsilon \mathbf{1} \quad (p \ge n).$$

Whence, for $k \ge \max\{k(p), m\}$, $p \ge n$, we infer

$$\pi_m |Ux| \le S(e - e_k) + \pi_m |Ux_k| \le S(e - e_k) + \varepsilon \mathbf{1},$$

$$\rho_n \pi_m |Ux| \le \rho_n \inf\{S(e - e_k) : \ k\ (p) \le k < k(p+1)\} + \varepsilon \mathbf{1} \le 2\varepsilon \mathbf{1}.$$

Summing up over n and m, we obtain $|Ux| \le 2\varepsilon \mathbf{1}$; consequently, $Ux = 0$. Since the element $x \in X$ is arbitrary, we have $U = 0$ and $S = 0$. ▷

4.3.11. *Under the hypotheses of Theorem 4.3.10, for every dominated operator $T : X \to Y$, there is a unique representation of T as the disjoint sum $T = T_i + T_{si}$, where T_i is a $*$-o-continuous operator and T_{si} is a dominated operator that has no $*$-o-continuous fragments.*

◁ The claim follows from 4.2.6 and 4.3.10. ▷

4.4. The Yosida–Hewitt-Type Theorems

We will consider the question about decomposition of a dominated operator into the order continuous and order singular parts. These results are conventionally called theorems of Yosida–Hewitt-type because of the classical fact on decomposition of a finitely additive measure into countably additive and finitely additive parts.

The decomposition in which singularity is understood as disjointness from the set of all order continuous operators is referred to as the *weak form of the Yosida–Hewitt decomposition.* Very often singularity means that an operator vanishes on a "huge" subset of its domain. In this case decomposition into the o-continuous and singular parts is called the *strong form of the Yosida–Hewitt decomposition.*

4.4.1. A dominated operator is called norm *order singular (bo-singular)* or *abnormal* if it is disjoint to each bo-continuous dominated operator. Denote by $M_{os}(X, Y)$ (respectively, $L^{\sim}_{os}(E, F)$ the set of all dominated (regular) bo-singular operator from X into Y from E into F. Then the above definition can be rewritten as follows:

$$T \in M_{os}(X, Y) \iff T \perp M_n(X, Y),$$
$$T \in L^{\sim}_{os}(E, F) \iff S \perp L^{\sim}_n(E, F).$$

Let X be a decomposable lattice-normed space and let Y be a Banach–Kantorovich space. Then, for an operator $T \in M(X, Y)$, the following are equivalent:

(1) *T is norm order singular;*

(2) *$|T|$ is order singular;*

(3) *T has no nonzero bo-continuous fragments.*

◁ The equivalence (2) ⇔ (1) follows from 4.3.2. The implication (1) ⇒ (3) is obvious and so it remains to observe that (3) ⇒ (2). If S is the projection of $|T|$ onto the band $L^{\sim}_n(E, F)$ then, in view of decomposability of the dominant norm (Theorem 4.2.6), there is a fragment T_0 of T such that $|T_0| = S$. Since T has no nonzero bo-continuous fragments, we have $T = 0$ and $S = 0$. ▷

4.4.2. Now the weak form of the Yosida–Hewitt decomposition can be easily deduced from the decomposability of dominant norm.

Theorem. *Let X be a decomposable lattice-normed space and let Y be a Banach–Kantorovich space. Then every operator $T \in M(X, Y)$ has a unique representation in the form $T = T_n + T_{os}$, where $T_n \in M_n(X, Y)$, and $T_{os} \in M_{os}(X, Y)$. Moreover,*

$$|T| = |T_n| + |T_{os}|, \ |T_n| = |T|_n, \ |T_{os}| = |T|_{os}.$$

A decomposition of T with the properties indicated above is unique.

◁ According to 3.2.3 (2) the positive operator $|T|$ has a unique representation in the form $|T| = |T|_n + |T|_{os}$, where $|T|_n \in L^{\sim}_n$ and $|T|_{os} \in L^{\sim}_{os}$. Using Theorem 4.2.6

we may find $T_1, T_2 \in M(X,Y)$ such that $|T_1| = |T|_n$, $|T_2| = |T|_{os}$, and $T = T_1+T_2$. It follows from 4.3.2 and 4.4.1 that $T_1 \in M_n(X,Y)$ and $T_2 \in M_{os}(X,Y)$. Thus, we may put $T_n := T_1$ and $T_{os} := T_2$. Uniqueness follows from 4.2.7. ▷

A similar result is valid for the σ-continuous fragments.

4.4.3. The following formulas for calculating the order continuous and order σ-continuous parts of a dominated operators can be easily deduced from 3.2.4 (2), 3.2.5 (4), 4.2.2, 4.2.4 (1):

(1) $T_n = bo\text{-}\lim_{A\in\mathscr{I}} \pi_A T \quad (\mathscr{I} := \mathscr{I}_d(E))$,

(2) $\pi_{[e]} T_{n\sigma} = bo\text{-}\lim_{A\in\mathscr{I}} \pi_A T \quad (\mathscr{I} := \mathscr{I}_{\sigma d}([e]),\ e \in E_+)$.

The fragments can be also calculated as pointwise limits:

(3) $T_n x = bo\text{-}\lim_{A\in\mathscr{I}} \pi_A T x \quad (\mathscr{I} := \mathscr{I}_d(E),\ x \in X)$,

(4) $T_{n\sigma} x = bo\text{-}\lim_{A\in\mathscr{I}} \pi_A T x \quad (\mathscr{I} := \mathscr{I}_{\sigma d}(\{e\}^{\perp\perp}),\ e \in E_+,\ |x| \in \{e\}^{\perp\perp})$.

The limits are taken over the decreasing net of order-dense ideals $A \subset E$.

Suppose now that E is a vector lattice with the projection property. Introduce two directed sets, Γ and Δ. Let Γ be the set of all partitions of unity in the Boolean algebra $\mathfrak{P}(X)$ which we order by "refinement." More precisely, if γ and γ' are partitions of unity in $\mathfrak{P}(X)$, then $\gamma \le \gamma'$ means that, for every $\pi' \in \gamma'$, there exists a $\pi \in \gamma$ such that $\pi \le \pi'$. Let the set Δ consist of all upward-directed subsets $\delta \subset \mathfrak{P}(X)$ such that $\sup \delta = I_X$. The order in Δ is introduced as above:

$$\delta \le \delta' \ \Leftrightarrow\ (\forall \pi' \in \delta')(\exists \pi \in \delta)\ \pi \le \pi'.$$

4.4.4. Now, consider the following formulas:

(1) $T_n x := bo\text{-}\lim_{\gamma\in\Gamma} bo\text{-}\sum_{\pi\in\gamma} T \circ \pi x \quad (x \in X)$;

(2) $|T_n| e := \inf_{\gamma\in\Gamma} \sum_{\pi\in\gamma} |T| \circ \pi e \quad (e \in E_+)$;

(3) $T_n x := bo\text{-}\lim_{\delta\in\Delta} bo\text{-}\lim_{\pi\in\delta} T \circ \pi x \quad (x \in X)$;

(4) $|T_n| e := \inf_{\delta\in\Delta} \sup_{\pi\in\delta} |T| \circ \pi e \quad (e \in E_+)$.

Theorem. *Let (X,E) be a decomposable LNS, let (Y,F) be a Banach–Kantorovich space, let E have the projection property. Then, for every $T \in M(X,Y)$, each of the formulas (1) and (3) well defines a dominated operator $T_n : X \to Y$. Moreover, $T_n \in M_n(X,Y)$, $T - T_n \in M_{os}(X,Y)$, and, the exact dominant can be calculated by (2) or (4).*

$\lhd$ We confine exposition to considering only the first pair of formulas. The first formula is an easy consequence of the second. Indeed, if (2) is valid then for any partition of unity $\gamma \in \Gamma$ we have

$$f_\gamma := \left| o\text{-}\sum_{\pi\in\gamma} T\circ\pi \right| = bo\text{-}\sum_{\pi\in\gamma} |T|\circ\pi \to |T_n|,$$

since the net $(f_\gamma)_{\gamma\in\Gamma}$ is decreasing and $\inf_{\gamma\in\Gamma} f_\gamma = o\text{-}\lim_{\gamma\in\Gamma} f_\gamma$. Thus, (1) follows from 4.2.4 (1) and 4.2.7. It remains to prove that $S_n e = f := \inf_{\gamma\in\Gamma} o\text{-}\sum_{\pi\in\gamma} S\circ\pi e$ for a positive operator $S : E \to F$ and $e \in E_+$.

Take an order-dense ideal $A \subset E$ of the form

$$A := \sum_{\pi\in\gamma} \pi(E) := \operatorname{lin}\left(\bigcup_{\pi\in\gamma} \pi(E)\right),$$

where $\gamma \in \Gamma$. Then $\pi_A S = o\text{-}\sum_{\pi\in\gamma} S\circ\pi$ in view of 4.2.3 (4) and 4.2.4 (2). By 3.2.4 (2) we may conclude $S_n e \le f$. To prove the reverse inequality, suppose that A is an order-dense ideal and choose a partition of unity $\gamma \in \Gamma$ such that $\pi e \in A$ for all $\pi \in \gamma$. According to 3.1.3 (1) $\pi_A Se \ge \pi_A S\pi e = S\pi e$; therefore, $\pi_A Se \ge f$. Since A is arbitrary, we obtain $Se \ge f$. $\rhd$

A similar result is valid for the σo-continuous fragment T_σ of an operator $T \in M(X,Y)$. Moreover, T_σ can be calculated by the formula

$$T_\sigma = bo\text{-}\lim_{(\pi_k)\in\Gamma_\sigma} \sum_{k=1}^{\infty} T\circ\pi_k x \quad (x \in X),$$

where Γ_σ is the subset of Γ consisting of countable partitions of unity. The operator $T - T_\sigma$ is σo-singular in the sense that it is disjoint from all sequentially bo-continuous dominated operators or, which is equivalent, has no sequentially bo-continuous fragments. The exact dominant satisfies $|T_\sigma| = |T|_\sigma$ and $|T - T_\sigma| = |T| - |T_\sigma|$.

4.4.5. We now present a strong form of the Yosida–Hewitt Theorem for dominated operators. An operator $T : X \to Y$ is called *singular* if there exists an order-dense ideal $X_0 \subset X$ on which T vanishes. From 4.1.5 it follows immediately that a dominated operator T is singular if and only if so is its exact dominant $|T|$. We denote the set of all dominated (regular) operators from X into Y (from E into F) by $M_s(X,Y)$ (respectively, $L_s^\sim(E,F)$). The *null ideal* $\mathscr{N}_T$ of the operator T is introduced by the formula

$$\mathscr{N}_T := \{x \in X : (\forall u \in X)(|u| \le |x| \Rightarrow Tu = 0)\}$$

Let E and F be vector lattices, with F order complete. For a positive operator $S : E \to F$, the following hold:

(1) $S \in L_n^{\sim}(E, F) \Leftrightarrow S \perp L_s^{\sim}(E, F)$;

(2) $S \in L_n^{\sim}(E, F) \Leftrightarrow (\forall 0 \leq U \leq S)(\mathcal{N}_U \in \mathfrak{B}(E))$.

$\lhd$ We have only to verify (2), since (1) was already proven in 3.2.3 (1). If S is order continuous and $0 \leq U \leq S$ then U is also order continuous, so that $\mathcal{N}_U$ is a band. Suppose that the right-hand side of the desired equivalence is valid. Let (g_α) be an increasing net in E with $g = \sup_\alpha g_\alpha$ and let $f := \sup_\alpha Sg_\alpha$. Check that $f = Sg$. To this end, for $0 < \varepsilon < 1$, denote $e_\alpha := (\varepsilon g - g_\alpha)^+$ and $S_\alpha := \pi_{e_\alpha} S$ (see 3.1.3 (3) and 3.1.4 (5) for the definitions). Since (e_α) is decreasing, (S_α) is also decreasing and $U := \inf_\alpha S_\alpha \leq S$. Also $U(\varepsilon g - g_\alpha)^- = 0$ or $(\varepsilon g - g_\alpha)^- \in \mathcal{N}_U$, because $S_\alpha(\varepsilon g - g_\alpha)^- = 0$. Moreover, $o\text{-}\lim(\varepsilon g - g_\alpha)^- = (1 - \varepsilon)g$, and hence, since by our hypothesis $\mathcal{N}_U$ is a band, $g \in \mathcal{N}_U$. Now observe that the following simple relations

$$
0 \leq Se_\alpha = S_\alpha e_\alpha \leq Sg,
$$

$$
0 \leq g - g_\alpha = (1 - \varepsilon)g + (\varepsilon g - g_\alpha) \leq (1 - \varepsilon)g + (\varepsilon g - g_\alpha)^+
$$

imply

$$
0 \leq Sg - f \leq S(g - g_\alpha) \leq (1 - \varepsilon)Sg + Se_\alpha \leq (1 - \varepsilon)Sg + S_\alpha g.
$$

From this, taking into consideration that $o\text{-}\lim S_\alpha g = Ug = 0$, we deduce $0 \leq Sg - f \leq (1 - \varepsilon)Sg$ for all $0 < \varepsilon < 1$. Hence, $f = Sg$ and the result follows. $\rhd$

4.4.6. Theorem. *If the hypotheses of Theorem 4.2.6 are met then, for an operator $T \in M(X, Y)$, the following are equivalent:*

(1) $T \in M_n(X, Y)$;

(2) $T \perp M_s(X, Y)$;

(3) *if* $U \in M(X, Y)$ *and* $|U| \leq |T|$ *then* $\mathcal{N}_U$ *is a band.*

$\lhd$ (1) $\Rightarrow$ (3): If $T \in M_n(X, Y)$ and $|U| \leq |T|$ then, in view of 4.3.2, $|U| \in L_n(E, F)$. According to 4.4.5 (2), $\mathcal{N}_{|U|}$ is a band in E and it remains to take account of the obvious equality (see 2.1.2)

$$
\mathcal{N}_U = h(\mathcal{N}_{|U|}) := \{x \in X : \ |x| \in \mathcal{N}_{|U|}\}.
$$

(3) $\Rightarrow$ (2): Suppose that $|U| \leq |T|$ for some singular $U \in M(X, Y)$. Then $\mathcal{N}_U$ is a band and, at the same time, $X = \mathcal{N}_U^{\perp\perp} = \mathcal{N}_U$. Thus, $U = 0$.

(2) $\Rightarrow$ (1): Suppose that $0 \leq S \leq |T|$ and the operator S is singular. According to 4.2.6, we may choose an operator $U \in M(X, Y)$, for which $|U| = S$. It is clear that $U \perp T$, since U is singular. But then $S \perp |T|$, i.e., $S = 0$. This means that $|T| \perp L_s(E, F)$ and, in view of 4.4.5 (1) and 4.3.2, we obtain $T \in M_n(X, Y)$. $\rhd$

4.4.7. To formulate our results for the strong form of the Yosida–Hewitt decomposition we need some new definitions and additional facts. A vector lattice E has the *Egorov property* if for every $0 \le e \in E$ and for every double sequence $e_{n,k} \in E$, we have that $0 \le e_{n,k} \nearrow_k f$ (for all $n \in \mathbb{N}$) implies the existence of a sequence $0 \le d_m \nearrow f$ with the relation $d_m \le e_{n,j(m,n)}$ holding for m, n arbitrary and $j(m,n)$ suitable.

A vector lattice is said to be *weakly σ-distributive*, if for every $e \in E$ and for every double sequence $(f_{m,n})_{m,n\in\mathbb{N}}$ in E_+ such that if $(f_{m,n})_{n\in\mathbb{N}}$ increases and o-converges to e for each $m \in \mathbb{N}$, there exists a net $(e_\varphi)_{\varphi\in\mathbb{N}^\mathbb{N}}$ with the properties $\sup_\varphi e_\varphi = e$ and $e_\varphi \le e_{m,\varphi(m)}$ for all $m \in \mathbb{N}$. The set $\mathbb{N}^\mathbb{N}$ is considered with the natural pointwise ordering: $\varphi_1 \le \varphi_2 \Leftrightarrow (\forall n \in \mathbb{N})\varphi_1(n) \le \varphi_2(n)$. It is obvious that a vector lattice with the Egorov property is weakly σ-distributive. The converse is false.

We also recall that a vector lattice is said to have the *countable sup property,* if whenever an arbitrary subset D has a supremum, then there exists an at most countable subset C of D with the same supremum.

4.4.8. *Theorem.* *The following are valid:*

(1) *a countable intersection of σ-order-dense ideals in a vector lattice with the Egorov property is also a σ-order-dense ideal;*

(2) *a countable intersection of σ-order-dense ideals in a weakly σ-distributive vector lattice is an order-dense ideal;*

(3) *a countable intersection of order-dense ideals in a vector lattice with both Egorov and countable sup property is also an order-dense ideal.*

$\lhd$ First we prove (1) and (2). Let (G_k) be a countable family of σ-order-dense ideals in a vector lattice E and $G = \bigcap_{k=1}^\infty G_k$. Take a nonzero $e \in E_+$. Each finite intersection of G_k is a σ-order-dense ideal. Thus, for any $n \in \mathbb{N}$ we may choose an increasing sequence $(e_{n,k})$ such that $e = \sup_k e_{n,k}$ and $e_{n,k} \in \bigcap_{l=1}^n G_l$. Using the Egorov property, we may find an increasing sequence $(e_m) \subset E_+$ such that $e = \sup e_m$ and for all m and n there is a suitable index $l(m,n)$ with the property $e_m \le d_{n,l(m,n)}$. This implies that $e_m \in G$ for all m, and G is a σ-order-dense ideal. In the case when E is weakly σ-distributive we may find an increasing net $(e_\varphi) \subset E_+$ such that $e = \sup e_\varphi$ and $e_\varphi \le e_{m,\varphi(m)}$ for all $m \in \mathbb{N}$. This implies that $e_\varphi \in G$ for all $\varphi \in \mathbb{N}^\mathbb{N}$, and G is an order-dense ideal.

Now, to prove (3) it is sufficient to observe that an order-dense ideal in a vector lattice with the countable sup property is σ-order-dense ideal. $\rhd$

4.4.9. Let E and F be vector lattices with F order complete.

(1) *If F has the countable sup property then every essentially positive σ-order continuous operator $T : E \to F$ is order continuous.*

$\lhd$ Assume $0 \le e_\tau \nearrow e \in E$ and set $f = \sup_\tau T(e_\tau)$. By the countable sup property of F, we find an increasing sequence $d_n := e_{\tau(n)}$ such that $f = \sup_n T(d_n)$.

Assume that $d_n \nearrow e$ is false. Then we find $h \in E$ and τ_0 such that $h \ge d_n$ for all n and $g := e_{\tau_0} \vee h - h > 0$. Since $e_{\tau_0} \vee h - e_{\tau_0} \vee d_n \le h - d_n$, we deduce $e_{\tau_0} \vee d_n - d_n \ge g > 0$ and $T(e_{\tau_0} \vee d_n) \ge T(d_n) + T(g)$. It follows that

$$\sup_n(T(d_n) + T(g)) = f + T(g) \le \sup_n T(e_{\tau_0} \vee d_n) \le \sup_\tau T(e_\tau) = f.$$

This is a contradiction since the last formula implies that $T(g) = 0$ for $g > 0$ despite of the strict positivity of T. Thus, we have proven that $d_n \nearrow e$ and, consequently, by σ-order continuity of T, we came to the sought relation $f = T(e)$. $\rhd$

(2) *If E has the Egorov property and F has the countable sup property then for every $e \in E_+$ the infimum is attained in the formula* (see 3.2.5 (1)):

$$\pi_A T_{n\sigma} e = \inf\{\pi_G Te : G \in \mathscr{I}_{\sigma d}(A)\}.$$

$\lhd$ Let A be the ideal $\mathscr{I}(e)$ generated by e. Because of the countable sup property of F, in the formula it suffices to take a countable set of σ-order-dense ideals $G(k) \subset \mathscr{I}(e)$:

$$T_{n\sigma} e = \pi_A T_{n\sigma} e = \inf_{k \in \mathbb{N}} \pi_{G(k)} Te.$$

Put $G := \bigcap_{k=1}^\infty G(k)$. According to 4.4.8 (1) G is also a σ-order-dense ideal. Therefore, $T_{n\sigma} e \le \pi_G Te \le \pi_{G(k)} Te$. By passing to infimum we deduce that $T_{n\sigma} e \le \pi_G Te \le \inf_k \pi_{G(k)} Te = T_{n\sigma} e$, whence $T_{n\sigma} e = \pi_G Te$. $\rhd$

(3) *If E weakly σ-distributive and F has the countable sup property then for every $e \in E_+$ there exists an order-dense ideal $G \in \mathscr{I}_{\sigma d}(A)$ such that $T_{n\sigma} e = \pi_G Te$.*

4.4.10. Theorem. *Let E be a vector lattice with the Egorov property and F be an order complete vector lattice with the countable sup property. Then*

(1) $L^\sim(E, F) = L^\sim_{n\sigma}(E, F) \oplus L^\sim_{s\sigma}(E, F)$;

(2) $L^\sim(E, F) = L^\sim_n(E, F) \oplus L^\sim_s(E, F)$.

$\lhd$ (1): Of course, what we have to prove reads as follows:

$$L^\sim_{n\sigma} = (L^\sim_{n\sigma})^\perp.$$

It suffices to show that

$$(L^\sim_{n\sigma})^\perp \subset L^\sim_{s\sigma},$$

since the reverse inclusion is obvious.

If $0 \le T \in (L^{\sim}_{n\sigma})^{\perp}$, then $T_{n\sigma} = 0$. Thus, for an arbitrary $e \in E_{+}$, in view of 4.4.9 (2), there is some σ-order-dense ideal $G \subset \mathscr{I}(e)$ such that $\pi_G Te = T_{n,\sigma}e = 0$. By definition of the projection $\pi_G T$ we have $G \subset \mathscr{N}_T$. Therefore, $\mathscr{N}_T$ is a σ-order-dense ideal, whence $T \in L^{\sim}_{s\sigma}$.

(2): Repeating the idea of the proof for (1) but using 4.4.9 (3) instead of 4.4.9 (2), we may obtain $(L^{\sim}_{n\sigma})^{\perp} \subset L^{\sim}_{s}$. Using this inclusion and taking into account that $L^{\sim}_{n} \subset L^{\sim}_{n\sigma}$, we have

$$L^{\sim} = L^{\sim}_{n} \oplus (L^{\sim}_{n})^{\perp} = L^{\sim}_{n\sigma} \oplus (L^{\sim}_{n\sigma})^{\perp} = L^{\sim}_{n} \oplus [L^{\sim}_{n\sigma} \cap (L^{\sim}_{n})^{\perp}] \oplus L^{\sim}_{s\sigma}.$$

Therefore, it is sufficient to prove that $L^{\sim}_{n\sigma} \cap (L^{\sim}_{n})^{\perp} \subset L^{\sim}_{s}$, since the inclusion $L^{\sim}_{s\sigma} \subset L^{\sim}_{s}$ is trivial.

Take $0 \le T \in L^{\sim}_{n\sigma} \cap (L^{\sim}_{n})^{\perp}$ denote by T_0 its restriction to $\mathscr{N}_T^{\perp}$. So, we obtain an essentially positive operator $T_0 \in L^{\sim}_{n\sigma}(\mathscr{N}_T^{\perp}, F)$. By virtue of 4.4.9 (1) T_0 is order continuous. Let $S \in L^{\sim}_{n}(E, F)$ be the smallest positive extension of T_0 to E. The inequality $0 \le S \le T \in (L^{\sim}_{n})^{\perp}$ implies $S \in (L^{\sim}_{n})^{\perp}$ and consequently $S = 0$. The latter says that $\mathscr{N}_T^{\perp} = \{0\}$. In other words $\mathscr{N}_T^{\perp}$ is an order-dense ideal and $T \in L^{\sim}_{s}$. $\triangleright$

4.4.11. So, if the hypotheses of the Theorem 2.4.6 are met then $M_n(X, Y) = M_s(X, Y)^{\perp}$, and $M_s(X, Y)$ is an order-dense ideal in $M_n(X, Y)^{\perp}$. In connection with the Yosida–Hewitt Theorem, of interest is the equality

$$M_s(X, Y) = M_n(X, Y)^{\perp} = M_{os}(X, Y)$$

or, which is the same, the representation

$$M(X, Y) = M_n(X, Y) \oplus M_s(X, Y).$$

Theorem. *Let (X, E) be a decomposable LNS, let E possess the Egorov property, let (Y, F) be a Banach–Kantorovich space, and let F be a K-space satisfying the countable sup condition. Then the above representation is valid.*

$\triangleleft$ The claim follows from 4.2.6 and 4.4.10 (2). $\triangleright$

4.4.12. A vector measure μ is *countably additive* (*completely additive*) if

$$\mu\left(\bigvee_{\gamma \in \Gamma} a_\gamma\right) = \sum_{\gamma \in \Gamma} \mu(a_\gamma)$$

for each countable (arbitrary) family $(a_\gamma)_{\gamma \in \Gamma}$ of pairwise disjoint elements of $\mathscr{A}$. Denote by $\operatorname{dca}(\mathscr{A}, Y)$ and $\operatorname{dao}(\mathscr{A}, Y)$ the respective sets of countably additive vector measure and completely additive vector measure.

(1) It can be easily checked that

$$\mu \in \mathrm{dca}(\mathscr{A}, Y) \Longleftrightarrow I_\mu \in M_\sigma(S(\mathscr{A}), Y),$$
$$\mu \in \mathrm{dao}(\mathscr{A}, Y) \Longleftrightarrow I_\mu \in M_n(S(\mathscr{A}), Y).$$

(2) The sets $\mathrm{dca}(\mathscr{A}, Y)$ and $\mathrm{dao}(\mathscr{A}, Y)$ are bands in $\mathrm{da}(\mathscr{A}, Y)$, so that

$$\mathrm{da}(\mathscr{A}, Y) = \mathrm{dca}(\mathscr{A}, Y) \oplus \mathrm{dca}(\mathscr{A}, Y)^\perp, \quad \mathrm{da}(\mathscr{A}, Y) = \mathrm{dao}(\mathscr{A}, Y) \oplus \mathrm{dao}(\mathscr{A}, Y)^\perp.$$

(3) Every $\mu \in \mathrm{da}(\mathscr{A}, Y)$ has a unique representation in the form $\mu = \mu_n + \mu_{os}$, where $\mu_n \in \mathrm{dca}(\mathscr{A}, Y)$ and $\mu_{os} \in \mathrm{dca}(\mathscr{A}, Y)^\perp$. Moreover,

$$|\mu| = |\mu_n| + |\mu_{os}|, \; |\mu_n| = |\mu|_n, \; |\mu_{os}| = |\mu|_{os}.$$

(4) Let $\mathscr{I}$ be an ideal in $\mathscr{A}$ and $\nu \in \mathrm{ba}_+(\mathscr{A}, F)$. Then there exists a unique measure $\pi_{\mathscr{I}}\nu \in \mathrm{ba}_+(\mathscr{A}, Y)$ such that $\nu = \pi_{\mathscr{I}}\nu$ on $\mathscr{I}$ and $\nu(a) = 0$ for every $a \in \mathscr{A}$ satisfying $a \wedge b = 0$ for all $b \in \mathscr{I}$.

(5) Let $\mathscr{I}$ be an ideal in $\mathscr{A}$ and let Y be *bo*-complete. For every $\mu \in \mathrm{da}(\mathscr{A}, Y)$ there exists a unique measure $\pi_{\mathscr{I}}\mu \in \mathrm{da}(\mathscr{A}, Y)$ such that $|\pi_{\mathscr{I}}\mu| = \pi_{\mathscr{I}}|\mu|$ and $|\mu - \pi_{\mathscr{I}}\mu| = |\mu| - \pi_{\mathscr{I}}|\mu|$.

(6) If $\nu \in \mathrm{ba}_+(\mathscr{A}, F)$ then

$$\nu_n(a) = \inf\big\{\sup\{\nu(a_\gamma) : a_\gamma \nearrow a\}\big\} \quad (a \in \mathscr{A}),$$
$$\nu_c(a) = \inf\big\{\sup\{\nu(a_n) : a_n \nearrow a\}\big\} \quad (a \in \mathscr{A}).$$

4.5. Extension of Dominated Operators

In this section we present a method of extending sequentially order continuous dominated operators which essentially coincides with the classical Daniell–Stone construction of the Lebesgue integral. The method works if and only if the underlying target vector lattice possesses the weak σ-distributivity property.

4.5.1. A vector lattice F is said to be *weakly (σ, ∞)-distributive*, if whenever $(f_{m,\alpha})_{\alpha \in \mathrm{A}(m)}$ $(m \in \mathbb{N})$ is an order-bounded sequence of nets in F such that if $(f_{m,\alpha})_{\alpha \in \mathrm{A}(m)}$ decreases to zero for each $m \in \mathbb{N}$, then

$$\inf\left\{\sup_{m \in \mathbb{N}} f_{m,\varphi(m)} : \varphi \in \prod_{m \in \mathbb{N}} A(m)\right\} = 0.$$

A vector lattice F is said to be *weakly σ-distributive* if whenever $(f_{m,n})_{m,n\in\mathbb{N}}$ is an order-bounded double sequence in F such that if $(f_{m,n})_{n\in\mathbb{N}}$ decreases and o-converges to zero for each $m \in \mathbb{N}$, then

$$\inf\left\{\sup_{m\in\mathbb{N}} f_{m,\varphi(m)} : \varphi \in \mathbb{N}^{\mathbb{N}}\right\} = 0.$$

As usual, $\mathbb{N}^{\mathbb{N}}$ denotes the set of all functions $\varphi : N \to \mathbb{N}$.

If the decreasing sequences $(f_{m,n})_{n\in\mathbb{N}}$ in this definition are replaced by countable nets or countable directed sets then we obtain formally new formulations. But these two are equivalent to the above (presuming the countable choice axiom). Thus, any weakly (σ,∞)-distributive lattice is weakly σ-distributive.

Theorem. *For an arbitrary K_σ-space (K-space) F the following are equivalent:*

(1) *F is weakly σ-distributive (weakly (σ,∞)-distributive);*

(2) *for every $0 \le f \in F$ the Boolean algebra of fragments $\mathfrak{E}(f)$ is weakly σ-distributive (weakly (σ,∞)-distributive);*

(3) *for every $0 \le f \in F$ the Stone space of the Boolean algebra $\mathfrak{E}(f)$ have the property: the union of every sequence of closed nowhere-dense G_δ-sets is nowhere-dense (each meager set is nowhere-dense).*

4.5.2. In order to apply the hypothesis of weak σ-distributivity to the operator extension problem, introduce the following concept. Take a double sequence $(f_{n,m})_{n,m\in\mathbb{N}}$ in F and denote $\Phi := \mathbb{N}^{\mathbb{N}}$. We say that the double sequence σ-depresses a nonempty set $B \subset F$ towards zero if: (a) $(f_{n,m})_{m\in\mathbb{N}}$ decreases to zero for every $n \in \mathbb{N}$, and (b) for every $\varphi \in \Phi$, there is a $b \in B$ such that $b \le \sup_{n\in\mathbb{N}} f_{n,\varphi(n)}$. A nonempty set $B \subset F$ is said to be *σ-depressed towards zero* if there is an order-bounded double sequence which depresses B towards zero.

(1) *If F is a weakly σ-distributive K_σ-space and B is a nonempty set which is σ-depressed towards zero, then $\inf\{b^+ : b \in B\} = 0$.*

◁ The proof is immediate from the definitions. ▷

(2) *If B and C are σ-depressed towards zero, then so is $B+C := \{b+c : b \in B,\ c \in C\}$.*

◁ If $(f_{n,m})_{n,m\in\mathbb{N}}$ and $(g_{n,m})_{n,m\in\mathbb{N}}$ depress towards zero B and C respectively then $\left(f_{n,m} + g_{n,m}\right)_{n,m\in\mathbb{N}}$ depresses $B + C$ towards zero. ▷

(3) *If B is σ-depressed towards zero, and every $b \in B$ dominates some $c \in C$ then C is σ-depressed towards zero. If $(f_n)_{n\in\mathbb{N}}$ decreases to zero in F, then $B := \{f_n : n \in \mathbb{N}\}$ is σ-depressed towards zero.*

◁ Follows immediately from the given definition. ▷

(4) *Let (B_n) be a sequence of nonempty sets in F, each σ-depressed towards zero. Then for every $f \in F_+$ the set*

$$B := \left\{ \sup_{n \in \mathbb{N}} \left(f \wedge \sum_{m=1}^{n} b_m \right) : (\forall m \in \mathbb{N})\, b_m \in B_m \right\}$$

is σ-depressed towards zero.

$\lhd$ For each $n \in \mathbb{N}$ choose a double sequence $(f_{n,m,l})_{m,l \in \mathbb{N}}$ depressing B_n towards zero. Put

$$w_{m,l} := \sup_{n \leq m} 2^n f_{n,n-m,l}, \quad v_{m,l} := f \wedge w_{m,l} \quad (m, l \in \mathbb{N}).$$

Then the sequences $(w_{m,l})_{l \in \mathbb{N}}$ and $(v_{m,l})_{l \in \mathbb{N}}$ decrease to zero for each $m \in \mathbb{N}$ and the second net is order-bounded. Given $\varphi \in \Phi$, we may choose, for each $k \in \mathbb{N}$, an element $b_k \in B_k$ such that $b_k \leq \sup\{f_{k,m,\varphi(k+m)} : m \in \mathbb{N}\}$. Since $w_{m+k,l} \geq 2^k f_{k,m,l}$ for $k, m, l \in \mathbb{N}$, we may estimate

$$b_k \wedge f \leq \inf_{m \in \mathbb{N}} \left(f \wedge 2^k w_{k+m,\varphi(k+m)} \right) \leq \sup_{m \in \mathbb{N}} \left(f \wedge 2^k w_{m,\varphi(m)} \right) =: u_k.$$

It is easy to check that $u_k = f \wedge 2^{n-k} u_n$ $(k \leq n)$. Summarizing the indicated properties of u_k, we deduce

$$\begin{aligned} f \wedge \sum_{k=1}^{n} b_k &\leq f \wedge \sum_{k=1}^{n} b_k^+ = f \wedge \sum_{k=1}^{n} (b_k^+ \wedge f) = f \wedge \sum_{k=1}^{n} (b_k \wedge f)^+ \\ &\leq f \wedge \sum_{k=1}^{n} u_k^+ = f \wedge \sum_{k=1}^{n} u_k = f \wedge \sum_{k=1}^{n} (f \wedge 2^{n-k} u_k) \\ &= f \wedge \sum_{k=1}^{n} (2^{n-k} u_n) \leq f \wedge 2^{n+1} u_n = f \wedge \sup_{m \in \mathbb{N}} 2^{n+1} f \wedge 2 w_{m,\varphi(m)} \\ &= \sup_{m \in \mathbb{N}} v_{m,\varphi(m)}. \end{aligned}$$

Since n is arbitrary, we have

$$\sup_{m \in \mathbb{N}} v_{m,\varphi(m)} \geq \sup_{m \in \mathbb{N}} \left(f \wedge \sum_{k=1}^{n} b_k \right) \in B.$$

Thus, B is depressed towards zero by $(v_{m,l})_{l \in \mathbb{N}}$ and the proof is complete. $\rhd$

4.5.3. Theorem. *Let F and G be K_σ-spaces with F weakly σ-distributive. Let E be a massive sublattice in G and let Y be a sequentially bo-complete lattice-normed space over F. Suppose that a linear operator $T : E \to Y$ has a sequentially G-continuous dominant $S : E \to F$. Then there exist a sequentially o-closed sublattice $\hat{E} \subset G$ and a unique pair of sequentially o-continuous operators $\hat{T} : \hat{E} \to Y$ and $\hat{S} : \hat{E} \to F$ such that $E \subset \hat{E}$, $\hat{T}$ is an extension of T, S is an extension of S, and $\hat{S}$ is a dominant of $\hat{T}$.*

$\lhd$ The proof is given below in 4.5.4–4.5.9. $\rhd$

4.5.4. Denote by $E^\uparrow$ the set of all $x \in G$ such that $x = \sup_n x_n$ for an increasing sequence $(x_n)_{n\in\mathbb{N}} \subset E$. It is evident that the set $E^\uparrow$ is closed under addition and lattice operations. Given $x \in E^\uparrow$, define

$$S^\uparrow x := \sup\{Sx_n : n \in \mathbb{N}\}.$$

The least upper bound on the left-hand side exists, since by the massiveness assumption x is dominated by some $u \in E$, and therefore, $(Sx_n)_{n\in\mathbb{N}}$ is bounded above by Su. If $(x'_n)_{n\in\mathbb{N}}$ is another sequence in E with $\sup x'_n = x$ then $o\text{-}\lim_{n,m}(x'_n - x_m) = 0$ in G and, by G-continuity of S, we have $o\text{-}\lim Sx'_n = o\text{-}\lim Sx_n$. Thus, the operator $S^\uparrow : E^\uparrow \to F$ is defined correctly.

(1) *The operator $S^\uparrow$ is increasing, additive, agrees with S on E, and $S^\uparrow u = \sup\{S^\uparrow u_n : n \in \mathbb{N}\}$ for an arbitrary sequence $(u_n)_{n\in\mathbb{N}}$ in $E^\uparrow$ increasing to $u \in E^\uparrow$.*

If x and x_n are as above then the sequence $(Tx_n)_{n\in\mathbb{N}}$ is bo-fundamental, since

$$|Tx_n - Tx_m| \le S(|x_n - x_m|) \le S^\uparrow(|x - x_\alpha|) + S^\uparrow(|x - x_m|) \xrightarrow{(o)} 0.$$

If $(x'_m)_{m\in\mathbb{N}}$ is another sequence in E increasing to $x \in E^\uparrow$, then

$$|Tx_n - Tx'_m| \le S^\uparrow(x - x_n) + S^\uparrow(x - x'_m) \xrightarrow{(o)} 0.$$

Thus, the relation $T^\uparrow x := bo\text{-}\lim Tx_n$ correctly defined an operator $T^\uparrow : E^\uparrow \to Y$.

(2) *The operator $T^\uparrow$ is additive, positively homogeneous, and agrees with T on E.*

Replacing increasing sequences by decreasing sequences in above definitions we may introduce $E^\downarrow$ and operators $S^\downarrow : E^\downarrow \to F$ and $T^\downarrow : E^\downarrow \to Y$. It can be easily verified that

$$|T^\downarrow x| \le S^\downarrow(|x|) \quad (x \in E^\downarrow),$$
$$|T^\uparrow x| \le S^\uparrow(|x|) \quad (x \in E^\uparrow).$$

Moreover, $E^\downarrow = -E^\uparrow$, $T^\downarrow = -(-T)^\uparrow$, and the operators $T^\downarrow$ and $T^\uparrow$ coincide on $E^\downarrow \cap E^\uparrow$.

4.5.5. Now define $\hat{E} \subset G$ to be the set of such $x \in G$ that the set

$$B(x) := \{Su - Sv : u \in E^{\uparrow}, v \in E^{\downarrow}, v \le x \le u\}$$

is σ-depressed towards zero in F. It can be easily observed that $\hat{E} = -\hat{E}$ and $\lambda\hat{E} = \hat{E}$ for $\lambda > 0$, since $B(x) = B(-x)$ and $B(\lambda x) = \lambda B(x)$. Moreover, the following is true:

(1) *The set $\hat{E}$ is a vector sublattice in G.*

$\lhd$ Let $x_1, x_2 \in \hat{E}$ and $x = x_1 + x_2$. Then the set $B(x_j)$ is σ-depressed towards zero for $j := 1, 2$. It follows easily from the definition of sets $B(x)$, $B(x_1)$, $B(x_2)$ that every member of $B(x_1) + B(x_2)$ dominates some member of $B(x)$. Thus, in view of 4.5.2 (2, 3), $B(x)$ is σ-depressed towards zero and $x \in \hat{E}$.

Now suppose that $x \in \hat{E}$. If $u \in E^{\uparrow}$, $v \in E^{\downarrow}$, and $v \le x \le u$. Then $u^+ \in E^{\uparrow}$, $v^+ \in E^{\downarrow}$, $v^+ \le x^+ \le u^+$, and $S(u^+ - v^+) \le S(u - v)$. It follows that every member of $B(x)$ dominates some member of $B(x^+)$. Thus, $B(x^+)$ is σ-depressed towards zero and $x^+ \in \hat{E}$. $\rhd$

If $x \in \hat{E}$ then $\inf B(x) = 0$ according to 4.5.2 (1). Therefore, the expressions $\inf\{Su : u \in E^{\uparrow}, x \le u\}$ and $\sup\{Sv : v \in E^{\downarrow} v \le x\}$ have equal values, say w. Thus, we may define an operator $\hat{S} : \hat{E} \to F$ by

$$\hat{S}x := \inf\{Su : u \in E^{\uparrow}, x \le u\} = \sup\{Sv : v \in E^{\downarrow} v \le x\} \quad (x \in \hat{E}).$$

It can be easily seen that $B(x)$ is σ-depressed towards zero if and only if there exists $w \in F$ such that the sets $B^+(x) := \{Su - w : u \in E^{\uparrow} u \ge x\}$ and $B^-(x) := \{-Su + w : v \in E^{\downarrow} v \le x\}$ are both σ-depressed towards zero. Moreover, in this case $\hat{S}x = w$.

(2) *The operator $\hat{S}$ is linear, positive and agrees with S on E.*

4.5.6. *The set $\hat{E}$ is sequentially order-closed sublattice in G and $\hat{S} : \hat{E} \to F$ is sequentially order continuous.*

$\lhd$ Take an increasing sequence $(x_n)_{n\in\mathbb{N}}$ in $\hat{E}_+$ with $x_1 = 0$ and let $x := \sup_{n\in\mathbb{N}} x_n$ exist in G. Fix on some $e \in E$ such that $x \le e$, and put $f := Se$. Let $w := \sup_{n\in\mathbb{N}} Sx_n$ and note that $w \le f$. We will show that the set $B(x)$ is σ-depressed towards zero. It is convenience to split the proof into two steps.

(1) For each $n \in \mathbb{N}$, denote

$$B_n := \{Su - \hat{S}(x_{n+1} - x_n) : u \in E^{\uparrow}, u \ge x_{n+1} - x_n\}.$$

It follows from 4.5.5 that for each $n \in \mathbb{N}$ the element $\hat{S}(x_{n+1} - x_n)$ exists and B_n is σ-depressed towards zero. Let B be defined as in 4.5.2 (4). Then B is σ-depressed towards zero. By definition each $b \in B$ is expressed as $b = \sup_{n\in\mathbb{N}} \left(f \wedge \sum_{m=1}^{n} b_m\right)$ where $b_m = Su_m - \hat{S}(x_{m+1} - x_m)$ with $u_m \geq x_{m+1} - x_m$. Denote

$$u := \sup_{n\in\mathbb{N}} \left(e \wedge \sum_{m=1}^{n} u_m \right).$$

Observe that $u \in E^{\uparrow}$ and $u \geq x$, since

$$u \geq e \wedge \sum_{m=1}^{n} u_m \geq e \wedge \sum_{m=1}^{n} (x_{m+1} - x_m) \geq e \wedge x_{n+1} = x_{n+1}$$

for every $n \in \mathbb{N}$. Now taking 4.5.4 (1) into consideration, we deduce

$$\begin{aligned} Su &= \sup_{n\in\mathbb{N}} S\left(e \wedge \sum_{m=1}^{n} u_m \right) \leq \sup_{n\in\mathbb{N}} \left(Se \wedge S\left(\sum_{m=1}^{n} u_m \right)\right) \\ &= \sup_{n\in\mathbb{N}} \left(Se \wedge \sum_{m=1}^{n} Su_m \right) = \sup_{n\in\mathbb{N}} \left(f \wedge \left(\sum_{m=1}^{n} b_m + \sum_{m=1}^{n} \hat{S}(x_{m+1} - x_m) \right)\right) \\ &= \sup_{n\in\mathbb{N}} \left(f \wedge \left(\sum_{m=1}^{n} b_m + \hat{S}x_{n+1} \right)\right) \leq \sup_{n\in\mathbb{N}} \left(f \wedge \sum_{m=1}^{n} b_m \right) + w = b + w. \end{aligned}$$

Thus, every $b \in B$ dominates some element of $B^{+}(x) := \{Su - w : u \in E^{\uparrow}\ u \geq x\}$ and, in view of 4.5.2 (2), $B^{+}(x)$ is σ-depressed to zero.

(2) For each $n \in \mathbb{N}$, denote

$$B_1 := \{w - \hat{S}x_n : n \in \mathbb{N}\}, \quad B_{n+1} := \{-Su + \hat{S}x_n : u \in E^{\downarrow},\ u \leq x_n\}.$$

By 4.5.5 B_n is σ-depressed to zero for each $n \in \mathbb{N}$. Again B, defined as in 4.5.2 (4), is σ-depressed to zero. But each $b \in B$ can be represented as

$$b := \sup_{n\in\mathbb{N}} \left(f \wedge \left(w - \hat{S}x_k + \sum_{m=1}^{n} (-Su_m + \hat{S}x_m) \right)\right)$$

where $k \in \mathbb{N}$ and $u_m \in E^{\downarrow}$, $u \leq x_n$ for every $n \in \mathbb{N}$. Since $-Su_m + \hat{S}x_m \geq 0$ we have

$$\begin{aligned} b &\geq f \wedge (w - \hat{S}x_k + (-Su_m + \hat{S}x_m)) = f \wedge (w - Su_m) \\ &= w \wedge (w - Su_m) = w + 0 \wedge S(-u_m) \geq w + S(0 \wedge (-u_m)) = w - S(0 \vee u_m). \end{aligned}$$

But $0 \vee u_m \in E^{\downarrow}$ and $0 \vee u_m \leq x_m \leq x$. Thus, every $b \in B$ dominates some element of the set $B^{-}(x) := \{w - Sv : v \in E^{\downarrow},\ v \leq x\}$ and $B^{-}(x)$ is σ-depressed towards zero.

According to 4.5.2 (2) $B^{+}(x) + B^{-}(x)$ is also σ-depressed towards zero. Since $B(x) = B^{+}(x) + B^{-}(x)$ we obtain that $x \in \hat{E}$. Moreover, $\hat{S}x = w$. ▷

4.5.7. Let $\Phi := \mathbb{N}^{\mathbb{N}}$ and define an order relation in Φ by

$$\varphi_1 \leq \varphi_2 \Leftrightarrow (\forall n \in \mathbb{N})\, \varphi_1(n) \leq \varphi_2(n).$$

Clearly, Φ is directed upward. We say that a net $(y_\varphi)_{\varphi\in\Phi}$ in Y is Φ-fundamental (Φ-converges to $y \in Y$), if there is a double sequence $(f_{m,n})_{m,n\in\mathbb{N}}$ in F such that $(f_{m,n})_{n\in\mathbb{N}}$ decreases to zero for each $m \in \mathbb{N}$ and

$$|y_{\varphi'} - y_{\varphi'}| \leq \sup_{m\in\mathbb{N}} f_{m,\varphi(m)}$$

for all $\varphi, \varphi', \varphi'' \in \Phi$ with $\varphi' \geq \varphi$ and $\varphi'' \geq \varphi$ (respectively,

$$|y - y_\varphi| \leq \sup_{m\in\mathbb{N}} f_{m,\varphi(m)}$$

for all $\varphi \in \Phi$).

Let $\hat{F}$ be an order completion of a σ-distributive vector lattice F. Then there exist a lattice-normed space $\hat{Y}$ over $\hat{F}$ and isometric embedding $\imath : Y \to \hat{Y}$ such that

(1) *if $(y_\varphi)_{\varphi\in\Phi}$ is a Φ-fundamental net in Y, then the net $(\imath y_\varphi)$ is bo-convergent to some element in $\hat{Y}$;*

(2) *every member of $\hat{Y}$ is the Φ-limit of a net $(y_\varphi)_{\varphi\in\Phi}$ in Y.*

$\lhd$ Denote by $\Phi(Y)$ the set of all Φ-fundamental nets in Y. Define some equivalence relation $\sim$ in $\Phi(Y)$ by putting $(y_\varphi) \sim (z_\varphi)$ if and only if there exists an order-bounded double sequence $(f_{m,n})$ in F such that the sequence $(f_{m,n})_{n\in\mathbb{N}}$ decreases to zero for every $m \in \mathbb{N}$ and

$$|y_{\varphi'} - z_{\varphi''}| \leq \sup_{m\in\mathbb{N}} f_{m,\varphi(m)}$$

for all $\varphi' \geq \varphi$ and $\varphi'' \geq \varphi$. Let $\hat{Y} := \Phi(Y)/\sim$ be the factor set and $\imath$ be the mapping that sends each $y \in Y$ to the coset of the constant net $y_\varphi = y$ ($\varphi \in \Phi$). If an element $\hat{y} \in \hat{Y}$ is the coset of a Φ-fundamental net (y_φ), then we set $|\hat{y}| := o\text{-lim}|y_\varphi|$. It follows from the definitions and σ-distributivity assumption that

$$o\text{-}\lim_{\varphi'} |\hat{y} - y_{\varphi'}| = o\text{-}\lim_{\varphi'} o\text{-}\lim_{\varphi} |y_\varphi - y_{\varphi'}| = 0,$$

which proves (1). Claim (2) paraphrases (1). $\rhd$

4.5.8. Now we will extend T to $\hat{E}$. Take $x \in \hat{E}$ and let a double sequence $(f_{m,n})_{m,n\in\mathbb{N}}$ σ-depresses $B(x)$ to zero. Then for every $\varphi \in \Phi$ there exist $u_\varphi \in E^\uparrow$ and $v \in E^\downarrow$ such that $v \leq x \leq u$ and

$$S^\uparrow(u_\varphi - v_\varphi) \leq \sup_{m\in\mathbb{N}} f_{m,\varphi(m)}.$$

Since $|T^\uparrow u| \leq \hat{S}(|u|)$ for $u \in E^\uparrow$, we have

$$|T^\uparrow u_{\varphi'} - T^\downarrow u_{\varphi''}| \leq \hat{S}(|u_{\varphi'} - u_{\varphi''}|)$$
$$\leq \hat{S}(u_{\varphi'} - x) + \hat{S}(u_{\varphi''} - x) \leq \sup_{m\in\mathbb{N}} f_{m,\varphi(m)}$$

for all $\varphi' \geq \varphi$ and $\varphi'' \geq \varphi$. From this it follows that the net $(T^\uparrow u_\varphi)_{\varphi\in\Phi}$ is Φ-fundamental. In a similar manner the nets $(T^\uparrow u_\varphi)_{\varphi\in\Phi}$ and $(T^\downarrow v_\varphi)_{\varphi\in\Phi}$ are equivalent. According to 4.5.7 these nets have the same Φ-limit in $\hat{Y}$ which is also their *bo*-limit because of the weak σ-distributivity assumption. Therefore, we may define

$$\hat{T}x := \lim_{\varphi\in\Phi} T^\uparrow u_\varphi = \lim_{\varphi\in\Phi} T^\downarrow v_\varphi.$$

Linearity of $\hat{T} : \hat{E} \to \hat{Y}$ and the inequality $|\hat{T}x| \leq \hat{S}(|x|)$ $(x \in \hat{E})$ are obvious. So, it remains to prove that $\hat{T}(\sigma(E)) \subset Y$.

Of course, we could use *bo*-completion from 2.2.11 instead of Φ-completion from 4.5.7. But it was the sole use of the axiom of choice. Thus, the proof of Theorem 4.5.3 runs by the countable choice axiom.

4.5.9. A subset in G is called *monotone* if it contains order limits of all monotone sequences in this set.

Let E_0 be the intersection of all monotone subsets in G including E, so that E_0 is the smallest monotone subset in G including E. Denote

$$E_1 := \{x \in \sigma(E) : \tilde{T}x \in Y\}.$$

If the sequence $(x_n) \subset E_1$ increases to some $x \in G$ then $x \in \sigma(E)$, and $(\hat{T}x_n)$ is fundamental, since $|\hat{T}x_n - \hat{T}x| \leq \hat{S}(|x_n - x|)$ and $\hat{S}$ is sequentially *o*-continuous. (see 4.5.7). But Y is sequentially *bo*-complete and $\hat{T}x = \lim \hat{T}x_n$ lie in Y. Thus $x \in E_1$, so that E_1 is a monotone subset including E. In particular, $E_0 \subset E_1 \subset \sigma(E)$. It can be shown that the smallest monotone subset containing a vector sublattice is itself a vector sublattice. Therefore, $E_0 = \sigma(E)$ and $E_1 = \sigma(E)$. Thus, $\hat{T}(\sigma(E)) \subset Y$ which completes the proof of Theorem 4.5.3.

Thus, the Daniell scheme works and results in sequentially order continuous extension to the "Baire completion" if F is weakly σ-distributive. The converse is also true: weak σ-distributivity of F is not only sufficient but also necessary for Theorem 4.5.3 to be valid. There are similar interrelations between possibility of extension to the "Borel completion" and (σ, ∞)-distributivity. We state only one result in this direction.

4.5.10. Theorem. *Let G and F be K-spaces with F weakly (σ,∞)-distributive. Let E be a massive sublattice in G, and let Y be a bo-complete lattice-normed space over F. Suppose that a linear operator $T : E \to Y$ is G-continuous. Then there exist a vector sublattice $\widetilde{E} \subset G$ and an extension $\widetilde{T} : \widetilde{E} \to Y$ of T to $\widetilde{E}$ such that the following hold:*

(1) *$\widetilde{E}$ is a sequentially o-closed sublattice in G containing $E^{\uparrow}$;*

(2) *$\widetilde{T}$ is a sequentially order continuous dominated operator;*

(3) *$|\widetilde{T}| : \widetilde{E} \to F$ is an extension of $|T|$;*

(4) *if a net $(x_\alpha) \subset E^{\uparrow}$ increases to $x \in E^{\uparrow}$ then $\widetilde{T}x = \lim_\alpha \widetilde{T}x_\alpha$.*

◁ We confine exposition to a rough sketch. Denote $S := |T|$ and observe that $S : E \to F$ is G-continuous (4.3.8). Given $x \in E^{\uparrow}$, we put by definition

$$S^{\uparrow}x := \sup\{Sx_\alpha : \alpha \in \mathrm{A}\}.$$

The operator $S^{\uparrow} : E^{\uparrow} \to F$ is correctly defined because E is massive in G and S is G-continuous. Moreover, $S^{\uparrow}y = \sup\{S^{\uparrow}y_\alpha : \alpha \in \mathrm{A}\}$ for an arbitrary net $(y_\alpha)_{\alpha\in\mathrm{A}}$ in $E^{\uparrow}$ increasing to $y \in E^{\uparrow}$. If x and x_α are as above then we let $T^{\uparrow}x := o\text{-}\lim Tx_\alpha$. This relation correctly defined an additive and positively homogeneous operator $T^{\uparrow} : E^{\uparrow} \to Y$.

We say that the sequence of nets $(f_{n,\alpha})_{\alpha\in\mathrm{A}(n)}$ (σ,∞)-depresses a nonempty set $B \subset F$ towards zero if: (a) $(f_{n,\alpha})$ decreases to zero for every $n \in \mathbb{N}$, and (b) for every $\varphi \in \Phi := \prod_{n\in\mathbb{N}} \mathrm{A}(n)$, there is a $b \in B$ such that $b \le \sup_{n\in\mathbb{N}} f_{n,\varphi(n)}$. A nonempty set $B \subset F$ is said to be (σ,∞)-*depressed towards zero* if there is an order-bounded sequence of nets which depresses B towards zero.

Now let $\widetilde{E} \subset G$ comprise such $x \in G$ that the set

$$B(x) := \{Sy - Sz : y \in E^{\uparrow},\ z \in E^{\downarrow}\ z \le x \le y\}$$

is (σ,∞)-depressed towards zero in F. Then the set $\widetilde{E}$ is a sequentially o-closed vector sublattice in G. Then the set $\widetilde{E}$ is a sequentially o-closed vector sublattice in G; moreover, there exists an extension $\widetilde{T} : \widetilde{E} \to Y$ with the required properties (2)–(4). Further details may be carried out in a similar way. ▷

4.6. Comments

4.6.1. (1) The notion of dominated operator (4.1.1) appeared in the second half of the 1930s in the papers of L. V. Kantorovich [154, 157, 158, 161]. The notion was motivated by two reasons: a theoretical motif was the development of the general theory of operations in semiordered spaces (see [153, 155–157]) and an applied stimulus was tied with the approximate methods of analysis (see [154, 158, 161]).

Speaking about the latter, it is worthwhile to use the words by L. V. Kantorovich himself [154]:

"In proving existence for a solution to various classes of functional equations, the method of successive approximations is of rather frequent use in analysis; in this case, the proof for convergence of the approximations rests on the fact that a given equation may be dominated by some equation of a simple form. Proofs of this sort occur in the theory of integral and differential equations.

Invoking semiordered spaces and operations in these spaces allows us to develop a rather easy but complete theory of functional equations of these type in abstract form."

(2) Without additional conditions like decomposability, we cannot guarantee existence of an exact dominant for a dominated operator (4.1.2). Simple examples demonstrate that a nonzero dominated operator can have even disjoint dominants. At the same time, decomposability conditions can be weakened to some extent. We point out three of such possibilities.

(2.1) *Suppose that X is r-decomposable in the following sense: for every $x \in X$ and $0 \leq e \leq |x|$, there exists an r-fundamental sequence (x_n) in X such that $|x_n| \xrightarrow{(r)} e$ and $|x - x_n| \xrightarrow{(r)} |x| - e$. Then each dominated operator from X into Y has an exact dominant.*

(2.2) *Let E be a vector lattice with the principal projection property and let X be d-decomposable. Then each dominated operator from X into Y has an exact dominant.*

(2.3) *Suppose that X is o-decomposable in the following sense: for every $x \in X$ and every $0 \leq e \leq |x|$ there exists an o-fundamental sequence (x_n) in X such that $|x_n| \xrightarrow{(o)} e$ and $|x - x_n| \xrightarrow{(o)} |x| - e$. Then each order continuous (see 2.3.1) dominated operator from X into Y has an exact dominant.*

(3) Various classes of dominated operators were studied rather independently. Alongside bounded operators in normed spaces, much attention was paid to regular operators, see [23, 60, 162, 163, 336, 388, 409]. Operators with abstract norm (4.1.3 (3)) were introduced by L. V. Kantorovich [163], see [60, 228] and the bibliography therein. Dominated operators in the sense of 4.1.3 (4) were studied by many authors in generality of various degree and under various names, e.g., see [55, 78, 115-117, 188, 237, 238, 296, 336]. In particular, the formula of 4.1.3 (4) is contained in [55, 238]. The operators of 4.1.3 (5) possess a peculiarity that, after the Boolean-valued representation of the domain and target (see 8.3.3) these operators become bounded in the sense of the theory of normed spaces. This fact provides a possibility of studying generalized orthomorphisms by interpreting, in a suitable Boolean-valued model, results of the theory of bounded operators in normed spaces (see A. G. Kusraev and S. S. Kutateladze [212]). We also point out that operators with abstract norm (4.1.3) can be represented as bounded functionals in a suitable Boolean-valued model (see E. I. Gordon [106]).

(4) The general formula of 4.1.5, for calculating an exact dominant, and its corollary 4.1.6 were obtained by A. G. Kusraev and V. Z. Strizhevskiĭ [218]. The improvement 4.1.8, connected with the possibility of taking the supremum over disjoint partitions, is proven by S. A. Malyugin [181]. Subdominated operators appeared for the first time in the paper [157] by L. V. Kantorovich. Assertion 4.1.11 (1) is also presented there.

(5) Consider a subdominated operator $T : X \to Y$. Denote

$$\Omega := \{S : Y \to F : \ |Ay| \le |y| \ (y \in Y)\}(= \partial|\cdot|).$$

For each $S \in \Omega$, the operator $S \circ T$ is dominated and the following formula holds:

$$[T]e = \sup\{|S \circ T|e : \ S \in \Omega\} \quad (e \in E_+).$$

$\lhd$ By the Hahn–Banach–Kantorovich Theorem the following equality holds:

$$|y| = \sup\{|Ay| : \ A \in \Omega\} \quad (y \in Y).$$

From 4.1.11 (1) it follows that the operator $S \circ T$ is dominated and $[A \circ T]e = |A \circ T|e$ $(e \in E_+)$. It remains to apply the formula for $[T]$ (4.1.10) and $[A \circ T]$ (4.1.5) and use associativity of suprema. $\rhd$

4.6.2. **(1)** The fact that the space of dominated operators $M(X, Y)$, where X is decomposable and Y is *bo*-complete, is a *bo*-complete LNS is known since the second half of the 1930s, see [157, 158, 161, 163]; however, the question about decomposability has remained open. Decomposability of the space $M(X, Y)$ (Theorem 4.2.6) was established by A. G. Kusraev and V. Z. Strizhevskiĭ [218] in 1987, see also [181]. E. V. Kolesnikov observed (see [181]) that the decomposability hypothesis for X in Theorem 4.2.6 can be somewhat weakened. More precisely, the following assertions are valid:

(1.1) If X is r-decomposable (see 4.6.1 (1)) and Y is a BKS then $M(X, Y)$ is a BKS.

(1.2) If X is o-decomposable (see 2.1.3 (3)) and Y is a BKS then $M_n(X, Y)$ is a BKS.

4.6.3. **(1)** Theorems 4.3.2 and 4.3.3 are established by A. G. Kusraev and V. Z. Strizhevskiĭ [218] for the case of order complete norm lattices, see also [197, 181]. The result on extending an o-continuous regular operator from a vector lattice onto its Dedekind completion (which the proof of 4.3.3 is referred to) is established by A. I. Veksler [376]. Formula 4.3.5 (3) was mentioned in [199]. Theorem 4.3.7 is valid in a sequential version as well: A dominated operator is sequentially o-continuous if and only if it is σ-additive (i.e., the condition of complete additivity in 4.3.6 is satisfied only for sequences). For a positive operator, this fact was established by A. G. Pinsker, see [163, Theorem VII.2.45].

(2) Almost integral operators were introduced by G. Ya. Lozanovskiĭ [253]. Theorem 2.5.10 is essentially due to A. G. Kusraev [199]. V. Z. Strizhevskiĭ [362] obtained the following result:

Theorem. *Let E and F be order complete vector lattices and let $E_n^\sim$ separate the points of E. If E is diffuse then every lattice homomorphism is disjoint from the band $J(E,F)$. If F is diffuse then every Maharam operator is disjoint from the band $J(E,F)$.*

4.6.4. **(1)** The question of decomposing a measure, functional, or operator into order continuous and order singular parts was studied from various points of view and has many different versions of stating and solving, e.g., see [23, 213–215, 228, 326, 409] and the bibliography in [60]. The formulas 4.4.3 (1–4) were obtained by E. V. Kolesnikov [178]; and 4.4.3 ($1'$–$4'$) are due to A. G. Kusraev and S. A. Malyugin [213]. Theorem 4.4.5 (2) belongs to C. D. Aliprantis and O. Burkinshaw [21], see also [23, 258, 388] and the survey [60]. Theorem 4.4.6 was proven in [199]; it is an easy consequence of Theorems 4.2.6 and 4.4.5 (2).

(2) As another corollary of Theorem 4.2.6, we indicate a version of a result by C. D. Aliprantis and O. Burkinshaw [23, Theorem 4.10] for dominated operators [199]:

Theorem. *Let X be a decomposable LNS, and let Y be a BKS. Then the following assertions are equivalent:*

(i) *every dominated operator from X into Y is order continuous;*

(ii) *the null ideal of every dominated operator from X into Y is a band;*

(iii) *every nonzero dominated operator from X into Y has nonzero carrier (= the disjoint complement of the null ideal).*

(3) It is well known that the strong form of the Yosida–Hewitt Theorem is not valid, even in the case of functionals, without some extra assumptions. For example, in the vector lattice $E := C[0,1]$ there are no nontrivial order continuous functionals (cf. [388]) but the Riemann integral gives an example of a nonsingular functional.

The generalized Yosida–Hewitt Theorems 4.4.10 and 4.4.11 were obtained by A. V. Bukhvalov and M. Ya. Yakubson (see [33, 228]; as regards Theorem 2.5.8, see [33]). The scalar case of Theorem 4.4.10 was established by W. A. J. Luxemburg [258]. For details about the Yosida–Hewitt Theorem and its applications, see [60, 239].

(4) We observe that Theorems 4.4.10 and 4.4.11 can be slightly improved. Namely, it suffices to require that E possesses an order-dense ideal E_0 that admits decomposition into arbitrarily many bands with the Egorov property. An instance of such a vector lattice is provided by a vector lattice E possessing an order-dense ideal with sufficiently many o-continuous functionals. To prove this fact, we may take the order completion of such an ideal, which also has a total set of o-continuous functionals and is realizable as an ideal space (see 1.4.11).

(5) The Egorov property was introduced by W. A. J. Luxemburg and A. C. Zaanen [263]. I. I. Shamaev [343, 347] gave an interesting characterization for the Egorov property. A K_σ-space E is said to be *weakly $\sigma\sigma$-distributive* (in [343] this property is called weak σ-distributivity of countable type) if for every order-bounded double sequence $(e_{n,k}) \in E$ such that the sequence $(e_{n,k})_{k\in\mathbb{N}}$ is decreasing for each $n \in \mathbb{N}$ there is an increasing sequence (φ_n) of mappings $\varphi : \mathbb{N} \to \mathbb{N}$ with

$$\bigvee_{n\in\mathbb{N}} \bigwedge_{k\in\mathbb{N}} e_{n,k} = \bigwedge_{k\in\mathbb{N}} \bigvee_{n\in\mathbb{N}} e_{n,\varphi_k(n)}.$$

A weakly σ-distributive K_σ-space is weakly $\sigma\sigma$-distributive. The converse is true when E has the countable sup property but fails in general. If M is the set of all strictly increasing mappings $\mathbb{N} \to \mathbb{N}$ then the vector lattice $l^\infty(M)$ is weakly σ-distributive but not weakly $\sigma\sigma$-distributive [343]. The following fact was established by I. I. Shamaev [343, 347].

Theorem. *For an arbitrary K_σ-space E the following are equivalent:*

(i) *E possesses the Egorov property;*

(ii) *E is weakly $\sigma\sigma$-distributive;*

(iii) *for every $e \in E_+$ the Stone space of the principal ideal $E(e)$ possesses the property: every sequence of closed nowhere-dense sets of type G_δ is contained in a single nowhere-dense set of type G_δ.*

(6) A vector lattice E is said to have the *diagonal property* if, for any double sequence $(e_{n,k}) \in E$ with the order limits $e_n := o\text{-}\lim_{k\to+\infty} e_{n,k}$ $(n \in \mathbb{N})$ and $e := o\text{-}\lim_{n\to+\infty} e_n$, we may choose a strictly increasing sequence of indices $(k_n)_{n\in\mathbb{N}}$ with $e = o\text{-}\lim_{n\to+\infty} e_{n,k_n}$. Every K_σ-space with the diagonal property is $\sigma\sigma$-distributive; the space l^∞ is a $\sigma\sigma$-distributive K_σ-space without the diagonal property [343]. A universally complete K-space have the diagonal property if and only if it is $\sigma\sigma$-distributive [343]. It is well known that a regular K-space have the diagonal property [163, 388]. D. A. Vladimirov [382] proved that the converse assertion is weaker than the continuum hypothesis.

(7) The countable sup property of F is essential for the validity of Theorem 4.4.9. This is illustrated by the following example (see [33]). Take $E = L^\infty(0,1)$ and, using the Brothers Kreĭn–Kakutani Theorem, realize E as the space $C(Q)$ for an extremal compact space Q. Let T be the identical embedding of $E := C(Q)$ into $F := l^\infty(Q)$. We claim that $T \in (L_n^\sim(E,F))^\perp \setminus L_s^\sim(E,F)$. It is clear that T does not belong to $L_s^\sim(E,F)$. To show that $T \in (L_n^\sim(E,F))^\perp$, let T_n be the order continuous part of T and f_0 the constant function 1 on Q. Proving $T_n(f_0) = 0$ we get that $T_n = 0$. Let $\{U_\alpha\}$ be a base for neighborhoods of an arbitrary point $q \in Q$. Set $F_\alpha = Q \setminus U_\alpha$ and $\alpha_1 \geq \alpha_2$ if $U_{\alpha_1} \subseteq U_{\alpha_2}$. By complete regularity, given α, we find $f_\alpha \in C(Q)$ with $f_\alpha(q) = 0$ and $f_\alpha = 1$ over F_α. As a consequence, we may affirm that $f_\alpha \uparrow f_0$ in $C(Q)$ (Q has no isolated points since $C(Q)$ is order isomorphic to $L^\infty(0,1)$) and by [23, Theorem 4.6] $T_n(f_0) \leq \sup_\alpha T(f_\alpha)$ (in $\ell^\infty(Q)$), hence $T_n(f_0)(q) = 0$. Since q is arbitrary, we are done.

4.6.5. The extension method of Section 4.5 is essentially the classical Daniell–Stone method for constructing the Lebesgue integral.

(1) For positive sequentially G-continuous operators with values in a regular K-space the scheme was implemented for the first time and the corresponding extension result was established in [160] by L. V. Kantorovich; see also [163]. Later an analogous result was obtained by E. J. McShane [280]. The result was extended to weakly (σ,∞)-distributive K_σ-spaces by K. Matthes in [277] (the local $\aleph$-regularity condition from [277] is equivalent to (σ,∞)-distributivity). It was J. D. M Wright [405] who proved that a σ-complete vector lattice admits σ-continuous extension of "preintegrals" (= positive G-continuous operators) if and only if it is weakly σ-distributive. Another proofs of this fact were given by D. H. Fremlin [101] and B. Riečan [329, 328]. In Section 4.5 we follow D. H. Fremlin's approach and use his concept of a set σ-depressed towards zero [101].

(2) It seems likely that weak σ-distributivity condition for Boolean algebras appeared for the first time in J. von Neumann's *Lectures on Continuous Geometry* (see [294]). This work provides examples of a weakly σ-distributive Boolean algebra (the algebra of Lebesgue measurable subsets of the real line modulo zero-measure sets) and a Boolean algebra which is not weakly σ-distributive (the same algebra modulo meager sets). In [265] weak σ-distributivity is one of the characteristic properties of Boolean algebras with measure, see also [167]. For the general concept of $(\mathfrak{n},\mathfrak{m})$-distributivity see in [352]. The equivalence relation (1) $\Leftrightarrow$ (3) in 4.5.1 was established independently by J. Kelley and J. C. Oxtoby (see [167]) for Boolean algebras with the countable chain condition and by Z. T. Dikanova [77] for universally complete vector lattices with the countable chain condition. The case of a σ-complete (an order complete) vector lattice is a straightforward generalization.

(3) Let $\mathscr{A}$ be a subalgebra of a complete Boolean algebra $\mathbb{B}$. A countably additive measure $\mu : \mathscr{A} \to Y$ is called *extendable* if there is a countably additive measure $\hat{\mu} : \sigma(\mathscr{A}) \to Y$ with $\hat{\mu}(a) = \mu(a)$ for all $a \in \mathscr{A}$. A lattice-normed space (or vector lattice) Y is said to have the *measure extension property* if each countable additive measure $\mu : \mathscr{A} \to Y$ is extendable. Analogously, we say that Y has the *operator extension property* if Theorem 4.5.3 is valid for all E, G, and T. J. D. M. Wright [402] also proved that a σ-complete vector lattice has the measure extension property if and only if it is weakly σ-distributive. This result remains valid for dominated measures with values in a sequentially *bo*-complete lattice-normed space. More precisely, the following result is true ([214]):

Theorem. *For an order σ-complete vector lattice F the following are equivalent:*

(i) *F is weakly σ-distributive;*

(ii) *F has the measure extension property;*

(iii) *F has the operator extension property;*

(iv) *every sequentially bo-complete lattice-normed space Y over F has the measure extension property;*

(v) *every sequentially bo-complete lattice-normed space Y over F has the operator extension property.*

(4) In [406] J. D. M. Wright also proved that an order σ-complete vector lattice F is weakly (σ, ∞)-distributive if and only if each F-valued Baire measure on each compact space can be extended to a regular $\widehat{F}$-valued Borel measure ($\widehat{F}$ is an order completion of F). In this connection we recall one more result.

Say that Y has the strong operator extension property if Theorem 4.5.10 is valid for all E, G, and T.

Let $\mathscr{A}$ be a subalgebra of a complete Boolean algebra $\mathbb{B}$. Say that a finitely additive measure $\mu : \mathscr{A} \to Y$ is smooth if $\mu(a) = bo\text{-}\lim \mu(a_\xi)$ for every net $(a_\xi) \subset \mathscr{A}$ increasing to $a \in \mathscr{A}$ in $\mathbb{B}$. A countably additive measure $\mu : \mathscr{A} \to Y$ is called *Borel extendable* if there is a countably additive measure $\widehat{\mu} : \widehat{\mathscr{A}} \to Y$ such that (a) $\widehat{\mathscr{A}}$ is a σ-subalgebra of $\mathbb{B}$ containing $\mathscr{A}^{\uparrow}$; (b) $\widehat{\mu}$ is an extension of μ; (c) $\widehat{\mu}(\bigvee a_\xi) = bo\text{-}\lim \mu(a_\xi)$ for every increasing net $(a_\xi) \subset \mathscr{A}$. A lattice-normed space (or vector lattice) Y is said to have the *strong measure extension property* if each smooth measure $\mu : \mathscr{A} \to Y$ is Borel extendable. The following result was established in [214].

Theorem. *For an order complete vector lattice F the following are equivalent:*

(i) *F is weakly (σ, ∞)-distributive;*

(ii) *F has the strong measure extension property;*

(iii) *F has the strong operator extension property;*

(iv) *every bo-complete lattice-normed space Y over F has the strong measure extension property;*

(v) *every bo-complete lattice-normed space Y over F has the strong operator extension property.*

(5) In [277] Matthes established that a σ-complete vector lattice E is (σ,∞)-distributive if and only if the order completion of E is also weakly (σ,∞)-distributive. A similar assertion for weakly σ-distributive vector lattices is false. In [406] J. D. M. Wright (assuming the continuum hypothesis) constructed a σ-complete vector lattice which is weakly (σ,∞)-distributive but whose order completion is not weakly σ-distributive. S. A. Malyugin informed the author that such a vector lattice can be constructed without employing the continuum hypothesis. Thus, weak (σ,∞)-distributivity is a strictly stronger condition than weak σ-distributivity.

(6) Theorems 4.5.3 and 4.5.10 are proved by a constructive method. There are several extension results for dominated operators that are "pure existence theorems" by nature. We state two such results. The first is an easy consequence of the Hahn–Banach–Kantorovich Theorem, while the second was proved (see A. G. Kusraev and S. A. Malyugin [213, 217]) by the Boolean-valued representation method using the Lipecki–Plachky–Thomsen characterization of extreme extensions [248] (compare the Comments on Chapter 3). Let E and F be vector lattices with F order complete and let E_0 be a massive sublattice of E. Denote by $\varepsilon^+(S_0) \subset L^\sim(E,F)$ the set of all positive extensions to the whole E of a positive operator $S_0 : E_0 \to F$.

(6.1) Theorem. *Let (X_0, E_0) be a lattice-normed space. Suppose that $T_0 \in M(X_0, F)$, while $S_0 : E_0 \to F$ is a dominant of T_0, and $S \in \varepsilon^+(S_0)$. Then there exists a linear extension T of T_0 to the whole X such that $|T| \le S$.*

(6.2) Theorem. *Let (Y, F) be a Banach–Kantorovich space. Suppose that $T_0 \in M(E_0, Y)$ and S is an extreme point of the set $\varepsilon^+(|T_0|)$. Then there exists a unique dominated operator $T : E \to Y$ such that T is an extension of T_0 and $|T| = S$.*

Chapter 5

Disjointness Preserving Operators

In the current chapter, we study disjointness preserving operators in vector lattices and lattice-normed spaces. We will concentrate mainly on decomposition and analytical representation of disjointness preserving operators. Somewhat more general class comprises n-disjoint operators (5.2.1). A dominated operator is n-disjoint if and only if it is representable as a sum of n disjointness preserving dominated operators (5.2.7). Thus, the space of n-disjoint operators has a rather simple structure similar to that of disjointness preserving operators.

The simplest representatives of the classes of disjointness preserving operators are band preserving operators. Simplicity of these operators notwithstanding, the question about their order-boundedness is far from triviality. All band preserving operators in a universally complete vector lattice are regular if and only if the vector lattice under study is locally one-dimensional (5.1.2) and a universally complete vector lattice is locally one-dimensional if and only if its base is σ-distributive (5.1.5). There exist nondiscrete locally one-dimensional vector lattices (5.1.6, 5.1.7).

A disjointness preserving operator defines a ring homomorphism between the Boolean algebras of bands, called the shadow of an operator (5.2.2). In turn, the shadow generates a so-called shift operator (5.3.2) which is a lattice homomorphism defined on a certain order-dense ideal of the universal completion of the domain vector lattice (5.3.1). Both are closely related with the initial disjointness preserving operator and concentrate, in a sense, its multiplicative properties. Using these simplest types of operators we may construct weighted shift operators, i.e. the composites $W \circ S \circ w$ of two orthomorphisms w and W and a shift operator S (5.3.4, 5.3.9). An arbitrary disjointness preserving operator is representable as the strongly disjoint sum of weighted shift operators (5.3.6, 5.3.10). Every weighted shift dominated operator admits multiplicative representation, i.e., it can be represented as a composite of a generalized shift operator and the operator of multiplication by an operator-valued weight (5.4.4, 5.4.9). This fact allows us to construct one of the analytical representations of general disjointness preserving operators (5.4.5, 5.4.10).

The notion of decomposable operator (5.5.5) is dual to that of disjointness preserving operator. An order continuous dominated operator is decomposable if and only if its exact dominant is a Maharam operator; i.e., if the dominant is order continuous and preserves order intervals (5.5.6). Decomposable operators admit a simple analytical description (5.5.9). In particular, an analog of the Radon–Nikodým Theorem holds for them in which the role of the integral is played by the exact dominant of the operator (5.5.10).

5.1. Band Preserving Operators

In this section we handle the problem: under what conditions all band preserving operators in a universally complete vector lattice are automatically order-bounded. Band preserving operators in lattice-normed space are also considered.

5.1.1. Let G be an arbitrary universally complete vector lattice with a fixed order-unity $\mathbf{1}$. We introduce some multiplication in G that makes G into a commutative ordered algebra with unity $\mathbf{1}$. A subset $\mathscr{E} \subset G$ is said to be *locally linearly independent* if whenever $e_1, \dots, e_n \in \mathscr{E}$, $\lambda_1, \dots, \lambda_n \in \mathbb{R}$, and π is a band projection in G with $\pi(\lambda_1 e_1 + \dots + \lambda_n e_n) = 0$ we have $\pi \lambda_k e_k = 0$ for all $k := 1, \dots, n$. A maximal locally linearly independent set in G is called a *local Hamel basis* for G. Using the Kuratowski–Zorn Lemma, we may easily deduce the existence of a local Hamel basis for G.

A locally linearly independent set $\mathscr{E}$ in G is a local Hamel basis for G if and only if for every $x \in G$ there exists a partition of unity $(\pi_\xi)_{\xi \in \Xi}$ in $\mathfrak{P}(G)$ such that $\pi_\xi x$ is a finite linear combination of elements of $\pi_\xi \mathscr{E}$ for each $\xi \in \Xi$. Such representation of $\pi_\xi x$ is unique in the band $\pi_\xi(G)$.

An element $e \in G_+$ is called *locally constant* with respect to $f \in G_+$ if $e = \sup_{\xi \in \Xi} \lambda_\xi \pi_\xi f$ for some numeric family $(\lambda_\xi)_{\xi \in \Xi}$ and a family $(\pi_\xi)_{\xi \in \Xi}$ of pairwise disjoint band projections.

For each universally complete vector lattice G the following are equivalent:

(1) *all elements of G_+ are locally constant with respect to $\mathbf{1}$;*

(2) *all elements of G_+ are locally constant with respect to an arbitrary order-unity $e \in G$;*

(3) *$\{\mathbf{1}\}$ is a local Hamel basis for G;*

(4) *every local Hamel basis of G consists of pairwise disjoint members.*

$\triangleleft$ Obviously, (2) $\Rightarrow$ (1). To prove the converse, note that for an arbitrary $x \in G$ we may choose a partition of unity $(\pi_\xi)_{\xi \in \Xi}$ such that for each $\xi \in \Xi$ both $\pi_\xi x$ and

$\pi_\xi e$ are multiples of $\pi_\xi \mathbf{1}$. So, $\pi_\xi x$ is a multiple of $\pi_\xi e$. A similar argument shows that $\{\mathbf{1}\}$ is a local Hamel basis if and only if so is $\{f\}$ for every order-unity $f \in G$. Thus, if (4) holds and $\mathscr{E}$ is a local Hamel basis for G then $f := \sup\{e : e \in \mathscr{E}\}$ exists and $\{f\}$ is a local Hamel basis for G. It follows that (4) $\Rightarrow$ (3); moreover, (3) $\Rightarrow$ (1) is obvious. To complete the proof, we had to show (1) $\Rightarrow$ (4). If (4) fails then we may choose a nonzero band projection π and a local Hamel basis with two members e_1 and e_2 such that both πe_1 and πe_2 are nonzero multiples of $\{\pi \mathbf{1}\}$. Consequently, $\pi(\lambda_1 e_1 + \lambda_2 e_2) = 0$ for some $\lambda_1, \lambda_2 \in \mathbb{R}$ and we arrive at the contradictory conclusion that $\{e_1, e_2\}$ is not locally linearly independent. $\rhd$

A universally complete vector lattice G is called *locally one-dimensional* if G satisfies the equivalent conditions (1–4) of the above proposition.

5.1.2. Theorem. *Let G be a universally complete vector lattice. Then the following are equivalent:*

(1) *G is locally one-dimensional;*

(2) *every band preserving operator $T : G \to G$ is order-bounded.*

$\lhd$ (1) $\Rightarrow$ (2): First observe that a linear operator $T : G \to G$ is band preserving if and only if $\pi T = T\pi$ for every band projection π in G. Assume that T is band preserving and put $\rho := T\mathbf{1}$. Since an arbitrary $e \in G_+$ can be expressed as $e = \sup_{\xi \in \Xi} \lambda_\xi \pi_\xi \mathbf{1}$, we deduce

$$\pi_\xi Te = T(\pi_\xi e) = T(\lambda_\xi \pi_\xi \mathbf{1}) = \lambda_\xi \pi_\xi T(\mathbf{1}) = \pi_\xi(e) T(\mathbf{1}) = \pi_\xi e \rho,$$

whence $Te = \rho e$. It follows that T is a multiplication operator in G which is obviously order-bounded.

(2) $\Rightarrow$ (1): Assume that (1) is false. Fix a local Hamel basis $\mathscr{E}$ in G. According to 5.1.1 (4) we may choose $e_1, e_2 \in \mathscr{E}$ that are not disjoint. Then the band projection $\pi := [e_1] \wedge [e_2]$ is nonzero. For an arbitrary $x \in G$ there exists a partition of unity $(\pi_\xi)_{\xi \in \Xi}$ such that $\pi_\xi x$ is a finite linear combination of elements of $\mathscr{E}$. Assume the elements of $\mathscr{E}$ have been labelled so that $\pi_\xi x = \lambda_1 \pi_\xi e_1 + \lambda_2 \pi_\xi e_2 + \cdots$. Define Tx to be a unique element in G with $\pi_\xi Tx := \lambda_1 \pi \pi_\xi e_2$. It is easy to check that T is a well defined linear operator from G into itself.

Take $x, y \in G$ with $x \perp y$ and let $(\pi_\xi)_{\xi \in \Xi}$ be a partition of unity such that both $\pi_\xi x$ and $\pi_\xi y$ are finite linear combination of elements from $\mathscr{E}$. Refining the partition of unity if necessary we may also require that at least one of the elements $\pi_\xi x$ and $\pi_\xi y$ equals to zero for each $\xi \in \Xi$. If $\pi_\xi y \neq 0$ then $\pi_\xi x = 0$, so the corresponding $\lambda_1 e_1 = 0$. If $\pi \pi_\xi \neq 0$ then $\lambda_1 = 0$, and in any case $\pi_\xi Tx = 0$. It follows that $Tx \perp y$ and T is band preserving.

If T were order-bounded then T would be presentable as $Tx = ax$ $(x \in G)$ for some $a \in G$. In particular, $Te_2 = ae_2$ and, since $Te_2 = 0$ by definition, we have $0 = [e_2]|a| \geq \pi|a|$. Thus $\pi e_2 = T(\pi e_1) = a\pi e_1 = 0$, contradicting the definition of π. $\rhd$

5.1.3. A subset of a Boolean algebra whose supremum is unity is called a *cover of the algebra*. A σ-complete Boolean algebra B is called *σ-distributive* if for every double sequence $(b_{n,m})_{n,m\in\mathbb{N}}$ in B the following conditions hold:

$$\bigvee_{n\in\mathbb{N}} \bigwedge_{m\in\mathbb{N}} b_{n,m} = \bigwedge_{\varphi\in\mathbb{N}^{\mathbb{N}}} \bigvee_{n\in\mathbb{N}} b_{n,\varphi(n)}.$$

Let B be an arbitrary Boolean algebra and let C be a cover of B. A subset C_0 of the algebra B is said to be *refined* from C if, for each $c_0 \in C_0$, there exists a $c \in C$ such that $c_0 \leqslant c$. An element $b \in B$ is called refined from C if the set $\{b\}$ is refined from C, i.e., $b \leqslant c$ for some element $c \in C$. If $(C_n)_{n\in\mathbb{N}}$ is a sequence of covers of the algebra B and an element $b \in B$ is refined from each of the covers C_n $(n \in \mathbb{N})$, then we say that b is refined from the sequence $(C_n)_{n\in\mathbb{N}}$. We also refer to a cover, all elements of which are refined from the sequence $(C_n)_{n\in\mathbb{N}}$, as refined from the sequence.

Let B be a complete Boolean algebra. The following are equivalent:

(1) *the algebra B is σ-distributive;*

(2) *from every sequence of countable covers of B, it is possible to refine a (possibly, uncountable) cover;*

(3) *from every sequence of finite covers of B, it is possible to refine a (possibly, infinite) cover;*

(4) *from every sequence of countable partitions of unity in B, it is possible to refine a (possibly, uncountable) partition;*

(5) *from every sequence of finite partitions of unity in B, it is possible to refine a (possibly, infinite) partition;*

$\lhd$ A proof of the equivalence (1) $\Leftrightarrow$ (2) can be found in [352, Theorem 19.3]. The implication (2) $\Rightarrow$ (3) is obvious. Moreover, (3) implies that a cover is refined from every sequence of two-element partitions of B. The latter is a paraphrase of the following equivalent definition of σ-distributivity: for every sequence (b_n) in B it holds

$$\bigvee_{\varepsilon\in\{1,-1\}^{\mathbb{N}}} \bigwedge_{n\in\mathbb{N}} \varepsilon(n)b_n = 1$$

where $1b_n = b_n$ and $(-1)b_n$ is the complement of b_n. The equivalences (2) $\Leftrightarrow$ (4) and (3) $\Leftrightarrow$ (5) are easy consequences of the Exhaustion Principle (see 1.1.6). $\rhd$

5.1.4. Let Q be the Stone space of the base of G, and $\operatorname{Clop}(Q)$ be the Boolean algebra of all clopen subsets of Q (see 1.1.7). Say that a function $e \in C_\infty(Q)$ is *refined* from a cover C of the Boolean algebra $\operatorname{Clop}(Q)$ if, for every two points $q', q'' \in Q$ satisfying the equality $e(q') = e(q'')$, there exists an element $U \in C$

such that $q', q'' \in U$. If $(C_n)_{n\in\mathbb{N}}$ is a sequence of covers of the algebra $\mathrm{Clop}(Q)$ and a function e is refined from each of the covers C_n $(n \in \mathbb{N})$, then we say that the function e is refined from the sequence $(C_n)_{n\in\mathbb{N}}$.

To each sequence of finite covers of $\mathrm{Clop}(Q)$, *there is a function in* $C(Q)$ *refined from it.*

$\vartriangleleft$ Let $(C_n)_{n\in\mathbb{N}}$ be a sequence of finite covers of the algebra $\mathrm{Clop}(Q)$. By induction, it is not difficult to construct a sequence of partitions $P_m = \{U_1^m, U_2^m, \dots, U_{2^m}^m\}$ of the algebra $\mathrm{Clop}(Q)$ possessing the following properties:

(1) for every $n \in \mathbb{N}$ there is a number $m \in \mathbb{N}$ such that the partition P_m is refined from the cover C_n;

(2) $U_n^m = U_{2j-1}^{m+1} \vee U_{2j}^{m+1}$ for all $m \in \mathbb{N}$ and $j \in \{1, 2, \dots, 2^m\}$. Given a number $m \in \mathbb{N}$, define the two-valued function $\chi_m \in C(Q)$ as follows:

$$\chi_m := \sum_{i=1}^{2^{m-1}} \chi(U_{2i}^m),$$

where $\chi(U)$ is the characteristic function of a subset $U \subset Q$. Since the series $\sum_{m=1}^{\infty} \frac{1}{3^m}\chi_m$ is uniformly convergent, its sum e belongs to $C(Q)$. We will show that the function e is refined from $(C_n)_{n\in\mathbb{N}}$. By property (1) of the sequence $(P_m)_{m\in\mathbb{N}}$, it is sufficient for this to establish that the function e is refined from $(P_m)_{m\in\mathbb{N}}$.

Assume the contrary and consider the smallest number $m \in \mathbb{N}$ for which the function e is not refined from the partition P_m. In this case, there are two points $q', q'' \in Q$ that satisfy the equality $e(q') = e(q'')$ and belong to distinct elements of P_m. Since the function e is refined from the partition P_{m-1} (for $m > 1$), from property (2) of the sequence $(P_m)_{m\in\mathbb{N}}$ it follows that the points q' and q'' belong to adjacent elements of P_m, i.e. elements of the form U_j^m and U_{j+1}^m, where $j \in \{1, \dots, 2^m - 1\}$. For definiteness, suppose that q' belongs to an element with even subscript and q'' with odd one, i.e., $\chi_m(q') = 1$ and $\chi_m(q'') = 0$. Therefore, taking into account the fact that $\chi_k(q') = \chi_k(q'')$ for all $k \in \{1, \dots, m-1\}$, we have:

$$e(q') - e(q'') = \frac{1}{3^m} + \sum_{k=m+1}^{\infty} \frac{1}{3^k}(\chi_i(q') - \chi_k(q'')) \geqslant \frac{1}{3^m} - \sum_{k=m+1}^{\infty} \frac{1}{3^k} = \frac{1}{2 \cdot 3^m} > 0,$$

which contradicts the equality $e(q') = e(q'')$. $\vartriangleright$

5.1.5. *Theorem.* *A universally complete vector lattice is locally one-dimensional if and only if its base is* σ*-distributive.*

$\vartriangleleft$ Let G be a universally complete K-space and let Q be the Stone space of its base. Suppose that G is locally one-dimensional and consider an arbitrary sequence

$(P_n)_{n\in\mathbb{N}}$ of finite partitions of the Boolean algebra Clop(Q). According to 5.1.3, in order to prove σ-distributivity of the base of G, it is sufficient to refine a cover of Clop(Q) from $(P_n)_{n\in\mathbb{N}}$. In view of Proposition 5.1.4, one can refine a function $e \in C_\infty(Q)$ from the sequence $(P_n)_{n\in\mathbb{N}}$. Since G is locally one-dimensional, there exists a partition $(U_\xi)_{\xi\in\Xi}$ of the algebra Clop(Q) such that the function e is constant on each of the sets U_ξ. Show that the partition $(U_\xi)_{\xi\in\Xi}$ is refined from the sequence $(P_n)_{n\in\mathbb{N}}$. To this end, we fix arbitrary indices $\xi \in \Xi$ and $n \in \mathbb{N}$ and establish that the set U_ξ is refined from the partition P_n. We may assume that $U_\xi \neq \varnothing$. Let q_0 be an element of U_ξ. Finiteness of the partition P_n allows us to find an element U of it such that $q_0 \in U$. It remains to observe that $U_\xi \subset U$. Indeed, if $q \in U_\xi$ then $e(q) = e(q_0)$ and, since the function e is refined from P_n, the points q and q_0 belong to the same element of the partition P_n, i.e., $q \in U$.

Now, assume that the base of G is σ-distributive and consider an arbitrary function $e \in C_\infty(Q)$. According to 5.1.1 (1), it is sufficient to construct a partition $(U_\xi)_{\xi\in\Xi}$ of the algebra Clop(Q) such that the function e is constant on each of the sets U_ξ. For every natural n and every integer m, denote by U_m^n the interior of the closure of the set of all points $q \in Q$ for which $\frac{m}{n} \leqslant e(q) < \frac{m+1}{n}$ and define $P_n := \{U_m^n : m \in \mathbb{Z}\}$. Due to 5.1.3 (4), from the sequence $(P_n)_{n\in\mathbb{N}}$ of countable partitions of the algebra Clop(Q), we may refine some partition $(U_\xi)_{\xi\in\Xi}$. It is not difficult to become convinced that the resultant partition was desired. ▷

5.1.6. Thus, the question about existence of a purely nonatomic locally one-dimensional K-space is reduced to existence of a purely nonatomic σ-distributive complete Boolean algebra. In the next two subsections will be constructed such an algebra.

A Boolean algebra B is called *σ-inductive* if every decreasing sequence of nonzero elements of B admits a nonzero lower bound. A subalgebra B_0 of a Boolean algebra B is said to be *dense* if, for every nonzero element $b \in B$, there exists a nonzero element $b_0 \in B_0$ such that $b_0 \leqslant b$.

If a σ-complete Boolean algebra contains a σ-inductive dense subalgebra then it is σ-distributive.

◁ Let B be a σ-complete Boolean algebra and let B_0 be a σ-inductive dense subalgebra of B. Consider an arbitrary sequence $(C_n)_{n\in\mathbb{N}}$ of countable covers of B, denote by C the set of all elements in B that are refined from $(C_n)_{n\in\mathbb{N}}$, and assume by way of contradiction that C is not a cover of B. Then there exists a nonzero element $b \in B$ that is disjoint with all elements of C.

By induction, we construct sequences $(b_n)_{n\in\mathbb{N}}$ and $(c_n)_{n\in\mathbb{N}}$ as follows. Let c_1 be an element of C_1 such that $b \wedge c_1 \neq 0$. Since B_0 is dense, there is an element $b_1 \in B_0$ such that $0 < b_1 \leqslant b \wedge c_1$. Suppose that the elements b_n and c_n are already constructed. Let c_{n+1} be an element of C_{n+1} such that $b_n \wedge c_{n+1} \neq 0$. As b_{n+1} we take an arbitrary element of B_0 that satisfies the inequalities $0 < b_{n+1} \leqslant b_n \wedge c_{n+1}$.

Thus, we have constructed sequences $(b_n)_{n\in\mathbb{N}}$ and $(c_n)_{n\in\mathbb{N}}$ such that $b_n \in B_0$, $b_n \leqslant c_n \in C_n$ and $0 < b_{n+1} \leqslant b_n \leqslant b$ for all $n \in \mathbb{N}$. Due to the fact that B_0 is σ-inductive, it contains an element b_0 which satisfies $b_0 \leqslant b_n$ for all $n \in \mathbb{N}$. In view of the inequalities $b_0 \leqslant c_n$, the element b_0 is refined from $(C_n)_{n\in\mathbb{N}}$, i.e., belongs to C. On the other hand, $b_0 \leqslant b$, which contradicts disjointness of b with all elements of C. ▷

5.1.7. As is known, for every Boolean algebra B, there exists a complete Boolean algebra $\overline{B}$ that contains B as a dense subalgebra (see [352; Section 35]). Such an algebra $\overline{B}$ is unique to within isomorphism and called a *completion* of B. Obviously, a completion of a purely nonatomic Boolean algebra is purely nonatomic. In addition, due to 5.1.6, a completion of a σ-inductive algebra is σ-distributive. Therefore, in order to prove existence of a purely nonatomic σ-distributive complete Boolean algebra, it is sufficient to present an arbitrary purely nonatomic σ-inductive Boolean algebra. Examples of such algebras are readily available. For the sake of completeness, we present here one of the simplest constructions.

Let B be the Boolean algebra of all subsets of $\mathbb{N}$ and let I be the ideal of B consisting of all finite subsets of $\mathbb{N}$. Then the factor algebra B/I is purely nonatomic and σ-inductive.

◁ Pure nonatomicity of the algebra B/I is obvious. In order to prove that the algebra is σ-inductive, it is sufficient to consider an arbitrary decreasing sequence $(b_n)_{n\in\mathbb{N}}$ of infinite subsets of $\mathbb{N}$ and construct an infinite subset $b \subset \mathbb{N}$ such that the difference $b\backslash b_n$ is finite for each $n \in \mathbb{N}$. We can easily obtain the desired set $b = \{m_n : n \in \mathbb{N}\}$ with the help of induction by letting $m_1 := \min b_1$ and $m_{n+1} := \min\{m \in b_{n+1} : m > m_n\}$. ▷

5.1.8. Now, introduce band preserving operators in lattice-normed spaces. Let E and F be vector sublattices of a vector lattice G, and let X and Y be lattice-normed spaces over E and F, respectively. As in 2.1.2, we call elements $x \in X$ and $y \in Y$ *disjoint* and write $x \perp y$ whenever $|x| \wedge |y| = 0$. The operator T is said to be *band preserving* if $x \perp y$ implies $Tx \perp y$ for all $x \in X$ and $y \in Y$. Clearly, if $|Y|$ is a minorant subset in G_+ then T is band preserving if it satisfies either of the following equivalent conditions:

(a) $|x| \perp g$ implies $|Tx| \perp g$ for all $x \in X$ and $g \in G$;

(b) $\{|Tx|\}^{\perp\perp} \subset \{|x|\}^{\perp\perp}$ for all $x \in X$, where $\{\cdot\}^{\perp\perp}$ is calculated in G.

Obviously, the last definition agrees with the notion of band preserving operator in vector lattices (see 3.3.2). In the rest of this section we assume that E and F are order-dense ideals in a K-space G. Recall also that we identify the Boolean algebras $\mathfrak{P}(G)$, $\mathfrak{P}(E)$, and $\mathscr{P}(X)$ whenever X is decomposable.

(1) *Let X be decomposable. A dominated operator $T : X \to Y$ is band preserving if and only if $|T| \in \mathrm{Orth}(E, F)$. In particular every dominated band preserving operator is bo-continuous.*

$\lhd$ Since a band preserving operator T preserves disjointness, by virtue of 4.1.11 (2) we have

$$|T|e := \sup\{|Tx| : |x| \leq e^+\} \quad (e \in E).$$

From this it follows that $e \perp f$ implies $|T|e \perp f$ for every $f \in F$, because $|x| \perp f$ for all $x \in X$ with $|x| \leq e$. Thus, $|T|$ is band preserving. The converse is trivial. $\rhd$

(2) *If X and Y are decomposable then a dominated operator $T : X \to Y$ is band preserving if and only if $T\pi = \pi T$ for all $\pi \in \mathfrak{P}(G)$.*

$\lhd$ If T is dominated and band preserving then, in view of (1) and 2.1.3

$$|\pi T(\pi^\perp x)| = \pi|T(\pi^\perp x)| \leq \pi|T||(\pi^\perp x)| = \pi\pi^\perp|T|(|x|) = 0,$$

whenever $x \in X$ and $\pi \in \mathfrak{P}(G)$. This implies $\pi T \pi^\perp = 0$ or $\pi T = \pi T \pi^\perp$. Substituting $\pi^\perp$ for π in the latter identity we obtain $T\pi = \pi T \pi^\perp$, so that $T\pi = \pi T$. The converse is obvious. $\rhd$

(3) *Let X be a bo-complete LNS, and $T \in \mathrm{Orth}(X, Y)$. For every orthomorphism $a \in \mathrm{Orth}(E)$ the product $aT(x)$ is defined in Y and the equality $T(ax) = aT(x)$ holds.*

$\lhd$ It follows from (2) and the Freudenthal Spectral Theorem. $\rhd$

We will say that an operator $T : X \to Y$ is *semibounded* whenever the following condition holds: if a sequence (x_n) in X *br*-converges to zero then $\inf\{|T(x_n)| : n \in \mathbb{N}\} = 0$.

5.1.10. Theorem. *The following properties of a band preserving operator T from a BKS into an LNS are equivalent:*

(1) *T is dominated;*

(2) *T is order-bounded;*

(3) *T is semibounded.*

$\lhd$ The implications (1) $\Rightarrow$ (2) $\Rightarrow$ (3) are obvious, (2) $\Rightarrow$ (1) follows from 4.1.11. It remains to show that (3) $\Rightarrow$ (1). Assume that a LNS X is order complete and an operator $T : X \to Y$ is band preserving and semibounded. Fix an arbitrary positive element $e \in G$ and prove that the set $\{|Tx| : |u| \leqslant e\}$ is order-bounded in F. We split the proof into two steps.

(a) Show first that the set $\{|Tu| : |u| \leqslant e\}$ is order-bounded in the universally complete K-space G. Without loss of generality, we may assume that $G = C_\infty(Q)$, where Q is an extremal compact space. Denote by D the totality of those points $q \in Q$, for which $\sup\{|Tu|(q) : |u| \leqslant e\} = \infty$. Assume that the set $\{|Tu| : |u| \leqslant e\}$ is not bounded in $C_\infty(Q)$. Then, according to [163; Chapter XIII, Theorem 2.32], the clopen set $U := \operatorname{int}\operatorname{cl} D$ is nonempty. For each natural n and each point $q \in U \cap D$, consider an element $u_n^q \in \mathscr{U}$ satisfying the conditions $|u_n^q| \leqslant e$ and $|Tu_n^q|(q) > n$. Denote by U_n^q a clopen subset of Q such that $q \in U_n^q \subset U$ and $|Tu_n^q|(p) \geqslant n$ for all $p \in U_n^q$. It is clear that, for each $n \in \mathbb{N}$ the relation $\sup_{q \in U \cap D} U_n^q = U$ holds in the Boolean algebra $\operatorname{Clop}(Q)$. In view of the Exhaustion Principle, there exists a family $(V_n^q)_{q \in U \cap D}$ of pairwise disjoint elements of $\operatorname{Clop}(Q)$ such that $V_n^q \subset U_n^q$ for all $q \in U \cap D$, and $\sup_{q \in U \cap D} V_n^q = U$. We know that the sum $o\text{-}\sum_{q \in U \cap D} \langle V_n^q \rangle u_n^q$ exists in the BKS X. Denote this sum by u_n. For all $n \in \mathbb{N}$ and $q \in U \cap D$, we have

$$\langle V_n^q \rangle |Tu_n| = |T \langle V_n^q \rangle u_n| = |T \langle V_n^q \rangle u_n^q| = \langle V_n^q \rangle |Tu_n^q| \geqslant n\chi_{V_n^q}.$$

After passing to the supremum over $q \in U \cap D$, we obtain $|Tu_n| \geqslant n\chi_U$ for all $n \in \mathbb{N}$; which, together with the inequalities $|u_n| \leqslant e$, yields a contradiction with semiboundedness of T.

(b) Denote by f the upper envelope of the set $\{|Tu| : |u| \leqslant e\}$ in the K-space G and show that $f \in F$. Without loss of generality, we may assume that $f > 0$ on some comeager subset of Q. Then, according to 1.4.2 (4), the set of all points $q \in Q$, for which $0 < \sup\{|Tu|(q) : |u| \leqslant e\} = f(q) < \infty$, is comeager in Q. For any such point q, consider an element $u_q \in \mathscr{U}$ satisfying the conditions $|u_q| \leqslant e$ and $|Tu_q|(q) > f(q)/2$. By repeating the idea of step (a) and "mixing up" the elements u_q in an appropriate way, we may construct an element $u \in \mathscr{U}$ such that $|Tu| \geqslant f/2$; whence the containment $f \in F$ follows directly. ▷

5.2. n-Disjoint Operators

The main goal of the present section is to describe the order ideal that is generated by disjointness preserving operators (= d-homomorphisms) in the space of dominated operators. For this purpose a new class of n-disjoint operators is introduced.

5.2.1. Let X and Y be lattice-normed spaces over vector lattices E and F respectively. An operator $T : X \to Y$ is said to be *n-disjoint* if, for all $n+1$ pairwise disjoint elements $x_0, \dots, x_n \in X$, the infimum of the set $\{|Tx_k| : k := 0, 1, \dots, n\}$ equals zero; symbolically:

$$(\forall x_0, x_1 \dots, x_n \in X)\, x_k \perp x_l \, (k \neq l) \;\Rightarrow\; |Tx_0| \wedge \dots \wedge |Tx_n| = 0.$$

Observe that an operator $T : X \to Y$ is disjointness preserving (or, which is the same, d-homomorphism) if it is 1-disjoint, i.e., $x_1 \perp x_2$ implies $Tx_1 \perp Tx_2$ for all $x_1, x_2 \in X$.

(1) *Let X be decomposable and let F be order complete. A dominated operator $T \in M(X, Y)$ is n-disjoint if and only if its exact dominant $|T|$ is an n-disjoint operator from E into F.*

$\triangleleft$ Sufficiency is obvious. Suppose that the operator T is n-disjoint. Take pairwise disjoint elements $e_0, \dots, e_n \in E_+$ and assign $f_k := \sup\{|Tu| : |u| \le e_k\}$. If $|u_k| \le e_k$ then $u_k \perp u_l$ $(k \ne l)$; therefore, $|Tu_0| \wedge \dots \wedge |Tu_n| = 0$. Passing to the supremum over $u_0, \dots, u_n$ in the last equality, we obtain $f_0 \wedge \dots \wedge f_n = 0$. If $|x_1| + \dots + |x_m| \le e_k$ then $\sum_{l=1}^m |Tx_l| \in \{f_k\}^{\perp\perp}$; consequently, $|T|e_k \in \{f_k\}^{\perp\perp}$ according to 4.1.5. Thus,

$$|T|e_0 \wedge \dots \wedge |T|e_n \in \{f_0\}^{\perp\perp} \cap \dots \cap \{f_k\}^{\perp\perp} = \{0\};$$

hence, $|T|e_0 \wedge \dots \wedge |T|e_n = 0$. $\triangleright$

(2) *A positive operator $S : E \to F$ is n-disjoint if and only if*

$$S\left(\bigvee_{k=0}^{n} x_k\right) = \bigvee_{k=0}^{n} S(x_0 \vee \dots \vee x_{k-1} \vee x_{k+1} \vee \dots \vee x_n)$$

for all $x_0, \dots, x_n \in X$ with $x_{-1} := x_n$ and $x_{n+1} := x_0$.

$\triangleleft$ Let S be n-disjoint and take arbitrary $x_0, \dots, x_n \in X$. If $y_k := x_0 \vee \dots \vee x_n - x_0 \vee \dots \vee x_{k-1} \vee x_{k+1} \vee \dots \vee x_n$ then $y_k \ge 0$ and $y_k \perp y_l$ $(k \ne l)$. Thus, $Sy_0 \wedge \dots \wedge Sy_n = 0$, which is equivalent to the required identity. Conversely, take a pairwise disjoint collection $x_0, \dots, x_n \in X$. Without loss of generality we may assume that x_k are positive. Then

$$\sum_{k=0}^{n} Sx_k = S\left(\sum_{k=0}^{n} x_k\right) = S\left(\bigvee_{k=0}^{n} x_k\right)$$

$$= \bigvee_{k=0}^{n} S(x_0 \vee \dots \vee x_{k-1} \vee x_{k+1} \vee \dots \vee x_n) = \bigvee_{k=0}^{n} \sum_{\substack{l=0,\dots,n \\ l \ne k}} Sx_l,$$

whence we deduce

$$\bigwedge_{k=0}^{n} Sx_k = -\bigvee_{k=0}^{n} (-Sx_k) = -\bigvee_{k=0}^{n} \left(\sum_{\substack{l=0,\dots,n \\ l \ne k}} Sx_l - \sum_{l=0}^{n} Sx_l \right)$$

$$= -\bigvee_{k=0}^{n} \sum_{\substack{l=0,\dots,n \\ l \ne k}} Sx_l + \sum_{l=0}^{n} Sx_l = 0. \ \triangleright$$

(3) *Under the hypotheses of* (1) *a dominated operator* $T : X \to Y$ *is disjointness preserving if and only if its exact dominant* $|T| : E \to F$ *is a lattice homomorphism. In particular, every positive disjointness preserving operator from* E *to* F *is a lattice homomorphism.*

$\lhd$ In view of (1) we need only to prove the second part. For all $x, y \in E$ the elements $x - x \wedge y$ and $y - x \wedge y$ are positive and disjoint. If $0 \leq T \in L^{\sim}(E, F)$ is disjointness preserving then $T(x - x \wedge y)$ and $T(y - x \wedge y)$ are also positive and disjoint. Thus, $0 = (Tx - z) \wedge (Ty - z) = T(x) \wedge T(y) - z$, where $z = T(x \wedge y)$. $\rhd$

5.2.2. Suppose that X and Y are decomposable lattice-normed spaces and $Y_T := T(X)^{\perp\perp}$. The *shadow* of an operator $T : X \to Y$ is the mapping $\operatorname{shdw} := \operatorname{shdw}(T) : \mathscr{P}(X) \to \mathscr{P}(Y_T)$ defined by the formula $\operatorname{shdw}(\pi) = [T\pi(X)]$. In other words, $\operatorname{shdw}(\pi)$ is the band projection onto $(T\pi(X))^{\perp\perp}$. Clearly, $\operatorname{shdw}(\pi) = \sup\{[T\pi x] : x \in X\}$.

(1) *Let* E *and* F *have the projection property. A linear operator* $T : X \to Y$ *is disjointness preserving if and only if its shadow* $\operatorname{shdw}(T)$ *is a Boolean homomorphism.*

$\lhd$ Sufficiency is trivial; prove necessity. Without loss of generality, we may assume that $Y_T = Y$. Assume that a linear operator $T : X \to Y$ is disjointness preserving. Let $\pi_1, \pi_2 \in \mathscr{P}(Y)$ and $\pi_1 \wedge \pi_2 = \pi_1 \circ \pi_2 = 0$. Then

$$\operatorname{shdw}(\pi_1) \wedge \operatorname{shdw}(\pi_2) = \bigvee_{x_1 \in X} [T\pi_1 x_1] \wedge \bigvee_{x_2 \in X} [T\pi_2 x_2] = \bigvee_{x_1, x_2 \in X} [T\pi_1 x_1] \wedge [T\pi_2 x_2] = 0,$$

i.e., $\operatorname{shdw}(\pi_1) \perp \operatorname{shdw}(\pi_2)$. Moreover,

$$\operatorname{shdw}(\pi_1) \vee \operatorname{shdw}(\pi_2) = \bigvee_{u, v \in X} [T\pi_1 u] \vee [T\pi_2 v]$$

$$= \bigvee_{u, v \in X} [T(\pi_1 u + \pi_2 v)] = \bigvee_{x \in X} [T(\pi_1 + \pi_2) u] = \operatorname{shdw}(\pi_1 \vee \pi_2).$$

It follows that $(\operatorname{shdw}(\pi^{\perp}) = \operatorname{shdw}(\pi)^{\perp}$ for every $\pi \in \mathscr{P}(X)$ and $\operatorname{shdw}(\pi_1) \vee \operatorname{shdw}(\pi_2) = \operatorname{shdw}(\pi_1 \vee \pi_2)$ for disjoint $\pi_1, \pi_2 \in \mathscr{P}(X)$. Thus, shdw is a Boolean homomorphism, since for arbitrary $\pi_1, \pi_2 \in \mathscr{P}(X)$ we have

$$\begin{aligned} \operatorname{shdw}(\pi_1 \vee \pi_2) &= \operatorname{shdw}\big((\pi_1 \backslash \pi_2) \vee (\pi_1 \wedge \pi_2) \vee (\pi_2 \backslash \pi_1)\big) \\ &= \operatorname{shdw}(\pi_1 \backslash \pi_2) \vee \operatorname{shdw}(\pi_1 \wedge \pi_2) \vee \operatorname{shdw}(\pi_2 \backslash \pi_1) \\ &= \big(\operatorname{shdw}(\pi_1 \backslash \pi_2) \vee \operatorname{shdw}(\pi_1 \wedge \pi_2)\big) \\ &\vee \big(\operatorname{shdw}(\pi_1 \wedge \pi_2) \vee \operatorname{shdw}(\pi_2 \backslash \pi_1)\big) \\ &= \operatorname{shdw}(\pi_1) \vee \operatorname{shdw}(\pi_2). \ \rhd \end{aligned}$$

(2) *Let F be order complete and T be dominated. Then the shadows of T and $|T|$ coincide.*

$\lhd$ Denote $h := \mathrm{shdw}(T)$ and $h' := \mathrm{shdw}(|T|)$. Of course, coincidence of the functions $h : \mathscr{P}(X) \to \mathscr{P}(Y)$ and $h' : \mathscr{P}(X) \to \mathscr{P}(Y)$ is understood with the identifications $\mathscr{P}(X) = \mathfrak{P}(E)$ and $\mathscr{P}(Y) = \mathfrak{P}(F)$. The inequality $h(\pi) \leqslant h'(\pi)$ $(\pi \in \mathfrak{P}(E))$ is obvious. To prove the reverse inequality, it is sufficient to observe, that for $e \in E$, $\pi \in \mathfrak{P}(E)$, and $u_1, \dots, u_n \in X$ the inequality $|u_1| + \dots + |u_n| \leqslant \pi e$ implies

$$[|Tu_1| + \dots + |Tu_n|] = [|T\pi u_1| + \dots + |T\pi u_n|] = [T\pi u_1] \vee \dots \vee [T\pi u_n] \leqslant h(\pi),$$

and to use 4.1.5. $\rhd$

Let $h : \mathscr{P}(X) \to \mathscr{P}(Y)$ be a ring homomorphism. We say that the mapping $T : X \to Y$ is *h-o-continuous* whenever $h\text{-}\lim_{\alpha \in \mathrm{A}} x_\alpha = x$ (see 2.2.10) implies $o\text{-}\lim_{\alpha \in \mathrm{A}} Tx_\alpha = Tx$ for every net $(x_\alpha)_{\alpha \in \mathrm{A}}$ in X and every $x \in X$.

(3) Theorem. *Let E and F be K-spaces. Every disjointness preserving operator $T : E \to F$ is h-o-continuous, where h is the shadow of T.*

$\lhd$ Since the shadow of $|T|$ coincides with the shadow of T (see (2)), we may assume that the operator T is positive. To prove h-o-continuity of T, it is sufficient to consider a net $(e_\alpha)_{\alpha \in \mathrm{A}}$ in E with $h\text{-}\lim_{\alpha \in \mathrm{A}} e_\alpha = 0$ and to show that $o\text{-}\lim_{\alpha \in \mathrm{A}} Te_\alpha = 0$. According to 1.3.7 (2), o-convergence of Te_α to zero will be established if we prove that $o\text{-}\lim_{\alpha \in \mathrm{A}} [Te][(Te_\alpha - Te/n])^+] = 0$ for all $e \in E$ and $n \in \mathbb{N}$. The latter relation can be obtained as follows:

$$[Te]\big[(T(e_\alpha - e/n))^+\big] = [Te][T((e_\alpha - e/n)^+)] \leqslant h([e])h([(e_\alpha - e/n)^+]) \xrightarrow{(o)} 0. \ \rhd$$

(4) *Every disjointness preserving dominated operator from a BKS into an LNS is h-o-continuous, where h is its shadow.*

$\lhd$ The claim follows from (2) and (3). $\rhd$

5.2.3. Theorem. *Let X and Y be decomposable and let E and F be order complete. Suppose that an operator $T \in M(X, Y)$ is disjointness preserving. Then for every dominated operator $S \in M(X, Y)$ the following are equivalent:*

(1) $S \in \{T\}^{\perp\perp}$;

(2) $Sx \in \{T(x)\}^{\perp\perp}$ $(x \in X)$;

(3) $S\pi = \mathrm{shdw}(\pi)S$ $\big(\pi \in \mathscr{P}(X)\big)$;

(4) $|S|\pi = \mathrm{shdw}(\pi)|S|$ $\big(\pi \in \mathscr{P}(X)\big)$.

◁ (1) ⇒ (2): Obvious.

(2) ⇒ (3): Fix arbitrary elements $x \in X$ and $\pi \in \mathscr{P}(X)$. From (2) it follows that $S\pi x$ and $S\pi^{\perp}x$ are disjoint. Consequently, there exist a projection $\rho \in \mathscr{P}(Y)$ such that $S\pi x = \rho Sx$ and $S\pi^{\perp}x = \rho^{\perp}Sx$. Denote $h := \operatorname{shdw}(\pi)$. In order to ensure the equality $\rho Sx = h(\pi)Sx$, it is sufficient to show that $\rho[Sx] = h(\pi)[Sx]$. The relations $\rho[Sx] = [S\pi x] \leqslant h(\pi)$ imply the inequality $\rho[Sx] \leqslant h(\pi)[Sx]$. We may establish similarly that $\rho^{\perp}[Sx] \leqslant h(\pi^{\perp})[Sx]$. The two last inequalities directly imply the equality $\rho[Sx] = h(\pi)[Sx]$.

(3) ⇒ (4): This follows from 5.2.2 (2).

(4) ⇒ (1): Since $S \in \{T\}^{\perp\perp}$ means that $|S| \in \{|T|\}^{\perp\perp}$, we may assume $S := |S|$ and $T := |T|$. Suppose $V \perp T$ for some $V \leq S$ and put $U := V + T$. By the proven implication (1) ⇒ (3) $U\pi = \operatorname{shdw}(\pi)U$ $\big(\pi \in \mathscr{P}(X)\big)$. In particular, U is a lattice homomorphism, as for disjoint $e_1, e_2 \in E_+$ we have $Ue_1 \wedge Ue_2 = \operatorname{shdw}([e_1])U(e_1) \wedge \operatorname{shdw}([e_2])U(e_2) = 0$. According to 3.3.3 $V = \rho U$ and $T = \rho' U$ for some orthomorphisms $0 \leq \rho, \rho' \leq I_F$. Disjointness of T and V implies that

$$\begin{aligned} 0 &= \inf\{\rho U e_1 + \rho' U e_2 : e = e_1 + e_2\} \\ &\geq \inf\{\rho f_1 + \rho' f_2 : f_1 + f_2 = Ue\} = (\rho \wedge \rho')Ue \geq 0 \end{aligned}$$

for every $e \in E_+$. Since $(\rho \wedge \rho')f = (\rho f) \wedge (\rho' f)$ for $f \in F_+$ we obtain $Ve \wedge Te = (\rho Ue) \wedge (\rho' Ue) = 0$. But then for every $e_1, e_2 \in E_+$ we deduce $Ve_1 \wedge Te_2 = Ve_1 \wedge T(e_1 + e_2)$ and $Ve_1 \wedge Te_2 = V(e_1 + e_2) \wedge Te_2$, so that $Ve_1 \perp Te_2$. Thus, if $\pi := [e]$ then $[Ve] \perp \operatorname{shdw}(\pi)$ and simultaneously $Ve = V\pi e = \operatorname{shdw}(\pi)Ve$, i.e., $[Ve] \leq \operatorname{shdw}(\pi)$. It follows that $V = 0$. ▷

5.2.4. Our next goal is to prove that a dominated n-disjoint operator is representable as a sum of disjointness preserving operator. To this end, we need some auxiliary constructions.

Let T and S be order-bounded operators from E to F. Define the set of order-bounded operators $\mathscr{X}(S)$ and the mapping $|\!\cdot\!| : \mathscr{X}(S) \to \operatorname{Orth}(F)$ as follows:

$$\begin{aligned} \mathscr{X}(S) &:= \{T \in L^{\sim}(E, F) : (\exists \rho \in \operatorname{Orth}(F)_+)\,(|T| \leq \rho \circ S)\}, \\ |T| &:= \inf\{\rho \in \operatorname{Orth}(F)_+ : |T| \leq \rho \circ S\} \quad (T \in \mathscr{X}(S)). \end{aligned}$$

Observe that $\mathscr{X}(S)$ coincide with $\mathscr{X}(p)$ as defined in 3.3.9 if $p(x) := S(|x|)$ $(x \in E)$.

(1) *The triple $(\mathscr{X}(S), |\!\cdot\!|, \operatorname{Orth}(F))$ is a Banach–Kantorovich lattice with an order semicontinuous M-norm.*

◁ The fact that $(\mathscr{X}(S), |\!\cdot\!|, \operatorname{Orth}(F))$ is a Banach–Kantorovich space follows from 3.3.9 (2). Monotonicity of the norm $|\!\cdot\!|$ is seen from its definition. In particular,

$|T_1| \vee |T_2| \leq |T_1 \vee T_2|$ for positive T_1, T_2. But the reverse inequality is also true, as $T_1 \leq |T_1|S$ and $T_2 \leq |T_2|S$, whence $T_1 \vee T_2 \leq (|T_1| \vee |T_2|)S$. If (T_α) is an increasing family of positive operators in $\mathscr{L}(S)$ with $T = \sup_\alpha T_\alpha$ then the equivalences hold

$$|T| \leq \gamma \Leftrightarrow (\forall \alpha)\, |T_\alpha| \leq \gamma \Leftrightarrow \sup |T\alpha| \leq \gamma.$$

Thus, $|T| = \sup_\alpha |T\alpha|$. ▷

The space $\mathscr{L}(S)$ has a natural module structure over $\mathrm{Orth}(F)$ given by $\rho T := \rho \circ T$. At the same time, according to 2.1.8 $\mathscr{L}(S)$ admits the structure of a module over $\mathrm{Orth}(\mathrm{Orth}(F))$. These two structures are identical in the following sense.

(2) *For every $\rho \in \mathrm{Orth}(F)$ there exists a unique $\bar{\rho} \in \mathrm{Orth}(\mathrm{Orth}(F))$ such that $\rho \circ T = \bar{\rho} * T$ for all $T \in \mathscr{L}(S)$.*

◁ According to 3.3.2 (3) the f-algebras $\mathrm{Orth}(F)$ and $\mathrm{Orth}(\mathrm{Orth}(F))$ are isomorphic. An isomorphism is given by assigning to $\rho \in \mathrm{Orth}(F)$ the orthomorphism $\bar{\rho} : \sigma \mapsto \rho \circ \sigma$ ($\sigma \in \mathrm{Orth}(F)$). Given $T \in \mathscr{L}(S)$ and $\rho \in \mathrm{Orth}(F)$ the product $\rho * T$ is defined by $|\bar{\rho}*T| = \bar{\rho}*|T| = \rho \circ |T|$ and $|T - \bar{\rho}*T| = |T| - \bar{\rho}*|T| = \rho^\perp \circ |T|$. By virtue of 3.3.9 $\rho \circ T$ satisfies the same identities: $|T - \rho \circ T| = \rho^\perp \circ |T|$, $|\rho \circ T| = \rho \circ |T|$. Therefore, $\bar{\rho} * T = \rho \circ T$, since disjoint decomposition is unique by 2.1.2 (3). ▷

(3) *Let X be an arbitrary vector space and let $p_1, \dots, p_n : \mathrm{X} \to F$ be sublinear operators. Then the following holds:*

$$\partial(p_1 \vee \dots \vee p_n) = \bigcup_{\substack{\sigma_1,\dots,\sigma_n \in \mathrm{Orth}(F)_+ \\ \sigma_1 + \dots + \sigma_n = I_F}} \left\{ \partial(\sigma_1 \circ p_1) + \dots + \partial(\sigma_n \circ p_n) \right\}.$$

◁ The proof can be found in [209]. ▷

(4) *Let $T \in L^\sim(E, F)$ and $p(x_1, \dots, x_n) := T(x_1 \vee \dots \vee x_n)$. Then p is a sublinear operator from E^n to F and*

$$\partial p = \left\{ (T_1, \dots, T_n) : 0 \leq T_k \in L^\sim(E, F),\ \sum_{k=1}^n T_k = T \right\},$$

where $(T_1, \dots, T_n)$ denotes the linear operator $(x_1, \dots, x_n) \mapsto (T_1 x_1 + \dots + T_n x_n)$.

◁ If $\mathscr{U}$ is the right-hand side of the desired equality then obviously $\partial p \supset \mathscr{U}$. Conversely, suppose that $0 \leq T_k \in L^\sim(E, F)$ and $T_1 x_1 + \dots + T_n x_n \leq T(x_1 \vee \dots \vee x_n)$ for all $x_k \in E$ ($k = 1, \dots, n$). Substituting $x_l = 0$ for $l \neq k$ and $x_k \leq 0$ in this inequality we get $T_k \geq 0$. Putting $x_1 = \dots = x_n$ we arrive at $T = T_1 + \dots + T_n$. ▷

5.2.5. Theorem. *Let E and F be vector lattices with F order complete, and let T be an order-bounded operator from E to F. The following are equivalent:*

(1) *T is an n-disjoint operator;*

(2) *for every collection of $n+1$ positive operators $T_0, \dots, T_n \in L^{\sim}(E, F)$ with $|T| = T_0 + \dots + T_n$ there exist collections of operators $\{T_{k,l} : k, l := 0, 1, \dots, n\}$ and $\{\sigma_l : l = 0, 1, \dots, n\}$ such that*

$$0 \le \sigma_l \in \mathrm{Orth}(F), \quad 0 \le T_{k,l} \in L^{\sim}(E, F), \quad T_{k,k} = 0,$$
$$\sum_{l=1}^{n} \sigma_l = I_F, \quad \sum_{k=1}^{n} \sigma_l T_{k,l} = |T|, \quad \sum_{l=1}^{n} \sigma_l T_{k,l} = T_k \quad (k, l = 0, 1, \dots, n);$$

(3) *for a collection of $n+1$ pairwise disjoint operators $T_0, \dots, T_n \in L^{\sim}(E, F)$ with $|T| = T_0 + \dots + T_n$ there exists a collection of orthomorphisms $\sigma_0, \dots, \sigma_n \in \mathrm{Orth}(F)_+$ such that $\sigma_0 + \dots + \sigma_n = I_F$ and $\sigma_k \circ T_k = 0$ $(k = 0, 1, \dots, n)$;*

(4) *for a collection of $n+1$ pairwise disjoint operators $T_0, \dots, T_n \in L^{\sim}(E, F)$ with $|T| = T_0 + \dots + T_n$ there exists a partition of unity $\pi_0, \dots, \pi_n$ in $\mathfrak{P}(F)$ such that $\pi_k \circ T_k = 0$ $(k = 0, \dots, n)$;*

(5) *T is a norm-n-decomposable element of the Banach–Kantorovich lattice $\mathscr{L}(T)$.*

◁ Without loss of generality, we may assume that $T \ge 0$.

(1) ⇔ (2): Define the operators $p, p_0, p_1, \dots, p_n : E^{n+1} \to F$ by

$$p(x_0, x_1, \dots, x_n) := T(x_0 \vee x_1 \vee \dots \vee x_n),$$
$$p_k(x_0, x_1, \dots, x_n) := T(x_0 \vee \dots \vee x_{k-1} \vee x_{k+1} \vee \dots \vee x_n).$$

It is easy to check that $p, p_0, \dots, p_n$ are increasing sublinear operators. In view of 5.2.1 (2) $p = p_0 \vee p_1 \vee \dots \vee p_n$. The latter is equivalent to $\partial p = \partial(p_0 \vee p_1 \vee \dots \vee p_n)$ by virtue of 3.1.9 (1). Using 5.2.4 (4), we deduce

$$\partial p = \left\{ (T_0, T_1, \dots, T_n) : 0 \le T_k \in L^{\sim}(E, F), \sum_{k=0}^{n} T_k = T \right\},$$
$$\partial p_l = \left\{ (T_{0,l}, T_{1,l}, \dots, T_{n,l}) : 0 \le T_{k,l} \in L^{\sim}(E, F), T_{k,k} = 0, \sum_{k=0}^{n} T_k = T \right\}.$$

It remains to apply 5.2.4 (3).

(2) ⇒ (3): Assume that $T = T_0 + \dots + T_n$ with pairwise disjoint $T_0, \dots, T_n \in L^{\sim}(E, F)_+$. Let $T_{k,l}$ and σ_l be as in (2). Then $\sigma_l T_{k,l} \perp T_l$ $(k \ne l)$ and, by summing the left-hand side over $k = 0, \dots, n$, we arrive at $\sigma_l T \perp T_l$. But $\sigma_k T_k \le T_k$ and $\sigma_k T_k \le \sigma_k T$, so that $\sigma_k T_k = 0$.

(3) $\Leftrightarrow$ (4) $\Leftrightarrow$ (5): These equivalences follow from Definitions 2.1.9 and 5.2.4 (2).

(5) $\Rightarrow$ (1): Take a collection $e_0, \dots, e_n$ of pairwise disjoint elements in E. Put $\pi_k := \pi_{e_k}$ $(k = 1, \dots, n)$ and $\pi_0 := (\pi_1 + \dots + \pi_n)^{\perp}$ where the band projections $\pi_1, \dots, \pi_n$ in $L^{\sim}(E, F)$ are defined as in 3.1.4 (5). If $T_k := \pi_k T$ then $T_k \perp T_l$ $(k \neq l)$ according to 3.1.4 (2) and $T_k(e_k) = T(e_k)$ in view of 3.1.3 (1). Moreover, $T = T_0 + T_1 + \dots + T_n$, so that $|T_0| \wedge \dots \wedge |T_n| = 0$ by Definition 2.1.9. But $|Te_k| = |T_k e_k| \leq T_k(|e_k|) \leq |T_k| T(|e_k|) \leq |T_k| f$ where $f := T(|e_0|) \vee \dots \vee T(|e_n|)$. $\rhd$

5.2.6. Let X be a lattice-ordered module over a lattice-ordered ring A. An element $x \in X$ is called *A-discrete* or *module-discrete* whenever for any $y \in X$ with $0 \leq y \leq |x|$ there exists $a \in A$ such that $0 \leq a \leq \mathbf{1}_A$ and $y = a|x|$. In other words $x \in X$ is A-discrete if and only if the equality of order intervals holds: $[0, |x|] = [0, \mathbf{1}_A]|x|$. Obviously, for the vector lattices E and F, with F order complete, $L^{\sim}(E, F)$ is a lattice-ordered $\mathrm{Orth}(F)$-module. Theorem 3.3.3 claims that an operator $T \in L^{\sim}(E, F)$ is an $\mathrm{Orth}(F)$-discrete if and only if T is disjointness preserving.

(1) *Let X be a decomposable br-complete lattice-normed lattice over a K_σ-space E. Assume that X has the strong Freudenthal property. Then an arbitrary element $x \in X$ is norm-indecomposable if and only if it is* Orth(F)*-discrete.*

$\lhd$ Assume that $x \in X$ is $\mathrm{Orth}(F)$-discrete and $|x| = u + v$ for order disjoint elements $u, v \in X_+$. By assumption there exist positive orthomorphisms $\rho, \sigma \in \mathrm{Orth}(F)$ such that $u = \rho|x|$ and $v = \sigma|x|$. Since X admits a compatible structure of a module over $\mathrm{Orth}(F)$ we may write

$$
\begin{aligned}
|u| \wedge |v| &= |\rho|x|| \wedge |\sigma|x|| = \rho|x| \wedge \sigma|x| \\
&= (\rho \wedge \sigma)|x| = |(\rho \wedge \sigma)|x|| = |(\rho|x|) \wedge (\sigma|x|)| = |u \wedge v| = 0.
\end{aligned}
$$

Let x be a positive norm-indecomposable element. If e is a fragment of x then by definition $|e| \wedge |x - e| = 0$ and $|x - e| = |x| - |e|$ in view of 2.1.2 (2). Thus, e is a fragment of x if and only if $|e|$ is a fragment of $|x|$. If π is the projection onto the band $|e|$ then $|\pi x - e| = \pi|x - e| = \pi(|x| - |e|) = 0$, so that $e = \pi x$. It follows that every finite-valued element $x_0 := \sum_{k=1}^{n} \lambda_k e_k$ in X where e_k is a fragment of x is representable as $x_0 = \sigma x$ with $\pi_k = [e_k]$, $\sigma := \sum_{k=1}^{n} \lambda_k \pi_k \in \mathrm{Orth}(F)$. Using the Freudenthal Spectral Theorem and br-completeness of X, we conclude that x is discrete. $\rhd$

(2) *Assume that the hypothesis of (1) are satisfied. Then every norm-n-decomposable element is a sum of n pairwise norm disjoint module discrete elements.*

$\lhd$ It follows from (1) and 2.1.10. $\rhd$

5.2.7. Theorem. *Suppose that X is decomposable and F is order complete. A dominated operator $T : X \to Y$ is n-disjoint if and only if T is representable as a sum of n disjointness preserving dominated operators.*

$\triangleleft$ Apply 5.2.1 (1), 5.2.5 (5), 2.1.10, and decomposability of the exact dominant (see 4.2.6). $\triangleright$

Denote by $H := H(E, F)$ the o-ideal generated in $L^{\sim}(E, F)$ by lattice homomorphisms and assign $M_H(X, Y) := \{T \in M(X, Y) : |T| \in H\}$. A positive operator preserves disjointness if and only if it is a lattice homomorphism. Therefore, from 3.3.3, 5.2.1 (1) and 5.2.7 it follows that if F is order complete then H consists of all order-bounded operators n-disjoint for some $n \in \mathbb{N}$. In addition, if X is decomposable then $M_H(X, Y)$ consists of all dominated operators n-disjoint for some $n \in \mathbb{N}$. At the same time, these spaces consist of finite sums of disjointness preserving operators. Thus, the space M_H has a rather simple structure modulo d-homomorphisms.

We close this section with the boundedness criteria for disjointness preserving operators. As we have seen in 5.1.10 a band preserving operator is order-bounded if and only if it is semibounded. This is not true for general disjointness preserving operators but remains valid for a disjointness preserving operator defined on a vector lattice.

5.2.8. Theorem *Let T be a disjointness preserving linear operator on a vector lattice E with values in a lattice-normed space Y. Then the following are equivalent:*

(1) *T is dominated;*

(2) *T is order-bounded;*

(3) *T is semibounded.*

$\triangleleft$ The implications (1) $\Rightarrow$ (2) $\Rightarrow$ (3) are obvious, (2) $\Rightarrow$ (1) follows from 4.1.11. It remains to show that (3) $\Rightarrow$ (2). We will split the proof into several elementary steps.

(1): Let us first observe that the claim is true if E possesses the strong Freudenthal property. This step is not necessary but in this case the proof becomes particularly simple and clear. Indeed, assume that an operator T enjoys condition (3). Take arbitrary elements $x, y \in E$ with $|x| \leqslant |y|$, and denote by $\mathscr{S}$ the set

$$\left\{ \sum_{k=1}^{n} \pi_k \lambda_k |y| : \pi_k \in \mathfrak{P}(E),\ |\lambda_k| \leqslant 1 \right\}.$$

It is not difficult to become convinced that $|Ts| \leqslant |Ty|$ for all $s \in \mathscr{S}$. Moreover, by the strong Freudenthal property, there exists a sequence $(s_n)_{n \in \mathbb{N}}$ in $\mathscr{S}$ that is relatively y-uniformly convergent to x. Condition (3) together with the relations $|Tx| \leqslant |Tx - Ts_n| + |Ty|$ $(n \in \mathbb{N})$ now yields the desired inequality $|Tx| \leqslant |Ty|$.

The general case is considered in (2)–(6).

(2): If $e_1, e_2 \in E$ then $|Te_1| \wedge |Te_2| \leq |T(e_1 \wedge e_2)|$.

$\lhd$ Indeed, $(e_1 - e_1 \wedge e_2) \perp (e_2 - e_1 \wedge e_2)$, so that $T(e_1 - e_1 \wedge e_2) \perp T(e_2 - e_1 \wedge e_2)$. Hence

$$\begin{aligned} |Te_1| \wedge |Te_2| - |T(e_1 \wedge e_2)| &= (|Te_1| - |T(e_1 \wedge e_2)|) \wedge (|Te_2| - |T(e_1 \wedge e_2)|) \\ &\leq |T(e_1 - e_1 \wedge e_2)| \wedge |T(e_2 - e_1 \wedge e_2)| = 0. \ \rhd \end{aligned}$$

Take $x, y \in E$ with $0 \leq y \leq x$ and prove that $|Ty| \leq |Tx|$, so that T is certainly order-bounded.

(3): If $u_\lambda := |T((y - \lambda x)^+)|$ and $v_\lambda := |T((y - \lambda x)^-)|$ then for each $\lambda \in [0, 1]$

$$u_\lambda \wedge v_\lambda \geq |Ty| - |Tx|; \quad 0 \leq \mu \leq \lambda \ \Rightarrow \ u_\lambda \wedge v_\mu = 0.$$

$\lhd$ If $\lambda \in [0, 1]$ then $(y - \lambda x)^+ \perp (y - \lambda x)^-$, so that

$$\begin{aligned} |T((y - \lambda x)^+)| \vee |T((y - \lambda x)^-)| &= |T((y - \lambda x)^+)| + |T((y - \lambda x)^-)| \\ &\geq |T((y - \lambda x)^+) - T((y - \lambda x)^-)| = |T(y - \lambda x)| \\ &\geq |Ty| - \lambda |Tx| \geq |Ty| - |Tx|. \end{aligned}$$

Note also that if $0 \leq \mu \leq \lambda$ then $(y - \lambda x)^+ \perp (y - \mu x)^-$, so that $u_\lambda \wedge v_\mu = 0$. $\rhd$

(4): For a collection of reals $0 = \lambda_0 < \lambda_1 < \lambda_2 < \cdots < \lambda_n = 1$ the inequality holds:

$$u_{\lambda_1} \vee \bigvee_{i=2}^{n} (u_{\lambda_{i-1}} \wedge v_{\lambda_i}) \geq \bigvee_{i=1}^{n} (u_{\lambda_i} \wedge u_{\lambda_i}).$$

$\lhd$ Put

$$f_k = v_{\lambda_1} \vee \bigvee_{i=2}^{k} (u_{\lambda_{i-1}} \wedge v_{\lambda_i}) \vee u_{\lambda_k} \quad (k = 2, \dots, n),$$

and observe that

$$\begin{aligned} f_k \wedge v_{\lambda_{k+1}} &= (v_{\lambda_1} \wedge v_{\lambda_{k+1}}) \vee \bigvee_{i=2}^{k} (u_{\lambda_{i-1}} \wedge v_{\lambda_i} \wedge v_{\lambda_{k+1}}) \vee (u_{\lambda_k} \wedge v_{\lambda_{k+1}}) \\ &\leq v_{\lambda_1} \vee \bigvee_{i=2}^{k+1} (u_{\lambda_{i-1}} \wedge v_{\lambda_i}). \end{aligned}$$

Moreover, using (3), we deduce

$$\begin{aligned} f_k \wedge u_{\lambda_{k+1}} &= (v_{\lambda_1} \wedge u_{\lambda_{k+1}}) \vee \bigvee_{i=2}^{k} (u_{\lambda_{i-1}} \wedge v_{\lambda_i} \wedge u_{\lambda_{k+1}}) \vee (u_{\lambda_k} \wedge u_{\lambda_{k+1}}) \\ &= u_{\lambda_k} \wedge u_{\lambda_{k+1}} \leq u_{\lambda_{k+1}}. \end{aligned}$$

We see then that

$$f_k \wedge (u_{\lambda_{k+1}} \wedge v_{\lambda_{k+1}}) = (f_k \wedge u_{\lambda_{k+1}}) \wedge (f_k \wedge u_{\lambda_{k+1}}) \leq f_{k+1}.$$

It follows from a simple induction argument that

$$f_n \geq \bigwedge_{i=1}^{n} (u_{\lambda_i} \vee v_{\lambda_i}),$$

and as $u_{\lambda_n} = 0$ we arrive at the desired relation. ▷

(5): Prove that $|Tz| \geq |Ty| - |Tx|$ if

$$z := (y - \lambda_1 x)^- \vee \bigvee_{k=2}^{n} \big((y - \lambda_{k-1} x)^+ \wedge (y - \lambda_k x)^-\big).$$

◁ If $2 \leq k, l \leq n$ and $k \neq l$, then

$$(y - \lambda_1 x)^-, (y - \lambda_{k-1} x)^+ \wedge (y - \lambda_k x)^- \perp (y - \lambda_{l-1} x)^+ \wedge (y - \lambda_l x)^-.$$

Applying successively of (2), (4) and (3) we deduce

$$|Tz| = |T(y - \lambda_1 x)^-| \vee \bigvee_{i=2}^{n} |T((y - \lambda_{i-1} x)^+ \wedge (y - \lambda_i x)^-)|$$

$$\geq v_{\lambda_1} \vee \bigvee_{i=2}^{n} (u_{\lambda_{i-1}} \wedge v_{\lambda_i}) \geq |Ty| - |Tx|. \ ▷$$

(6): Now note that if $\delta = \max\{\lambda_k - \lambda_{k-1} : k = 1, \ldots, n\}$ then

$$(y - \lambda_{k-1} x)^+ \wedge (\lambda_k x - y)^+ \leq (y - \lambda_{k-1} x + \lambda_k x - y)^+ \leq \delta x,$$
$$(y - \lambda_1 x)^+ \leq \lambda_1 x = (\lambda_1 - \lambda_0) x \leq \delta x,$$

so that $0 \leq z \leq \delta x$. If for each $m \in \mathbb{N}$ we choose a partition of $[0, 1]$ with $\delta < 1/m$ and let z_m be the corresponding z, then $z_m \to 0$ relatively uniformly so that by hypothesis $\bigwedge_{m=1}^{\infty} |Tz_m| = 0$. Proposition (5) now shows that $|Ty| - |Tx| \leq 0$ as required. ▷

5.3. Weight-Shift-Weight Factorization

The main result of the present section is representation of an arbitrary disjointness preserving operator as a strongly disjoint sum of operators admitting weight-shift-weight factorization.

5.3.1. Throughout this section, E and F are order-dense ideals of some universally complete vector lattices $\mathscr{E}$ and $\mathscr{F}$. In the spaces $\mathscr{E}$ and $\mathscr{F}$, we fix order-unities $\mathbf{1}_{\mathscr{E}}$ and $\mathbf{1}_{\mathscr{F}}$ and consider the multiplication that makes the spaces f-algebras with unities $\mathbf{1}_{\mathscr{E}}$ and $\mathbf{1}_{\mathscr{F}}$, respectively. We recall that orthomorphisms in the K-spaces under consideration are multiplication operators and we identify them with the corresponding multipliers. For every $f \in \mathscr{E}$, there exists a unique element $g \in \mathscr{E}$ such that $fg = [f]\mathbf{1}_{\mathscr{E}}$. We denote such an element g by $1/f := \mathbf{1}_{\mathscr{E}}/f$. The product $e(1/f)$ is denoted by e/f for brevity. The ideal of the K-space $\mathscr{E}$ which is generated by the element $1 := \mathbf{1}_{\mathscr{E}}$ is denoted by $\mathscr{E}(1) := \mathscr{E}(\mathbf{1})$. Observe that some notions in this section depend on a specific choice of the unities $\mathbf{1}_{\mathscr{E}}$ and $\mathbf{1}_{\mathscr{F}}$.

Let $h : \mathfrak{P}(E) \to \mathfrak{P}(F)$ be a Boolean homomorphism. Denote by $\mathscr{E}(1,h)$ the set of elements $x \in \mathscr{E}$ representable as

$$x = o\text{-}\sum_{n=1}^{\infty} \pi_n x_n$$

where (x_n) is an arbitrary sequence in $\mathscr{E}(1)$ and (π_n) is a countable partition of unity in $\mathfrak{P}(\mathscr{E})$ such that $(h(\pi_n))$ is a partition of unity in $\mathfrak{P}(\mathscr{F})$. Of course, $\mathscr{E} = \mathscr{E}(1,h)$ if and only if h is sequentially order continuous.

(1) *The set $\mathscr{E}(1,h)$ is an order ideal in $\mathscr{E}$ and $\mathscr{E}(1)$ is relatively uniformly dense in $\mathscr{E}(1,h)$.*

$\lhd$ It is clear that $\mathscr{E}(1,h)$ is a subspace in $\mathscr{E}$. If x is as in the definition of $\mathscr{E}(1,h)$ and $|y| \le |x|$ then $\pi_n|y| \le \pi_n|x_n| \in \mathscr{E}(1)$, so that $y_n := \pi_n y_n \in \mathscr{E}(1)$. Moreover, $y = o\text{-}\sum_{n=1}^{\infty} \pi_n y_n$ whence $y \in \mathscr{E}(1,h)$. Furthermore, using the same notation, denote $z_n = \sum_{k=1}^{\infty} \pi_k x_k$ and $e = o\text{-}\sum_{n=1}^{\infty} n\pi_n y_n$. Then $z_n \in \mathscr{E}(1)$, $e \in \mathscr{E}(1,h)$, and $|x - z_n| \le (1/n)e$ $(n \in \mathbb{N})$. $\rhd$

Let D be a subset of E, and let Y be an LNS. We say that an operator $T : E \to Y$ is *wide on the set* D whenever $T(D)^{\perp\perp} = T(E)^{\perp\perp}$. If d is a positive element of $\mathscr{E}$, then we say that an operator T is *wide at the element* d whenever it is wide on the set $\{e \in E : e \text{ is a fragment of } d\}$.

(2) *Suppose that $T : E \to Y$ is a disjointness preserving operator, and $h : \mathfrak{P}(E) \to \mathfrak{P}(Y)$ is its shadow. Then T is wide on the set $D \subset E$ if and only if every element $e \in E$ is the h-limit of some net in the order ideal E_D generated by D.*

$\triangleleft$ First observe that the operator T is wide on the set D if and only if it is wide on the ideal E_D generated by D. Assume T to be wide on E_D, consider an arbitrary element $e \in E$, and show that $h\text{-}\lim_{\pi\in\Pi} \pi e = e$, where $\Pi = \{\pi \in \mathrm{Pr}(E) : \pi e \in E_D\}$. For every $n \in \mathbb{N}$ and $d \in E_D$, assign $\pi_n := [(n|d| - |e|)^+]$. Obviously, $\pi_n \in \Pi$. Since

$$|d - \pi_n d| = (\pi_n)^\perp |d| \leqslant (\pi_n)^\perp |e|/n \leqslant |e|/n$$

for all $n \in \mathbb{N}$, we have $r\text{-}\lim_{n\to\infty} \pi_n d = d$. Using r-continuity of the operator T and taking account of the equality $T(\pi_n d) = h(\pi_n)Td$, we arrive at the relation $\sup_{n\in\mathbb{N}} h(\pi_n) \geqslant [Td]$. Since the element $d \in E_D$ was chosen arbitrarily, we conclude by (2) that $\sup_{\pi\in\Pi} h(\pi) = h(1)$ and, consequently, $h\text{-}\lim_{\pi\in\Pi} \pi e = e$.

Conversely, consider an arbitrary element $e \in E$ and let e be the h-limit of some net $(e_\alpha)_{\alpha\in\mathrm{A}}$ of elements in E_D. In view of 5.2.2 (4), we have $o\text{-}\lim_{\alpha\in\mathrm{A}} T(e_\alpha) = Te$. Thus, the set $T(E_D)$ is o-dense in $T(E)$ and T is wide on $T(E_D)$. $\triangleright$

(3) *Let $T \in L^\sim(E, F)$ be a disjointness preserving operator and $h :=$ $\mathrm{shdw}(T)$. Then $E \subset \mathscr{E}(1, h)$ if and only if T is wide on* 1.

$\triangleleft$ Follows immediately from (2). $\triangleright$

5.3.2. *For every ring homomorphism $h : \mathrm{Pr}(\mathscr{E}) \to \mathrm{Pr}(\mathscr{F})$, there exists a unique regular operator $S : \mathscr{E}(1, h) \to \mathscr{F}$ such that the shadow of S is equal to h and $S(\mathbf{1}_{\mathscr{E}}) = h(\mathbf{1})\mathbf{1}_{\mathscr{F}}$. Furthermore, the operator S is positive.*

$\triangleleft$ For the sake of convenience, assume that $h(1) = 1$. We split the construction of the operator S into three steps.

(1): Define the operator S on the set of step-elements of $\mathscr{E}$ by letting

$$S\left(\sum_{i=1}^{n} \lambda_i \pi_i \mathbf{1}_{\mathscr{E}}\right) := \sum_{i=1}^{n} \lambda_i h(\pi_i) \mathbf{1}_{\mathscr{F}}$$

for arbitrary $\lambda_1, \dots, \lambda_n \in \mathbb{R}$ and $\pi_1, \dots, \pi_n \in \mathrm{Pr}(\mathscr{E})$.

(2): Extend the operator S onto $\mathscr{E}(1)$. To this end, fix an arbitrary element $e \in \mathscr{E}(1)$ and choose a sequence $(e_n)_{n\in\mathbb{N}}$ of step-elements in $\mathscr{E}$ so that it r-converges to e with regulator $\mathbf{1}_{\mathscr{E}}$. It is easy to verify that the sequence $(Se_n)_{n\in\mathbb{N}}$ is r-fundamental (with regulator $\mathbf{1}_{\mathscr{F}}$). Assign $Se := r\text{-}\lim_{n\to\infty} Se_n$.

(3): Finally, extend S onto the entire set $\mathscr{E}(1, h)$. Every element $e \in \mathscr{E}(1, h)$ can be represented as the mixing $o\text{-}\sum_{n\in\mathbb{N}} \pi_n e_n$ of elements $e_n \in \mathscr{E}(1)$ by an h-partition $(\pi_n)_{n\in\mathbb{N}}$. Assign $Se := o\text{-}\sum_{n\in\mathbb{N}} h(\pi_n) Se_n$.

It is easy to verify that the definition of S is sound at each of the steps. Obvious positivity of S ensures its regularity. In order to prove uniqueness of S, it is sufficient to observe that, at step 3, the sequence $\left(\sum_{n=1}^{m} \pi_n e_n\right)_{m\in\mathbb{N}}$ is r-convergent to e with regulator $o\text{-}\sum_{n\in\mathbb{N}} n\pi_n |e_n| \in \mathscr{E}(1, h)$. $\triangleright$

5.3.3. The operator S, whose existence is asserted in 5.3.2, is called the *shift by* h and denoted by S_h. Let E be an order-dense ideal of $\mathscr{E}$ and F be an order-dense ideal of $\mathscr{F}$. We say that an operator $S : E \to F$ is a *shift operator*, if there exists a ring homomorphism $h : \mathfrak{P}(\mathscr{E}) \to \mathfrak{P}(\mathscr{F})$ such that $E \subset \mathscr{E}(1, h)$ and $S = S_h$ on E. It is clear that, in this case, the homomorphism h is the shadow of S. Observe that the notion of shift and that of a shift operator depend on the choice of unities $\mathbf{1}_{\mathscr{E}}$ and $\mathbf{1}_{\mathscr{F}}$ in the K-spaces $\mathscr{E}$ and $\mathscr{F}$.

Theorem. *A linear operator $S : E \to F$ is a shift operator if and only if it satisfies the following conditions:*

(1) *S is disjointness preserving;*

(2) *S is regular;*

(3) *S takes fragments of $\mathbf{1}_{\mathscr{E}}$ into fragments of $\mathbf{1}_{\mathscr{F}}$;*

(4) *S is wide at $\mathbf{1}_{\mathscr{E}}$.*

$\lhd$ Necessity of (1)–(3) is obvious and necessity of (4) follows from 5.3.1 (3). Let us show sufficiency. Suppose that the operator S satisfies conditions (1)–(4), denote the shadow of S by h and assign $\Pi := \{\pi \in \mathfrak{P}(\mathscr{E}) : \pi\mathbf{1}_{\mathscr{E}} \in E\}$. Proposition 5.3.1 (3) implies the equality $[S(\pi\mathbf{1}_{\mathscr{E}})] = h(\pi)$ for each $\pi \in \Pi$, which, together with condition (3), yields $S(\pi\mathbf{1}_{\mathscr{E}}) = S_h(\pi\mathbf{1}_{\mathscr{E}})$. The same proposition ensures the inclusion $E \subset \mathscr{E}(1, h)$. In view of 5.2.2 (3) and 5.3.1 (1), we now conclude that $S = S_h$ on E. $\rhd$

5.3.4. Say that a linear operator $T : E \to F$ admits a *weight-shift-weight factorization* if there exist order-dense ideals $E' \subset \mathscr{E}$ and $F' \subset \mathscr{F}$, orthomorphisms $w : E \to E'$ and $W : F' \to F$, and a shift operator $S : E' \to F'$ such that $T = W \circ S \circ w$. The composite $W \circ S \circ w$ is called a *WSW-representation* of T, and the operators W, S, and w are respectively called the *outer weight*, *shift*, and *inner weight* of the representation $W \circ S \circ w$.

An operator can have different *WSW*-factorizations. However, the constituents of these factorizations for a given operator T are connected with one another. Two main aspects of this connection are presented in the following proposition.

Let $T : E \to F$ admit a *WSW*-factorization $T = W \circ S \circ w$. Assign $\rho := [\operatorname{im}(T)]$.

(1) *Denote the shift of T by S_T. Then S_T extends $\rho \circ S$ and the equality $W \circ S \circ w = W \circ S_T \circ w$ holds.*

$\lhd$ The claim follows readily from 5.3.2 and 3.3.3. $\rhd$

(2) *Identify w and W with the corresponding elements of $\mathscr{E}$ and $\mathscr{F}$ and assign $W_T := o\text{-}\lim_{\pi \in \Pi} T\pi(\mathbf{1}_{\mathscr{E}}/w) \in \mathscr{F}$, where $\Pi = \{\pi \in \mathfrak{P}(\mathscr{E}) : \pi(\mathbf{1}_{\mathscr{E}}/w) \in E\}$. Then $\rho W = W_T$ and $W \circ S \circ w = W_T \circ S \circ w$.*

$\lhd$ By the obvious equality $T \circ [w]^\perp = 0$, we do not restrict generality on assuming that $[w] = [1]$. Then

$$o\text{-}\lim_{\pi\in\Pi} T\pi(\mathbf{1}_{\mathscr{E}}/w) = o\text{-}\lim_{\pi\in\Pi} WS_T w\pi(\mathbf{1}_{\mathscr{E}}/w) = o\text{-}\lim_{\pi\in\Pi} WS_T\pi\mathbf{1}_{\mathscr{E}} = \big(\sup_{\pi\in\Pi} h(\pi)\big)W,$$

where h is the shadow of T. Since $\rho = h(1)$, it is sufficient to show the relation $\sup_{\pi\in\Pi} h(\pi) = h(1)$. From $E \subset \operatorname{dom}(S_T \circ w)$ it follows that $w(E) \subset \operatorname{dom}(S_T) = \mathscr{E}(1, h)$ and, hence, $E \subset \mathscr{E}(1/w, h)$. It remains to employ Proposition 5.3.1 (1). $\rhd$

(3) *The inner weight in a WSW-representation can be chosen positive. If $W \circ S \circ w$ is a WSW-representation of an operator T with w positive, then the operators T^+, T^-, and $|T|$ admit the following WSW-representations: $T^+ = W^+ \circ S \circ w$, $T^- = W^- \circ S \circ w$, and $|T| = |W| \circ S \circ w$.*

$\lhd$ Consider an arbitrary WSW-factorization $W \circ S \circ w$. Identifying the orthomorphism w with an element of $\mathscr{E}$, denote the projection $[w^+] \in \mathfrak{P}(E)$ by π and assign $\rho := \big[S(\pi\mathbf{1}_{\mathscr{E}})\big]$. Then

$$W \circ S \circ w = W \circ S \circ \big(\pi|w| - \pi^\perp|w|\big)$$
$$= W \circ \big(\rho \circ S \circ |w| - \rho^\perp \circ S \circ |w|\big) = (\rho W - \rho^\perp W) \circ S \circ |w|.$$

The remaining is obvious. $\rhd$

5.3.5. Theorem. *Let w be an arbitrary positive element of $\mathscr{E}$. A linear operator $T : E \to F$ admits a WSW-factorization with inner weight w if and only if T is disjointness preserving, regular, and wide at the element $\mathbf{1}_{\mathscr{E}}/w$.*

$\lhd$ Necessity ensues from 5.3.4 (2). Let us prove sufficiency. Suppose that a disjointness preserving operator $T : E \to F$ is wide at $\mathbf{1}_{\mathscr{E}}/w$. Without loss of generality, we may assume that the operator T is positive. Assign $\Pi := \big\{\pi \in \Pr(\mathscr{E}) : \pi(\mathbf{1}_{\mathscr{E}}/w) \in E\big\}$ and denote by W the orthomorphism of multiplication by $\sup_{\pi\in\Pi} T\pi(\mathbf{1}_{\mathscr{E}}/w) \in \mathscr{F}$. Consider the composite $(\mathbf{1}_{\mathscr{F}}/W) \circ T \circ (\mathbf{1}_{\mathscr{E}}/w)$ as an operator from $w[E]$ into $\mathscr{F}$ and denote it by S. In accordance with Theorem 5.3.3, it is sufficient to show that the operator S satisfies conditions (1)–(4) presented in the statement of the theorem. Verification of the conditions causes no difficulties, so that S is a shift operator and we obtain the desired WSW-factorization $W \circ S \circ w$ for T. $\rhd$

Operators S and T are called *strongly disjoint* if $Su \perp Tv$ for all $u, v \in X$. Let X and Y be LNSs and let $(T_\xi)_{\xi\in\Xi}$ be a family of linear operators from X into Y. Say that an operator $T : X \to Y$ decomposes into the *strongly disjoint sum* of T_ξ (and write $T = \bigoplus_{\xi\in\Xi} T_\xi$), whenever the operators T_ξ are strongly disjoint and, for every $x \in X$, the relation $Tx = o\text{-}\sum_{\xi\in\Xi} T_\xi x$ holds. It is easy to observe that the strongly disjoint sum of (T_ξ) is disjointness preserving if and only if so is each summand T_ξ.

5.3.6. Theorem. *Let $T : E \to F$ be a disjointness preserving regular operator. Then there exists a partition of unity $(\rho_\xi)_{\xi\in\Xi}$ in the algebra $\mathfrak{P}(F)$ such that, for each $\xi \in \Xi$, the composite $\rho_\xi \circ T$ admits a WSW-factorization with inner weight $\mathbf{1}_{\mathscr{E}}/e_\xi$, where e_ξ is a positive element of E. In this case, the operator T decomposes into the strongly disjoint sum*

$$T = \bigoplus_{\xi\in\Xi} W \circ \rho_\xi S \circ (\mathbf{1}_{\mathscr{E}}/e_\xi),$$

where S is the shift of T and $W : \mathscr{F} \to \mathscr{F}$ is the orthomorphism of multiplication by o-$\sum_{\xi\in\Xi} \rho_\xi T e_\xi$.

$\lhd$ By applying the Exhaustion Principle (see 1.1.6) to the relation

$$\sup\{[Te] : e \in E_+\} = [\mathrm{im}(T)],$$

we obtain a disjoint family $(\rho_\xi)_{\xi\in\Xi}$ in the algebra $\mathfrak{P}(F)$ and a family $(e_\xi)_{\xi\in\Xi}$ of positive elements in E such that $\sup_{\xi\in\Xi} \rho_\xi[Te_\xi] = [\mathrm{im}(T)]$. After adding the band projection onto $[\mathrm{im}(T)]^\perp$ to the family $(\rho_\xi)_{\xi\in\Xi}$ and the zero element to the family $(e_\xi)_{\xi\in\Xi}$, we make $(\rho_\xi)_{\xi\in\Xi}$ into a partition of unity while $\sup_{\xi\in\Xi} \rho_\xi[Te_\xi] = [\mathrm{im}(T)]$. By Theorem 5.3.5, for each $\xi \in \Xi$, the composite $\rho_\xi \circ T$ admits a WSW-factorization with inner weight $\mathbf{1}_{\mathscr{E}}/e_\xi$. If S is the shift of T then the shift of $\rho_\xi \circ T$ is equal to $\rho_\xi S$ (see 3.3.3); thus, using Proposition 5.3.4 (2), we conclude that $\rho_\xi \circ T = \rho_\xi T e_\xi \circ \rho_\xi S \circ (\mathbf{1}_{\mathscr{E}}/e_\xi)$. $\rhd$

5.3.7. To obtain WSW-factorization representations for operators in Banach–Kantorovich spaces which are analogous to those for operators in K-spaces we need some auxiliary facts. Let X be an LNS over E, let X_0 be a vector subspace of X, and let Y be an LNS over F. Consider a linear operator $T_0 : X_0 \to Y$ and a disjointness preserving positive operator $S : E \to F$ with the shadow $h : \mathfrak{P}(E) \to \mathfrak{P}(F)$. Denote by hX_0 the LNS of all elements of X that are h-approximated by X_0.

Assume that $|T_0u_0| \leqslant S|u_0|$ ($|T_0u_0| = S|u_0|$) for all $u_0 \in X_0$. Then there exists a unique linear extension $T : hX_0 \to Y$ of the operator T_0 such that $|Tu| \leqslant S|u|$ (respectively, $|Tu| = S|u|$) for all $u \in hX_0$.

$\lhd$ First, we prove the assertion about extension with the inequality preserved. If $\pi \in \mathfrak{P}(X)$ and $u_0 \in X_0$ are such that $\pi u_0 = 0$, then $h(\pi)T_0u_0 = 0$, since $h(\pi)|T_0u_0| \leqslant h(\pi)S|u_0| = S\pi|u_0| = 0$. This fact implies that the following definition of an operator $\bar{T}_0 : d_0(X_0) \to Y$ is sound:

$$\bar{T}_0\left(\sum_{k=1}^{n} \pi_k u_k\right) := \sum_{k=1}^{n} h(\pi_k)T_0 u_k$$

where $u_h \in X$ and $\pi_k \in \mathscr{P}(X)$ are pairwise disjoint. Evidently, $\bar{T}_0$ and satisfies the inequality $|\bar{T}_0 u| \leqslant S|u|$ for all $u \in d_0(X_0)$. In view of Proposition 2.2.10 (3), for every $u \in hX_0$, there exists a net $(u_\alpha)_{\alpha \in \mathrm{A}}$ in $d_0(X_0)$ that is h-convergent to u. From the inequality $|\bar{T}_0 u_\alpha - \bar{T}_0 u_\beta| \leqslant S|u_\alpha - u_\beta|$ and h-o-continuity of S (see 5.2.2 (3)) it follows that the net $(\bar{T}_0 u_\alpha)_{\alpha \in \mathrm{A}}$ is o-fundamental. Since the LNS Y is bo-complete, it contains a bo-limit of the net. Obviously, the limit depends only on u and, therefore, can be denoted by Tu. It is not difficult to become convinced that the operator $T : hX_0 \to Y$ thus obtained is the desired one. Uniqueness of this extension is ensured by its h-o-continuity inherited from S.

Assume now that $|T_0 u_0| = S|u_0|$ for all $u_0 \in X_0$. In view of what was proven above, there exists an extension $T : hX_0 \to Y$ of the operator T_0 such that $|Tu| \leqslant S|u|$ for all $u \in hX_0$. For every $u_0 \in X_0$ and $\pi \in \mathrm{Pr}(X)$, the relations

$$S|u_0| = |Tu_0| = |T\pi u_0| + |T\pi^\perp u_0| \leqslant S|\pi u_0| + S|\pi^\perp u_0| = S|u_0|$$

and the inequalities $|T\pi u_0| \leqslant S|\pi u_0|$ and $|T\pi^\perp u_0| \leqslant S|\pi^\perp u_0|$ imply $|T\pi u_0| = S|\pi u_0|$. Since $u_0 \in X_0$ and $\pi \in \mathscr{P}(X)$ were chosen arbitrarily, we have $|Tu| = S|u|$ for all $u \in d_0 X_0$. The equality $|Tu| = S|u|$ for all $u \in hX_0$ is now deduced from the fact that $d_0 X_0$ is h-dense in X. ▷

5.3.8. An operator $S : X \to Y$ is called a *shift operator* if there exists a shift operator $s : E \to F$ such that $|Su| = s|u|$ for all $u \in X$. Obviously, $s = |S|$, i.e., the operator s is the exact dominant of S.

(1) *An operator $S : X \to Y$ is a shift operator if and only if there exist a shift operator $s : E \to F$ and an F-isometric embedding $\imath : sX \to Y$ such that $S = \imath \circ s_X$, where $s_X : X \to sX$ is the norm transformation of X by means of s (see 2.2.13 (2)).*

◁ Only necessity requires proving. An elementary verification shows that the formula

$$\imath\left(\sum_{k=1}^{n} \rho_k s_X x_k\right) = \sum_{k=1}^{n} \rho_k S x_k \quad (x_k \in X,\ \rho_k \in \mathscr{P}(Y))$$

soundly defines a function $\imath : sX \to Y$ that is the desired isometry. ▷

The following description of shift operators generalizes criterion 5.3.3 to LNSs.

(2) Theorem. *An operator $S : X \to Y$ is a shift operator if and only if S satisfies the following:*

(a) *S is disjointness preserving;*

(b) *S is bounded;*

(c) *if $u \in X$ and $|u|$ is a fragment of $\mathbf{1}_{\mathscr{E}}$ then $|Su|$ is a fragment of $\mathbf{1}_{\mathscr{F}}$;*

(d) *S is wide at $\mathbf{1}_{\mathscr{E}}$.*

$\lhd$ The proof is omitted, since the claim is not used in the sequel. More details can be found in [121, 123]. $\rhd$

5.3.9. Let X be a BKS over an order-dense ideal $E \subset \mathscr{E}$ and let Y be a BKS over an order-dense ideal $F \subset \mathscr{F}$. We say that a linear operator $T : X \to Y$ admits a *weight-shift-weight factorization* if there exist a BKS X' over an order-dense ideal $E' \subset \mathscr{E}$, a BKS Y' over an order-dense ideal $F' \subset \mathscr{F}$, orthomorphisms $w : X \to X'$ and $W : Y' \to Y$, and a shift operator $S : X' \to Y'$ such that $T = W \circ S \circ w$, i.e., the diagram

$$\begin{array}{ccc} X & \xrightarrow{\ T\ } & Y \\ {\scriptstyle w}\big\downarrow & & \big\uparrow{\scriptstyle W} \\ X' & \xrightarrow{\ S\ } & Y' \end{array}$$

is commutative. As in the case of an operator in K-spaces, the composite $W \circ S \circ w$ is called a *WSW-representation* of T and the operators W, S and w are respectively called the *outer weight*, the *shift*, and the *inner weight* of the representation $W \circ S \circ w$. In order to avoid confusion, we call a weighted shift representation *scalar* or *vector*, referring to Definitions 5.3.3 or 5.3.9, respectively. By analogous reasons, we speak about scalar or vector *WSW*-factorizations. A vector *WSW*-factorization $W \circ S \circ w$ of an operator $T : X \to Y$ will be called *semivector* if w is a *scalar orthomorphism*, i.e., X and X' are order-dense ideals of the same BKS over $\mathscr{E}$ and the orthomorphism w acts by the rule $u \mapsto eu$ for some fixed orthomorphisms $e \in \mathrm{Orth}(E, E')$.

(1) *If $\bar{W} \circ \bar{S} \circ \bar{w}$ is a vector WSW-representation of T then $|T|$ admits a scalar WSW-factorization $W \circ |\bar{S}| \circ |\bar{w}|$ such that $0 \leqslant W \leqslant |\bar{W}|$.*

$\lhd$ Clearly, $|T| \leqslant |\bar{W}| \circ |\bar{S}| \circ |\bar{w}|$. According to 3.3.3 there exists $\pi \in \mathrm{Orth}(F)$ with $|T| = \pi |\bar{W}| \circ |\bar{S}| \circ |\bar{w}|$, so that we may put $W := \pi |\bar{W}|$. $\rhd$

(2) **Theorem.** *Assume that $T : X \to Y$ is a dominated operator and its exact dominant $|T| : E \to F$ admits a scalar WSW-factorization $W \circ S \circ w$ with positive weights W and w. Then T admits a semivector WSW-factorization $\bar{W} \circ \bar{S} \circ \bar{w}$ such that $|\bar{W}| = W$, $|\bar{S}| = S$, and $\bar{w}$ is the orthomorphism of multiplication by w.*

$\lhd$ Suppose that $W \circ S \circ w$ is a scalar *WSW*-factorization of $|T|$, where $w : E \to E'$, $S : E' \to F'$ and $W : F' \to F$. Let mX be the universal completion of X, let X' be the ideal $\{u \in mX : |u| \in E'\}$ of the BKS mX, and let $\bar{w} : X \to X'$ be the orthomorphism of multiplication by w. Denote by Y' the *bo*-completion of the norm transformation of X' by means of S (see 2.2.13 (2)) and consider the corresponding operator of norm transformation $\bar{S} : X' \to Y'$. Now, we are to construct an orthomorphism $\bar{W} : Y' \to Y$.

Assign $Y_0' := (\bar{S} \circ \bar{w})(X)$ and define a linear operator $\bar{W}_0 : Y_0' \to Y$ by letting $\bar{W}_0(\bar{S}\bar{w}u) := Tu$. Such a definition is sound, since the equality $\bar{S}\bar{w}u_1 = \bar{S}\bar{w}u_2$ implies

$$\begin{aligned} |Tu_1 - Tu_2| &\leqslant |T||u_1 - u_1| = WSw|u_1 - u_1| \\ &= WS|\bar{w}u_1 - \bar{w}u_2| = W|\bar{S}\bar{w}u_1 - \bar{S}\bar{w}u_2| = 0. \end{aligned}$$

Denote $\rho := [\mathrm{im}(T)]$. Since $\rho \leqslant \big[(\bar{S} \circ \bar{w})(X)\big]$ and $\bar{w}(X) = \big\{v' \in Y' : |v'| \in w(E)\big\}$, the operator $\rho \circ \bar{S}$ is wide on the ideal $w(E) \subset E'$. It follows from the definitions that then the shadow of the restriction of $\rho \circ \bar{S}$ onto the set $\bar{w}(X)$ coincide with the shadow of $\rho \circ \bar{S}$. Therefore, the set $Y_0' = (\rho \circ \bar{S})\big(\bar{w}(X)\big)$ approximates $(\rho \circ \bar{S})(X')$. The latter set, by the definition of the norm transformation 2.2.13 (2), approximates the set $\rho(SX')$, which in turn approximates $\rho(Y')$. Thus, by virtue of 2.2.10 (2), the set Y_0' approximates $\rho(Y')$. Obviously, $|\bar{W}_0 v_0'| \leqslant W|v_0'|$ for all $v_0' \in Y_0'$. According to 5.3.7, the operator $\bar{W}_0$ admits a (unique) linear extension $\bar{W}_1 : \rho[Y'] \to Y$ such that $|\bar{W}_1 v'| \leqslant W|v'|$ for all $v' \in Y'$. Then the composite $\bar{W}_1 \circ \rho : Y' \to Y$ satisfies the inequality $|\bar{W}_1 \circ \rho| \leqslant W$ and, consequently, it is an orthomorphism. Thus, we have constructed a *WSW*-factorization $(\bar{W}_1 \circ \rho) \circ \bar{S} \circ \bar{w}$ of the operator T. However, the equality $|\bar{W}_1 \circ \rho| = W$ is not be guaranteed and the operator $\bar{W}_1 \circ \rho$ should be rectified.

For all positive $e \in E$, we have

$$\begin{aligned} |\bar{W}_1 \circ \rho|Swe &= \sup\big\{|\bar{W}_1 \rho v'| : v' \in Y',\ |v'| = Swe\big\} \\ &\geqslant \sup\big\{\rho|\bar{W}_0 v_0'| : v_0' \in Y_0',\ |v_0'| = Swe\big\} \\ &= \sup\big\{\rho|\bar{W}_0 \bar{S}\bar{w}u| : u \in X,\ |\bar{S}\bar{w}u| = Swe\big\} \\ &= \sup\big\{|Tu| : S\,w|u| = Swe\big\} \\ &\geqslant \sup\big\{|Tu| : |u| = e\big\} = |T|e = WSwe, \end{aligned}$$

whence $|\bar{W}_1 \circ \rho|Swe = WSwe$ by the inequality $|\bar{W}_1 \circ \rho| \leqslant W$. Thus, $W \circ S \circ w$ and $|\bar{W}_1 \circ \rho| \circ S \circ w$ are two *WSW*-factorizations of the operator $|T|$. Hence, according to Proposition 5.3.4 (2), the equality $|\bar{W}_1 \circ \rho| = \rho W$ holds. To ensure the equality $|\bar{W}| = W$, we define $\bar{W} := \bar{W}_1 \circ \rho + \bar{W}_2 \circ \rho^{\perp}$, where $\bar{W}_2 : X \to Y$ is an arbitrary dominated operator with $|\bar{W}_2| = W$. ▷

(3) *A linear operator $T : X \to Y$ is a vector weighted shift operator if and only if it is dominated and its exact dominant $|T| : E \to F$ is a scalar weighted shift operator.*

◁ Follows from (1) and (2). ▷

5.3.10. *Theorem.* *Suppose that X and Y are Banach–Kantorovich spaces over order-dense ideals $E \subset \mathscr{E}$ and $F \subset \mathscr{F}$, and $T : X \to Y$ is a disjointness* *preserving bounded operator. Then there exists a partition of unity $(\rho_\xi)_{\xi\in\Xi}$ in the algebra $\mathscr{P}(Y)$ such that, for each $\xi \in \Xi$, the composition $\rho_\xi \circ T$ admits WSW-factorization with inner weight of norm $\mathbf{1}_{\mathscr{E}}/e_\xi$, where e_ξ is a positive element of E.*

For each $\xi \in \Xi$, assign $E_\xi := \{e/e_\xi : e \in E\}$ and $X_\xi := \{u \in mX : |u| \in E_\xi\}$, and denote by $w_\xi : X \to X_\xi$ the scalar orthomorphism of multiplication by $\mathbf{1}_{\mathscr{E}}/e_\xi$. Then there exist a BKS Y' over $\mathscr{F}$, strongly disjoint shift operators $S_\xi : X_\xi \to Y'$ ($\xi \in \Xi$), and an orthomorphism $W : Y' \to mY$ such that the operators T and $|T|$ decompose into the following strongly disjoint sums:

$$T = \bigoplus_{\xi\in\Xi} W \circ S_\xi \circ w_\xi, \quad |T| = \bigoplus_{\xi\in\Xi} |W| \circ |S_\xi| \circ |w_\xi|.$$

$\lhd$ Consider an arbitrary disjointness preserving bounded operator $T : X \to Y$. By Theorem 5.3.6, there exists a partition of unity $(\rho_\xi)_{\xi\in\Xi}$ in the algebra $\mathfrak{P}(F)$ such that, for each $\xi \in \Xi$, the composition $\rho_\xi \circ |T|$ is a weighted shift operator and, moreover, admits a *WSW*-factorization with inner weight $\mathbf{1}_{\mathscr{E}}/e_\xi$, where e_ξ is a positive element of E. Define BKSs X_ξ and orthomorphisms $w_\xi : X \to X_\xi$ in the same way as in the statement of the theorem under proof. By Theorem 5.3.7 (2), for each $\xi \in \Xi$, there exist a BKS Y_ξ over an order-dense ideal $F_\xi \subset \rho_\xi[\mathscr{F}]$, a shift operator $S_\xi : X_\xi \to Y_\xi$, and an orthomorphism $W_\xi : Y_\xi \to \rho_\xi(Y)$ such that $\rho_\xi \circ T = W_\xi \circ S_\xi \circ w_\xi$ and $\rho_\xi \circ |T| = |W_\xi| \circ |S_\xi| \circ |w_\xi|$. In order to complete the proof, it remains to construct the desired BKS Y' and "glue" the orthomorphisms W_ξ together to obtain a single orthomorphism W.

Assign $Y_0' := \bigoplus_{\xi\in\Xi} Y_\xi$ and denote by Y' a universal completion of the BKS Y_0'. Naturally identifying Y_ξ and $\rho_\xi[Y_0']$, we regard S_ξ as an operator from X_ξ into Y'. For each element $v_0' = (v_\xi)_{\xi\in\Xi} \in Y_0'$, assign $W_0(v') := o\text{-}\sum_{\xi\in\Xi} W_\xi(v_\xi) \in m\mathscr{V}$. Due to 5.3.7, the orthomorphism $W_0 : Y_0' \to mY$ admits a unique extension to an orthomorphism $W : Y' \to mY$. $\rhd$

5.4. Multiplicative Representation

In this section we give some analytical representations of disjointness preserving operators constructed with the help of such operations as a continuous change of variable and the pointwise multiplication by a real-valued function.

5.4.1. Throughout the section, P and Q are extremal compact spaces, E and F are order-dense ideals in the universally complete vector lattices $\mathscr{E} := C_\infty(P)$ and $\mathscr{F} := C_\infty(Q)$, respectively. The symbol $\mathbf{1}_M$ denotes the function on a set M which is identically equal to unity. Denote by $C_0(Q, P)$ the totality of all continuous functions $s : Q_0 \to P$ defined on various clopen subsets $Q_0 \subset Q$.

(1) *A mapping h : $\mathrm{Clop}(P) \to \mathrm{Clop}(Q)$ is a ring homomorphism if and only if there exists a function $s \in C_0(Q, P)$ such that $h(U) = s^{-1}(U)$ for all $U \in \mathrm{Clop}(P)$. For every ring homomorphism h, such a function s is unique.*

$\lhd$ The claim follows directly from the Sikorski Theorem 1.1.5 (1). $\rhd$

The relation $h(U) = s^{-1}(U)$ is called the *representation* of the ring homomorphism h by means of the function s.

(2) *Let E and E' be order-dense ideals of $C_\infty(Q)$. A mapping $W : E \to E'$ is an orthomorphism if and only if there exists a function $w \in C_\infty(Q)$ such that $W(e) = we$ for all $e \in E$. For every orthomorphism W, such a function w is unique.*

$\lhd$ See 3.3.2 (5). $\rhd$

The relation $W(e) = we$ is called the *representation* of W by means of w.

5.4.2. Given arbitrary $s \in C_0(Q, P)$ and $e \in C_\infty(P)$, the function $e \bullet s : Q \to \overline{\mathbb{R}}$ is defined as follows:

$$(e \bullet s)(q) := \begin{cases} e(s(q)), & \text{if } q \in \operatorname{dom}(s), \\ 0, & \text{if } q \in Q \setminus \operatorname{dom}(s). \end{cases}$$

Obviously, the function $e \bullet s$ is continuous but, in general, does not belong to $C_\infty(Q)$, since it can assume infinite values on a set with nonempty interior. The totality of all functions $e \in C_\infty(P)$ for which $e \bullet s \in C_\infty(Q)$ is denoted by $C_s(P)$.

(1) *Let $h : \mathfrak{P}(C_\infty(P)) \to \mathfrak{P}(C_\infty(Q))$ be a ring homomorphism, and let $C_h(P) := \mathscr{E}(\mathbf{1}_P, h)$ be the order-dense ideal of $\mathscr{E} := C_\infty(P)$ defined in 5.3.1. Then $C_h(P) = C_s(P)$, where $s \in C_0(Q, P)$ represents h by means of a formula $h(U) = s^{-1}(U)$.*

$\lhd$ The claim follows from the definition of $\mathscr{E}(1, h)$ from 5.3.1 and 5.4.1 (1). $\rhd$

(2) *A mapping $S : E \to F$ is a shift operator if and only if there exists a function $s \in C_0(Q, P)$ such that $Se = e \bullet s$ for all $e \in E$.*

$\lhd$ Sufficiency can be easily established with the help of Theorem 5.3.3. Let us show necessity. Suppose that $S : E \to F$ is a shift operator and $h : \mathfrak{P}(E) \to \mathfrak{P}(F)$ is its shadow which is identified with the corresponding homomorphism from $\operatorname{Clop}(P)$ to $\operatorname{Clop}(Q)$. Consider the representation $h(U) = s^{-1}(U)$ of h by means of a mapping $s \in C_0(Q, P)$. According to Proposition (1), the equality $C_h(P) = C_s(P)$ holds. Since the operators $e \mapsto e \bullet s$ acts from $C_s(P)$ to $C_\infty(Q)$ and $S_h : C_h(P) \to C_\infty(Q)$ have the same shadow h and satisfy the equalities $\mathbf{1}_P \bullet s = S_h(\mathbf{1}_P) = h(1)\mathbf{1}_Q$, they coincide in view of Proposition 5.3.2. Therefore, $Se = S_h e = e \bullet s$ for all $e \in E$. $\rhd$

5.4.3. The function s in the representation 5.4.2 (2) of the shift operator S is not unique in general. Indeed, assume that the compact space P contains two different nonisolated points p_1 and p_2, assign $E := \{e \in C_\infty(P) : e(p_1) = e(p_2) = 0\}$ and consider the functions $s_1, s_2 : Q \to P$ identically equal to p_1 and p_2, respectively. Then $e \bullet s_1 = e \bullet s_2 = 0$ for all $e \in E$. The following propositions clarifies the question about uniqueness of a representation of a shift operator $S : E \to F$. Assign $Q_0 := \operatorname{supp}(\operatorname{im} S) := \operatorname{cl} \bigcup_{e \in E} \operatorname{supp}(Se)$, where $\operatorname{supp}(f) := \operatorname{cl}\{q \in Q : f(q) \neq 0\}$.

(1) *If functions $s_1, s_2 \in C_0(Q, P)$ satisfy the equalities $Se = e \bullet s_1 = e \bullet s_2$ for all $e \in E$ then $Q_0 \subset \mathrm{dom}(s_1) \cap \mathrm{dom}(s_2)$ and $s_1 = s_2$ on Q_0.*

$\lhd$ Denote by D the set of all points in P, at which some functions in E are nonzero. Obviously, the set $s_1^{-1}(D)$ is dense in Q_0; therefore, it is sufficient to establish the equality $s_1 = s_2$ on this set. Take an arbitrary point $q \in s_1^{-1}(D)$ and assume to the contrary that $s_1(q) \neq s_2(q)$. Since $s_1(q) \in D$, there exists a function $e \in E$ that satisfies the relations $e(s_1(q)) \neq 0$ and $e(s_2(q)) = 0$, which contradicts the equality $e \bullet s_1 = e \bullet s_2$. $\rhd$

(2) *There exists a unique function $s \in C(Q_0, P)$ such that $Se = e \bullet s$ for all $e \in E$. Furthermore, if s is such a function then $h(U) = s^{-1}(U)$ is a representation of the shadow h of the operator S.*

$\lhd$ Such a function s exists according to Proposition 5.4.2 (2), and its uniqueness follows from (1). The fact that s represents the shadow of S ensues from the proof of Proposition 5.4.2 (2). $\rhd$

If a function s satisfies the conditions stated in assertion (2) then the relation $Se = e \bullet s$ is called the *representation* of the shift operator S by means of the function s.

5.4.4. (1) Theorem. *A mapping $T : E \to F$ admits a WSW-representation if and only if there exist functions $s \in C_0(Q, P)$, $w \in C_\infty(P)$, and $W \in C_\infty(Q)$ such that $we \bullet s \in C_\infty(Q)$ and $Te = W(we \bullet s)$ for all $e \in E$.*

$\lhd$ The claim readily follows from Propositions 5.4.1 (2) and 5.4.2 (2). $\rhd$

Simple examples show that the constituents of the representation $Te = W(we \bullet s)$ of a weighted shift operator T are not unique. However, omitting certain details, we may say that the function s is unique and W is uniquely determined by the choice of w. Let $T : E \to F$ be a disjointness preserving regular operator. Assign $Q_0 := \mathrm{supp}(\mathrm{im}(T))$.

(2) *Let functions $s_1, s_2 \in C_0(Q, P)$, $w_1, w_2 \in C_\infty(P)$ and $W_1, W_2 \in C_\infty(Q)$ be such that $Te = W_1(w_1 e \bullet s_1) = W_2(w_2 e \bullet s_2)$ for all $e \in E$. Then $Q_0 \subset \mathrm{dom}(s_1) \cap \mathrm{dom}(s_2)$ and $s_1 = s_2$ on Q_0. If, in addition, $w_1 = w_2$ then $W_1 = W_2$ on Q_0.*

$\lhd$ Follows immediately from Proposition 5.3.4 due to 5.4.1 (2) and 5.4.2 (2). $\rhd$

(3) *Let a positive function $w \in C_\infty(P)$ be such that the operator T is wide at $1/w$ (see 5.3.1). Then there exist unique functions $s \in C(Q_0, P)$ and $W \in C_\infty(Q)$ such that $W = 0$ outside Q_0 and $Te = W(we \bullet s)$ for all $e \in E$. Furthermore, $\mathrm{supp}(W) = s^{-1}(\mathrm{supp}(w)) = Q_0$, $Se = e \bullet s$ is a representation of the shift S of the operator T, and $h(U) = s^{-1}(U)$ is a representation of its shadow h.*

$\lhd$ Such functions s and W exist by virtue of Theorem 5.3.5 and (1); their uniqueness follows from (1). Connection of the function s with the shift and shadow of the operator T follows from Propositions 5.3.4 (1) and 5.4.3 (2). $\rhd$

If s, w, and W satisfy the conditions stated in assertion (2), then the relation $Te = W(we \bullet s)$ is called the *representation* of the weighted shift operator T by means of the functions s, w, and W.

(4) *If $Te = W(we \bullet s)$ is a representation of T then the operators T^+, T^-, and $|T|$ admit the following representations: $T^+e = W^+(we \bullet s)$, $T^-e = W^-(we \bullet s)$, and $|T|e = |W|(we \bullet s)$.*

5.4.5. Theorem. *Let E be an order-dense ideal of $C_\infty(P)$, let F be an order-dense ideal of $C_\infty(Q)$, and let $T : E \to F$ be a disjointness preserving regular operator. Consider the representation $h(U) = s^{-1}(U)$ of the shadow h of the operator T by means of a function $s \in C_0(Q, P)$. Then there exist a family $(w_\xi)_{\xi\in\Xi}$ of positive functions in $C_\infty(P)$ and a family $(W_\xi)_{\xi\in\Xi}$ of pairwise disjoint functions in $C_\infty(Q)$ such that $1/w_\xi \in E$ for all $\xi \in \Xi$ and*

$$Te = o\text{-}\sum_{\xi\in\Xi} W_\xi(w_\xi e \bullet s) \quad (e \in E).$$

$\lhd$ The assertion stated is a reformulation of Theorem 5.3.6 with account taken of Proposition 5.4.4 (3). $\rhd$

Observe that the functions $w_\xi e \bullet s$ in the above representation, being continuous functions from Q into $\overline{\mathbb{R}}$, need not belong to $C_\infty(Q)$, while the products $W_\xi(w_\xi e \bullet s)$ do belong to $C_\infty(Q)$.

If $Te = o\text{-}\sum_{\xi\in\Xi} W_\xi(w_\xi e \bullet s)$ is a representation of the operator T then the operators T^+, T^-, and $|T|$ admit the following representations:

$$T^+e = o\text{-}\sum_{\xi\in\Xi} W_\xi^+(w_\xi e \bullet s),$$

$$T^-e = o\text{-}\sum_{\xi\in\Xi} W_\xi^-(w_\xi e \bullet s),$$

$$|T|e = o\text{-}\sum_{\xi\in\Xi} |W_\xi|(w_\xi e \bullet s).$$

5.4.6. Now we proceed to constructing analytical representations for operators in Banach–Kantorovich spaces which are analogous to above results for operators in K-spaces. If $\mathscr{X}$ and $\mathscr{Y}$ are ample Banach bundles over Q, $u \in C_\infty(Q, \mathscr{X})$ and $w \in C_\infty(Q, B(\mathscr{X}, \mathscr{Y}))$ then the section $w \otimes u : q \in Q \mapsto w(q)u(q) \in Y_q$ is continuous and defines a unique element of $C_\infty(Q, \mathscr{Y})$, see 2.4.6 and 2.4.13. This element will be denoted by $w \bar{\otimes} u$. As was indicated in 2.4.6 $w \bar{\otimes} u$ can be identified with the maximal extension of $w \otimes u$.

Let $\mathscr{X}$ and $\mathscr{Y}$ be ample Banach bundles over Q, let E and E' be order-dense ideals in $C_\infty(Q)$. An operator $W : E(\mathscr{X}) \to E'(\mathscr{Y})$ is an orthomorphism if and only if there exists a section $w \in C_\infty(Q, B(\mathscr{X}, \mathscr{Y}))$ such that the representation $Wu = w\bar{\otimes}u$ holds for all $u \in E(\mathscr{X})$. Moreover, $|W|(e) = |w| \cdot e$ $(e \in E)$ and the function w with these properties is unique.

$\lhd$ Let $W : E(\mathscr{X}) \to F(\mathscr{Y})$ be such that $|W| \in \mathrm{Orth}(E, F)$. Then the representation $|W|(e) = ge$ $(e \in E)$ holds with a suitable function $g \in C_\infty(Q)$. Denote by D the set of all $q \in Q$ such that $|g(q)| < \infty$ and $e(q) \neq 0$ for some $e \in E$. Put $E_1 := E \cap C(Q)$. Define an mapping $w_0 : q \in D \mapsto w(q) \in B(\mathscr{X}(q), \mathscr{Y}(q))$ by the following rule: for arbitrary $q \in D$ and $x \in \mathscr{X}(q)$ take a section $u \in E_1(\mathscr{X})$ with $u(q) = x$, and put $w_0(q)x := (Wu)(q)$. This definition is sound and the operator $w_0(q) : \mathscr{X}(q) \to \mathscr{Y}(q)$ is bounded, since the identities

$$\|(Wu)(q)\| = |Wu|(q) \leqslant (|W||u|)(q) = (g|u|)(q) = g(q)\|u(q)\|$$

are valid for all $q \in D$ and $u \in E_1(\mathscr{X})$. According to Theorem 2.4.13 we have $w_0 \in C(D, B(\mathscr{X}, \mathscr{Y}))$. If $w := \mathrm{ext}(w_0) \in C_\infty(Q, B(\mathscr{X}, \mathscr{Y}))$. then it follows from the definitions that $Wu = w\bar{\otimes}u$ for all $u \in E_1(\mathscr{X})$. The set $E_1(\mathscr{X})$ is a *bo*-dense ideal in the Banach–Kantorovich space $E(\mathscr{X})$ and the operators W and $u \mapsto w\bar{\otimes}u$ are *bo*-continuous. Thus, W and $u \mapsto w\bar{\otimes}u$ coincide on $E(\mathscr{X})$.

Assume that the sections $w_1, w_2 \in C_\infty(Q, B(\mathscr{X}, \mathscr{Y}))$ met the relation $w_1\bar{\otimes}u = w_2\bar{\otimes}u$ for all $u \in E(\mathscr{X})$. Denote by D_0 the set of all $q \in Q$ such that $e(q) \neq 0$ for some $e \in E$. Put $D := D_0 \cap \mathrm{dom}(w_1) \cap \mathrm{dom}(w_2)$ and take $q \in D$ and $x \in \mathscr{X}(q)$. Since there exists a section $u \in E(\mathscr{X})$ with $u(q) = x$, we deduce

$$w_1(q)x = (w_1\bar{\otimes}u)(q) = (w_2\bar{\otimes}u)(q) = w_2(q)x.$$

It follows that $w_1 = w_2$, because D is dense in Q.

To prove the identity $|W|(e) = |w|e$ observe that, by virtue of 2.4.2 (5), we have $|w| = \sup\{|w\bar{\otimes}u| : u \in C(Q, \mathscr{X}), |u| \leqslant 1\}$. Therefore, for all $e \in E_+$ we may deduce:

$$\begin{aligned} |W|e &= \sup_{|u| \leqslant e} |Wu| = \sup_{|u| \leqslant 1} |W(eu)| \\ &= \sup_{|u| \leqslant 1} |w\bar{\otimes}(eu)| = \sup_{|u| \leqslant 1} |w\bar{\otimes}u|e = |w|e. \ \rhd \end{aligned}$$

The relation $Wu = w\bar{\otimes}u$ we will call the *representation* of W *by means of the section* w.

5.4.7. (1) *Let $\mathscr{X}$ and $\mathscr{Y}$ be continuous Banach bundles over Q with $\mathscr{Y}$ ample. Assume that X and Y are decomposable subspaces of lattice-normed spaces $E(\mathscr{X})$ and $E(\mathscr{Y})$, respectively. If the set X is an approximating set in $E(\mathscr{X})$ then a mapping $I : X \to Y$ is an isometric embedding of lattice-normed space if and only if there exists an isometric imbedding $\imath$ of the bundle $\mathscr{X}$ into $\mathscr{Y}$ such that $I(u) = \imath\bar{\otimes}u$ for all $u \in X$.*

$\lhd$ Sufficiency is obvious. To prove necessity suppose that the operator $I : X \to Y$ is an isometric embedding. In view of 5.3.7 there exists an isometric embedding $\bar{I} : C_\infty(Q, \mathscr{X}) \to C_\infty(Q, \mathscr{Y})$ extending I. Let $\overline{\mathscr{X}}$ be an ample hull of $\mathscr{X}$. Identify $C_\infty(Q, \mathscr{X})$ and $C_\infty(Q, \overline{\mathscr{X}})$ according to 2.4.11 (3). By virtue of 5.4.6 we have the representation $I(u) = \imath\bar{\otimes}u$ of the orthomorphism I by means of some section $\imath \in C_\infty\big(Q, B(\overline{\mathscr{X}}, \mathscr{Y})\big)$. It is easy to check that $\imath$ is an isometric imbedding of $\overline{\mathscr{X}}$ into $\mathscr{Y}$. For $q \in Q$ denote the restriction of $\imath(q)$ onto $\mathscr{X}(q)$ by the same symbol $\imath(q)$. By Definition 2.4.9 $\imath \in \mathrm{Hom}_Q(\mathscr{X}, \mathscr{Y})$, and so $\imath$ is an isometric embedding of $\mathscr{X}$ into $\mathscr{Y}$. Obviously $I(u) = \imath\bar{\otimes}u$ for all $u \in X$. $\rhd$

(2) *Let $\mathscr{X}$ and $\mathscr{Y}$ be ample Banach bundles over Q and let E be an order-dense ideal in $C_\infty(Q)$. A linear operator $I : E(\mathscr{X}) \to E(\mathscr{Y})$ is an isometric embedding (isometry) if and only if there exist an isometric embedding (isometry) $\imath$ of $\mathscr{X}$ into (onto) $\mathscr{Y}$ such that $I(u) = \imath\bar{\otimes}u$ for all $u \in E(\mathscr{X})$.*

5.4.8. Let $\mathscr{X}$ be a Banach bundle over P and s be an arbitrary mapping from a subset $\mathrm{dom}(s) \subset Q$ to P. Then the composite $\mathscr{X} \circ s$ is a Banach bundle over $\mathrm{dom}(s)$. Moreover, if u is a section of $\mathscr{X}$ over $D \subset P$ then $u \circ s$ is a section of $\mathscr{X} \circ s$ over $s^{-1}(D)$. For an arbitrary set $\mathscr{U}$ of sections of $\mathscr{X}$, denote by $\mathscr{U} \circ s$ the set of sections $\{u \circ s : u \in \mathscr{U}\}$ of a bundle $\mathscr{X} \circ s$. The bundle $\mathscr{X} \circ s$, extended by the zero stalks on $Q \setminus \mathrm{dom}(s)$, will be denoted by $\mathscr{X} \bullet s$. More precisely, the Banach bundle $\mathscr{X} \bullet s$ over Q is defined by

$$(\mathscr{X} \bullet s)(q) := \begin{cases} \mathscr{X}\big(s(q)\big), & \text{if } q \in \mathrm{dom}(s), \\ \{0\}, & \text{if } q \in Q \setminus \mathrm{dom}(s). \end{cases}$$

If u is a section of $\mathscr{X}$ over $D \subset P$ then we denote by $u \bullet s$ the section

$$(u \bullet s)(q) := \begin{cases} u\big(s(q)\big), & \text{if } q \in s^{-1}[D], \\ 0, & \text{if } q \in Q \setminus \mathrm{dom}(s) \end{cases}$$

of the bundle $\mathscr{X} \bullet s$ over $s^{-1}(D) \cup (Q \setminus \mathrm{dom}(s))$.

Let $\mathscr{C}_{\mathscr{X}}$ be a continuity structure in the Banach bundle $\mathscr{X}$. Then the set $\mathscr{C}_{\mathscr{X}} \bullet s := \{u \bullet s : u \in \mathscr{C}_{\mathscr{X}}\}$ is a continuity structure in the Banach bundle $\mathscr{X} \bullet s$, so that $\mathscr{X} \bullet s$ is considered as a continuous Banach bundle (2.4.3).

(1) *Let $\mathscr{X}$ be an ample Banach bundle over P, $s \in C_0(Q, P)$, and $u \in C_\infty(P, \mathscr{X})$. If $|u| \in C_s(P)$ then $u \bullet s \in C_\infty(Q, \mathscr{X} \bullet s)$.*

$\lhd$ Indeed, the domain of $u \bullet s$ coincides with $\mathrm{dom}(|u| \bullet s)$ which is dense in Q due to $|u| \in C_s(P)$. At the same time, if the section $u \bullet s$ has a limit at $q \in Q$, then $q \in \mathrm{dom}|u \bullet s| = \mathrm{dom}(|u| \bullet s) = \mathrm{dom}(u \bullet s)$. $\rhd$

(2) *Let $\mathscr{X}$ be an ample Banach bundle over P, let $S : E \to F$ be a shift operator. Let $\mathscr{U}$ be the norm transformation of X by means of S and denote by $\hat{S}$ the operator of norm transformation (see 2.2.13 (2)). Let $Se = e \bullet s$ be the representation of S by means of some mapping $s \in C_0(Q, P)$. Then there exists an isometric embedding $\imath : \mathscr{U} \to F(\mathscr{X} \bullet s)$ such that $\imath\hat{S}u = u \bullet s$ for all $u \in E(\mathscr{X})$.*

$\lhd$ Define an operator $\imath_0 : \hat{S}(\mathscr{U}) \to F(\mathscr{X} \bullet s)$ by putting $\imath_0\hat{S}u := u \bullet s$ for all $u \in \mathscr{U}$. The definition is sound since $|u \bullet s| = |u| \bullet s = S|u| = |\hat{S}u|$ $(u \in \mathscr{U})$, and $|\imath_0(v)| = |v|$ for all $v \in \hat{S}(\mathscr{U})$. In view of 5.3.7 $\imath_0$ extends to a desired isometric embedding $\imath : \mathscr{U} \to F(\mathscr{X} \bullet s)$. $\rhd$

(3) *Let $\mathscr{X}$ and $\mathscr{Y}$ be ample Banach bundles over P and Q, respectively. A linear operator $S : E(\mathscr{X}) \to F(\mathscr{Y})$ is a shift operator if and only if there exist a mapping $s \in C_0(Q, P)$ and a isometric embedding $\imath$ of the bundle $\mathscr{X} \bullet s$ into $\mathscr{Y}$ such that $Su = \imath \otimes (u \bullet s)$ for all $u \in E(\mathscr{X})$. Moreover, $|S|e = e \bullet s$ for all $e \in E$.*

$\lhd$ Sufficiency is easily verified by employing Theorem 5.3.8 (2). Necessity follows from (2), (3), and 5.4.7 (1). $\rhd$

5.4.9. Theorem. *Let $\mathscr{X}$ and $\mathscr{Y}$ be ample Banach bundles over P and Q, respectively. A linear operator $T : E(\mathscr{X}) \to F(\mathscr{Y})$ admits WSW-representation if and only if there exist a positive function $w \in C_\infty(P)$, a mapping $s \in C_0(Q, P)$, and a section $W \in C_\infty(Q, B(\overline{\mathscr{X} \bullet s}, \mathscr{Y}))$, where $\overline{\mathscr{X} \bullet s}$ is an ample hull of $\mathscr{X} \bullet s$, such that $Tu = W\bar{\otimes}(wu \bullet s)$ for all $u \in E(\mathscr{X})$ and $|T|e = |W|(we \bullet s)$ for all $e \in E$.*

$\lhd$ Sufficiency follows immediately from 5.4.6 and 5.4.8 (3). Let a linear operator $T : E(\mathscr{X}) \to F(\mathscr{Y})$ admit WSW-factorization. Then by Theorem 5.3.9 (2) there exist a Banach–Kantorovich space $\mathscr{V}'$ over an order-dense ideal $F' \subset C_\infty(Q)$, a scalar orthomorphism $\bar{w} : E(\mathscr{X}) \to C_\infty(P, \mathscr{X})$ generated by a positive orthomorphism $w : E \to C_\infty(P)$, a shift operator $\bar{S} : (wE)(\mathscr{X}) \to \mathscr{V}'$, and an orthomorphism $\bar{W} : \mathscr{V}' \to F(\mathscr{Y})$ such that $T = \bar{W} \circ \bar{S} \circ \bar{w}$ and $|T| = |\bar{W}| \circ |\bar{S}| \circ |\bar{w}|$. Using Theorem 2.4.10, we may assume that $\mathscr{V}' = F'(\mathscr{Z})$, where $\mathscr{Z}$ is an ample Banach bundle over Q. According to 5.4.8 (3) there are a mapping $s \in C_0(Q, P)$ and an isometric imbedding $\imath$ of the bundle $\mathscr{X} \bullet s$ into $\mathscr{Z}$ such that $\bar{S}u = \imath \otimes (u \bullet s)$ for all $u \in (wE)(\mathscr{X})$. By virtue of 2.4.11 (2) the homomorphism $\imath$ can be extended to an isometric embedding $\bar{\imath}$ of the bundle $\overline{\mathscr{X} \bullet s}$ into $\mathscr{Z}$. In view of 5.4.6 the operator $(v \mapsto \bar{W}(\bar{\imath} \otimes v)) : F'(\overline{\mathscr{X} \bullet s}) \to F(\mathscr{Y})$ has the representation $v \mapsto W\bar{\otimes}v$, where $W \in C_\infty(Q, B(\overline{\mathscr{X} \bullet s}, \mathscr{Y}))$. It is easy to verify that the functions w, s and W possess the desired properties. $\rhd$

5.4.10. Theorem. *Let $\mathscr{X}$ and $\mathscr{Y}$ be ample Banach bundles over P and Q, respectively. Let $T : E(\mathscr{X}) \to F(\mathscr{Y})$ be a bounded disjointness preserving operator* *and let* $s \in C_0(Q,P)$ *be its shift function. Then there exist a family* $(w_\xi)_{\xi\in\Xi}$ *of positive functions in* $C_\infty(P)$, *a pairwise disjoint family* $(Q_\xi)_{\xi\in\Xi}$ *in* $\mathrm{Clop}(Q)$, *and a section* $W \in C_\infty(Q, B(\overline{\mathscr{X} \bullet s}, \mathscr{Y}))$ *such that* $(\mathrm{supp}(W)) = \mathrm{cl}\bigcup_{\xi\in\Xi} Q_\xi = \mathrm{dom}(s) = \mathrm{supp}(\mathrm{im}(T))$, $1/w_\xi \in E$ *for all* $\xi \in \Xi$ *and*

$$Tu = \bigoplus_{\xi\in\Xi} W \otimes (w_\xi u \bullet s)|_{Q_\xi} \quad (u \in E(\mathscr{X})).$$

$\lhd$ The claim follows from Theorems 5.3.10 and 5.4.9. $\rhd$

5.5. Decomposable Operators

In this section we consider several results concerning an analytical description for the dual classes of disjointness preserving operators and decomposable operators acting in the spaces of continuous vector-functions. We give an independent representation though some results may be deduced from Section 5.3 or can be generalized to the case of Banach bundles.

5.5.1. Throughout this section X and Y are normed spaces with normed duals X' and Y'; furthermore, E and F are order-dense ideals in universally complete vector lattices $\mathscr{E} := C_\infty(P)$ and $\mathscr{F} := C_\infty(Q)$, respectively. Recall (see 4.1.3 (3)) that $L_A(X,E)$ is the Banach–Kantorovich space of operators $T : X \to E$ with *abstract norm* $|T| = \sup\{Tx : \|x\| \le 1\}$. For the definitions of $E(X)$, $E_s(X)$, and $\mathfrak{M}_Q(X,Y')$ see Section 2.3.

(1) *To each operator with abstract norm* $T : X \to E$ *there is a unique* $u_T \in E_s(X')$ *satisfying*

$$Tx = \langle x, u_T\rangle \quad (x \in X).$$

The mapping $T \mapsto u_T$ *is a linear isometry between the Banach–Kantorovich spaces* $L_A(X,E)$ *and* $E_s(X')$.

$\lhd$ If $e := |T|$ then, for every $x \in X$, the function $Tx \in C_\infty(Q)$ takes a finite value at each point of $Q_0 := \{t \in Q : e(t) < +\infty\}$ since $|Tx| \le e\|x\|$. The last estimate also implies that, for $t \in Q_0$, the functional $v(t) : x \mapsto (Tx)(t)\ \ (x \in X)$ is bounded and $\|v(t)\| \le e(t)$. This gives rise to the mapping $v : Q_0 \to X'$ continuous in the weak topology $\sigma(X',X)$. Let u_T denote the coset of v. Then $Tx = \langle x, u_T\rangle$ for all $x \in X$. In particular, the following supremum exists: $\sup\{\langle x, u_T\rangle| : \|x\| \le 1\} = e$. Hence, $u_T \in E_s(X')$ and $|u_T| = |T|$. We thus see that $T \mapsto u_T$ is an isometry from $L_A(X,E)$ to $E_s(X')$. Clearly, this mapping is linear and surjective. $\rhd$

Consider $T \in L_A(X\widehat{\otimes}Y, E)$, where $X\widehat{\otimes}Y$ is the projective tensor product of X and Y. It is an easy matter to show that the bilinear operator $b := T\otimes : X \times Y \to E$ has the abstract norm

$$|b| := \sup\{|b(x,y)| : \|x\| \le 1,\ \|y\| \le 1\},$$

with $|b| = |T|$. Denote by $\mathscr{B}_A(X \times Y, E)$ the set of all bilinear operators $b : X \times Y \to E$ with abstract norm. We further let $\mathscr{B}(X \times Y)$ denote the set of all bilinear forms on $X \times Y$. Since the isometric isomorphy $(X \widehat{\otimes} Y)' \simeq \mathscr{B}(X \times Y)$ is available, from (1) we derive the following proposition:

(2) *To every $b \in \mathscr{B}_A(X \times Y, E)$ there is a unique $u_b \in E_s(\mathscr{B}(X \times Y))$ such that*

$$b(x, y) = \langle x \otimes y, u_b \rangle \quad (x \in X,\ y \in Y).$$

The mapping $b \mapsto u_b$ is a linear isometry between $\mathscr{B}_A(X \times Y, E)$ and $E_s(\mathscr{B}(X \times Y))$.

5.5.2. Let $\Phi : E \to mF$ be a lattice homomorphism. Assume that $F = \Phi(E)^{\perp\perp}$ is a K-space. In the universal completion mF, we fix the multiplicative structure that is uniquely determined by choice of unity. Assign $F' := \{f \in mF : f \cdot \Phi(E) \subset F\}$. Let $\mathscr{L}^{\Phi}(E, F)$ be the set of all regular operators $S : E \to F$ such that $S \in \{\Phi\}^{\perp\perp}$, where S is regarded as an operator from E into mE. Assign

$$M^{\Phi}(E(X), F_s(Y')) := \{T \in M(E(X), F_s(Y')) : |T| \in \mathscr{L}^{\Phi}(E, F)\}.$$

(1) *Every disjointness preserving sequentially o-continuous regular operator $T : E \to F$ admits a WSW-representation with an arbitrary order-unity $w \in C_{\infty}(P)$ as an inner weight.*

$\lhd$ If T is sequentially o-continuous then its shadow $\operatorname{shdw}(T)$ is also sequentially o-continuous and, by Definition 5.1.3, $\mathscr{E} = \mathscr{E}(1, h)$. According to 5.3.1 (3) T is wide at 1. Now, the claim follows from 5.3.5. $\rhd$

(2) *Assume that $\Phi : E \to F$ be an order continuous lattice homomorphism. Then there exist a continuous open mapping φ from a closed subset of Q into P and a function $\alpha \in C_{\infty}(Q)$ such that $\Phi e = \alpha \cdot e \bullet \varphi$ $(e \in E)$.*

$\lhd$ The representation $\Phi e = \alpha \cdot e \bullet \varphi$ follows from (1) and 5.4.4 (1). If T is order continuous then $\operatorname{shdw}(T)$ is also order continuous and then φ is an open mapping; see [352]. $\rhd$

(3) *If the substitution operator $\varphi^* : e \mapsto e \bullet \varphi$ acts from E into F and is order continuous then for any Banach space X the mapping $\varphi^*_X : u \mapsto u \bullet \varphi$ is a unique dominated operator from $E(X)$ to $F(X)$ such that $|s^*_X| = s^*$ and $\varphi^*_X(x \otimes e) = x \otimes \varphi^* e$ $(x \in X,\ e \in E)$.*

$\lhd$ The operator φ^* is order continuous if and only if the mapping φ is open; moreover, the inverse image of every meager subset of Q under φ is a meager subset of P; see [352]. Therefore, the mapping $u \mapsto u \bullet \varphi$ acts from $E(X)$ into $C_{\infty}(Q, X)$ and is uniquely defined on $X \otimes E$ by the formula

$$\varphi^*_X\left(\sum_{k=1}^{n} x_k \otimes e_k\right) := \sum_{k=1}^{n} x_k \otimes \varphi^* e_k.$$

Straightforward verification shows that this operator is dominated. From Theorem 4.3.3 it is clear that φ_X^* admits a unique extension onto $E(X)$. The equality $|\varphi_X^*| = \varphi^*$ follows from 4.3.5 (3). $\triangleright$

5.5.3. Theorem. *Suppose that Φ is an order continuous lattice homomorphism from E into mF. For every operator $T \in M^\Phi(E(X), F_s(Y'))$ there exists a unique (to within equivalence) operator-function $K \in \mathfrak{M}_Q(X, Y')$ satisfying $|K| \in F'$ and providing the representation*

$$Tu = K(\alpha \cdot \varphi_X^*(u)) \quad (u \in E(X)).$$

The correspondence $T \mapsto K$ defines a linear isometry between the Banach–Kantorovich spaces $M^\Phi(E(X), F_s(Y'))$ and $F_s'(\mathscr{L}(X, Y'))$.

$\triangleleft$ According to 5.5.2 (2), the operator Φ admits the multiplicative representation $\Phi e = \alpha\varphi^* e$ $(e \in E)$. Take a $T \in M^\Phi(E(X), F_s(Y'))$ and assign $Se = S_{x,y}(e) := \langle y, T(x \otimes e)\rangle$, where $x \in X$, $y \in Y$, and $e \in E$. The element $x \otimes e \in E(X)$ is determined by the formula $x \otimes e : t \mapsto e(t)x$, $|e(t)| < +\infty$. Observe that

$$|Se| \leq |T(x \otimes e)| \|y\| \leq |T|(|e|) \cdot \|x\| \|y\|.$$

Since $|T| \in \mathscr{L}^\Phi$, we also have $S \in \mathscr{L}^\Phi$, because $\mathscr{L}^\Phi$ is a band. Denote by λ the isomorphism from $\mathscr{L}^\Phi(E, F)$ onto F' that was presented in 3.3.4 (2). Assign $b(x, y) := \lambda(S_{x,y})$. From the definitions it is clear that the mapping $(x, y) \mapsto b(x, y)$ is a bilinear operator from $X \times Y$ into F'. Moreover,

$$|b(x, y)| = \lambda(|S_{x,y}|) \leq \lambda(|T|) \|x\| \|y\| \quad (x \in X,\ y \in Y),$$

i.e., b is an operator with abstract norm and $|b| \leq \lambda(|T|)$. By the definition of λ, we have

$$\langle y, T(x \otimes e)\rangle = b(x, y)\Phi e \quad (x \in X,\ y \in Y,\ e \in E).$$

Taking account of what was said, as well as of the formula 4.3.5 (3), we may write the chain

$$\begin{aligned} |T|e &= \sup\Big\{\sum |T(x(k) \otimes c(k))| : (x, c) \in \mathscr{U}\Big\} \\ &= \sup\Big\{\sup_{\|y\| \leq 1} \sum \langle y, T(x(k) \otimes c(k))\rangle : (x, c) \in \mathscr{U}\Big\} \\ &= \sup\Big\{\sup_{\|y\| \leq 1} \sum b(x(k), y)\Phi(c(k)) : (x, c) \in \mathscr{U}\Big\} \\ &\leq \sup_{\|y\| \leq 1} \sup\Big\{\sum |b|\Phi(c(k)) \|x(k)\| \|y\| : (x, c) \in \mathscr{U}\Big\} \\ &\leq \sup\Big\{|b|\Phi\Big(\sum c(k)\Big)\Big\} \\ &= |b|\Phi(e), \end{aligned}$$

where the set $\mathscr{U}$ is constituted by all pairs (x, c) in which $x : \{1, \dots, n\} \to X$, $c : \{1, \dots, x_n\} \to E_+$, and $n \in \mathbb{N}$; moreover, $\|x(k)\| \le 1$ $(k := 1, \dots, n)$, $c(k) \perp c(l)$ $(k \neq l)$, and $\sum_{k=1}^{n} c(k) = e$. From the estimates made it is clear that $\lambda(|T|) \le |b|$. Taking account of the reverse inequality established above, we obtain $|b| = \lambda(|T|)$. Now apply Theorem 5.5.1 (2), according to which there exists an operator-function $K \in \mathfrak{M}_Q(X, Y')$ such that $|b| = |K|$ and the following hold:

$$\langle y, T(x \otimes e)\rangle = b(x, y)\Phi(e) = \langle y, Kx\rangle\alpha\varphi^* e = \langle y, K(\alpha\varphi_X^*(x \otimes e))\rangle.$$

Since the choice of $y \in Y$ is arbitrary,

$$T(x \otimes e) = K(\alpha\varphi_X^*(x \otimes e)) \quad (x \in X,\ e \in E).$$

Using this fact, linearity of T and K, and the definition of φ_X^* (see 5.5.2 (3)), we deduce that

$$Tu = K(\alpha\varphi_X^* u) \quad (u \in X \otimes E).$$

Since the operators T and φ_X^* are *bo*-continuous and $X \otimes E$ is *bo*-dense in $E(X)$, the above representation on $X \otimes E$ remains valid on the entire $E(X)$. Uniqueness of K is clear from the following reasoning. If $L \in \mathfrak{M}_Q(X, Y')$ represents T too, then $L(\alpha\varphi_X^* u) = K(\alpha\varphi_X^* u)$ for $u \in E(X)$. In particular,

$$(L(t)x - K(t)x)\alpha(t)e(\varphi(t)) = 0 \quad (x \in X,\ e \in E).$$

Since $\Phi(E)$ is an order-dense ideal in $C_\infty(Q)$, we conclude that α is an order-unity in $C_\infty(Q)$ and $\operatorname{dom}(\varphi)$ is dense in Q; hence, $L(t)x = K(t)x$ for all $t \in Q$ with the possible exception of points in some meager set. It remains to observe that if $K \in \mathfrak{M}_Q(X, Y')$ and $|K| \in F'$ then the operator $Su = K(\alpha\varphi_X^* u)$ belongs to M^Φ, because

$$|Su| \le |K||\alpha\varphi_X^*(u)| \le \alpha|K|\varphi^*(|u|) = |K|\Phi(|u|) \quad (u \in E(X)). \ \rhd$$

5.5.4. Theorem. *Let Φ be an order continuous lattice homomorphism from E into F and suppose that an operator $T \in M(E(X), F_s(Y, Z))$ satisfies the condition $|T| \in \{\Phi\}^{\perp\perp}$. Then there exist an operator-function $K \in \mathfrak{M}_Q(X, Y')$, a function $\alpha \in C_\infty(Q)$, and a continuous mapping $\varphi : \operatorname{dom}(\varphi) \to P$, with $\operatorname{dom}(\varphi)$ a closed subset of Q, such that the following hold:*

$$\begin{aligned} Tu &= K(\alpha\varphi_X^* u) && (u \in E(X)), \\ |T|e &= |K|\alpha\varphi^* e && (e \in E), \\ \Phi e &= \alpha\varphi^* e && (e \in E). \end{aligned}$$

$\lhd$ Identify Y with a closed subspace of Z'. Then $F_s(Y, Z)$ is a *bo*-closed subspace of $F_s(Z')$. It is clear that $T \in M^\Phi(E(X), F_s(Z'))$, because $|T| \in \{\Phi\}^{\perp\perp}$ By Theorem 5.5.3, there exists a $K \in \mathfrak{M}_Q(X, Z')$ for which the required representations hold. $\rhd$

5.5.5. Consider LNSs (X, E) and (Y, F). A linear operator $T : X \to Y$ is called *decomposable* if, for all $x \in X$ and disjoint $y_1, y_2 \in Y$, the equality $Tx = y_1 + y_2$ implies existence of $x_1, x_2 \in X$ such that $x = x_1 + x_2$ and $T(B_X(|x_k|)) \perp y_l$ $(k, l := 1, 2,\ k \neq l)$. The last relation means that if $|u| \leq |x_k|$ then $Tu \perp y_l$ $(k \neq l)$. In particular, $Tx_k \perp y_l$ $(k \neq l)$; therefore, the coinciding elements $Tx_1 - y_1$ and $y_2 - Tx_2$ are disjoint and, hence, $Tx_k = y_k$ $(k = 1, 2)$. A positive operator $S : E \to F$ is called *positively decomposable* if, for all $e \in E_+$ and disjoint $f_1, f_2 \in F_+$, the equality $Se = f_1 + f_2$ implies existence of $e_1, e_2 \in E_+$ such that $e = e_1 + e_2$ and $Se_k = f_k$ $(k = 1, 2)$. Recall that the operator S is said to be *interval preserving* if $S([0, e]) = [0, Se]$ for each $e \in E_+$. An order continuous, interval preserving operator is called a *Maharam operator.*

5.5.6. Theorem. *Let X and Y be decomposable and let F be order complete. Given a dominated operator $T : X \to Y$, the following hold:*

(1) *T is decomposable if and only if $|T|$ is positively decomposable on the o-ideal generated by $|X|$;*

(2) *T is decomposable and bo-continuous if and only if $|T|$ is positively decomposable and o-continuous;*

(3) *if E is order complete then T is decomposable and bo-continuous if and only if $|T|$ is a Maharam operator.*

$\lhd$ Suppose that the operator $|T|$ is positively decomposable on the set $|X|$. Assume that the representation $Tx = y_1 + y_2$ holds for $x \in X$ and $y_1, y_2 \in Y$, $y_1 \perp y_2$. Take the projection π_1 onto the band $|y_1|^{\perp\perp}$ and assign $\pi_2 := \pi_1^{\perp}$ and $f_k := \pi_k |T|(|x|)$ $(k := 1, 2)$. Then $|T|(|x|) = f_1 + f_2$, $f_1 \perp f_2$, and, hence, there are $e_1, e_2 \in E_+$ such that $e_1 + e_2 = |x|$ and $f_k = |T|e_k$ $(k := 1, 2)$. Decomposability of X provides the representation $x = x_1 + x_2$ with $|x_k| = e_k$ $(k := 1, 2)$. If $u \in X$ and $|u| \leq |x_k|$ then $|Tu| \leq |T|(|u|) \leq f_k$; therefore, $|Tu| \perp f_l$ $(k \neq l)$. Since $|y_l| \leq f_l$, we have $Tu \perp y_l$ $(k \neq l)$. Thus, T is decomposable.

Conversely, let T be decomposable. Assume that $|T|e = f_1 + f_2$ for some $e := |x|$, $x \in X$, and disjoint $f_1, f_2 \in F_+$. Assign $y_k := \pi_k Tx$, where π is the projection onto the band $\{f_1\}^{\perp\perp}$ and $\pi_2 := \pi_1^{\perp}$. In view of decomposability of T, there are $x_1, x_2 \in X$ such that $x = x_1 + x_2$, $Tx_k = y_k$, and $T(B_X(|x_k|)) \perp y_l$ $(k \neq l)$. It follows that $e \leq c_1 + c_2$, $|T|c_k \perp f_l$ $(k \neq l)$, and $c_k = |x_k|$. If $e_k \leq c_k$ are such that $e = e_1 + e_2$ then $f_1 + f_2 = |T|e_1 + |T|e_2$ and $|T|e_k \perp f_l$ $(k \neq l)$; consequently, $|T|e_k = f_k$ $(k = 1, 2)$. Suppose now that $e \leq |x_1| + \dots + |x_n|$, where $x_1, \dots, x_n \in X$. Then, due to decomposability of X, we have $e = c_1 + \dots + c_n$ with $c_1, \dots, c_n \in |X|$. In view of what was proven above, there exist $e_{1k}, e_{2k} \in E_+$ $(k := 1, \dots, n)$ such that $c_k = e_{1k} + e_{2k}$, $\sum_{k=1}^{n} |T|e_{1k} = f_1$, and $\sum_{k=1}^{n} |T|e_{2k} = f_2$. It remains to assign $e_1 := e_{11} + \dots + e_{1n}$ and $e_2 := e_{21} + \dots + e_{2n}$. Thus, assertion (1) is completely proven. In order to prove (2), observe that an o-continuous positive operator $S : E \to F$ is decomposable if S is such on an order-dense ideal E_0 (in our case, E_0 is the ideal generated by $|X|$, thus being an order-dense ideal in $|X|^{\perp\perp}$, and $S := |T|$ vanishes on $|X|^{\perp}$).

Taking account of (2), in (3) it is sufficient to prove the following: an o-continuous positive operator $S : E \to F$ preserves intervals if and only if S is positively decomposable. If S preserves intervals then, obviously, S is positively decomposable. Suppose that S is positively decomposable. Without loss of generality, we may assume that S is essentially positive. Assign $|e|_S = S(|e|)$ $(e \in E)$. Then $|\cdot|_S$ is a d-decomposable F-valued norm in E, see 3.4.3. According to 2.1.3, there exist a complete Boolean algebra $\mathscr{B}$ of projections in E and an isomorphism $h : \mathfrak{P}(F_0) \to \mathscr{B}$, with $F_0 = |E|_S^{\perp\perp}$, such that $\pi|e|_S = |h(\pi)e|_S$ $(e \in E,\ \pi \in \mathfrak{P}(F_0))$. Moreover, $\mathscr{B}$ is an order-closed subalgebra of $\mathfrak{P}(E)$. Using 2.1.8, we may extend the isomorphism h onto $\Lambda := \mathrm{Orth}(F)$, thereby endowing E with the structure of a unital Λ-module. Hence it easily follows that S preserves intervals. $\triangleright$

5.5.7. Theorem. *Let X and Y be decomposable and let E and F be order complete. Suppose that an operator $T \in M(X,Y)$ is decomposable and bo-continuous. Assign $Y_T := T(X)^{\perp\perp}$ and $X_T := \{x \in X : |T|(|x|) = 0\}^{\perp}$. There exists a Boolean isomorphism h from $\mathscr{P}(Y_T)$ onto $\mathscr{P}(X_T)$ such that, for each bo-continuous $S \in M(X,Y)$, the following are equivalent:*

(1) $S \in \{T\}^{\perp\perp}$;

(2) $Sx \in T\big(B_X(|x|)\big)^{\perp\perp}$ $(x \in X)$;

(3) $\pi S = Sh(\pi)$ $\big(\pi \in \mathscr{P}(Y_T)\big)$.

$\triangleleft$ Without loss of generality, we may assume that $E = |X|$ (see 2.1.7 (3)). By Theorem 5.5.6 (3), $\Phi := |T|$ is a Maharam operator. The general properties of Maharam operators (see 3.4.3) ensure existence of a Boolean isomorphism h from $\mathscr{P}(F_\Phi)$ onto $\mathfrak{P}(E_\Phi)$ such that $\pi\Phi = \Phi h(\pi)$ for all $\pi \in \mathfrak{P}(F_\Phi)$. Denote by the same letter h the isomorphism from $\mathscr{P}(Y_T)$ onto an order-closed subalgebra of $\mathscr{P}(X_T)$; such an isomorphism exists, since the Boolean algebras $\mathfrak{P}(F_\Phi)$ and $\mathscr{P}(Y_T)$, as well as $\mathfrak{P}(E_\Phi)$ and $\mathscr{P}(X_T)$, are pairwise isomorphic. For completing the proof, it is sufficient to show that the required properties (1), (2), and (3) are respectively equivalent to the following:

(1′) $\Psi \in \{\Phi\}^{\perp\perp}$;

(2′) $\Psi e \in \{\Phi e\}^{\perp\perp}$ $(e \in E)$;

(3′) $\pi\Psi = \Psi h(\pi)$ $\big(\pi \in \mathfrak{P}(F_\Phi)\big)$,

where $\Psi := |S|$.

(1) $\Leftrightarrow$ (1′): It is evident.

(2) $\Leftrightarrow$ (2′): Suppose $Sx \in T\big(B_X(|x|)\big)^{\perp\perp}$ $(x \in X)$. If $x_1, \dots, x_n \in X$ are such that $|x| = |x_1| + \dots + |x_n|$, then $\sum_{k=1}^{n} T\big(B_X(|x_k|)\big)^{\perp\perp} \subset T\big(B_X(|x|)\big)^{\perp\perp}$. At

the same time $\sum_{k=1}^{n}|S(x_k)| \in \sum_{k=1}^{n} T(B_X(|x_k|))^{\perp\perp}$; therefore, $\sum_{k=1}^{n}|S(x_k)| \in \Phi(|x|)^{\perp\perp}$. Thus, we deduce that

$$\Psi(|x|) = \sup\left\{\sum_{k=1}^{n}|S(x_k)|\right\} \in \Phi(|x|)^{\perp\perp}.$$

Conversely, let $\Psi e \in \{\Phi e\}^{\perp\perp}$ for all $e \in E$. Then $|Sx| \in \{\Psi e\}^{\perp\perp} \subset \{\Phi e\}^{\perp\perp}$ and it remains to observe that

$$\begin{aligned}\{\Phi e\}^{\perp\perp} &= \sup\left\{\sum_{k=1}^{n}|Tx_k| : \sum_{k=1}^{n}|x_k| = |x|,\ n \in \mathbb{N}\right\}^{\perp\perp}\\ &= \bigvee\left\{|Ty|^{\perp\perp} : |y| \leq |x|\right\} = T(B_X(|x|))^{\perp\perp}.\end{aligned}$$

(3) $\Leftrightarrow$ (3′): If (3) is valid then the following equalities hold:

$$\begin{aligned}\pi\Psi e &= \sup\left\{\sum_{k=1}^{n}\pi|Sx_k| : \sum_{k=1}^{n}|x_k| = e,\ n \in \mathbb{N}\right\}\\ &= \sup\left\{\sum_{k=1}^{n}|\pi Sx_k| : \sum_{k=1}^{n}|x_k| = e,\ n \in \mathbb{N}\right\}\\ &= \sup\left\{\sum_{k=1}^{n}|Sh(\pi)x_k| : \sum_{k=1}^{n}|x_k| = e,\ n \in \mathbb{N}\right\}\\ &= \sup\left\{\sum_{k=1}^{n}|Sy_k| : \sum_{k=1}^{n}|y_k| = h(\pi)e,\ n \in \mathbb{N}\right\}\\ &= \Psi h(\pi)e.\end{aligned}$$

Conversely, assume that $\pi\Psi = \Psi h(\pi)$ for all $\pi \in \mathfrak{P}(F_\Phi))$. Then $|Sh(\pi)x| \leq \pi\Psi(|x|)$ and $|\pi^\perp Sh(\pi)x| = \pi^\perp|Sh(\pi)x| = 0$. Thus, $\pi^\perp Sh(\pi) = 0$ or $\pi Sh(\pi) = Sh(\pi) = 0$. Replacing π by $\pi^\perp$ in the last equality we obtain also $\pi Sh(\pi) - \pi S = 0$, whence $\pi S = \pi Sh(\pi)$. $\triangleright$

5.5.8. Considering the universal completion mE, fix the multiplicative structure that is uniquely determined by a choice of an order-unity $\mathbf{1}$. Suppose that an F-valued essentially positive Maharam operator Φ is defined on some order-dense ideal $\mathscr{D}(\Phi) \subset mE$. Denote by $L^1(\Phi)$ the maximal order-dense ideal of mE onto which we may extend Φ by o-continuity. We assume that $\mathscr{D}(\Phi) = L^1(\Phi)$. Let Φ_0 be the restriction of Φ onto $E_0 := E \cap L^1(\Phi)$. Denote by $\mathscr{L}_\Phi(E, F)$ the set of all o-continuous regular operators $S : E \to F$ such that the restriction of S onto E_0 belongs to the band $\{\Phi_0\}^{\perp\perp}$. It is easily seen that $\mathscr{L}_\Phi(E,F)$ is a band of $\mathscr{L}_n(E,F)$. Assign $E' := \{e' \in mE : e' \cdot E \subset L^1(\Phi)\}$.

Now, given Banach spaces X and Y, consider the LNS $E'_s(\mathscr{L}(X,Y'))$ defined as a factor set of the class $\mathfrak{M}_P(X,Y')$ of operator-functions, see 2.3.6. Take $K \in \mathfrak{M}_P(X,Y')$, $u \in E(X)$, and $y \in Y$. If $s, s_0 \in \operatorname{dom}(K) \cap \operatorname{dom}(u)$ then

$$\begin{aligned} &|\langle y, K(s)u(s)\rangle - \langle y, K(s_0)u(s_0)\rangle| \\ &\le |K|(s)\|u(s) - u(s_0)\| + |\langle y, (K(s) - K(s_0))u(s_0)\rangle|. \end{aligned}$$

Hence it is clear that the function $t \mapsto \langle y, K(t)u(t)\rangle$ is continuous on $\operatorname{dom}(K) \cap \operatorname{dom}(u)$; therefore, it admits an $\overline{\mathbb{R}}$-valued continuous extension onto the entire P. Denote by $\langle y, Ku\rangle$ the corresponding element of $C_\infty(P)$. From the inequality $|\langle y, Kx\rangle| \le |K|\|x\| \cdot \|y\|$ it follows that $|\langle y, Ku\rangle| \le |K| \cdot |u|$. If $|K| \in E'$ and $|u| \in E$ then $|K| \cdot |u| \in L^1(\Phi)$; thus, the element $\Phi(\langle y, Ku\rangle) \in F$ is defined.

Now we formulate and prove the main result of the present section.

5.5.9. Theorem. *For every dominated operator $T \in M_\Phi(E(X), F_s(Y'))$, there exists a unique (to within equivalence) operator-function $K \in \mathfrak{M}_P(X,Y')$ such that $|K| \in E'$ and the following holds:*

$$\langle y, Tu\rangle = \Phi(\langle y, Ku\rangle) \quad (u \in E(X),\ y \in Y).$$

The correspondence $T \mapsto K$ establishes a linear isometry of the Banach–Kantorovich spaces $M_\Phi(E(X), F_s(Y'))$ and $E'_s(\mathscr{L}(X,Y'))$.

$\lhd$ In the same way as in the proof of Theorem 5.5.3, for $x \in X$ and $y \in Y$ we define an operator $S := S_{x,y} : E \to F$ by $Se := \langle y, T(x \otimes e)\rangle$. It turns out again that $S \in \mathscr{L}_\Phi(E,F)$. Denote by λ the isomorphism from $\mathscr{L}_\Phi(E,F)$ onto E' discussed in the theorem of 3.4.8 and assign $b(x,y) := \lambda(S_{x,y})$. Then, due to the mentioned theorem, we have $\Phi(eb(x,y)) = \langle y, T(x \otimes e)\rangle$ for all $e \in E$, $x \in X$, and $y \in Y$. It is easy to see that the mapping $(x,y) \mapsto b(x,y)$ is a bilinear operator from $X \times Y$ into E'. Moreover, b is an operator with abstract norm and $|b| \le \lambda(|T|)$. At the same time, by formula 4.3.5 (3) we have

$$\begin{aligned} |T|e &= \sup\Big\{\sum |T(x(k) \otimes c(k))| : (x,c) \in \mathscr{U}\Big\} \\ &= \sup\Big\{\sup_{\|y\|\le 1} \sum \Phi\Big(c(k)b\big(x(k), y\big)\Big) : (x,c) \in \mathscr{U}\Big\} \\ &\le \sup_{\|y\|\le 1} \sup\Big\{\sum \Phi(c(k)|b|)\|x(k)\|\|y\| : (x,c) \in \mathscr{U}\Big\} \\ &\le \Phi\Big(\sum c(k)|b|\Big) \\ &= \Phi(e|b|). \end{aligned}$$

Here the set $\mathscr{U}$ is the same as in 5.5.3. It is clear that $\lambda(|T|) \leq |b|$, which, together with the above-mentioned reverse inequality, yields $|b| = \lambda(|T|)$. According to Proposition 5.5.1 (2), there exists a $K \in E'_s(\mathscr{L}(X, Y'))$ such that $|K| = |b|$ and $b(x, y) = \langle y, Kx \rangle$ $(x \in X,\ y \in Y)$. Taking the definition of b into account, we may write

$$\langle y, T(x \otimes e) \rangle = \Phi(eb(x, y)) = \Phi(e\langle y, Kx \rangle) = \Phi(\langle y, Kx \otimes e \rangle).$$

Thus, $\langle y, Tu \rangle = \Phi(\langle y, Ku \rangle)$ for all $u \in X \otimes E$. This means that the operators $T_1 u := \langle y, Tu \rangle$ and $T_2 u := \Phi(\langle y, Ku \rangle)$ from $E(X)$ into F coincide on $X \otimes E$. Moreover, these operators admit o-continuous dominants S_1 and S_2, respectively:

$$S_1 e := |T|(e)\|y\|, \quad S_2 e := \Phi(e|K|)\|y\| \quad (e \in E).$$

Since $X \otimes E$ is bo-dense in $E(X)$, we have $T_1 = T_2$ on the entire $E(X)$. It is also clear that $\lambda(|T|) = |K|$.

If there is another element $L \in E'_s(\mathscr{L}(X, Y'))$ that serves as a representing operator-function for T then $\Phi(\langle y, (K - L)u \rangle) = 0$ for all $u \in E(X)$. Assigning $u := x \otimes e$, we conclude that $\Phi(e\langle y, (K - L)x \rangle) = 0$ for all $e \in E$. Therefore, $\langle y, (K - L)x \rangle = 0$. Since $x \in X$ and $y \in Y$ are arbitrary, $K \sim L$. Thus, the correspondence $T \mapsto K$ defines a linear isometry from $M_\Phi(E(X), F_s(Y'))$ into $E'_s(\mathscr{L}(X, Y'))$. We will prove that it is surjective.

Take $K \in E'_s(\mathscr{L}(X, Y'))$, $u \in E(X)$, and $y \in Y$. As was mentioned in 5.5.8, we have the element $\Phi(\langle y, Ku \rangle) \in F$ defined correctly. Assign $S_u(y) := \Phi(\langle y, Ku \rangle)$. The operator $S_u : Y \to F$ is linear and $|S_u(y)| \leq \Phi(|K||u|)\|y\|$. Thus, S_u is an operator with abstract norm and, by Proposition 5.5.1 (2), there exists an element $v \in F_s(Y')$ such that $S_u(y) = \langle y, v \rangle$ $(y \in Y)$ and, moreover, $|v| = |S_u|$. Assign $T_u := v$. Then $\langle y, Tu \rangle = S_u(y) = \Phi(\langle y, Ku \rangle)$. This defines a linear operator $T : E(X) \to F_s(Y')$ that satisfies the relations

$$|Tu| = |S_u| = \sup_{\|y\| \leq 1} \Phi(\langle y, Ku \rangle) \leq \Phi(|K||u|).$$

Consequently, T admits an o-continuous dominant $S : e \mapsto \Phi(e|K|)$ and $S \in \mathscr{L}_\Phi(E, F)$. Therefore, $T \in M_\Phi(E(X), F_s(Y'))$. ▷

5.5.10. Theorem. *Let $\Phi : E \to F$ be a Maharam operator and let a dominated operator $T : E(X) \to F_s(Y, Z)$ be such that $|T| \in \{\Phi\}^{\perp\perp}$. Then there exists a unique (to within equivalence) operator-function $K \in \mathfrak{M}_P(X, Z')$ such that the following hold:*

$$\begin{aligned} &\langle Tu, z \rangle = \Phi(\langle z, Ku \rangle) && (u \in \mathscr{D}(K),\ z \in Z), \\ &|T|e = \Phi(e|K|) && (e \in \mathscr{D}(|K|)), \end{aligned}$$

where Z is a norming subspace of Y', $\mathscr{D}(|K|) := \{e \in E : e|K| \in E\}$, and $\mathscr{D}(K) := \{u \in E(X) : |u| \in \mathscr{D}(|K|)\}$.

$\triangleleft$ Arguing as in 5.5.4 and using Theorem 5.5.9, find an operator-function $K \in \mathfrak{M}_P(X, Z')$ satisfying

$$\langle z, Tu \rangle = \widetilde{\Phi}(\langle z, Ku \rangle) \quad (u \in E(X),\ z \in Z),$$
$$|T|e = \widetilde{\Phi}(e|K|) \quad (e \in E),$$

where $\widetilde{\Phi}$ is the extension of Φ onto $L^1(\Phi)$ by o-continuity. In these representations, we may write Φ instead of $\widetilde{\Phi}$ in case $u \in \mathscr{D}(K)$ and $e \in D(|K|)$. $\triangleright$

5.6. Comments

Disjointness preserving operators in vector lattices have attracted attention for a long time. B. Z. Vulikh [385–387] was one of the first who considered disjointness preserving operators. Later, disjointness preserving operators were studied by Yu. A. Abramovich, E. L. Arenson, W. Arendt, S. J. Bernau, A. E. Gutman, D. R. Hart, C. B. Huijsmans, A. K. Kitover, A. V. Koldunov, S. S. Kutateladze, P. T. N. McPolin, B. de Pagter, A. I. Veksler, A. W. Wickstead, A. C. Zaanen, and many others (see, for instance, [4, 7–12, 26, 27, 40–42, 121–123, 130, 136, 279, 310, 396, 408]). The theory of disjointness preserving operators is rich in results and covers such topics as boundedness, continuity, compactness, spectral properties, analytical representation, etc. Chapter 5 concentrates on boundedness, decomposition, and analytical representation.

5.6.1. (1) An orthomorphism is a band preserving operator that is order-bounded. In [396] A. W. Wickstead raised the question whether every band preserving operator must be order-bounded automatically. Existence of an unbounded band preserving operator was announced for the first time in [12: Theorem 1]. Later, it was clarified that the situation described in the paper is typical in a sense. Namely, Yu. A. Abramovich, A. I. Veksler, and A. V. Koldunov [11: Theorem 2.1] and P. T. N. McPolin and A. W. Wickstead [279: Theorem 3.2]) established that all band preserving operators in a universally complete K-space are automatically bounded if and only if this K-space is locally one-dimensional (Theorem 5.1.2). (The definitions of a locally one-dimensional K-space and a local Hamel basis, as well as the equivalence conditions (1)–(4) from 5.1.1, are presented in [279].)

(2) Thus, A. W. Wickstead's question about boundedness of band preserving operators was given an exhaustive answer modulo the structure of one-dimensional K-space. It was conjectured implicitly in [11] and [279] that the notions of locally one-dimensional and discrete K-space coincide. In [13] A. W. Wickstead fixed the conjecture as an open question. The negative solution of 5.1.6

and 5.1.7 was found by A. E. Gutman [122]: there exists a diffuse locally one-dimensional K-space (see also [121, 123]). He established this result by describing a locally one-dimensional K-space in terms of its base (Theorem 5.1.5).

(3) A. W. Wickstead's question admits different answers depending on spaces in which the operator in question acts. There are many results that guarantee automatic boundedness for a band preserving operator acting in concrete classes of vector lattices. According to [11, 12] every band preserving operator from a Banach lattice to a normed vector lattice is bounded. This claim remains valid if the domain Banach lattice is replaced by a relatively uniformly complete vector lattice [11]. In [279] a similar result is obtained for band preserving operators acting in a relatively uniformly complete vector lattice which is endowed with a locally convex locally solid topology.

(4) The problem of finding sufficient conditions for boundedness of disjointness preserving operators (see [199: 6.5, Problem 7]) remains actual for operators in lattice-normed spaces. Theorem 5.1.10 is due to A. E. Gutman [121, 123]. We call an operator semibounded (the term was introduced in [121]) (5.1.8) if it satisfy the McPolin–Wickstead condition of [279].

(5) In the case of a universally complete K-space, a band preserving order-unbounded operator can be constructed on using $\mathbf{V}^{(B)}$. Moreover, inside an appropriate $\mathbf{V}^{(B)}$ this problem reduces to existence of a discontinuous authomorphism of the group $(\mathscr{R}, +)$, i.e., an additive but nonlinear function from $\mathscr{R}$ to $\mathscr{R}$. Let E be a universally complete K-space and let $B := \mathfrak{B}(E)$. Take a Boolean algebra B such that $\mathbb{R}^\wedge \neq \mathscr{R}$, see 8.6.1 (7). Then $\mathscr{R}$ is an infinite-dimensional space over $\mathbb{R}^\wedge$ inside $\mathbf{V}^{(B)}$. By the Kuratowski–Zorn Lemma, there exist an $\mathbb{R}^\wedge$-linear but not $\mathscr{R}$-linear function $u : \mathscr{R} \to \mathscr{R}$ in the model $\mathbf{V}^{(B)}$. The operator $U_0 := u{\downarrow} : \mathscr{R}{\downarrow} \to \mathscr{R}{\downarrow}$ is linear, band preserving, but order-unbounded. If ι is an isomorphism of E onto $\mathscr{R}{\downarrow}$ then $U := \iota^{-1} \circ U_0 \circ \iota$ is a band preserving o-unbounded operator.

5.6.2. **(1)** The notion of n-disjoint operator in vector lattices was introduced by S. J. Bernau, C. B. Huijsmans, and B. de Pagter [42] and was adapted to operators in lattice-normed spaces by A. G. Kusraev [204]. Different characterizations of n-disjoint operators presented in 5.2.1 (2) and 5.2.5 were obtained by V. A. Radnaev [323, 324] employing Kutateladze's approach [219] to characterizing lattice homomorphisms. It should be emphasized that the equivalence (1) $\Leftrightarrow$ (5) in 5.2.5 gives a purely algebraic characterization of n-disjoint operators as metric n-decomposable elements of the associated Banach–Kantorovich space (5.2.4).

(2) Theorem 5.2.7 is due to S. J. Bernau, C. B. Huijsmans, and B. de Pagter [42] (see also [41]); an algebraic approach to the proof (2.1.10, 5.2.6, 5.2.7) was found by V. A. Radnaev [323, 324]. The method of proof in [42] makes it clear that the decomposition of an n-disjoint operator into the sum of n lattice homomorphisms is nonunique. V. A. Radnaev [323, 324] noticed that, first, the disjoint preserving operators $T_1, \dots, T_n$ in the decomposition $T = T_1 + \dots + T_n$ can be chosen pairwise disjoint; second, if $S_1, \dots, S_n$ is another disjoint collection of disjointness preserving operators with $T = S_1 + \dots + S_n$ then for every $k = 1, \dots, n$ there exist a partition of unity $\pi_{k1}, \dots, \pi_{kn}$ in $\mathfrak{P}(F)$ such that $S_k = \pi_{k1}T_1 + \dots + \pi_{kn}S_n$ for all $k = 1, \dots, n$. Decomposition of an operator into infinite series of lattice homomorphisms was considered in [344].

(3) A positive operator $T : E \to F$ is said to be a *local homomorphism* if there exists a partition of unity $(\pi_\xi)_{\xi \in \Xi}$ in $\mathfrak{P}(E)$ such that $T \circ \pi_\xi$ is a lattice homomorphism for every $\xi \in \Xi$. N. Kalton [147, 148] proved that each finite sum of lattice homomorphisms in an ideal space over a standard measure space is a local homomorphism. This result was generalized to operators in arbitrary order complete vector lattices by I. I. Shamaev [344].

(4) The shadow of an operator (5.2.2) as a Boolean homomorphism (without introducing the corresponding term) was first considered in [259] for lattices homomorphisms and in [199] for a disjointness preserving operators in lattice normed spaces. Theorem 5.2.3 is due to A. G. Kusraev [199]; for lattice homomorphisms the equivalence (1) $\Leftrightarrow$ (2) was proved by W. A. J. Luxemburg and A. Schep [261]; while the equivalence (1) $\Leftrightarrow$ (3) was proved by C. B. Huijsmans and B. de Pagter [136]. A. E. Gutman [123] found that some properties of disjointness preserving operators can be expressed in terms of its shadows; in particular, he proved 5.2.2 (3, 4).

(5) Yu. A. Abramovich's condition (R) [4: Theorem A] was the first paraphrase of boundedness for disjointness preserving operators weaker than sequential r-o-continuity. Later, this condition was also weakened: P. T. N. McPolin and A. W. Wickstead showed [279: Theorem 2.1] that, for a disjointness preserving operator in vector lattices to be bounded, it suffices that the operator under test be semibounded (see Theorem 5.2.8). The proof of the nontrivial implication (4) $\Rightarrow$ (1) in Theorem 5.2.8 repeats P. T. N. McPolin and A. W. Wickstead arguments. In [279] the case $Y = F$ is considered; however, the result remains valid for an operator with values in an arbitrary LNS.

(6) Attempts are unsuccessful at generalizing the Abramovich–McPolin–Wickstead criterion to the case of operators in lattice-normed spaces. A. E. Gutman studied four types of boundedness for this class of operators but it turned out that they are pairwise different. Thus the problem of finding sufficient conditions for boundedness remains open for disjointness preserving operators in LNSs. A small step in this direction is made in the following proposition.

Let X be a BKS over E and let Y be an LNS over F. A disjointness preserving operator $T : X \to Y$ is subdominated if and only if T is semibounded and, for every *positive element $e \in E$, the set $\{|Tu| : u \in X,\ |u| = e\}$ is order-bounded in F.*

Note that every semibounded disjointness preserving operator defined on a vector lattice obviously meets the hypotheses of the last proposition. This allows us to consider this proposition as a generalization of Theorem 5.2.8.

5.6.3. **(1)** In Section 5.3 we follow A. E. Gutman [121, 123]. The main idea here is that, since every disjointness preserving regular or dominated operator T is h-o-continuous with respect to the shadow h of T (5.2.2 (3, 4)), T can be uniquely extended to the h-closure of the domain of T (5.7.3). From this it follows immediately that the operator admits a WSW-factorization on principal ideals of the domain which can be also extended to the corresponding h-closure. These h-closures are sufficiently representative if the operator under consideration is wide at some set (5.3.1 (2, 3)). Now, an easy application of the Exhaustion Principle gives description for disjointness preserving operators through WSW-representation (Theorems 5.3.6 and 5.3.19).

(2) The domain $\mathscr{E}(1,h)$ of the shift S_h is maximally wide. More precisely, $\mathscr{E}(1,h)$ contains the domain of every regular operator S acting from an order-dense ideal of $\mathscr{E}$ into $\mathscr{F}$, having shadow h, and satisfying the equality $S(\mathbf{1}_{\mathscr{E}}) = h(1)\mathbf{1}_{\mathscr{F}}$.

Shift operators are abstract analogs of the composite mappings $f \mapsto f \circ s$. As regards, composition in spaces of measurable functions, see the survey paper [60].

(3) A linear operator $S : E \to \mathscr{F}$ defined on an order-dense ideal $E \subset \mathscr{E}$ is called *multiplicative* if $Se_1Se_2 = S(e_1e_2)$ for any two elements $e_1, e_2 \in E$ whose product belongs to E. Observe that the notion of multiplicative operator depends on the choice of the unities $\mathbf{1}_{\mathscr{E}}$ and $\mathbf{1}_{\mathscr{F}}$. The following fact was proved in [123].

Theorem. *Let E be an order-dense ideal of $\mathscr{E}$. A linear operator $S : E \to \mathscr{F}$ is a shift operator if and only if S is multiplicative.*

The study of multiplicative operators in vector lattices was initiated by B. Z. Vulikh [385–387] who proved that o-continuous shift operators in K-spaces with unity are multiplicative (see also [163]). The above theorem generalizes this result to the case of arbitrary shift operators in arbitrary K-spaces. There are some results describing multiplicative operators as extreme points of certain sets of operators (see [71, 90, 93, 318]).

(4) The idea of considering the shift of a disjointness preserving operator appears in different contexts. An analogous notion occurs, for instance in [9, 10, 130] and in many papers about isometries of L_p-spaces. The S-correspondence of a positive operator considered in [362] is also a functional analog of shift. If $\imath : \mathfrak{P}(E) \to \mathrm{Clop}(P)$ and $\imath' : \mathfrak{P}(F) \to \mathrm{Clop}(Q)$ are the Stone transforms then the

S-correspondence of $T \in L^{\sim}(E, F)$ is defined by

$$q \mapsto \bigcap \Big\{ \imath\Big([\mathscr{C}_{\pi \circ T}] \Big) : q \in \imath'(\pi),\ \pi \in \mathfrak{P}(F) \Big\} \quad (q \in Q).$$

(5) The criterion for *WSW*-representability stated in 5.3.5 is close to [10]. Some other criteria are presented also in [8–10, 121, 123]). The conventional notion of weighted shift operator does not contain an inner weight (see [8, 10–12, 199]). Involving an inner weight allows us to decompose an arbitrary bounded disjointness preserving operator in lattice normed spaces into the strongly disjoint sum of weighted shift operators (Theorems 5.3.6 and 5.3.10).

(6) Not every disjointness preserving regular operator admits a WSW-factorization. The corresponding example was given by Yu. A. Abramovich [4]. Let Q be an extremal compact space without isolated points. In this case, we may find an order-dense ideal $E \subset C_\infty(Q)$, a family $(e_\xi)_{\xi \in \Xi}$ in E, and a family $(q_\xi)_{\xi \in \Xi}$ in Q so that the following conditions be satisfied: the set $\{q_\xi : \xi \in \Xi\}$ is dense in Q, $e_\xi(q_\xi) = \infty$ for all $\xi \in \Xi$, and, for each $e \in E$, the numeric set $\big\{(e/e_\xi)(q_\xi) : \xi \in \Xi\big\}$ is bounded. Then the operator $T : E \to l^\infty(\Xi)$ acting by the rule $(Te)(\xi) = (e/e_\xi)(q_\xi)$ is disjointness preserving and regular (even positive), but T is not a weighted shift operator. Denoting by ρ_ξ the operator of multiplication by the characteristic function $\chi_{\{\xi\}}$, we obtain a partition of unity $(\rho_\xi)_{\xi \in \Xi}$ in the algebra $\mathfrak{P}\big(l^\infty(\Xi)\big)$ such that all fragments of the form $\rho_\xi \circ T$ are weighted shift operators. Theorem 5.3.6 says that all disjointness preserving regular operators has the same structure.

5.6.4. (1) The main results of Section 5.4 (Theorems 5.4.5 and 5.4.10) belong to A. E. Gutman [121, 123]. The facts presented in 5.4.1–5.1.4 just repeat Yu. A. Abramovich's results [4]. Theorem 5.4.5 interprets the decomposition in 5.3.6 of a disjointness preserving operator into the sum of weighted shift operators in terms of their functional representations. As is seen from the proofs, Theorem 5.4.10 leans upon the theory of ample Banach bundles from Section 2.4. Further development of this approach and extension of the multiplicative representation to operators acting in lattice-normed spaces of continuous or measurable sections can be found in the papers of A. E. Gutman [118–123].

(2) The global representation of 5.4.5 and 5.4.10 for a disjointness preserving operator, as well as the notions of the shift of an operator and the corresponding shift function, allows us to interpret the abstract properties of the operator in terms of its concrete functional representation or in terms of the properties of its shift function. Examples of similar interpretations can be found, for instance, in [8–10, 27, 152, 173, 174].

(3) The isometries of function spaces prove very often to be disjointness preserving operators. This phenomenon seems to be discovered by J. Lamperti

[232]; and so disjointness preserving operators (in spaces of measurable functions) are sometimes referred to as *Lamperti operators*. The literature on the theory of Lamperti operators is extensive; some aspects are reflected in [6, 9, 26, 36, 113, 114, 174, 358]; see also the survey [60].

5.6.5. (1) Section 5.5 follows the articles [199, 204]. The multiplicative representation 5.5.2 (2) was established by Yu. A. Abramovich [4]. It should be noted that the recent progress in the multiplicative representation of disjointness preserving operators stems from this work of Yu. A. Abramovich. Theorems 5.5.3 and 5.5.4 were proved in [199] (see also [204]). As is seen from the proof, these results are obtained by combining the representation method of Yu. A. Abramovich and the technique of dominated operators.

(2) The notion of a decomposable operator was introduced in [199]. The main results on decomposable operators, which are presented in 5.5.6, 5.5.7, 5.5.8, and 5.5.9, belong to A. G. Kusraev [199, 204]. The auxiliary result in 5.5.1 (2) belongs to G. N. Shotaev [349].

Chapter 6

Integral Operators

This chapter deals with different classes of dominated operators whose common feature is integrality. Speaking of integrality we mean the possibility of integral representation with respect to a vector-valued or scalar-valued measure. Given a σ-additive measure taking values in a *bo*-complete lattice-normed space, a Lebesgue-type integral of numerical functions or, more generally, of elements of a universally complete vector lattice can be constructed (6.1.1, 6.1.2, 6.1.3). This is fairly straightforward and all simple properties of the resultant integral as well as analog of the Lebesgue convergence results are easily deduced (6.1.4, 6.1.5, 6.1.6). This integral is used to obtain the Riesz–Markov-type representation theorem for dominated operators defined on a lattice of bounded continuous functions (6.2.6). The corresponding class of measures is *quasi-Radon measures.* A dominated measure and its least dominant are or are not quasi-Radon measures simultaneously (6.2.2).

The space of all integrable functions (elements) is some lattice-normed space that is neither *bo*-complete nor decomposable in general. This circumstance gives rise to new classes of measures: the space of integrable elements is decomposable if and only if the measure is *modular* (6.1.9 (3)) and it is *bo*-complete if and only if the measure is *ample* (6.1.9 (4)). A Radon–Nikodým-type theorem is also valid for ample measures (6.1.11). Ample measures are closely connected with the theory of Maharam operators. Maharam extension of a positive operator, described in Section 3.5, leads to a Maharam operator whose domain is too large in general. But for an order continuous operator the extended domain space admits functional representation, while the extended Maharam operator is represented as the integral with respect to some ample measure (6.3.6).

Classical integral operators defined by measurable kernels admit the following intrinsic characterization: a linear operator between ideal spaces is an integral operator if and only if it takes order-bounded sequences converging to zero in measure into sequences converging to zero almost everywhere (6.4.5). The same is true for dominated operators acting in the spaces of measurable vector-functions (6.4.10,

6.4.11) if integrality is suitably defined (6.4.6). But this fact relies upon the inheritance of integrality under domination. If some dominant of an operator acting in the spaces of measurable vector-functions is an integral operator then the operator itself admits weak integral representation (6.4.10 (2)). The converse is true too (6.4.7).

A broad class of operators arises from integration with respect to a family of measures depending on some measurable parameter. Such operators are called *pseudointegral* (6.5.1). It turns out that a positive operator admits pseudointegral representation if and only if it is order continuous, i.e. takes order-bounded sequences converging to zero almost everywhere into sequences converging to zero almost everywhere (6.5.4). From this fact, using the properties of dominated operators, the following criterion for the weak pseudointegrality is deduced: a dominated operator admits weak pseudointegral representation whenever it is order continuous (6.5.8).

The above-mentioned results imply assertions about the general form of various classes of dominated operators (6.1.7, 6.3.8, 6.3.11, 6.4.12, 6.5.9).

6.1. Vector Integration

The main goal of this section is to present some Lebesgue-type integration for measures taking values in lattice-normed spaces. A Radon–Nikodým-type theorem is also established for a specific class of ample measures.

6.1.1. Let G be a universally complete vector lattice with order-unity $\mathbf{1}$ and let (Y, F) be a sequentially *bo*-complete lattice-normed space over an order complete vector lattice F. Fix a subalgebra $\mathscr{A}$ in the complete Boolean algebra $\mathfrak{E}(\mathbf{1})$ of unit elements of G and a finitely additive measure $\mu : \mathscr{A} \to Y$ with the bounded vector variation $|\mu| : \mathscr{A} \to F$. Denote by $S(\mathscr{A})$ the vector sublattice of G comprising all $\mathscr{A}$-simple (finite-valued) elements, i.e. $x \in S(\mathscr{A})$ means that there is a representation $x = \sum_{k=1}^{n} \alpha_k e_k$, where $\{\alpha_1, \dots, \alpha_n\} \subset \mathbb{R}$ and $\{e_1, \dots, e_n\} \subset \mathscr{A}$ are pairwise disjoint. Put

$$I_\mu(x) := \int x\, d\mu := \sum_{k=1}^{n} \alpha_k \mu(e_k) \quad (x \in S(\mathscr{A})).$$

It is clear that this formula correctly defines some dominated linear operator $I_\mu : S(\mathscr{A}) \to Y$ and

(1) $|\int x\, d\mu| \le \int |x|\, d|\mu| \quad (x \in S(\mathscr{A}))$.

Consider the principal ideal $G(\mathbf{1})$ generated by $\mathbf{1}$ with the norm $\|x\| := \inf\{\lambda : |x| \le \lambda \mathbf{1}\}$, so that $G(\mathbf{1})$ is an AM-space (see 1.5.5). Let $C(\mathscr{A})$ be the closure of $S(\mathscr{A})$ in the AM-space $G(\mathbf{1})$.

(2) *The operator I_μ admits a unique dominated extension to $C(\mathscr{A})$ which is denoted by the same symbol. Moreover,* $|I_\mu| = I_{|\mu|}$.

$\lhd$ From (1) we immediately deduce that

$$\left|\int x\,d\mu\right| \leq \|x\| \cdot |\mu|(\mathbf{1}) \quad (x \in S(\mathscr{A})),$$

so that I_μ is *bo*-continuous. Now, to obtain a unique dominated extension of I_μ to $S(\mathscr{A})$ we apply Theorem 4.3.3, taking into consideration the sequential *bo*-completeness of Y. The inequality (1) preserves under passage to the limit. Thus, $|I_\mu| \leq I_{|\mu|}$ and it remains to check that $|I_\mu|e = I_{|\mu|}e$ for all $e \in \mathscr{A}$. Using Theorem 4.1.8 with $E := X$ and 4.2.9 (1), we derive

$$|I_\mu|e = \bigvee \left\{ \sum_{k=1}^{n} |I_\mu e_k| : e_k \wedge e_l = 0\,(k \neq l)\ \bigvee_{k=1}^{n} e_k = e \right\} = |\mu|(e) = I_{|\mu|}e. \ \rhd$$

(3) *For every dominated operator $T : C(\mathscr{A}) \to Y$ there is a unique dominated measure $\mu : \mathscr{A} \to Y$ such that*

$$Tx = \int x\,d\mu \quad (x \in C(\mathscr{A})).$$

The correspondence $T \mapsto \mu$ is a linear isometric isomorphism of the lattice-normed spaces $M(C(\mathscr{A}), Y)$ and $\mathrm{da}(\mathscr{A}, Y)$.

Moreover, $T \in M_G(C(\mathscr{A}), Y)$ ($T \in M_{\sigma G}(C(\mathscr{A}), Y)$) if and only if we have $\mu \in \mathrm{dao}(\mathscr{A}, Y)$ (respectively $\mu \in \mathrm{dca}(\mathscr{A}, Y)$).

Now we assume that $\mathscr{A}$ is a σ-subalgebra in $\mathfrak{C}(\mathbf{1})$. Consider a universally complete K_σ-space $E \subset G$ comprising all $\mathscr{A}$-valued resolutions of unity (spectral functions, see 1.4.3). The inclusion $E \subset G$ is understood by virtue of Theorem 1.4.4 which enables us to identify the K-spaces G and $\mathfrak{K}(\mathfrak{C}(\mathbf{1}))$.

6.1.2. For the sequel we need some technical facts.

(1) If a sequence $(x_n)_{n\in\mathbb{N}} \subset E$ decreases to zero then

$$(\forall\, 0 < \varepsilon \in \mathbb{R}) \ \bigwedge_{n=1}^{\infty} |\mu|(\mathbf{1} - e_\varepsilon^{x_n}) = 0.$$

$\lhd$ The proof is immediate from the σ-additivity of $|\mu|$ (4.4.12) and the suitable properties of spectral functions (1.3.8 (9)):

$$\bigvee_{n=1}^{\infty} e_\varepsilon^{x_n} = e_\varepsilon^{\inf_n x_n} = e_\varepsilon^0 = \mathbf{1}. \ \rhd$$

(2) If a sequence $(x_n)_{n\in\mathbb{N}} \subset S(\mathscr{A})$ decreases and $\inf_n x_n = 0$ in G, then $\inf_n \int x_n \, d|\mu| = 0$.

◁ Take a positive number α with $x_1 \le \alpha\mathbf{1}$. Then for every $\varepsilon > 0$ and $n \ge 1$ we have $x_n \le \varepsilon\mathbf{1} + \alpha(\mathbf{1} - e_\varepsilon^{x_n})$. From this it is seen that

$$\int x_n \, d|\mu| \le \varepsilon|\mu|(\mathbf{1}) + \alpha|\mu|\,(\mathbf{1} - e_\varepsilon^{x_n}).$$

Using (1), we obtain

$$\bigwedge_{n=1}^{\infty} \int x_n \, d|\mu| \le \varepsilon|\mu|(\mathbf{1}).$$

Since $\varepsilon > 0$ is arbitrary, the result follows. ▷

(3) *For every positive element $x \in E$ there is a sequence of positive finite-valued elements $(x_n)_{n\in\mathbb{N}} \subset S(\mathscr{A})$ with $x_n \nearrow x$.*

(4) *Let E be a universally σ-complete vector lattice with order-unity* $\mathbf{1}$ *and let $(x_n)_{n\in\mathbb{N}}$ be an unbounded sequence in E_+. Then there exists a unit element $e_\infty \in \mathfrak{E}(\mathbf{1})$ such that the sequence $([\mathbf{1} - e_\infty]x_n)_{n\in\mathbb{N}}$ is order-bounded and $te_\infty = \sup_n(te_\infty \wedge x_n)$ for every $0 < t \in \mathbb{R}$.*

◁ Put

$$\sigma'(\lambda) := \bigwedge_{n=1}^{\infty} e_\lambda^{x_n}, \quad \sigma(\lambda) := \bigvee_{\nu<\lambda} \sigma(\lambda), \quad e := \bigvee_{n=1}^{\infty} \sigma(n).$$

If $E_0 := [e]E$ then e is an order-unity in E_0 and the function $\lambda \mapsto \sigma(\lambda)$ $(\lambda \in \mathbb{R})$ is a resolution of unity in the Boolean algebra $\mathfrak{E}(e) = [0, e] \cap \mathfrak{E}(\mathbf{1})$, see 1.4.3 (1–3). Since E_0 is universally complete, there exists $x \in E_0$ with $\sigma(\lambda) = e_\lambda^x$ $(\lambda \in \mathbb{R})$. According to 1.3.8 (12) $\lambda \mapsto e \wedge e_\lambda^{x_n}$ is the spectral function of $[e]x_n$ in E_0. Denote $e_\infty := \mathbf{1} - e$, $y_n := [e_\infty]x_n$. It suffices to prove that $te_\infty = \sup_n(te_\infty \wedge y_n)$, since $\sup_n(te_\infty \wedge y_n) \le \sup_n(te_\infty \wedge x_n) \le te_\infty$. Put $\tau(\lambda) := \inf_n \left(e_\lambda^{y_n} \vee e_\lambda^{te}\right)$ and show that τ coincides with the spectral function of te_∞. If $\lambda < t$ then, using 1.3.8 (12), we derive

$$\begin{aligned}\tau(\lambda) &= \bigwedge_{n=1}^{\infty} e_\lambda^{y_n} \vee \mathbf{0} = \bigwedge_{n=1}^{\infty} e_\infty \wedge e_\lambda^{x_n} = e_\infty \wedge \bigwedge_{n=1}^{\infty} e_\lambda^{x_n} \\ &= e_\infty \wedge \sigma'(\lambda) \le e_\infty \wedge (\mathbf{1} - e_\infty) = \mathbf{0}.\end{aligned}$$

Similarly, if $\lambda > t$ then $\tau(\lambda) = \inf_n e_\lambda^{y_n} \vee e_\infty = e_\infty$. According to what was said after Theorem 1.3.8 we conclude that τ is the spectral function of $\sup_n(te_\infty \wedge y_n)$ and the result follows. ▷

6.1.3. Now define the integral for elements which can be approximated by $\mathscr{A}$-simple elements. We say that a positive element $x \in E$ is *integrable by* μ, or μ*-integrable* if there is an increasing sequence $(x_n)_{n\in\mathbb{N}}$ of positive elements in $S(\mathscr{A})$ o-converging in G to x and the supremum $\sup_{n\in\mathbb{N}} \int x_n\, d|\mu|$ existing in F. For such a sequence (x_n) the sequence of the integrals $(I_\mu(x_n))_{n\in\mathbb{N}}$ is bo-fundamental. Indeed, by applying 6.1.1 (1) we have

$$\left|\int x_n\, d\mu - \int x_m\, d\mu\right| \le \int |x_n - x_m|\, d|\mu| \le \bigvee_{k=1}^{\infty} \left\{\int x_k\, d|\mu|\right\} - \int x_p\, d|\mu| \xrightarrow[p\to\infty]{} 0,$$

where $p = \min\{m, n\}$. Now we may define the integral of x by putting

$$I_\mu(x) := \int x\, d\mu := bo\text{-}\lim_{n\to\infty} \int x_n\, d\mu.$$

To check the soundness of this definition, take one more sequence $(y_n)_{n\in\mathbb{N}} \subset S(\mathscr{A})$ increasing to x in G and assume that $\sup_n I_{|\mu|}(y_n)$ exists in F. Using 6.1.1 (1), 6.1.2 (2), and 1.3.2 (5) we deduce

$$\begin{aligned}\left|\int x_n\, d\mu - \int y_m\, d\mu\right| &\le \int |x_n - y_m|\, d|\mu| \\ &\le \int x_n \vee y_m\, d|\mu| - \int x_n \wedge y_m\, d|\mu| \\ = \bigvee_{k,l\in\mathbb{N}} &\int (x_n \vee y_m) \wedge (x_k \wedge y_l)\, d|\mu| - \int x_n \wedge y_m\, d|\mu| \\ &\le \bigvee_{k,l\in\mathbb{N}} \int x_k \wedge y_l\, d|\mu| - \int x_n \wedge y_m\, d|\mu|.\end{aligned}$$

Consequently,

$$bo\text{-}\lim_{m,n\to\infty} \left(\int x_n\, d\mu - \int y_m\, d\mu\right) = 0$$

and the soundness of our definition is established.

An element $x \in E$ is *integrable* ($= \mu$-integrable) if its positive part x^+ and negative part x^- are both integrable. Denote by $\mathscr{L}^1(\mu)$ the set of all integrable elements and, given $x \in \mathscr{L}^1(\mu)$, put

$$I_\mu(x) := \int x\, d\mu := \int x^+\, d\mu - \int x^-\, d\mu.$$

It can be easily checked, using 6.1.2 (2), that $\mathscr{L}^1(\mu)$ is an order-dense ideal in E and $I_\mu : \mathscr{L}^1(\mu) \to Y$ is a linear operator. Moreover, 6.1.1 (1) holds for all $x \in \mathscr{L}^1(\mu)$. Note also that the construction of the integral implies $\mathscr{L}^1(\mu) = \mathscr{L}^1(|\mu|)$.

Define in $\mathscr{L}^1(\mu)$ an F-valued seminorm

$$|x|_1 := \int |x|\, d|\mu| \quad (x \in \mathscr{L}^1(\mu)).$$

We say that two elements $x, y \in \mathscr{L}^1(\mu)$ are μ-equivalent if there is a unit element $e \in \mathfrak{E}(\mathbf{1})$ with $|\mu|(\mathbf{1} - e) = 0$ and $[e]x = [e]y$. The set $\mathscr{N}(\mu)$ of all elements that are μ-equivalent to zero is a sequentially o-closed order ideal in $\mathscr{L}^1(\mu)$. It follows from the definition of integral that $\mathscr{N}(\mu) = \{x \in \mathscr{L}^1(\mu) : |x|_1 = 0\}$. Define the K_σ-space $L^1(\mu)$ as the factor space of $\mathscr{L}^1(\mu)$ by σ-ideal $\mathscr{N}(\mu)$. The coset of an element $x \in \mathscr{L}^1(\mu)$ will be denoted by $\tilde{x}$. An F-valued norm in $\mathscr{L}^1(\mu)$ is introduced by setting $|\tilde{x}|_1 := |x|$ $(x \in \mathscr{L}^1(\mu))$. Thus, $(L^1(\mu), |\cdot|)$ is a lattice-normed space.

6.1.4. Monotone Convergence Theorem. *Assume that $(x_n)_{n\in\mathbb{N}}$ is an increasing sequence of μ-integrable elements and let the sequence $\big(I_{|\mu|}(x_n)\big)_{n\in\mathbb{N}}$ be order-bounded in F. Then there is $x \in \mathscr{L}^1(\mu)$ such that the identity $\tilde{x} = \bigvee_n \tilde{x}_n$ holds in $L^1(\mu)$ and*

$$\int x\, d\mu = bo\text{-}\lim_{n\to\infty} \int x_n\, d\mu.$$

$\lhd$ First we observe that if an increasing sequence $(y_n)_{n\in\mathbb{N}}$ of positive elements in $\mathscr{L}^1(\mu)$ has a least upper bound $y := \sup_n y_n$ in G and the sequence $I_{|\mu|}(y_n)$ has a least upper bound in F then $y \in \mathscr{L}^1(\mu)$ and

$$\int y\, d|\mu| = \bigvee_{n=1}^{\infty} \int y_n\, d|\mu|.$$

Indeed, for every $n \in \mathbb{N}$ take an increasing sequence of positive elements $(z_{n,k})_{k\in\mathbb{N}} \subset S(\mathscr{A})$ such that $y_n = \sup_k z_{n,k}$ $(n \in \mathbb{N})$. Since $y = \sup_k \sup_{n\le k} z_{n,k}$, employing the definition of integral with respect to the measure $|\mu|$ we deduce

$$\int y\, d|\mu| = \bigvee_{k=1}^{\infty} \int \bigvee_{n\le k} z_{n,k}\, d|\mu| \le \bigvee_{k=1}^{\infty} \int \bigvee_{n\le k} y_n\, d|\mu| = \bigvee_{n=1}^{\infty} \int y_n\, d|\mu|.$$

Since the reverse inequality is evident, the result follows.

According to 6.1.2 (4) there is a unit element $e \in \mathscr{A}$ such that $m(\mathbf{1} - e) = \sup_n (m(\mathbf{1} - e) \wedge x_n)$ $(m \in \mathbb{N})$. It follows from this and the above observation that

$$m|\mu|(\mathbf{1} - e) = \bigvee_{n=1}^{\infty} \int m(\mathbf{1} - e) \wedge x_n\, d|\mu| \le \bigvee_{n=1}^{\infty} \int x_n\, d|\mu|.$$

As $m \in \mathbb{N}$ is arbitrary and the integrals on the right-hand side are order-bounded, we conclude that $|\mu|(\mathbf{1} - e) = 0$. Put $x := \sup_n [e]x_n$. Then $\tilde{x} = \sup_n \tilde{x}_n$ and

$$\int x \, d|\mu| = \bigvee_{n=1}^{\infty} \int [e]x_n \, d|\mu| = \bigvee_{n=1}^{\infty} \int x_n \, d|\mu|.$$

Taking into consideration the fact that $I_{|\mu|}$ is a dominant of I_μ we deduce

$$\left| \int x \, d\mu - \int x_n \, d\mu \right| \leq \int |x - x_n| \, d|\mu| = \int x \, d|\mu| - \int x_n \, d|\mu| \xrightarrow[n \to \infty]{(o)} 0,$$

and the proof is complete. ▷

6.1.5. *Dominated Convergence Theorem.* *Let $(x_n)_{n \in \mathbb{N}}$ be a sequence of μ-integrable elements and let o-$\lim x_n = x$ in G. If $y \in \mathscr{L}^1(\mu)$ and $|x_n| \leq y$ $(n \in \mathbb{N})$ then $x \in \mathscr{L}^1(\mu)$ and*

$$\int x \, d\mu = bo\text{-}\lim_{n \to \infty} \int x_n \, d\mu.$$

◁ Since $\mathscr{L}^1(\mu)$ is an order-dense ideal in E and E is sequentially order-closed in G, we have $x \in \mathscr{L}^1(\mu)$. The assumption that o-$\lim x_n = x$ and the evident inequality $|x - x_n| \leq 2y$ imply the existence of a decreasing sequence $(y_n)_{n \in \mathbb{N}} \subset \mathscr{L}^1(\mu)$ such that $|x - x_n| \leq y_n$ and o-$\lim y_n = 0$. Thus, $y_1 - y_n \nearrow y_1$ and, using the Monotone Convergence Theorem, we have

$$\int y_1 \, d|\mu| = o\text{-}\lim_{n \to \infty} \int (y_1 - y_n) \, d|\mu|.$$

Consequently,

$$\left| \int x \, d\mu - \int x_n \, d\mu \right| \leq \int |x - x_n| \, d|\mu| \leq \int y_n \, d|\mu| \xrightarrow[n \to \infty]{} 0$$

and the result follows. ▷

6.1.6. The above results can be summarized as follows:

Theorem. *Let $\mu : \mathscr{A} \to Y$ be a dominated countably additive measure. There exist an order-dense ideal $\mathscr{L}^1(\mu) \subset E$ and a sequentially bo-continuous dominated operator $I_\mu : \mathscr{L}^1(\mu) \to Y$ such that*

(1) *$\mathscr{L}^1(\mu) \supset \mathscr{A}$;*

(2) *$I_\mu e = \mu(e)$ $(e \in \mathscr{A})$;*

(3) *if $L \supset \mathscr{A}$ and $Ie = \mu(e)$ $(e \in \mathscr{A})$ for some order-dense ideal $L \subset E$ and a bo-continuous dominated operator $I : E \to Y$ then $L \subset \mathscr{L}^1(\mu)$ and $Ix = I_\mu x$ $(x \in L)$;*

(4) $|I_\mu| = I_{|\mu|}$.

$\triangleleft$ The existence and sequential order continuity of I_μ defined on the order-dense ideal $\mathscr{L}^1(\mu) \subset E$ and satisfying (1) and (2) was justified in 6.1.1, 6.1.3 and 6.1.5. The condition (3) is deduced from 6.1.4 and 6.1.3 (3). Finally, (4) is an easy consequence of 6.1.1 (1), 6.1.2 (3), and sequential order continuity $I_{|\mu|}$. $\triangleright$

6.1.7. *Theorem.* *Let E_0 be an order-dense ideal in E containing order-unity and let $T : E_0 \to Y$ be a sequentially bo-continuous dominated linear operator. Then there exists a unique dominated σ-additive measure $\mu : \mathscr{A} \to Y$ such that $\mathscr{L}^1(\mu) \supset E_0$ and*

$$Tx = \int x\,d\mu, \quad |T|x = \int x\,d|\mu| \quad (x \in E_0).$$

$\triangleleft$ We have only to define $\mu(e) := Te$ $(e \in \mathscr{A})$ and observe that T and I_μ coincide on $S(\mathscr{A})$. Then T and I_μ coincide on $\mathscr{L}(\mu)$ by *bo*-continuity. $\triangleright$

6.1.8. Now we proceed to a natural question under what conditions the lattice-normed space $(L^1(\mu), |\cdot|)$ is *bo*-complete.

The space $(L^1(\mu), |\cdot|)$ is br-complete.

$\triangleleft$ The proof uses much the same arguments as the Riesz–Fisher Completeness Theorem for a scalar measure. We sketch the proof. If a sequence $(x_n)_{n\in\mathbb{N}}$ is *br*-fundamental then $|x_n - x_m| \le r_k f$ $(n, m \ge k)$ for some $f \in F$ and a numeric sequence $(r_n)_{n\in\mathbb{N}}$ converging to zero. Denote $L(f) := \{x \in L^1(\mu) : |x| \in F(f)\}$, where $F(f)$ is an order ideal generated by f. Then $(x_n) \subset L(f)$ and it suffices to show that $L(f)$ is a Banach space under the norm $\|x\| := \||x|\|_{F(f)}$. Suppose that the series $\sum_{n=1}^{\infty} x_n$ is absolutely convergent in $L(f)$. Then the series $\sum_{n=1}^{\infty} |x_n|$ converges in the *AM*-space $F(f)$ and also *r*-converges to the same limit. Put

$$t := \sum_{k=1}^{\infty} |x_k|, \quad \sigma_n := \sum_{k=1}^{n} x_k, \quad s_n := \sum_{k=1}^{n} |x_k|.$$

Clearly, (s_n) has positive entries, increases, and $\int s_n \le t$. Hence, by the Monotone Convergence Theorem, there exists a limit $y := o\text{-}\lim s_n$, with the resultant element y a member of $L^1(\mu)$. The inequalities $|\sigma_n| \le s_n \le y$ imply that the series $\sum_{k=1}^{\infty} x_k$ *o*-converges. For the sum x_0 the estimate holds: $|x_0| \le y$, whence $x_0 \in L^1(\mu)$. Appealing to the Dominated Convergence Theorem, conclude that $|\sigma_n - f_0| = \int |\sigma_n - x_0|\,d|\mu| \to 0$. $\triangleright$

To obtain further completeness results, some additional assumption are necessary. For example, sequential *bo*-completeness of $L^1(\mu)$ can be proved whenever F is regular. But the decomposability phenomenon involves essentially different properties of measures.

6.1.9. In the rest of this section we assume that Y is a Banach–Kantorovich space over a K-space F and $\mu : \mathscr{A} \to Y$ is a countably additive measure. Define the *null ideal* of μ by

$$\mathscr{N}(\mu) := \{a \in \mathscr{A} : (\forall\, a' \in \mathscr{A})\, a' \leq a \Rightarrow \mu(a') = 0\}.$$

Evidently $\mathscr{N}(\mu) = \mathscr{N}(|\mu|) = \{a \in \mathscr{A} : |\mu|(a) = 0\}$. Let $\tilde{\mathscr{A}}$ and ϕ denote the factor algebra $\mathscr{A}/\mathscr{N}(\mu)$ and the canonical factor mapping $\mathscr{A}/\mathscr{N}(\mu) \to \tilde{\mathscr{A}}$ respectively. There is a unique measure $\tilde{\mu} : \tilde{\mathscr{A}} \to Y$ such that $\tilde{\mu} \circ \phi = \mu$. Furthermore, $|\tilde{\mu}| = \widetilde{|\mu|}$.

Given a Boolean homomorphism $h : \mathbb{B} := \mathfrak{P}(F) \to \tilde{\mathscr{A}}$, we say that μ is *modular with respect to* h, or h-*modular* if $b\tilde{\mu}(\phi a) = \tilde{\mu}(h(b) \wedge \phi(a))$ for all $a \in \mathscr{A}$ and $b \in \mathbb{B}$. Clearly, the modularity of μ means that $b\mu(a) = \mu(b' \wedge a')$ for all $a' \in \phi(a)$ and $b' \in h(b)$.

Let $e := \bigvee\{b \in \mathbb{B} : (\forall\, a \in \mathscr{A})\, b\mu(a) = 0\}$. Then $e\mu(\mathscr{A}) = \{0\}$ and $\mu(\mathscr{A}) \subset (\mathbf{1} - e)Y$. Moreover, $b\mu(\mathscr{A}) = \{0\}$ if and only if $h(b) \in \mathscr{N}(\mu)$. Thus, h is injective on $[\mathbf{0}, \mathbf{1} - e]$. In the sequel we agree that $\mu(\mathscr{A})^{\perp\perp} = Y$ and in this event h is an isomorphic embedding of $\mathbb{B}$ into $\tilde{\mathscr{A}}$. An h-modular measure μ is said to be *ample* (with respect to h) if for any partition of unity $(b_\xi)_{\xi \in \Xi}$ in $\mathbb{B}$ and an arbitrary family $(a_\xi)_{\xi \in \Xi}$ in $\mathscr{A}$ there exists a unique (to within equivalence) element $a \in \mathscr{A}$ such that $b_\xi|\mu|(a \Delta a_\xi) = 0$ for all $\xi \in \Xi$. This condition is equivalent to $h(b_\xi) \wedge \phi(a) = h(b_\xi) \wedge \phi(a_\xi)$ $(\xi \in \Xi)$, since μ is h-modular.

(1) *A measure μ is modular with respect to a Boolean isomorphism h if and only if so is its exact dominant $|\mu|$.*

◁ Suppose that μ is h-modular and prove the identity $b\widetilde{|\mu|}(\phi(a)) = \widetilde{|\mu|}(h(b) \wedge \phi(a))$. It is equivalent to $b|\tilde{\mu}|(\phi(a)) = |\tilde{\mu}|(h(b) \wedge \phi(a))$, since $\widetilde{|\mu|} = |\tilde{\mu}|$. Now the desired identity is verified by means of the following simple calculations with the use of 4.2.9 (1):

$$\begin{aligned}
b|\tilde{\mu}|(\phi(a)) &= b\bigvee\left\{\sum_{k=1}^{n} \tilde{\mu}(\tilde{a}_k) : \tilde{a}_1 \vee \dots \vee \tilde{a}_n = \phi(a)\right\} \\
&= \bigvee\left\{\sum_{k=1}^{n} b\tilde{\mu}(\phi(a_k)) : a_1 \vee \dots \vee a_n = a\right\} \\
&= \bigvee\left\{\sum_{k=1}^{n} \tilde{\mu}(\phi(b') \wedge \phi(a_k)) : a_1 \vee \dots \vee a_n = a\right\} \\
&= \bigvee\left\{\sum_{k=1}^{n} \tilde{\mu}(\phi(c_k)) : c_1 \vee \dots \vee c_n = \phi(b') \wedge \phi(a)\right\} \\
&= |\mu|(h(b) \wedge \phi(a)),
\end{aligned}$$

where the finite sets $\{\tilde{a}_1, \dots, \tilde{a}_n\} \subset \tilde{\mathscr{A}}$, $\{a_1, \dots, a_n\} \subset \mathscr{A}$, and $\{c_1, \dots, c_n\} \subset \mathscr{A}$ are pairwise disjoint and $h(b) = \phi(b')$ for some $b' \in \mathscr{A}$. The converse follows from Proposition 6.1.11 (1) below. ▷

(2) *Let F obey the countable chain condition. Then every countably additive h-modular measure μ defined on a σ-algebra is ample.*

◁ For a countable partition of unity $(b_n) \subset \mathbb{B}$ and a sequence $(a_n) \subset \mathscr{A}$ put $a := \bigvee\{c_n \wedge a_n : n \in \mathbb{N}\}$, where (c_n) is a sequence of pairwise disjoint elements in $\mathscr{A}$ with $h(b_n) = \phi(c_n)$, $n \in \mathbb{N}$. Using the modularity and countable additivity of μ we derive:

$$
\begin{aligned}
b_m\mu(a) &= \sum_{n=1}^{\infty} b_m\mu(c_n \wedge a_n) = \sum_{n=1}^{\infty} b_m\tilde{\mu}(\phi(c_n) \wedge \phi(a_n)) \\
&= \sum_{n=1}^{\infty} \tilde{\mu}(\phi(c_m) \wedge \phi(c_n) \wedge \phi(a_n)) = \tilde{\mu}(h(b_m) \wedge \phi(a_m)) \\
&= b_m\tilde{\mu}(\phi(a_m)) = b_m\mu(a_m). \ \triangleright
\end{aligned}
$$

In the sequel we follow the common practice of identifying μ and $\tilde{\mu}$ when this leads to no confusion.

(3) *The lattice-normed space $L^1(\mu)$ is disjointly decomposable if and only if μ is a modular measure.*

◁ Suppose that μ is modular with respect to some Boolean isomorphism h. Prove that $b|x|_1 = |h(b)x|_1$ for all $b \in \mathbb{B}$ and $x \in L^1(\mu)$. We identify a unit element $h(b) \in \mathscr{A}$ with the band projection $[h(b)]$ in E and write $h(b)x$ instead of $[h(b)]x$. If $x = \tau_1 a_1 + \dots + \tau_n a_n$ with $\tau_1, \dots, \tau_n \in \mathbb{R}$ and $a_1, \dots, a_n \in \mathscr{A}$ pairwise disjoint, then $h(b)x = \tau_1 h(b) \wedge a_1 + \dots + \tau_n h(b) \wedge a_n$ and we may write

$$
I_\mu(h(b)x) = \sum_{k=1}^{n} \tau_k \mu(h(b) \wedge a_k) = \sum_{k=1}^{n} \tau_k b\mu(a_k) = bI_\mu(x).
$$

Now take an increasing sequence $(x_n) \subset S(\mathscr{A})$ such that $I_\mu(x) = bo\text{-}\lim_n I_\mu(x_n)$. Then $h(b)x = o\text{-}\lim_n h(b)x_n$, since the band projection $[h(b)]$ is order continuous. Moreover, $h(b)x_n \le x$ and also $h(b)x \in L^1(\mu)$, because $L^1(\mu)$ is an order ideal in E. Using the Dominated Convergence Theorem, we deduce

$$
bI_\mu(x) = b\big(bo\text{-}\lim_{n\to\infty} I_\mu(x_n)\big) = bo\text{-}\lim_{n\to\infty} I_\mu(h(b)x_n) = I_\mu(h(b)x).
$$

It follows that $L^1(\mu)$ is d-decomposable. The converse is evident. ▷

(4) *The lattice-normed space $L^1(\mu)$ is disjointly complete if and only if μ is an ample measure.*

◁ Suppose that μ is ample. Let $(b_\xi)_{\xi\in\Xi}$ be a partition of unity in $\mathfrak{P}(F)$ and let $(x_\xi)_{\xi\in\Xi}$ be a norm-bounded family in $L^1(\mu)$. Let $e_\xi(\cdot):\mathbb{R}\to\mathscr{A}$ be the spectral function of x_ξ and define

$$e(\lambda):=\bigvee_{\xi\in\Xi} h(b_\xi)\wedge e_\xi(\lambda)\quad(\lambda\in\mathbb{R}).$$

The function $\lambda\mapsto e(\lambda)$ is a resolution of unity, i.e. satisfies 1.4.3 (1–3). This is checked by straightforward calculation employing the associativity laws (1.1.2) and the infinite distributive laws (1.1.3). The only claim that needs clarification is the identity $e:=\bigwedge\{e(\lambda):\lambda\in\mathbb{R}\}=0$. Since $\big(h(b_\xi)\big)$ is a partition of unity in $\mathscr{A}/\mathscr{N}(\mu)$, it is sufficient to show that $h(b_\xi)\wedge e=0$ for all $\xi\in\Xi$. The due calculations are as follows:

$$h(b_\eta)\wedge e=\bigwedge_{n=1}^{\infty}h(b_\eta)\wedge e(-n)=\bigwedge_{n=1}^{\infty}\left(h(b_\eta)\wedge\bigvee_{\xi\in\Xi}h(b_\xi)e_\xi(-n)\right)$$
$$=\bigwedge_{n=1}^{\infty}\bigvee_{\xi\in\Xi}h(b_\eta)\wedge h(b_\xi)e_\xi(-n)=\bigwedge_{n=1}^{\infty}h(b_\eta)\wedge e_\eta(-n)=0.$$

By Theorem 1.4.4 there is an element x in the universal completion $mL^1(\mu)$ with $e(\lambda)=e^x_\lambda$ $(\lambda\in\mathbb{R})$. If $y\in L^1(\mu)$ is an upper bound of the family $(|x_\xi|)$ then $|x|\le y$, so that $x\in L^1(\mu)$. Applying 1.3.8 (12), obtain $e_\lambda^{h(b_\xi)x}=e_\lambda^{h(b_\xi)x_\xi}$ $(\lambda\in\mathbb{R})$, whence $h(b_\xi)x=h(b_\xi)x_\xi$. Thus we have verified that $L^1(\mu)$ is disjointly complete. Since the reverse is evident, the proof is complete. ▷

6.1.10. *Theorem.* *For a countably additive measure with values in a Banach–Kantorovich space the following are equivalent:*

(1) *μ is an ample measure;*

(2) *$|\mu|$ is an ample measure;*

(3) *$L^1(\mu)$ is a Banach–Kantorovich space;*

(4) *$L^1(\mu)$ is an order complete vector lattice and the operator $T:L^1(\mu)\to F$ defined by $T\tilde{x}=I_{|\mu|}(x)$ $\big(x\in\mathscr{L}^1(\mu)\big)$ is a Maharam operator.*

◁ (1) ⇒ (2): Follows from 6.1.9 (1).

(2) ⇒ (3): Follows from 6.1.8, 6.1.9 (4), and 2.2.3.

(3) ⇒ (4): We know that $L^1(\mu)$ is order σ-complete. The Dominated Convergence Theorem implies that the F-valued norm in $L^1(\mu)$ is sequentially order continuous, i.e. for every decreasing sequence $(x_n)\subset L^1(\mu)$ we have $o\text{-}\lim_n|x_n|=0$ whenever $o\text{-}\lim_n x_n=0$. Take an order-bounded set $M\subset L^1(\mu)$ and let $u\in L^1(\mu)$

be its upper bound. Without loss of generality we may assume that M contains all elements of the form $x_1 \vee \dots \vee x_n$ and bo-$\sum_{\xi\in\Xi} h(b_\xi)x_\xi$, where $\{x_1, \dots, x_n\} \subset M$, $(x_\xi) \subset M$, and (b_ξ) is a partition of unity in $\mathfrak{P}(F)$, since supplementing M with these elements, we will not change the set of upper bounds of M. Put $f := \sup\{|x| : x \in M\} \le |u|$ and choose a sequence $(x_n) \subset M$ with $f - |x_n| \le (1/n)f$. If $y_n := x_1 \vee \dots \vee x_n$ then $f - |y_n| \le (1/n)f$ $(n \in \mathbb{N})$ and (y_n) is increasing. If $y = \sup_n y_n$ then $|y| = f$. We now take an arbitrary $z \in M$ and verify that $z \le y$. To this end, observe that o-$\lim_n y_n \vee z = y \vee z \ge y$ and o-$\lim_n |y_n \vee z| = |y \vee z| \le f$, since $y_n \vee z \in M$. Thus, $f \le |y| \le |y \vee z| \le f$, so that $f = |y \vee z|$. Since the norm on $L^1(\mu)_+$ is additive, we have $f = |(y\vee z - y) + y| = |y\vee z - y| + |y| = |y \vee z - y| + f$, whence $|y \vee z - y| = 0$ and $y \vee z = y$. So, we have proved that $y = \sup(M)$ and $L^1(\mu)$ is order complete.

Consider a downward directed set D with $\inf(D) = 0$ and put $f := \inf\{|x| : x \in D\}$. Repeating the above arguments we may choose a decreasing sequence $(y_n) \subset L^1(\mu)$ (not in D in general) such that $f = \inf_n |y_n|$ and $0 = \inf_n y_n$. Since the norm is order σ-continuous, we obtain $f = 0$. Thus, T is order continuous, because $Tx = |x^+| - |x^-|$. According to 2.1.8 (4) $L^1(\mu)$ admits a compatible module structure over $\operatorname{Orth}(F)$. Therefore, if $0 \le f \le Tx$ for some $0 \le x \in L^1(\mu)$, then there is $\pi \in \operatorname{Orth}(F)$ such that $f = \pi Tx = \pi|x| = |\pi x| = T(\pi x)$, so that T possess the Maharam property.

(4) $\Rightarrow$ (1): This is an easy consequence of 3.4.3. $\triangleright$

6.1.11. Let $\nu : \mathscr{A} \to F$ and $\mu : \mathscr{A} \to Y$ be finitely additive measures. We say that μ is *absolutely continuous with respect to* ν and write $\mu \ll \nu$ if $|\mu(a)| \in \nu(A)^{\perp\perp}$ for all $a \in \mathscr{A}$. If $\mu \ll \nu$ then $\mathscr{N}(\nu) \subset \mathscr{N}(\mu)$ and a natural Boolean homomorphism $\varrho : \mathscr{A}/\mathscr{N}(\nu) \to \mathscr{A}/\mathscr{N}(\mu)$ can be defined by $\varrho\circ\phi' = \phi$, where ϕ' is a canonical factor mapping $\mathscr{A} \to \mathscr{A}/\mathscr{N}(\nu)$. Denote by χ_a the band projection in E corresponding to some unit element $a \in \mathscr{A}$.

(1) *Suppose that ν is modular with respect to h and $\mu \ll \nu$. Then μ is modular with respect to $h \circ \varrho$.*

$\triangleleft$ Without loss of generality we may assume that ν is positive. By hypothesis, for every $b \in \mathbb{B}$ we have $|\mu(h(b) \wedge \phi(a))| \in \nu(h(b) \wedge \phi'(a))^{\perp\perp}$, so that $b^\perp\mu(h(b) \wedge \phi(a)) = 0$ for all $b \in \mathbb{B}$ and $a \in \mathscr{A}$. From this we obtain $\mu(h(b) \wedge \phi(a)) = b\mu(h(b) \wedge \phi(a))$. Substituting $b^\perp$ for b, we arrive at $b\mu(h(b^\perp) \wedge \phi(a)) = 0$, whence $b\mu(\phi(a)) = b\mu(h(b) \wedge \phi(a))$. Hence, $b\mu(\phi(a)) = \mu(h(b) \wedge \phi(a))$ and the result follows. $\triangleright$

(2) Radon–Nikodým Theorem. *Let $\mu, \nu : \mathscr{A} \to F$ be countably additive measures with μ positive and ample. If ν is absolutely continuous with* *respect to μ then there exists $y \in \mathscr{L}^1(\mu)$ such that*

$$\nu(a) = \int \chi_a y \, d\mu \quad (a \in \mathscr{A}).$$

$\lhd$ First assume that $|\nu| \le \mu$. Define an operator $S_\nu : L^1(\mu) \to F$ by

$$S_\nu(\tilde{x}) := \int x \, d\mu \quad (x \in \mathscr{L}^1(\mu)).$$

Clearly, $S_\nu \ll T$, so that by Theorem 3.4.9 there is $\rho \in \operatorname{Orth}(L^1(\mu))$ such that $|\rho| \le I$ and $S_\nu(u) = T(\rho u)$ $(u \in L(\mu))$. The orthomorphism ρ is representable as $\rho(u) = \tilde{y}u$ for some $\tilde{y} \in L^1(\mu)$, $|y| \le \mathbf{1}$, whence

$$\nu(a) = T(\chi_a \tilde{y}) = \int \chi_a y \, d\mu \quad (a \in \mathscr{A}).$$

To handle the general case, put $\nu_n := \nu \wedge (n\mu)$ and $S_n := S_{\nu_n}$ $(n \in \mathbb{N})$. Then $\nu_n \nearrow \nu$, $S_n \nearrow S_\nu$ and by above proved there is an increasing sequence $(y_n) \in \mathscr{L}^1(\mu)$ with $\nu_n(a) = I_\mu(\chi_a y_n)$ $(a \in \mathscr{A})$. Since $I_\mu(y_n) = \nu_n(\mathbf{1}) \le \nu(\mathbf{1})$, we may apply the Monotone Convergence Theorem. Thus, the element $y := \sup_n y_n$ is contained in $\mathscr{L}^1(\mu)$ and $\nu(a) = I_\mu(\chi_a y)$ $(a \in \mathscr{A})$. $\rhd$

(3) *Let μ be a positive ample measure. Then the mapping $x \to \nu_x$, where ν_x is defined by $\nu_x(a) := I_\mu(\chi_a x)$ $(a \in \mathscr{A})$, is a lattice isomorphism of vector lattices $L^1(\mu)$ and $\{\mu\}^{\perp\perp}$.*

6.2. Integral Representation by Quasi-Radon Measures

In this section we establish an integral representation result for dominated operators on some lattices of continuous functions. Of vital importance in this connection is the class of quasi-Radon measures.

6.2.1. (1) We now specify the vector integral of the proceeding section for elements of some abstract K_σ-space. Take as a universally complete K-space G the vector lattice $Q^{\mathbb{R}}$ of all real-valued functions defined on a nonempty set Q. Let $\mathscr{A}$ be an algebra of subsets of Q, i.e. $\mathscr{A} \subset \mathscr{P}(Q)$. This algebra we identify with the isomorphic algebra of the characteristic functions $\{1_A := \chi_A : A \in \mathscr{A}\}$ so that $S(\mathscr{A})$ is the space of all $\mathscr{A}$-simple functions on Q, i.e., $f \in S(\mathscr{A})$ means that $f = \sum_{k=1}^n \alpha_k \chi_{A_k}$ for some $\alpha_1, \dots, \alpha_n \in \mathbb{R}$ and disjoint $A_1, \dots, A_n \in \mathscr{A}$. Let a measure μ be defined on $\mathscr{A}$ and take values in a bo-complete lattice-normed space Y over a K-space F. We suppose that $\mu \in \operatorname{da}(\mathscr{A}, Y)$. If $f \in S(\mathscr{A})$ then we put by definition

$$I_\mu := \int f \, d\mu = \sum_{k=1}^n \alpha_k \mu(A_k).$$

As was described in Section 6.1, the integral I_μ can be extended to the spaces of μ-summable functions $\mathscr{L}^1(\mu)$ for which the more informative notations $\mathscr{L}^1(Q,\mu)$ and $\mathscr{L}^1(Q,\mathscr{A},\mu)$ are also used. On identifying equivalent functions, we obtain the K_σ-space $L^1(\mu) := L^1(Q,\mu) := L^1(Q,\mathscr{A},\mu)$. Observe that order convergence in $G := Q^{\mathbb{R}}$ coincides with pointwise convergence, while order convergence in $L^1(\mu)$ is defined by almost everywhere convergence.

(2) In this section we fix the following notation: Q is a completely regular topological space; $\mathscr{T}$, $\mathscr{F}$, and $\mathscr{K}$ are respectively the collections of all open, closed and compact subsets of Q; $C_b(Q)$ is the space of all bounded continuous functions on Q; $\mathscr{M}(Q)$ and $\mathscr{M}_b(Q)$ are respectively the spaces of all Borel and bounded Borel functions on Q. For a family $\mathscr{D}$ of subsets of Q denote by $\sigma_0(\mathscr{D})$ $(\sigma(\mathscr{D}))$ the smallest subalgebra (respectively, σ-algebra), containing $\mathscr{D}$. In this event we say that $\sigma_0(\mathscr{D})$ is generated and $\sigma(\mathscr{D})$ is σ-generated by $\mathscr{D}$.

(3) Let $C \in \mathscr{A}$. Consider the families of sets $\mathscr{K}_C = \{K \in \mathscr{K} \cap \mathscr{A} : K \subset C\}$ and $\mathscr{F}_C = \{D \in \mathscr{F} \cap \mathscr{A} : D \subset C\}$ directed by inclusion. A measure $\mu : \mathscr{A} \to Y$ is said to be *Radon* (*quasi-Radon*) if for each $C \in \mathscr{A}$ (for every $C \in \mathscr{T} \cap \mathscr{A}$) the identity $\mu(C) = bo\text{-}\lim\{\mu(K) : K \in \mathscr{K}_C\}$ holds. A measure $\mu : \mathscr{A} \to Y$ is called *regular* (*quasiregular*) if for every $C \in \mathscr{A}$ (for every $C \in \mathscr{T} \cap \mathscr{A}$) the equality $\mu(C) = bo\text{-}\lim\{\mu(D) : D \in \mathscr{F}_C\}$ holds.

In the case of a compact space Q these two definitions are equivalent. Moreover, we may prove that $\mu \in \mathrm{da}(\mathscr{A}, Y)$ is a Radon measure (a regular measure) if and only if the vector variation of μ is a Radon measure (a regular measure). But in our considerations of particular interest is a similar result for quasi-Radon measure. To prove it, we need an auxiliary fact that can be easily deduced from the Birkhoff–Ulam Theorem.

(4) *Let Q be an extremal compact space. There exists an order σ-continuous lattice homomorphism β from the vector lattice of Borel functions $\mathscr{M}(Q, \mathscr{B}or(Q))$ onto $C_\infty(Q)$ such that h is the identity operator on $C_\infty(Q)$. Moreover, $\beta(\sup_\alpha f_\alpha) = \bigvee_\alpha \beta(f_\alpha)$ for every increasing order-bounded net (f_α) in $C_\infty(Q)$ (sup on the left-hand side denotes the pointwise supremum, while $\bigvee$ in the right-hand side means the least upper bound in $C_\infty(Q)$).*

◁ We need only to put

$$\beta(f) := \int_Q f\,d\varphi \quad (f \in \mathscr{M}(Q, \mathscr{B}or(Q))),$$

where $\varphi : \mathscr{B}or(Q) \to C_\infty(Q)$ is the Birkhoff–Ulam homomorphism of 1.2.6. ▷

6.2.2. Theorem. *Suppose that a measure $\mu \in \mathrm{da}(\mathscr{A}, Y)$ satisfies one of the following conditions:*

(1) *$\mathscr{A}$ is generated by its closed sets, i.e.* $\mathscr{A} = \sigma_0(\mathscr{F} \cap \mathscr{A})$;

(2) *μ is countably additive and $\mathscr{A}$ is σ-generated by its closed sets, i.e.* $\mathscr{A} = \sigma(\mathscr{F} \cap \mathscr{A})$.

Then μ is quasi-Radon measure (quasiregular measure) if and only if its exact dominant $|\mu|$ is quasi-Radon measure (quasiregular measure).

$\lhd$ We confine exposition to (2); the case (1) is settled along the same lines. Suppose that μ is a quasi-Radon measure but $|\mu|$ lacks this property. Then there is a set $U \in \mathscr{T} \cap \mathscr{A}$ such that

$$f := |\mu|(U) - \bigvee\{|\mu|(K) : K \in \mathscr{K} \cap \mathscr{A},\ K \subset U\} > 0.$$

If $K \in \mathscr{K} \cap \mathscr{A}$ and $K \subset U$ then $|\mu|(U \setminus K) \geq f$. By Definition 4.2.9 (1) there are $0 < \varepsilon_0 \in \mathbb{R}$, $0 < e_0 \in \mathfrak{G}(e)$, $e = |\mu|(U)$, and a finite collection $(C_i)_{i\in\mathbb{N}} \subset \mathscr{A}$ such that

$$\bigcup_{i=1}^{n} C_i = U \backslash K, \quad C_k \cap C_i = \varnothing\ (k \neq i), \quad \sum_{i=1}^{n} |\mu(C_i)| \geq \varepsilon_0 e_0.$$

Let ω_1 be the least uncountable ordinal. For some countable ordinal $\alpha_0 < \omega_1$ all C_i $(i = 1, \dots, n)$ are included to the Baire class $\mathscr{B}_{\alpha_0}(\sigma_0(\mathscr{F} \cap \mathscr{A}))$ over the algebra $\sigma_0(\mathscr{F} \cap \mathscr{A})$. The ordinal α_0 may be thought nonlimit. Each set from the Baire class $\mathscr{B}_\alpha = \mathscr{B}_\alpha(\sigma_0(\mathscr{F} \cap \mathscr{A}))$ is either a countable union or a countable intersection of some sets from the preceding Baire classes. Therefore, there are $\alpha_1 < \alpha_0$ and sequences $(C_{i,k})_{k\in\mathbb{N}}$ $(i = 1, \dots, n)$ in $\mathscr{B}_{\alpha_i}$ such that the sequence $(C_{i,k})_{k\in\mathbb{N}}$ is monotone and converges to C_i for all i. It can be also assumed that $C_{i,k} \subset U \backslash K$. for $i \leq n$ and $k \in \mathbb{N}$. Take an arbitrary $\delta > 0$. By additivity of μ, we may choose $e_1 \in \mathfrak{G}(e)$, $0 \leq e_1 \leq e_0$, and $k_1 \in \mathbb{N}$, with

$$\sum_{i=1}^{n} \left|\mu\left(C_{i,k_1} \backslash \bigcup_{j<i} C_{j,k_1}\right)\right| \geq \left(\varepsilon_0 - \frac{\delta}{4}\right) e_1.$$

Denote $C_i^1 := C_{i,k_1}$ $(i = 1, \dots, n)$. Repeating the above procedure we obtain a decreasing sequence of ordinals $\alpha_0 > \alpha_1 > \dots > \alpha_{m-1} > \alpha_m$, a sequence of elements $(e_k)_{k=1}^m \subset \mathfrak{G}(e)$, sequences of sets $(C_i^n)_{i=1}^n \subset \mathfrak{B}_{\alpha_k}$ $(k = 0, 1, \dots, m)$ such that $0 < e_m \leq e_{m-1} \leq \dots \leq e_1$ and

$$\sum_{i=0}^{n} \left|\mu\left(C_i^k \backslash \bigcup_{j<i} C_j^k\right)\right| \geq \left(\varepsilon_0 - \frac{\delta}{4} - \dots - \frac{\delta}{2^{k+1}}\right) e_k \quad (k = 0, 1, \dots, m).$$

Since ordinals are well-ordered, it follows that $\alpha_m = 0$ for some $m \in \mathbb{N}$. Observe that if

$$D_i := C_i^m \backslash \bigcup_{j<i} C_j^m \quad (i = 1, \dots, n), \quad D_0 := (U \backslash K) \backslash \bigcup_{i=1}^{n} D_i, \quad g := e_m$$

then $0 < g \le e$, $g \in \mathfrak{G}(e)$, and $\sum_{i=0}^{n}|\mu(D_i)| \ge (\varepsilon_0 - \delta/2)g$; moreover $D_0, D_1, \dots, D_n$ are pairwise disjoint elements of $\sigma_0(\mathscr{F} \cap \mathscr{A})$ with $D_0 \cup D_1 \cup \dots D_n = U \backslash K$. Without loss of generality we may assume that each D_i has the form $D_i = U_i \backslash V_i$, where $U_i, V_i \in \mathscr{T} \cap \mathscr{A}$ and $U_i \cup V_i \subset U \backslash K$ for all $i = 0, 1, \dots, n$. (This is because of the construction of the algebra $\sigma_0(\mathscr{F} \cap \mathscr{A})$.)

Now enumerate all U_i and V_i in a single finite sequence $(W_i)_{i=1}^m$ and denote $M = \{1, \dots, m\}$. For every $J \subset M$ put $H_J = \bigcap\{W_i : i \in M \backslash J\}$. Obviously, $H_J \cap H_{J'} = H_{J \cap J'}$. Since μ is a quasi-Radon measure, for each $\varepsilon' > 0$ there are a compact space $K_\varnothing \in \mathscr{K} \cap \mathscr{A}$, $K_\varnothing \subset H_\varnothing$, and a fragment $g_0 \in \mathfrak{E}(e)$, $0 < g_0 \le g$, such that for every $K' \in \mathscr{K} \cap \mathscr{A}$ with $K_\varnothing \subset K' \subset H_\varnothing$, the relations hold:

$$g_0|\mu(K' \backslash K_\varnothing)| \le \varepsilon' e, \quad g_i|\mu(H_\varnothing \backslash K_\varnothing)| \le \varepsilon' e.$$

(Here we mean the multiplication of the order ideal $F(e)$ with ring unity e.) For every $i \in M$ there are a compact space $K_{\{i\}} \in \mathscr{K} \cap \mathscr{A}$ and a fragment $g_i \in \mathfrak{G}(e)$ such that $K_{\{i\}} \subset H_{\{i\}} \backslash K_\varnothing$, $0 < g_i \le g_0$, and, for every $K' \in \mathscr{K} \cap \mathscr{A}$, with $K_{\{i\}} \subset K' \subset H_{\{i\}} \backslash K_\varnothing$, the relations hold:

$$g_i|\mu(K' \backslash K_{\{i\}})| \le \varepsilon' e, \quad g_i|\mu((H_{\{i\}} \backslash K_\varnothing) \backslash K_{\{i\}})| \le \varepsilon' e.$$

We may assume that $g_m \le g_{m-1} \le \dots \le g_1$. If $i \ne j$ then $K_{\{i\}} \cap K_{\{j\}} = H_\varnothing$; therefore, $g_0|\mu(K_{\{i\}} \cap K_{\{j\}})| \le \varepsilon' e$.

Assume that for some $k \le m$ the above construction is performed for each $J \subset M$, with card $J < k$. In particular, for $J \subset M$, card $J < k$, there is $K_J \in \mathscr{K} \cap \mathscr{A}$. Let $J \subset M$ and card $J = k$. Take $K_J \in \mathscr{K} \cap \mathscr{A}$ and $g_J \in \mathfrak{G}(e)$ so that

$$K_J \subset H_J \backslash \bigcup\{K_{J'} : J' \subset J\}, \quad 0 < g_J \le \bigwedge\{g_{J'} : \operatorname{card} J' < k\}$$

and for every $K' \in \mathscr{K} \cap \mathscr{A}$ with $K_J \subset K' \subset H_J \backslash \bigcup\{K_{J'} : J' \subset J\}$, the relations hold

$$g_J|\mu(K' \setminus K_J)| \le \varepsilon' e, \quad g_J|\mu((H_J \backslash \bigcup\{K_{J'} : J' \subset J\}) \backslash K_J)| \le \varepsilon' e.$$

All g_J with card $J \le k$ can be assumed linearly ordered. Moreover, for every two subsets $J \subset M$ and $J' \subset M$, $J \ne J'$, we have $K_J \cap K_{J'} \subset H_J \cap H_{J'} = H_{J \cap J'}$ provided that either $\operatorname{card} J = k$ and $\operatorname{card} J' \le k$, or $\operatorname{card} J \le k$ and $\operatorname{card} J' = k$. If $J'' \subset J \cap J'$ then by construction either $K_J \cap K_{J''} = \varnothing$, or $K_{J'} \cap K_{J''} = \varnothing$. This amounts to the fact that $\bar{g}|\mu(K')| \le \varepsilon' e$ for every $K' \in \mathscr{K} \cap \mathscr{A}$, $K' \subset K_J \cap K_{J'}$, with $\bar{g} = \bigwedge\{g_{J'} : \operatorname{card} J' \le k\}$. The construction is finished on the induction step $k = m - 1$. Thus we obtain the following:

For any function $\sigma : M \to \{0, 1\}$ denote $W^\sigma = \bigcup_{i=1}^m W_i^{\sigma(i)}$, where $W_i^0 = (U \backslash K) \backslash W_i$ and $W_i^1 = W_i$ for $i \in M$. We have proven that for every $\varepsilon' > 0$ there exist $\bar{g} \in \mathfrak{E}(e)$, $0 < \bar{g} \le g$, and $K_\sigma \in \mathscr{K} \cap \mathscr{A}$ $(\sigma \in \{0,1\}^M)$ such that $K_\sigma \subset U \backslash K$ and

$$\bar{g}|\mu(W^\sigma) - \mu(K_\sigma)| \le \varepsilon' \bar{g}, \quad \bar{g}|\mu(K')| \le \varepsilon' \bar{g} \qquad (*)$$
$$\big(K' \in \mathscr{K} \cap \mathscr{A},\ K' \subset K_\sigma \cap K_{\sigma'},\ \big(\sigma, \sigma' \in \{0,1\}^M,\ \sigma \ne \sigma'\big)\big).$$

This amounts to the inequalities

$$\bar{g}|\mu|(K_1) \ge \bar{g} \sum_\sigma \Big|\mu\Big(K_\sigma \backslash \bigcup_{\sigma' \ne \sigma} K_{\sigma'}\Big)\Big| + \bar{g}\Big|\mu\Big(\bigcup_{\sigma \ne \sigma'} K_\sigma \cap K_{\sigma'}\Big)\Big|$$
$$\ge \bar{g} \sum_\sigma |\mu(K_\sigma)| - \bar{g} \sum_\sigma \Big|\mu\Big(\bigcup_{\sigma' \ne \sigma} K_\sigma \cup K_{\sigma'}\Big)\Big|,$$

where $K_1 = \bigcup\{K_\sigma : \sigma \in \{0,1\}^M\}$. For every $\sigma \in \{0,1\}^M$, evaluate the following term in the right-hand side of the above inequality: $\bar{g}\Big|\mu\Big(\bigcup_{\sigma' \ne \sigma} K^\sigma \cap K^{\sigma'}\Big)\Big|$. To this end, denote $\mathscr{M}_\sigma := \{0,1\}^M \backslash \{\sigma\}$ and $L_{\sigma'} = K_\sigma \cap K_{\sigma'}$ and consider the identity

$$\mu\Big(\bigcup_{\sigma'} L_{\sigma'}\Big) = \sum_{k=1}^{l} (-1)^{k+1} \Big(\sum_{\sigma_1 < \dots < \sigma_k} \mu(L_{\sigma_1} \cap \dots \cap L_{\sigma_k})\Big).$$

Here $\sigma', \sigma_1, \dots, \sigma_k \in \mathscr{M}_\sigma$, $l = 2^m - 1$, and the finite set $\mathscr{M}_\sigma$ is linearly ordered somehow. On the right-hand side of the above identity we have $2^l - 1$ summands, each of which evaluated by the above formulas $(*)$. Hence, for sufficiently small ε' we obtain

$$\bar{g}|\mu|(K_1) \ge \bar{g} \sum_\sigma |\mu(W^\sigma)| - 2^{l+m} \varepsilon' \bar{g}$$
$$\ge \bar{g} \sum_{i=0}^{n} |\mu(D_i)| - 2^{l+m} \varepsilon' \bar{g} \ge (\varepsilon_0 - \delta)\bar{g}.$$

Thus, we have proven that for every $\delta > 0$, $e_1 \in \mathfrak{E}(e)$, and $K \in \mathscr{K} \cap \mathscr{A}$ with $0 < e_1 \le e_0$, $K \subset U$, there are $e_2 \in \mathfrak{G}(e)$ and $K_1 \in \mathscr{K} \cap \mathscr{A}$ such that $0 < e_2 \le e_1$, $K_1 \subset U \backslash K$, and $|\mu|(K_1) \ge (\varepsilon_0 - \delta)e_2$. First, put $K := \varnothing$ and find K_1 with the above property. Further, put $K := K_1$ and choose $K_2 \in \mathscr{K} \cap \mathscr{A}$ and $e_3 \in \mathfrak{G}(e)$, such that $K_2 \subset U \backslash K_1$, $0 < e_3 \le e_2$, and $|\mu|(K_2) \ge (\varepsilon_0 - \delta - \delta/2)e_3$. Continuing this process, we produce the sequences $(K_n)_{n \in \mathbb{N}}$ in $\mathscr{K} \cap \mathscr{A}$ and $(e_n)_{n \in \mathbb{N}}$ in $\mathfrak{G}(e)$ for which we have

$$K_n \subset \bigcup_{i=1}^{n-1} K_i, \quad 0 < e_{n+1} \le e_n,$$
$$|\mu|(K_n) \ge (\varepsilon_0 - \delta - \dots - \delta/2^{n-1})e_{n+1} \quad (n \in \mathbb{N}).$$

This amounts to the fact that for every $n \in \mathbb{N}$ the inequality $|\mu|(U) \geq n(\varepsilon_0 - 2\delta)e_{n+1}$ holds. So, we arrive at a contradiction, since $e_{n+1} \leq e = |\mu|(U)$.

Conversely, if $|\mu|$ is a quasi-Radon measure then it follows immediately from the inequalities

$$|\mu(C) - \mu(D)| \leq |\mu|(C \backslash D) \quad (C, D \in \mathscr{A}, D \subset C)$$

that μ is also a quasi-Radon measure. The proof is complete. $\triangleright$

The following two subsections give some easy corollaries to the above theorem.

6.2.3. *Assume that a measure $\mu \in \mathrm{da}(\mathscr{A}, Y)$ satisfies one of the two conditions of Theorem* 6.2.2. *Then the following definitions of a quasi-Radon measure μ are equivalent:*

(1) $\mu(C) = bo\text{-}\lim\{\mu(K) : K \in \mathscr{K}_C\}$ *for every* $C \in \mathscr{T} \cap \mathscr{A}$;

(2) *the identity (1) is valid for all* $C \in \sigma_0(\mathscr{F} \cap \mathscr{A})$;

(3) $|\mu|(C) = \bigvee\{|\mu|(K) : K \in \mathscr{K}_C\}$ *for every* $C \in \mathscr{T} \cap \mathscr{A}$;

(4) *the identity* (3) *is valid for all* $C \in \sigma_0(\mathscr{F} \cap \mathscr{A})$.

6.2.4. (1) *If we substitute $\mathscr{F}_C$ for $\mathscr{K}_C$ in Corollary* 6.2.3 *then each of the conditions* (1–4) *is equivalent to the quasiregularity of μ.*

(2) *If a measure $\mu \in \mathrm{da}(\mathscr{A}, Y)$ is quasi-Radon and meets at least one of the conditions* (1) *and* (2) *of Theorem* 6.2.2 *then μ is quasiregular.*

In this connection it is worth observing that a positive quasi-Radon measure is quasiregular. Moreover, the following is true:

(3) *Suppose that the vector variation $|\mu|$ of a measure $\mu \in \mathrm{da}(\mathscr{A}, Y)$ is quasiregular and*

$$|\mu|(Q) = \bigvee\{|\mu|(K) : K \in \mathscr{K} \cap \mathscr{A}\}.$$

Then μ and $|\mu|$ are quasi-Radon measures.

6.2.5. Theorem. *Let $\mu : \mathscr{A} \to Y$ be a dominated finitely additive measure. If μ is a quasi-Radon measure then its restriction to the algebra $\sigma_0(\mathscr{F} \cap \mathscr{A})$ is countably additive.*

$\triangleleft$ We denote the restriction of μ to the subalgebra $\sigma_0(\mathscr{F} \cap \mathscr{A})$ by the same letter. The vector variation of μ with respect to this subalgebra do not coincide with the restriction of $|\mu|$, so that we denote by $|\mu|_0$ the vector variation of the restricted measure. Ensure that $|\mu|_0$ is σ-additive and the result will follow. Take a sequence $(A_k)_{k \in \mathbb{N}}$ in $\sigma_0(\mathscr{F} \cap \mathscr{A})$ decreasing to the empty set and suppose that $\bigwedge_{n=1}^{\infty} |\mu|_0(A_n) > 0$. If $f := |\mu|_0(Q)$ then there are $0 < \varepsilon_0 \in \mathbb{R}$ and $0 < e_0 \in$

$\mathfrak{E}(e)$ with $|\mu|_0(A_n) \geq \varepsilon_0 e_0$ $(n \in \mathbb{N})$. By Theorem 6.2.2 $|\mu|_0$ is a quasi-Radon measure, so that we may find $K_1 \in \mathscr{K} \cap \mathscr{A}$ and $e_1 \in \mathfrak{G}(e_0)$ such that $K_1 \subset A_1$, $0 < e_1 \leq e_0$, and $e_1|\mu|_0(A_1 \backslash K_1) \leq (\varepsilon_0/4)e_0$. By induction, we produce sequences $(K_n)_{n\in\mathbb{N}} \subset \mathscr{K} \cap \mathscr{A}$ and $(e_n)_{n\in\mathbb{N}} \subset \mathfrak{G}(e)$ such that $0 \leq e_{n+1} \leq e_n$, $K_n \subset A_n$, and $e_n|\mu|_0(A_n \backslash K_n) \leq (\varepsilon_0/2^{n+1})e_0$ $(n \in \mathbb{N})$. Putting $K'_n = \bigcap_{l=1}^n K_l$, we arrive at $e_n|\mu|_0(A_n \backslash K'_n) \leq (\varepsilon_0/2)e_0$ $(n \in \mathbb{N})$. From this it follows that $|\mu|_0(K'_n) = |\mu|_0(A_n) - |\mu|_0(A_n \backslash K'_n) \geq (\varepsilon_0/2)e_n$. But then $K'_{n_0} = \varnothing$ for some $n_0 \in \mathbb{N}$, since $K'_n \searrow \varnothing$, This is a contradiction, and so the claim follows. $\triangleright$

Similar fact is valid for Radon measures without any additional restrictions, i.e., if $\mu \in \mathrm{da}(\mathscr{A}, Y)$ is a Radon measure then μ is σ-additive.

6.2.6. We now address the question of integral representation of dominated operators by quasi-Radon measures. Let Q be a completely regular topological space and let L be a sublattice of the vector lattice $C_b(Q)$. Denote by $\mathscr{T}(L)$ the weakest topology in which all functions in L are continuous. If $\mathscr{T}(L)$ coincides with the initial topology $\mathscr{T}$ of Q, then we say that L *generates* $\mathscr{T}$.

Theorem. *Assume that a vector sublattice $L \subset C_b(Q)$ contains the identically one function $\mathbf{1}$ and generates the topology of Q. Let $T : L \to Y$ be a dominated operator. The following are equivalent:*

(1) *there exists a unique quasi-Radon measure $\mu \in \mathrm{dca}(\mathscr{B}or(Q), Y)$ such that the representation holds*

$$Tf = \int f\, d\mu \quad (f \in L);$$

(2) $|T|(\mathbf{1}) = \bigvee \{\bigwedge\{|T|g : g \in L,\ g \geq \chi_K\} :\ K \in \mathscr{K}\}$.

$\triangleleft$ Suppose that (1) holds and take a net $(f_\alpha)_{\alpha\in\mathrm{A}}$ in L decreasing to zero pointwise. Fix some $\alpha_0 \in \mathrm{A}$ and $\varepsilon > 0$. Then for some $0 < M \in \mathbb{R}$ we have $0 \leq f_\alpha \leq M\mathbf{1}$ for all $\alpha \geq \alpha_0$. Given a compact set $K \subset Q$, denote $a_K := |T|(\mathbf{1}) - \bigwedge\{|T|f : f \in L,\ f \geq \chi_K\}$. Observe that

$$a_K = \bigvee\{|T|(\mathbf{1} - g) : g \geq \chi_K\} = \bigvee\{|T|(h) : h \leq \mathbf{1} - \chi_K\}.$$

Since the net (f_α) vanishes uniformly on the compact set K, we may choose $\alpha_1 \geq \alpha_0$ so that $f_\alpha - f_\alpha \wedge (\varepsilon\mathbf{1})$ vanishes on K whenever $\alpha \geq \alpha_1$. Moreover, $h_\alpha := M^{-1}(f_\alpha - f_\alpha \wedge (\varepsilon\mathbf{1})) \leq \mathbf{1}$, whence $h \leq \chi_K$ for the same α. Thus, $|T|(h_\alpha) \leq a_K$ and we may evaluate

$$0 \leq |T|(f_\alpha) = |T|\big(f_\alpha \wedge (\varepsilon\mathbf{1})\big) + |T|(Mh_\alpha) \leq \varepsilon|T|(\mathbf{1}) + Ma_K.$$

By Theorem 1.4.6 (1) the band $\{|T|(\mathbf{1})\}^{\perp\perp}$ in F is isomorphic to an order-dense ideal F' in $C_\infty(P)$, where P is an extremal compact space. We may assume that this isomorphism carries the element $|T|(\mathbf{1})$ into the identically one function on P. By hypothesis the net $(a_K)_{K\in\mathscr{K}}$ decreases to zero. Therefore, there exists a comeager set $P_0 \subset P$ such that the numerical net $a_K(p)$ converges to zero for all $p \in P_0$. From the above evaluation it follows that if a net $(f_\alpha)_{\alpha\in\mathrm{A}} \subset L$ decreases to zero then for each $p \in P_0$ the net $\omega_p(f_\alpha) := (|T|f_\alpha)(p)$ converges to zero. The positive functional ω_p then can be extended to an order σ-continuous functional $\tilde{\omega}_q : \mathscr{M}_b(Q) \to \mathbb{R}$ (see 4.5.3). Define an operator $V : \mathscr{M}_b(Q) \to \mathbb{R}^P$ as follows:

$$(Vf)(p) := \begin{cases} \tilde{\omega}_p(f), & q \in P_0, \\ 0, & p \notin P_0. \end{cases}$$

Clearly, $Vf \in \mathscr{M}_b(P)$ for every $f \in L$. Moreover, from the order σ-continuity of $\tilde{\omega}_p$ it follows that if $Vf_n \in \mathscr{M}_b(P)$ for a sequence $f_n \in \mathscr{M}_b(P)$ with $f := o\text{-}\lim_n f_n$ then $Vf \in \mathscr{M}_b(P)$. Whence we conclude that $Vf \in \mathscr{M}_b(P)$ for every $f \in \mathscr{M}_b(Q)$. Put $W = \beta \circ V$, where $\beta : \mathscr{M}_b(P) \to F$ is defined by 6.2.1 (4). Then $W : \mathscr{M}_b(P) \to F$ is an order σ-continuous extension of $|T|$.

The operator T is extended as follows: Take $f \in \mathscr{M}_b(Q)$ and a bounded net $(g_\alpha)_{\alpha\in\mathrm{A}}$ in L that increases to f pointwise. From the estimate $|Tg_\alpha - Tg_\beta| \leq W(|g_\alpha - g_\beta|)$ $(\alpha, \beta \in A)$ it follows that Tg_α is *bo*-fundamental. Thus, we may define $T_0 f = bo\text{-}\lim Tg_\alpha$. Let SC_b denote the cone of all bounded lower semicontinuous functions on P. Then T can be extended to a dominated operator $T_0 : M_0 \to Y$ with $M_0 = SC_b^\uparrow - SC_b^\uparrow$. Further extension goes on by transfinite induction up to the least uncountable ordinal ω_1. This extension preserves the inequality $|T_0 f| \leq W(|f|)$ $(f \in M_0)$. Suppose that for every ordinal $\beta < \alpha < \omega_1$ we have defined a vector sublattice $M_\beta \subset \mathscr{M}_b(P)$ and a linear operator $T_\beta : M_\beta \to Y$, such that the following relations are valid:

$$|T_\beta f| \leq W(|f|) \ \ (f \in M_\beta); \quad M_\beta \subset M_\gamma, \ \ T_\gamma|_{M_\beta} = T_\beta, \quad \beta < \gamma < \alpha.$$

If α is a limit ordinal then we put $M_\alpha := \bigcup\{M_\beta : \beta < \alpha\}$ and define a linear operator $T_\alpha : M_\alpha \to Y$ by the relation $T_\alpha|_{M_\beta} = T_\beta$ $(\beta < \alpha)$. If α is a nonlimit ordinal then we consider the set $M_{\alpha-1}^\uparrow$ (of the least upper bounds of countable bounded subsets of $M_{\alpha-1}$). Take an increasing sequence $(f_n)_{n\in\mathbb{N}}$ in $M_{\alpha-1}$ with $\sup_n f_n = f \in M_{\alpha-1}^\uparrow$. Using the above reasoning we may easily check that $(T_{\alpha-1} f_n)_{n\in\mathbb{N}}$ is a *bo*-fundamental sequence. Hence we may define $T_{\alpha-1}^\uparrow f := bo\text{-}\lim T_{\alpha-1} f_n$. By doing so we obtain an operator $T_{\alpha-1}^\uparrow : M_{\alpha-1}^\uparrow \to Y$ satisfying the inequality $|T_{\alpha-1}^\uparrow f| \leq Wf$ $(0 \leq f \in M_{\alpha-1}^\uparrow)$. Soundness of this definition follows from sequential order continuity of W. Let $M_\alpha := M_{\alpha-1}^\uparrow - M_{\alpha-1}^\uparrow$ and let $T_\alpha : M_\alpha \to Y$ be the extension of $T_{\alpha-1}^\uparrow$ by differences. It is easy to see that $\mathscr{M}_b(Q) = M_{\omega_1}$ and the operator $T_1 := T_{\omega_1}$ is a sequentially order continuous extension of T to the space $\mathscr{M}_b(Q)$. Moreover, $|T_1 f| \leq W(|f|)$ $(f \in \mathscr{M}_b(Q))$.

Now it is clear that if we define a measure μ by letting $\mu(C) := T_1(\chi_C)$ $\big(C \in \mathscr{B}or(Q)\big)$ then the integral representation 6.2.6 (1) holds. It should be proven that μ is a quasi-Radon measure. For every $U \in \mathscr{T}$ the characteristic function χ_U is lower semicontinuous and by the above construction

$$|\mu|(U) = |T|(\chi_U) = \bigvee\{|T|f : f \in L,\ 0 \le f \le \chi_U\}.$$

Fix $0 < \varepsilon \in \mathbb{R}$, $f \le \chi_U$, $f \in L$, and put $D := \{x \in Q : f(x) \ge \varepsilon\}$. Then

$$f \le \chi_D + \varepsilon\mathbf{1}, \quad |T|f \le |\mu|(D) + \varepsilon|\mu|(U).$$

This amounts to the following quasiregularity condition

$$|\mu|(U) = \bigvee\{|\mu|(D) : D \in \mathscr{F},\ D \subset U\}.$$

In a similar way, using 6.2.6 (2), we may deduce the relation $|\mu|(Q) = \bigvee\{|\mu|(K) : K \in \mathscr{K}\}$. By virtue of 6.2.4 (3) $|\mu|$ and μ are quasi-Radon measures.

Conversely, suppose that T admits the integral representation 6.2.6 (1) with a quasi-Radon $\mu \in \mathrm{dca}(\mathscr{B}or(Q), Y)$. If $K \in \mathscr{K}$ then

$$|\mu|(Q\backslash K) = \bigvee\{|\mu|(K') : K' \in \mathscr{K}, K' \subset U\backslash K\}.$$

Since for every $K' \subset U\backslash K$ there is a function $f \in L$ such that $f(K') = \{0\}$, $f(K) = \{1\}$, and $0 \le f \le 1$, we have $|\mu|(K) = \bigvee\{|T|f : f \in L,\ f \ge \chi_K\}$ and the desired identity 6.2.6 (2) holds. The proof is complete. $\triangleright$

6.2.7. Observe two corollaries to Theorem 6.2.7. Denote by $\mathrm{dqa}(\mathscr{A}, F)$ the space of countably additive quasiregular F-valued measures on $\mathscr{A}$.

(1) *Theorem.* *Let Q be an arbitrary compact space, and let F be a K-space. Then for every order-bounded linear operator $T : C(Q) \to F$ there is a unique order-bounded countably additive quasiregular measure $\mu : \mathscr{B}or(Q) \to F$ such that*

$$T(f) = \int\limits_Q f\,d\mu \quad (f \in C(Q)).$$

The correspondence $T \mapsto \mu$ is a linear and lattice isomorphism between the vector lattices $L^{\sim}(C(Q), F)$ and $\mathrm{dqa}(\mathscr{B}or(Q), F)$.

$\triangleleft$ Under the stated conditions, 6.2.2 (2) holds automatically; we may put $K := Q$, because Q is compact. $\triangleright$

Now we give a corollary concerning the measure extension problem. Let $C(\mathscr{A})$ denote the uniform closure of $S(\mathscr{A})$, and let $\mathscr{T}_0$ stand for the functionally open sets. An algebra $\mathscr{A} \subset \mathscr{P}(Q)$ is called *tight,* if the following conditions are met: (a) the vector lattice $L = C(\mathscr{A}) \cap C_b(Q)$ generates the topology $\mathscr{T}$; (b) for every $V \in \mathscr{A} \cap \mathscr{T}_0$ there is a function $\varphi \in C(\mathscr{A}) \cap C_b(Q)$ with $V = \{x \in Q : \varphi(x) > 0\}$.

(2) Theorem. *Assume that a quasi-Radon measure $\mu_0 \in \mathrm{da}(\mathscr{A}, Y)$ is defined on a tight algebra $\mathscr{A}$ with $\mathscr{A} = \sigma_0(\mathscr{F}_0 \cap \mathscr{A})$. Then there exists a unique quasi-Radon measure $\mu \in \mathrm{dca}(\mathscr{B}or(Q), Y)$ extending μ_0.*

$\lhd$ Consider the dominated operator $T : L \to Y$ defined on the vector lattice $L = C(\mathscr{A}) \cap C_b(Q)$ by $Tf = \int f\, d\mu_0$ $(f \in L)$. The operator $|T|$ meets 6.2.2 (2), since L separates compact subsets in Q. Thus, there is a unique quasi-Radon measure $\mu \in \mathrm{dca}(\mathscr{B}or(Q), Y)$ such that the integral representation 6.2.2 (1) holds. To ensure that μ is an extension of μ_0 it is sufficient to justify the identity $\mu_0(U) = \mu(U)$ provided that $U \in \mathscr{T}_0 \cap \mathscr{A}$. Since $\mathscr{A}$ is tight, there is a function $\varphi \in L$ with $U = \{x \in Q : \varphi(x) > 0\}$. Put $\varphi_n = (n\varphi) \wedge \mathbf{1}$ $(n \in \mathbb{N})$ and observe that $\varphi_n \nearrow \chi_U$. Using σ-additivity of μ and μ_0, we obtain $\mu_0(U) = bo\text{-}\lim \int \varphi_n\, d\mu_0 = bo\text{-}\lim \int \varphi_n\, d\mu = \mu(U)$, which completes the proof. $\rhd$

(3) Let Q be a σ-compact topological space and let $\mathscr{A} \subset \mathscr{P}(Q)$ be an algebra with $\mathscr{A} = \sigma_0(\mathscr{F}_0 \cap \mathscr{A})$. Then every measure $\mu_0 \in \mathrm{dca}(\mathscr{A}, Y)$ has a unique extension to a quasi-Radon measure $\mu \in \mathrm{dca}(\mathscr{B}or(Q), Y)$.

6.3. Functional Representation of Maharam's Extension

The Maharam extension of a positive operator presented in Section 3.5 result in a vector lattice $L^1(\Phi)$ on which the operator is well behaved. But the space $L^1(\Phi)$ has a complicated structure which is troublesome in applications. In this section we give an explicit description for $L^1(\Phi)$ in terms of measurable functions.

6.3.1. Let P be a σ-compact topological space and let $E_0 := C_0(P)$ be the vector lattice of compactly-supported continuous functions on P, i.e., $C_0(P) := \{f \in C(P, \mathbb{R}) : \mathrm{supp}(f) \text{ is a compact space}\}$. In this section A is a nonempty set, $\mathscr{A}$ is a σ-algebra of its subsets, and $\mathscr{N}$ is a σ-ideal in $\mathscr{A}$. Let $M(A, \mathscr{A}, \mathscr{N})$ be the space of cosets of measurable functions on A as defined in 1.4.7 (1). We will suppose that the measurable space $(\mathscr{A}, \mathscr{N})$ is of *countable type;* i.e., an arbitrary family $(\mathscr{A}_\alpha) \subset \mathscr{A} \backslash \mathscr{N}$ with $A_\alpha \cap A_\beta \in \mathscr{N}$ $(\alpha \neq \beta)$ is at most countable. In this event $M(A, \mathscr{A}, \mathscr{N})$ is an order complete vector lattice. In this section F is an order-dense ideal in $M(A, \mathscr{A}, \mathscr{N})$. A sequence $(A_n) \subset \mathscr{A}$ of pairwise disjoint sets is called a *partition of a measurable set* $A_0 \in \mathscr{A}$ if $\chi_{A_0} = \sup \chi_{A_n}$ in F, where χ_C always stands for the characteristic function of C.

Denote by V the vector lattice of functions of two variables $f : A \times P \to \mathbb{R}$ representable as $v(s,t) = o\text{-}\sum_{n=1}^{\infty} \chi_{A_n}(s) e_n(t)$, where (A_n) is a partition of A, (e_n) is an order-bounded sequence in E_0, and the infinite sum is understood for every $s \in P$ in the sense of order convergence in the vector lattice $M(A, \mathscr{A}, \mathscr{A}_0)$, i.e. almost everywhere on A. Extend the operator $\Phi : E_0 \to F$ to the vector lattice V by letting $\Phi v = \sum_{n=1}^{\infty} \chi_{A_n} \Phi e_n$. This definition is sound. Indeed, if $v(s,t) =$

$\sum_{n=1}^{\infty} \chi_{B_n}(s)d_n(t)$ for another partition (B_n) of A and sequence $(d_n) \subset E_0$, then the representation $v(s,t) = \sum_{n,k=1}^{\infty} \chi_{C_{n,k}}(s)g_{n,k}(t)$ is also true with $C_{n,k} := A_n \cap B_k$ and $g_{n,k} := e_n = d_k$, whenever $C_{n,k} \neq \varnothing$. Therefore,

$$\begin{aligned} o\text{-}\sum_{n=1}^{\infty} \chi_{A_n} \Phi e_n &= o\text{-}\sum_{n=1}^{\infty} o\text{-}\sum_{k=1}^{\infty} \chi_{C_{n,k}} \Phi g_{n,k} \\ &= o\text{-}\sum_{k=1}^{\infty} o\text{-}\sum_{n=1}^{\infty} \chi_{C_{n,k}} \Phi g_{n,k} = o\text{-}\sum_{k=1}^{\infty} \chi_{B_k} \Phi e_k. \end{aligned}$$

6.3.2. *If a sequence (v_n) in the vector lattice V decreases and $(v_n(s,t))$ converges to zero for all $(s,t) \in A \times P$, then $\inf \Phi v_n = 0$.*

$\lhd$ Suppose that a sequence (v_n) meets the hypotheses, and still $\inf \Phi v_n > \varepsilon \chi_A$ for some $0 < \varepsilon \in \mathbb{R}$ and $C \in \mathscr{A} \setminus \mathscr{N}$. Let $v_n = \sum_{k=1}^{\infty} \chi_{A_{n,k}} e_{n,k}$. Consider the sequence $v_n' := \chi_C v_n$ and a function $e \in (E_0)_+$, with $e_{1,k} \leq e$ for all $k \in \mathbb{N}$. Evidently, each function v_n' can be represented as $v_n' = \sum_{k=1}^{\infty} \chi_{C_{n,k}} e_{n,k} + \chi_{A_n} \varphi_n$, where for every $n \in \mathbb{N}$ the sequence $(C_{n,k})_{k \in \mathbb{N}} \subset \mathscr{A} \setminus \mathscr{N}$ is a partition of C, the function $\varphi_n : A \times P \to \mathbb{R}$ is bounded above by $\chi_A \otimes e$, and $C_n := C \setminus \bigcup_{k=1}^{\infty} C_{n,k} \in \mathscr{N}$. Moreover, the partitions of C can be chosen so that $(C_{n+1,k})_{k=1}^{\infty}$ is a refinement of $(C_{n,k})_{k=1}^{\infty}$. Take any element $s \in C \setminus \cup C_n$. There exist a decreasing sequence $C_{n,k_n(s)}$ such that $s \in \bigcap_{n=1}^{\infty} A_{n,k_n(s)}$. Then for the sequence $(e_{n,k_n(s)})$ in $C_0(P)$ we have

$$e_{n,k_n(s)}(t) = v_n(s,t) \geq v_{n+1}(s,t) = e_{n+1,k_{n+1}(s)}(t) \quad (t \in P).$$

Since $\lim_n v_n(s,t) = 0$, it follows that $(e_{n,k_n(s)})$ vanishes uniformly on the compact set $\operatorname{supp}(e)$. Thus, for every $m \in \mathbb{N}$ there is $n \in \mathbb{N}$ such that

$$\chi_{C_{n,k_n(s)}} e_{n,k_n(s)} \leq (\varepsilon/m) \cdot e \wedge \mathbf{1} \leq (\varepsilon/m) g,$$

where g is a positive compactly-supported function coinciding with unity on $\operatorname{supp}(e)$. From this we deduce

$$\begin{aligned} \varepsilon \chi_{C_{n,k_n(s)}} &= \chi_{C_{n,k_n(s)}}(\varepsilon \chi_C) \leq \varepsilon \chi_{C_{n,k_n(s)}} \Phi v_n \\ &= \varepsilon \chi_{C_{n,k_n(s)}} \Phi v_n' = \varepsilon \chi_{C_{n,k_n(s)}} \Phi e_{n,k_n(s)} \leq (\varepsilon/m) \Phi g, \end{aligned}$$

whence $\varepsilon \leq (\varepsilon/m)(\Psi g)(s) \to 0$. This is a contradiction, and the claim follows. $\rhd$

6.3.3. A set $D \subset A \times P$ is said to be *negligible* or, more precisely, *Φ-negligible* if for every $0 < \varepsilon \in \mathbb{R}$ there exist an increasing sequence of positive functions (v_n) in V such that $\sup_n v_n(s,t) \geq 1$ $((s,t) \in D)$ and $\Phi v_n \leq \varepsilon \mathbf{1}_A$ almost everywhere on A.

(1) *The countable union of negligible sets is negligible.*

◁ Suppose that D is the union of some sequence of negligible sets $D_n \subset A \times P$. Take $0 < \varepsilon \in \mathbb{R}$, $n \in \mathbb{N}$ and let $(v_{n,k})_{k\in\mathbb{N}}$ be an increasing sequence in V_+ such that $\sup_k (v_{n,k}(s,t) \geq 1$ for all $(s,t) \in D_n$ and $\Phi(v_{n,k}) \leq \varepsilon/2^n$. If $v_k := v_{1,k} \vee \dots \vee v_{k,k}$ then $\sup_{k\in\mathbb{N}} v_k(s,t) \geq 1$ for all $(s,t) \in D$ and

$$\Phi(v_k) \leq \sum_{l=1}^{k} v_{l,k} \leq \sum_{l=1}^{k} \frac{\varepsilon}{2^l} = \varepsilon \mathbf{1}_A,$$

whence the claim is clear. ▷

(2) *If a decreasing sequence $(v_n) \subset V$ vanishes almost everywhere on $A \times P$ then $\inf \Phi v_n = 0$.*

◁ Since v_1 is a bounded function, $\beta := \sup\{v_1(s,t) : (s,t) \in A \times P\} < \infty$. If $\beta = 0$ then there is nothing to prove, so that we assume $\beta > 0$. Let $D \subset A \times P$ be the negligible set on which the sequence (v_n) does not vanish. For a fixed $\varepsilon > 0$ choose an increasing sequence (w_n) in V such that $\sup_n w_n(s,t) \geq 1$ for $(s,t) \in D$ and $\Phi(w_n) \leq \varepsilon/\beta$ $(n \in \mathbb{N})$. The sequence $(u_n)_{n\in\mathbb{N}}$ with $u_n := (v_n - \beta w_n)^+$ is decreasing and $\lim_n u_n = 0$ everywhere on $A \times P$. By 6.3.2 $o\text{-}\lim_n \Phi(u_n) = 0$, whence

$$o\text{-}\lim_{n\to\infty} \Phi(v_n) - \beta o\text{-}\lim_{n\to\infty} \Phi(w_n) = o\text{-}\lim_{n\to\infty} \Phi(v_n - \beta w_n) \leq o\text{-}\lim_{n\to\infty} \Phi(u_n) = 0,$$

so that

$$0 \leq o\text{-}\lim_{n\to\infty} \Phi(v_n) \leq \beta o\text{-}\lim_{n\to\infty} \Phi(w_n) \leq \beta \frac{\varepsilon}{\beta} = \varepsilon. \ ▷$$

6.3.4. Now we will extend the "preintegral" $\Phi : V \to F$ to some larger function space. Unfortunately, we cannot apply the Extension Theorem 4.5.3 since F is not weakly σ-distributive. Nevertheless, the Daniell construction of the Lebesgue integral may be successfully carried out due to the specific properties of our "preintegral" Φ. Consider the set $V^\uparrow$ comprising the extended real-valued functions $f : A \times P \to \mathbb{R} \cup \{\pm\infty\}$ that are representable as almost everywhere limits of increasing sequences $(v_n) \subset V$ with $\sup_n \Phi(v_n) \in F$.

(1) *Every function $f \in V^\uparrow$ takes finite values almost everywhere.*

◁ Indeed, take a sequence $(v_n) \subset V_+$ with $v_n \nearrow f$ and $\sup \Phi v_n =: g \in F$. Put $D := \{f = +\infty\} := \{(s,t) \in A \times P : f(s,t) = +\infty\}$, $C_0 := \{g = +\infty\}$, and $C_m := \{m - 1 \leq g < m\} := \{s \in A : m - 1 \leq g(s) < m\}$ $(m \in \mathbb{N})$. Clearly, $\{f = +\infty\} \subset A \times P$, $C_m \in \mathscr{A}$ $(m = 0, 1, \dots)$ and $(C_m)_{m\in\mathbb{N}}$ is a partition of A. It suffices to prove that $D_k := D \cap (C_k \times P)$ is a negligible set for every $k \in \mathbb{N}$. Take an arbitrary $\varepsilon > 0$ and put $w_n = \frac{\varepsilon}{k} \chi_{C_k} v_n$. It is easy to see that

$w_n \in V$ and $\Phi(w_n) \le \frac{\varepsilon}{k}\chi_{C_k} g \le \varepsilon\mathbf{1}_A$. Moreover, for every $d = (s,t) \in D_k$ there is a number $n \in \mathbb{N}$, such that $v_n(d) > \frac{k}{\varepsilon}$ (since $v_n(d) \nearrow +\infty$). From this we deduce $w_n(d) = \frac{\varepsilon}{k}\chi_{C_k}(t)v_n(d) \ge \mathbf{1}$, so that $\sup w_n(d) \ge 1$ for each $d \in D_k$. ▷

For every $f \in V^{\uparrow}$ define $\Phi f := \sup \Phi v_n$, where $(v_n) \subset V$ and $v_n \nearrow f$. This soundly defines some mapping $V^{\uparrow} \to F$ that we will denote by the same symbol. It is an easy matter to verify that $V^{\uparrow}$ is closed under lattice operations, addition, and multiplication by positive scalars. Moreover the following is true:

(2) *If the sequence $(f_n) \subset V^{\uparrow}$ increases almost everywhere to some $f : A \times P \to \mathbb{R} \cup \{\pm\infty\}$ and $\Phi(f_n)$ is bounded above in F then $f \in V^{\uparrow}$ and $\lim_n \Phi(f_n) = \Phi(f)$.*

◁ If for every $n \in \mathbb{N}$ the sequence $(v_{n,k})_{k\in\mathbb{N}}$ in V increases and converges almost everywhere to f_n, then under the stated conditions the sequence $(v_n)_{n\in\mathbb{N}}$ of the functions $v_n := v_{1,n} \vee \dots \vee v_{n,n}$ converges almost everywhere to f and $\Phi(v_n)$ is bounded above, since $v_n \le f_n$. Thus, $f \in V^{\uparrow}$. Since $\Phi(v_n) \le \Phi(f_n) \le \Phi(f)$, we have $\Phi(f) = \sup_n \Phi(f_n)$. ▷

6.3.5. A function $f = A \times P \to \mathbb{R} \cup \{\pm\infty\}$ is said to be Φ-*summable* or Φ-*integrable* if

$$\Phi^* f := \inf\{\Phi h : h \ge f,\, h \in V^{\uparrow}\} = \sup\{-\Phi g : -g \le f,\, g \in V^{\uparrow}\} =: \Phi_* f.$$

In this event we set $\widehat{\Phi} f := \Phi^* f = \Phi_* f$. Denote by $\mathscr{L}^1(A \times P, \Phi)$ the set of all Φ-summable functions. Evidently, $\mathscr{L}^1(A \times P, \Phi) \supset V^{\uparrow} \supset V$.

A function $f = A \times P \to \mathbb{R} \cup \{\pm\infty\}$ is Φ-summable if and only if for every $0 < \varepsilon \in \mathbb{R}$ there exist $h \in V^{\uparrow}$ and $g \in -V^{\uparrow}$ such that $g \le f \le h$ and $\Phi(h-g) \le \varepsilon\mathbf{1}_A$.

◁ Since F has the countable sup property, we may find a decreasing sequence $(h_n) \subset V^{\uparrow}$ and an increasing sequence $(g_n) \subset -V^{\uparrow}$ with $g_n \le f \le h_n$ $(n \in \mathbb{N})$ and $\Phi f = \inf \Phi h_n = \sup \Phi g_n$. The last identities imply that there exist a partition $(C_n) \subset \mathscr{A}$ of A such that $\chi_{C_n}(\Phi h_n - \Phi f) \le (\varepsilon/2)\mathbf{1}_A$ and $\chi_{C_n}(\Phi f - \Phi g_n) \le (\varepsilon/2)\mathbf{1}_A$. Put

$$h = \sum_{n=1}^{\infty} \chi_{C_n} h_n, \quad g = \sum_{n=1}^{\infty} \chi_{C_n} g_n.$$

Clearly, $-g, h \in V^{\uparrow}$, $g \le f \le h$, and $\Phi h - \Phi g \le \varepsilon\mathbf{1}_A$. ▷

Using this proposition, it is easy to verify that $\mathscr{L}^1(\Phi)$ is an order σ-complete vector lattice and $\widehat{\Phi} : \mathscr{L}^1(A \times P, \Phi) \to F$ is an order σ-continuous operator. Moreover, if $(f_n) \subset \mathscr{L}^1(A \times P, \Phi)$ is an increasing sequence with $\sup \Phi f_n \in F$ then $\sup f_n \in \mathscr{L}^1(A \times P, \Phi)$. A stronger assertion will be proved in the next subsection.

Introduce the equivalence relation $f_1 \sim f_2 \Leftrightarrow \widehat{\Phi}(f_1 - f_2) = 0$ in $\mathscr{L}^1(A \times P, \Phi)$ and denote by $L^1(A \times P, \Phi)$ the factor vector lattice $\mathscr{L}^1(A \times P, \Phi)/\sim$. Let $\jmath$ be the natural embedding of E_0 into $\mathscr{L}^1(A \times P, \Phi)$; more precisely, $\jmath e$ is the coset of the function $\mathbf{1}_A \otimes e : (s,t) \mapsto e(t)\ (s,t) \in A \times P$. Denote by ϕ the factor mapping which sends every Φ-summable function f to its coset $\tilde{f}$. There is a unique operator $\Psi : L^1(A \times P, \Phi) \to F$ satisfying $\widehat{\Phi} = \Psi \circ \phi$. We will also denote $\widehat{\Phi} := \Psi$.

Recall that (see 3.5.2, 3.5.3) there exist an order complete Banach–Kantorovich lattice $L^1(\Phi)$, an essentially positive Maharam operator $\widetilde{\Phi} : L^1(\Phi) \to F$, and an isomorphic embedding $\imath : E_0 \to L^1(\Phi)$ such that $\iota(X)^{\perp\perp} = L^1(\Phi)$ and $\widetilde{\Phi} \circ \imath = \Phi$. Moreover, $L^1(\Phi)$ admits the structure of a lattice-ordered module over $\Lambda := \mathrm{Orth}(F)$ such that $\widetilde{\Phi}$ is a module homomorphism, i.e. a Λ-linear operator.

6.3.6. Theorem. *The following hold:*

(1) *$L^1(A \times P, \Phi)$ is an order complete vector lattice;*

(2) *$\widehat{\Phi} : L^1(A \times P, \Phi) \to F$ is an essentially positive Maharam operator;*

(3) *there exists a unique lattice isomorphism β from $L^1(A \times P, \Phi)$ onto $L^1(\Phi)$ such that $\beta \circ \jmath = \imath$ and $\widetilde{\Phi} \circ \beta = \widehat{\Phi}$.*

$\lhd$ It suffices to verify (3). Take an arbitrary $v \in V$, $v = \sum_{k=1}^{\infty} \chi_{C_n} e_n$, where $(C_n) \subset \mathscr{A}$ is a partition of A and $(e_n) \subset E_0$ is an order-bounded sequence. Let π_n be the projection in F defined as multiplication by the characteristic function χ_{C_n}. Put $\tilde{\beta}(v) := \sum_{k=1}^{\infty} \pi_k \imath(e_k)$. If $(v_n) \subset V$, $v_n \nearrow h$, and $\sup_n \Phi(v_n) \in F$, then we set $\tilde{\beta}(h) := \sup_n \tilde{\beta}(v_n)$. Finally, if $(h_n) \subset V^{\uparrow}$ and $h_n \searrow f$, $f \in \mathscr{L}^1(A \times P, \Phi)$, then we define $\tilde{\beta}(f) := \inf_n \tilde{\beta}(h_n)$. Observe that $\tilde{\beta}$ is a lattice homomorphism which maps the lattice V onto $\mathscr{M}_0$, the cone $V^{\uparrow}$ onto $\mathscr{M}_0^{\uparrow}$, and the vector lattice $\mathscr{L}^1(A \times P, \Phi)$ onto the K-space $L^1(\Phi) = \mathscr{M}_0^{\uparrow\downarrow}$. It follows from the definitions of $\widetilde{\Phi}$ and $\widehat{\Phi}$ that $\widetilde{\Phi} \circ \tilde{\beta} = \widehat{\Phi}$. Consequently, if $f_1 \sim f_2$ then $0 = \widehat{\Phi}(|f_1 - f_2|) = \widetilde{\Phi}(\tilde{\beta}|f_1 - f_2|)$; therefore, $\tilde{\beta}(|f_1 - f_2|) = 0$. Thus, we may define β on $L^1(A \times P, \Phi)$ by the equation $\beta(\tilde{f}) = \tilde{\beta}(f)$, where $\tilde{f}$ is the coset of a function f. Now, it is clear that $\widetilde{\Phi} \circ \beta = \widehat{\Phi}$ and β is the desired isomorphism.

It remains to establish uniqueness of β. Suppose that there is an isomorphism $\beta' : L^1(A \times P, \Phi) \to L^1(\Phi)$ with $\beta' \circ \jmath = \imath$ and $\widehat{\Phi} = \widetilde{\Phi} \circ \beta'$. In order to verify the identity $\beta = \beta'$ it suffices to check that $\beta'(\chi_C e) = \pi_C \imath(e)$, where $e \in (E_0)_+$, $C \in \mathscr{A}$, and $\pi_C \in \mathfrak{B}(F)$ is the projection corresponding to χ_C. Observe that $\widetilde{\Phi}(\pi_C^{\perp} \beta'(\chi_C e)) = \pi_C^{\perp} \widetilde{\Phi} \beta'(\chi_C e) = \chi_{A \setminus C} \chi_C \Phi e = 0$. Hence, $\beta'(\chi_C e) = \pi_C \beta'(\chi_C e) \leq \pi_C \beta'(e) = \pi_C \imath(e)$. At the same time $\widetilde{\Phi} \beta'(\chi_C e) = \chi_C \Phi e = \pi_C \Phi e = \widetilde{\Phi}(\pi_C \imath(e))$. From this we derive $\beta'(\chi_C e) = \pi_C \imath(e)$, since $\widetilde{\Phi}$ is essentially positive. $\rhd$

6.3.7. A set $D \subset A \times P$ is called *Φ-summable* if $\chi_D \in L^1(A \times P, \Phi)$. A set is *$\Phi$-measurable* if it is representable as the union of a countable family of Φ-summable sets.

(1) *If $B \subset P$ is a Baire set and $C \in \mathscr{A}$ then the rectangle $D := B \times C$ is Φ-measurable.*

$\lhd$ First we observe that every rectangle of the form $D := A \times \{e > 1\}$ with $e \in (E_0)_+$ is Φ-summable. Indeed, $\chi_D = \chi_A \otimes \sup_n \big(e \wedge (1+1/n)\mathbf{1}_P - e\wedge \mathbf{1}\big)/n \in V^{\uparrow}$. Now, the standard measure theoretic arguments show that $A \times B$ is measurable for every Baire set $B \subset P$. Since P is a σ-compact space, the rectangle $C \times P$ is also measurable for all $C \in \mathscr{A}$. It remains to note that $C \times B = (C \times P) \cap (A \times B)$. $\rhd$

Denote by $\mathscr{A} \otimes \mathscr{B}$ the σ-algebra generated by the rectangles $C \times B$ where $B \subset P$ is an arbitrary Baire set and $C \in \mathscr{A}$. It follows from (1) that $\mathscr{A} \otimes \mathscr{B}$ consists of measurable sets.

(2) *For every Φ-measurable set $D \subset A \times P$ there exists a set $D' \in \mathscr{A} \otimes \mathscr{B}$ such that the symmetric difference $D \triangle D'$ is negligible.*

$\lhd$ Denote by $\mathscr{D}$ the subalgebra in the Boolean algebra of unit elements $\mathfrak{E}\big(L^1(\widetilde{\Phi})\big)$ which is generated by cosets of the form $\phi(B \times C)$. Take $x \in C_0(P)$ and let $C := \{p \in P : x(p) > 0\}$. A coset $\phi(v) \in L^1(\widetilde{\Phi})$ is disjoint from $\jmath(x)$ if and only if v may be chosen disjoint from the function $(s,t) \mapsto x(s)$ $(t \in P)$. Therefore, $\jmath(x)$ and $\phi(v)$ are disjoint if and only if so are $\phi(\chi_C)$ and $\phi(v)$. This means that $\jmath(x)$ and $\phi(\chi_C)$ generates the same band in $L^1(\widetilde{\Phi})$.

Now, it follows from 3.5.10 (1) that $\mathfrak{E}\big(L^1(\widetilde{\Phi})\big) = \mathscr{D}^{\downarrow\uparrow}$. Since the vector lattice F has the countable chain property, the formula $\mathfrak{E}\big(L^1(\widetilde{\Phi})\big) = \mathscr{D}^{\downarrow\uparrow}$ is also true. Taking into account the identity $\phi(\mathscr{A} \otimes \mathscr{B}) = \mathfrak{E}\big(L^1(\widetilde{\Phi})\big)$, we conclude that there exists some $D' \in \mathscr{A} \otimes \mathscr{B}$ with $\phi(D) = \phi(D')$, or $\phi(D \triangle D') = 0$. Hence, $\hat{\Phi} \circ \phi(\chi_{D\triangle D'}) = 0$, whence $D \triangle D'$ is negligible. $\rhd$

6.3.8. Denote by $\mathscr{L}^0(A \times P, \Phi)$ the space of almost everywhere finite measurable functions on $A \times P$ with respect to the σ-algebra $\mathscr{A} \otimes \mathscr{B}$. Introduce the factor space $L^0(A \times P, \Phi) := \mathscr{L}^0(A \times P, \Phi)/\sim$, where $\sim$ is the equivalence relation in 6.3.5. Clearly, $\mathscr{L}^0(A \times P, \Phi)$ is a K_σ-space. Moreover, $L^1(A \times P, \Phi)$ is an order-dense ideal in $L^0(A \times P, \Phi)$ and the latter is the universal completion of the former.

Theorem. *For any operator $S \in \{\Phi\}^{\perp\perp}$ there exists a unique (up to equivalence) Φ-summable function $K_S \in \mathscr{L}^1(A \times P, \Phi)$ such that $Se = \widehat{\Phi}(\phi(K_S)\jmath e)$ $(e \in E_0)$. Moreover, the correspondence $S \mapsto K_S$ induces a linear and order isomorphism from $\{\Phi\}^{\perp\perp}$ onto $L^1(A \times P, \Phi)$.*

$\lhd$ Observe that for an arbitrary $u \in L^0(A \times P, \Phi)$ we have

$$u \cdot \jmath\big(C_0(P)\big) \subset L^1(A \times P, \Phi) \Leftrightarrow u \in L^1(A \times P, \Phi)$$

Then the claim follows from 6.3.6, 3.4.5, and 3.5.5 (1). $\rhd$

6.3.9. Assume now that P is a compact space, so that a weak order-unity $\mathbf{1} := \phi(1_{A\times P})$ belongs to $L^1(A \times P, \Phi)$. Then we are able to define an F-valued measure $\varphi : \mathscr{A} \otimes \mathscr{B} \to F$ by letting $\varphi(D) := \widehat{\Phi}(D)$. According to 6.1.7 φ is a unique σ-additive measure with $\mathscr{L}^1(A \times P, \Phi) = \mathscr{L}^1(\varphi)$ and

$$\widehat{\Phi}(\phi f) = \int\limits_{A\times P} f\, d\varphi \quad (f \in \mathscr{L}^1(\varphi)).$$

If π is a band projection in F then there is $C \in \mathscr{A}$ with $\pi\tilde{f} = \widetilde{\chi_C f}$, $\tilde{f} \in F$. Define some band projection $h(\pi)$ in $L^1(\varphi)$ as follows: if $g \in \mathscr{L}^1(\varphi)$ then $h(\pi)\tilde{g}$ is the coset of the function $(s,t) \mapsto \chi_C(s)g(s,t)$. It is clear that h is a Boolean homomorphism from $\mathfrak{P}(F)$ into $\mathfrak{P}(L^1(\varphi))$.

(1) *Under the above assumptions φ is ample with respect to h.*

$\triangleleft$ Follows from 6.3.6 and 6.1.10. $\triangleright$

Summing up (1) and 6.3.6, we obtain the following:

(2) *Let $\Phi : C(P) \to F$ be a positive operator. Then there exists a unique ample measure $\varphi : \mathscr{A} \otimes \mathscr{B} \to F$ such that*

$$\Phi g = \iint\limits_{A\times P} 1_A \otimes g\, d\varphi \quad (g \in C(P)).$$

(3) *For every order-bounded operator $S \in \{\Phi\}^{\perp\perp}$ there exists a unique (up to φ-equivalence) φ-integrable function $K_S \in \mathscr{L}^1(\varphi)$ such that*

$$Sg = \iint\limits_{A\times P} K_S 1_A \otimes g\, d\varphi \quad (g \in C(P)).$$

(4) *The correspondence $S \mapsto \phi(K_S)$ is a linear and lattice isomorphism from $\{\Phi\}^{\perp\perp}$ onto $L^1(\varphi)$.*

6.3.10. Let μ be a regular Borel measure on a compact space P, and let $L^0(P,\mu)$ be the vector lattice of cosets of real μ-measurable functions on P. Assume that E is an order-dense ideal in $L^0(P,\mu)$ containing the identically one function 1_P.

(1) *If Φ admits an order continuous extension to E then every μ-negligible set is Φ-negligible.*

$\triangleleft$ Take a μ-negligible set $Z \subset P$. For every $k \in \mathbb{N}$ choose an increasing sequence $(f_{n,k})_{n\in\mathbb{N}}$ in $C(P)_+$ such that $\int_P f_{n,k}\, d\mu \leq 1/k$ $(n,k \in \mathbb{N})$ and $g_k(t) \geq 1$ $(t \in Z)$, where $g_k(t) = \sup_n f_{n,k}(t)$ $(t \in P)$. Substituting, if necessary, $f_{n,k} \wedge f_{n,k+1}$ for $f_{n,k+1}$ we may assume that $f_{n,k} \geq f_{n,k+1}$. By the Monotone Convergence Theorem

$$\int_P g_k \, d\mu = \lim_{n\to\infty} \int_P f_{n,k} \leq \frac{1}{k}.$$

Thus $(g_k)_{k\in\mathbb{N}}$ is a decreasing sequence of μ-summable functions and $\int_P g_k \, d\mu \to 0$ as $k \to \infty$, so that (g_k) converges to zero almost everywhere. Since Φ is order continuous on E and almost everywhere convergence in E is order convergence, we have $\inf_{k\in\mathbb{N}} \Phi(g_k) = 0$. From this we deduce that for every $\varepsilon > 0$ there is a countable partition $(C_k)_{k\in\mathbb{N}}$ of the set A such that $\Phi(g_k)(s) \leq 1$ for all $s \in C_k$. Now define $v_n \in V$ by

$$v_n := \sum_{k=1}^{\infty} \chi_{C_k} \otimes f_{n,k}, \quad \chi_{C_k} \otimes f_{n,k} : (s,t) \mapsto f_{n,k}(t)\chi_{C_k}(s) \quad (t \in P,\ s \in A).$$

Using the definition in 6.3.1 we infer

$$\Phi(v_n) = \sum_{k=1}^{\infty} \chi_{C_k} \Phi(f_{n,k}) \leq \sum_{k=1}^{\infty} \chi_{C_k} \Phi(g_k) \leq \varepsilon 1_A.$$

If $s \in C_m$ then $\sup_n v_n(s,t) = \sup_n f_{n,m}(t) = g_m(t) \geq 1$ for $t \in Z$. Since every s is contained in some C_k, we obtain that $\sup_n v_n(s,t) \geq 1$ for all $t \in Z$ and $s \in A$. $\triangleright$

(2) *If Φ admits an order continuous extension to E then the function $\bar{g} := 1_A \otimes g : (s,t) \mapsto g(t)1_P(s) = g(t)$ is Φ-negligible for every μ-negligible function g.*

$\triangleleft$ If g is μ-negligible then there is a decreasing sequence (g_n) of lower semicontinuous functions with $|g| \leq g_n \leq 1_P$ $(n \in \mathbb{N})$ and $\int_P g_n \, d\mu \xrightarrow[n\to\infty]{} 0$. Hence, (g_n) converges to zero μ-almost everywhere. Moreover, (g_n) is contained in E and o-converges to zero there. By assumption, $(\Phi(g_n))$ o-converges to zero. It follows that $\Phi^*(\bar{g}) = 0$ provided that $\bar{g}_n \in V^{\uparrow}$ and $\Phi(g_n) = \widehat{\Phi}(\bar{g_n})$.

Take an increasing net of continuous functions $(f_\alpha) \subset C(P)$ and put $f(t) = \sup_\alpha f_\alpha(t)$ $(t \in P)$. Without loss of generality we may set $f \leq 1_P$. Then there is an increasing sequence $(\alpha(n))_{n\in\mathbb{N}}$ such that $f(t) = \sup_n f_{\alpha(n)}(t)$ μ-almost everywhere. In other words, the sequence $(f_{\alpha(n)})$ o-converges to f. Again by order continuity,

$$\Phi(f) = \sup \Phi(f_{\alpha(n)}) = \Phi\big(\sup f_{\alpha(n)}\big) \leq \Phi(1_P) \in F.$$

At the same time $\Phi(f_n) = \widehat{\Phi}(\bar{f}_n) \to \widehat{\Phi}(\bar{g})$ by definition. Moreover, according to (1) the identity $f(t) = \sup_n f_{\alpha(n)}$ is valid Φ-almost everywhere. $\triangleright$

6.3.11. Theorem. *For an arbitrary order continuous positive operator $T : E \to F$ there exists a unique ample measure $\varphi : \mathscr{A} \otimes \mathscr{B} \to F$ such that*

$$T\tilde{e} = \iint 1_A \otimes e(s,t)\, d\varphi(s,t) \quad (\tilde{e} \in E).$$

Moreover, for every order-bounded operator $S \in \{T\}^{\perp\perp}$ there is a unique (up to φ-equivalence) φ-measurable function $K := K_S$ such that

$$S\tilde{e} = \iint_{A\times P} K(s,t) 1_A \otimes e(s,t)\, d\varphi(s,t) \quad (\tilde{e} \in E).$$

The correspondence $S \mapsto K_S$ defines a linear and lattice isomorphism from $\{T\}^{\perp\perp}$ onto the order-dense ideal $L_\varphi(E) := \{g \in L^0(A \times P, \varphi) : g \cdot \jmath(E) \subset L^1(A \times P, \varphi)\}$ in $L^0(A \times P, \varphi)$.

$\lhd$ Let Φ be the restriction of T to the lattice $C(P)$. Apply 6.3.9 to Φ and find an ample measure $\varphi : \mathscr{B} \otimes \mathscr{A} \to F$ with the representation 6.3.9 (2) valid. According to 6.3.10 (2) the integral on the right-hand side of 6.3.9 (2) is correctly defined for $g \in \mathscr{L}^0(P, \mu)$ with $\tilde{g} \in E$. By o-continuity we thus have

$$\widehat{\Phi}\tilde{g} = \iint_{A\times P} 1_A \otimes g\, d\mu \quad (\tilde{g} \in E),$$

where $\tilde{g}$ denotes the μ-equivalence class of g.

For every $K \in L_\varphi(E)$ the formula

$$S_K(\tilde{g}) = \iint_{A\times P} K(s,t) 1_A \otimes e(s,t)\, d\varphi(s,t) \quad (\tilde{g} \in E)$$

correctly defines some operator $S_K \in \{T\}^{\perp\perp}$. Moreover, the mapping $\beta : K \mapsto S_K$ is a linear and order isomorphism from $L_\varphi(E)$ into $\{T\}^{\perp\perp}$. Show that this mapping is surjective. By virtue of the Freudenthal Spectral Theorem it is sufficient to prove that β covers the Boolean algebra of unit elements $\mathfrak{E}(T)$. Let $\mathscr{K}$ denote the set of all functions $K \in L_\varphi(E)$ representable as $K = \sum_{k=1}^n \chi_{C_k} \otimes \chi_{B_k}$ for arbitrary $B_n \in \mathscr{B}$ and disjoint $C_n \in \mathscr{A}$. Given such a K, we may check by simple calculations that $\beta(K) = \sum_{k=1}^n [\chi_{C_k}] \circ T \circ [\chi_{B_k}]$. Hence $\beta(\mathscr{K})$ covers $\mathscr{A}(T)$ (see 3.5.9 for the definition) and $\mathfrak{E}(T) = (\beta(\mathscr{K}))^{\uparrow\downarrow\uparrow}$ by 3.5.11 (3). Denote $\mathscr{K}_1 := \{g \in \mathscr{L}^0(P \otimes A) : 0 \le g \le 1_{A\times P}\}$. It remains to prove that $\beta(\mathscr{K}_1) \supset (\beta(\mathscr{K}))^{\uparrow}$ and $\beta(\mathscr{K}_1) \supset (\beta(\mathscr{K}))^{\downarrow}$, since in this event $\beta(\mathscr{K}_1)$ covers $\mathfrak{E}(T)$. We confine exposition to the former relation; the latter is proved in a similar way.

Take an increasing net $(\beta(K_\alpha))$ with $K_\alpha \in \mathscr{K}_1$ and $S := \sup_\alpha \beta(K_\alpha)$. Denote $L := \sup_\alpha \tilde{K}_\alpha$, where $\tilde{K}_\alpha$ is the coset of K_α and the supremum is taken in $L^0(A \times P, \varphi)$. Since the vector lattice $L^0(A \times P, \varphi)$ satisfies the countable chain condition, there is a sequence of indexes $(\alpha(n))_{n\in\mathbb{N}}$ such that $L = \sup_n \tilde{K}_{\alpha(n)}$. Put $K(s,t) := \sup_n K_{\alpha(n)}(s,t)$ $(s \in A,\ t \in P)$. Observe that $K \in \mathscr{K}_1$, $L = \tilde{K}$ and $\beta(K) = \sup_n \beta(K_{\alpha(n)})$ by the Monotone Convergence Theorem. At the same time $\tilde{K}_\alpha \le \tilde{K}$ implies $\beta(K_\alpha) \le \beta(K)$, so that $S \le \beta(K) \le S$. Hence, $S \in \beta(\mathscr{K}_1)$. ▷

6.3.12. Suppose now that P is a σ-compact space. Then there is an increasing sequence of compact sets $P_n \subset P$ with $P = \bigcup_{n\in\mathbb{N}} P_n$. Let $\mathscr{B}_n$ and π_n denote respectively the Baire σ-algebra of P_n and the band projection in F corresponding to multiplication by χ_{P_n}. For every $n \in \mathbb{N}$, in view of 6.3.9, there is a unique ample measure $\varphi_n : \mathscr{A} \otimes \mathscr{B}_n \to F$ such that

$$\pi_n \widehat{\Phi}(\phi g) = o\text{-}\lim_{n\to\infty} \iint\limits_{A\times P} g(s,t)\, d\varphi_n(s,t) \quad (g \in \mathscr{L}^1(\varphi)).$$

Thus, all that was said in 6.3.10 and 6.3.11 can be carried out in the general case of a σ-compact space P. Of course, we may define the measure

$$\varphi(D) := \bigvee_{n=1}^{\infty} \varphi_n(D \cap A \times P_n) \quad (D \in \mathscr{A} \times \mathscr{B})$$

taking values in $F \cup \{+\infty\}$. A little should be added to 6.1 for developing the theory of $F \cup \{+\infty\}$-valued measures along the same lines. We pass over these possibilities and proceed to integral representation by means of real-valued measures.

6.4. Integral Operators

In the present section, we study the question of integral representation for a dominated operator between spaces of measurable vector-functions. Our approach grounds on the technique of dominated operators.

6.4.1. Throughout this section $(A, \mathscr{A}, \lambda)$ and $(B, \mathscr{B}, \mu)$ are σ-finite measure spaces; $(\Omega, \Sigma, \nu) = (A \times B, \mathscr{A} \otimes B, \lambda \otimes \mu)$ is their product; and E and F denote some ideal spaces over $(B, \mathscr{B}, \mu)$ and $(A, \mathscr{A}, \lambda)$, respectively. An operator $S : E \to F$ is called *integral* if there exists a measurable function $K : A \times B \to \overline{\mathbb{R}}$ such that for every $x \in E$ the value $y = Sx$ is the function

$$y(s) = \int\limits_B K(s,t)\, x(t)\, d\mu(t).$$

The integral is understood to be the usual Lebesgue integral. The definition presumes the following two conditions to be satisfied:

(1) For every $x \in E$ the integrand is summable for almost all $s \in A$; i.e.,

$$\int |K(s,t)\,x(t)|\,d\mu(t) < \infty$$

for almost all $s \in A$.

(2) For every $x \in E$ the function $y(\cdot)$ belongs to the space F.

The function $K(s,t)$ is referred to as the *kernel* of the integral operator S. In addition, we say that S admits *integral representation* or S is an integral operator with kernel K and write

$$Sx(s) = \int_B K(s,t)\,x(t)\,d\mu(t) \quad (x \in E).$$

The set of all integral operators from E to F is denoted by $\mathscr{I}(E,F)$.

(3) *Let $S : E \to L^0$ be an integral operator. If $x_n \to 0$ in measure and $|x_n| \le x \in E$ $(n \in \mathbb{N})$ then $Sx_n \to 0$ almost everywhere.*

$\lhd$ This is an obvious corollary to the Lebesgue Dominated Convergence Theorem by virtue of (1). $\rhd$

Alongside with S we consider the integral operator with kernel $|K(\cdot,\cdot)|$:

$$(\bar{S}x)(s) = \int_B |K(s,t)|\,x(t)\,d\mu(t).$$

By (1), $\bar{S}x$ is a function defined for all $x \in E$ and finite almost everywhere, i.e., $\bar{S}x \in L^0(A,\mathscr{A},\lambda)$. Thus, the operator $\bar{S}$ acts always from E into L^0. But $\bar{S}$ may fail to act from E into F. In this connection, we give the following definition:

An operator S is called a *regular integral operator* from E into F if the operator $\bar{S}$ with kernel $|K(s,t)|$ acts from E into F. It is evident that if $\bar{S}$ acts from E to F then S acts from E to F. The converse is true only for regular integral operators. Property (1) shows that every integral operator is a regular integral operator if considered as acting from E into $L^0(A,\lambda)$.

The set of all regular integral operators is denoted by $\mathscr{I}^{\sim}(E,F)$.

6.4.2. Theorem. *An integral operator S is a regular integral operator from E into F if and only if S is o-bounded from E into F. Moreover, if $K(\cdot,\cdot)$ is the kernel of S then the modulus $|S|$ (in the sense of 3.1.2) is also an integral operator with kernel $|K(\cdot,\cdot)|$, i.e.,*

$$(|S|x)(s) = \int |K(s,t)|\,x(t)\,d\mu(t), \quad x \in E.$$

◁ We confine exposition to the case of separable measure spaces. The general case is settled in [162].

It suffices to establish that if an integral operator S acts from E into L^0 then the desired integral representation for $|S|$ holds for all $x \in E$, $x \geq 0$. Fix such a function x and consider the set $M = \{y : |y| \leq x\}$ involved in 3.1.2 (5) for calculating the modulus. Formula 3.1.2 (5) asserts that

$$|S|x = \sup\{Sy : y \in M\} =: f;$$

but the supremum on the right-hand side is that in $L^0(A, \mathscr{A}, \lambda)$. By virtue of the separability of $L^0(A, \mathscr{A}, \lambda)$, there exists a countable everywhere dense set $(y_n)_{n=1}^{\infty}$ in M. For an arbitrary element $y \in M$ there is a sequence y_{n_k} such that $y_{n_k} \to y$ in measure. Since the set M is o-bounded in E, we have $Sy_{n_k} \to Sy$ almost everywhere by 6.4.1 (3). Therefore,

$$Sy \leq \sup_{k \in \mathbb{N}} Sy_{n_k} \leq \sup Sy_n,$$

which immediately implies

$$\sup\{Sy : y \in M\} = \sup\{Sy_n : n \in \mathbb{N}\}.$$

Putting $y := y_s$ in the last identity with $y_s(t) := \operatorname{sign}(K(s,t))\, x(t)$, we obtain

$$\int |K(s,t)|\, x(t)\, d\mu(t) = \int K(s,t)\, y_s(t)\, d\mu(t) = (|Sx|)(s)$$
$$= \sup\left\{\int K(s,t)\, y_n(t)\, d\mu(t) : n \in \mathbb{N}\right\} \quad \text{almost everywhere,}$$

since for every $s \in A$ the inequality $|y_s| \leq x$ is satisfied. ▷

6.4.3. Define the *dual space* E' (see 3.4.7) as

$$E' := \left\{ y \in L^0(B, \mu) : \int |xy|\, d\mu < \infty \ (\forall x \in E) \right\}.$$

It is clear that the dual space is an ideal space. (If $E = L^p$ $(1 \leq p \leq \infty)$ then $E' = L^{p'}$ $(p' := \frac{p}{p-1})$, if $1 < p < \infty$; $p' := \infty$, if $p = 1$; $p' := 1$, if $p = \infty$.) But it may happen that $E' = \{0\}$: for instance, if $E = L^0([0,1], dx)$ or $E = L^p([0,1], dx)$, $0 < p < 1$.)

Given $x' \in E'$, we may construct some linear functional $\varphi_{x'}$ on E by the formula

$$\varphi_{x'}(x) = \int x(t) x'(t)\, d\mu(t) \quad (x \in E).$$

By the Lebesgue Dominated Convergence Theorem it is obvious that $\varphi_{x'} \in E_n^\sim$. It is an easy consequence of 3.4.8 that the mapping $x' \mapsto \varphi_{x'}$ is an order and linear isomorphism of the K-spaces E' and $E_n^\sim$. Thus, $E_n^\sim$ is exactly the set of those functionals on E which admit integral representation.

Fix $x' \in E'$ and $y \in F$. Denote by $x' \otimes y$ the rank-one operator

$$(x' \otimes y)(x) = \left(\int x(t)\, x'(t)\, d\mu(t) \right) y \quad (x \in E),$$

that obviously belongs to $L_n^\sim(E, F)$ and is a regular integral operator with kernel $K(s,t) = y(s)\, x'(t)$. Recall that $J(E, F)$ denotes the band in the K-space $L_n^\sim(E, F)$ which is generated by all operators of the form $x' \otimes y$ $(x' \in E', y \in F)$ (see 4.3.9).

(1) Theorem. $\mathscr{I}^\sim(E, F) = J(E, F)$.

$\lhd$ The proof is contained in 6.4.4 (1–3). $\rhd$

Before launching into proof, observe two corollaries to the above theorem.

(2) *The set of regular integral operators from E into F is a band.*

(3) *An operator $S : E \to L^0$ is an integral operator if and only if there exists an integral operator $V \geq 0$ such that $|S| \leq V$.*

6.4.4. (1) *Every continuous linear operator $S : L^1(B, \mu) \to L^\infty(A, \lambda)$ is an integral operator.*

$\lhd$ Let G be the set of functions of the form

$$L(s,t) = \sum_{k=1}^{n} r_k \chi_{A_k}(s)\, \chi_{B_k}(t)$$

where $r_1, \dots, r_n \in \mathbb{R}$, $A_1, \dots, A_n \in \mathscr{A}$, $B_1, \dots, B_n \in \mathscr{B}$ with $\mu(B_k), \lambda(A_k) < \infty$, and $B_k \cap B_l = \varnothing$ $(k \neq l)$. Put

$$\varphi(L) = \sum_{k=1}^{n} r_k \int_{A_k} S(\chi_{B_k})\, d\lambda.$$

It is easy to see that the linear functional φ is defined on the linear subspace $G \subset L^1(\Omega, \nu)$ correctly. Then

$$|\varphi(L)| \leq \|S\| \sum_{k=1}^{n} r_k \lambda(A_k)\, \mu(B_k) = \|S\|\, \|L\|_{L^1(\Omega,\nu)},$$

whence φ is continuous on G. Since G is dense in $L^1(\Omega,\nu)$ in norm, we may extend φ by continuity to a bounded functional over $L^1(\Omega,\nu)$ for which we preserve the previous notation. Then there exists a function $K \in L^\infty(\Omega,\nu)$ such that

$$\varphi(L) = \int_\Omega L(s,t)\,K(s,t)\,d\nu\,(s,t) \quad \big(L \in L^1(\Omega)\big).$$

Taking the definition of φ into account, we conclude that Sx admits some integral representation from 6.4.1 with kernel K for finite functions $x \in L^1(B,\mu)$. Since $K \in L^\infty(\Omega,\nu)$, we infer that the representation can be extended by continuity to the whole $L^1(B,\mu)$. $\triangleright$

Without loss of generality we may assume that $\mathbf{1}_B \in E'$ and $\mathbf{1}_A \in F$ or, what is the same, $L^\infty(A,\lambda) \subset F$ and $L^\infty(B,\mu) \subset E'$. The general case reduces easily to this case by decomposition of measure spaces.

(2) *Let V_n be a (rank-one) integral operator with kernel $n\mathbf{1}_\Omega = n\mathbf{1}_{A\times B}$ and $0 \le S \le V_n$. Then S is an integral operator.*

$\triangleleft$ It is obvious that

$$\|Sx\|_{L^\infty} \le \|V_n(|x|)\|_{L^\infty} = n\int |x(t)|\,d\mu(t) = n\|x\|_{L^1}$$

for all $x \in L^1(B,\mu)$; therefore, we may extend S to a continuous linear operator from $L^1(B,\mu)$ into $L^\infty(A,\lambda)$ and apply (1). $\triangleright$

(3) *For an arbitrary $0 \le S \in J(E,F)$ the sequence $(S\wedge V_n)_{n\in\mathbb{N}}$ increases and o-converges to S.*

$\triangleleft$ First we prove the sought relation for $S = x'\otimes y$, where $x' \in E'_+$ and $y \in F_+$. It is clear that

$$(S\wedge V_n)(x)(s) = \int x(t)\,(x'(t)\,y(s) \wedge n\mathbf{1}_{A\times B}(s,t))\,d\mu(t)$$

(this follows, for example, from Theorem 6.4.2, but can be checked straightforwardly). Since the sequence $\big(x'(t)\,y(s)\wedge n\mathbf{1}_{A\times B}(s,t)\big)_{n\in\mathbb{N}}$ increases and converges pointwise to $x'(t)\,y(s)$, we need only to use Beppo Levy's Theorem and the formula $(\sup_n S\wedge V_n)x = \sup\{(S\wedge V_n)x : n \in \mathbb{N}\}$.

We now show that $\{V_1\}^{\perp\perp} = J(E,F)$. Indeed, take an arbitrary operator $0 \le U \in J(E,F)$ with $V_1 \wedge U = 0$ and put $V := x'\otimes y$, where $x' \in E'_+$ and $y \in F_+$. Then $\sup(V\wedge V_n) = V$ by the above. Hence,

$$U\wedge V = U\wedge \sup(V\wedge V_n) = \sup(U\wedge V\wedge V_n) = 0,$$

for $0 \le U \wedge V_n \le n(U \wedge V_1) = 0$. By the definition of $J(E, F)$ we have $U = 0$.

By Theorem 3.1.2, there exists an operator $V := \sup(S \wedge V_n) \le S$ in the K-space $L^{\sim}(E, F)$. It is clear that

$$\sup(V_n \wedge V) = \sup[(V_n \wedge S) \wedge V] = \sup(V_n \wedge S) \wedge V = V.$$

Put $U := (S - V) \wedge V_1 \ge 0$ and observe that

$$(V_n \wedge V) + U = (V_n + U) \wedge (V + U) \le V_{n+1} \wedge S$$

whence,

$$V = \sup(S \wedge V_n) = \sup(S \wedge V_{n+1}) \ge \sup(V \wedge V_n) + U = V + U.$$

Consequently, $U \le 0$. Thus, $U = 0$, i.e., $(S - V) \wedge V_1 = 0$. But we have just proven that V_1 is a weak order-unity in $J(E, F)$; therefore, $S = V$. $\triangleright$

(4) Now we pass to proving Theorem 6.4.3 (1).

$\triangleleft$ Take an operator $S \in J(E, F)$. Since $S = S^+ - S^-$, we may assume that $S \ge 0$. We have to show that S admits integral representation. By (2), for $S_n := S \wedge V_n$ there exists a kernel $K_n(s, t)$, i.e.,

$$(S_n x)(s) = \int K_n(s, t)\, x(t)\, d\mu(t) \quad (x \in E).$$

Since the sequence (S_n) is increasing, (K_n) is also an increasing sequence by Theorem 6.4.2. According to (3) $\sup_n S_n = S$. Put $K(s, t) = \sup_n K_n(s, t)$. Since $\sup_n S_n x = Sx$ $(x \in E_+)$, by Beppo Levy's Theorem, we infer that S is an integral operator with kernel $K(s, t)$.

Let S be a regular integral operator. Show that $S \in J(E, F)$. By the regularity of S, we have $S \in L_n^{\sim}(E, F)$. Without loss of generality we may assume that $S \ge 0$. Assign $K_n(s, t) := K(s, t) \wedge n\mathbf{1}_\Omega(s, t)$ and let S_n be the integral operator with kernel K_n. Then for $x \in E_+$ we have

$$(S_n x)(s) \le \left(n \int x(t)\, d\mu(t)\right) \mathbf{1}_A(s) = (V_n x)(s).$$

Since $V_n \in J(E, F)$ and $0 \le S_n \le V_n$, we obtain $S_n \in J(E, F)$, for every band is an ideal. By Beppo Levy's Theorem, $\sup_n S_n = S$, whence $S \in J(E, F)$ by the definition of band. $\triangleright$

6.4.5. Bukhvalov Theorem. *Let $S : E \to F$ be a linear operator. The following are equivalent:*

(1) *S is an integral operator;*

(2) *if $0 \leq x_n \leq x \in E$ $(n \in \mathbb{N})$ and $x_n \to 0$ in measure then $Sx_n \to 0$ almost everywhere;*

(3) *the operator S satisfies the following conditions:* (a) *if $\mu(B_n) \to 0$ $(B_n \in \mathscr{B})$ and $\chi_{B_n} \leq x \in E$ $(n \in \mathbb{N})$ then $S(\chi_{B_n}) \to 0$ almost everywhere;* (b) *if $0 \leq x_n \leq x \in E$ $(n \in \mathbb{N})$ and $x_n \to 0$ almost everywhere then $Sx_n \to 0$ almost everywhere.*

$\lhd$ (1) $\Leftrightarrow$ (2) follows from 4.3.10 and 6.4.3 (1); (2) $\Leftrightarrow$ (3) is straightforward. $\rhd$

6.4.6. Now we turn to the question of finding some conditions for a dominated linear operator in lattice-normed spaces of measurable vector-functions to admit integral representation. First of all we introduce the corresponding conceptions of integrality.

(1) Take Banach spaces X and Y and let $Z \subset Y'$ be a norming subspace. Consider a Z-weakly measurable operator-function $K : \Omega \to \mathscr{L}(X, Z')$ and a measurable vector-function $u : B \to X$. The vector-function $(s,t) \mapsto K(s,t)u(t)$ $((s,t) \in A \times B)$ is Z-weakly measurable. Suppose that, for all $z \in Z$ and almost all $s \in B$, the integral

$$w(s,z) := \int_B \langle z, K(s,t)u(t) \rangle \, d\mu(t)$$

is defined and, moreover, the linear functional $z \mapsto w(s,z)$ is continuous for almost all $s \in A$. Then the vector-function $v : s \mapsto w(s,\cdot) \in Z'$ is Z-weakly measurable. Denote the coset of this vector-function by $\bar{v}$. Assign $T\bar{u} := \bar{v}$. If, for each $\bar{u} \in E(X)$, there exists a $T\bar{u}$ and if $|T\bar{u}| \in F$, then a linear operator $T : E(X) \to F_s(Z')$ appears. In this case we say that T is the *weakly integral operator with kernel* K and take the liberty of writing

$$\langle z, Tu \rangle(s) = \int_B \langle z, K(s,t)u(t) \rangle \, d\mu(t) \quad \text{a.e.} \quad \big(u \in E(X)\big).$$

If the kernel K belongs to $\mathfrak{M}_\nu(X, Z')$ (see 2.3.11) and the integral operator S with kernel $|K| \in L^0(\nu)$ acts from E into F, then the weakly integral operator T is dominated; moreover, S is its exact dominant (see 6.4.7 (1) below).

(2) Using the same notation as in (1), take a simply measurable operator-function $K : \Omega \to \mathscr{L}(X, Y)$. If $u : B \to X$ is a measurable vector-function then the vector-function $(s,t) \mapsto K(s,t)u(t)$ is measurable too. Suppose that, for each $\bar{u} \in E(X)$, we have $K(s,\cdot)u(\cdot) \in L^1(\mu, Y)$ for almost all $s \in B$ and that the measurable vector-function

$$v(s) := \int_B K(s,t)u(t)\,d\mu(t)$$

satisfies $\bar{v} \in F(Y)$. Then a linear operator $T : E(X) \to F(Y)$ can be defined by the formula $T\bar{u} = \bar{v}$. This operator is called a *strongly integral operator with kernel* K. Again, we take the liberty of writing

$$(Tu)(s) = \int_B K(s,t)u(t)\,d\mu(t) \text{ a.e. } \quad \big(u \in E(X)\big).$$

If the kernel K belongs to $\mathfrak{M}_\nu^s(X,Y)$ (see 2.3.12) and the integral operator S with kernel $|K|$ acts from E into F, then T is a dominated operator and S is its exact dominant.

6.4.7. Consider a weakly (strongly) integral operator T acting from $E(X)$ into $F_s(Y,Z)$ $\big($into $F(Y)\big)$ with kernel K. We say that T is *regular* if $K \in \mathfrak{M}_\nu(X,Z')$ and the measurable function $|K| \in L^0(\nu)$ is a kernel of an integral operator from E into F.

(1) *The weakly (strongly) integral operator T with kernel K is dominated if and only if it is regular. In this case, $|T|$ is the integral operator with kernel $|K|$.*

$\lhd$ If $K \in \mathfrak{M}_\nu(X,Z')$ then $|\langle z, K(s,t)x\rangle| \le |K|(s,t)\|z\|\|x\|$; therefore,

$$|\langle z, Tu\rangle| \le \|z\| S_K\big(|u|\big),$$

where S_K is the integral operator with kernel $|K|$. Hence it is clear that $|T| \le S_K$. Conversely, assume that T is a dominated operator. Then, for $x \in X$, $z \in Z$, and $e \in E$, the inequality

$$\left| \int_B \langle z, K(s,t)x\rangle e(t)\,d\mu(t) \right| \le |T|(e)\|x\|\|z\|$$

holds. Therefore, for the integral operator $S_{x,z}$ with kernel $\langle z, K(s,t)x\rangle$ we have $|S_{x,z}| \le |T|$ whenever $\|z\| \le 1$ and $\|x\| \le 1$. If $S := \sup\{|S_{x,z}| : \|x\| \le 1, \|z\| \le 1\}$ then, according to 6.4.3 (2), S is an integral operator from E into F. It remains to observe that $S \le |T|$ and $S = S_K$. $\rhd$

(2) *Let T be a dominated weakly (strongly) integral operator. If a sequence $(u_n)_{n\in\mathbb{N}}$ in $E(X)$ is such that $|u_n| \le e \in E$ $(n \in \mathbb{N})$ and u_n converges to zero in measure, then Tu_n converges to zero almost everywhere. In particular, T is bo-continuous.*

$\lhd$ According to (1), the exact dominant $|T|$ is an integral operator from E to F. Hence, the required assertion follows from the estimate $|Tu_n| \le |T|(|u_n|)$ in view of 6.4.1 (3). $\rhd$

6.4.8. Recall that an operator $S : X \to F$ is called an operator with abstract norm (in symbols $S \in L_A(X, F)$ if there exists $|S| := \sup\{|Sx| : \|x\| \le 1\}$ in F.

(1) *To each operator with abstract norm $S : X \to F$ there is a unique $u_S \in F_s(X')$ satisfying*

$$Tx = \langle x, u_T\rangle \quad (x \in X).$$

The mapping $T \mapsto u_T$ is a linear isometry between the Banach–Kantorovich spaces $L_A(X, F)$ and $F_s(X')$.

$\lhd$ If $f := |S|$ then $S(X) \subset \{f\}^{\perp\perp}$ and we may assume without loss of generality that $0 < f(t) < \infty$ for every $s \in A$. Let U_f denote the operator in $L^0(A, \lambda)$ defined by $g \mapsto f^{-1}g$. Then U_f is an invertible order isomorphism and $U_f^{-1} = U_{f^{-1}}$. If $T := U_f \circ S$ then obviously $T \in L_A(X, L^\infty(A, \lambda))$ and $|T| = U_f(f) = \mathbf{1}$. Let ρ be a lifting of the space $L^\infty(A, \lambda)$, see 1.4.8. For each $s \in A$ the functional $\varphi_s : x \mapsto (\rho \circ Tx)(s)$ is contained in X', since $|Tx| \le \|x\|$. Thus, the vector-function $u : s \mapsto \varphi_s$ is $\sigma(X', X)$-measurable and $\langle x, u(s)\rangle = \varphi_s(x) = \rho(Tx)(s)$. Moreover, $\|u(s)\| = \sup\{\|\rho(Tx)(s)\|; \|x\| \le 1\} = 1$, so that $|u| = \mathbf{1}$. Denote by u_S the coset of the vector-function fu. Then $u_S \in F_s(X)$, $|u_S| = f$, and $Sx = U_f^{-1}Tx = f\langle x, u(\cdot)\rangle = \langle x, f(\cdot)u(\cdot)\rangle = \langle x, fu\rangle = \langle x, u_S\rangle$. $\rhd$

Let E, Φ, E', and $\mathscr{L}_\Phi(E, F)$ be the same as in 5.5.8, but require in addition that $mE = L^0(\mu)$. Take a Y-weakly measurable operator-function $K \in \mathfrak{M}_\nu(X, Y')$. If $u \in E(X)$ then, for each $y \in Y$, the function $\langle y, Ku\rangle : t \mapsto \langle y, K(t)u(t)\rangle$ is measurable. Moreover, $\langle y, Ku\rangle \in L^1(\Phi)$ whenever $|K| \in E'$, and the element $\Phi(\langle y, Ku\rangle) \in F$ is defined.

(2) Theorem. *For every dominated operator $T \in \mathscr{L}_\Phi(E(X), F_s(Y'))$, there exists a unique (to within equivalence) operator-function $K \in \mathfrak{M}_\nu(X, Y')$ such that $|K| \in E'$ and*

$$\langle y, Tu\rangle = \Phi(\langle y, Ku\rangle) \quad (u \in E(X),\ y \in Y).$$

The correspondence $T \mapsto K$ is a linear isometry of the Banach–Kantorovich spaces $\mathscr{L}_\Phi(E(X), F_s(Y'))$ and $E'_s(\mathscr{L}(X, Y'))$.

$\lhd$ The proof repeats the reasoning of 5.5.10 with the only difference that we should use (1) instead of Theorem 5.5.1 (1). $\rhd$

(3) *Every dominated operator $T : L^1(\mu, X) \to L^\infty(\lambda)_s(Y')$ admits weak integral representation.*

$\lhd$ The operator $|T| : L^1(\mu) \to L^\infty(\lambda)$ is integral. Let $L(\cdot,\cdot) \in L^\infty(\nu)$ be its kernel. Consider the operator $\Phi : L^{1,\infty}(\nu) \to L^\infty(\lambda)$ defined by the formula

$$(\Phi e)(s) = \int\limits_B L(s,t)e(s,t)\,d\mu(t) \quad \big(e \in L^{1,\infty}(\nu)\big),$$

where $L^{1,\infty}(\nu)$ is the space of measurable functions of two variables $e(\cdot,\cdot)$ such that the function $|e|(s) := \int_B |e(s,t)|\,d\mu(t)$ $(s \in A)$ belongs to $L^\infty(A,\lambda)$. Obviously, Φ is a Maharam operator. By Theorem 4.3.3 there exists a unique dominated operator $\widetilde{T} : L^{1,\infty}(\nu, X) \to L^\infty(\lambda)_s(Y')$ such that $|\widetilde{T}| = \Phi$ and $gTf = \widetilde{T}(g \otimes f)$, where $g \in L^\infty(\lambda)$, $f \in L^1(\mu, X)$, and $(g \otimes f)(s,t) = g(s)f(t)$. According to (2), the following representation holds:

$$\langle y, \widetilde{T}u\rangle = \Phi\big(\langle y, \widetilde{K}u\rangle\big) \quad \big(u \in L^{1,\infty}(\nu, X),\ y \in Y\big),$$

where $\widetilde{K} \in \mathfrak{M}_\nu(X, Y')$ and $|\widetilde{K}| = \mathbf{1}$. Put $K(s,t) := \widetilde{K}(s,t)L(s,t)$ $(s \in A,\ t \in B)$. From the definitions of $\widetilde{T}$ and Φ it is clear that we obtain the desired integral representation for $u := \mathbf{1} \otimes f$. $\rhd$

6.4.9. *Let Y be a dual Banach space possessing the Radon–Nikodým property. Every dominated operator $T : L^1(\mu, X) \to L^\infty(\lambda, Y)$ admits strong integral representation.*

$\lhd$ Denote by $\mathscr{F}$ the set of functions $u : \Omega \to \mathbb{R}$ of the form $u = \sum_{k=1}^n \varphi_k \otimes \psi_k$, where $\varphi_k \in L^1(B,\mu)$, $\psi_k \in L^1(A,\lambda)$, and $\varphi_k \otimes \psi_k : (s,t) \mapsto \varphi_k(t)\psi_k(s)$ $(k := 1,\dots,n)$. For any fixed $x \in X$, define the operator $G_x : \mathscr{F} \to Y$ by the formula

$$G_x(u) := \sum_{k=1}^n \int\limits_A \psi_k(s)T(x \otimes \varphi_k)(s)\,d\lambda(s).$$

The operator $|T| : L^1(\mu) \to L^\infty(\lambda)$ is integral and its kernel L belongs to $L^\infty(\nu)$. Employing the Fubini Theorem, it is easy to derive the inequality

$$\|G_x(u)\| \le \Phi(|u|) := \int\limits_\Omega L(s,t)|u(s,t)|\,d\nu(s,t) \quad (u \in \mathscr{F}).$$

By Theorem 4.3.3 the operator G_x admits a unique dominated extension onto $L^1(\nu)$; we denote this extension by the same symbol G_x. In our case, $|G_x| \le \Phi$.

In particular, the operator $G_x : L^1(\nu) \to Y$ is bounded. Without loss of generality, we may assume that the measure ν is finite. Since Y possesses the Radon–Nikodým property, there exists a $g_x \in L^\infty(\nu, Y)$ such that

$$G_x(u) = \int_\Omega g_x u\, d\nu, \quad |G_x|e = \int_\Omega |g_x| e\, d\nu \quad \big(e \in L^1(\mu),\ u \in L^1(\mu X)\big).$$

Assign $Ux := g_x$. We obtain a linear operator $U : X \to L^\infty(\nu, Y)$ that satisfies the following conditions:

$$\int_A \psi T(x \otimes \varphi)\, d\lambda = \int_\Omega (Ux)\varphi \otimes \psi\, d\nu \quad \big(\varphi \in L^1(\lambda),\ \psi \in L^1(\mu)\big),$$

$$|U| := \sup_{\|x\| \le 1} |Ux| \le L.$$

Using 4.3.5 (3), as well as the formula for calculating the supremum of a set of regular operators, we may show that $\sup\{|G_x| : \|x\| \le 1\} = \Phi$; therefore, $|U| = L$. Now let ρ be a lifting of the space $L^\infty(\nu, Y)$ associated with that of $L^\infty(\nu)$. Assign $K(s,t)x := (\rho Ux)(s,t)$ $\big((s,t) \in \Omega\big)$. As is seen, the operator-function $K : \Omega \to \mathscr{L}(X,Y)$ is simply measurable and $|K| = L$. The definitions of U and K imply the following representation:

$$\int_A \psi T(x \otimes \varphi)\, d\lambda = \int_A \psi(t) \left(\int_B K(s,t)(x \otimes \varphi)(s)\, d\mu(s) \right) d\lambda(t).$$

Hence we obtain

$$(Tu)(t) = \int_B K(s,t)u(s)\, d\mu(s) \quad \big(u \in X \otimes L^1(\mu)\big).$$

The same equality remains valid for all $u \in L^1(\mu, X)$, since $X \otimes L^1(\mu)$ is dense in $L^1(\mu, X)$, the operator T is o-continuous, and passage to the limit is possible under the Bochner integral sign. ▷

6.4.10. Theorem. *For every dominated operator T from $E(X)$ to $F_s(Y,Z)$, the following are equivalent:*

(1) *T admits weak integral representation;*

(2) *some dominant of T admits integral representation;*

(3) *if a sequence $(u_n)_{n\in\mathbb{N}}$ in $E(X)$ is such that $|u_n| \le e \in E$ $(n \in \mathbb{N})$ and $u_n \to 0$ in measure, then $Tu_n \to 0$ almost everywhere;*

(4) *T is sequentially o-continuous and if a sequence of measurable sets $A_n \in \mathscr{A}$ is such that $\lambda(A_n) \to 0$ then $T(u\chi_{A_n}) \to 0$ almost everywhere for each function $u \in E(X)$.*

◁ The implication (1) ⇒ (2) follows from 6.4.7 (1), and (2) ⇒ (3) from 6.4.7 (2); it is obvious that (3) ⇒ (4). It remains to prove validity of (4) ⇒ (1). Taking it into account that $*$-convergence of an o-bounded sequence in $L^0(\mu)$ coincides with convergence in measure, from (4) we easily deduce $*$-o-continuity of the operator T. By Theorem 4.3.10 the operator $|T|$ is almost integral. According to 6.4.3 (1), $|T|$ is an integral operator; let $L(\cdot,\cdot) \in L^0(\nu)$ be its kernel. There exists a countable partition (Ω_n) of the set Ω into disjoint ν-measurable subsets such that $L_n := \chi_{\Omega_n} L \in L^\infty(\nu)$ for all $n \in \mathbb{N}$. Denote by S_n the integral operator with kernel L_n. It is clear that $(S_n)_{n\in\mathbb{N}}$ is a sequence of positive pairwise disjoint operators from E into F. Moreover, S_n can be also regarded as a positive operator from $L^1(\mu)$ into $L^\infty(\lambda)$. Since the exact dominant is decomposable (see 4.2.6), there exists a sequence of pairwise disjoint operators $T_n : E(X) \to F_s(Z')$ such that $|T_n| = S_n$ $(n \in \mathbb{N})$ and $Tu = bo\text{-}\sum_{k=1}^\infty T_n u$ $\big(u \in E(X)\big)$. The restriction of T_n onto $L^1(\mu, X) \cap E(X)$ can be extended to a dominated operator from $L^1(\mu, X)$ into $L^\infty(\lambda)_s(Z')$ with the exact dominant S_n preserved (see Theorem 4.3.3). We denote such an extension by the same symbol T_n. According to 6.4.8 (3), the operator T_n admits weak integral representation with kernel $K_n : \Omega \to \mathscr{L}(X, Z')$, for which $|K_n| = L_n$. Now define an operator-function $K : \Omega \to \mathscr{L}(X, Z')$ so that the restriction of K onto Ω_n coincide with K_n. Then K is a Z-weakly measurable function, $|K| = L$, and, moreover, for all $z \in Z$ and $u \in W := E(X) \cap L^1(\mu, X)$, the following equality holds almost everywhere:

$$\langle z, Tu\rangle(s) = \int\limits_A \big\langle z, K(s,t)u(t)\big\rangle \, d\mu(t).$$

This representation is valid for all $u \in E(X)$, since W is bo-dense in $E(X)$, the operator T is bo-continuous, and passage to the limit is possible under the integral sign. ▷

6.4.11. Theorem. *Let Y be a dual Banach space possessing the Radon–Nikodým property. A dominated operator $T : E(X) \to F(Y)$ admits strong integral representation if and only if one of the conditions (2)–(4) of Theorem 6.4.10 is satisfied.*

◁ The proof follows the same scheme as in 6.4.10, but we should use 6.4.9 instead of 6.4.8 (3). ▷

6.4.12. Theorems 6.4.10 and 6.4.11 allow us to obtain results about the general from of some classes of dominated operators. Recall that $M_J(X, Y)$ is the band of almost integral operators from X into Y (see 4.3.9). Denote by EF the o-ideal of $L^0(\nu)$ constituted by the kernels of regular integral operators from E into F.

(1) Theorem. *The mapping sending a weakly measurable operator-function to the corresponding weakly integral operator establishes a linear isometry of the Banach–Kantorovich spaces $EF_s(\mathscr{L}(X,Y'))$ and $M_J(E(X),F_s(Y'))$.*

(2) Theorem. *Let Y be a dual Banach space possessing the Radon–Nikodým property. The mapping, sending a simply measurable operator-function to the corresponding strongly integral operator, establishes a linear isometry of the Banach–Kantorovich spaces $EF_s(\mathscr{L}(X,Y))$ and $M_J(E(X),F(Y))$.*

(3) Theorem. *Let Y be the same as in (2) and $G := L^{p,\infty}(\nu\otimes\lambda)$, $1\le p<\infty$. Then the BKSs $G_s(\mathscr{L}(X,Y))$ and $M(L^1(\nu,X),L^p(\lambda,Y))$ are linearly isometric in the sense of Theorem (2).*

6.5. Pseudointegral Operators

The main result of the present section is a criterion for pseudointegrality of a dominated operator in spaces of measurable vector-functions.

6.5.1. Let, as above, $(A,\mathscr{A},\lambda)$ and $(B,\mathscr{B},\mu)$ be spaces with complete σ-finite measures; let E and F be ideal spaces on $(B,\mathscr{B},\mu)$ and $(A,\mathscr{A},\lambda)$, respectively, and let (Ω,Σ,ν) be the product of these measure spaces. By a *representing measure* or a *measure kernel* we mean a positive countably additive function $m:\Sigma\to\mathbb{R}\cup\{+\infty\}$ satisfying the following conditions: (a) there exists a countable increasing set of measurable sets $B_n\in\mathscr{B}$ and, for every $n\in\mathbb{N}$, a countable set of pairwise disjoint measurable sets $A_{n,k}\in\mathscr{A}$ such that $B=\bigcup_{n=1}^{\infty}B_n$, $A=\bigcup_{k=1}^{\infty}A_{n,k}$ $(n\in\mathbb{N})$, and $m(A_{n,k}\times B_n)<\infty$ $(n,k\in\mathbb{N})$; (b) $m(A_0\times B_0)=0$ whenever $A_0\times B_0\in\Sigma$ and either $\lambda(A_0)=0$ or $\mu(B_0)=0$.

Take a representing measure $m:\Sigma\to\mathbb{R}\cup\{+\infty\}$. For $e\in L^0(\mu)$, denote $F(e):=F(m,e):=\{f\in L^0(\lambda):e\otimes f\in L^1(m)\}$, where $(e\otimes f)(s,t):=e(t)f(s)$ $(s\in A,\ t\in B)$. Let $\mathscr{D}(m)$ be constituted by $e\in L^0(\mu)$ such that $F(e)$ is an order-dense ideal of $L^0(\lambda)$. Then $\mathscr{D}(m)$ is an order-dense ideal of $L^0(\mu)$.

(1) *There exists a unique order continuous operator $\Phi_m:\mathscr{D}(m)\to L^0(\lambda)$ such that*

$$\int\limits_A f(s)(\Phi_m e)(s)\,d\lambda(s)=\iint\limits_\Omega e(s)f(t)\,dm(s,t)$$

for all $e\in\mathscr{D}(m)$ and $f\in F(m,e)$. The operator Φ_m is positive if and only if so is m.

$\triangleleft$ Indeed, for any fixed $e \in E$, the right-hand side of the indicated equality defines an o-continuous functional φ_e on $F(m,e)$. By Theorem 3.4.8 there exists a unique element $g \in F(m,e)' := \{f' \in L^0(\lambda) : f' \cdot F(m,e) \subset L^1(\lambda)\}$ such that

$$\varphi_e(f) = \int_A g(s)f(s)\,d\lambda(s).$$

It remains to assign $\Phi_m e := g$ and observe that if $m \geq 0$ and $e \geq 0$ then $\varphi_e \geq 0$ and $\Phi_m e \geq 0$. Order continuity of Φ_m follows from the Dominated Convergence Theorem. $\triangleright$

An operator $S : E \to F$ is called *pseudointegral* if there exists a representing measure $m : \Sigma \to \mathbb{R} \cup \{+\infty\}$ such that $E \subset \mathscr{D}(m)$ and $Se = \Phi_m e$ $(e \in E)$. In view of the above proposition every pseudointegral operator is order continuous.

(2) Suppose that $(B, \mathscr{B})$ is a *standard Borel space;* i.e., $(B, \mathscr{B})$ is Borel isomorphic to a Borel subset of a complete separable metric space. In this event the representing measure m admits disintegration: $dm(s,t) = dm_s(t)d\lambda(s)$. More precisely, there exists a mapping $s \mapsto m_s$ $(s \in A)$, with each m_s a Borel measure on B, such that the following conditions are satisfied: (1) if $B_0 \in \mathscr{B}$ and $\mu(B_0) = 0$ then $m_s(B_0) = 0$ for λ-almost all s; (2) for every $B_0 \in \mathscr{B}$, the function $s \mapsto m_s(B_0)$ is Borel; (3) if $A_0 \in \mathscr{A}$, $B_0 \in \mathscr{B}$, and $m(A_0 \times B_0) < +\infty$ then

$$m(A_0 \times B_0) = \int_{A_0} m_s(B_0)\,d\lambda(s).$$

A mapping $s \mapsto m_s$ with this properties is called a *disintegrated kernel.* A pseudointegral representation under the stated additional assumption takes the form

$$(Tu)(s) = \int_B e(t)\,dm_s(t).$$

6.5.2. *Let $\mathscr{I}(\Phi_m)$ be the order ideal in $L^{\sim}(\mathscr{D}(m), L^0(\lambda))$ generated by the operator Φ_m. There exists a linear and order isomorphism Λ from $\mathscr{I}(\Phi)$ onto $L^{\infty}(m)$ such that, for each $S \in \mathscr{I}(\Phi_m)$, the following representation holds:*

$$\int_A (Se)(s)f(s)\,d\lambda(s) = \iint_\Omega \Lambda(S)(s,t)f(s)e(t)\,dm(s,t)$$

$$\big(e \in \mathscr{D}(m),\ f \in F(m,e)\big).$$

$\triangleleft$ Suppose that $e_1, \dots, e_n \in E_+$ and $f_1, \dots, f_n \in F(m,e)_+$. If $0 \le S \le \Phi_m$ and $z = \sum_{k=1}^n e_k \otimes f_k$ then

$$\widetilde{S}z := \sum_{k=1}^n \int_A S(e_k) f_k \, d\lambda \le \iint_\Omega z(s,t) \, dm(s,t).$$

This implies that there exists a unique bounded positive functional $\widetilde{S}$ on $L^1(m)$, for which $\widetilde{S}(e \otimes f) = \int_A (Se) f \, d\mu$ $\big(e \in E,\ f \in F(m,e)\big)$. The mapping $S \mapsto \widetilde{S}$ is an algebraic and order isomorphism of $\mathscr{I}(\Phi)$ onto $L^1(m)'$. It remains to observe that $L^1(m)' \simeq L^\infty(m)$. $\triangleright$

6.5.3. Now suppose that B is a σ-compact topological space and μ is a regular Borel measure on B.

Theorem. *Let $T : E \to F$ be an order continuous positive operator. Then there exists a representing measure m such that*

$$\int_A g(s)(Te)(s) \, d\lambda(s) = \iint_\Omega g(s)e(t) \, dm(s,t) \quad \big(e \in E,\ g \in F(m,e)\big).$$

Moreover, for every operator $S \in \{T\}^{\perp\perp}$ there is a m-measurable function $K(s,t)$ such that

$$\int_A g(s)(Se)(s) \, d\lambda(s) = \iint_\Omega K(s,t) g(s) e(t) \, dm(s,t) \quad \big(e \in E,\ F(Km,e)\big).$$

$\triangleleft$ Let (B_n) be a sequence of compact subspaces with $B = \bigcup_{n=1}^\infty B_n$ and let E_n be the band in E generated by B_n, i.e. $E_n = [B_n]E$, where $[B_n]$ denotes the band projection $\tilde{e} \mapsto \widetilde{\chi_{B_n} e}$. Denote $T_n := T \circ [B_n]$. According to Theorem 6.3.11 there exists a unique modular measure $\varphi_n : \mathscr{A} \otimes \mathscr{B}_n \to F$ such that

$$T_n e = \iint e(t) \, d\varphi_n(s,t) \quad (\tilde{e} \in E_n).$$

Given $n \in \mathbb{N}$, choose a partition $(A_{n,k})_{k \in \mathbb{N}}$ of A such that $[A_{n,k}]\varphi_n(D) \in L^1(\lambda)$ for all $D \in \mathscr{A} \otimes \mathscr{B}_n$ and $n, k \in \mathbb{N}$. Define a real-valued measure

$$m_n(D) := \int \varphi_n(D) \, d\lambda(s) \quad (D \in \mathscr{A} \otimes \mathscr{B}_n).$$

There is a unique positive countably additive numerical measure m on $\mathscr{A} \otimes \mathscr{B}$ that agrees with m_n on $\mathscr{A} \otimes \mathscr{B}_n$. Clearly $m\,(A_{n,k} \times B_n) = m_n\,(A_{n,k} \times B_n) < \infty$. If $B' \subset B_n$ and $A' \subset A_{n,k}$ then for $D = A' \times B'$, by modularity of φ, we have

$$m(D) = \int_A \varphi(A' \times B')\,d\lambda = \int_{A'} \varphi(A \times B')\,d\lambda.$$

Now it is evident that $\lambda(A') = 0$ implies $m(D) = 0$. The fact that $m(B') = 0$ implies $m(D) = 0$ was proved in 6.3.10 (1). Hence, m is a representing measure. Now, if $e \in E$ and $f \in F(m, e)$ then

$$\int_\Omega f(s)e(t)\,dm_n(s,t) = \int_A f(s) \left(\int_\Omega e(t)\,d\varphi_n(s,t) \right) d\lambda(s).$$

This identity is evident for $e := 1_C$ and $f := 1_D$, needs some simple calculation if e and f are step-functions, and is deduced with the help of the Monotone Convergence Theorem for arbitrary $e \in E$ and $f \in F(m, e)$. For the operator we have the representation

$$\int g(s)[B_n](Te)(s)\,d\lambda(s) = \iint g(s)e(t)\,dm_n(s,t) \quad \big(e \in E,\ g \in F(m,e),\ n \in \mathbb{N}\big).$$

Employing again the Monotone Convergence Theorem we obtain the desired representation for T. The assertion concerning S is deduced by similar arguments on using the second part of Theorem 6.3.11. $\rhd$

6.5.4. Sourour Theorem. *A positive operator from E into F admits pseudointegral representation if and only if it is order continuous.*

$\lhd$ See 6.5.1 (1) and 6.5.3. $\rhd$

6.5.5. Given an order continuous operator $\Phi : \mathscr{D}(\Phi) \to F$, where $\mathscr{D}(\Phi)$ is an order-dense ideal in $L^0(\lambda)$, we introduce the space $\mathscr{L}_\Phi(E, F)$ as in 3.2.4; the latter is a band of $L_n(E, F)$.

Theorem. *Let Φ be an order continuous operator from $\mathscr{D}(\Phi)$ into F. There exists a linear and order isomorphism Λ from $\mathscr{L}_\Phi(E, F)$ onto some order-dense ideal of the space $L^0(m)$ such that, for each $S \in \mathscr{L}_\Phi(E, F)$, the following representation holds:*

$$\int_A (Se)(t)f(t)\,d\lambda(t) = \iint_\Omega \Lambda(S)(s,t)f(s)e(t)\,dm(s,t)$$

whenever $e \in E$ and $f \in F(km, e)$ with $k := \Lambda(S)$.

◁ According to 6.5.3, there exists a representing measure m, for which $\mathscr{D}(\Phi) = \mathscr{D}(m)$ and $\Phi = \Phi_m$. Suppose that $0 \le S \in \mathscr{L}_\Phi(E, F)$, assign $E_0 := E \cap \mathscr{D}(m)$, and denote by S_0 and Φ_0 the restrictions of the respective operators S and Φ onto E_0. Put $S_n := S_0 \wedge (n\Phi_0)$. By Proposition 6.5.2 the operator S_n is pseudointegral and admits $k_n m$ as a representing measure, where $k_n \in L^\infty(m)$. The sequence (k_n) is increasing and bounded almost everywhere. Indeed, by Theorem 6.5.4 the operator S admits a pseudointegral representation with some representing measure m'. Consequently,

$$\iint\limits_{\Omega} f(s)e(t)\, dm'(s,t) = \lim_{n\to\infty} \iint\limits_{\Omega} k_n(s,t)f(s)e(t)\, dm(s,t)$$

for all $e \in E_0$ and $f \in F(m', e)$. Therefore, the sequence (k_n) is bounded and $m' = km$, where $k = \sup k_n$. Assign $\Lambda(S) := k$. It is easy that Λ defines a linear and order isomorphism. ▷

6.5.6. Now consider a linear operator $T : E(X) \to F_s(Y, Z)$, where X and Y are Banach spaces. Take a Z-weakly m-measurable operator-function $K : \Omega \to \mathscr{L}(X, Z')$, a strongly μ-measurable vector-function $u : B \to X$, and some order-dense ideal F_u in $L^0(\lambda)$. Suppose that for all $z \in Z$ and $g \in F_u$ the following integral exists:

$$w(z, g, u) := \int\limits_{\Omega} \langle z, K(s,t)u(t)\rangle g(s)\, dm(s,t).$$

For a fixed $z \in Z$, the functional $w(z, \cdot, u)$ on F_0 is order continuous; therefore, there exists a function $v(z, \cdot, u) \in L^0(\lambda)$ such that

$$\int v(z,s)g(s)\, d\lambda(s) = w(z, g, u) \quad (g \in F_0).$$

Next, observe that the operator $U : z \mapsto v(z, \cdot, u)$ $(z \in Z)$ is linear. If it has the form $U_z(s) = \langle z, Tu\rangle(s)$ (for the above operator $T : E(X) \to F_s(Y, Z)$), then T is said to be the *weakly pseudointegral operator with representing measure m and kernel K*. In this case we take the liberty of writing

$$\int\limits_A \langle z, Tu\rangle(s)g(s)\, d\lambda(s) = \iint\limits_{\Omega} \langle z, K(s,t)u(t)\rangle g(s)\, dm(s,t).$$

If $K \in \mathfrak{M}_m(X, Z')$ and the operator $S : E \to F$ acts by the formula

$$\int\limits_A (Se)(s)\, d\lambda(s) = \iint\limits_{\Omega} |K|(s,t)e(t)g(s)\, dm(t,s),$$

then S is a dominant for T (moreover, S is the exact dominant of T, see 6.5.7). A weakly pseudointegral operator T is called *regular* if $K \in \mathfrak{M}_m(X, Z')$ and the pseudointegral operator with representing measure $|K|m$ acts from E into F.

If $(A, \mathscr{A})$ is a standard Borel space then the weakly pseudointegral operator has the form

$$\langle z, Tu\rangle(s) = \int_A \langle z, K(s,t)u(t)\rangle \, dm_s(t).$$

6.5.7. Theorem. *The weakly pseudointegral operator $T : E(X) \to F_s(Y, Z)$ with kernel K and representing measure m is dominated if and only if T is regular. In this case, $|T|$ is the pseudointegral operator with representing measure $|K|m$.*

$\lhd$ Assume that a linear operator T admits weak pseudointegral representation 6.5.6. If $K \in \mathfrak{M}_m(X, Z')$ and $S : E \to F$ is the pseudointegral operator with kernel $|K|$ and representing measure m, then

$$\int_A \langle Tu, z\rangle(s) g(s) \, d\lambda(s) \le \|z\| \int_A S(|u|)(s) g(s) \, d\lambda$$

for all $g \in F_u := F(|K|m, |u|)$. Whence we deduce $\langle Tu, z\rangle \le S(|u|)\|z\|$, and passage to the supremum over all $\|z\| \le 1$ yields $|Tu| \le S(|u|)$.

Conversely, assuming that a weakly pseudointegral operator T is dominated, prove that its kernel K belongs to $\mathfrak{M}_m(X, Z')$. Let Φ be the pseudointegral operator from $\mathscr{D}(m)$ into $L^0(\lambda)$ generated by a representing measure m. For $x \in X$ and $z \in Z$, define the operator $S_{x,z}$ from E into $L^0(\lambda)$ by the formula

$$\int_A S_{x,z}(e)(s) g(s) \, d\lambda(s) = \iint_\Omega \langle z, K(s,t)x\rangle e(t) g(s) \, dm(s,t),$$

where $e \in E$ and $g \in F(e)$. It is clear that $S_{x,z} \in \mathscr{L}_\Phi(E, F)$. Since T is dominated, we have the estimate

$$\int_A S_{x,z}(e)(s) g(s) \, d\lambda(s) \le \|x\|\|z\| \cdot \int_A |T|(e)(s) g(s) \, d\lambda(s)$$

for $e \in E_+$ and $g \in F(e)$. Hence it is clear that $S_{x,z} \le |T|$ whenever $\|x\| \le 1$ and $\|z\| \le 1$. Let S_0 be the supremum in $L^\sim(E, F)$ of the bounded set $\{S_{x,z} : \|x\| \le 1, \ \|z\| \le 1\}$. Since $\mathscr{L}_\Phi(E, F)$ is a band, $S_0 \in \mathscr{L}_\Phi(E, F)$. By Theorem 6.5.5

$$L := \sup\left\{\langle z, Kx\rangle : \|x\| \le 1, \ \|z\| \le 1\right\}$$

exists in the space $L^0(m)$ and the following representation holds:

$$\int\limits_A S(e)(s)g(s)\,d\lambda(s) = \iint\limits_\Omega L(s,t)e(t)g(s)\,dm(s,t).$$

Thus, $K \in \mathfrak{M}_m(X, Z')$, $|K| = L$, and $S \le |T|$. While proving sufficiency, we have seen that $|T| \le S$; hence, $S = |T|$, i.e., $|T|$ is the pseudointegral operator with kernel $|K|$ and representing measure m. $\rhd$

6.5.8. *Theorem.* *For every dominated operator $T : E(X) \to F_s(Y, Z)$, the following are equivalent:*

(1) *the operator T is order continuous;*

(2) *the operator $|T|$ admits pseudointegral representation;*

(3) *the operator T admits weak pseudointegral representation.*

$\lhd$ (1) $\Leftrightarrow$ (2): This follows from Theorems 4.3.2 and 6.5.3.

(2) $\Rightarrow$ (3): Let m be a representing measure of the operator $|T|$. According to 6.5.5, there exists an isomorphism Λ from $\{|T|\}^{\perp\perp}$ onto an ideal subspace $L \subset L^0(m)$. For $x \in X$ and $z \in Z$, the operator $S := S_{x,z} : e \mapsto \langle T(x \otimes e), z\rangle$ acts from E into F, is regular, and belongs to $\{|T|\}^{\perp\perp}$. Assign $b(x,z) := \Lambda(S_{x,z})$. We know that b is a bilinear operator with abstract norm and $|b| = \Lambda(|T|)$, see 6.4.8 (1) and 5.5.1 (2). By Theorem 6.4.8 (1) there exists a Z-weakly measurable operator-function $K \in \mathfrak{M}_m(X, Z')$ such that $|b| = |K|$ and

$$b(x,z) = \langle z, Kx\rangle \quad (x \in X,\ z \in Z).$$

For any operator $S \in \{|T|\}^{\perp\perp}$, in view of what was said above, we may write:

$$\int\limits_B (Se)(s)g(s)\,d\lambda(s) = \iint\limits_\Omega \Lambda(S)e(t)g(s)\,d\nu(s,t).$$

Hence, according to the definition of the operator b, we have:

$$\int\limits_B \big\langle z, T(x \otimes e)\big\rangle(s)g(s)\,d\lambda(s) = \iint\limits_\Omega \big\langle z, K(s,t)x\big\rangle e(t)g(s)\,d\nu(s,t).$$

Taking account of the fact that $X \otimes E$ is *bo*-dense in $E(X)$, after appropriate passage to the limit under the integral sign, we obtain a required pseudointegral representation for T.

(3) $\Rightarrow$ (2): This is an immediate consequence of 6.5.7 and 6.5.1 (1). $\rhd$

6.5.9. Let G be the ideal of $L^0(m)$ consisting of functions k such that the pseudointegral operator with representing measure km acts from E into F. Assign

$$M_\Phi(E(X), F_s(Y')) := \{T \in M(E(X), F_s(Y')) : |T| \in \mathscr{L}_\Phi(E, F)\}.$$

Theorem. *To each $K \in G_s(\mathscr{L}(X, Y'))$, there corresponds the weakly pseudointegral operator $T \in M(E(X), F_s(Y'))$ with kernel K and representing measure m. The mapping $K \mapsto T$ establishes a linear isometry between the BKSs $G_s(\mathscr{L}(X, Y'))$ and $M_\Phi(E(X), F_s(Y'))$.*

$\lhd$ Only the first part of the assertion requires proving, the rest of the claims follows immediately from 6.5.3, 6.5.5, 6.5.7, and 6.5.8. Take a $K \in G_s(\mathscr{L}(X, Y'))$. For any fixed $u \in E(X)$ and $y \in Y$, the functional

$$\varphi : f \mapsto \iint_\Omega \langle y, K(s,t)u(s)\rangle f(t)\, dm(s,t) \quad (f \in F(m, |u|))$$

is order continuous. Consequently, the following representation holds:

$$\varphi(f) = \int_A f(s)g(s)\, d\lambda(s) \quad (f \in F(m, |u|)),$$

where g is a uniquely defined element of $L^0(\mu)$. Consider the operator $S : Y \to L^0(\mu)$ defined by the formula $Sy := g$. If Ψ is the pseudointegral operator with representing measure $|K|m$ then

$$\int_A f(s)g(s)\, d\lambda(s) \le \|y\| \iint_\Omega |K|(s,t)|u|(s)f(t)\, dm(s,t)$$

$$= \|y\| \int_B \Psi(|u|)(s)f(s)\, d\lambda(s) \quad (f \in F(m, |u|)).$$

Hence it is clear that $Sy := g \le \Psi(|u|)\|y\|$. Therefore, S is an operator with abstract norm and $|S| \in F$. By the representation theorem for operators with abstract norm (see 6.4.8 (1)), there exists a $v \in F_s(Y')$ such that $Sy = \langle y, v\rangle$ $(y \in Y)$. Now define the operator T by the equality $Tu := v$. From the definitions we have:

$$\int_A \langle y, Tu\rangle(s)f(s)\, d\lambda(s) = \int_A g(s)f(s)\, d\lambda(s)$$

$$= \varphi(g) = \iint_\Omega \langle y, K(s,t)u(t)\rangle f(s)\, dm(s,t)$$

and the result follows. $\rhd$

6.6. Comments

6.6.1. (1) Some integration theory of scalar-valued function with respect to a σ-additive measure with values in a Stone algebra (say F) has been developed by J. D. M. Wright [404]. This theory (including all convergence theorems) remains valid when F is an arbitrary K_σ-space, see [402]. The construction in 6.1.1–6.1.7 has been performed in [214] and essentially repeats J. D. M. Wright's [404] considerations in a more general situation: integrable objects are elements of some K_σ-space as in [163, 388] and the vector lattice F is replaced by a lattice-normed space.

(2) It can be easily observed that the Radon–Nikodým Theorem for measures with values in an (order complete) vector lattices fails in general: if μ is a σ-additive (finite) measure then the $\mathbb{R}^2$-valued measures $\mu_1 := (1,0)\mu$ and $\mu_2 := (0,1)\mu$ are absolutely continuous with respect to one another, but $\mu_1(A) = (1,0)\mu(A) \neq (0,1)I_\mu(f\chi_A) = I_{\mu_2}(f\chi_A)$ provided that $\mu(A) \neq 0$. J. D. M. Wright has shown in [400] that the Radon–Nikodým Theorem is true for special class of ample measures. By definition [400] μ is ample if the space $L^2(\mu)$ is a Kaplansky–Hilbert module (see Section 7.4). J. D. M. Wright deduced his Radon–Nikodým Theorem [400; Theorem 4.1] (that essentially coincides with 6.1.11 (2)) from the following auxiliary fact [400; Lemma 4.2] (which is immediate from Kaplansky's Theorem [166; Theorem 5], see 7.5.7 (2)).

Let μ be an ample $C(Q)$-valued measure and let $T : L^2(\mu) \to C(Q)$ be a norm-bounded module homomorphism. Then there exists a unique $g \in L^2(\mu)$ such that

$$Tf = \int fg\,d\mu \quad (f \in L^2(\mu)).$$

(3) By [400] a measure $\mu : \mathscr{B} \to C(Q)$ (with Q an extremal compact space) is modular with respect to an algebra homomorphism $\pi : C(Q) \to L^\infty(\mu)$ if

$$\int \pi(a) f\,d\mu = a \int f\,d\mu \quad (a \in C(Q),\ f \in L^1(\mu)).$$

Equivalence of this definition to that in 6.1.9 follows from 6.1.8, 6.1.8 (3), and 2.1.8. According to 6.1.9 (4), 2.2.3, and 7.4.4 the definitions of ample measure in [400] and 6.1.9 are also equivalent.

6.6.2. (1) The main results of Section 6.2 (Theorems 6.2.2 and 6.2.6) as well as the concept of quasi-Radon measure are due to S. A. Malyugin [273] and stem from Wright's theory of Stone-algebra-valued measures [400, 402, 404–406]. The key idea in the proof of Theorem 6.2.6 is to use the Birkhoff–Ulam homomorphism (in the form 6.1.1 (4) or like) was employed for the first time by J. D. M. Wright [404].

Other Riesz-type representation theorems for positive and order-bounded operators are presented in [43, 103, 171, 172, 214, 247, 314]. In this connection we must mention the two classical works by A. A. Markov [276] and A. D. Alexandrov [16] which are inspirational sources for studying Riesz-type representations and related topics for more than half of a century (see also [34, 112]).

(2) Theorem 6.2.7 (1) was proved by J. D. M. Wright [404; Theorem 4.1]. In this result μ cannot be chosen regular rather than quasiregular. Moreover, J. D. M. Wright [406; Theorem T] obtained the following elegant characterization of order complete vector lattices for which this choice is always possible.

Theorem. *Let F be an order complete vector lattice. Then the following are equivalent:*

(i) *F is weakly (σ, ∞)-distributive;*

(ii) *each F-valued Baire measure on every compact space can be extended to a regular F-valued Borel measure;*

(iii) *every F-valued quasiregular Borel measure on every compact space is regular.*

An analogous result for measures with values in $F \cup \{\infty\}$ was obtained in [314].

(3) Another peculiarity of Theorem 6.2.7 (1) is that it cannot be proven by the Daniell extension method. The Daniell construction fails since the Baire measure may be irregular. The following fact is also due to J. D. M. Wright [402]. We say that F has the *measure extension property* if, for every set Ω and each algebra $\mathscr{A}$ of subsets of Ω each countably additive measure $\mu : \mathscr{A} \to F$ has some σ-additive extension to the σ-algebra $\sigma(\mathscr{A})$ generated by $\mathscr{A}$.

Theorem. *Let F be an order σ-complete vector lattice. Then the following conditions are equivalent:*

(i) *F is weakly σ-distributive;*

(ii) *F has the measure extension property;*

(iii) *every F-valued Baire measure on every compact space is regular.*

(4) Speaking of the measure extension problem, we should mention the so-called extreme extension of measures. This direction stems from the classical result by A. Horn and A. Tarski [133]: *A finitely additive positive real-valued measure defined on a subalgebra of a Boolean algebra admits a finitely additive positive extension to the whole algebra.* Let $\mathscr{A}$ be a Boolean algebra and let $\mathscr{A}_0$ be its subalgebra. Let $\mathrm{Ch}(\varepsilon_F^+(\mu_0))$ denote the set of *extreme extensions,* i.e. the collection of all positive finitely additive extensions of a measure $\mu_0 : \mathscr{A}_0 \to F$ to the whole algebra $\mathscr{A}$. The following characterization of an extreme extension of a real-valued measure was given by D. Plachky [320]:

$$\mu \in \mathrm{Ch}\left(\varepsilon_{\mathbb{R}}^+(\mu_0)\right) \Leftrightarrow (\forall\, a \in \mathscr{A}) \inf\{\mu(a \triangle a_0) : a_0 \in \mathscr{A}_0\} = 0.$$

This result was generalized to measures with values in an order complete vector lattice in [248]. Such results are somehow connected with the Buck–Phelps characterization of extreme points, see [132]. An operator version of the Buck–Phelps result was obtained by S. S. Kutateladze [223]; see also [209].

(5) Let $\mathscr{A}$ and $\mathscr{A}_0$ be the same as in (4), with F an order complete vector lattice and Y a Banach–Kantorovich space over F. Observe two more results of [217]. The first claims existence of an extreme extension and, in the case of measures with values in an order complete vector lattice, it was proved in [248]; the second guarantees existence of simultaneous extension and, for scalar-valued measures, it was obtained in [126].

(5.1) *Theorem.* *Let $\mu_0 \in \mathrm{da}(\mathscr{A}_0, Y)$. Then $\mathrm{Ch}\,(\varepsilon_F^+(|\mu_0|)) \neq \varnothing$ and for each extreme extension $\nu : \mathscr{A} \to F$ of $|\mu_0|$ there exists a unique measure $\mu \in \mathrm{da}(\mathscr{A}, Y)$ such that μ extends μ_0 and $|\mu| = \nu$.*

Denote by ϱ and ϱ_0 the restriction mappings from $\mathrm{ba}(\mathscr{A}, F)$ onto $\mathrm{ba}(\mathscr{A}_0, F)$ and from $\mathrm{da}(\mathscr{A}, F)$ onto $\mathrm{da}(\mathscr{A}_0, F)$, respectively.

(5.2) *Theorem.* *There exist an order continuous lattice homomorphism ε_0 from $\mathrm{ba}(\mathscr{A}_0, F)$ onto an order-closed sublattice of $\mathrm{ba}(\mathscr{A}, F)$ and an Orth-linear operator ε from $\mathrm{da}(\mathscr{A}_0, Y)$ into $\mathrm{ba}(\mathscr{A}, Y)$ such that $|\varepsilon\mu| = \varepsilon_0|\mu|$, $\varrho \circ \varepsilon$ is the identity mapping in $\mathrm{da}(\mathscr{A}, F)$, and $\varrho_0 \circ \varepsilon_0$ is the identity mapping onto $\mathrm{da}(\mathscr{A}_0, F)$.*

6.6.3. (1) The material of Section 6.3 is mainly taken from the article by E. V. Kolesnikov and A. G. Kusraev [180]. After the first unusual step (extension of a given "preintegral" to the lattice V in 6.3.1) the construction follows the classical Daniell scheme, see [250, 348]. The main results remain valid if $M(A, \mathscr{A}, \mathscr{N})$ satisfies the *local countable chain condition* (is of countable type locally), i.e. there is a partition of unity (π_α) in $\mathfrak{P}(F)$ such that π_α satisfies the countable chain condition for all α. Evidently $M(A, \mathscr{A}, \mathscr{N})$ satisfy the local countable chain condition if and only if so is $(\mathscr{A}, \mathscr{N})$ The latter means that there is a partition (A_α) of A satisfying: (a) $(\mathscr{A} \cap A_\alpha, \mathscr{N} \cap A_\alpha)$, where $\mathscr{B} \cap A_\alpha = \{B \cap A_\alpha : B \in \mathscr{B}\}$, obeys the countable chain condition for all α; (b) if $C \subset A$ and $C \cap \mathrm{A}_\alpha \in \mathscr{A}$ for all α then $C \in \mathscr{A}$; (c) if $C \subset A$ and $C \cap \mathrm{A}_\alpha \in \mathscr{N}$ for all α then $C \in \mathscr{N}$.

(2) In [398] A. W. Wickstead developed some integration theory for vector-valued functions by Stone-algebra-valued functions very close to that in Section 6.3. Let Q be a Stone space (= an extremal compact space) and let $\mathscr{B}$ be a σ-algebra of subsets of a nonempty set Ω. Given a countably additive positive measure $\mu : \mathscr{B} \to C(Q)$, the integral of a simple function $f := o\text{-}\sum_{n=1}^{\infty} \chi_{A_n} e_n$ with $A_1, \dots, A_n \in \mathscr{B}$ and $e_1, \dots, e_n \in C_\infty(Q)$ in [398] is defined by

$$I_\mu := \int_\Omega f\, d\mu := o\text{-}\sum_{n=1}^{\infty} e_n \mu(A_n).$$

Further, using I_μ as a preintegral, the Lebesgue integral is developed along the lines of the Daniell scheme. Thus, integrable functions are defined on Ω, taking their values in $C_\infty(Q)$. On the contrary, our simple functions (the lattice V) can be identified with the functions defined on Q and taking values in $M(\Omega, \mathscr{B}, \mathscr{N})$ (see 6.5.1 and 3.5.6). This moderate difference actually turns out essential, leading to a simpler and more flexible construction.

(3) As was mentioned in 3.6.4 the theory of Sections 3.4, 3.5 and 6.3 stems from the D. Maharam fundamental papers [264–269]. In particular, variants of Theorems 6.3.6, 6.3.8, 6.3.9, and 6.3.11 were obtained in [268, 269]. The main difference is that in Maharam's representation is performed on the Stone spaces of vector lattices E and F, while our results are formulated in the initial terms of a locally σ-compact space Q and measurable structure $(A, \mathscr{A}, \mathscr{N})$.

(4) In [339] I. E. Schochetman studied integral operators in the spaces of measurable sections of Lebesgue spaces. The relevant definitions can be easily extended to the spaces $E(\mathscr{X})$. It would be interesting to obtain necessary and sufficient conditions for integrality of (dominated) linear operators in the spaces of type $E(\mathscr{X})$ as well as to study these integral operators in the spirit of [128, 183, 188].

6.6.4. The main references to the topic of the classical integral operators, related to the theme of Section 6.4, are [128, 183, 188]; see also [162, 228].

(1) The question about integrality of a linear operator was posed by J. von Neumann. In the fundamental article [292] he solved the problem of finding all operators in $L^2(0,1)$ unitarily equivalent to some selfadjoint integral operator (see the monograph [183] by V. B. Korotkov) and posed the problem of finding necessary and sufficient conditions for a given operator in $L^2(0,1)$ to admit the integral representation of 6.4.1 [292, p. 4].

(2) Theorem 6.4.5 was established by A. V. Bukhvalov [56, 57]. The proof presented is very close to the original. The whole history, references, related ideas and results can be found in [59, 60], see also [228, 409, 410]. We only mention two preceding results by S. I. Zhdanov [412] and L. Lessner [235, 236] that are close to Bukhvalov's criteria, the article by A. Schep [338] in which Theorem 6.4.5 was re-proved, and a new proof of the implication (1) $\Rightarrow$ (2) in Theorem 6.4.5 found by L. Weis [392].

(3) The criteria for weak and strong integrality of a dominated operator in the spaces of measurable vector-functions (Theorems 6.4.10 and 6.4.11), as well as the general form of dominated operators (three theorems in 6.4.12), were obtained by A. G. Kusraev [199, 198]. For linear operators acting from the space of measurable vector-functions into an arbitrary Banach space, i.e., in case $F = \mathbb{R}$

similar results were obtained by V. G. Navodnov [290, 291]; for compact and weakly compact operators, see also the paper by Kevin [168].

(4) In spite of the fact that Theorem 6.4.2 is natural and easy for separable measure spaces, its proof is very involved in the general case. W. A. J. Luxemburg and A. C. Zaanen [263] gave a proof that grounds on approximating the kernel $K(\cdot,\cdot)$ with finite-rank kernels. Another proof presented in [162; Theorem XI.1.2] is based on a Yu. I. Gribanov's result [115]. All subtlety of the theorem lies in the fact that, under some conditions on the set of functions in the domain of an integral operator, the supremum of the values of the operator on this set calculated in the K-space L^0 coincides with the pointwise supremum. In the case when E and F are KB-spaces on $[0,1]$, Theorem 6.4.3 (1) was proven by G. Ya. Lozanovskiĭ in [253]. The general form was obtained in [56]. Theorem 6.4.8 (1) is due to A. V. Bukhvalov [55]; for 6.4.8 $(2,3)$ see [199, 204].

6.6.5. **(1)** Pseudointegral operators (6.5.1) were introduced by W. Arveson [28] in connection with operator algebras in L^2. Pseudointegral operators were later considered by H. Fakhoury [94] (operators in L^1) and N. J. Kalton [147] (operators in L^p with $0 < p \leq 1$). Different aspects of pseudointegral operators are reflected in [147–150, 344–347, 359–358, 393-395].

(2) In Section 6.5 we mainly follow [204]. Theorems 6.5.3 and 6.5.4 were proven by Sourour [361] but we pursue another our approach. The results 6.5.2 and 6.5.5 can be also deduced from [359, 361]. The main results 6.5.7, 6.5.8, and 6.5.9 on pseudointegrality of dominated operators in the spaces of measurable vector-functions are obtained by K. T. Tibilov [373].

(3) It follows from 4.6.3 (2) that in an ideal spaces over a diffuse measure space integral operators are disjoint from all lattice homomorphisms and all Maharam operators. In this connection an interesting example of a pseudointegral operator was constructed in [345]. Let Γ be a compact group of the unit disk and let ν be an arbitrary diffuse measure disjoint from the Haar measure μ (generated by the Lebesgue measure on $[0,1]$). For every $t \in \Gamma$ define the Borel measure ν_t on Γ by $\nu_t(B) := \nu(t+B)$. The relation $(Tf)(t) = \int_\Gamma f(s)\,d\nu_t(s)$ defines a pseudointegral operator acting from $L^\infty(\mu)$ to $L^\infty(\mu)$. I. I. Shamaev [345] has proved that the operator T is disjoint from all integral operators, all lattice homomorphisms, and all Maharam operators in $L^\infty(\mu)$.

Chapter 7

Operators in Spaces with Mixed Norm

In the present chapter, we study various classes of linear operators acting in spaces with mixed norm and defined in mixed terms of norm and order. If (X, E) is a lattice-normed space and E is a norm lattice of X then X can be endowed with a mixed norm so that X becomes a normed space, and even a Banach space in case the lattice-normed space is br-complete (7.1.1 and 7.1.2). The dual of a space with mixed norm is a space with mixed norm too (7.1.4); furthermore, the canonical embedding into the second dual preserves the vector norm (7.1.5). A more general result states that the space of dominated operators between spaces with mixed norm is itself a space with mixed norm if some natural conditions are met (7.1.9). Passage to the dual of an operator commutes rather often with the taking of the exact dominant of this operator (7.1.10).

Various classes of operators under study in functional analysis are often defined in mixed terms that employ norm and order. Using the (positively homogeneous) functional calculus in Banach lattices, we introduce the class of (p, q)-summing operators in spaces with mixed norm (7.2.3). The set of (p, q)-summing operators acting in fixed spaces with mixed norm is a Banach space (7.2.4). A bounded operator is (p, q)-summing if and only if the dual operator is (q', p')-summing (7.2.6). Note that $(1, 1)$-summing operators can be characterized in terms of convergent series (7.2.7) and, under some additional requirements, this class of operators coincides with the class of dominated operators (7.2.8). Particular cases of the notion of (p, q)-summing operator are presented by (p, q)-convex operators (7.2.11 (1)), (p, q)-concave operators (7.2.11 (2)), and (p, q)-regular operators (7.2.11 (3)).

Lattice-normed spaces provide reasonable grounds for constructing some isometric classification of spaces with mixed norm. The key point in this respect is the presence of a complete Boolean algebra of projections in a Banach space as well as a special geometric property of the unit ball which is related to the algebra. The property is called $\mathscr{B}$-cyclicity (7.3.3) or $(\mathscr{B}, p)$-cyclicity with $1 \leq p < +\infty$ (7.3.3). A Banach space is linearly isometric to a bo-complete space with mixed norm whose norm lattice is an order complete AM-space with unity (AL^p-space, $1 \leqslant p < \infty$), if and only if it is $(\mathscr{B}, \infty)$-cyclic $\big((\mathscr{B}, p)$-cyclic$\big)$ with respect to some complete (Badé-complete) Boolean algebra $\mathscr{B}$ of projections (7.3.2 and 7.3.4).

An interesting class of Banach–Kantorovich spaces arises as a generalization of Hilbert spaces by allowing the inner product to take values in a Stone algebra instead of the complex numbers (7.4.3). If the space is complete under the vector-valued norm defined by the Stone-algebra-valued inner product then it is called a Kaplansky–Hilbert module (7.4.5). An orthonormal set and a basis are defined in a Kaplansky–Hilbert module just as it is done in a Hilbert space (7.4.6). Kaplansky–Hilbert modules with a basis are called homogeneous. Unlike Hilbert space, not every Kaplansky–Hilbert module has a basis, but every Kaplansky–Hilbert module splits into the direct sum of homogeneous bands (7.4.7). In spite of this essential dissimilarity, these two objects have common features (7.4.9, 7.4.10, 7.5.7 (2)). Moreover, every homogeneous Kaplansky–Hilbert module is unitarily equivalent to the space of continuous function defined on comeager sets of an extremal compact space and taking values in a Hilbert space (7.4.12).

The space of all bounded endomorphisms of a Kaplansky–Hilbert module is a C^*-algebra; moreover, it is an AW^*-algebra (7.5.7), i.e. it meets additionally the axioms of a Baer $*$-ring (7.5.1). Conversely, every type I AW^*-algebra is $*$-isomorphic to such algebra (7.5.11). The endomorphism algebra is homogeneous if so is the underlying module. Any AW^*-algebra admits a Stone-algebra-valued norm, so that the original norm is a mixed norm and the algebra itself is a Banach–Kantorovich space (7.5.5). Finally, every homogeneous AW^*-algebra is representable as the space of continuous functions defined on comeager sets of an extremal compact space and taking values in the algebra of bounded operators in a Hilbert space with the strong operator topology; an arbitrary type I AW^*-algebra splits into the direct sum of such algebras (7.4.11, 7.4.12).

7.1. Spaces with Mixed Norm

In this section, we introduce spaces with mixed norm and study their simplest properties. We also consider the interrelation between the notions of dominated operator and mixed norm.

7.1.1. Recall that a normed (Banach) lattice is a vector lattice E that is simultaneously a normed (Banach) space whose norm is monotone in the following sense: if $|x| \leqslant |y|$ then $\|x\| \leqslant \|y\|$ $(x, y \in E)$, see Section 1.5. If (X, E) is a lattice-normed space with E a norm lattice of X then, by definition, $|x| \in E$ for every $x \in X$, and we may introduce some *mixed norm* in X by the formula

$$|\!|\!| x |\!|\!| := \big\| |x| \big\| \quad (x \in X).$$

In this situation, the normed space $(X, ⦀\cdot⦀)$ is called a *space with mixed norm*. In view of the inequality $\big||x| - |y|\big| \leqslant |x - y|$ and monotonicity of the norm in E we have

$$\big\||x| - |y|\big\| \leqslant ⦀x - y⦀ \quad (x, y \in X),$$

so that the vector norm $|\cdot|$ is a norm continuous operator from $(X, ⦀\cdot⦀)$ into E. All the notions introduced in Chapter 2 for a lattice-normed space make sense for a space with mixed norm, including decomposability, *br*-completeness, *d*-completeness, *bo*-completeness, etc.

A *Banach space with mixed norm* is a pair (X, E) in which E is a Banach lattice and X is a *br*-complete lattice-normed space with E-valued norm. The following proposition justifies this definition.

7.1.2. *Let E be a Banach lattice. Then $(X, ⦀\cdot⦀)$ is a Banach space if and only if the lattice-normed space (X, E) is complete with respect to relative uniform convergence.*

$\lhd$ $\Leftarrow$ Take a fundamental sequence $(x_n) \subset X$. Without loss of generality, we may assume that $⦀x_{n+1} - x_n⦀ \leqslant 1/n^3$ $(n \in \mathbb{N})$. Assign

$$e_n := |x_1| + \sum_{k=1}^{n} k|x_{k+1} - x_k| \quad (n \in \mathbb{N}).$$

Then we may estimate

$$\|e_{n+l} - e_n\| = \left\| \sum_{k=n+1}^{n+l} k|x_{k+1} - x_k| \right\|$$

$$\leqslant \sum_{k=n+1}^{n+l} k⦀x_{k+1} - x_k⦀ \leqslant \sum_{k=n+1}^{n+l} \frac{1}{k^2} \xrightarrow[n,l\to\infty]{} 0.$$

Thus, the sequence (e_n) is fundamental and hence it has a limit $e := \lim_{n\to\infty} e_n$. Since $e_{n+k} \geqslant e_n$ $(n, k \in \mathbb{N})$, we have $e = \sup e_n$. If $n \geqslant m$ then

$$m|x_{n+l} - x_n| \leqslant \sum_{k=n+1}^{n+l} k|x_{k+1} - x_k| \leqslant e_{n+l} - e_n \leqslant e;$$

consequently, $|x_{n+l} - x_n| \leqslant (1/m)e$. This means that the sequence (x_n) is *br*-fundamental. By *br*-completeness, the limit $x := br\text{-}\lim_{n\to\infty} x_n$ exists. It is clear that $\lim_{n\to\infty} ⦀x - x_n⦀ = 0$.

$\Rightarrow$ Suppose now that a sequence $(x_n) \subset X$ is *br*-fundamental; i.e., $|x_n - x_m| \leqslant \lambda_k e$ $(m, n, k \in \mathbb{N},\ m, n \geqslant k)$, where $0 \leqslant e \in E$ and $\lim_{k\to\infty} \lambda_k = 0$. Then $⦀x_n - x_m⦀ \leqslant \lambda_k \|e\| \to 0$ as $k \to \infty$; consequently, the limit $x := \lim_{n\to\infty} x_n$ exists. By continuity of the vector norm, we have $|x - x_n| \leqslant \lambda_k e$ $(n \geq k)$; therefore, $x = br\text{-}\lim x_n$. $\rhd$

7.1.3. *Let E be a Banach lattice with an order continuous norm. Then the following assertions are valid:*

(1) $(X, |\!|\!| \cdot |\!|\!|)$ *is a Banach space if and only if* $(X, |\cdot|)$ *is bo-complete;*

(2) *if X is decomposable then* $(X, |\!|\!| \cdot |\!|\!|)$ *is a Banach space if and only if* $(X, |\cdot|)$ *is a Banach–Kantorovich space.*

$\lhd$ Assertion (2) is an obvious consequence of (1). Sufficiency in (1) follows from 7.1.2. It remains to show that if $(X, |\!|\!| \cdot |\!|\!|)$ is a Banach space then $(X, |\cdot|)$ is *bo*-complete. Let a net $(x_\alpha) \subset X$ be such that $|x_\alpha - x_\beta| \leqslant e_\gamma$ for $\alpha, \beta \geqslant \alpha(\gamma)$ and suppose that e_γ decreases and *o*-converges to zero. Since $|\!|\!|x_\alpha - x_\beta|\!|\!| \leqslant \|e_\gamma\|$ and $\|e_\gamma\| \to 0$ (due to condition (A)), (x_α) is fundamental with respect to the mixed norm. Consequently, $\lim_\alpha |\!|\!|x - x_\alpha|\!|\!| = 0$ for some $x \in X$. Taking continuity of the vector norm into account, we can pass to the limit over β in the mixed norm; therefore, $|x - x_\alpha| \leqslant e_\gamma$ for $\alpha \geqslant \alpha(\gamma)$. This means that $x = bo\text{-}\lim_\alpha x_\alpha$. $\rhd$

7.1.4. Theorem. *If (X, E) is a decomposable space with mixed norm then (X', E') is a decomposable bo-complete space with mixed norm. The least dominant $|x'| \in E'$ of a functional $x' \in X'$ serves as its vector norm; in particular,*

$$\langle x, x' \rangle \leqslant \big\langle |x|, |x'| \big\rangle \quad (x \in X,\ x' \in X').$$

$\lhd$ A continuous functional $x' \in X'$ is bounded on the set $\{x \in X : |x| \leqslant e\}$ for each $e \in E_+$. Hence, in view of 4.1.11 (1), it follows that x' is dominated and

$$\big\langle e, |x'| \big\rangle = \sup \big\{ \langle x, x' \rangle : |x| \leqslant e,\ x \in X \big\}. \tag{2}$$

By Theorem 4.2.6, $M(X, \mathbb{R})$ is a Banach–Kantorovich space. Consequently, X' is a Banach–Kantorovich space too, since $X' := \big\{ x^* \in M(X, \mathbb{R}) : |x^*| \in E' \big\}$ and E' is an order ideal of $L_r(E, \mathbb{R})$. It remains to observe that the norm in X' is mixed:

$$\begin{aligned}
|\!|\!| |x'| |\!|\!| &= \sup \big\{ \big\langle e, |x'| \big\rangle : e \in E_+,\ \|e\| \leqslant 1 \big\} \\
&= \sup \Big\{ \sup \big\{ \langle x, x' \rangle : |x| \leqslant e,\ x \in X \big\} : e \in E_+,\ \|e\| \leqslant 1 \Big\} \\
&= \sup \big\{ \langle x, x' \rangle : x \in X,\ |\!|\!| |x| |\!|\!| \leqslant 1 \big\} \\
&= \|x'\|. \ \rhd
\end{aligned}$$

7.1.5. *Let (X, E) be a decomposable space with mixed norm. The canonical embedding of X into the second dual X'' preserves the vector norm. More precisely, if $\varkappa$ and λ are the canonical embeddings $X \to X''$ and $E \to E''$, respectively, then* $|\varkappa(x)| = \lambda(|x|)$ $(x \in X)$.

◁ Observe first that, according to 7.1.4, we have

$$\langle x', \varkappa(x)\rangle = \langle x, x'\rangle \leqslant \langle |x|, |x'|\rangle = \langle |x'|, \lambda(|x|)\rangle$$

for $x \in X$ and $x' \in X$. Consequently, $|\varkappa(x)| \leqslant \lambda(|x|)$. Take an arbitrary functional $e' \in E'$. By the Hahn–Banach Theorem, for a fixed $x \in X$ there is a functional $x' \in X$ such that $\langle x, x'\rangle = \langle |x|, e'\rangle$ and $\langle u, x'\rangle \leqslant \langle |u|, e'\rangle$ for all $u \in X$. Using 7.1.4 again, we conclude that

$$\langle e', \lambda(|x|)\rangle = \langle |x|, e'\rangle = \langle x, x'\rangle = \langle x', \varkappa(x)\rangle \leqslant \langle |x'|, |x(x)|\rangle \leqslant \langle e', |\varkappa(x)|\rangle.$$

Thus, $\lambda(|x|) \leqslant |\varkappa(x)|$. ▷

7.1.6. We say that the mixed norm in a lattice-normed space (X, E) is *bo-continuous* (with respect to the vector norm $|\cdot|$) if $x_\alpha \xrightarrow{(bo)} 0$ implies $|\!|\!|x_\alpha|\!|\!| \to 0$. It is easy to see that, for a normed lattice $X = E$, *bo*-continuity of the mixed norm is equivalent to order continuity of the lattice norm as defined in 1.5.3. Let E_0 denote the order ideal in E generated by $|X| := \{|x| : x \in X\}$.

(1) *Let E have the principal projection property and (X, E) be decomposable. Then the mixed norm is bo-continuous if and only if the norm in E_0 is order continuous.*

◁ Sufficiency is evident. To prove the necessity take a decreasing net (e_α) with $\inf_\alpha e_\alpha = 0$. According to 4.1.4 we may choose $x_\alpha \in X$ such that $|x_\alpha| = e_\alpha$. Then the net (x_α) *bo*-converges to zero and, by hypothesis $\lim_\alpha \|e_\alpha\| = \lim_\alpha |\!|\!|x_\alpha|\!|\!| = 0$. ▷

Suppose that the space X itself is a vector lattice. The vector norm in X is called *order continuous* if $\inf_\alpha x_\alpha = 0$ implies $\inf_\alpha |x_\alpha| = 0$ for every decreasing net $(x_\alpha) \subset X$. We say that the mixed norm is *mixed order continuous* if the vector norm is order continuous and the norm of E_0 is order continuous.

(2) *Suppose that (X, E) is a decomposable lattice-normed lattice and E_0 is order complete. Then the mixed norm in X is order continuous if and only if it is mixed order continuous.*

◁ Sufficiency is again obvious. Assume that the mixed norm is order continuous. If $e \leq |x_\alpha|$ for some $e \in E$ and decreasing net $(x_\alpha) \subset X$ with $\inf_\alpha x_\alpha = 0$ then $\|e\| \leq |\!|\!|x_\alpha|\!|\!| \to 0$. Thus, $\inf_\alpha |x_\alpha| = 0$ and the vector norm is order continuous. Next, take a decreasing sequence $(e_n) \subset E_0$ with $\inf_n e_n = 0$ and $(e_n - e_{n+1}) \perp e_{n+1}$ $(n \in \mathbb{N})$. By 4.1.4 there is $x_1 \in X$ with $|x_1| = e_1 = e_2 + (e_1 - e_2)$. By decomposability there exist $x_2 \in X$ such that $|x_2| = e_2$ and $0 \leq x_2 \leq x_1$. By induction we may find a decreasing sequence (x_n) in X with $|x_n| = e_n$. It follows that $\inf_n x_n = 0$ and by hypothesis $\lim_\alpha \|e_\alpha\| = \lim_\alpha |\!|\!|x_\alpha|\!|\!| = 0$. In view of 1.5.3 (4) the norm of E_0 is order continuous. ▷

7.1.7. Theorem. *For a decomposable bo-complete space with mixed norm (X, E), the following assertions are equivalent:*

(1) *the mixed norm in X is bo-continuous;*

(2) *if $\varkappa$ is the canonical embeddings $X \to X''$ then $\varkappa(X)$ is a decomposable subspace in (X'', E'').*

$\triangleleft$ By virtue of 7.1.6 (1) and 1.5.3 (5) it sufficient to show that $\varkappa(X)$ is decomposable in X'' if and only if $\lambda(E)$ is an order ideal in E'', where $E := E_0$ and λ is the canonical embedding $E \to E''$.

If $\lambda(E)$ is an order ideal in E'' and the representation $\lambda(|x|) = e_1 + e_2$ holds for some $x \in X$ and $0 \leq e_1, e_2 \in E$ then $e_1, e_2 \in \lambda(E)$. By the decomposability assumption there exists $x_1, x_2 \in X$ such that $x = x_1 + x_2$ and $\lambda(|x_k|) = e_k$ $(k = 1, 2)$. Conversely, suppose that $\varkappa(X)$ is decomposable in X''. Take $e \in E$ and $e'' \in E''$ with $0 \leq e'' \leq e$. From 2.1.7 (3) it follows that $e'' = |x|$ for some $x \in X$. According to 7.1.8 $\varkappa(X)$ is a submodule of the $\mathscr{L}(E'')$-module X''. Therefore, there exists an orthomorphism $0 \leq \pi \in \mathscr{L}(E'')$ such that $e'' = \pi e = |\pi x|$. Thus, $e'' \in \lambda(E)$, since $\pi x \in \varkappa(X)$. $\triangleright$

7.1.8. Theorem. *For a decomposable bo-complete space with mixed norm (X, E), the following are equivalent:*

(1) *the mixed norm in X is bo-continuous;*

(2) *every closed bo-ideal in X is a bo-band;*

(3) *every bounded functional on X is bo-continuous;*

(4) *the norm in E_0 is order continuous.*

$\triangleleft$ (1) $\Rightarrow$ (2): If $X \in (A)$ then every closed *bo*-ideal is *bo*-closed, and the claim follows from 2.1.6 (3).

(2) $\Rightarrow$ (3): The zero ideal of each bounded functional $f \in X'$ is a closed *bo*-ideal, and it remains to apply 4.4.6.

(3) $\Rightarrow$ (4): Without loss of generality we may assume that $E = |X|^{\perp\perp}$. Take a positive functional $\varphi \in E'$. By Theorem 4.4.2, we have the decomposition $\varphi = \varphi_n + \varphi_s$, where φ_n is *bo*-continuous and φ_s is *bo*-singular. Assume that $\varphi_s \neq 0$. Since the functional $x \mapsto \varphi_s(|x|)$ $(x \in X)$ is sublinear, continuous, and nonzero, by the Hahn–Banach Theorem there exists a nonzero functional $f \in X'$ such that $f(x) \leqslant \varphi_s(|x|)$ $(x \in X)$. Hence, it is clear that $0 \neq |f| \leqslant \varphi_s$. By condition (3), the functional f is *bo*-continuous; consequently, $|f|$ is *o*-continuous according to 4.3.2. This contradicts *bo*-singularity of φ_s; hence, we must have $\varphi_s = 0$ and $\varphi = \varphi_n$. Thus, $E' = E'_+ - E'_+ \subset E'_n$; therefore, the norm in E is order continuous according to 1.5.3 (8).

(4) $\Rightarrow$ (1): This is obvious. $\triangleright$

7.1.9. Now consider spaces with mixed norm which consist of dominated operators. Let (X, E) and (Y, F) be spaces with mixed norm. Denote by $\mathscr{M}(X, Y)$ the set of all linear operators from X into Y which admit bounded dominants. In other words, $T \in \mathscr{M}(X, Y)$ means that T is linear and $|Tx| \leqslant S(|x|)$ $(x \in X)$ for some $0 \leqslant S \in \mathscr{L}(E, F)$. The dominant norm $\mu(T)$ of the operator $T \in \mathscr{M}(X, Y)$ is defined by the formula

$$\mu(T) := \inf\{\|S\| : S \in \mathscr{L}(E, F) \cap \operatorname{maj}(T)\}.$$

In the special case $X = E$ and $Y = F$, we evidently obtain the space of regular operators $(\mathscr{R}(E, F), \rho) := (\mathscr{M}(X, Y), \mu)$. Observe that if E and F are Banach lattices then $\mathscr{M}(X, Y) = M(X, Y)$, since $\mathscr{R}(E, F) = L^\sim(E, F)$ in this case.

(1) *If X is decomposable and F is order complete then the norm μ is mixed: $\mu(T) = \||T|\|$. If, moreover, E and F are Banach lattices and Y is a BKS then $(\mathscr{M}(X, Y), \mathscr{R}(E, F))$ is a decomposable bo-complete space with mixed norm.*

$\lhd$ The claim follows from 4.1.2 and 4.2.6. $\rhd$

A linear operator $T : X \to Y$ is called *predominated* if there exists an operator $0 \leqslant S \in \mathscr{L}_r(E, F'')$, called a *predominant*, such that $|Tx| \leqslant S(|x|)$ $(x \in X)$. Assign $\mu^p(T) = \inf\{\|S\|\}$, where the infimum is taken over all S's indicated. The space of all predominated operators is denoted by $\mathscr{M}^p(X, Y)$.

(2) *If X is decomposable then the norm μ^p is mixed: $\mu^p(T) = \||T|\|$, since the lattice F'' is order complete and the least predominant of $|T|$ exists in $\mathscr{R}(E, F'')$.*

7.1.10. We conclude this section by considering several simple properties of dominated operators. We start with the following question: When does passing to the dual of an operator commute with the taking of the exact dominant of this operator? Let (X, E) and (Y, F) be decomposable spaces with mixed norm and take an operator $T \in \mathscr{M}(X, Y)$. Then the dual $T' \in \mathscr{L}(Y', X')$ is defined. In case F is order complete, the operator $|T|$ exists.

(1) *If $T \in M(X, Y)$ then $T' \in M(Y', X')$ and $|T'| \leqslant |T|'$.*

$\lhd$ For arbitrary $y' \in Y'$ and $e \in E_+$,

$$\langle e, |T'y'|\rangle = \langle e, |y' \circ T|\rangle \leqslant \langle |T|e, |y'|\rangle = \langle e, |T|'|y'|\rangle.$$

It follows that $|T'y'| \leqslant |T|'(|y'|)$ $(y' \in Y')$; therefore, $T \in M(Y', X')$ and $|T'| \leqslant |T|'$. $\rhd$

The inequality between the exact dominants may be strict. However, this is impossible for an order continuous dominated operator. More precisely, the following proposition holds.

(2) *Suppose that E and F are Banach lattices with E'_n and F'_n point-separating. Suppose further that F is order complete. Let (X, E) and (Y, F) be decomposable Banach spaces with mixed norm. If $T \in M_n(X, Y)$ then $T'(Y'_n) \subset X'_n$ and*

$$|T'|f' = |T|'f' \quad (f' \in F'_n).$$

◁ The desired inclusion is an obvious consequence of the *bo*-continuity of T. By Theorem 4.3.2, $|T|$ is *o*-continuous; therefore, $|T|'(F'_n) \subset E'_n$. Let T^* and $|T|^*$ be the restrictions of the operators T' and $|T|'$ onto Y'_n and F'_n, respectively. Since F'_n is an *o*-ideal in F' and $Y'_n := \{y' \in Y : |y'| \in F'_n\}$; therefore, $|T^*|$ is the restriction of $|T'|$ onto F'_n. This follows readily from 2.2.2.

Hence, in view of (1), we have $|T^*| \leqslant |T|^*$. It is clear that the operator $|T|^*$ is *o*-continuous. Then T^* is *bo*-continuous too, since it has an *o*-continuous dominant. Now, for the operator $S := T^* : Y'_n \to X'_n$ we have $S'\big((X'_n)'_n\big) \subset (Y'_n)'_n$ and, in view of what was already proven, $|S^*| \leqslant |S|^*$, i.e., $|T^{**}| \leqslant |T^*|^*$. It is clear that the restriction of T^{**} onto X coincides with T. Since E is an *o*-ideal in $(E'_n)'_n$ and $X = \{x'' \in X'' : |x''| \in E\}$, $|T|$ is the restriction of $|T^{**}|$ onto E. Thus, $|T| \leqslant |T^*|^*\big|_E$. Now if $e \in E_+$ and $f' \in F'_{n+}$ then

$$\langle e, |T|'f'\rangle = \langle |T|e, f'\rangle \leqslant \langle |T^*|^*e, f'\rangle = \langle e, |T^*|f'\rangle.$$

This means that $|T|^* \leqslant |T^*|$, which yields $|T^*| = |T|^*$ in view of what was obtained above. ▷

(3) *In the setting of Proposition (2), suppose further that X and Y satisfy condition (A). If $T \in M_n(X, Y)$ then $T' \in M_n(Y', X')$ and $|T'| = |T|'$.*

7.1.11. Suppose that E and F are Banach lattices. The following assertions hold:

(1) *Every dominated operator T is bounded and $\|T\| \leqslant \mu(T)$.*

(2) *If $T \in \mathscr{M}(X, Y)$ and $U \in \mathscr{M}(Y, Z)$ then $UT \in \mathscr{M}(X, Z)$ and $\mu(U \circ T) \leqslant \mu(U) \cdot \mu(T)$.*

(3) *If $T \in \mathscr{M}(X, Y)$ then $T' \in \mathscr{M}(Y', X')$ and $\mu(T') \leqslant \mu(T)$.*

(4) *Let X and Y be decomposable. An operator $T \in \mathscr{L}(X, Y)$ satisfies $T' \in \mathscr{M}(Y', X')$ if and only if $T \in \mathscr{M}^p(X, Y)$; in this case, $\mu^p(T) = \mu(T')$.*

◁ Take a bounded operator $T : X \to Y$. Let $T' \in \mathscr{M}(Y', X')$ and $U := |T'|$. Then U' is a dominant for T'' (see 7.1.9 (1)). Denote by S the restriction of U' onto E. It is clear that $S \in \mathscr{L}_+(E, F'')$ and, in view of 7.1.5, $|Tx| \leqslant S(|x|)$ $(x \in X)$. This means that $T \in \mathscr{M}(X, Y)$; furthermore, $\mu^p(T) \leqslant \|S\| \leqslant \|U'\| = \|U\| = \mu(T')$. Conversely, suppose that $T \in \mathscr{M}^p(X, Y)$ and an operator $S \in \mathscr{L}_+(E, F'')$ serves as a predominant for T. As above, taking account of 7.1.4 and 7.1.5, we conclude that the restriction of S' onto F' is a dominant for T'. Therefore, $T' \in M(Y', X')$ and $\mu(T') \leq \|S'\| = \|S\|$. Hence, $\mu(T') \leqslant \mu^p(T)$ and, finally, $\mu^p(T) = \mu(T')$. ▷

(5) *If there is a positive projection with norm 1 from F'' onto the image of F under the canonical embedding $F \to F''$, then $\mathscr{M}(X,Y) = \mathscr{M}^p(X,Y)$ and $\mu = \mu^p$. If, furthermore, X and Y are decomposable then, for every bounded operator $T : X \to Y$, the containments $T \in \mathscr{M}(X,Y)$ and $T' \in \mathscr{M}(Y',X')$ are equivalent and, moreover, $\mu(T) = \mu(T')$.*

$\lhd$ Let $\pi : F'' \to F$ be a positive projection such that $\|\pi\| = 1$. Take a bounded operator $T : X \to Y$. If an operator $S \in \mathscr{L}_+(E,F'')$ is a predominant for $T \in L(X,Y)$ then $\pi \circ S$ is a dominant for T and $\|\pi \circ S\| \leqslant \|S\|$. Therefore, $T \in \mathscr{M}^p(X,Y)$ implies $T \in \mathscr{M}(X,Y)$ and $\mu^p(T) \geqslant \mu(T)$. The reverse inequality is obvious. The second part of the assertion follows from the first and (4). $\rhd$

7.2. Summing Operators

Various classes of operators studied in functional analysis are defined in mixed terms of norm and order. A considerable part of the corresponding problems falls naturally within the theory of LNSs. Furthermore, new relations and possibilities arise. In this section, we consider a small fragment connected with the notion of summing operator.

7.2.1. Let E be a Banach lattice and let $f : \mathbb{R}^n \to \mathbb{R}$ be a continuous positively homogeneous function. Since E is uniformly complete, we may apply 1.5.7, so that there is a mapping $\hat{f} : E^n \to E$ continuous positively homogeneous and satisfying the condition

$$\hat{f}(t_1 e, \dots, t_n e) = e f(t_1, \dots, t_n) \quad (t_1, \dots, t_n \in \mathbb{R}^n,\ e \in E_+).$$

The element $\hat{f}(e_1, \dots, e_n)$ belongs to the ideal generated by $|e_1|, \dots, |e_n|$.

Take a function $f_{p,n} : (t_1, \dots, t_n) \mapsto (\sum_{k=1}^n |t_k|^p)^{\frac{1}{p}}$. (For $p = \infty$, we define $f_{p,n}(t_1, \dots, t_n) = \max\{|t_k| : k := 1, \dots, n\}$.) The function is positively homogeneous and continuous. Let $Q_{p,n} := \hat{f}_{p,n}$ be the corresponding mapping from E_n into E. We use a more intricate but expressive notation:

$$Q_{p,n}(e_1, \dots, e_n) =: \left(\sum_{k=1}^n |e_k|^p \right)^{\frac{1}{p}} \quad (e_1, \dots, e_n \in E).$$

Since $h : \mathscr{H}(\mathbb{R}^l) \to E$ is a lattice homomorphism with $h(dx_j) = e_j$ $(j := 1, \dots, l)$, it is easy to see that

$$\left(\sum_{k=1}^n |e_k|^1 \right)^1 = \sum_{k=1}^n |e_k|, \quad \left(\sum_{k=1}^n |e_k|^\infty \right)^{\frac{1}{\infty}} = \bigvee_{k=1}^n |e_k|.$$

The following explicit description for this element can be taken as a definition:

(1) *For all* $e_1, \dots, e_n \in E$ *and reals* $1 \le p, q \le \infty$ *with* $1/p + 1/q = 1$ *the following is valid:*

$$\left(\sum_{k=1}^{n} |e_k|^p\right)^{\frac{1}{p}} = \sup\left\{\sum_{k=1}^{n} \lambda_k e_k : \left(\sum_{k=1}^{n} |\lambda_k|^q\right)^{\frac{1}{q}} \le 1\right\}.$$

$\lhd$ Put $u := |e_1| \vee \dots \vee |e_n|$ and observe that the set

$$M := \left\{\sum_{k=1}^{n} \lambda_k e_k : \left(\sum_{k=1}^{n} |\lambda_k|^q\right)^{\frac{1}{q}} \le 1\right\}$$

is precompact in the AM-space $E(u)$. According to the Arzelà–Ascoli Theorem for a precompact set $M \subset C(Q)$ the supremum exists pointwise and defines a continuous function $(\sup M)(t) = \sup\{f(t) : f \in M\}$. By virtue of the Brothers Kreĭn–Kakutani Theorem 1.5.6 (2) the supremum $e := \sup M$ exists in $E(u)$. Moreover, $e = (\sum_{k=1}^{n} |e_k|^p)^{\frac{1}{p}}$ in $E(u)$ and we are done, since supremum of M in E coincides with e. $\rhd$

(2) *Let* $S : E \to F$ *be a positive operator and* $e_1, \dots, e_n \in E$. *Then*

$$\left(\sum_{k=1}^{n} |Se_k|^p\right)^{\frac{1}{p}} \le S\left(\sum_{k=1}^{n} |e_k|^p\right)^{\frac{1}{p}}.$$

If, in addition, S *preserves suprema (=* S *is order continuous lattice homomorphism) then we have equality instead of inequality.*

$\lhd$ Since S is linear and increasing, we use (1) and proceed as follows:

$$\begin{aligned}\left(\sum_{k=1}^{n} |Se_k|^p\right)^{\frac{1}{p}} &= \sup\left\{S\left(\sum_{k=1}^{n} \lambda_k e_k\right) : \left(\sum_{k=1}^{n} |\lambda_k|^q\right)^{\frac{1}{q}} \le 1\right\} \\ &\le S\left(\sup\left\{\sum_{k=1}^{n} \lambda_k e_k : \left(\sum_{k=1}^{n} |\lambda_k|^q\right)^{\frac{1}{q}} \le 1\right\}\right) \\ &= S\left(\sum_{k=1}^{n} |e_k|^p\right)^{\frac{1}{p}}.\end{aligned}$$

The second assertion follows immediately from (1). $\rhd$

7.2.2. We need two more properties of the mapping $Q_{p,n}$.

(1) *Let E be a Banach lattice, $e_1, \dots, e_n \in E$, $e'_1, \dots, e'_n \in E'$, and $1 \le p \le \infty$. Then the Hölder inequality holds:*

$$\sum_{k=1}^{n} |\langle e_k, e'_k \rangle| \le \left\langle \left(\sum_{k=1}^{n} |e_k|^p \right)^{\frac{1}{p}}, \left(\sum_{k=1}^{n} |e'_k|^q \right)^{\frac{1}{q}} \right\rangle.$$

$\lhd$ First we assume that $E = C(Q)$ for a compact topological space Q. Then $\mu_k := |e'_k|$ $(k = 1, \dots, n)$ are Radon measures absolutely continuous with respect to $\mu := \sum_{k=1}^{n} \mu_k$. By the Radon–Nikodým Theorem we may find functions $g_k \in L^1(Q, \mu)$ such that $d\mu_k = g_k \, d\mu$ $(k = 1, \dots, n)$. Using the Hölder inequality we obtain

$$\sum_{k=1}^{n} |\langle e_k, e'_k \rangle| \le \sum_{k=1}^{n} \langle |e_k|, |\mu_k| \rangle = \sum_{k=1}^{n} \int |e_k| \, d\mu_k$$
$$= \sum_{k=1}^{n} \int |e_k| g_k \, d\mu \le \int \left(\sum_{k=1}^{n} |e_k|^p \right)^{\frac{1}{p}} \left(\sum_{k=1}^{n} {g_k}^q \right)^{\frac{1}{q}} d\mu.$$

It remains to note that

$$\left(\sum_{k=1}^{n} {g_k}^q \right)^{\frac{1}{q}} \mu = \left(\sum_{k=1}^{n} |e'_k|^q \right)^{\frac{1}{q}},$$

since $L^1(Q, \mu)$ is isomorphic to the band in $C(Q)'$ generated by μ.

In the case of a general Banach lattice E we set $u := |e_1| + \dots + |e_n|$ and note that $e_k \in E(u)$. Let J be an isomorphism from $C(Q)$ onto $E(u)$ and $f_k := J^{-1} e_k$. From the above we deduce

$$\sum_{k=1}^{n} |\langle e_k, e'_k \rangle| = \sum_{k=1}^{n} |\langle f_k, J' e'_k \rangle| \le \left\langle \left(\sum_{k=1}^{n} |f_k|^p \right)^{\frac{1}{p}}, \left(\sum_{k=1}^{n} |J' e'_k|^q \right)^{\frac{1}{q}} \right\rangle =: t.$$

Since J preserves suprema and J' is positive, we may apply 7.2.1 (2) to derive:

$$t \le \left\langle \left(\sum_{k=1}^{n} |f_k|^p \right)^{\frac{1}{p}}, J' \left(\sum_{k=1}^{n} |e'_k|^q \right)^{\frac{1}{q}} \right\rangle = \left\langle \left(\sum_{k=1}^{n} |J f_k|^p \right)^{\frac{1}{p}}, \left(\sum_{k=1}^{n} |e'_k|^q \right)^{\frac{1}{q}} \right\rangle. \rhd$$

(2) *Let E be a Banach lattice, $e_1, \dots, e_n \in E$ and $1 \le p \le \infty$. Then*

$$\left\| \left(\sum_{k=1}^{n} |e_k|^p \right)^{\frac{1}{p}} \right\| = \sup \left\{ \sum_{k=1}^{n} \langle e_k, e'_k \rangle : \left\| \left(\sum_{k=1}^{n} |e'_k|^q \right)^{\frac{1}{q}} \right\| \le 1 \right\}.$$

$\triangleleft$ By virtue of 3.1.2 (8) we obtain

$$\left\| \left(\sum_{k=1}^{n} |e_k|^p \right)^{\frac{1}{p}} \right\| = \sup \left\{ \left\langle \left(\sum_{k=1}^{n} |e_k|^p \right)^{\frac{1}{p}}, e' \right\rangle : e' \in E'_+, \|e'\| \le 1 \right\}$$

$$= \sup_{e'} \sup \left\{ \sum_{l=1}^{m} \left\langle \sum_{k=1}^{n} \lambda'_k e_k, f'_l \right\rangle : f'_l \in E'_+, e' = \sum_{l=1}^{m} f'_l, \left(\sum_{k=1}^{n} |\lambda'_k|^q \right)^{\frac{1}{q}} \le 1 \right\}$$

$$= \sup \sup \sum_{k=1}^{n} \left\langle e_k, \sum_{l=1}^{m} \lambda'_k f'_l \right\rangle .$$

It remains to observe that if $e'_k := \sum_{l=1}^{m} \lambda'_k f'_l$ then

$$\left\| \left(\sum_{k=1}^{n} |e'_k|^q \right)^{\frac{1}{q}} \right\| \le \left\| \sum_{l=1}^{m} f'_l \right\| = \|e'\| \le 1. \ \triangleright$$

7.2.3. Suppose that $1 \le q \le p \le +\infty$. A linear operator $T : X \to Y$ is called (p, q)-*summing* if there exists a number $\varkappa > 0$ such that, for every finite set $\{x_1, \dots, x_n\} \subset X$, the following inequality holds:

$$\left\| \left(\sum_{k=1}^{n} |Tx_k|^p \right)^{\frac{1}{p}} \right\| \le \varkappa \left\| \left(\sum_{k=1}^{n} |x_k|^q \right)^{\frac{1}{q}} \right\| .$$

Denote by $\sigma_{pq}(T)$ the infimum of the set of all $\varkappa$'s indicated. Let $\mathfrak{S}_{p,q}(X, Y)$ be the space of all (p, q)-summing operators from X into Y. We can see from the definition that every summing operator T is bounded and $\|T\| \le \sigma_{pq}(T)$. It is also clear that σ_{pq} is a norm in $\mathfrak{S}_{p,q}(X, Y)$. The triangle inequality for σ_{pq} follows from sublinearity of the mapping $Q_p : (f_1, \dots, f_n) \mapsto (\sum_{k=1}^{n} |f_k|^p)^{\frac{1}{p}}$ acting from F_n

into F. The norm σ_{pq} can be calculated by the following formulas:

$$\sigma_{pq}(T) = \sup\left\{ \left\| \left(\sum_{k=1}^{n} |Tx_k|^p \right)^{\frac{1}{p}} \right\| : \ x_k \in X, \ n \in \mathbb{N}, \ \left\| \left(\sum_{k=1}^{n} |x_k|^q \right)^{\frac{1}{q}} \right\| \leq 1 \right\},$$

$$\sigma_{pq}(T) = \sup\left\{ \frac{\left\| \left(\sum\limits_{k=1}^{n} |Tx_k|^p \right)^{\frac{1}{p}} \right\|}{\left\| \left(\sum\limits_{k=1}^{n} |x_k|^q \right)^{\frac{1}{q}} \right\|} : \ x_k \in X, \ n \in \mathbb{N}, \ \sum_{k=1}^{n} |x_k| \neq 0 \right\}.$$

In case $p = q = 1$, we use the term a *summing* operator and write σ instead of σ_{11} and $\mathfrak{S}$ instead of $\mathfrak{S}_{11}$.

7.2.4. Theorem. *Let X and Y be spaces with mixed norm and suppose that Y is a Banach space. Then $(\mathfrak{S}_{p,q}(X,Y), \sigma_{pq})$ is a Banach space.*

$\triangleleft$ Let (T_n) be a fundamental sequence in $\mathfrak{S}_{pq}(X,Y)$. Then (T_n) is also fundamental in the weaker norm of the space $\mathscr{L}(X,Y)$. The space is complete; consequently, there exists an operator $T \in \mathscr{L}(X,Y)$ such that $\lim_{n\to\infty} \|T - T_n\| = 0$. For an $\varepsilon > 0$, choose a number $k_0 \in \mathbb{N}$ so that $\sigma_{pq}(T_n - T_m) \leq \varepsilon$ for all $n, m \geq k_0$. Then, for each finite set $x_1, \dots, x_l \in X$ we have

$$\left\| \left(\sum_{k=1}^{l} |x_k|^q \right)^{\frac{1}{q}} \right\| \leq 1 \Rightarrow \alpha_{nm} := \left\| \left(\sum_{k=1}^{l} |T_n x_k - T_m x_k|^p \right)^{\frac{1}{p}} \right\| \leq \varepsilon.$$

Observe that the operator $Q_{p,l}$ is continuous, since the following inequality holds due to sublinearity:

$$\left| Q_{p,l}(f_1, \dots, f_l) - Q_{p,l}(g_1, \dots, g_l) \right| \leq Q_{p,l}(f_1 - g_1, \dots, f_l - g_l).$$

Taking it into account that a vector norm is a continuous operator from Y into F, we conclude

$$\lim_{m\to\infty} \left(\sum_{k=1}^{l} |T_n x_k - T_m x_k|^p \right)^{\frac{1}{p}} = \left(\sum_{k=1}^{l} |T_n x_k - T x_k|^p \right)^{\frac{1}{p}}.$$

Thus,

$$\lim_{m\to\infty} \alpha_{nm} = \left\| \left(\sum_{k=1}^{l} |T_n x_k - T x_k|^p \right)^{\frac{1}{p}} \right\| \leq \varepsilon.$$

Arbitrariness of the choice of $x_1, \dots, x_n$ implies $\sigma_{pq}(T_n - T) \leq \varepsilon$ for $n \geq k_0$. Hence, it is clear that $T \in \mathfrak{S}_{p,q}(X,Y)$ and $\sigma_{pq}(T_n - T) \to 0$ as $n \to \infty$. $\triangleright$

7.2.5. *If $T \in \mathfrak{S}(X,Y)$, $S \in \mathscr{M}(X,Y_1)$, and $U \in \mathscr{M}(X_1,X)$, then $STU \in \mathfrak{S}_{pq}(X_1,Y_1)$ and $\sigma_{pq}(STU) \le \mu(S)\sigma_{pq}(T)\mu(U)$.*

$\lhd$ Let S_0 and U_0 be arbitrary dominants of the operators S and U respectively. Then, by applying 7.2.1 (2) twice for arbitrary $x_1,\dots,x_n \in X_1$, we obtain:

$$\left\| \left(\sum_{k=1}^n |STVx_k|^p \right)^{\frac{1}{p}} \right\| \le \left\| \left(\sum_{k=1}^n (S_0|TVx_k|)^p \right)^{\frac{1}{p}} \right\|$$

$$\le \|S_0\|\sigma_{pq}(T) \left\| \left(\sum_{k=1}^n |Vx_k|^p \right)^{\frac{1}{p}} \right\| \le \|S_0\|\sigma_{pq}(T)\|U_0\| \left\| \left(\sum_{k=1}^n |x_k|^p \right)^{\frac{1}{p}} \right\|.$$

Thus, $STU \in \mathfrak{S}_{pq}(X_1,Y_1)$ and $\sigma_{pq}(STU) \le \|S_0\|\sigma_{pq}(T)\|U_0\|$. Taking the infimum over S_0 and U_0, we come to the desired inequality. $\rhd$

7.2.6. Theorem. *Let X and Y be Banach spaces with decomposable mixed norms. A bounded operator $T : X \to Y$ is (p,q)-summing if and only if the dual T' of T is (q',p')-summing. Furthermore, $\sigma_{pq}(T) = \sigma_{q'p'}(T')$.*

$\lhd$ Suppose that $T \in \mathfrak{S}_{p,q}(X,Y)$ and take a finite n-tuple $y'_1,\dots,y'_n \in Y'$. Consecutively applying formulas 7.2.2 (2), 7.1.4, and 7.2.2 (1), we may write down the following chain:

$$\left\| \left(\sum_{k=1}^n |T'y'_k|^{q'} \right)^{\frac{1}{q'}} \right\| = \sup\left\{ \sum_{k=1}^n \langle e_k, |T'y'_k|\rangle : \ e_k \in E_+, \ \left\| \left(\sum_{k=1}^n |e_k|^q \right)^{\frac{1}{q}} \right\| \le 1 \right\}$$

$$= \sup\left\{ \sum_{k=1}^n \sup\{\langle x, T'y'_k\rangle : \ |x| \le e_k\} : \ e_k \in E_+, \ \left\| \left(\sum_{k=1}^n |e_k|^q \right)^{\frac{1}{q}} \right\| \le 1 \right\}$$

$$\le \sup\left\{ \sum_{k=1}^n \langle |Tx_k|, |y'_k|\rangle : \ x_k \in X, \ \left\| \left(\sum_{k=1}^n |x_k|^q \right)^{\frac{1}{q}} \right\| \le 1 \right\}$$

$$\le \sup\left\{ \left\langle \left(\sum_{k=1}^n |Tx_k|^p \right)^{\frac{1}{p}}, \left(\sum_{k=1}^n |y'_k|^{p'} \right)^{\frac{1}{p'}} \right\rangle : \left\| \left(\sum_{k=1}^n |x_k|^q \right)^{\frac{1}{q}} \right\| \le 1 \right\}$$

$$\le \sup\left\{ \left\| \left(\sum_{k=1}^n |Tx_k|^p \right)^{\frac{1}{p}} \right\| \left\| \left(\sum_{k=1}^n |y'_k|^{p'} \right)^{\frac{1}{p'}} \right\| : \left\| \left(\sum_{k=1}^n |x_k|^q \right)^{\frac{1}{q}} \right\| \le 1 \right\}$$

$$= \sigma_{pq}(T) \left\| \left(\sum_{k=1}^n |y'_k|^{p'} \right)^{\frac{1}{p'}} \right\|.$$

Thus, $T' \in \mathfrak{S}_{q',p'}(Y', X')$ and $\sigma_{q'p'}(T') \le \sigma_{pq}(T)$. Conversely, assume that T' is a (q', p')-summing operator. Then, in view of what was proven, T'' is (p, q)-summing and $\sigma_{pq}(T'') \le \sigma_{q'p'}(T')$. Since the canonical embeddings $X \to X''$ and $Y \to Y''$ preserve vector norms (see 7.1.5); therefore, $T \in \mathfrak{S}_{p,q}(X, Y)$, and from the definition of σ_{pq} (see 7.2.3) it is clear that $\sigma_{pq}(T) \le \sigma_{pq}(T'')$. Thus, $\sigma_{pq}(T) = \sigma_{q'p'}(T')$. $\rhd$

7.2.7. Theorem. *A linear operator $T : X \to Y$ is summing if and only if, for every sequence $(v_n) \subset X$, convergence in norm of the series $\sum_{k=1}^{\infty} |v_k|$ implies convergence in norm of the series $\sum_{k=1}^{\infty} |Tv_k|$.*

$\lhd$ Necessity is obvious. In order to prove sufficiency, assume that $T \notin \mathfrak{S}(X, Y)$. Then, for each $n \in \mathbb{N}$, we may choose a finite set $\{v_{n1}, \dots, v_{nl(n)}\} \subset X$ such that

$$\left\| \sum_{j=1}^{l(n)} |v_{nj}| \right\| \le 1, \quad \left\| \sum_{j=1}^{l(n)} |Tv_{nj}| \right\| \ge 2^n.$$

Construct a new sequence $(u_k) \subset X$ by letting $u_k := (1/2^n) v_{nj}$ for $k := l(1) + \dots + l(n-1) + j$ $(n \in \mathbb{N},\ j := 1, \dots, l(n))$. Then the series $\sum_{k=1}^{\infty} |u_k|$ converges in norm, since

$$\left\| \sum_{k=l(1)+\dots+l(m-1)+1}^{l(m)+\dots+l(m+s)} |u_k| \right\| \le \sum_{n=m}^{m+s} \frac{1}{2^n} \left\| \sum_{j=1}^{l(n)} |v_{nj}| \right\| \le \sum_{n=m}^{m+s} \frac{1}{2^n}.$$

At the same time, for the element

$$s_n := \sum_{k=l(1)+\dots+l(n-1)+1}^{l(1)+\dots+l(n)} |Tu_k| = \frac{1}{2^n} \sum_{j=1}^{l(n)} |Tv_{nj}|,$$

we have $\|s_n\| \ge 1$. Consequently, the series $\sum_{k=1}^{\infty} |Tu_k|$ diverges. A contradiction. $\rhd$

7.2.8. Theorem. *Let X be decomposable, and let one of the following conditions be satisfied:*

(1) *$E = X$ and there is a projection with norm 1 from F'' onto the image of F under the canonical embedding $F \to F''$ (i.e., F enjoys Property (B) and Property (C));*

(2) *F is an order complete AM-space with unity.*

Then $\mathscr{M}(V, W) = \mathfrak{S}(V, W)$ and $\mu = \sigma$.

$\lhd$ Take a $T \in \mathfrak{S}(V, W)$ and denote by $U(e)$ the set of sums of the form $a(\theta) := \sum_{k=1}^{n} |Tv_k|$, where $e \in E^+$, $\theta := \{v_1, \dots, v_n\} \subset X$, and $\sum_{k=1}^{n} |v_k| \le e$.

It is clear that $\|a(\theta)\| \le \sigma(T)\|e\|$, i.e., $\mathscr{U}(e)$ is norm-bounded subset in F. If (1) is true, the set $\mathscr{U}(e)$ is directed upward and the norm in F is order semicontinuous.

If (2) is true then norm-boundedness is equivalent to order-boundedness. In both cases, $T \in \mathscr{M}(X,Y)$. In view of 4.1.5,

$$\mu(T) = \sup_{\substack{\|e\|=1 \\ e\geq 0}} \| |T|e \| = \sup_{\substack{\|e\|=1 \\ e\geq 0}} \sup_{a\in U(e)} \|a\| = \sigma(T)$$

and the result follows. ▷

7.2.9. Corollary. *Assume that one of the conditions (1) or (2) in 4.2.8 is satisfied. Then, for a linear operator $T : X \to Y$, the following are equivalent:*

(1) *T is a dominated operator;*

(2) *T is a summing operator;*

(3) *for every sequence $(x_n) \subset X$, convergence of the series $\sum_{k=1}^{\infty} |x_k|$ in E implies convergence of the series $\sum_{k=1}^{\infty} |Tx_k|$ in F.*

7.2.10. If the hypotheses of Theorem 4.2.8 are met then the following formula holds:

$$\mu(T) = \sup \left\{ \left\| \sum_{k=1}^{n} |Tx_k| \right\| \ : \ x_k \in X,\ n \in \mathbb{N},\ \left\| \sum_{k=1}^{n} |x_k| \right\| \leq 1 \right\}.$$

7.2.11. We indicate some particular cases of the class $\mathfrak{G}_{pq}(X,Y)$ which may be encountered in the literature on functional analysis.

(1) *(p,q)-convex operators.* This term is conventionally used for operators of the class $\mathfrak{G}_{p,q}(X,Y)$ in case $Y = F$ and $E = \mathbb{R}$ (i.e., X is a Banach space). An operator $T : X \to F$ is called *(p,q)-convex* if there exists a number $\varkappa > 0$ such that, for all $x_1, \dots, x_n \in X$, the following inequality holds:

$$\left\| \left(\sum_{k=1}^{n} |Tx_k|^p \right)^{\frac{1}{p}} \right\| \leq \varkappa \left(\sum_{k=1}^{n} \|x_k\|^q \right)^{\frac{1}{q}}.$$

The terms a *p-superadditive* or *dominated* operator are used instead of a $(p,1)$-convex or (∞,∞)-convex operator, respectively.

(2) *(p,q)-concave operators.* The class of these operators coincides with $\mathfrak{G}_{p,q}(X,Y)$ in case $E = X$ and $F = \mathbb{R}$ (i.e., Y is a Banach space). Thus, an operator $T : E \to Y$ is called *(p,q)-concave* if there exists a number $\varkappa > 0$ such that, for all $e_1, \dots, e_n \in X$, the following holds:

$$\left(\sum_{k=1}^{n} \|Te_k\|^p \right)^{\frac{1}{p}} \leq \varkappa \left\| \left(\sum_{k=1}^{n} |e_k|^q \right)^{\frac{1}{q}} \right\|.$$

For the particular cases of (∞,q)-concave and $(1,1)$-concave operators, the respective terms *q-superadditive* and *summing* operators are used.

(3) *(p, q)-regular operators.* This class of operators arises when two Banach lattices are considered, i.e., when $X = E$ and $Y = F$. Hence, an operator $T : E \to F$ is called *(p, q)-regular* if there exists a number $\varkappa > 0$ such that, for all $e_1, \dots, e_n \in E$, the following inequality holds:

$$\left\| \left(\sum_{k=1}^{n} |Te_k|^p \right)^{\frac{1}{p}} \right\| \leq \varkappa \left\| \left(\sum_{k=1}^{n} |e_k|^q \right)^{\frac{1}{q}} \right\|.$$

Theorems 7.2.6 and 7.2.9 imply the following corollaries.

(4) *An operator $T : X \to F$ (with X a Banach space) is (p, q)-convex if and only if T' is a (q', p')-concave operator. An operator $T : E \to Y$ (with Y a Banach space) is (p, q)-concave if and only if T' is a (p', q')-convex operator.*

(5) *If there is a positive projection with norm 1 from F'' onto the image of F under the canonical embedding then an operator $T : E \to F$ is $(1, 1)$-regular if and only if T is regular.*

7.3. Isometric Classification

In this section, we are going to show that lattice-normed spaces provide reasonable grounds for constructing isometric classification of spaces with mixed norm. To this end, we chose a small number of natural questions connected with norms take values AM- and AL^p-spaces.

7.3.1. We now turn to the following natural question: Which Banach spaces are linearly isometric to Banach spaces with mixed norm? We confine our study to the case in which the norm lattice is an AM- or an AL^p-space. Below, we present several results in this direction. We start with the necessary definitions.

Let X be a normed space. Suppose that $\mathscr{L}(X)$ has a complete Boolean algebra of norm-one projections $\mathscr{B}$ which is isomorphic to B. In this event we will identify the Boolean algebras $\mathscr{B}$ and B, writing $B \subset \mathscr{L}(X)$. Say that X is a *normed B-space* if $B \subset \mathscr{L}(X)$ and for every partition of unity $(b_\xi)_{\xi \in \Xi}$ in B the two conditions hold:

(1) If $b_\xi x = 0$ $(\xi \in \Xi)$ for some $x \in X$ then $x = 0$;

(2) If $b_\xi x = b_\xi x_\xi$ $(\xi \in \Xi)$ for $x \in X$ and a family $(x_\xi)_{\xi \in \Xi}$ in X then $\|x\| \leq \sup\{\|b_\xi x_\xi\| : \xi \in \Xi\}$.

Conditions (1) and (2) amount to the respective conditions (1′) and (2′):

(1′) To each $x \in X$ there corresponds the greatest projection $b \in B$ such that $bx = 0$;

(2′) If x, (x_ξ), and (b_ξ) are the same as in (2) then $\|x\| = \sup\{\|b_\xi x_\xi\| : \xi \in \Xi\}$.

From (2′) it follows in particular that

$$\left\|\sum_{k=1}^{n} b_k x\right\| = \max_{k=1,\dots,n} \|b_k x\|$$

for $x \in X$ and pairwise disjoint projections $b_1, \dots, b_n$ in B.

Given a partition of unity (b_ξ), we refer to $x \in X$ satisfying the condition $(\forall \xi \in \Xi)\, b_\xi x = b_\xi x_\xi$ as a *mixing* of (x_ξ) by (b_ξ). If (1) holds then there is a unique mixing x of (x_ξ) by (b_ξ). In these circumstances we naturally call x the *mixing* of (x_ξ) by (b_ξ). Condition (2) maybe paraphrased as follows: The unit ball U_X of X is closed under mixing or is *mix-complete*.

7.3.2. Theorem. *For a Banach space X the following are equivalent:*

(1) *X is a decomposable space with mixed norm whose norm lattice is an AM-space with unity;*

(2) *X is a Banach B-space.*

$\lhd$ (1) $\Rightarrow$ (2): Appeal to the appropriate definitions and 2.1.3.

(2) $\Rightarrow$ (1): Suppose that X is a Banach B-space and $J : B \to \mathscr{B}$ is a corresponding isomorphism of B onto the Boolean algebra of projections $\mathscr{B}$. Denote by E the ideal generated by $\bar{1}$ in the universally complete K-space of all B-valued resolutions of unity (cf. 1.4.3). Take the finite-valued element $d := \sum_{i=1}^{n} \lambda_i b_i \in E$, where $\lambda_1, \dots, \lambda_n \in \mathbb{R}$, the family $\{b_1, \dots, b_n\}$ is a partition of unity in B, and λb stands for the spectral function $e : \mu \mapsto e(\mu) \in B$ equal to the zero of B for $\mu \le \lambda$ and equal to the unity of B for $\mu > \lambda$. Put $J(\alpha) := \sum_{i=1}^{n} \lambda_i J(b_i)$ and note that $J(\alpha)$ is a bounded linear operator in X. Calculating the norm of $J(\alpha)$, obtain

$$\begin{aligned}\|J(\alpha)\| &= \sup_{\|x\|\le 1} \|J(\alpha)x\| = \sup_{\|x\|\le 1} \sup_{l=1,\dots,n} \{\|\pi_l x\| |\lambda_l|\}\\ &= \sup_{l=1,\dots,n} \sup\{\|\pi_l x\|\, |\lambda_l| : \|x\| \le 1\} = \max\{|\lambda_1|, \dots, |\lambda_n|\}.\end{aligned}$$

On the other hand, the norm $\|\alpha\|_\infty$ of a member α of the AM-space E coincides with $\max\{|\lambda_1|, \dots, |\lambda_n|\}$ too. Hence, J is a linear isometry of the subspace E_0 of finite-valued members of E to the algebra of bounded operators $\mathscr{L}(X)$. It is also clear that $J(\alpha\beta) = J(\alpha) \circ J(\beta)$ for all α, $\beta \in E_0$. Since E_0 is norm dense in E and $\mathscr{L}(X)$ is a Banach algebra; therefore, we may extend J by continuity to an isometric isomorphism of E onto a closed subalgebra of $\mathscr{L}(X)$. Assigning $x\alpha := \alpha x := J(\alpha)x$ for $x \in X$ and $\alpha \in E$, we make X into a unital E-module so that

$$\|\alpha x\| \le \|x\|\, \|\alpha\|_\infty \quad (\alpha \in E,\ x \in X).$$

Moreover, $\alpha U_X + \beta U_X \subset U_X$ for $|\alpha| + |\beta| \le \mathbf{1}$. Define the mapping $p : X \to E_+$ by the formula

$$p(x) := \inf\{\alpha \in E_+ : x \in \alpha U_X\} \quad (x \in X),$$

with the infimum taken in the K-space E. If $p(x) = \mathbf{0}$ then to $\varepsilon > \mathbf{0}$ there are a partition of unity $(\pi_\xi) \subset B$ and a family $(\alpha_\xi) \subset E_+$ such that $\pi_\xi \alpha_\xi \le \varepsilon \mathbf{1}$ and $x \in \alpha_\xi U_X$ for all ξ. But then $\pi_\xi x \in \pi_\xi \alpha_\xi U_X \subset \varepsilon U_X$. Since the unit ball U_X is closed under mixing; therefore, $x = \operatorname{mix}(\pi_\xi x_\xi) \in \varepsilon U_X$. The arbitrary choice of $\varepsilon > \mathbf{0}$ implies $x = \mathbf{0}$. If $x \in \alpha U_X$ and $y \in \beta U_X$ for some $\alpha, \beta \in E_+$, then, putting $\gamma := \alpha + \beta + \varepsilon \mathbf{1}$, we may write down

$$x + y = \gamma\,(\gamma^{-1}x + \gamma^{-1}y) \in \gamma\,(\gamma^{-1}\alpha U_X + \gamma^{-1}\beta U_X) \subset \gamma U_X.$$

Consequently, $p(x+y) \le \alpha + \beta + \varepsilon\mathbf{1}$; and taking the infimum over α, β, and ε yields $p(x+y) \le p(x) + p(y)$. Furthermore, granted $\pi \in B$ and $x \in X$, observe the equalities

$$\begin{aligned}\pi p(x) &= \inf\{\pi\alpha : \mathbf{0} \le \alpha \in E,\ x \in \alpha U_X\} \\ &= \inf\{\alpha \in E_+ : \pi x \in \alpha U_X\} = p(\pi x).\end{aligned}$$

But then, for $\alpha = \sum \lambda_i \pi_i$, with $\{\pi_1, \dots, \pi_n\}$ a partition of unity in B, we see that

$$p(\alpha x) = \sum \pi_i p(\lambda_i x) = \sum_{i=1}^{n} \pi_i |\lambda_i| p(x) = |\alpha| p(x).$$

Hence, $p(\alpha x) = |\alpha| p(x)$ for all $\alpha \in E$. Therefore, (X, p, E) is a decomposable lattice-normed space.

Show now that the norm of X is a mixed norm; i.e., $\|x\| = \|p(x)\|_\infty \quad (x \in X)$. Take $0 \ne x \in X$ and put $y = x/\|x\|$. Then $y \in U_X$ and $p(y) \le \mathbf{1}$. Consequently, $p(x) \le \|x\|\mathbf{1}$ or $\|p(x)\|_\infty \le \|x\|\|\mathbf{1}\|_\infty = \|x\|$. Conversely, given $\varepsilon > 0$, we may find a partition of unity $(\pi_\xi)_{\xi\in\Xi}$ in $\mathfrak{Pr}(E)$ and a family $(\alpha_\xi)_\xi \subset E_+$ such that $\pi_\xi \alpha_\xi \le p(x) + \varepsilon\mathbf{1} \le (\|p(x)\|_\infty + \varepsilon)\mathbf{1}$ and $x \in \alpha_\xi U_X \quad (\xi \in \Xi)$. Whence $\pi_\xi x_\xi \in \pi_\xi \alpha_\xi U_X \subset (\|p(x)\|_\infty + \varepsilon)\pi_\xi \mathbf{1} U_X \subset (\|p(x)\|_\infty + \varepsilon) U_X$. Consequently, $\|\pi_\xi x_\xi\| \le \|p(x)\|_\infty + \varepsilon$. Considering the arbitrary choice of $\varepsilon > 0$ together with 7.3.1 (2), we deduce $\|x\| \le \|p(x)\|_\infty$. ▷

7.3.3. A normed B-space X is *B-cyclic* if we may find in X a mixing of each norm-bounded family by any partition of unity in B. Taking 7.3.2 into consideration, we may assert that X is a B-cyclic normed space if and only if, given a partition of unity $(b_\xi) \subset B$ and a family $(x_\xi) \subset U_X$, we may find a unique element $x \in U_X$ such that $b_\xi x = b_\xi x_\xi$ for all ξ.

(1) Theorem. *A Banach space is linearly isometric to a bo-complete space with mixed norm, whose norm lattice is an order complete AM-space with unity, if and only if it is $\mathscr{B}$-cyclic with respect to some complete Boolean algebra $\mathscr{B}$ of projections.*

$\triangleleft$ In view of 7.3.2 it is sufficient to observe that a Banach B-space is B-cyclic if and only if it is disjointly complete as a lattice-normed space. $\triangleright$

A Boolean algebra $\mathscr{B}$ is called *Badé-complete* if it is order complete and, for every increasing net of projections $(b_\alpha) \subset \mathscr{B}$, from $b = \sup b_\alpha$ it follows that $\langle bx, x' \rangle = \lim_\alpha \langle b_\alpha x, x' \rangle$ whenever $x \in X$ and $x' \in X'$. We say that a set $C \subset X$ is *$\mathscr{B}$-bounded* if bounded is the following set of reals:

$$\left\{ \left\| \sum_{k=1}^{n} b_k x_k \right\| : x_1, \dots, x_n \in C,\ b_1, \dots, b_n \in \mathscr{B},\ n \in \mathbb{N}, b_k \circ b_l = 0 \quad (k \neq l) \right\}.$$

A Banach space X is called *$(\mathscr{B}, p)$-cyclic* $(1 \leq p < \infty)$ if the following conditions are satisfied:

(a) there is a complete Boolean algebra $\mathscr{B}$ of norm-one projections in X;

(b) $\|bx + b^\perp y\|^p = \|bx\|^p + \|b^\perp y\|^p$ for all $b \in \mathscr{B}$ and $x, y \in X$;

(c) for each $\mathscr{B}$-bounded family in X, there exists a mixing by an arbitrary partition of unity (with the same index set) in $\mathscr{B}$;

(2) *If a Banach space X is $(\mathscr{B}, p)$-cyclic and $\mathscr{B}$ is Badé-complete then, for every decreasing net $(b_\alpha) \subset \mathscr{B}$, from $\inf_\alpha b_\alpha = 0$ it follows that $\lim_n \|b_\alpha x\| = 0$ for all $x \in X$.*

$\triangleleft$ Let (b_α) be a decreasing net of projections; suppose that $\inf b_\alpha = 0$ but $\|b_\alpha x\|$ does not converge to zero. Without loss of generality, we may assume that $\|b_\alpha x\| > \varepsilon > 0$ for all α. Since the Boolean algebra is Badé-complete, the net $b_\alpha x$ converges to zero in the weak topology $\sigma(X, X')$. By the Mazur Theorem, there exists a convex combination $y = \sum_{k=1}^{n} \lambda_k b_{\alpha(k)} x$, $\lambda_k \in \mathbb{R}_+$, $\sum_{k=1}^{n} \lambda_k = 1$, such that $\|y\| < (\varepsilon/2)$. If $b_\alpha \leq b_{\alpha(k)}$, $k = 1, \dots, n$, then $b_\alpha y = b_\alpha x$ and, by condition (b) in the definition of $(\mathscr{B}, p)$-cyclicity, we have:

$$\varepsilon^p < \|b_\alpha x\|^p \leq \|b_\alpha y\|^p + \left\|b_\alpha^\perp y\right\|^p = \|y\|^p < (\varepsilon/2)^p.$$

A contradiction is obtained. Hence, $\lim \|b_\alpha x\| = 0$. $\triangleright$

(3) *Under the hypotheses of Proposition* (1), *there is a unique mixing $x = \operatorname{mix}_{\xi \in \Xi} (b_\xi x_\xi)$. Moreover, $x = \sum_{\xi \in \Xi} b_\xi x_\xi$ and*

$$\|x\|^p = \sum_{\xi \in \Xi} \|b_\xi x_\xi\|^p.$$

7.3.4. Theorem. *A Banach space is linearly isometric to a bo-complete space with mixed norm whose norm lattice is an AL^p-space with $1 \le p < +\infty$ if and only if the space is $(\mathscr{B}, p)$-cyclic with respect to some Badé-complete Boolean algebra of projections.*

$\triangleleft$ We only need to prove sufficiency. Suppose that X is a $(\mathscr{B}, p)$-cyclic Banach space with $1 \le p < \infty$ and let $\mathscr{B}$ be a Badé-complete Boolean algebra of projections in X. If $b \in \mathscr{B}$ is a nonzero projection then $\langle bx, x' \rangle \neq 0$ for suitable $x \in X$ and $x' \in X'$. At the same time, the function $b \mapsto \langle bx, x' \rangle$ $(b \in \mathscr{B})$ is additive and, due to Badé-completeness, it is o-continuous as well. Thus, there exists a point-separating set of o-continuous measures on $\mathscr{B}$. Let Z be a universally complete K-space of all resolutions of unity in the algebra $\mathscr{B}$. Then the base of Z is isomorphic to $\mathscr{B}$; consequently, there exists an order-dense ideal in Z, on which an essentially positive o-continuous functional Φ is defined, see 1.4.10. Denote by $L^1(\Phi)$ the greatest order-dense ideal onto which Φ can be extended by o-continuity. Assign $L^\infty(\Phi) := \bigcup_{n=1}^{\infty}[-n\mathbf{1}, n\mathbf{1}]$; i.e., $L^\infty(\Phi)$ is the o-ideal in Z generated by the unity $\mathbf{1} \in Z$. The mapping that associates with an element $z \in L^1(\Phi)$ the functional $\alpha \mapsto \Phi(\alpha z)$ $\big(\alpha \in L^\infty(\Phi)\big)$, is a linear and order isomorphism of $L^1(\Phi)$ onto $L^\infty(\Phi)'_n$, see 3.4.8. In the same way as in 7.3.2, we may establish that X can be endowed with the structure of a faithful unital module over the ring $L^\infty(\Phi)$; furthermore, $\|\alpha x\| \le \|\alpha\| \cdot \|x\|$ $\big(\alpha \in L^\infty(\Phi),\ x \in X\big)$.

Now, take an arbitrary element $x \in X$ and define a function $\varphi_x : b \mapsto \|bx\|^p$ $(b \in \mathscr{B})$. From the $(\mathscr{B}, p)$-cyclicity condition (see 7.3.3 (b)) it is clear that φ_x is additive and o-continuous. Given $\alpha \in L^\infty(\Phi)$, assign

$$\Phi_x(\alpha) := \int\limits_{-\|\alpha\|}^{\|\alpha\|} \lambda d\varphi_x(e_\lambda^\alpha),$$

where (e_λ^α) is the spectral function of the element α, and the integral is defined as the r-limit of the integral sums

$$\sum_{n=-k}^{k-1} l_n \varphi_x(e_{\lambda_{n+1}}^\alpha - e_{\lambda_n}^\alpha) \quad l_n \in [\lambda_n, \lambda_{n+1}),$$

over refining partitions of the interval $[-\|\alpha\|, \|\alpha\|]$ of the real line:

$$-\|\alpha\| =: \lambda_{-k} < \lambda_{-k+1} < \dots < \lambda_{k-1} < \lambda_k = \|\alpha\|.$$

Then Φ_x is a positive o-continuous functional on $L^\infty(\Phi)$; consequently, there exists a unique positive element $z \in L^1(\Phi)$ such that $\Phi_x(\alpha) = \Phi(\alpha z)$ $(\alpha \in L^\infty(\Phi))$. Define $|x| := \sqrt[p]{z}$. Introduce an AL^p-space $L^p(\Phi)$ by the following formulas:

$$L^p(\Phi) := \big\{z \in Z : |z|^p \in L^1(\Phi)\big\}, \quad \|z\|_p := \big(\Phi(|z|^p)\big)^{\frac{1}{p}} \quad \big(z \in L^p(\Phi)\big).$$

Thus, a mapping $|\cdot|: X \to L^p(\Phi)$ is defined so that

$$\|bx\|^p = \Phi(b|x|^p) \quad (b \in \mathscr{B},\ x \in X).$$

Using the last relation, show that $\|\cdot\|$ is a decomposable norm and X is a space with mixed norm. First of all, observe that $\|x\| = \big\||x|\big\|_p$ $(x \in X)$. In particular, if $|x| = 0$ then $x = 0$. Next, for every partition of unity $\{b_1, \dots, b_n\} \subset \mathscr{B}$, every set $\{x_1, \dots, x_n\} \subset X$, and every element $b \in \mathscr{B}$, we have (see 7.3.3 (b)):

$$\Phi\left(b\left|\sum_{k=1}^n b_k x_k\right|^p\right) = \left\|\sum_{k=1}^n b b_k x_k\right\|^p = \sum_{k=1}^n \|b b_k x_k\|^p = \Phi\left(b\sum_{k=1}^n b_k|x_k|^p\right).$$

Since the projection $b \in \mathscr{B}$ is arbitrary and the elements $b_k|x_k|$ are pairwise disjoint in $L^p(\Phi)$, we have

$$\left|\sum_{k=1}^n b_k x_k\right| = \left(\sum_{k=1}^n b_k|x_k|^p\right)^{\frac{1}{p}} = \sum_{k=1}^n b_k|x_k|.$$

Hence, by letting $\alpha := \lambda_1 b_1 + \dots + \lambda_n b_n$ and $x_k := \lambda_k x$ $(k := 1, \dots, n)$, we obtain $|\alpha x| = |\alpha| \cdot |x|$. The latter equality is true for all $\alpha \in L^\infty(\Phi)$. Indeed, if $\alpha_n \to \alpha$ in $L^\infty(\Phi)$ then we have

$$\Phi(b|\alpha_n x|^p) = \|b\alpha_n x\|^p \to \|b\alpha x\|^p = \Phi(b|\alpha x|^p)$$

for all $x \in X$ and $b \in \mathscr{B}$. Therefore,

$$\Phi(b|\alpha x|^p) = \Phi\big(b(|\alpha||x|)^p\big) = \lim \Phi\big(b(|\alpha_n||x|)^p\big).$$

Since $b \in \mathscr{B}$ was chosen arbitrarily, $|\alpha x| = |\alpha||x|$.

Now, take $x, y \in X$ and numbers $0 \le \alpha, \beta \in \mathbb{R}$ such that $\alpha + \beta \le 1$. Employing the triangle inequality for the norm $\|\cdot\|_p$ and the Hölder inequality for finite sums, we obtain

$$\begin{aligned}
\Phi(b|\alpha x + \beta y|^p) &\le \big(\Phi(b|\alpha x|^p)\big)^{\frac{1}{p}} + \big(\Phi(b|\beta y|^p)\big)^{\frac{1}{p}} \\
&= \left(\alpha^{1-\frac{1}{p}}\big(\alpha\Phi(b|x|^p)\big)^{\frac{1}{p}} + \beta^{1-\frac{1}{p}}\big(\beta\Phi(b|y|^p)\big)^{\frac{1}{p}}\right)^p \\
&\le \big(\alpha\Phi(b|x|^p) + \beta\Phi(b|y|^p)\big)\big(\alpha^{(1-\frac{1}{p})p'} + \beta^{(1-\frac{1}{p})p'}\big) \\
&\le \Phi\big(b(\alpha|x|^p + \beta|y|^p)\big).
\end{aligned}$$

Using again arbitrariness of $b \in \mathscr{B}$, we infer that $|\alpha x + \beta y|^p \le \alpha|x|^p + \beta|y|^p$. If $\alpha := \alpha_1 b_1 + \dots + \alpha_n b_n$ and $\beta := \beta_1 b_1 + \dots + \beta_n b_n$, with $\alpha_k, \beta_k \in \mathbb{R}_+$, $\alpha_k + \beta_k = 1$ $(k := 1, \dots, n)$, then

$$|\alpha x + \beta y|^p = \left|\sum_{k=1}^n b_k(\alpha_k x + \beta_k y)\right|^p = \sum_{k=1}^n b_k|\alpha_k x + \beta_k x|$$

$$\le \sum_{k=1}^n b_k(\alpha_k|x|^p + \beta_k|y|^p) = \alpha|x|^p + \beta|y|^p.$$

As above, using uniform approximation by finite-valued elements of $L^p(\Phi)$, we may say that the inequality is true for $\alpha, \beta \in L^\infty(\Phi)_+$ whenever $\alpha + \beta \le \mathbf{1}$. Now, assign $\gamma := |x| + |y| + 2\varepsilon\mathbf{1}$, $\alpha := \gamma^{-1}(|x| + \varepsilon\mathbf{1})$, and $\beta := \gamma^{-1}(|y| + \varepsilon\mathbf{1})$, $\varepsilon > 0$. It is clear that $\alpha \ge 0$, $\beta \ge 0$, and $\alpha + \beta \le \mathbf{1}$. Moreover, if $x_1 := (|x| + \varepsilon\mathbf{1})^{-1}x$ and $y_1 := (|y| + \varepsilon\mathbf{1})^{-1}y$ then $|x_1| \le \mathbf{1}$ and $|y_1| \le \mathbf{1}$; consequently,

$$|\gamma^{-1}(x + y)| = |\alpha x_1 + \beta y_1| \le (\alpha|x|^p + \beta|y|^p)^{\frac{1}{p}} \le \mathbf{1}.$$

Thus, $|x + y| \le \gamma$. Passing to the limit as $\varepsilon \to 0$, we obtain the triangle inequality for $|x|$.

Thus, $(X, |\cdot|, L^p(\Phi))$ is an LNS with decomposable norm and, furthermore, $\big\||x|\big\|_p = \|x\|$ $(x \in X)$. In view of 7.1.2, X is br-complete. In order to prove disjoint completeness, take a family $(x_\xi)_{\xi\in\Xi}$ that is vector norm-bounded in X. If $\Theta \subset \Xi$ is a finite set and $(b_\nu)_{\nu\in\Theta}$ is a partition of unity in $\mathscr{B}$, then

$$\left\|\sum_{\nu\in\Theta} b_\nu x_\nu\right\|^p = \sum_{\nu\in\Theta} \|b_\nu x_\nu\|^p = \sum_{\nu\in\Theta} \Phi(b_\nu|x_\nu|^p) \le \Phi(e^p),$$

where $e := \sup_{\xi\in\Xi}|x_\xi|$. Hence, it is clear that the family (x_ξ) is $\mathscr{B}$-bounded; therefore, there exists a mixing $\operatorname{mix}(\pi_\xi x_\xi)$ by an arbitrary partition of unity (π_ξ). Finally, $(X, |\cdot|, L^p(\Phi))$ is a BKS and the original norm in X is a mixed norm. ▷

7.3.5. Theorem. *A Banach space Y is linearly isometric to $L^p(\mu, X)$, $1 \leqslant p < \infty$, for some finite measure μ and Banach space X, if and only if there is a closed subspace Z in Y and a Badé-complete Boolean algebra $\mathscr{B}$ of projections in Y with the following properties:*

(1) *Y is $(\mathscr{B}, p)$-cyclic;*

(2) *X and Z are linearly isometric;*

(3) *the set of all sums of the form $\sum_{k=1}^n \pi_k z_k$, with $\pi_1, \dots, \pi_n \in \mathscr{B}$, $z_1, \dots, z_n \in Z$, and $n \in \mathbb{N}$, is dense in Y;*

(4) *for every* $0 \neq \pi \in \mathscr{B}$, *the function* $z \mapsto \|\pi z\|$ *is a positive constant on the sphere* $\{z \in Z : \|z\| = 1\}$.

$\lhd$ Necessity of these properties is checked directly. To proof sufficiency, suppose that Y, Z and $\mathscr{B}$ satisfy (1)–(4). By virtue of Theorem 7.3.4 we may assume that Y is a space with mixed norm, its norm lattice is $L^p(\mu)$ for some measure μ, and the relation

$$\|\pi y\| = \left(\int \pi |y|^p \, d\mu\right)^{\frac{1}{p}} \quad (y \in Y, \pi \in \mathscr{B})$$

holds. Put $e := |z|/\|z\|$, $z \in Z$. By hypothesis e does not depend on z, since

$$\int \pi e^p \, d\mu = \int \pi (|z|/\|z\|)^p \, d\mu = (\|\pi z\|/\|z\|)^p = \text{const}$$

for each $0 \neq \pi \in \mathscr{B}$. From this we may also see that $\pi e \neq 0$ for every nonzero π; thus, e^p is an order-unity in $L^1(\mu)$. Put $\nu := e^p \mu$ and $|y|_0 := |y|/e$, $y \in Y$. Clearly, ν is a finite measure and $(Y, L^p(\nu))$ is a *bo*-complete lattice-normed space, since $y \mapsto |y|_0$ is an $L^p(\nu)$-valued decomposable norm in Y. Moreover,

$$\|\pi y\|^p = \int \pi |y|_0^p \, d\nu \quad (y \in Y, \ \pi \in \mathscr{B})$$

and in particular

$$\left(\frac{\|\pi z\|}{\|z\|}\right)^p = \frac{1}{\|z\|} \int \pi |z|_0^p \, d\nu = \nu(\pi).$$

Let J denote a linear isometry from Z onto X. Take $y = \sum_{k=1}^n \pi_k z_k$ with $z_1, \dots, z_n \in Z$ and $\pi_1, \dots, \pi_n \in \mathscr{B}$ pairwise disjoint and define

$$J(y) := \sum_{i=1}^n \pi_i J(z_i) \in L^p(\nu),$$

where $\pi_i x$ denotes the vector-function with constant value x on π_i and vanishing in $\pi_i^\perp$. Then, using the fact that Y is $\mathscr{B}_p$-cyclic, we may derive

$$\|J(y)\| = \left(\int |J(y)|_0^p \, d\nu\right)^{\frac{1}{p}} = \left(\sum_{i=1}^n \nu(\pi_i)\|z_i\|^p\right)^{\frac{1}{p}}$$

$$= \left(\sum_{i=1}^n \|\pi_i z_i\|^p\right)^{\frac{1}{p}} = \left\|\sum_{i=1}^n \pi_i z_i\right\| = \|y\|.$$

It follows that J is a linear isometry of dense subspaces, which can be extended by continuity up to an isometry of spaces Y and $L^p(\nu, X)$. It is also clear that J preserves the vector norm, i.e., J is an isometry also in the sense of vector norms. $\rhd$

Observe a corollary to Theorem 7.3.5. Let μ and ν be finite measures. Denote by $L^{p,q}(\mu \otimes \nu)$ the space of measurable functions u of two variables having finite mixed norm

$$\|u\|_{pq} := \left(\int \left(\int |u(s,t)|^q d\mu(s) \right)^{\frac{p}{q}} d\nu(t) \right)^{\frac{1}{p}}.$$

Equivalent functions are conventionally identified.

7.3.6. Theorem. *A Banach space Y is linearly isometric to the space $L^{p,q}(\mu \otimes \nu)$ for some finite measures μ and ν if and only if there are Badé-complete Boolean algebras $\mathscr{A}$ and $\mathscr{B}$ of projections in Y and an element $e \in Y$ such that the following are satisfied:*

(1) *Y is $(\mathscr{B}, p)$-cyclic;*

(2) *for all $\pi \in \mathscr{B}$, disjoint projections $\rho, \sigma \in \mathscr{A}$, and numbers α, β, the equality $\|\alpha\pi\rho e + \beta\pi\sigma e\|^q = \|\alpha\pi\rho e\|^q + \|\beta\pi\sigma e\|^q$ holds;*

(3) *the linear span of the set $\{\pi\rho e : \pi \in \mathscr{B},\ \rho \in \mathscr{A}\}$ is dense in Y;*

(4) *for every $0 \neq \pi \in \mathscr{B}$, the function $\rho \mapsto \|\pi\rho e\|/\|\rho e\|$, $0 \neq \rho \in \mathscr{A}$, is a positive constant.*

$\lhd$ Denote by Z the closure of the linear span of the set $\{\rho e : \rho \in \mathscr{A}\}$. It follows from (3) that the set

$$\left\{ \sum_{i=1}^{n} \pi_i z_i : \pi_1, \dots, \pi_n \in \mathscr{B}, z_1, \dots, z_n \in Z,\ n \in \mathbb{N} \right\}$$

is dense in Y. Let $0 \neq \pi \in \mathscr{B}$ and $z = \sum_{i=1}^{n} \alpha_i \rho_i e$, where $\rho_1, \dots, \rho_n \in \mathscr{A}$ are pairwise disjoint and $\alpha_1, \dots, \alpha_n \in \mathbb{R}$. Using (2) and (4), we may deduce

$$\frac{\|\alpha_1 \pi \rho_1 e\|^q}{\|\alpha_1 \rho_1 e\|^q} = \dots = \frac{\|\alpha_n \pi \rho_n e\|^q}{\|\alpha_n \rho_n e\|^q} = \frac{\sum\limits_{i=1}^{n} \|\alpha_i \pi \rho_i e\|^q}{\sum\limits_{i=1}^{n} \|\alpha_i \rho_i e\|^q} = \frac{\|\pi z\|^q}{\|z\|^q}.$$

Now, we see that the function $z \mapsto \|\pi z\|$ is a positive constant on the unit sphere of Z whatever the projections $0 \neq \pi \in \mathscr{B}$ may be. Let Z_+ be the closure of the conic hull of $\{\rho e : \rho \in \mathscr{A}\}$. Then the space Z, ordered by the cone Z_+, is a Banach lattice. Moreover, Z is an AL^q-space with weak order-unity e. Indeed, if $z_1 = \sum_{i=1}^{n} \alpha^i \rho_i e$

and $z_2 = \sum_{j=1}^{n} \beta^j \sigma_j e$, with $0 \leq \alpha^1, \dots, \beta^m \in \mathbb{R}$, and $\rho_1, \dots, \rho_n, \sigma_1, \dots, \sigma_m \in \mathscr{A}$ pairwise disjoint then, in view of (2),

$$\|z_1 + z_2\|^q = \sum_{i=1}^{n} \|\alpha^i \rho_i e\|^q + \sum_{j=1}^{n} \|\beta^j \sigma_j e\|^q = \|z_1\|^q + \|z_2\|^q.$$

It should be also noted that elements of the form z_1 and z_2 constitute a dense subset in the cone Z_+ and arbitrary $u_1, u_2 \in Z$ are disjoint if and only if there exists a projection $\rho \in \mathscr{A}$ such that $\rho u_1 = u_1$ and $\rho u_2 = 0$. In view of Theorem 1.5.6 (3) Z is isometrically and latticially isomorphic to the Banach lattice $L^q(\mu)$ for a suitable finite measure μ. It remains to apply Theorem 7.3.5 with $X := L^q(\mu)$ and observe that the Banach lattices $L^p(\nu, L^q(\mu))$ and $L^{p,q}(\mu \times \nu)$ are isometrically isomorphic. ▷

7.4. Kaplansky–Hilbert Modules

In this section we introduce the class of AW^*-modules as *bo*-complete Banach modules with mixed norm, consider some important structural properties, and establish that every such module may be constructed, up to isomorphism, by forming direct sums and "smearing" of a Hilbert space over an extremal compact space.

7.4.1. We recall some preliminaries concerning complex algebras. Note also that by an *algebra* we always mean a unital associative algebra. An *involutive algebra* or $*$-*algebra* A is a complex algebra with *involution*, i.e. a mapping $x \mapsto x^*$ $(x \in A)$ satisfying the conditions:

(1) $x^{**} = x \quad (x \in A)$;

(2) $(x + y)^* = x^* + y^* \quad (x, y \in A)$;

(3) $(\lambda x)^* = \lambda^* x^* \quad (\lambda \in \mathbb{C},\ x \in A)$;

(4) $(xy)^* = y^* x^* \quad (x, y \in A)$.

An element x of an involutive algebra A is called *hermitian* if $x^* = x$. An element x of A is called *normal* if $x^* x = x x^*$. A hermitian element p is a *projection* whenever p is an idempotent, i.e. $p^2 = p$. The symbol $\mathfrak{P}(A)$ stands for the set of all projections of an involutive algebra A. Two projections $p, q \in \mathfrak{P}(A)$ are called *orthogonal* if $pq = 0$. A projection p is *central* if $px = xp$ for all $x \in A$. Denote the set of all central projections by $\mathfrak{P}_c(A)$.

A scalar $\lambda \in \mathbb{C}$ is a *spectral value* of x, if $\lambda - x$ is not invertible in A. The set of all spectral values of x is called the *spectrum* of x and denoted by $\mathrm{Sp}(x)$. An element x of a $*$-algebra A is called *positive* if x is hermitian and $\mathrm{Sp}(x) \subset \mathbb{R}_+$. The set of all positive elements of A is denoted by A_+.

If $(A, *)$ and $(B, *)$ are involutive algebras and $\mathscr{R} : A \to B$ is a multiplicative linear operator, then $\mathscr{R}$ is called a *$*$-representation* of A in B whenever $\mathscr{R}(x^*) = \mathscr{R}(x)^*$ for all $x \in A$. If $\mathscr{R}$ is also an isomorphism then $\mathscr{R}$ is a *$*$-isomorphism* of A and B. In the presence of norms in the algebras, the naturally understood terms "isometric $*$-representation" and "isometric $*$-isomorphism" are in common parlance.

7.4.2. A norm $\|\cdot\|$ on an algebra A is *submultiplicative* if

$$\|xy\| \leq \|x\|\,\|y\| \quad (x, y \in A).$$

A *Banach algebra* A is an algebra furnished with a submultiplicative norm making A into a Banach space. An *involutive Banach algebra* is a Banach algebra which is also an involutive algebra and its involution satisfies the condition

$$\|x^*\| = \|x\| \quad (x \in X).$$

If A is an involutive Banach algebra satisfying

$$\|xx^*\| = \|x\|^2 \quad (x \in A)$$

then A is called a *C^*-algebra*. The spectrum of an element of a C^*-algebra is a nonempty compact subset of $\mathbb{C}$. Let $C(\mathrm{Sp}(x), \mathbb{C})$ denote the C^*-algebra of complex continuous functions on $\mathrm{Sp}(x)$.

(1) Spectral Theorem. *Let x be a normal element of a C^*-algebra A, with $\mathrm{Sp}(x)$ the spectrum of x. There is a unique isometric $*$-representation $\mathscr{R}_x : C(\mathrm{Sp}(x), \mathbb{C}) \to A$ such that $x = \mathscr{R}_x(\imath)$, where $\imath$ is the identity mapping on $\mathrm{Sp}(x)$.*

The representation $\mathscr{R}_x : C(\mathrm{Sp}(x), \mathbb{C}) \to A$ is called the *continuous functional calculus* (for a normal element x of A). The element $\mathscr{R}_x(f)$ with $f \in C(\mathrm{Sp}(x), \mathbb{C})$ is usually denoted by $f(x)$. In particular, for every positive $x \in A$ the square root $\sqrt{x}$ is defined, since $\mathrm{Sp}(x) \subset \mathbb{R}_+$, and for each normal $x \in A$ the modulus can be defined as $|x| := \sqrt{x^* x}$.

(2) *Let x be a normal element of a C^*-algebra A and $f \in C(\mathrm{Sp}(x), \mathbb{C})$. Then $(g \circ f)(x) = g(f(x))$ for each $g \in C(\mathrm{Sp}\, f(x), \mathbb{C})$.*

(3) *An element x of a C^*-algebra A is positive if and only if $x = y^* y$ for some $y \in A$. The set A_+ of all positive elements is an ordering cone and so (A, A_+) is an ordered vector space.*

Treating a $*$-algebra A as an ordered vector space, we always imply the order that is conventionally induced by A_+.

7.4.3. Suppose that Λ is an order complete complex AM-space (see 1.3.11 and 1.5.5) with strong order-unity $\mathbf{1}$. According to the Brothers Kreĭn–Kakutani Theorem 1.5.6 (2) and Theorem 1.5.9 Λ is linearly isometric and order isomorphic to the space of continuous functions $C(Q)$ on some extremal compact space Q. Therefore, A can be endowed with some multiplication and involution so that Λ becomes a commutative C^*-algebra. Such C^*-algebra is often called a Stone algebra. Thus, a *Stone algebra* is a commutative C^*-algebra (with unity) which is order complete vector lattice with respect to the ordering 7.4.2 (3). An element $e \in \Lambda$ is a projection if and only if it is a fragment of $\mathbf{1}$. Moreover, the isomorphism $\Lambda \to C(Q)$ defines a bijection between the set of fragments of $\mathbf{1}$ and the set of characteristic functions of clopen sets in Q, so that the Boolean algebras $\mathfrak{E}(\mathbf{1}) := \mathfrak{E}(A)$ coincides with the set of all projections $\mathfrak{P}(\Lambda)$ and is isomorphic to $\operatorname{Clop}(Q)$. A hermitian element $p \in \Lambda$ is a projection if and only if the multiplicative operator $x \mapsto px$ is a band projection. Given a complete Boolean B there exists a unique (up to $*$-isomorphism) Stone algebra Λ such that B and $\mathfrak{P}(\Lambda)$ are isomorphic. Each of these algebras will be denoted by $\mathscr{S}(B)$.

(The same symbol $\mathscr{S}(B)$ we used in 1.2.1 (2) to denote the Stone space of B. We hope that this liberty will not lead to confusions, since the meaning is always clear from the context.)

Let Λ be a Stone algebra and consider a unitary Λ-module X. The mapping $\langle \cdot \mid \cdot \rangle : X \times X \to \Lambda$ is a *Λ-valued inner product*, if for all $x, y, z \in X$ and $a \in \Lambda$ the following are satisfied:

(1) $\langle x \mid x \rangle \geq \mathbf{0}$; $\langle x \mid x \rangle = \mathbf{0} \Leftrightarrow x = \mathbf{0}$;

(2) $\langle x \mid y \rangle = \langle y \mid x \rangle^*$;

(3) $\langle ax \mid y \rangle = a\langle x \mid y \rangle$;

(4) $\langle x + y \mid z \rangle = \langle x \mid z \rangle + \langle y \mid z \rangle$.

Using a Λ-valued inner product, we may introduce the norm in X by the formula

(5) $|\!|\!| x |\!|\!| := \sqrt{\| \langle x|x \rangle \|} \quad (x \in X)$,

and the vector norm

(6) $|x| := \sqrt{\langle x|x \rangle} \quad (x \in X)$.

Employing the continuous functional calculus 7.4.2 (1, 2) we may deduce from the properties (2) and (3) that $|\lambda x| = |\lambda||x|$ for all $\lambda \in \Lambda$ and $x \in X$. The fact that $|\cdot|$ satisfies the triangle inequality results as usual from the Cauchy–Bunyakovskiĭ–Schwarz inequality

(7) $\langle x \mid y \rangle \le |x|\,|y|$.

On taking norms in (7) and using submultiplicativity and monotonicity of the norm in Λ, we further obtain the numerical version of the Cauchy–Bunyakovskiĭ–Schwarz inequality

(8) $\|\langle x \mid y \rangle\| \le |\!|\!|x|\!|\!|\, |\!|\!|y|\!|\!|$.

It follows from (5) and (6) that

(9) $|\!|\!|x|\!|\!| = \big\| |x| \big\| \quad (x \in X)$,

since $\|a\| = \|(\sqrt{a})^2\| = \|\sqrt{a}\|^2$ for every positive $a \in \Lambda$. Therefore, the formula (5) defines a mixed norm on X (cf. 7.1.1).

7.4.4. Let X be a Λ-module with an inner product $\langle \cdot \mid \cdot \rangle : X \times X \to \Lambda$. If X is complete with respect to the mixed norm $|\!|\!| \cdot |\!|\!|$, it is called a *C^*-module* over Λ.

Theorem. *Let X be a C^*-module. The pair $(X, |\!|\!| \cdot |\!|\!|)$ is a B-cyclic Banach space if and only if $(X, |\cdot|)$ is a Banach–Kantorovich space over $\Lambda := \mathscr{S}(B)$.*

◁ Note that 7.4.4 (6) gives a decomposable norm since $|bx| = b|x|$ $(x \in X,\ b \in B)$. By Theorem 7.1.2, the normed space $(X, |\!|\!| \cdot |\!|\!|)$ is complete if and only if $(X, |\cdot|)$ is br-complete. Furthermore, it is clear that the B-cyclicity of $(X, |\!|\!| \cdot |\!|\!|)$ amount to the disjoint completeness of $(X, |\cdot|)$. The above remarks justify 2.2.3, so completing the proof. ▷

7.4.5. A *Kaplansky–Hilbert module* or an *AW^*-module* over Λ is a unitary C^*-module satisfying each of the equivalent conditions of Theorem 7.4.4. According to 2.2.3 X is a Kaplansky–Hilbert module over Λ if and only if it is an C^*-module over Λ and enjoys the following two properties:

(1) let x be an arbitrary element in X, and let $(e_\xi)_{\xi \in \Xi}$ be a partition of unity in $\mathfrak{P}(\Lambda)$ with $e_\xi x = 0$ for all $\xi \in \Xi$; then $x = 0$;

(2) let $(x_\xi)_{\xi \in \Xi}$ be a norm-bounded family in X, and let $(e_\xi)_{\xi \in \Xi}$ be a partition of unity in $\mathfrak{P}(\Lambda)$; then there exists an element $x \in X$ such that $e_\xi x = e_\xi x_\xi$ for all $\xi \in \Xi$.

The element of (2) is the bo-sum of the family $(e_\xi x_\xi)_{\xi \in \Xi}$, see 2.2.1. According to 7.4.3 (7) the inner product is bo-continuous in each variable. In particular,

(3) $\left\langle bo\text{-}\sum_{\xi \in \Xi} e_\xi x_\xi \,\middle|\, y \right\rangle = bo\text{-}\sum_{\xi \in \Xi} \langle e_\xi x_\xi \mid y \rangle$

for every bounded family $(x_\xi)_{\xi \in \Xi}$ in X and partition of unity $(e_\xi)_{\xi \in \Xi}$ in $\mathfrak{P}(\Lambda)$.

Let X be a Kaplansky–Hilbert module over Λ. A Kaplansky–Hilbert submodule of X is a bo-closed submodule $X_0 \subset X$. Theorems 7.1.2 and 2.2.3 imply that X_0

is a Kaplansky–Hilbert submodule if and only if X_0 is a submodule in the conventional algebraic sense closed in norm topology and containing all sums of the form $bo\text{-}\sum_{\xi\in\Xi} e_\xi x_\xi$, where $(x_\xi)_{\xi\in\Xi}$ is a bounded family in X_0 and $(e_\xi)_{\xi\in\Xi}$ is a partition of unity in $\mathfrak{P}(\Lambda)$. The intersection of any number of Kaplansky–Hilbert submodules is again a Kaplansky–Hilbert submodule. Thus, for each nonempty subset $M \subset X$ there exists a smallest Kaplansky–Hilbert submodule containing M; it is called the Kaplansky–Hilbert *submodule generated by* M.

(4) *The orthogonal complement* $M^\perp := \{x \in X : (\forall y \in M)\, \langle x \,|\, y\rangle = 0\}$ *of any nonempty subset* $M \subset X$ *is a Kaplansky–Hilbert submodule of* X.

A Kaplansky–Hilbert module over Λ is called *faithful* if for every $a \in \Lambda$ the condition $(\forall x \in X)\, ax = 0$ implies that $a = 0$. It is clear that the faithfulness of X amounts to the condition $|X|^{\perp\perp} = \Lambda$. In the sequel we restrict our attention to faithful Kaplansky–Hilbert modules over Λ.

7.4.6. Suppose that X is a unitary Kaplansky–Hilbert module over a Stone algebra Λ. A subset $\mathscr{E}$ of X is said to be *orthonormal* if

(1) $\langle x \,|\, y\rangle = \mathbf{0}$ for all distinct $x, y \in \mathscr{E}$;

(2) $\langle x \,|\, x\rangle = \mathbf{1}$ for every $x \in \mathscr{E}$.

An orthonormal set $\mathscr{E} \subset X$ is a *basis* for X provided that

(3) the condition $(\forall e \in \mathscr{E})\langle x \,|\, e\rangle = \mathbf{0}$ implies $x = \mathbf{0}$.

Say that a Kaplansky–Hilbert module X is λ-*homogeneous,* if λ is a cardinal and X has a basis of cardinality λ. Granted $\mathbf{0} \neq b \in B$, denote by $\varkappa(b)$ the least cardinal γ such that a Kaplansky–Hilbert module bX over $b\Lambda$ is γ-homogeneous. If X is homogeneous then $\varkappa(b)$ is defined for all $\mathbf{0} \neq b \in B$. It is convenient to assume that $\varkappa(\mathbf{0}) = 0$. We shall say that a Kaplansky–Hilbert module X is *strictly* γ-*homogeneous* if X is homogeneous and $\gamma = \varkappa(b)$ for all nonzero $b \in B$. A Kaplansky–Hilbert module is said to be (strictly homogeneous) if it is λ-homogeneous (strictly λ-homogeneous) for some cardinal λ.

If γ is a finite cardinal then the property of γ-homogeneity and strict γ-homogeneity of a AW^*-module are equivalent. Denote by $|M|$ the cardinality of M; i.e., a cardinal number bijective with M.

7.4.7. **(1)** *Suppose that* X *is a Kaplansky–Hilbert module over* Λ. *The mapping* $\varkappa$ *preserves suprema of nonempty sets, i.e.* $\varkappa(\sup(D)) = \sup(\varkappa(D))$ *for every* $D \subset B$.

$\triangleleft$ Put $\bar{b} := \sup D$. By definition $\varkappa$ is increasing: $b_1 \leq b_2 \Rightarrow \varkappa(b_1) \leq \varkappa(b_2)$. Therefore, $\sup_{b\in D} \varkappa(b) \leq \varkappa(\bar{b})$. Prove the reverse inequality. For an arbitrary nonzero $b \in B$ the set of cardinals $\{\varkappa(b') : \mathbf{0} \neq b' \leq b\}$ has the least element, say $\gamma := \varkappa(b_0)$. Obviously, $b_0 \neq \mathbf{0}$ and $\varkappa(b_0) = \varkappa(b')$ for all nonzero $b' \leq b_0$.

Thus the set D' of all $b \in B$, with bX strictly homogeneous, minorizes D. By the Exhaustion Principle (see 1.1.6) there exists a partition $(b_\xi)_{\xi\in\Xi}$ of the element $\bar{b}$ such that $b_\xi X$ is a strictly $\varkappa(b_\xi)$-homogeneous Kaplansky–Hilbert module over $b_\xi\Lambda$. Let $\mathscr{E}_\xi := (e_{\gamma,\xi})_{\gamma<\varkappa(b_\xi)}$ be a basis in $b_\xi X$. Put $\lambda := \sup_{\xi\in\Xi} \varkappa b_\xi$ and $e_{\gamma,\xi} = 0$ for $\varkappa(b_\xi) \leq \gamma < \lambda$. Now, define a set $\mathscr{E} := (e_\gamma)_{\gamma\in\lambda}$ by putting

$$e_\gamma := bo\text{-}\sum_{\xi\in\Xi} e_{\gamma,\xi} \quad (\gamma \in \lambda).$$

The set $\mathscr{E}$ is orthonormal, since

$$\langle e_\gamma \,|\, e_\beta\rangle = \left\langle bo\text{-}\sum_{\xi\in\Xi} e_{\gamma,\xi} \,\middle|\, bo\text{-}\sum_{\eta\in\Xi} e_{\beta,\eta}\right\rangle = bo\text{-}\sum_{\xi,\eta\in\Xi} \langle e_{\gamma,\xi} \,|\, e_{\beta,\eta}\rangle$$

$$= bo\text{-}\sum_{\xi,\eta\in\Xi} \langle b_\xi e_{\gamma,\xi} \,|\, b_\eta e_{\beta,\eta}\rangle = bo\text{-}\sum_{\xi,\eta\in\Xi} b_\xi b_\eta \langle e_{\gamma,\xi} \,|\, e_{\beta,\eta}\rangle =: e,$$

where $e = 0$ if $(\gamma,\xi) \neq (\beta,\eta)$ and $e = \mathbf{1}$ if $(\gamma,\xi) = (\beta,\eta)$. The set $\mathscr{E}$ is in fact a basis in $\bar{b}X$. Indeed, if $x \in X$ and $\langle x \,|\, e_\gamma\rangle = \mathbf{0}$ for all $\gamma \in \lambda$, whence $\langle x \,|\, e_{\gamma,\xi}\rangle = \mathbf{0}$ for all $\xi \in \Xi$ and $\gamma < \varkappa(b_\xi)$. Thus, $b_\xi x \perp \mathscr{E}_\xi$, whence $b_\xi x = \mathbf{0}$. By 7.4.5 (1) $x = 0$. Since $|\mathscr{E}| \leq \lambda$ we have by definition of $\varkappa$ that $\varkappa(b) \leq \lambda \leq \sup_{b\in D} \varkappa(b)$. ▷

(2) *Theorem.* *Let X be a Kaplansky–Hilbert module over Λ. Then there exists a partition of unity $(b_\xi)_{\xi\in\Xi}$ in $\mathfrak{P}(\Lambda)$ such that $b_\xi X$ is a strictly $\varkappa(b_\xi)$-homogeneous Kaplansky–Hilbert module over $b_\xi\Lambda$.*

◁ Assume that $B := \mathfrak{P}(\Lambda)$ and B' consists of the elements $b' \in B$ with homogeneous $b'X$. Extend $\varkappa$ from B' to the whole B by letting $\varkappa(b) := \sup\{\varkappa(b');\ b' \in B',\ b' < b\}$ This definition is sound due to Proposition (1). In the same way as in the proof of (1) we may choose a partition of unity (b_ξ) satisfying the desired properties. ▷

7.4.8. Now, we will give two examples of Kaplansky–Hilbert modules. Denote by $C_\#(Q,H)$ the part of $C_\infty(Q,H)$ that consists of vector-functions z satisfying $|z| \in C(Q)$ (cf. 2.3.3).

(1) *Suppose that Q is an extremal compact space, and H is a Hilbert space of dimension λ. The space $C_\#(Q,H)$ is a λ-homogeneous Kaplansky–Hilbert module over the algebra $\Lambda := C(Q,\mathbb{C})$.*

◁ Let $(\cdot\,|\,\cdot)$ stand for the inner product of H. Introduce some Λ-valued inner product in $C_\#(Q,H)$ as follows. Take continuous vector-functions $u : \mathrm{dom}(u) \to H$ and $v : \mathrm{dom}(v) \to H$. The function $q \mapsto \langle u(q)|v(q)\rangle$ $(q \in \mathrm{dom}(u) \cap \mathrm{dom}(v))$ is continuous and admits a unique continuation $z \in C(Q)$ to the whole of Q. If x and y are the cosets containing vector-functions u and v then assign $(x \,|\, y) := z$. Clearly,

$(\cdot\,|\,\cdot)$ is a Λ-valued inner product and $|x| = \sqrt{(x\,|\,x)}$ $(x \in C_{\#}(Q,H))$. The space $C_{\#}(Q,H)$ is disjointly complete, since $C_{\#}(Q,H)$ is a Banach–Kantorovich space. Moreover, $C_{\#}(Q,H)$ is a Banach space whose norm satisfies the conditions

$$\|x\| = \||x|\|_{\infty} = \sqrt{\|(x\,|\,x)\|_{\infty}} \quad (x \in C_{\#}(Q,H)).$$

Suppose that $\mathscr{E}$ is a basis for H. Given $e \in \mathscr{E}$, introduce the vector-function $\bar{e} : q \mapsto e \quad (q \in Q)$ and put $\overline{\mathscr{E}} := \{\bar{e} : e \in \mathscr{E}\}$. It is easy to note that $\overline{\mathscr{E}}$ is a basis for $C_{\#}(Q,H)$. Summarizing, conclude that $C_{\#}(Q,H)$ is a λ-homogeneous Kaplansky–Hilbert module, with $\lambda = \dim(H)$. $\triangleright$

Let $\mathscr{E}$ be a nonempty set and denote by $l_2(\mathscr{E},\Lambda)$ the set of all families $(a_e)_{e\in\mathscr{E}}$ of elements of Λ such that $\sum_{e\in\mathscr{E}} |a_e|^2$ is o-summable in Λ. Define a Λ-valued inner product in $l_2(\mathscr{E},\Lambda)$ as

$$\langle u\,|\,v\rangle := o\text{-}\sum_{e\in\mathscr{E}} u_e v_e^* \quad (u,v \in l_2(\mathscr{E},\Lambda)).$$

It is easy to check (in the same way as in the scalar case $\Lambda := \mathbb{C}$) that the sum exists for all $u,v \in l_2(\mathscr{E},\Lambda)$ and the mapping $\langle\cdot\,|\,\cdot\rangle$ meets the conditions 7.4.3 (1–4). The Λ-valued norm 7.4.3 (6) has the form

$$|u| := \sqrt{o\text{-}\sum_{e\in\mathscr{E}} |a_e|^2} \quad (u \in l_2(\mathscr{E},\Lambda)).$$

(2) *For any nonempty set $\mathscr{E}$ with $\lambda := |\mathscr{E}|$ the space $l_2(\mathscr{E},\Lambda)$ is a λ-homogeneous Kaplansky–Hilbert module over Λ.*

$\triangleleft$ Using Theorem 2.2.3 it is easy to prove that $(l_2(\mathscr{E},\Lambda), |\cdot|)$ is a Banach–Kantorovich space. Thus, $l_2(\mathscr{E},\Lambda)$ is a Kaplansky–Hilbert module over Λ. For each $e \in \mathscr{E}$ denote by $\langle e\rangle$ the family $(u_c)_{c\in\mathscr{E}}$ such that $u_c = 0$ for $c \neq e$ and $u_e = e$. It is an easy matter to check that the set $\langle\mathscr{E}\rangle := \{\langle e\rangle : e \in \mathscr{E}\}$ is a basis in $l_2(\mathscr{E},\Lambda)$. $\triangleright$

(3) *For a Stone algebra Λ and a cardinal number λ there exists a λ-homogeneous module over Λ.*

$\triangleleft$ Immediate from (1) or (2). $\triangleright$

7.4.9. Assuming that X is an arbitrary Kaplansky–Hilbert module over a Stone algebra Λ we prove two key properties of orthonormal sets.

(1) *Let $\mathscr{E}$ be an orthonormal subset in X, and let X_0 be the Kaplansky–Hilbert submodule generated by $\mathscr{E}$. If $(a_e)_{e\in\mathscr{E}}$ is a family in Λ such that $\sum_{e\in\mathscr{E}} |a_e|^2$ is o-convergent then there exist an element $x_0 \in X_0$ with*

$$|x_0|^2 = o\text{-}\sum_{e\in\mathscr{E}} |a_e|^2, \quad \langle x_0\,|\,e\rangle = a_e \quad (e \in \mathscr{E}).$$

$\lhd$ Let Θ be the set of all finite subsets of $\mathscr{E}$. Given $\theta \in \Theta$, put

$$s_\theta := \sum_{e\in\theta} a_e e, \quad \sigma_\theta := \sum_{e\in\theta} |a_e|^2, \quad \sigma := o\text{-}\sum_{e\in\mathscr{E}} |a_e|^2, \quad \delta_\theta := \sigma - \sigma_\theta.$$

Take $\theta, \theta_1, \theta_2 \in \Theta$ with $\theta \subset \theta_1 \cap \theta_2$ and denote by θ' and $\theta_1 \triangle \theta_2$ the complement of θ and the symmetric difference of θ_1 and θ_2, respectively. Since the set $\mathscr{E}$ is orthonormal, we may write

$$|s_{\theta_1} - s_{\theta_2}|^2 = \Big| \sum_{e\in\theta_1\triangle\theta_2} a_e e \Big|^2 = \sum_{e\in\theta_1\triangle\theta_2} |a_e|^2 \le \sum_{e\in\theta'} |a_e|^2 = \sigma - \sigma_\theta = \delta_\theta.$$

By hypothesis $(\delta_\theta)_{\theta\in\Theta}$ decreases to zero, so that $(s_\theta)_{\theta\in\Theta}$ is *bo*-fundamental. By Definition 7.4.5 there exists the *bo*-limit of $(s_\theta)_{\theta\in\Theta}$ in X. Denote

$$x_0 := bo\text{-}\lim_{\theta\in\Theta} s_\theta := bo\text{-}\sum_{e\in\mathscr{E}} a_e e.$$

Now, taking 7.4.5 (3) into consideration, we deduce $\langle x_0 \,|\, e\rangle = \sum_{c\in\mathscr{E}} a_c \langle c \,|\, e\rangle = a_e$ for all $e \in \mathscr{E}$. Moreover, $|s_\theta|^2 = \sigma_\theta$, and passing to the limits, we obtain $|x_0|^2 = \sigma$. $\rhd$

(2) *Let X_0 be a homogeneous Kaplansky–Hilbert submodule of X with a basis $\mathscr{E}$. Then $\mathscr{E}$ and X_0 have the same orthogonal complements in X.*

$\lhd$ Given an element $y \in X$ with $\langle y \,|\, e\rangle = 0$ $(e \in \mathscr{E})$, we have to show that $\langle y \,|\, x\rangle = 0$ for all $x \in X_0$. Take an arbitrary $x \in X_0$ and put $a_e := \langle x \,|\, e\rangle$. Employing the notation from (1) we observe that $\sigma_\theta \le |x|^2$ for all $\theta \in \Theta$. Thus, the family $(a_e)_{e\in\mathscr{E}}$ is *bo*-summable and, in view of 7.4.9, there exists $x_0 \in X_0$ with $\sigma = |x_0|^2$ and $\langle x_0 \,|\, e\rangle = a_e$ $(e \in \mathscr{E})$. Evidently $\langle x - x_0 \,|\, e\rangle = 0$ for every $e \in \mathscr{E}$ and, by 7.4.6 (3), $x - x_0 = 0$. The orthogonal complement $\{y\}^\perp$ is a Kaplansky–Hilbert module by 7.4.5 (4). Thus, $\{y\}^\perp$ contains the Kaplansky–Hilbert submodule generated by $\mathscr{E}$, since it contains $\mathscr{E}$, hence contains $x_0 = x$. $\rhd$

7.4.10. Observe some important corollaries to Propositions 7.4.9 (1, 2). Let $\mathscr{E}$ be a basis for a homogeneous Kaplansky–Hilbert module X and $x \in X$. The family $\widehat{x} := (\widehat{x}_e)_{e\in\mathscr{E}}$ in $\Lambda^{\mathscr{E}}$, given by the identity $\widehat{x}_e := \langle x \,|\, e\rangle$, is the *Fourier coefficient family* of x with respect to $\mathscr{E}$ or the *Fourier transform* of x (relative to $\mathscr{E}$).

(1) *The Fourier coefficient family of an element $x \in X$ is square bo-summable; moreover, the identity holds $|x|^2 = o\text{-}\sum_{e\in\mathscr{E}} |\langle x \,|\, e\rangle|^2$.*

(2) Riesz–Fisher Isomorphism Theorem. *Let X be a homogeneous Kaplansky–Hilbert module over Λ with a basis $\mathscr{E}$. The Fourier transform $\mathscr{F} : x \mapsto \widehat{x}$ (relative to $\mathscr{E}$) is an isometric isomorphism of X onto $l_2(\mathscr{E}, \Lambda)$. The inverse Fourier* *transform, the Fourier summation* $\mathscr{F}^{-1} : l_2(\mathscr{E}, \Lambda) \to X$, *acts by the rule* $\mathscr{F}^{-1}(x) := bo\text{-}\sum_{e\in\mathscr{E}} x_e e$ *for* $x := (x_e)_{e\in\mathscr{E}} \in l_2(\mathscr{E}, \Lambda)$. *Moreover, the Fourier transform preserves inner product or, in other words, for all* $x, y \in X$ *the Parseval identity holds:*

$$\langle x \mid y\rangle = o\text{-}\sum_{e\in\mathscr{E}} \widehat{x}_e \widehat{y}_e^{\,*}.$$

◁ By (1), the Fourier transform acts into $l_2(\mathscr{E}, \Lambda)$. By Proposition 7.4.9 (1), $\widehat{\cdot}$ is an epimorphism. By 7.4.6 (3), $\widehat{\cdot}$ is a monomorphism. Evidently, $\mathscr{F}^{-1}\widehat{x} = x$ for $x \in X$ and $\widehat{\cdot} \circ \mathscr{F}^{-1}(u) = u$ for $u \in l_2(\mathscr{E}, \Lambda)$. The equality

$$|x|^2 = o\text{-}\sum_{e\in\mathscr{E}} |\widehat{x}|^2 = |\widehat{x}_e|_2^2 \quad (x \in X)$$

follows from (1). At the same time

$$\langle x \mid y\rangle = \left\langle bo\text{-}\sum_{e\in\mathscr{E}} \widehat{x}_e e \,\middle|\, bo\text{-}\sum_{e\in\mathscr{E}} \widehat{y}_e e \right\rangle = o\text{-}\sum_{e,e'\in\mathscr{E}} \widehat{x}_e \widehat{y}_{e'}^{\,*} \langle e \mid e'\rangle = o\text{-}\sum_{e\in\mathscr{E}} \widehat{x}_e \widehat{y}_e^{\,*}. \ \triangleright$$

(3) *Any two* λ*-homogeneous Kaplansky–Hilbert modules over a Stone algebra are isomorphic.*

◁ By (2) all λ-homogeneous Kaplansky–Hilbert modules over Λ are isomorphic to $l_2(\mathscr{E}, \Lambda)$ with $|\mathscr{E}| = \lambda$. ▷

(4) *Let* X *be a Kaplansky–Hilbert module over* Λ, *with* X_0 *a Kaplansky–Hilbert submodule of* X *and* $X_1 := X_0^{\perp}$. *Then* $X = X_0 \oplus X_1$.

◁ For a homogeneous Kaplansky–Hilbert module the claim follows from (2). According to 7.4.7 (2) in an arbitrary Kaplansky–Hilbert module we can choose a partition of unity $(b_\xi)_{\xi\in\Xi}$ in $B := \mathfrak{P}(\Lambda)$ such that $b_\xi X$ is homogeneous Kaplansky–Hilbert module over $b_\xi\Lambda$. For an arbitrary $x \in X$ we may decompose $b_\xi x$ into a sum $b_\xi x = y_\xi + z_\xi$ of elements $y_\xi \in b_\xi X_0$ and $z_\xi \in b_\xi X_1$. Taking $y := bo\text{-}\sum_{\xi\in\Xi} y_\xi$ and $z := bo\text{-}\sum_{\xi\in\Xi} z_\xi$ we obtain a desired decomposition $x = y + z$. ▷

7.4.11. Take some cardinal λ. Granted $b \in B$ and $\beta \in \mathrm{On}$, denote by $b(\beta)$ the set of all partitions of b having the form $(b_\alpha)_{\alpha\in\beta}$. Define the $[\mathbf{0}, b]$-valued function d on $b(\beta)$ by the formula

$$d(u, v) := \left(\bigvee_{\alpha\in\beta} u_\alpha \wedge v_\alpha\right)^*, \quad \big(u = (u_\alpha),\ \ v = (v_\alpha) \in b(\beta)\big).$$

It is the matter of direct verification that the function d is a Boolean-valued metric, i.e. the following hold:

(1) $d(u, v) = \mathbf{0} \Leftrightarrow u = v$;

(2) $d(u, v) = d(v, u)$;

(3) $d(u, v) \leq d(u, w) \vee d(w, v)$.

Granted $\gamma \in \mathrm{On}$, write $b(\beta) \simeq b(\gamma)$ if there is a bijection $\imath$ between $b(\beta)$ and $b(\gamma)$ which preserves the Boolean metric; i.e., $d(\imath(u), \imath(v)) = d(u, v)$. We call the Boolean algebra B and its Stone space *λ-stable* provided that $\lambda \leq \alpha$ for all nonzero $b \in B$ and each ordinal α with $b(\lambda) \simeq b(\alpha)$. A nonzero element $b \in B$ is λ-stable by definition whenever $[\mathbf{0}, b]$ is a λ-stable Boolean algebra.

7.4.12. Theorem. *To each Kaplansky–Hilbert module X there is a family of nonempty extremal compact spaces $(Q_\gamma)_{\gamma \in \Gamma}$, with Γ a set of cardinals, such that Q_γ is γ-stable for all $\gamma \in \Gamma$ and the following unitary equivalence holds:*

$$X \simeq \sum_{\gamma \in \Gamma}^{\oplus} C_\#(Q_\gamma, l_2(\gamma)).$$

If some family $(P_\delta)_{\delta \in \Delta}$ of extremal compact spaces satisfies the above conditions then $\Gamma = \Delta$ and P_γ is homeomorphic with Q_γ for all $\gamma \in \Gamma$.

$\triangleleft$ The representation follows from 7.4.7 (2), 7.4.8 (1), and 7.4.10 (3). The uniqueness of the family $(Q_\gamma)_{\gamma \in \Gamma}$ and Γ-stability of the compact space (Q_γ) will be proved in Section 8.4. $\triangleright$

7.5. AW^*-Algebras

In this section we introduce the class of AW^*-algebras and prove that the set of all bounded endomorphisms of an AW^*-module is an AW^*-algebra of type I; moreover, every AW^*-algebra of type I arises in this way.

7.5.1. Consider an involutive algebra A. Granted a nonempty set $M \subset A$, define the *right annihilator* $M^\perp$ of M and the *left annihilator* ${}^\perp M$ of M by

$$M^\perp := \{y \in A : (\forall x \in M)\, xy = \mathbf{0}\};$$
$${}^\perp M := \{x \in A : (\forall y \in M)\, xy = \mathbf{0}\}.$$

General properties of annihilators imply that the inclusion-ordered sets of all right and all left annihilators are order complete lattices. The mapping $K \mapsto K^* := \{x^* : x \in K\}$ is an isotone bijection between these lattices for $(M^\perp)^* = {}^\perp(M^*)$ and $({}^\perp M)^* = (M^*)^\perp$.

A *Baer $*$-algebra* we call an involutive algebra A provided that, for each nonempty $M \subset A$, there is some p in $\mathfrak{P}(A)$ satisfying $M^\perp = pA$. Clearly, this amounts to the condition that each left annihilator has the form ${}^\perp M = Aq$ for an appropriate projection q. To each left annihilator L in a Baer $*$-algebra there is a unique projection $q_L \in A$ such that $x = xq_L$ for all $x \in L$ and $q_L y = \mathbf{0}$ whenever $y \in L^\perp$. The mapping $L \mapsto q_L$ is an isomorphism between the poset of left annihilators and the poset of all projections. The inverse isomorphism has the form $q \mapsto {}^\perp(\mathbf{1} - q) \quad (q \in \mathfrak{P}(A))$. An analogous claim is true for right annihilators. This implies in particular that the poset $\mathfrak{P}(A)$ is an order complete lattice. The mapping $p \mapsto p^\perp := \mathbf{1} - p \quad (p \in \mathfrak{P}(A))$ satisfies the conditions:

$$p^{\perp\perp} = p, \ p \wedge p^\perp = \mathbf{0}, \ p \vee p^\perp = \mathbf{1},$$
$$(p \wedge q)^\perp = p^\perp \vee q^\perp, \ (p \vee q)^\perp = p^\perp \wedge q^\perp,$$
$$p \leq q \Rightarrow p \vee (p^\perp \wedge q) = q.$$

In other words, $(\mathfrak{P}(A), \wedge, \vee, \perp)$ is a *orthomodular lattice.*

7.5.2. An *AW^*-algebra* is a C^*-algebra presenting a Baer $*$-algebra. More explicitly, an AW^*-algebra is a C^*-algebra whose every right annihilator has the form pA, with p a projection. An element $z \in A$ is said to be *central* if it commutes with every member of A; i.e. $(\forall x \in A)\, xz = zx$. The *center* of an AW^*-algebra A is the set $\mathscr{Z}(A)$ comprising central elements. Clearly, $\mathscr{Z}(A)$ is a commutative AW^*-subalgebra of A, with $\lambda\mathbf{1} \in \mathscr{Z}(A)$ for all $\lambda \in \mathbb{C}$. If $\mathscr{Z}(A) = \{\lambda\mathbf{1} : \lambda \in \mathbb{C}\}$ then the AW^*-algebra A is called an *AW^*-factor.*

A C^*-algebra A is an AW^*-algebra if and only if the following conditions are satisfied:

(1) *Every orthogonal family in $\mathfrak{P}(A)$ has a supremum;*

(2) *Every maximal commutative $*$-subalgebra A_0 of A is a Stone algebra.*

The space $B(H)$ of all bounded linear endomorphisms of a complex Hilbert space H exhibits an example of an AW^*-algebra. Recall that the structure of a Banach algebra in $B(H)$ results from the conventional addition and composition of operators and the routine operator norm. The involution in $B(H)$ is the taking of the adjoint of an operator. Observe also that every commutative AW^*-algebra is just a Stone algebra.

7.5.3. Spectral Theorem. *To each hermitian element a of an AW^*-algebra A there is a unique resolution of unity $\lambda \mapsto e_\lambda \quad (\lambda \in \mathbb{R})$ in $\mathfrak{P}(A)$ such that*

$$a = \int_{-\|a\|}^{\|a\|} \lambda \, de_\lambda.$$

Moreover, an element $x \in A$ commutes with a if and only if $xe_\lambda = e_\lambda x$ for all $\lambda \in \mathbb{R}$.

◁ Recall that the term "resolution of identity" (or "spectral function") in $\mathfrak{P}(A)$ means the same as in the case of a Boolean algebra; i.e., we call so every function $\lambda \mapsto e_\lambda$ ($\lambda \in \mathbb{R}$) that satisfies 1.4.3 (1–3) (cf. 1.3.8 (1–3)). Every maximal commutative $*$-subalgebra of A which contains a is a complex K-space by 7.5.2 (2). Hence, the sought representation ensue from the Freudenthal Theorem 1.3.9 (1). The claim about commuting follows from the fact that the element a and the set $\{e_\lambda : \lambda \in \mathbb{R}\}$ generate the same maximal $*$-subalgebra. ▷

7.5.4. A Banach algebra A is *B-cyclic* with respect to a complete Boolean algebra B of projections of A provided that A is a B-cyclic Banach space in the sense of 7.3.3 and every member of B is a *multiplicative projection*, i.e.

$$\pi(xy) = \pi(x)\pi(y) = x\pi y = \pi(x)y \quad (x, y \in A;\ \pi \in B).$$

The definition of *B-cyclic involutive algebra* appears on requiring additionally that every member of B is *$*$-preserving*, i.e.

$$\pi(x^*) = (\pi x)^* \quad (x \in A,\ \pi \in B).$$

Finally, the definition of *B-cyclic C^*-algebra* is evident.

Recall that we consider only unital algebras. With this in mind, let $\mathbf{1}$ be the unity of A and identify each projection $b \in B$ with the element $b\mathbf{1}$. If A is involutive then $b\mathbf{1}$ is a central projection of A in the sense of 7.4.1. In this event we write $B \subset \mathfrak{P}_c(A)$.

Take B-cyclic algebras A_1 and A_2. A bounded operator $\Phi : A_1 \to A_2$ is a *B-homomorphism* whenever Φ is multiplicative ($\Phi(xy) = \Phi(x) \cdot \Phi(y)$) and commuting with every projection in B ($b \circ T = T \circ b$ for all $b \in B$).

If A_1 and A_2 are involutive algebras and some B-homomorphism Φ is $*$-preserving, i.e. $\Phi(x^*) = \Phi(x)^*$ ($x \in A_1$); then Φ is a *$*$-B-homomorphism*. Hence, A_1 and A_2 are B-isomorphic whenever there is an isomorphism acting from A_1 to A_2 and commuting with projections in B. If a B-isomorphism is $*$-preserving then we call it a *$*$-B-isomorphism*.

7.5.5. Theorem. *An AW^*-algebra A is a B-cyclic C^*-algebra for every order-closed subalgebra B of the complete Boolean algebra $\mathfrak{P}_c(A)$. In particular, every AW^*-algebra is a bo-complete space with mixed norm.*

◁ Let U denote the unit ball of A. It suffices to demonstrate that to every partition of unity $(b_\xi)_{\xi \in \Xi} \subset B$ and every family $(a_\xi)_{\xi \in \Xi} \subset U$ there is a unique element a in U satisfying $b_\xi a_\xi = b_\xi a$ for all $\xi \in \Xi$. Assume first that a_ξ is hermitian for all $\xi \in \Xi$. Then the family $(b_\xi a_\xi)$ consists of pairwise commuting hermitian elements since $(b_\xi a_\xi) \cdot (b_\eta a_\eta) = (b_\xi b_\eta) \cdot (a_\xi a_\eta) = 0$ for $\xi \neq \eta$.

Denote by A_0 the maximal commutative $*$-subalgebra of A that includes $(b_\xi a_\xi)$. By 7.5.3 (2), A_0 is a complex K-space of bounded elements. Hence, A_0 contains the element $a = o\text{-}\sum_{\xi\in\Xi} b_\xi a_\xi$, where o-summation is done in A_0. Clearly, $b_\xi a_\xi = b_\xi a$ for all $\xi \in \Xi$. On the other hand, $-\mathbf{1} \le a_\xi \le \mathbf{1}$ implies that $-\mathbf{1} \le a \le \mathbf{1}$, and so $\|a\| \le 1$.

Uniqueness is now in order. Assume that for some hermitian element $d \in A$ we have $b_\xi d = 0$ for all $\xi \in \Xi$. By 1.3.8 (12),

$$e_\lambda^{b_\xi d} = b_\xi^\perp \vee e_\lambda^d = \mathbf{1} = e_\lambda^{\mathbf{1}} \quad (\lambda \in \mathbb{R},\ \lambda > 0),$$

$$e_\lambda^{b_\xi d} = b_\xi \wedge e_\lambda^d = \mathbf{0} = e_\lambda^{\mathbf{0}} \quad (\lambda \in \mathbb{R},\ \lambda \le 0).$$

The equalities $b_\xi^\perp \vee e_\lambda^d = \mathbf{1}$ and $b_\xi \wedge e_\lambda^d = \mathbf{0}$ are equivalent with the respective inequalities $e_\lambda^d \ge b_\xi$ and $e_\lambda^d \le b_\xi^\perp$. Therefore, $e_\lambda^d = \mathbf{1}$ for $\lambda > 0$ and $e_\lambda^d = \mathbf{0}$ for $\lambda \le 0$; i.e. the spectral function of d is that of the zero element. Consequently, $d = 0$.

In the general case of arbitrary $a_\xi \in U$, use the presentation $a_\xi = u_\xi + iv_\xi$, where i stands for the imaginary unity, and u_ξ and v_ξ are uniquely determined hermitian elements of U. The above shows that there are hermitian elements $u, v \in U$ satisfying $b_\xi u = b_\xi u_\xi$ and $b_\xi v = b_\xi v_\xi$ for all $\xi \in \Xi$. Observe that $a = u + iv$ is a sought element. Indeed, $b_\xi a = b_\xi a_\xi$ for all $\xi \in \Xi$. Moreover, the hermitian elements $a_\xi^* a_\xi$ belong to U, and $b_\xi a^* a = b_\xi a_\xi^* a_\xi$ for all $(\xi \in \Xi)$. Only one element meets these conditions. Since so is a^*a, we see that $a^*a \in U$. Whence $a \in U$, for $\|a\|^2 = \|a^*a\| \le 1$. $\triangleright$

7.5.6. We will now introduce the classification of AW^*-algebras into types. The type of such an algebra is determined from the structure of its lattice of projections. Recall the relevant definitions. Take an AW^*-algebra A. Clearly, the formula

$$q \le p \Leftrightarrow q = qp = pq \quad (q, p \in \mathfrak{P}(X))$$

specifies some order $\le$ on the set of projections. We call projections p and q *equivalent* and write $p \sim q$ if there is an element x in A satisfying $x^*x = p$ and $xx^* = q$. In this case x is a *partial isometry* with initial projection p and final projection q. The relation $\sim$ is in fact an equivalence over $\mathfrak{P}(A)$.

A projection $\pi \in A$ is called:

(a) *abelian* if the algebra $\pi A \pi$ is commutative;

(b) *finite*, if for every projection $\rho \in A$ from $\pi \sim \rho \le \pi$ it follows that $\rho = \pi$;

(c) *infinite*, if π is not finite;

(d) *purely infinite*, if π does not contain nonzero finite projections.

As usual, the phrase "a projection π contains a projection ρ" stands for the inequality $\rho \le \pi$.

An algebra A has *type I*, if each nonzero projection in A contains nonzero abelian projection. An algebra A has *type II*, if A does not contain nonzero abelian projections and each nonzero projection in A contains a nonzero finite projection. An algebra A has *type III*, if the unity of A is a purely infinite projection. An algebra A is *finite* if the unity of A is a finite projection. We say that an algebra A is λ-*homogeneous* if there exists a set $\mathscr{P}$ of pairwise orthogonal equivalent abelian projection with $\sup \mathscr{P} = \mathbf{1}$ and $|\mathscr{P}| = \lambda$.

7.5.7. Our next goal is to demonstrate that the space $B_\Lambda(X)$ of all continuous Λ-linear operators acting in a Kaplansky–Hilbert module X is an AW^*-algebra. The structure of a Banach algebra in $B_\Lambda(X)$ results from the conventional multiplication by the elements of Λ, addition and composition of operators, and the operator norm. We will prove that the involution in $B_\Lambda(X)$ can be defined as the taking of the adjoint of an operator in the sense of Λ-valued inner product. To do this we need some auxiliary facts

(1) *Let X and Y be Kaplansky–Hilbert modules over Λ and let $T : X \to Y$ be a continuous Λ-linear operator. Then T is dominated and bo-continuous. In addition, the kernel of T is a Kaplansky–Hilbert submodule of X.*

$\lhd$ If U is the unit ball in X then $U = \{x \in X : |x| \leq \mathbf{1}\}$. Since any set of type $\{x \in X : |x| \leq a\}$, $a \in \Lambda$, is contained in $a'U$ for some $a' \in \Lambda$, the operator T is subdominated and hence dominated, by virtue of 4.1.11. To prove the *bo*-continuity of T it is sufficient to verify, in view of 4.3.7, that T is completely additive. Take a *bo*-summable family $(x_\alpha)_{\alpha\in\mathrm{A}}$ of pairwise disjoint elements. Let b_α be the projection of $\mathbf{1}$ onto the band generated by $|x_\alpha|$ and $x := bo\text{-}\sum_{\alpha\in\mathrm{A}} b_\alpha x_\alpha$. Then by Λ-linearity we have $b_\alpha Tx = T(b_\alpha x_\alpha) = Tx_\alpha$. Summing these formulas with respect to α, we obtain $Tx = bo\text{-}\sum_{\alpha\in\mathrm{A}} Tx_\alpha$, and the proof is complete. $\rhd$

(2) *Let X be a Kaplansky–Hilbert module over Λ and let $f : X \to \Lambda$ be a continuous Λ-linear operator. Then there exists a unique element $y \in X$ such that*

$$f(x) = \langle x \,|\, y\rangle \quad (x \in X).$$

$\lhd$ In view of (1) $X_0 := \ker(f)$ is a Kaplansky–Hilbert submodule in X. By 7.4.10 (4) $X = X_0 \oplus X_0^\perp$ and $X_0^\perp$ is a Kaplansky–Hilbert module over Λ by 7.4.5 (4). By virtue of 2.1.7 (3) we may choose an element $y_0 \in X_0^\perp$ with $\langle y_0 \,|\, y_0\rangle = \mathbf{1}$. (We always assume that the Kaplansky–Hilbert module under consideration is faithful.) Observe that any element of the form $f(x)y_0 - f(y_0)x$, $x \in X$, is contained in X_0. Thus $\langle f(x)y_0 - f(y_0)x \,|\, y_0\rangle = 0$, or $f(x)\langle y_0 \,|\, y_0\rangle = \langle x \,|\, f(y_0)^* y_0\rangle$. It remains to set $y := f(y_0)^* y_0$. $\rhd$

Let X and Y be Kaplansky–Hilbert modules over Λ and let $T : X \to Y$ be a Λ-linear operator. A Λ-linear operator $T^* : Y \to X$ is said to be the *adjoint* of T if $\langle Tx \,|\, y\rangle = \langle x \,|\, T^* y\rangle$ for all $x \in X$ and $y \in Y$.

(3) *A Λ-linear operator $T : X \to Y$ is continuous if and only if T has an adjoint.*

$\lhd$ Suppose T is continuous. For fixed $y \in Y$ the Λ-linear operator $x \mapsto \langle Tx \,|\, y\rangle$ is continuous and, by (2), can be represented as $\langle Tx \,|\, y\rangle = \langle x \,|\, u\rangle$ $(x \in X)$ with $u \in X$. Put by definition $T^*y := u$. It can be easily verifies that the so defined operator T^* has the desired properties.

Now, let T has an adjoint T^*. If $|\!|\!| x |\!|\!| \leq 1$ then $\langle Tx \,|\, y\rangle = \langle x \,|\, T^*y\rangle$; thus, the set $\{\langle Tx \,|\, y\rangle : |\!|\!| x |\!|\!| \leq 1\}$; is bounded for every $y \in Y$. By the Uniform Boundedness Principle T is continuous. $\rhd$

7.5.8. Theorem. *Let X be a Kaplansky–Hilbert module over a Stone algebra Λ. Then the algebra $B_\Lambda(X)$ of continuous Λ-linear operators on X is an AW^*-algebra of type I with center isomorphic to Λ. Moreover, if X is λ-homogeneous, so is $B_\Lambda(X)$.*

$\lhd$ We split the proof into several steps, denoting for short $\mathscr{A} := B_\Lambda(X)$.

(1) *The center of $\mathscr{A}$*: Multiplication of X by any element of Λ gives rise to a central element of $\mathscr{A}$. Conversely, take a central operator $T \in \mathscr{A}$. The fact that T commutes with the operator $x \mapsto \langle x \,|\, y\rangle z$ for fixed $y, z \in X$ can be written as $\langle Tx \,|\, y\rangle z = \langle x \,|\, y\rangle Tz$. Put $x = y$ and, using 2.1.7 (3), choose $x \in X$ with $|x| = \mathbf{1}$. Then $Tz = \langle Tx \,|\, x\rangle z$, so that T is simply the multiplication by $\langle Tx \,|\, x\rangle$.

(2) *The Baer property*: Observe that the right annihilator of each element of $\mathscr{A}$ coincides with the annihilator of its range. Therefore, for a given subset $M \subset X$ we have to prove that the annihilator of M has the form $\pi\mathscr{A}$ for some projection π. Since the kernel of each bounded Λ-linear operator is a Kaplansky–Hilbert submodule (Proposition 7.5.7 (1)), the annihilator of M coincides with the annihilator of the Kaplansky–Hilbert submodule, say X_0, generated by M. According to 7.4.10 (4) we have the orthogonal decomposition $X = X_0 \oplus X_0^\perp$. Define π to be zero on X_0 and the identity on $X_0^\perp$. Then π is a projection and ${}^\perp(X_0) = \pi\mathscr{A}$.

(3) *Axioms of a C^*-algebra*: We already know that $\mathscr{A}$ is a Banach algebra with involution (see Proposition 7.5.7 (3)). Take any $T \in \mathscr{A}$ and prove that the norm satisfy the two required identities (see 7.4.2): $\|T\| = \|T^*\|$ and $\|T^*T\| = \|T\|^2$. Using the Cauchy–Bunyakovskiĭ–Schwarz inequality 7.4.3 (7) we obtain

$$|Tx|^2 = \langle Tx \,|\, Tx\rangle = \langle T^*Tx \,|\, x\rangle \leq |x||T^*Tx|.$$

Taking norms in the latter and employing the definition of the operator norm we deduce

$$\|T\|^2 := \sup\{\|Tx\|^2 : |\!|\!| x |\!|\!| \leq 1\} \leq \|T^*T\| \leq \|T\|\|T^*\|,$$

whence $\|T\| \leq \|T^*\|$. Since T is also the adjoint of T^*, the reverse inequality also holds. Thus, the first identity holds $\|T\| = \|T^*\|$, and the inequalities $\|T\|^2 \leq \|T^*T\| \leq \|T\|\|T^*\| = \|T\|^2$ imply the second identity.

(4) *Abelian projections*: Take an element $u \in X$ with $|u|$ a projection in Λ and define the operator $\rho_u \in \mathscr{A}$ by $\rho_u x := \langle x \,|\, u\rangle u$. Then ρ_u is hermitian:

$$\langle \rho_u x \,|\, y\rangle = \langle x \,|\, u\rangle\langle u \,|\, y\rangle = \langle x \,|\, \rho_u y\rangle.$$

Moreover, ρ_u is a projection, since $\rho_u^2 x = \langle x \,|\, u\rangle\langle u \,|\, u\rangle u = \langle x \,|\, u\rangle u$. The last identity is true because the projection $|u|$ has the following property: $|u|u = [\,|u|\,]u = u$. For any operators $S, T \in \mathscr{A}$ we compute

$$(\rho_u S \rho_u)(\rho_u T \rho_u)x = \langle x \,|\, u\rangle\langle Su \,|\, u\rangle\langle Tu \,|\, u\rangle u = (\rho_u T \rho_u)(\rho_u S \rho_u)x,$$

so that ρ_u is an abelian projection. If π is an arbitrary nonzero projection in $\mathscr{A}$ then we may choose an element $u \in X$ with $\pi u = u$ and $|u|$ a projection in Λ. In this event $\pi\rho_u x = \langle x \,|\, u\rangle u = \rho_u x$ and $\rho_u \pi x = \langle \pi x \,|\, u\rangle u = \langle x \,|\, \pi u\rangle u = \rho_u x$, whence $\rho_u \leq \pi$. Hence $\mathscr{A}$ is an AW^*-algebra of type I.

(5) *Homogeneity*: Suppose that X is homogeneous and has basis $\mathscr{E}$ with $|\mathscr{E}| = \lambda$. Given arbitrary $e, d \in \mathscr{E}$, define operators π_e and π_{ed} by

$$\pi_e x := \langle x \,|\, e\rangle e, \quad \pi_{ed} x := \langle x \,|\, e\rangle d \quad (x \in X).$$

By (4) π_e is an abelian projection; moreover $\pi_e \circ \pi_d = 0$, whenever $e \neq d$. If a nonzero projection $\pi \in \mathscr{A}$ is orthogonal to all π_e, $e \in \mathscr{E}$, then there is a nonzero element $x \in X$ with $0 = \pi_e x = \langle x \,|\, e\rangle e$ and $\langle x \,|\, e\rangle = 0$. This contradiction proves that $\sup_{e\in\mathscr{E}} \pi_e = I_X$. Since $\pi_{ed} \circ \pi_{de} = \pi_e$, it follows that π_e and π_d are equivalent. This shows that $\mathscr{A}$ is λ-homogeneous. ▷

7.5.9. Suppose that Q is some extremal compact space, H is a Hilbert space, and $B(H)$ is the space of bounded linear endomorphisms of H.

Denote by $\mathfrak{C}(Q, B(H))$ the set of all operator-functions $u : \mathrm{dom}(u) \to B(H)$, defined on the comeager sets $\mathrm{dom}(u) \subset Q$ and continuous in the strong operator topology.

If $u \in \mathfrak{C}(Q, B(H))$ and $h \in H$, then the vector-function $uh : q \mapsto u(q)h$ ($q \in \mathrm{dom}(u)$) is continuous thus determining a unique element $\widetilde{uh} \in C_\infty(Q, H)$ from the condition $uh \in \widetilde{uh}$ (cf. 2.3.6). Introduce an equivalence on $\mathfrak{C}(Q, B(H))$ by putting $u \sim v$ if and only if u and v agree on $\mathrm{dom}(u) \cap \mathrm{dom}(v)$. If $\tilde{u}$ is a coset of the operator-function $u : \mathrm{dom}(u) \to B(H)$ then $\tilde{u}h := \widetilde{uh}$ $(h \in H)$ by definition.

Denote by $SC_\infty(Q, B(H))$ the set of all cosets $\tilde{u}$ such that $u \in \mathfrak{C}(Q, B(H))$ and the set $\{|\tilde{u}h| : \|h\| \leq 1\}$ is bounded in $C_\infty(Q)$.

Since $\lvert \tilde{u}h \rvert$ agrees with the function $q \mapsto \|u(q)h\|$ $(q \in \mathrm{dom}(u))$ on some comeager set, the inclusion $\tilde{u} \in SC_\infty(Q, B(H))$ means that the function $q \mapsto \|u(q)\|$ $(q \in \mathrm{dom}(u))$ is continuous on a comeager set. Hence, there are an element $\lvert \tilde{u} \rvert \in C_\infty(Q)$ and a comeager set $Q_0 \subset Q$ satisfying $\lvert \tilde{u} \rvert(q) = \|u(q)\|$ $(q \in Q_0)$. Moreover, $\lvert \tilde{u} \rvert = \sup\{\lvert \tilde{u}h \rvert : \|h\| \le \mathbf{1}\}$, where the supremum is taken over $C_\infty(Q)$. We naturally furnish $SC_\infty(Q, B(H))$ with the structure of a $*$-algebra and a unitary $C_\infty(Q)$-module by means of the operations

$$
\begin{aligned}
(u+v)(q) &:= u(q) + v(q) \quad (q \in \mathrm{dom}(u) \cap \mathrm{dom}(v)),\\
(uv)(q) &:= u(q) \circ v(q) \quad (q \in \mathrm{dom}(u) \cap \mathrm{dom}(v)),\\
(av)(q) &:= a(q)v(q) \quad (q \in \mathrm{dom}(a) \cap \mathrm{dom}(v)),\\
u^*(q) &:= u(q)^* \quad (q \in \mathrm{dom}(u)),
\end{aligned}
$$

with $u, v \in \mathfrak{C}(Q, B(H))$ and $a \in C_\infty(Q)$. Furthermore, we note the following

$$
\begin{gathered}
\lvert \tilde{u} + \tilde{v} \rvert \le \lvert \tilde{u} \rvert + \lvert \tilde{v} \rvert,\\
\lvert \tilde{u}\tilde{v} \rvert \le \lvert \tilde{u} \rvert \cdot \lvert \tilde{v} \rvert,\\
\lvert a\tilde{v} \rvert = |a| \lvert \tilde{v} \rvert, \ \lvert \tilde{u} \cdot \tilde{u}^* \rvert = \lvert \tilde{u} \rvert^2.
\end{gathered}
$$

If $\tilde{u} \in SC_\infty(Q, B(H))$ and the element $\tilde{x} \in C_\infty(Q, H)$ is determined by a continuous vector-function $x : \mathrm{dom}(x) \to H$ then we may define $\tilde{u}\tilde{x} := \widetilde{ux} \in C_\infty(Q, H)$, with $ux : q \mapsto u(q)x(q)$ $(q \in \mathrm{dom}(u) \cap \mathrm{dom}(x))$; since the last function is continuous. We also have

$$
\lvert \tilde{u}x \rvert \le \lvert \tilde{u} \rvert \cdot \lvert x \rvert \quad (x \in C_\infty(Q, H)).
$$

It follows in particular that

$$
\lvert \tilde{u} \rvert = \sup \left\{ \lvert \tilde{u}x \rvert : x \in C_\infty(Q, H), \lvert x \rvert \le \mathbf{1} \right\}.
$$

Denote the operator $x \mapsto \tilde{u}x$ by $S_{\tilde{u}}$.

We now introduce the following normed $*$-algebra

$$
\begin{gathered}
SC_\#(Q, B(H)) := \{v \in SC_\infty(Q, B(H)) : \lvert v \rvert \in C(Q)\},\\
\|v\| = \|\lvert v \rvert\|_\infty \quad (v \in SC_\#(Q, B(H))).
\end{gathered}
$$

7.5.10. Theorem. *To each operator $U \in \mathrm{End}(C_\#(Q, H))$ there is a unique element $u \in SC_\#(Q, B(H))$ satisfying $U = S_u$. The mapping $U \mapsto u$ is a $*$-B-isomorphism of $\mathrm{End}(C_\#(Q, H))$ onto $A := SC_\#(Q, B(H))$. In particular, A is a λ-homogeneous algebra. Moreover, if Q is a λ-stable compact space then A is a strictly λ-homogeneous AW^*-algebra, with $\lambda = \dim(H)$.*

◁ First of all note that the operator S_u obeys the inequality $|S_u x| \le |u| \cdot |x|$ for all $x \in C_\#(Q,H)$. Consequently, for each $u \in SC_\#(Q,B(H))$ we see that S_u acts in $C_\#(Q,H)$, $C(Q)$ as a bounded linear operator. Moreover,

$$\|S_u\| = \sup_{\|x\|\le 1} \| |S_u x| \|_\infty = \sup_{|x|\le 1} \sup_{q\in Q} |ux|(q) = \sup_{q\in Q} |u|(q) = \|u\|.$$

Clearly, $S_{u^*} = S_u{}^*$ for all $u \in SC_\#(Q,B(H))$. Therefore, the mapping $u \mapsto S_u$ is a $*$-B-isomorphic embedding of $SC_\#(Q,B(H))$ into $\mathrm{End}(C_\#(Q,H))$. Prove that this embedding is surjective. The mapping $U \in \mathrm{End}(C_\#(Q,H))$ is a *dominated operator*; i.e., U obeys the inequality $|Ux| \le f \cdot |x|$ for all $x \in C_\#(Q,H)$, where $f := \sup\{|Ux| : |x| \le 1\} \in C(Q)$. By Theorem 5.5.4 there is an operator-function $u : \mathrm{dom}(u) \to B(H)$, satisfying the conditions: (1) the function $q \mapsto \langle u(q)h|g\rangle$ $(q \in \mathrm{dom}(u))$ is continuous for all $g,h \in H$; (2) there is a function $\varphi \in C_\infty(Q)$ such that $\|u(q)\| \le \varphi(q)$ $(q \in \mathrm{dom}(u))$; (3) $Ux = \tilde{u}x$ for all $x \in C_\#(Q,H)$ and $|u| = f$. Thus, $U = S_{\tilde{u}}$ and we are left with justifying only that u is continuous in the strong operator topology. Recalling the definition of least upper bound in the K-space $C_\infty(Q)$, we may observe that $\|u(q)\| = |u|(q)$ $\ (q \in Q_0)$, where Q_0 is some comeager subset of Q. Therefore, substituting $Q_0 \cap \mathrm{dom}(u)$ for $\mathrm{dom}(u)$ if need be, we may assume that $q \mapsto \|u(q)\|$ $\ (q \in \mathrm{dom}(u))$ is a continuous function. Together with the above condition (1) this implies the continuity of u in the strong operator topology; i.e., $u \in SC_\#(Q,B(H))$. The rest of the theorem ensues from 5.3.4 (3). ▷

We say that the families of nonempty compact sets $(Q_\gamma)_{\gamma\in\Gamma}$ and $(P_\delta)_{\delta\in\Delta}$ are *congruent* provided that $\Gamma = \Delta$, and Q_γ and P_γ are homeomorphic to one another for all $\gamma \in \Gamma$.

7.5.11. Theorem. *Let A be a type I AW^*-algebra with center Λ. Then there is a Kaplansky–Hilbert module X over Λ such that A is $*$-isomorphic to $B_\Lambda(X)$.*

◁ For diversity, we shall give a short Boolean-valued proof in 8.4.14. ▷

7.5.12. Theorem. *To every type I AW^*-algebra A there exists a family of nonempty extremal compact spaces $(Q_\gamma)_{\gamma\in\Gamma}$ such that the following conditions are met:*

(1) *Γ is a set of cardinals and Q_γ is γ-stable for each $\gamma \in \Gamma$;*

(2) *There is a $*$-B-isomorphism:*

$$A \simeq \sum_{\gamma\in\Gamma}^{\oplus} SC_\#(Q_\gamma, B(l_2(\gamma))).$$

This family is unique up to congruence.

◁ The representation (2) follows from Theorems 7.4.12, 7.5.10, and 7.5.11. The uniqueness of the family $(Q_\gamma)_{\gamma\in\Gamma}$ and Γ-stability of the compact (Q_γ) will be proved in Section 8.4.11. ▷

7.6. Comments

7.6.1. (1) Banach spaces with mixed norm in the sense of 7.1.1 were introduced in [199]. Recall that if E and F are ideal spaces then $E[F]$ is also an ideal space and even a Banach–Kantorovich space, see 2.3.9. If, in addition, E is a Banach lattice then $E[G]$ is conventionally endowed with norm $|||K||| := \| |K| \|$ and is called a space with mixed norm; see [61, 162]. Clearly, $E[F] = L^{p,q}(\mu \otimes \nu)$ for $E = L^q(\mu)$ and $E = L^p(\nu)$. This justifies the term "Banach space with mixed norm" in 7.1.1.

(2) The material of Section 7.1 is taken from [199, 205]. In case $X = E$ and $|\cdot| = |\cdot|$, a space with mixed norm is just a normed lattice and the results presented in 7.1.3–7.1.8 become a portion of the theory of normed lattices [23, 162, 336, 341, 409]. In particular, the equivalence of Theorem 7.1.7 becomes the Andó Theorem [25], see [162, 231, 336].

(3) Order properties of taking the adjoint operator were studied by J. Synnatzchke [364, 365]. The inequality in 7.1.10 (1) can be strict, see [23]. It follows from 7.1.10 (2) that the adjoint to the modulus of an order continuous regular operator and the modulus of the adjoint to the original operator agree on the order continuous linear functionals. This result is due to J. Synnatzchke [364, 365] and U. Krengel [190]. Concerning the Banach lattice prototypes of other results from 7.1.9–7.1.11, see [336, 341].

7.6.2. (1) The class of (p,q)-summing operators was introduced by A. G. Kusraev in [199, 205]. In the same papers, the main results 7.2.4–7.2.9 were established. The classes of (p,q)-convex and (p,q)-concave operators (7.2.11 (1, 2)) were introduced by J. L. Krivine [191] for $p = q$ and B. Maurey [278] in the general case, see [242, 341]. Finally, (p,q)-regular operators (7.2.10 (3)) were introduced by A. V. Bukhvalov [58]. As regards connections with other classes of operators, see the survey [60].

(2) Note that $(1,1)$-concave operators were first considered by V. L. Levin [237, 238] as the so-called summing operators; see [60, 239] for further references. The class of summing operators and the dual class of reqular (order-bounded) operators are discussed in the monographs by V. L. Levin [239] and H. H. Schaefer [336]. In a more general situation of a Banach space with regular cone the classes of these operators were introduced and investigated by V. T. Khudalov [169, 170]. A particular case of Corollary 7.2.8 in which $X = E$ and $Y = F$ was also established by V. L. Levin; see [239]. An interesting class of lattice-summing operators was studied by Nielsen and Szulga [297] and by L. P. Yanovskiĭ [407].

7.6.3. (1) In [240] J. Lindenstrauss and L. Tzafriri wrote: "There are many classes of special Banach spaces appearing in analysis which have not yet been studied in detail from the point of view of the geometric theory of Banach spaces.

We are sure that such a study, if made, will give new insight into the specific areas of analysis in which the spaces appear, and also bring to light several natural and interesting problems in areas which do not seem to be a priori closely connected with Banach space theory."

As an example of a poorly-studied class, we mention the class of classical Banach spaces with mixed norm: $L^{p,q}(\mu \otimes \nu)$ or $L^p(\mu, X)$, where $1 \leqslant p, q \leqslant \infty$, $p \neq q$, X is a Banach space, and μ and ν are finite or σ-finite measures. In Section 7.3, we sketched an approach to the indicated problems which is based on the theory of lattice-normed spaces. In general, it seems interesting and useful to develop the geometric theory of Banach spaces with mixed norm (7.1.1) in the spirit of [73, 76, 231, 241, 242] (Problem 20 in [199]).

(2) The main results, 7.3.2, 7.3.4, 7.3.5, and 7.3.6, were obtained by A. G. Kusraev in [199, 205]. Earlier, G. Takeuti [371] obtained a version of Theorem 7.3.3 (1) while studying Boolean-valued representations of C^*-algebras. Actually, he also employed the notion of B-cyclic Banach space but did not use the term. Theorem 7.3.4 can be considered as an analog of the classical Bohnenblust Theorem [47] in the class of Banach spaces with mixed norm.

(3) W. G. Badé [31] initiated the study of operator algebras generated by Boolean algebras of projections, see also [32] and [88]. In this connection complete (σ-complete) Boolean algebras (7.3.3) were introduces in [31]. Moreover, in [31] W. G. Badé introduced the so-called cyclic Banach spaces (other than B-cyclic Banach spaces of 7.3.3). Let $\mathscr{B}$ be a σ-complete Boolean algebra of bounded projections in a Banach space X. If there exists an element $e \in X$ such that the closed linear span of $\{\pi e : \pi \in \mathscr{B}\}$ coincides with X then X is called *cyclic Banach spaces,* see [336]. A. I. Veksler [379] have proved that a cyclic Banach space is isomorphic to a Banach lattice with order continuous norm and weak order-unity. A close result was independently proved in [336]; see also [242].

(4) W. G. Badé [31] also proved the following reflexivity theorem: *A continuous linear operator T on a Banach space X belongs to the strongly closed algebra generated by a σ-complete Boolean algebra $\mathscr{B}$ of projections if and only if T leaves invariant each $\mathscr{B}$-invariant subspace of X.* This result was extended to the setting of locally convex spaces in [82] by P. G. Dodds and B. de Pagter and in [85] by P. G. Dodds and W. Ricker. About further development see [83, 84] and the literature therein.

7.6.4. (1) J. von Neumann started the study of involutive operator algebras, inspired by the mathematical problems of quantum mechanics, cf. [293–295]. This traditional connection with theoretical physics is still alive (cf., for instance, [52]). However, the present-day theory of involutive topological algebras contains several rather abstract fields of research, raising many subtle mathematical problems. To enter this field, the reader may consult [29, 79, 80, 146, 286, 333, 356, 363, 366].

(2) The study of C^*-algebras was originated by I. M. Gelfand and M. A. Naĭmark in 1943. The principal structural properties of C^*-algebra are connected with positivity. As regards the preliminaries of Baer involutive algebras, consult [43]. See [29, 79, 225, 284] for the details of C^*-algebras and [80, 333, 366] for the details of von Neumann algebras.

(3) The theory of Kaplansky–Hilbert modules was developed by I. Kaplansky in [166]. The definition of 7.4.3, 7.4.5 (1, 2), and 7.4.6 except strict homogeneity, as well as the results of 7.4.8 (3), 7.4.9, and 7.4.10 are taken from [166]. The concepts of strict λ-homogeneity (7.4.6) and λ-stability (7.4.11) were introduced in [201]. The examples of Kaplansky–Hilbert modules of 7.4.8 (1) and 7.4.8 (2) were constructed in [201] and [166], respectively. One more example can be found in [400]: it is $L^2(\mu)$ for a modular ample measure (see also 6.6.1 (2)). As to Theorem 7.4.12, see 8.6.4 (3).

(4) General C^*-modules have other interesting applications. In particular, in C^*-index theory, they play the role of Hilbert spaces, see [356] and the references therein. To enter this field, the reader may consult [96, 233, 316].

7.6.5. (1) The modern structure theory of AW^*-algebras originated with the articles by I. Kaplansky [164–166]. These objects appear naturally by way of algebraization of the theory of von Neumann operator algebras. The results of 7.5.7 and 7.5.6 are taken from [166]. Theorems 7.5.10 and 7.5.11 are due to A. G. Kusraev [201].

(2) In [400] J. D. M. Wright proved an important spectral theorem for Kaplansky–Hilbert modules. To state it, we will use the following notation: $\mathscr{M}(Q,\mathbb{C})$ symbolizes the K_σ-space of all complex Borel functions on Q, while $C(Q,\mathbb{C})$ is the space of complex continuous functions on Q, with A_{sa} standing for the selfadjoint part of a complex algebra A.

Theorem. *Let X be a Kaplansky–Hilbert module over a Stone algebra Λ and let A be a commutative AW^*-subalgebra of $B_\Lambda(X)$ with unity. Let Q be a compact space and let $\varphi : C(Q,\mathbb{C}) \to A$ be a $*$-homomorphism. Then there exists a unique quasiregular Borel measure $\mu : \mathscr{B}or(Q) \to A_{\mathrm{sa}}$ with the following properties:*

(i) $\varphi(f) = \int_Q f\,d\mu \quad (f \in C(Q,\mathbb{C}))$;

(ii) *the formula*

$$\widehat{\varphi}(f) = \int\limits_Q f\,d\mu \quad (f \in \mathscr{M}(Q,\mathbb{C}))$$

defines a $*$-homomorphism from $\mathscr{M}(Q,\mathbb{C})$ into A;

(iii) *μ is a projection-valued (spectral) measure;*

(iv) *if the A_{sa}-valued measure μ_x is defined as follows*

$$\mu_x(C) := \langle \mu(C)x \,|\, x \rangle \quad (C \in \mathscr{B}or(Q)),$$

then μ is a positive quasiregular Borel measure and

$$\left\langle \left(\int_Q f \, d\mu \right) x \,\middle|\, x \right\rangle = \int_Q f \, d\mu_x$$

for all $f \in \mathscr{M}(Q, \mathbb{C})$ and $x \in X$.

Chapter 8

Applications of Boolean-Valued Analysis

The term *Boolean-valued analysis* signifies the technique of studying properties of an arbitrary mathematical object by means of comparison between its representations in two different set-theoretic models whose construction utilizes distinct Boolean algebras. As these models, we usually take the classical sets in the shape of the *von Neumann universe* and a specially-trimmed *Boolean-valued universe* in which the conventional set-theoretic concepts and propositions acquire nonstandard interpretations. Thus to understand the subject of this chapter, we need some information from the theory of Boolean-valued models. We will use the Boolean-valued technique of ascending and descending as presented in [206] and [212]. A concise presentation of Boolean-valued tools is given in the Appendix.

It turns out that Boolean-valued analysis is inseparable from the concepts of the theory of ordered vector spaces and, above all, with that of K-space. Sections 8.1 and 8.2 constitute a Boolean-valued introduction to the theory of vector lattices. The fundamental contribution of Boolean-valued analysis to this theory is Gordon's Theorem (8.1.2) which we may read as follows: *Every universally complete K-space is an interpretation of the reals in an appropriate Boolean-valued model.*

Moreover, every Archimedean vector lattice embeds in a Boolean-valued model, becoming a vector sublattice of the reals viewed as such over some dense subfield of the reals (8.1.6). According to these facts and fundamental principles of Boolean-valued models, each theorem about the reals within Zermelo–Fraenkel set theory has an analog in the original K-space interpreted as the Boolean-valued reals. Translation of theorems is carried out by appropriate general operations of Boolean-valued analysis.

This *Boolean-valued transfer principle for K-space* is applied to establish some important structure properties of vector lattices: representability by means of continuous functions (8.2.5 (1–3)) and spectral functions (8.2.3, 8.2.4 (1, 3)), the Freudenthal Spectral Theorem (8.2.4 (2)), the spectral integral (8.2.6, 8.2.7, 8.2.11, 8.2.12, 8.2.13), the functional calculus (8.2.8, 8.2.10, 8.2.14, 8.2.15), etc.

The main topic of Section 8.3 is Banach spaces in Boolean-valued universes. Theorems 8.3.1, 8.3.2 and 8.3.5 (2) also rank among the principal results of the current chapter. The first of them claims that an arbitrary Banach space inside a Boolean-valued model may be interpreted as a universally complete lattice-normed space. The second declares that a lattice-normed space may be realized as a dense subspace of a Banach space viewed a vector space over some field. Finally, the third theorem means essentially that a Banach space X results from the procedure of bounded descent from a Boolean-valued model if and only if X includes a complete Boolean algebra of norm one projections which possesses the cyclicity property. In other words, X is an order complete lattice-normed space and the norm of X is a mixed norm. This fact serves as a starting point for the approach to involutive algebras which we pursue in the next section.

The Boolean-valued approach to studying Banach algebras, presented in Section 8.4, bases on the following observation: If the center of an algebra is properly qualified and perfectly located then it becomes a one dimensional subalgebra after immersion in a suitable Boolean-valued model. This might lead to a simpler algebra. On the other hand, the Boolean-valued transfer principle implies that the scope of the formal theory of the initial algebra is the same as that of its Boolean-valued representation. Theorems 8.4.1 and 8.4.2 (1) elaborate this claim for a general Banach algebra and a C^*-algebra.

Further exposition focuses on analysis of AW^*-algebras and Kaplansky–Hilbert modules which transform into AW^*-factors and Hilbert spaces in a Boolean-valued model (Theorems 8.4.4 (2) and 8.3.8). The dimension of a Hilbert space in such a model is a Boolean-valued cardinal referred naturally to as the Boolean dimension of the Kaplansky–Hilbert module depicting the space (8.4.7 and 8.4.8).

A rather delicate effect reveals itself here, the so-called *cardinal shift*: standard cardinals may glue together when embedded into a Boolean-valued model. In other words, the bases of isomorphic Kaplansky–Hilbert modules may differ in cardinality (8.4.12 (4)). This also implies that every type I AW^*-algebra decomposes into the direct sum of homogeneous subalgebras in many ways. I. Kaplansky conjectured the fact as far back as in 1953.

Leaning on results about the Boolean-valued immersion of Kaplansky–Hilbert modules and AW^*-algebras, we further expose functional representations of these objects. To put it more precisely, we prove that every Kaplansky–Hilbert module is unitarily equivalent to the direct sum of homogeneous Kaplansky–Hilbert modules consisting of continuous vector-functions with range in a Hilbert space (8.4.13). An analogous representation holds for an arbitrary type I AW^*-algebra on replacing continuous vector-function with operator-valued functions continuous in the strong operator topology (8.4.14).

In Section 8.5 deals with the so-called cyclically compact operators in Banach spaces. This notion is closely related with the Boolean-valued interpretation of the conventional compactness (8.5.1, 8.5.2, 8.5.3) and seems to deserve an independent study. We confine exposition to interpretation of the general form of a compact operator in Hilbert space and the Fredholm Alternative (8.5.6, 8.5.9).

8.1. Real Numbers in Boolean-Valued Models

Boolean-valued analysis stems from the assertion due to D. Scott that the image of every field of reals in a Boolean-valued model represents a universally complete K-space (of measurable functions). Depending on which Boolean algebra B (the algebra of measurable sets, open regular sets, or projections in a Hilbert space) forms the base for constructing a Boolean-valued model $\mathbf{V}^{(B)}$, we obtain different K-spaces (the spaces of measurable functions, semicontinuous functions, or selfadjoint operators). Thereby there appears a remarkable opportunity for transferring all the treasure-trove of knowledge about real numbers to a profusion of classical objects of analysis. This will constitute the topic of the section.

8.1.1. By a field of reals we mean each algebraic system that satisfies the axioms of an Archimedean ordered field (with distinct zero and unity) and the axiom of completeness. Recall the following two well-known assertions:

(1) *There exists a field of reals* $\mathbb{R}$ *that is unique to within isomorphism.*

(2) *If* $\mathbf{P}$ *is an Archimedean ordered field then there is an isomorphic embedding* h *of the field* $\mathbf{P}$ *into* $\mathbb{R}$ *such that the image* $h(\mathbf{P})$ *is a subfield of* $\mathbb{R}$ *containing the subfield of rational numbers. In particular,* $h(\mathbf{P})$ *is dense in* $\mathbb{R}$.

Successively applying the Transfer and Maximum Principles of Boolean-valued analysis to (1), we find an element $\mathscr{R} \in \mathbf{V}^{(B)}$ for which $[\![\, \mathscr{R}$ is a field of reals $]\!] = \mathbf{1}$. Moreover, if an arbitrary $\mathscr{R}' \in \mathbf{V}^{(B)}$ satisfies the condition $[\![\, \mathscr{R}'$ is a field of reals $]\!] = \mathbf{1}$ then it also satisfies $[\![$ the ordered fields $\mathscr{R}$ and $\mathscr{R}'$ are isomorphic$]\!] = \mathbf{1}$. In other words, there exists a field of reals $\mathscr{R}$ in the model $\mathbf{V}^{(B)}$ and such a field is unique to within isomorphism. We call $\mathscr{R}$ the *reals* in $\mathbf{V}^{(B)}$.

Note also that $\varphi(x)$, formally presenting the expressions of the axioms of an Archimedean ordered field x, is bounded; therefore, $[\![\, \varphi(\mathbb{R}^{\wedge})\,]\!] = \mathbf{1}$, i.e., $[\![\, \mathbb{R}^{\wedge}$ is an Archimedean ordered field$\,]\!] = \mathbf{1}$. "Pulling" assertion (2) through the Transfer Principle, we conclude that $[\![\, \mathbb{R}^{\wedge}$ is isomorphic to a dense subfield of the field $\mathscr{R}\,]\!] = \mathbf{1}$. In this regard, we further assume that $\mathscr{R}$ is the reals in the model $\mathbf{V}^{(B)}$ and $\mathbb{R}^{\wedge}$ is a dense subfield of $\mathscr{R}$. It is easy to note that the elements $0 := 0^{\wedge}$ and $1 := 1^{\wedge}$ are the zero and unity of the field $\mathscr{R}$.

Denoting implication and equivalence in the sequel, we use $\Rightarrow$ and $\Leftrightarrow$ outside $\mathbf{V}^{(B)}$ and $\rightarrow$ and $\leftrightarrow$ inside $\mathbf{V}^{(B)}$.

Now, consider the descent $\mathscr{R}{\downarrow}$ of the algebraic system $\mathscr{R}$ (see A.9 and A.16). In other words, we consider the descent of the underlying set of the system $\mathscr{R}$ together with descended operations and order. For simplicity, we denote the operations and order in $\mathscr{R}$ and $\mathscr{R}\downarrow$ by the same symbols $+$, $\cdot$, and $\leq$. In more detail, we introduce addition, multiplication, and order in $\mathscr{R}\downarrow$ by the formulas

$$\begin{gathered} z = x + y \Leftrightarrow [\![z = x + y]\!] = \mathbf{1}, \\ z = x \cdot y \Leftrightarrow [\![z = x \cdot y]\!] = \mathbf{1}, \\ x \leq y \Leftrightarrow [\![x \leq y]\!] = \mathbf{1} \\ (x,\ y,\ z \in \mathscr{R}\downarrow). \end{gathered}$$

Also, we may introduce multiplication by the usual reals in $\mathscr{R}\downarrow$ by the rule

$$y = \lambda x \Leftrightarrow [\![\lambda^{\wedge} x = y]\!] = \mathbf{1} \quad (\lambda \in \mathbb{R},\ x, y \in \mathscr{R}\downarrow).$$

8.1.2. Gordon Theorem. *Let $\mathscr{R}$ be the reals in the model $\mathbf{V}^{(B)}$. Then $\mathscr{R}\downarrow$ (with the descended operations and order) is a universally complete K-space with order-unity 1. Moreover, there exists an isomorphism χ of the Boolean algebra B onto the base $\mathfrak{P}(\mathscr{R}\downarrow)$ such that the following equivalences hold:*

$$\begin{gathered} \chi(b)x = \chi(b)y \Leftrightarrow b \leq [\![x = y]\!], \\ \chi(b)x \leq \chi(b)y \Leftrightarrow b \leq [\![x \leq y]\!] \end{gathered}$$

for all $x, y \in \mathscr{R}\downarrow$ and $b \in B$.

$\lhd$ We omit any elementary verification of the fact that $\mathscr{R}\downarrow$ is a vector space over $\mathbb{R}$ and an ordered set. Show that the operations and order in $\mathscr{R}\downarrow$ agree and the necessary exact bounds exist. Take elements $x, y \in \mathscr{R}\downarrow$ such that $x \leq y$. It means that

$$\mathbf{V}^{(B)} \models \text{“}x \text{ and } y \text{ are reals and } x \leq y.\text{”}$$

Let $u := x + z$, $v := y + z$, $x' := \lambda x$, and $y' := \lambda y$, where $z \in \mathscr{R}\downarrow$ and $\lambda \in \mathbb{R}$, $\lambda \geq 0$. By the definition of the operations and order in $\mathscr{R}\downarrow$, we have $\mathbf{V}^{(B)} \models$ “x', y', u, and v are reals; moreover, $u = x + z$, $v = y + z$, $x' = \lambda^{\wedge} x$, and $y' = \lambda^{\wedge} y$.” the inequality $\lambda \geq 0$ implies $\mathbf{V}^{(B)} \models \lambda^{\wedge} \geq 0^{\wedge} = 0$. Using the requested properties of numbers inside $\mathbf{V}^{(B)}$, we obtain $\mathbf{V}^{(B)} \models$ “$u \leq v$ and $x' \leq y'$.” Thereby $u \leq v$ and $x' \leq y'$.

Suppose that a set $A \subset \mathscr{R}\downarrow$ is bounded above by an element $y \in \mathscr{R}\downarrow$. By definition, it means that $[\![x \leq y]\!] = \mathbf{1}$ for every element $x \in A$. Then $\mathbf{V}^{(B)} \models$ “$A\uparrow$ is a set of numbers bounded above by the number y” or, in view of A.10 (1),

$$[\![(\forall x \in A\uparrow)(x \leq y)]\!] = \bigwedge_{x \in A} [\![x \leq y]\!] = \mathbf{1}.$$

The completeness of $\mathscr{R}$ yields

$$[\![(\exists a \in \mathscr{R})(a = \sup(A\uparrow))]\!] = \mathbb{1}.$$

Employing the Maximum Principle, we find $a \in \mathbf{V}^{(B)}$ such that $[\![a \in \mathscr{R}]\!] = [\![a = \sup(A\uparrow)]\!] = \mathbb{1}$. Thereby $a \in \mathscr{R}\!\downarrow$ and if $z \in \mathscr{R}\!\downarrow$ is an upper bound of A then, as was already shown, $[\![z$ is an upper bound of $A\uparrow]\!] = \mathbb{1}$; therefore, $[\![a \leq z]\!] = \mathbb{1}$ or $a \leq z$. Consequently, a is the supremum of the set A in $\mathscr{R}\!\downarrow$. Incidentally, we have established that $a = \sup(A)$ if and only if $[\![a = \sup(A\uparrow)]\!] = \mathbb{1}$. In particular, given arbitrary $x, x_1, x_2 \in \mathscr{R}\!\downarrow$, we have $x = x_1 \vee x_2$ if and only if

$$[\![x = \sup\{x_1, x_2\} = x_1 \vee x_2]\!] = \mathbb{1},$$

since $[\![\{x_1, x_2\}\uparrow = \{x_1, x_2\}]\!] = \mathbb{1}$. Of course, an analogous assertion is valid for greatest lower bounds. Now, take an arbitrary set $A \subset \mathscr{R}\!\downarrow$ of positive pairwise disjoint elements. We may see from the above remarks and A.10 that

$$[\![(\forall x_1 \in A\uparrow)(\forall x_2 \in A\uparrow) x_1 \wedge x_2 = 0]\!] = \bigwedge_{x_1, x_2 \in A} [\![x_1 \wedge x_2 = 0]\!] = \mathbb{1}.$$

Hence, the numeric set $A\uparrow$ (inside $\mathbf{V}^{(B)}$) consists of pairwise disjoint positive elements. For such a set we have only the following two possibilities: either $[\![A\uparrow = \{0\}]\!] = \mathbb{1}$ and then $A \subset A\uparrow\downarrow = \{0\}$, or $[\![A\uparrow = \{0, a\}]\!] = \mathbb{1}$ for some $0 \leq a \in \mathscr{R}\!\downarrow$ (by the Maximum Principle) and then $[\![\sup(A\uparrow) = a]\!] = \mathbb{1}$. As was mentioned above, the latter relation is equivalent to the equality $a = \sup A$. Now, we may conclude that $\mathscr{R}\!\downarrow$ is a universally complete K-space. Recalling that $1 := 1^{\wedge}$ is the unity of the field $\mathscr{R}$ inside $\mathbf{V}^{(B)}$ and employing the formulas of A.9 (2) and A.5, we find

$$\mathbb{1} = [\![(\forall x \in \mathscr{R})(x \wedge 1 = 0 \rightarrow x = 0)]\!] = \bigwedge_{x \in \mathscr{R}\downarrow} [\![x \wedge 1 = 0]\!] \Rightarrow [\![x = 0]\!].$$

Hence, we see that $[\![x \wedge 1 = 0]\!] \leq [\![x = 0]\!]$ for each $x \in \mathscr{R}\!\downarrow$. If $x \wedge 1 = 0$ then $[\![x \wedge 1 = 0]\!] = \mathbb{1}$ and so $[\![x = 0]\!] = \mathbb{1}$, i.e., $x = 0$. Thereby 1 is the unity of the K-space $\mathscr{R}\!\downarrow$.

Now, introduce some mapping $\chi : B \rightarrow \mathfrak{P}(\mathscr{R}\!\downarrow)$. Take an arbitrary element $b \in B$ and put $\chi(b)x := \operatorname{mix}\{bx, b^*0\}$ for $x \in \mathscr{R}\!\downarrow$. In other words, the element $\chi(b)x \in \mathscr{R}\!\downarrow$ is uniquely determined by the following relations (see A.7 (3)):

$$b \leq [\![\chi(b)x = x]\!], \quad b^* \leq [\![\chi(b)x = 0]\!].$$

It implies that $\pi := \chi(b) : \mathscr{R}\!\downarrow \to \mathscr{R}\!\downarrow$ is an extensional mapping. Indeed, the following inequalities hold for $x, y \in \mathscr{R}\!\downarrow$ (see A.5 (3)):

$$[\![x = y]\!] \wedge b \leq [\![x = y]\!] \wedge [\![x = \pi x]\!] \wedge [\![y = \pi y]\!] \leq [\![\pi x = \pi y]\!],$$
$$b^* \leq [\![\pi x = 0]\!] \wedge [\![\pi y = 0]\!] \leq [\![\pi x = \pi y]\!].$$

If $\rho := \pi\!\uparrow$ then $[\![\rho : \mathscr{R} \to \mathscr{R}]\!] = \mathbf{1}$ by A.10 (4) and $\rho = \operatorname{mix}\{bI_{\mathscr{R}}, b^*0\}$. Since $\mathbf{0}$ and $I_{\mathscr{R}}$ are idempotent positive linear mappings from $\mathscr{R}$ to $\mathscr{R}$, so is π. Moreover, $[\![(\forall x \in \mathscr{R}_+)\rho x \leq x]\!] = \mathbf{1}$; therefore, $\pi x \leq x$ for all $x \in \mathscr{R}\!\downarrow_+$. Thus, $\pi = \chi(b)$ is a band projection. Since ρ is positive, we have $[\![x \leq y \to \rho x \leq \rho y]\!] = \mathbf{1}$ for $x, y \in \mathscr{R}\!\downarrow$ and hence

$$[\![x \leq y]\!] \leq [\![\rho x \leq \rho y]\!] = [\![\pi x \leq \pi y]\!].$$

Assume $\pi x \leq \pi y$. Then

$$b = [\![\pi x \leq \pi y]\!] \wedge [\![\pi x = x]\!] \wedge [\![\pi y = y]\!] \leq [\![x \leq y]\!].$$

Conversely, if we assume that $b \leq [\![x \leq y]\!]$ then $b \leq [\![\pi x \leq \pi y]\!]$. Moreover,

$$b^* \leq [\![\pi x = 0]\!] \wedge [\![\pi y = 0]\!] \wedge [\![0 \leq 0]\!] \leq [\![\pi x \leq \pi y]\!];$$

consequently, $[\![\pi x \leq \pi y]\!] = \mathbf{1}$ or $\pi x \leq \pi y$.

Thereby we have established the second of the required equivalences. The first ensues from that by virtue of the formula $u = v \Leftrightarrow u \leq v \wedge v = u$.

It remains to demonstrate that the mapping χ is an isomorphism between the Boolean algebras B and $\mathfrak{P}(\mathscr{R}\!\downarrow)$. Take an arbitrary band projection $\pi \in \mathfrak{P}(\mathscr{R}\!\downarrow)$ and put $b := [\![\pi\!\uparrow = I_{\mathscr{R}}]\!]$. The fact that a band projection is extensional (and hence the ascent $\pi\!\uparrow$ of π is well-defined) follows from the above-established equivalences, because

$$\begin{aligned} c = [\![x = y]\!] &\Rightarrow \chi(c)x = \chi(c)y \Rightarrow \pi\chi(c)x = \pi\chi(c)y \\ &\Rightarrow \chi(c)\pi x = \chi(c)\pi y \Rightarrow c \leq [\![\pi x = \pi y]\!]. \end{aligned}$$

Since π is idempotent, $\pi\!\uparrow$ as well is an idempotent mapping in $\mathscr{R}$; i.e., either $\pi\!\uparrow = I_{\mathscr{R}}$ or $\pi = \mathbf{0}$. Hence, we derive $b^* = [\![\pi \neq I_{\mathscr{R}}]\!] = [\![\pi = 0]\!]$ and thereby $\pi\!\uparrow = \operatorname{mix}\{bI_{\mathscr{R}}, b^*(\mathbf{0})\}$. The mixing is unique; therefore, $\pi\!\uparrow = \chi(b)\!\uparrow$, i.e., $\pi = \chi(b)$. Thus, χ is a bijection between B and $\mathfrak{P}(\mathscr{R}\!\downarrow)$.

Let $b_1, b_2 \in B$ and $\rho_k := \chi(b_k)$ $(k := 1, 2)$. Recalling that $\rho_k = \operatorname{mix}\{b_k I_{\mathscr{R}}, b_k^* \mathbf{0}\}$, we derive

$$[\![\chi(b_1 \wedge b_2)\!\uparrow = I_{\mathscr{R}}]\!] = b_1 \wedge b_2 = [\![\rho_1 = I_{\mathscr{R}} \wedge \rho_2 = I_{\mathscr{R}}]\!] = [\![\rho_1 \circ \rho_2 = I_{\mathscr{R}}]\!],$$
$$[\![\chi(b_1 \wedge b_2)\!\uparrow = 0]\!] = (b_1 \wedge b_2)^* = [\![\rho_1 = \mathbf{0} \vee \rho_2 = \mathbf{0}]\!] = [\![\rho_1 \circ \rho_2 = \mathbf{0}]\!].$$

Thus,

$$[\![\chi(b_1 \wedge b_2)\!\uparrow = \rho_1 \circ \rho_2 = (\chi(b_1) \wedge \chi(b_2))\!\uparrow]\!] = \mathbf{1}$$

and hence

$$\chi(b_1 \wedge b_2) = \chi(b_1) \wedge \chi(b_2).$$

In particular, $0 = \chi(b) \wedge \chi(b^*)$ for $\chi(0) = 0$. Given elements $\rho := \chi(b)\!\uparrow$ and $\rho' := \chi(b^*)$, we have $[\![\rho, \rho' \in \{\mathbf{0}, I_{\mathscr{R}}\}$; $\rho = \mathbf{0}$ or $\rho' = \mathbf{0}$; and ρ and ρ' do not vanish simultaneously$]\!] = \mathbf{1}$. Hence, we see that $[\![\rho + \rho' = I_{\mathscr{R}}]\!] = \mathbf{1}$ and thereby $\chi(b) + \chi(b^*) = I_{\mathscr{R}\downarrow}$. Taking stock of the above, we conclude that χ preserves greatest lower bounds and complements; i.e., χ is an isomorphism. $\triangleright$

8.1.3. *The universally complete K-space $\mathscr{R}\!\downarrow$ is a faithful f-algebra with ring unity* 1; *moreover, for every $b \in B$ the projection $\chi(b)$ is the operator of multiplication by the order-unity $\chi(b)1$.*

$\triangleleft$ The multiplicative structure on $\mathscr{R}\!\downarrow$ was defined in 8.1.1. As in 8.1.2, we establish that $\mathscr{R}\!\downarrow$ is a faithful f-algebra. Take $x \in \mathscr{R}\!\downarrow$ and $b \in B$. By the definition of the projection $\chi(b)$, we have $b \leq [\![\chi(b)x = x]\!]$ and $b^* \leq [\![\chi(b^*)x = 0]\!]$. Applying these relations to $x := 1$ and appealing to the definition of multiplication in $\mathscr{R}\!\downarrow$, we obtain $b \leq [\![x = x \cdot 1 = x \cdot \chi(b)1]\!]$ and $b^* \leq [\![0 = x \cdot 0 = x \cdot \chi(b)1]\!]$. Thereby

$$[\![\chi(b)x = x \cdot \chi(b)1]\!] \geq [\![\chi(b)x = x]\!] \wedge [\![x = x \cdot \chi(b)1]\!] \geq b.$$

In a similar way, $b^* \leq [\![\chi(b)x = \chi(b)1 \cdot x]\!]$. Hence, $[\![\chi(b)x = x \cdot \chi(b)1]\!] = \mathbf{1}$. $\triangleright$

We see from the above that the mapping $b \mapsto \chi(b)1$ $(b \in B)$ is a Boolean isomorphism between B and the algebra $\mathfrak{E}(\mathscr{R}\!\downarrow)$ of order-unities. This isomorphism is denoted by the same letter χ. Thus, depending on the context, $x \mapsto \chi(b)x$ is either a band projection or the operator of multiplication by the order-unity $\chi(b)$.

8.1.4. Henceforth, $\mathscr{R}$ denotes the reals in the model $\mathbf{V}^{(B)}$. We will clarify the meaning of the exact bounds and order limits in the K-space $\mathscr{R}\!\downarrow$.

(1) *Let $(b_\xi)_{\xi\in\Xi}$ be a partition of unity in B and let $(x_\xi)_{\xi\in\Xi}$ be a family in $\mathscr{R}\!\downarrow$. Then*

$$\operatorname*{mix}_{\xi\in\Xi}(b_\xi x_\xi) = o\text{-}\sum_{\xi\in\Xi} \chi(b_\xi) x_\xi.$$

$\triangleleft$ If $x := \operatorname{mix}_{\xi\in\Xi}(b_\xi x_\xi)$ then $b_\xi \leq [\![x = x_\xi]\!]$ $(\xi \in \Xi)$ (see A.7 (3)). According to 8.1.2, $\chi(b_\xi)x_\xi = \chi(b_\xi)x$ for all $\xi \in \Xi$. Summing the last relations over ξ, we arrive at what was required. $\triangleright$

(2) *The following equivalences hold for a nonempty set $A \subset \mathscr{R}\!\downarrow$ and arbitrary $a \in \mathscr{R}$ and $b \in B$:*

$$b \leq [\![a = \sup(A\!\uparrow)]\!] \Leftrightarrow \chi(b)a = \sup \chi(b)(A),$$
$$b \leq [\![a = \inf(A\!\uparrow)]\!] \Leftrightarrow \chi(b)a = \inf \chi(b)(A).$$

◁ We will prove only the first equivalence. The equality

$$\chi(b)a = \sup\{\chi(b)x : x \in A\}$$

holds if and only if $b \le [\![x \le a]\!]$ for all $x \in A$ and for each $y \in \mathscr{R}{\downarrow}$ the relation $(\forall x \in A)(b \le [\![x \le y]\!])$ implies $b \le [\![a \le y]\!]$ (see 8.1.2).

Using the rules for calculating the truth-values for quantifiers (see A.4), we can represent the conditions in question in the following equivalent form:

$$b \le [\![(\forall x \in A{\uparrow})\, x \le a]\!],$$
$$b \le [\![(\forall y \in \mathscr{R})\, (A{\uparrow} \le y \to a \le y)]\!].$$

This system of inequalities is equivalent to the formula $b \le [\![a = \sup(A{\uparrow})]\!]$. ▷

(3) *Let* A *be an upward-directed set and* $s : \mathrm{A} \to \mathscr{R}{\downarrow}$ *be a net in* $\mathscr{R}{\downarrow}$. *Then* $\mathrm{A}^{\wedge}$ *is directed upward and* $\sigma := s{\uparrow} : \mathrm{A}^{\wedge} \to \mathscr{R}$ *is a net in* $\mathscr{R}$ *(inside* $\mathbf{V}^{(B)}$*); moreover,*

$$b \le [\![x = \lim \sigma]\!] \Leftrightarrow \chi(b)x = o\text{-}\lim \chi(b) \circ s$$

for arbitrary $x \in \mathscr{R}{\downarrow}$ *and* $b \in B$.

◁ The assertion "A is an upward-directed set" is a bounded formula. By virtue of the Restricted Transfer Principle A.8 (4), we have $\mathbf{V}^{(B)} \models$ "$\mathrm{A}^{\wedge}$ is an upward-directed set." The equality $\chi(b)x = o\text{-}\lim \chi(b) \circ s$ means that there exists a net $d : \mathrm{A} \to \mathscr{R}{\downarrow}$ for which the following system of conditions is compatible:

$$\alpha \le \beta \to d(\alpha) \le d(\beta) \quad (\alpha, \beta \in \mathrm{A}), \quad \inf_{\alpha \in \mathrm{A}} d(\alpha) = 0,$$
$$|\chi(b)x - \chi(b)s(\alpha)| \le d(\alpha) \quad (\alpha \in \mathrm{A}).$$

Taking account of the easy formula $[\![s(\mathrm{A}){\uparrow} = \sigma(\mathrm{A}^{\wedge})]\!] = \mathbf{1}$ and putting $\delta := d{\uparrow}$, we see that the indicated system of conditions is equivalent to the following system of inequalities:

$$b \le [\![\inf \sigma(\mathrm{A}^{\wedge}) = 0]\!],$$
$$b \le [\![(\forall \alpha, \beta \in \mathrm{A}^{\wedge})\, (\alpha \le \beta \to \sigma(\alpha) \le \sigma(\beta))]\!],$$
$$b \le [\![(\forall \alpha \in \mathrm{A}^{\wedge})\, (|x - \sigma(\alpha)| < \delta(\alpha))]\!],$$

whose short form is just the relation $b \le [\![x = \lim \sigma]\!]$. ▷

(4) *Suppose that* A *and* $\sigma \in \mathbf{V}^{(B)}$ *are such that* $[\![$ A *is directed upward and* $\sigma : \mathrm{A} \to \mathscr{R}]\!] = \mathbf{1}$. *Then* $\mathrm{A}{\downarrow}$ *is an upward-directed set and hence the mapping* $s := \sigma{\downarrow} : \mathrm{A}{\downarrow} \to \mathscr{R}{\downarrow}$ *is a net in* $\mathscr{R}{\downarrow}$. *Moreover,*

$$b \le [\![x = \lim \sigma]\!] \Leftrightarrow \chi(b)x = o\text{-}\lim \chi(b) \circ s$$

for arbitrary $x \in \mathscr{R}{\downarrow}$ *and* $b \in B$.

◁ The proof is similar to that of (3). ▷

(5) *Let f be a mapping from a nonempty set Ξ to $\mathscr{R}\!\downarrow$ and $g := f\!\uparrow$. Then*

$$b \le [\![x = \sum_{\xi \in \Xi^\wedge} g(\xi)]\!] \Leftrightarrow \chi(b)x = \sum_{\xi \in \Xi} \chi(b) f(\xi)$$

for arbitrary $x \in \mathscr{R}\!\downarrow$ and $b \in B$.

◁ First of all observe that the required equivalence holds for a finite set $\Xi_0 \subset \Xi$. Afterwards, apply (3) to the net $s : \mathscr{P}_{\mathrm{fin}}(\Xi) \to \mathscr{R}\!\downarrow$, where $\mathscr{P}_{\mathrm{fin}}(\Xi)$ is the set of finite subsets of Ξ and $s(\theta) := \sum_{\xi \in \theta} f(\xi)$, and employ the relation $[\![\mathscr{P}_{\mathrm{fin}}(\Xi)^\wedge = \mathscr{P}_{\mathrm{fin}}(\Xi^\wedge)]\!] = \mathbf{1}$ (see A.8 (6)). ▷

8.1.5. *The following relations hold for every element $x \in \mathscr{R}\!\downarrow$:*

$$e_x := \chi([\![x \neq 0]\!]), \quad e_\lambda^x = \chi([\![x < \lambda^\wedge]\!]) \quad (\lambda \in \mathbb{R}).$$

◁ A real number t is distinct from zero if and only if the supremum of the set $\{1 \wedge (n|t|) : n \in \omega\}$ is equal to 1. Consequently, for $x \in \mathscr{R}\!\downarrow$ the Transfer Principle yields $b := [\![x \neq 0]\!] = [\![1 = \sup A]\!]$, where $A \in \mathbf{V}^{(B)}$ is determined by the formula $A := \{1 \wedge (n|x|) : n \in \omega^\wedge\}$. If $C := \{1 \wedge (n|x|) : n \in \omega\}$ then we prove that $[\![C\!\uparrow = A]\!] = \mathbf{1}$ using the second formula of A.10 (1) and the representation $\omega^\wedge = (\iota\omega)\!\uparrow$ of A.11. Hence, $[\![\sup(A) = \sup(C\!\uparrow)]\!] = \mathbf{1}$. Employing 8.1.4 (2), we derive

$$b = [\![\sup(C\!\uparrow) = 1]\!] = [\![\sup(C) = 1]\!] = [\![e_x = 1]\!].$$

On the other hand, $[\![e_x = 0]\!] = [\![e_x = 1]\!]^* = b^*$. Now, according to 8.1.2, we may write down

$$\chi(b)e_x = \chi(b)1 = \chi(b), \quad \chi(b^*)e_x = 0 \Rightarrow \chi(b)e_x = e_x.$$

Finally, $\chi(b) = e_x$.

Take $\lambda \in \mathbb{R}$ and put $y := (\lambda 1 - x)^+$. Since $[\![\lambda^\wedge = \lambda 1]\!] = \mathbf{1}$, we have $[\![y = (\lambda^\wedge - x)^+]\!] = \mathbf{1}$. Consequently, $e_\lambda^x = e_y = \chi([\![y \neq 0]\!])$. It remains to observe that inside $\mathbf{V}^{(B)}$ the number $y = (\lambda^\wedge - x) \vee 0$ is distinct from zero if and only if $\lambda^\wedge - x > 0$, i.e., $[\![y \neq 0]\!] = [\![x < \lambda^\wedge]\!]$. ▷

8.1.6. Theorem. *Let E be an Archimedean vector lattice, let $\mathscr{R}$ be the reals in the model $\mathbf{V}^{(B)}$, and let $\jmath$ be an isomorphism of B onto the base $\mathfrak{B}(E)$. Then there exists an element $\mathscr{E} \in \mathbf{V}^{(B)}$ satisfying the following conditions:*

(1) $\mathbf{V}^{(B)} \models$ *"$\mathscr{E}$ is a vector sublattice of $\mathscr{R}$ considered as a vector lattice over $\mathbb{R}^\wedge$";*

(2) *$E' := \mathscr{E}{\downarrow}$ is a vector sublattice of $\mathscr{R}{\downarrow}$ invariant under each band projection $\chi(b)$ $(b \in B)$ and such that every set of positive pairwise disjoint elements in it has a supremum;*

(3) *there is an o-continuous lattice isomorphism $\iota : E \to E'$ such that $\iota(E)$ is a massive sublattice in $\mathscr{R}{\downarrow}$;*

(4) *for every $b \in B$ the band projection in $\mathscr{R}{\downarrow}$ onto $\{\iota(\jmath(b))\}^{\perp\perp}$ coincides with $\chi(b)$.*

$\triangleleft$ Assign $d(x,y) := \jmath^{-1}(\{|x-y|\}^{\perp\perp})$. Let $\mathscr{E}$ be the Boolean-valued realization of the B-set (E,d) and $E' := \mathscr{E}{\downarrow}$ (see A.12 (1, 2)). By A.12 (2), without loss of generality we may assume that $E \subset E'$, $d(x,y) = [\![x \neq y]\!]$ $(x, y \in E)$, and $E' = \operatorname{mix} E$. Further, furnish E' with a vector lattice structure. To this end, take a number $\lambda \in \mathbb{R}$ and elements $x, y \in E'$ of the form $x := \operatorname{mix}(b_\xi x_\xi)$ and $y := \operatorname{mix}(b_\xi y_\xi)$, where $(x_\xi) \subset E$, $(y_\xi) \subset E$, and (b_ξ) is a partition of unity in B, and define

$$
\begin{aligned}
x + y &:= \operatorname{mix}(b_\xi(x_\xi + y_\xi)),\\
\lambda x &:= \operatorname{mix}(b_\xi(\lambda x_\xi)),\\
x \leq y &\Leftrightarrow x = \operatorname{mix}(b_\xi(x_\xi \wedge y_\xi)).
\end{aligned}
$$

Inside $\mathbf{V}^{(B)}$, define the addition $\oplus$, multiplication $\odot$, and order $\sqsubseteq$ in the set $\mathscr{E}$ as the ascents of the corresponding objects in E'. More precisely, the operations $\oplus : \mathscr{E} \times \mathscr{E} \to \mathscr{E}$ and $\odot : \mathscr{E} \times \mathbb{R}^\wedge \to \mathscr{E}$ and the predicate $\sqsubseteq \subset \mathscr{E} \times \mathscr{E}$ are defined by the relations

$$
\begin{gathered}
[\![x \oplus y = x + y]\!] = \mathbf{1},\\
[\![\lambda \odot x = \lambda x]\!] = \mathbf{1} \quad (x, y \in E',\ \lambda \in \mathbb{R}),\\
[\![x \sqsubseteq y]\!] = \bigvee \{ [\![x = x']\!] \wedge [\![y = y']\!] : x', y' \in E',\ x' \leq y' \}.
\end{gathered}
$$

Thus, we may claim that $\mathscr{E}$ is a vector lattice over the field $\mathbb{R}^\wedge$ and, in particular, a lattice-ordered group inside $\mathbf{V}^{(B)}$. Also, it is clear that the Archimedean axiom is valid on $\mathscr{E}$, since E' is an Archimedean lattice.

Note that if $x \in E_+$ then $\{x\}^{\perp\perp} = d(x,0) = [\![x \neq 0]\!]$, i.e., $\{x\}^{\perp} = [\![x = 0]\!]$. Consequently, we have

$$
[\![x = 0]\!] \vee [\![y = 0]\!] = \{x\}^{\perp} \vee \{y\}^{\perp} = \mathbf{1}_B
$$

for disjoint $x, y \in E$. Hence, we easily derive that $[\![\mathscr{E} \text{ is linearly ordered}]\!] = \mathbf{1}$, for

$$
[\![(\forall x \in \mathscr{E})(\forall y \in \mathscr{E})\ (|x| \wedge |y| = 0 \to x = 0 \vee y = 0)]\!] = \mathbf{1}.
$$

It is well known that an Archimedean linearly ordered group is isomorphic to an additive subgroup of the reals. Applying this assertion to $\mathscr{E}$ inside $\mathbf{V}^{(B)}$, without loss of generality we may assume that $\mathscr{E}$ is an additive subgroup of the field $\mathscr{R}$. Furthermore, we suppose that $1 \in \mathscr{E}$, since otherwise $\mathscr{E}$ could be replaced by the isomorphic group $e^{-1}\mathscr{E}$ with $0 < e \in \mathscr{E}$. The multiplication $\odot$ represents a continuous $\mathbb{R}^\wedge$-bilinear mapping from $\mathbb{R}^\wedge \times \mathscr{E}$ to $\mathscr{E}$. Let $\beta : \mathscr{R} \times \mathscr{R} \to \mathscr{R}$ be its extension by continuity. Then β is $\mathscr{R}$-bilinear and $\beta(1,1) = 1^\wedge \odot 1 = 1$. Consequently, β coincides with the usual multiplication in $\mathscr{R}$; i.e., $\mathscr{E}$ is a vector sublattice of the field $\mathscr{R}$ considered as a vector lattice over $\mathbb{R}^\wedge$. Thereby $E' \subset \mathscr{R}{\downarrow}$.

The fact that E' is massive in $\mathscr{R}{\downarrow}$ obviously ensues from the fact that $[\![\mathscr{E}$ is dense in $\mathscr{R}]\!] = \mathbf{1}$. Prove that E is minorizing in E'.

It follows from the properties of the isomorphism χ (see 8.1.2) that

$$\chi(b)\iota x = 0 \Leftrightarrow \jmath(b) \leq \{x\}^\perp \Leftrightarrow x \in \jmath(b^\perp),$$

whatever $b \in B$ and $x \in E_+$ might be. Hence, $\chi(b)$ is the band projection onto the band in $\mathscr{R}{\downarrow}$ generated by the set $\iota(\jmath(b))$. Moreover, if $\chi(b)x = 0$ for all $x \in E_+$ then $b = \{0\}$. Thus, for every $b \in B$ we may find a positive element $y \in E$ for which $y = \chi(b)y$. Now, take $0 < z \in E'$. The representation $z = o\text{-}\sum_{\xi\in\Xi} \chi(b_\xi)x_\xi$ is valid, where (b_ξ) is a partition of unity in B and $(x_\xi) \subset E_+$. We see that $\chi(b_\xi)x_\xi \neq 0$ at least for one index ξ. Let $\pi := \chi(b_\xi) \circ \chi([\![x_\xi \neq 0]\!])$ and y be a strictly positive element in E such that $y = \pi y$. Then for $x_0 := y \wedge x_\xi$ we have

$$0 < x_0 \leq \pi x_\xi \leq \chi(b_\xi)x_\xi \leq z$$

and $x_0 \in E$. Thereby E is minorizing in E'. $\triangleright$

8.1.7. The element $\mathscr{E} \in \mathbf{V}^{(B)}$ arising in Theorem 8.1.6 is called the *Boolean-valued realization* of the vector lattice E. Thus, the Boolean-valued realizations of Archimedean vector lattices are vector sublattices of the reals $\mathscr{R}$ considered as a vector lattice over the field $\mathbb{R}^\wedge$.

Now, we indicated some corollaries to 8.1.2 and 8.1.6, with the same notations B, E, E', $\mathscr{E}$, ι, and $\mathscr{R}$.

(1) *For every $x' \in E'$ there exist a family $(x_\xi) \subset E$ and a partition of unity (π_ξ) in $\mathfrak{P}(\mathscr{R}{\downarrow})$ such that*

$$x' = o\text{-}\sum_{\xi\in\Xi} \pi_\xi \iota x_\xi.$$

(2) *For arbitrary $x \in \mathscr{R}{\downarrow}$ and $\varepsilon > 0$ there is $x_\varepsilon \in E'$ such that $|x - x_\varepsilon| \leq \varepsilon\mathbf{1}$.*

◁ This is a consequence of the fact that $[\![\mathscr{E} \text{ is dense in } \mathscr{R}]\!] = \mathbb{1}$. ▷

(3) *If $h : \mathscr{E} \to \mathscr{R}{\downarrow}$ is a lattice isomorphism and for every $b \in B$ the band projection onto the band in $\mathscr{R}{\downarrow}$ generated by the set $h(\jmath(b))$ coincides with $\chi(b)$ then there exists $a \in \mathscr{R}{\downarrow}$ such that $hx = a \cdot \iota(x)$ $(x \in E)$.*

◁ Indeed, if $E_0 := \operatorname{im} \iota$ and $h_0 := h \circ \iota^{-1}$ then the isomorphism $h_0 : E_0 \to \mathscr{R}{\downarrow}$ is extensional; therefore, for $\tau := h_0{\uparrow}$ we have $[\![$ the mapping $\tau : \mathscr{E} \to \mathscr{R}$ is isotonic, injective, and additive $]\!] = \mathbb{1}$. Consequently, h_0 is continuous and has the form $\tau(\alpha) = a \cdot \alpha$ $(\alpha \in \mathscr{R})$, where a is a fixed element in $\mathscr{R}{\downarrow}$. Hence, we derive that $h_0(y) = a \cdot y$ $(y \in E_0)$ or $h(x) = a \cdot \iota(x)$ $(x \in E)$. ▷

(4) *If there exists an order-unity $\mathbf{1}$ in E then the isomorphism ι is uniquely determined by the extra requirement that $\iota\mathbf{1} = 1$.*

(5) *If E is a K-space then $\mathscr{E} = \mathscr{R}$, $E' = \mathscr{R}{\downarrow}$, and $\iota(E)$ is an order-dense ideal of the K-space $\mathscr{R}{\downarrow}$. Moreover, $\iota^{-1} \circ \chi(b) \circ \iota$ is the band projection onto $\jmath(b)$ for every $b \in B$.*

◁ If E is order complete then so is the lattice E'. From 8.1.4 (2) we see that the order completeness of E' is equivalent to the axiom of existence of exact bounds for bounded sets in $\mathscr{E}$. By 8.1.1, $\mathscr{E} = \mathscr{R}$ and $E' = \mathscr{R}{\downarrow}$. Let $e \in E_+$, $y \in \mathscr{R}{\downarrow}$, and $|y| \le \iota e$. Since $\iota(E)$ is a massive sublattice in $\mathscr{R}{\downarrow}$, we have $y^+ = \sup \iota(A)$, where $A := \{x \in E_+ : \iota x \le y^+\}$. But the set A is bounded in E by the element e; therefore, $\sup A \in E$ and $y^+ = \iota(\sup A) \in \iota E$. Similarly, $y^- \in \iota(E)$ and, finally, $y \in \iota(E)$. ▷

(6) *The image $\iota(E)$ coincides with the whole $\mathscr{R}{\downarrow}$ if and only if E is a universally complete K-space.*

◁ If E is a K-space then $\mathscr{E} = \mathscr{R}$ by (5) and, hence, $\mathscr{R}{\downarrow} = \mathscr{E}{\downarrow} = \operatorname{mix} \iota(E)$. However, for the universally complete K-space E we have $\operatorname{mix} \iota(E) = \iota(E)$. The converse is obvious. ▷

(7) *Universally complete K-spaces are isomorphic if and only if their bases are isomorphic.*

◁ If E and F are universally complete K-spaces and the Boolean algebras $\mathfrak{B}(E)$ and $\mathfrak{B}(F)$ are isomorphic then E and F are isomorphic to the same K-space $\mathscr{R}{\downarrow}$ by (6). On the other hand, if h is an isomorphism from E onto F then the mapping $K \mapsto h(K)$ $(K \in \mathfrak{B}(E))$ is an isomorphism between the bases. ▷

(8) *Let E be a universally complete K-space with unity $\mathbf{1}$. Then we may uniquely define the multiplication in E so as to make E into an f-algebra and $\mathbf{1}$, into a ring unity.*

◁ By (6) and (4), we may assume that $E = \mathscr{R}{\downarrow}$ and $\mathbf{1} = 1$. The existence of the required multiplication in E follows from 8.1.3. Assume that there is another multiplication $\odot : E \times E \to E$ in E and $(E, +, \odot, \le)$ is a faithful f-algebra with unity **1**. The faithfulness of the f-algebra implies that $\odot$ is an extensional mapping. But then the ascent $\times := \odot\!\uparrow$ is a multiplication in $\mathscr{R}$. By virtue of uniqueness of the multiplicative structure in $\mathscr{R}$, we conclude that $\times = \cdot$. Hence, we derive that $\odot$ coincides with the original multiplication in E (see 8.1.3). $\rhd$

8.1.8. Interpreting the concept of a convergent numeric net inside $\mathbf{V}^{(B)}$ and employing 8.1.4 (3) and 8.1.7 (5), we obtain useful tests for o-convergence in a K-space E with unity **1**.

Theorem. *Let $(x_\alpha)_{\alpha\in \mathrm{A}}$ be an order-bounded net in E and $x \in E$. The following are equivalent:*

(1) *the net (x_α) o-converges to the element x;*

(2) *for every number $\varepsilon > 0$ the net $(e_\varepsilon^{y(\alpha)})_{\alpha\in\mathrm{A}}$ of unit elements, where $y(\alpha) := |x - x_\alpha|$, o-converges to **1**;*

(3) *for every number $\varepsilon > 0$ there exists a partition of unity $(\pi_\alpha)_{\alpha\in\mathrm{A}}$ in the Boolean-valued algebra $\mathfrak{P}(E)$ such that*

$$\pi_\alpha|x - x_\beta| \le \varepsilon\mathbf{1} \quad (\alpha, \beta \in \mathrm{A},\ \beta \ge \alpha);$$

(4) *for every number $\varepsilon > 0$ there exists an increasing net $(\rho_\alpha)_{\alpha\in\mathrm{A}} \subset \mathfrak{P}(E)$ of projections such that*

$$\rho_\alpha|x - x_\beta| \le \varepsilon\mathbf{1} \quad (\alpha, \beta \in \mathrm{A},\ \beta \ge \alpha).$$

$\lhd$ Without loss of generality we may assume that E is an order-dense ideal of the universally complete K-space $\mathscr{R}{\downarrow}$ (see 8.1.7 (5)).

(1) $\Leftrightarrow$ (2): It suffices to consider the case $y_\alpha = x_\alpha$ ($\alpha \in \mathrm{A}$), i.e., $(x_\alpha) \subset E_+$ and $x_\alpha \xrightarrow{(o)} 0$.

Let σ be the modified ascent of the mapping $s : \alpha \to x_\alpha$. Then $[\![\sigma$ is a net in $\mathscr{R}_+]\!] = \mathbf{1}$. By 8.1.4 (3), o-$\lim s = 0$ if and only if $[\![\lim\sigma = 0]\!] = \mathbf{1}$. We may rewrite the last equality in equivalent form:

$$\mathbf{1} = [\![(\forall\varepsilon \in \mathbb{R}^\wedge)(\varepsilon > 0 \to (\exists\alpha \in \mathrm{A}^\wedge)(\forall\beta \in \mathrm{A}^\wedge)\,(\beta \ge \alpha \to x_\beta < \varepsilon))]\!].$$

Calculating the Boolean truth-values for the quantifiers, we find another equivalent form

$$(\forall\varepsilon > 0)(\exists(b_\alpha)_{\alpha\in\mathrm{A}} \subset B)\ \Big(\bigvee_{\alpha\in\mathrm{A}} b_\alpha = \mathbf{1} \wedge (\forall\beta \in \mathrm{A})\ (\beta \ge \alpha \to [\![x_\beta < \varepsilon^\wedge]\!] \ge b_\alpha)\Big)$$

which in turn amounts to the following:

$$(\forall\varepsilon > 0)\Big(\bigvee_{\alpha\in\mathrm{A}}\ \bigwedge_{\substack{\beta\in\mathrm{A}\\ \beta\ge\alpha}} [\![x_\beta < \varepsilon^\wedge]\!] = \mathbf{1}\Big).$$

Since $\chi([\![x_\beta < \varepsilon^\wedge]\!]) = e_\varepsilon^{x_\beta}$ (see 8.1.5), we see from the above that $x_\alpha \xrightarrow{(o)} x$ if and only if

$$\liminf_{\alpha\in\mathrm{A}} e_\varepsilon^{x_\alpha} = \bigvee_{\alpha\in\mathrm{A}} \bigwedge_{\substack{\beta\in\mathrm{A}\\ \beta\geq\alpha}} e_\varepsilon^{x_\beta} = \mathbf{1}$$

for every $\varepsilon > 0$, i.e., $e_\varepsilon^{x_\alpha} \xrightarrow{(o)} \mathbf{1}$ for every $\varepsilon > 0$.

(1) $\Leftrightarrow$ (3): Arguing as in (1) $\Rightarrow$ (2), we find that the relation $o\text{-}\lim x_\alpha = x$ is equivalent to the following:

$$(\forall\varepsilon > 0)(\exists(c_\alpha)_{\alpha\in\mathrm{A}} \subset B)\Big(\bigvee_{\alpha\in\mathrm{A}} c_\alpha = 1 \wedge (\forall\beta\in\mathrm{A})(\beta\geq\alpha \Rightarrow c_\alpha \leq [\![|x_\alpha - x| \leq \varepsilon^\wedge]\!])\Big).$$

By virtue of the Exhaustion Principle for Boolean algebras, there exist a partition of unity $(d_\xi)_{\xi\in\Xi}$ in B and a mapping $\delta : \Xi \to \mathrm{A}$ such that $d_\xi \leq c_{\delta(\xi)}$ $(\xi\in\Xi)$. Put $b_\alpha := \bigvee\{d_\xi : \alpha = \delta(\xi)\}$ if $\alpha \in \delta(\Xi)$ and $b_\alpha = 0$ if $\alpha \notin \delta(\Xi)$. We see that $(b_\alpha)_{\alpha\in\mathrm{A}}$ is a partition of unity and $b_\alpha \leq c_\alpha$ $(\alpha\in\mathrm{A})$. Thus, if $x_\alpha \to x$ then for every $\varepsilon > 0$ there is a partition of unity (b_α) such that

$$b_\alpha \leq [\![|x - x_\beta| \leq \varepsilon^\wedge]\!] \quad (\alpha,\beta\in\mathrm{A},\ \beta\geq\alpha).$$

As follows from 8.1.2, the latter means that

$$\pi_\alpha|x - x_\beta| \leq \varepsilon\mathbf{1} \quad (\alpha,\beta\in\mathrm{A},\ \beta\geq\alpha),$$

where $\pi_\alpha := \chi(b_\alpha)$. Since (π_α) is a partition of unity in $\mathfrak{P}(E)$, necessity is proven.

To prove sufficiency, observe that if the indicated conditions are satisfied and $a := \limsup|x_\alpha - x|$ then

$$\pi_\alpha a \leq \bigvee_{\beta\geq\alpha} |x_\beta - x| \leq \varepsilon\pi_\alpha\mathbf{1}$$

for all $\alpha\in\mathrm{A}$. Consequently,

$$0 \leq a = \sum \pi_\alpha a \leq \varepsilon\sum\pi_\alpha 1 = \varepsilon\mathbf{1}.$$

Since $\varepsilon > 0$ is arbitrary, we have $a = 0$ and $o\text{-}\lim x_\alpha = x$.

(3) $\Leftrightarrow$ (4): We only have to put $\rho_\alpha := \bigvee\{\pi_\beta : \beta\in\mathrm{A},\ \alpha\leq\beta\}$ in (3). $\rhd$

8.1.9. Let $\mathscr{C}$ be the field of complex numbers in the model $\mathbf{V}^{(B)}$. Then the algebraic system $\mathscr{C}{\downarrow}$ represents the complexification of the K-space $\mathscr{R}{\downarrow}$. In particular, $\mathscr{C}{\downarrow}$ is a universally complete complex K-space and a complex algebra.

$\lhd$ Since $\mathbb{C} = \mathbb{R} \oplus i\mathbb{R}$ is equivalent to a bounded formula, we have $[\![\, \mathbb{C}^\wedge = \mathbb{R}^\wedge \oplus \mathbb{R}^\wedge \,]\!] = \mathbf{1}$ (see A.8 (4)), where i is the imaginary unity and the element $i^\wedge$ is denoted by the same letter i. From 8.1.1 we see that $[\![\, \mathbb{C}^\wedge$ is a dense subfield of the field $\mathscr{C} \,]\!] = \mathbf{1}$ and, in particular, $[\![\, i$ is the imaginary unity of the field $\mathscr{C} \,]\!] = \mathbf{1}$. If $z \in \mathscr{C}{\downarrow}$ then z is a complex number inside $\mathbf{V}^{(B)}$; therefore,

$$[\![\, (\exists! x \in \mathscr{R})(\exists! y \in \mathscr{R})\, z = x + iy \,]\!] = \mathbf{1}.$$

The Maximum Principle implies that there is a unique pair of elements $x, y \in \mathbf{V}^{(B)}$ such that

$$[\![\, x, y \in \mathscr{R} \,]\!] = [\![\, z = x + iy \,]\!] = \mathbf{1}.$$

Hence, we obtain $x, y \in \mathscr{R}{\downarrow}$, $z = x + iy$, and thereby $\mathscr{C}{\downarrow} = \mathscr{R}{\downarrow} \oplus i\mathscr{R}{\downarrow}$. Appealing to 8.1.2 and 8.1.4, complete the proof. $\rhd$

8.2. Boolean-Valued Analysis of Vector Lattices

In this section, we show that the most important structure properties of vector lattices such as representability by means of function spaces, the spectral theorem, the functional calculus, etc. are the images of properties of the reals in an appropriate Boolean-valued model.

8.2.1. We start with several useful remarks to be used below without further specification. Take a K_σ-space E. By Theorem 8.1.6, we may assume that E is a sublattice of the universally complete K-space $\mathscr{R}{\downarrow}$, where, as usual, $\mathscr{R}$ is the field of the reals in the model $\mathbf{V}^{(B)}$ and $B := \mathfrak{B}(E)$. Moreover, the ideal $\widehat{E} := I(E)$ generated by the set E in $\mathscr{R}{\downarrow}$ is an order-dense ideal of $\mathscr{R}{\downarrow}$ and an o-completion of E. The unity $\mathbf{1}$ of the lattice E is also a unity in $\mathscr{R}{\downarrow}$. The exact bounds of countable sets in E are inherited from $\mathscr{R}{\downarrow}$. In more detail, if the least upper (greatest lower) bound x of a sequence $(x_n) \subset E$ exists in $\mathscr{R}{\downarrow}$ then x is also the least upper (greatest lower) bound in E, provided that $x \in E$. Thus, it does not matter whether the o-limit (o-sum) of a sequence in E is calculated in E or $\mathscr{R}{\downarrow}$, provided the result belongs to E. The same is true for the r-limit and r-sums. In particular, we may claim that if $x \in E$ then the trace e_x and the spectral function (characteristic) e_λ^x of an element x calculated in $\mathscr{R}{\downarrow}$ are an element of $B := \mathfrak{C}(E)$ and a mapping from $\mathbb{R}$ to B respectively.

8.2.2. As a first easy application we prove the properties of every spectral function of 1.3.8 (1–12). According to the remarks of 8.2.1, without loss of generality we may assume that the K_σ-space under consideration coincides with $\mathscr{R}{\downarrow}$. But then the required relations can be easily derived from the elementary properties of numbers with the help of 8.1.5. We confine exposition to (2), (6), and (8).

$\lhd$ First of all observe that $\mathbf{P}^\wedge$ is a dense subfield of the field $\mathscr{R}$ inside $\mathbf{V}^{(B)}$ Take $x \in \mathscr{R}{\downarrow}$ and consider the two formulas $\varphi(x) := (\exists t \in \mathbf{P}^\wedge)\,(x < t)$ and $\psi(x) := (\forall t \in \mathbf{P}^\wedge)\,(x < t)$. For a real number x the formula $\varphi(x)$ is true and $\psi(x)$ is false. Consequently, the Transfer Principle implies $[\![\varphi(x)]\!] = \mathbf{1}$ and $[\![\psi(x)]\!] = \mathbf{0}$. Calculating the Boolean truth-values for the quantifiers by the rules of A.8 (1) yields

$$\bigvee_{t \in \mathbf{P}} [\![x < t^\wedge]\!] = \mathbf{1}, \quad \bigwedge_{t \in \mathbf{P}} [\![x < t^\wedge]\!] = \mathbf{0}$$

which is equivalent to (2) by 8.1.5.

Take positive elements $x, y \in \mathscr{R}{\downarrow}$ and a number $0 < t \in \mathbf{P}$. Then x, y, and $t^\wedge$ are reals in the model $\mathbf{V}^{(B)}$. Make use of the following property of numbers:

$$x \geq 0 \wedge y \geq 0 \to (xy < t^\wedge \leftrightarrow (\exists r, s \in \mathbf{P}_+^\wedge)(x < r \wedge y < s \wedge rs = t)).$$

Employing again the Transfer Principle and the rules of A.8 (1) for calculating the Boolean truth-values, we arrive at the relation

$$[\![xy < t^\wedge]\!] = \bigvee_{\substack{0 \leq r, s \in \mathbf{P} \\ rs = t}} [\![x < r^\wedge]\!] \bigwedge [\![y < s^\wedge]\!].$$

Hence, the required equality (6) ensues if we apply χ to both sides of the preceding equality (see 8.1.5).

Now, let A be a set in the considered K_σ-space. Then $A{\uparrow}$ is some set of reals inside $\mathbf{V}^{(B)}$ and the formula $\inf(A) < t \Leftrightarrow (\exists a \in A{\uparrow})(a < t)$ holds. Employing 8.1.4 (2) and A.10 (1), we may write down the following chain of equivalences:

$$\begin{aligned} x = \inf(A) &\Leftrightarrow [\![x = \inf(A{\uparrow})]\!] = \mathbf{1} \Leftrightarrow [\![(\forall t \in \mathbf{P}^\wedge) \\ &(x < t \leftrightarrow \inf(A{\uparrow}) < t)]\!] = \mathbf{1} \Leftrightarrow (\forall t \in \mathbf{P})[\![x < t^\wedge]\!] \\ &= [\![(\exists a \in A{\uparrow})(a < t^\wedge)]\!] \Leftrightarrow (\forall t \in \mathbf{P})[\![x < t^\wedge]\!] = \bigvee_{a \in A} [\![a < t^\wedge]\!]. \end{aligned}$$

Appealing to 8.1.5, complete the proof of (8). $\rhd$

8.2.3. Thus, to each element of a K_σ-space with unity there corresponds the spectral function; moreover, the operations transform in a rather definite way. This circumstance suggests that an arbitrary K_σ-space with unity can be realized as some space of "abstract spectral functions." This was done in 1.4.3 and 1.4.4. Now, we will pursue the Boolean-valued approach to the same problem.

Theorem. *Let B be a complete Boolean algebra. The set $\mathfrak{R}(B)$ with introduced operations and order represents a universally complete K-space. The mapping sending an element $x \in \mathscr{R}{\downarrow}$ to the resolution of unity $t \mapsto [\![x < t^{\wedge}]\!]$ $(t \in \mathbb{R})$ is an isomorphism between the K-spaces $\mathscr{R}{\downarrow}$ and $\mathfrak{R}(B)$.*

$\lhd$ Denote the indicated mapping from $\mathscr{R}{\downarrow}$ to $\mathfrak{R}(B)$ by the letter h. By Theorem 1.3.8, h preserves the operations and order. Moreover, h is one-to-one, since the equality $h(x) = h(y)$ means

$$[\![x < t^{\wedge}]\!] = [\![y < t^{\wedge}]\!] \quad (t \in \mathbb{R})$$

or (see A.8 (1))

$$[\![(\forall t \in \mathbb{R}^{\wedge})\,(x < t \leftrightarrow y < t)]\!] = \mathbf{1}$$

and thereby is equivalent to coincidence of two numbers x and y inside $\mathbf{V}^{(B)}$. By virtue of Gordon's Theorem 8.1.2, it remains to establish that h is surjective. Take an arbitrary resolution of unity e in the Boolean algebra B. Let $\beta := (t_n)_{n\in\mathbb{Z}}$ be a partition of the real axis; i.e., $t_n < t_{n+1}$ $(n \in \mathbb{Z})$, $\lim_{n\to\infty} t_n = \infty$, and $\lim_{n\to-\infty} t_n = -\infty$. The disjoint sum

$$\bar{x}(\beta) := \sum_{n\in\mathbb{Z}} t_{n+1}(\chi(e(t_{n+1})) - \chi(e(t_n)))$$

exists in the universally complete K-space $\mathscr{R}{\downarrow}$; here χ is the isomorphism of B onto $\mathfrak{E}(\mathscr{R}{\downarrow})$ (see 8.1.2 and 8.1.3). Denote by the letter A the set of all elements $\bar{x}(\beta)$. Every element of the form

$$x(\beta) := \sum_{n\in\mathbb{Z}} t_n(\chi(e(t_{n+1})) - \chi(e(t_n)))$$

is a lower bound of A. Therefore, there exists $x := \inf A := \inf\{\bar{x}(\beta)\}$. It is easy to observe that

$$e_\lambda^{\bar{x}(\beta)} = \bigvee\{\chi(e(t_n)) : t_n < \lambda\}.$$

Hence, by 1.3.8 (9), we infer

$$e_\lambda^x = \bigvee_{a\in A} e_\lambda^a = \bigvee_{t\in\mathbb{R},\, t<\lambda} \chi(e(t)) = \chi(e(\lambda)) \quad (\lambda \in \mathbb{R}).$$

Thereby $h(x) = e$ (see 8.1.5). $\rhd$

8.2.4. We derive several important corollaries to the just-proven theorem.

(1) *A universally complete K-space E with unity $\mathbf{1}$ is isomorphic to the K-space $\mathfrak{K}(B)$, where $B = \mathfrak{C}(\mathbf{1})$. The isomorphism is established by the mapping $x \mapsto ((e^x_\lambda)_{\lambda\in\mathbb{R}})$ $(x \in E)$.*

$\triangleleft$ It suffices to compare 8.1.8 (6) and 8.2.3. $\triangleright$

(2) Freudenthal Spectral Theorem. *Let E be an arbitrary K_σ-space with unity $\mathbf{1}$. Every element $x \in E$ admits the representation*

$$x = \int\limits_{-\infty}^{\infty} \lambda de^x_\lambda,$$

where the integral is understood to be the limit with regulator $\mathbf{1}$ of the integral sums

$$x(\beta) := \sum_{n\in\mathbb{Z}} \tau_n(e^x_{t_{n+1}} - e^x_{t_n}), \quad t_n < \tau_n < t_{n+1},$$

as $\delta(\beta) := \sup_{n\in\mathbb{Z}}(t_{n+1} - t_n) \to 0$.

$\triangleleft$ We may assume that $\mathscr{R}\!\downarrow$ is a universal completion of E and $E \subset \mathscr{R}\!\downarrow$. Let $x \in E$, $\beta := (t_n)_{n\in\mathbb{Z}}$ be a partition of $\mathbb{R}$, and $t_n < \tau_n < t_{n+1}$ $(n \in \mathbb{Z})$. Denote $b_n := e_{t_{n+1}} - e_{t_n}$. Then

$$\begin{aligned} b_n &= [\![t_n^\wedge \leq x < t_{n+1}^\wedge]\!] \wedge [\![t_n^\wedge \leq \tau_n^\wedge < t_{n+1}^\wedge]\!] \wedge [\![t_{n+1}^\wedge - t_n^\wedge \leq \delta(\beta)^\wedge]\!] \\ &\leq [\![|x - \tau_n^\wedge| \leq \delta(\beta)^\wedge]\!]. \end{aligned}$$

Taking the equality $x(\beta) := \mathrm{mix}_{n\in\mathbb{Z}}(b_n\tau_n^\wedge)$ into account, we derive

$$[\![|x - x(\beta)| \leq \delta(\beta)^\wedge]\!] = \mathbf{1} \quad \text{or} \quad |x - x(\beta)| \leq \delta(\beta)\mathbf{1}.$$

It remains to recall the remarks of 8.2.1. $\triangleright$

(3) *For an arbitrary σ-algebra B, the set $\mathfrak{K}(B)$ (with the structure defined as in 1.4.3) is a universally complete K_σ-space with unity. Conversely, every universally complete K_σ-space E with unity is isomorphic to $\mathfrak{K}(B)$, where $B = \mathfrak{C}(E)$.*

$\triangleleft$ Let $\widehat{B}$ be an o-completion of the σ-algebra B. According to 8.2.3, $\mathfrak{K}(\widehat{B})$ is a universally complete K-space. The set $\mathfrak{K}(B)$ is contained in $\mathfrak{K}(\widehat{B})$. Moreover, from 1.3.8 (4–7) and 8.2.4 we may see that the vector lattice structure and the exact bounds of countable sets in $\mathfrak{K}(B)$ are inherited from $\mathfrak{K}(\widehat{B})$. Consequently, $\mathfrak{K}(B)$ is a K_σ-space with unity. The same arguments imply that every countable set of pairwise disjoint elements in $\mathfrak{K}(B)$ is bounded.

Now, take an arbitrary K_σ-space E with unity and a universal completion $\widehat{E}$ of E. If $B = \mathfrak{C}(E)$ and $\widehat{B} := \mathfrak{C}(\widehat{E})$ then $\widehat{B}$ is an o-completion of B. By (1), the spaces $\widehat{E}$ and $\mathfrak{K}(\widehat{B})$ are isomorphic; moreover, $\mathfrak{K}(B)$ is the image of the subspace E by (2). $\triangleright$

8.2.5. From 8.2.3 and 8.2.4 we may immediately derive some results on function realization of vector lattices.

(1) Theorem. *Let Q be the Stone space of a σ-algebra B. The vector lattices $C_\infty(Q)$ and $\mathfrak{K}(B)$ are isomorphic. In particular, $C_\infty(Q)$ is a universally complete K_σ-space with unity for every quasiextremal compact space Q.*

$\lhd$ Take $e \in \mathfrak{K}(B)$. Let G_t be a clopen set in Q corresponding to the element $e(t) \in B$. By 1.4.1, there exists a unique continuous function $\hat{e} : Q \to \overline{\mathbb{R}}$ such that

$$\{\hat{e} < t\} \subset G_t \subset \{\hat{e} \leq t\} \quad (t \in \mathbf{R}).$$

It follows from the relations 1.3.8 (2) that the closed set $\bigcap\{G_t : t \in \mathbb{R}\}$ has empty interior and the open set $\bigcup\{G_t : t \in \mathbb{R}\}$ is dense in Q. Hence, the function is finite everywhere, except possibly the points of a nowhere-dense set; therefore, $\hat{e} \in C_\infty(Q)$.

It is easy to check that the mapping $e \mapsto \hat{e}$ is the sought isomorphism. $\rhd$

(2) Theorem. *Let Q be the Stone space of a complete Boolean algebra B, and let $\mathscr{R}$ be the reals in the model $\mathbf{V}^{(B)}$. The vector lattice $C_\infty(Q)$ is isomorphic to the universally complete K-space $\mathscr{R}{\downarrow}$. The isomorphism is established by assigning to an element $x \in \mathscr{R}{\downarrow}$ the function $\hat{x} : Q \to \overline{\mathbb{R}}$ by the formula*

$$\hat{x}(q) := \inf\{t \in \mathbb{R} : [\![x < t^\wedge]\!] \in q\} \quad (q \in Q).$$

$\lhd$ The proof is immediate from (1) and 8.2.3. $\rhd$

(3) Theorem. *Let E be an Archimedean vector lattice and Q be the Stone space of the base $\mathfrak{B}(Q)$. Then E is isomorphic to a minorizing sublattice $E_0 \subset C_\infty(Q)$. Moreover, E is an order-dense ideal of $C_\infty(Q)$ (coincides with $C_\infty(Q)$) if and only if E is a K-space (a universally complete K-space).*

$\lhd$ See (2), 8.1.6, and 8.1.7 (5, 6). $\rhd$

8.2.6. In the sequel, we need the concept of integral with respect to a spectral measure. Suppose that (Ω, Σ) is a measure space; i.e., Ω is a nonempty set and Σ is a fixed σ-algebra of subsets of Ω. A *spectral measure* is defined to be a σ-continuous Boolean homomorphism μ from Σ into the Boolean σ-algebra B. More precisely, a mapping $\mu : \Sigma \to B$ is a spectral measure if $\mu(\Omega - A) = \mathbf{1} - \mu(A)$ $(A \in \Sigma)$ and

$$\mu\left(\bigcup_{n=1}^{\infty} A_n\right) = \bigvee_{n=1}^{\infty} \mu(A_n)$$

for each sequence (A_n) of elements of Σ.

Let $B := \mathfrak{E}(E)$ be the Boolean algebra of unit elements of a K_σ-space E with a fixed unity $\mathbf{1}$. Take a measurable function $f : \Omega \to \mathbb{R}$. Given an arbitrary partition of the real axis

$$\beta := (\lambda_k)_{k\in\mathbb{Z}}, \quad \lambda_k < \lambda_{k+1} \ (k \in \mathbb{Z}), \quad \lim_{n\to\pm\infty} \lambda_n = \pm\infty,$$

assign $A_k := f^{-1}([\lambda_k, \lambda_{k+1}))$ and compose the integral sums

$$\underline{\sigma}(f,\beta) := \sum_{-\infty}^{\infty} \lambda_k \mu(A_k), \quad \overline{\sigma}(f,\beta) := \sum_{-\infty}^{\infty} \lambda_{k+1}\mu(A_k),$$

where the sums are calculated in E. It is clear that

$$\underline{\sigma}(f,\beta) \le \sum_{-\infty}^{\infty} f(t_k)\mu(A_k) \le \overline{\sigma}(f,\beta)$$

for every choice of $t_k \in A_k$ $(k \in \mathbb{Z})$. Also, it is evident that $\underline{\sigma}(f,\beta)$ increases and $\overline{\sigma}(f,\beta)$ decreases as we refine the partition β. If there exists an element $x \in E$ such that $\sup\{\underline{\sigma}(f,\beta)\} = x = \inf\{\overline{\sigma}(f,\beta)\}$, where the exact bounds are calculated over all partitions $\beta := (\lambda_k)_{k\in\mathbb{Z}}$ of the real axis, then we say that the function f is *integrable with respect to the spectral measure* μ or *the spectral integral* $I_\mu(f)$ *exists;* in this event we write

$$I_\mu(f) := \int_T f\,d\mu := \int_T f(t)\,d\mu(t) := x.$$

8.2.7. *The spectral integral $I_\mu(f)$ exists for every bounded measurable function f. If E is a universally complete K_σ-space then every almost everywhere finite measurable function is integrable with respect to each spectral measure.*

◁ Note that $A_k \cap A_l = \varnothing$ $(k \ne l)$ and $\bigcup_{k\in\mathbb{Z}} A_k = \Omega$; therefore, $(\mu(A_k))_{k\in\mathbb{Z}}$ is a resolution of unity in the Boolean algebra B. Putting $\delta := \sup_{k\in\mathbb{Z}}\{\lambda_{k+1} - \lambda_k\}$, we may write down

$$0 \le \overline{\sigma}(f,\beta) - \underline{\sigma}(f,\beta) \le \sum_{k\in\mathbb{Z}} \delta\mu(A_k) = \delta\mathbf{1}.$$

Consequently, a measurable function f is integrable with respect to μ if and only if $\overline{\sigma}(f,\beta)$ and $\underline{\sigma}(f,\beta)$ exist at least for one partition β. If f is bounded then the sums $\overline{\sigma}(f,\beta)$ and $\underline{\sigma}(f,\beta)$ contain at most finitely many nonzero summands. If E is a universally complete K_σ-space and a measurable function f is arbitrary then the indicated sums also make sense, since in this case they involve at most countably many pairwise disjoint elements. ▷

8.2.8. Theorem. *Let $E := \mathscr{R}{\downarrow}$ and let μ be a spectral measure with values in $B := \mathfrak{E}(E)$ Then for every measurable function f the integral $I_\mu(f)$ is a unique element of the K-space E satisfying the condition*

$$[\![I_\mu(f) < \lambda^\wedge]\!] = \mu(\{f < \lambda\}) \quad (\lambda \in \mathbb{R}).$$

$\lhd$ Take an arbitrary number $\lambda \in \mathbb{R}$ and a partition of the real axis $\beta := (\lambda_k)_{k\in\mathbb{Z}}$ such that $\lambda_0 = \lambda$. If $b := [\![I_\mu(f) < \lambda^\wedge]\!]$ then

$$b = [\![(\exists t \in \mathbb{R}^\wedge)\,(I_\mu(f) < t \wedge t < \lambda^\wedge)]\!].$$

By the Mixing Principle, there exist a partition $(b_\xi)_{\xi\in\Xi}$ of the element b and a family $(t_\xi)_{\xi\in\Xi} \subset \mathbb{R}$ such that $t_\xi < \lambda$ and $b_\xi \leq [\![I_\mu(f) \leq t_\xi^\wedge]\!]$ for all ξ. Hence, applying 8.1.2, we derive

$$b_\xi \underline{\sigma}(f, \beta) \leq t_\xi b_\xi < \lambda b_\xi \quad (\xi \in \Xi)$$

and further

$$\lambda_k b_\xi \mu(A_k) \leq t_\xi b_\xi \mu(A_k) < \lambda b_\xi \mu(A_k) \quad (\xi \in \Xi,\ k \in \mathbb{Z}).$$

For $k \geq 1$ we have $\lambda_k > \lambda$; therefore, $b_\xi \mu(A_k) = 0$. Thereby

$$b = \bigvee_{\xi\in\Xi} b_\xi \leq \bigwedge_{k=1}^{\infty} \mu(A_k)^* = \mu\left(\Omega - \bigcup_{k=1}^{\infty} A_k\right) = \mu(\{f < \lambda\}).$$

On the other hand, $b^* = [\![I_\mu(f) \geq \lambda^\wedge]\!]$ and, by 8.1.2, we again infer that $\lambda b^* \leq b^* I_\mu(f) \leq b^* \overline{\sigma}(f, \beta)$ or

$$\lambda b^* \mu(A_k) \leq b^* \lambda_k \mu(A_k) \quad (k \in \mathbb{Z}).$$

For $k < 0$ we have $\lambda_k < \lambda$; therefore, $b^* \mu(A_k) = 0$. Consequently,

$$b^* \leq \bigwedge_{k=-1}^{-\infty} \mu(A_k)^* = \mu\left(\Omega - \bigcup_{k=-1}^{-\infty} A_k\right) = \mu(\{f \geq \lambda\}).$$

This implies $b \geq \mu(\{f < \lambda\})$ and we finally obtain $b = \mu(\{f < \lambda\})$.

Assume that

$$[\![x < \lambda^\wedge]\!] = \mu(\{f < \lambda\}) \quad (\lambda \in \mathbb{R})$$

for some $x \in \mathscr{R}{\downarrow}$. Then by what was established above we have $[\![x < \lambda^\wedge]\!] = [\![I_\mu(f) < \lambda^\wedge]\!]$ for all $\lambda \in \mathbb{R}$. This is equivalent to the relation

$$[\![(\forall \lambda \in \mathbb{R}^\wedge)\,(x < \lambda \leftrightarrow I_\mu(f) < \lambda]\!] = \mathbf{1}.$$

Hence, recalling that $\mathbb{R}^\wedge$ is dense in $\mathscr{R}$, we obtain the equality $[\![x = I_\mu(f)]\!] = \mathbf{1}$ or $x = I_\mu(f)$. $\rhd$

8.2.9. *Take a measurable function $f : \Omega \to \mathbb{R}$ and a spectral measure $\mu : \Sigma \to B := \mathfrak{E}(E)$, where E is some K-space. If the integral $I_\mu(f) \in E$ exists then $\lambda \mapsto \mu(\{f < \lambda\})$ $(\lambda \in \mathbb{R})$ coincides with the spectral function of the element $I_\mu(f)$.*

$\triangleleft$ We have only to compare 8.2.8 with 8.1.5. $\triangleright$

8.2.10. Theorem. *Let E be a universally complete K_σ-space, and let $\mu : \Sigma \to B_0 := \mathfrak{E}(E)$ be some spectral measure. The spectral integral $I_\mu(\cdot)$ represents a sequential o-continuous (linear, multiplicative, and latticial) homomorphism from the f-algebra $\mathscr{M}(\Omega, \Sigma)$ of measurable functions into E.*

$\triangleleft$ Without loss of generality we may assume that $E \subset \mathscr{R}{\downarrow}$ and $\mathscr{R}{\downarrow}$ is a universal completion of E (see 8.1.7). Here $\mathscr{R}$ is the field of the reals in $\mathbf{V}^{(B)}$, where B is a completion of the algebra B_0. It is obvious that the operator I_μ is linear and positive. Prove its sequential o-continuity. Take a decreasing sequence $(f_n)_{n \in \mathbb{N}}$ of measurable functions such that $\lim_{n\to\infty} f_n(t) = 0$ for all $t \in \Omega$, and let $x_n := I_\mu(f_n)$ $(n \in \mathbb{N})$ and $0 < \varepsilon \in \mathbb{R}$. If we assign $A_n := \{t \in \Omega : f_n(t) < \varepsilon\}$ then $\Omega = \bigcup_{n=1}^\infty A_n$. By Proposition 8.2.9, we may write down

$$o\text{-}\lim_{n\to\infty} e_\varepsilon^{x_n} = o\text{-}\lim_{n\to\infty} \mu(A_n) = \bigvee_{n=1}^{\infty} \mu(A_n) = \mathbf{1}.$$

Appealing to the test for o-convergence 8.1.9 (2), we obtain $o\text{-}\lim_{n\to\infty} x_n = 0$. Further, given arbitrary measurable functions $f, g : \Omega \to \mathbb{R}$, we derive from 1.3.8 (8) and 8.2.9 that

$$e_\lambda^{I(f\vee g)} = \mu(\{f \vee g < \lambda\}) = \mu(\{f < \lambda\}) \wedge \mu(\{g < \lambda\}) = e_\lambda^{I(f)} \wedge e_\lambda^{I(g)} = e_\lambda^{I(f)\vee I(g)}$$

(with $I := I_\mu$); consequently, $I(f \vee g) = I(f) \vee I(g)$. It means that I_μ is a lattice homomorphism. In a similar way, for $f \geq 0$ and $g \geq 0$ it follows from 1.3.8 (6) and 8.2.9 that

$$\begin{aligned} e_\lambda^{I(f\cdot g)} &= \mu(\{f \cdot g < \lambda\}) = \mu\Big(\bigcup_{\substack{r,s\in\mathbb{Q}_+\\ rs=\lambda}} \{f < r\} \cap \{g < s\}\Big) \\ &= \bigvee_{\substack{r,s\in\mathbb{Q}_+\\ rs=\lambda}} \mu(\{f < r\}) \wedge \mu(\{g < s\}) = \bigvee_{\substack{r,s\in\mathbb{Q}_+\\ rs=\lambda}} e_r^{I(f)} \wedge e_s^{I(g)} = e_\lambda^{I(f)\cdot I(g)} \end{aligned}$$

for $\lambda > 0$, $\lambda \in \mathbb{Q}$, with $\mathbb{Q}$ the *rationals*. Thus, $I(f \cdot g) = I(f) \cdot I(g)$. The validity of the latter equality for arbitrary functions f and g ensues from the above-established properties of the spectral integral:

$$\begin{aligned} I(f \cdot g) &= I(f^+g^+) + I(f^-g^-) - I(f^+g^-) - I(f^-g^+) \\ &= I(f)^+I(g)^+ + I(f)^-I(g)^- - I(f)^+I(g)^- - I(f)^-I(g)^+ \\ &= I(f) \cdot I(g). \ \triangleright \end{aligned}$$

8.2.11. *Let $e_1, \dots, e_n : \mathbb{R} \to B$ be a finite collection of spectral functions with values in a σ-algebra B. Then there exists a unique B-valued spectral measure μ defined on the Borel σ-algebra $\mathscr{B}or(\mathbb{R}^n)$ of the space $\mathbb{R}^n$ such that*

$$\mu\left(\prod_{k=1}^{n}(-\infty, \lambda_k)\right) = \bigwedge_{k=1}^{n} e_k(\lambda_k)$$

for all $\lambda_1, \dots, \lambda_n \in \mathbb{R}$.

$\lhd$ Without loss of generality we may assume that $B = \mathrm{Clop}(Q)$, where Q is the Stone space of B. According to 1.4.1 (1), there are continuous functions $x_k : Q \to \overline{\mathbb{R}}$ such that $e_k(\lambda) = \mathrm{cl}\{x_k < \lambda\}$ for all $\lambda \in \mathbb{R}$ and $k := 1, \dots, n$. Put $f(t) = (x_1(t), \dots, x_n(t))$ if all $x_k(t)$ are finite and $f(t) = \infty$ if $x_k(t) = \pm\infty$ at least for one index k. Thereby we have defined a continuous mapping $f : Q \to \mathbb{R}^n \cup \{\infty\}$ (the neighborhood filterbase of the point ∞ is composed of the complements to various balls with center the origin). It is clear that f is measurable with respect to the Borel algebras $\mathscr{B}or(Q)$ and $\mathscr{B}or(\mathbb{R}^n)$. Let $\mathrm{Clop}_\sigma(Q)$ and φ be the same as in 1.2.6.

Define the mapping $\mu : \mathscr{B}or(\mathbb{R}^n) \to B$ by the formula

$$\mu(A) := \varphi(f^{-1}(A)) \quad (A \in \mathscr{B}or(\mathbb{R}^n)).$$

It is obvious that μ is a spectral measure. If $A := \prod_{k=1}^{n}(-\infty, \lambda_k)$ then

$$f^{-1}(A) = \bigcap_{k=1}^{n}\{x_k < \lambda_k\},$$

and hence $\mu(A) = e_1(\lambda_1) \wedge \dots \wedge e_n(\lambda_n)$. If ν is another spectral measure with the same properties as μ then the set $\mathscr{B} := \{A \in \mathscr{B}or(\mathbb{R}^n) : \nu(A) = \mu(A)\}$ is a σ-algebra containing all sets of the form

$$\prod_{k=1}^{n}(-\infty, \lambda_k) \quad (\lambda_1, \dots, \lambda_n \in \mathbb{R}).$$

Hence, $\mathscr{B} = \mathscr{B}or(\mathbb{R}^n)$. $\rhd$

8.2.12. Now, take an ordered collection of elements $x_1, \dots, x_n$ in a K_σ-space E with unity $\mathbf{1}$. Let $e^{x_k} : \mathbb{R} \to B := \mathfrak{C}(\mathbf{1})$ denote the spectral function of the element x_k. According to the above-proven assertion, there exists a spectral measure $\mu : \mathscr{B}or(\mathbb{R}^n) \to B$ such that

$$\mu\left(\prod_{k=1}^{n}(-\infty, \lambda_k)\right) = \bigwedge_{k=1}^{n} e^{x_k}(\lambda_k).$$

We may see that the measure μ is uniquely determined by the ordered collection $\mathfrak{x} := (x_1, \dots, x_n) \in E^n$. For this reason, we write $\mu_{\mathfrak{x}} := \mu$ and say that $\mu_{\mathfrak{x}}$ is the spectral measure of the collection $\mathfrak{x}$. The following notations are accepted for the integral of a measurable function $f : \mathbb{R}^n \to \mathbb{R}$ with respect to the spectral measure $\mu_{\mathfrak{x}}$:

$$\hat{\mathfrak{x}}(f) := f(\mathfrak{x}) := f(x_1, \dots, x_n) := I_\mu(f).$$

If $\mathfrak{x} = (x)$ then we also write $\hat{x}(f) := f(x) := I_\mu(f)$ and call $\mu_x := \mu$ the spectral measure of x. Recall that the space $\mathscr{B}(\mathbb{R}^n, \mathbb{R})$ of all Borel functions in $\mathbb{R}^n$ is a universally complete K_σ-space and a faithful f-algebra.

8.2.13. Theorem. *The spectral measures of a collection $\mathfrak{x} := (x_1, \dots, x_n)$ and the element $f(\mathfrak{x})$ maintain the relation*

$$\mu_{f(\mathfrak{x})} = \mu_{\mathfrak{x}} \circ f^{\leftarrow},$$

where $f^{\leftarrow} : \mathscr{B}or(\mathbb{R}) \to \mathscr{B}or(\mathbb{R}^n)$ is the homomorphism acting by the rule $A \mapsto f^{-1}(A)$. In particular,

$$(f \circ g)(\mathfrak{x}) = g(f(\mathfrak{x}))$$

for measurable functions $f \in \mathscr{B}(\mathbb{R}^n, \mathbb{R})$ and $g \in \mathscr{B}(\mathbb{R}, \mathbb{R})$ whenever $f(\mathfrak{x})$ and $g(f(\mathfrak{x}))$ exist.

$\lhd$ By 8.2.9, we have

$$\mu_{f(\mathfrak{x})}(-\infty, t) = e_t^{f(\mathfrak{x})} = [\![f(\mathfrak{x}) < t]\!] = \mu_{\mathfrak{x}} \circ f^{-1}(-\infty, t)$$

for every $t \in \mathbb{R}$. Hence, the spectral measures $\mu_f(\mathfrak{x})$ and $\mu_{\mathfrak{x}} \circ f^{\leftarrow}$ defined on $\mathscr{B}or(\mathbb{R})$ coincide on the intervals of the form $(-\infty, t)$. Reasoning in a standard manner, we then conclude that the measures coincide everywhere. To prove the second part, it suffices to observe that $(g \circ f)^{\leftarrow} = f^{\leftarrow} \circ g^{\leftarrow}$ and apply what was established above twice. $\rhd$

8.2.14. Theorem. *For every ordered collection $\mathfrak{x} := (x_1, \dots, x_n)$ of a universally complete K_σ-space E, the mapping*

$$\hat{\mathfrak{x}} : f \mapsto \hat{\mathfrak{x}}(f) \quad (f \in \mathscr{B}(\mathbb{R}^n, \mathbb{R}))$$

is a unique sequentially o-continuous homomorphism of the f-algebra $\mathscr{B}(\mathbb{R}^n, \mathbb{R})$ into E satisfying the conditions

$$\hat{\mathfrak{x}}(dt_k) = x_k \quad (k := 1, \dots, n),$$

where $dt_k : (t_1, \dots, t_n) \mapsto t_k$ stands for the kth coordinate function on $\mathbb{R}^n$.

◁ As was established in 8.2.10, the mapping $f \mapsto \hat{\mathfrak{x}}(f)$ is a sequentially o-continuous homomorphism of f-algebras. Theorem 8.2.13 yields the equalities

$$\mu_{dt_k(\mathfrak{x})} = \mu_{\mathfrak{x}} \circ (dt_k)^{\leftarrow} = \mu_{x_k}.$$

Consequently, the elements $\hat{\mathfrak{x}}(dt_k) = dt_k(\mathfrak{x})$ and x_k coincide, for they have the same spectral function. If $h : \mathscr{B}(\mathbb{R}^n, \mathbb{R}) \to E$ is another homomorphism of f-algebras with the same properties as $\hat{\mathfrak{x}}(\cdot)$ then h and $\hat{\mathfrak{x}}(\cdot)$ coincide on all polynomials. Afterwards, we infer that h and $\hat{\mathfrak{x}}(\cdot)$ coincide on the whole $\mathscr{B}(\mathbb{R}^n, \mathbb{R})$ due to o-continuity. ▷

8.2.15. *Theorem.* *An element $x \in E$ has the form $x = f(\mathfrak{x})$ with some $\mathfrak{x} \in E^n$ and $f \in \mathscr{B}(\mathbb{R}^n, \mathbb{R})$ if and only if $\operatorname{im}(\mu_x) \subset \operatorname{im}(\mu_{\mathfrak{x}})$.*

◁ Necessity follows from 8.2.13. Sufficiency is left to the reader as an exercise. ▷

8.3. Boolean-Valued Banach Spaces

In this section we discuss the Boolean-valued transfer principle for lattice-normed spaces. The interpretation of a Banach space inside an arbitrary Boolean-valued model is a Banach–Kantorovich space. Conversely, the maximal extension of any lattice-normed space, having been embedded into a suitable Boolean-valued model, becomes a Banach space. A possibility thus appears of transferring theorems on Banach spaces to analogous results on lattice-normed spaces by means of Boolean-valued methods. As in Section 8.1, B is a fixed complete Boolean algebra and $\mathbf{V}^{(B)}$ is the Boolean-valued universe constructed over B. Let $\mathscr{R}$ and $\mathscr{C}$ be the fields of real and complex numbers inside $\mathbf{V}^{(B)}$. Denote by $\oplus$ and $\odot$ the addition and multiplication in the fields $\mathscr{R}$ and $\mathscr{C}$.

8.3.1. *Theorem.* *Let $(\mathscr{X}, \rho)$ be a Banach space in the model $\mathbf{V}^{(B)}$. Assign $X := \mathscr{X}\!\downarrow$ and $|\cdot| := \rho\!\downarrow(\cdot)$. The following hold:*

(1) *$(X, |\cdot|, \mathscr{R}\!\downarrow)$ is a universally complete Banach–Kantorovich space;*

(2) *the space X can be endowed with the structure of a faithful unital module over the ring $\Lambda = \mathscr{C}\!\downarrow$ so that*

(a) $(\lambda \mathbf{1})x = \lambda x \quad (\lambda \in \mathbb{C},\, x \in X)$;

(b) $|ax| = |a|\,|x| \quad (a \in \mathscr{C}\!\downarrow,\, x \in X)$;

(c) $b \le [\![x = 0]\!] \Leftrightarrow \chi(b)x = 0 \quad (b \in B,\, x \in X)$,

where χ is an isomorphism of B onto $\mathfrak{P}(X)$.

◁ We denote the addition in $\mathscr{X}$, $\mathscr{C}$, and $\mathscr{R}$ by the same symbol $\oplus$. Let $\odot$ denote the external composition law $\mathscr{C} \times \mathscr{X} \to \mathscr{X}$ of the complex vector space $\mathscr{X}$, as well as the multiplication in $\mathscr{R}$ and in $\mathscr{C}$. Assign $+ := \oplus\!\downarrow$ and $\cdot := \odot\!\downarrow$. This means that

$$x + y = z \Leftrightarrow [\![x \oplus y = z]\!] = \mathbf{1} \quad (x, y, z \in X);$$

$$a \cdot x = y \Leftrightarrow [\![a \odot x = y]\!] = \mathbf{1} \quad (a \in \Lambda,\ x, y \in X).$$

From simple properties of the descent, it follows that $(X, +)$ is an abelian group. For instance, commutativity of the $+$ operation is deduced as follows. Inside the model, we have $[\![\oplus \circ \imath = \oplus]\!] = \mathbf{1}$, where $\imath : \mathscr{X} \times \mathscr{X} \to \mathscr{X} \times \mathscr{X}$ is a permutation of coordinates. Therefore, $\jmath = \imath\!\downarrow$ is a permutation of coordinates in $X \times X$, and

$$+ \circ \jmath = \oplus\!\downarrow \circ\, \imath\!\downarrow = (\oplus \circ \imath)\!\downarrow = \oplus\!\downarrow = +.$$

Given arbitrary $b \in B$ and $x \in X$, assign $\chi(b)x := \operatorname{mix}\{bx, b^*0\}$, where 0 is the neutral element of the group $(X, +)$. In other words, $\chi(b)x$ is a unique element of X for which $[\![\chi(b)x = x]\!] \geq b$ and $[\![\chi(b)x = 0]\!] \geq b^*$. A mapping $\chi(b) : X \to X$ is thus defined; moreover, $\chi(b)$ is additive and idempotent. Let $\mathfrak{P} := \{\chi(b) : b \in B\}$. Then $\mathfrak{P}$ is a complete Boolean algebra and χ is a Boolean isomorphism. Taking account of the fact that, inside the model $\mathbf{V}^{(B)}$, the axioms of a vector space are valid for $\mathscr{X}$, we may write

$$\begin{aligned}
a \cdot (x + y) = a \odot (x + y) &= a \odot x + a \odot y = a \cdot x + a \cdot y, \\
(a + b) \cdot x = (a + b) \odot x &= a \odot x + b \odot x = a \cdot x + b \cdot x, \\
(ab) \cdot x = (ab) \odot x &= a \odot (b \odot x) = a \cdot (b \cdot x), \\
\mathbf{1} \cdot x = \mathbf{1} \odot x &= x \quad (a, b \in \Lambda;\ x, y \in X).
\end{aligned}$$

Since $\mathbf{V}^{(B)}$ is separated, from these relations it follows that the $+$ and $\cdot$ operations determine the structure of a unital Λ-module in X. Letting $\lambda x := (\lambda 1) \cdot x$ $(\lambda \in \mathbb{C},\ x \in X)$, we obtain the structure of a complex vector space in X; moreover, equality (a) is valid. Since, in the model $\mathbf{V}^{(B)}$, the following hold:

$$\begin{aligned}
\chi(b) = 1 &\Rightarrow \chi(b) \odot x = x, \\
\chi(b) = 0 &\Rightarrow \chi(b) \odot x = 0,
\end{aligned}$$

for $b \leq [\![x = 0]\!]$ we have by definition

$$\begin{aligned}
b \leq [\![\chi(b) \odot x = x]\!] \wedge [\![x = 0]\!] &\leq [\![\chi(b) \cdot x = 0]\!], \\
b^* \leq [\![\chi(b) \odot x = 0]\!] &= [\![\chi(b) \cdot x = 0]\!].
\end{aligned}$$

Hence, $[\![\chi(b) \cdot x = 0]\!] = \mathbf{1}$, i.e., $\chi(b)x = 0$, which implies (c). Now turn to Banach properties of the space $(\mathscr{X}, \rho)$. Subadditivity and homogeneity of the norm ρ can be written as follows:

$$\rho \circ \oplus \leq \oplus \circ (\rho \times \rho), \quad \rho \circ \odot = \odot \circ (|\cdot| \times \rho),$$

where $\rho \times \rho : (x, y) \mapsto \big(\rho(x), \rho(y)\big)$ and $|\cdot| \times \rho : (a, x) \mapsto \big(|a|, \rho(x)\big)$. Taking account of the rules of descending composition, for $p := |\!\cdot\!|$ we have

$$p \circ + \leq + \circ (p \times p), \quad p \circ \cdot = \cdot \circ \big(|\cdot| \times p\big).$$

This means that the operator $|\!\cdot\!| : X \to \operatorname{Re} \Lambda$ is a vector seminorm and condition (b) is satisfied. If $|x| = 0$ for some $x \in X$, from $[\![\, \rho(x) = |x| \,]\!] = \mathbb{1}$ we have $[\![\, \rho(x) = 0 \,]\!] = \mathbb{1}$; hence, $[\![\, x = 0 \,]\!] = \mathbb{1}$, i.e., $x = 0$. Thus, $|\!\cdot\!|$ is a vector norm. Decomposability ensues from property (b). Indeed, assume that $c := p(x) = c_1 + c_2 \quad (x \in X;\ c_1, c_2 \in \Lambda^+)$. There exist $a_1, a_2 \in \Lambda^+$ such that $a_k c = c_k \ \ (k := 1, 2)$ and $a_1 + a_2 = 1$. (Assign $a_k := c_k \big(c + (1 - e_c)\big)^{-1}$, where e_c is the trace of the element c.) If $x_k := a_k \cdot x$ $(k := 1, 2)$ then $x = x_1 + x_2$ and $|x_k| = |a_k x| = a_k |x| = c_k$.

It remains to prove *bo*-completeness for X. Take a *bo*-fundamental net $s : \mathrm{A} \to X$. If $\bar{s}(\alpha, \beta) := s(\alpha) - s(\beta) \quad (\alpha, \beta \in \mathrm{A})$, then $\lim |\!\cdot\!| \circ \bar{s}(\alpha, \beta) = 0$. Let $\sigma : \mathrm{A}^\wedge \to \mathscr{X}$ be the modified ascent of s, and $\bar{\sigma}(\alpha, \beta) := \sigma(\alpha) - \sigma(\beta) \ \ (\alpha, \beta \in \mathrm{A}^\wedge)$. Then $\bar{\sigma}$ is the modified ascent of $\bar{s}$, and $\rho \circ \bar{\sigma}$ is the modified ascent of $|\!\cdot\!| \circ s$. Then, due to 8.1.4 (3), $[\![\, \lim \rho \circ \bar{\sigma} = 0 \,]\!] = \mathbb{1}$, i.e., $\mathbf{V}^{(B)} \models$"σ is a fundamental net in $\mathscr{X}$." Since $\mathscr{X}$ is a Banach space inside $\mathbf{V}^{(B)}$, by the Maximum Principle there is an element $x \in X$ such that $[\![\, \lim \rho \circ \sigma_0 = 0 \,]\!] = \mathbb{1}$, where $\sigma_0 : \mathrm{A}^\wedge \to \mathscr{X}$ is defined by the formula $\sigma_0(\alpha) := \sigma(\alpha) - x \quad (\alpha \in \mathrm{A}^\wedge)$. The modified descent of σ_0 is presented by the net $s_0 : \alpha \mapsto s(\alpha) - x \quad (\alpha \in \mathrm{A})$. Consequently, according to 8.1.4 (3), we have o-$\lim |\cdot| \circ s_0 = 0$, i.e., o-$\lim |s(\alpha) - x| = 0$. $\triangleright$

A universally complete Banach–Kantorovich space $\mathscr{X}{\downarrow} := (\mathscr{X}, \rho){\downarrow} := (\mathscr{X}{\downarrow}, \rho{\downarrow})$ is called the *descent of the Banach space* $(\mathscr{X}, \rho)$.

8.3.2. Theorem. *For every lattice-normed space (X, p), there exists a unique (to within a linear isometry) Banach space $\mathscr{X}$ inside $\mathbf{V}^{(B)}$, with $B \simeq \mathfrak{B}\big(|X|^{\perp\perp}\big)$, for which the descent $\mathscr{X}{\downarrow}$ is the universal completion of X.*

$\triangleleft$ Without loss of generality, we assume that $E = p(X)^{\perp\perp} \subset mE = \mathscr{R}{\downarrow}$ and $B = \mathfrak{B}(E)$. Assign

$$d(x, y) := p(x - y)^{\perp\perp} \quad (x, y \in X).$$

It is easy to verify that d is a B-metric in the set X. If we endow the field $\mathbb{C}$ with the discrete B-metric d_0, then the operations of addition, $+ : X \times X \to X$, and multiplication, $\cdot : \mathbb{C} \times X \to X$, are nonexpanding mappings. The vector norm p is nonexpanding too. All these assertions are almost evident. For instance, as regards multiplication, whenever $\alpha, \beta \in \mathbb{C}$ and $x, y \in X$, we have

$$\begin{aligned} d(\alpha x, \beta y) = p(\alpha x - \beta y)^{\perp\perp} &\leq \big(|\alpha| p(x - y)\big)^{\perp\perp} \vee \big(|\alpha - \beta| p(y)\big)^{\perp\perp} \\ &\leq d(x, y) \vee d_0(\alpha, \beta). \end{aligned}$$

Let $\mathscr{X}_0$ be the Boolean-valued representation of the B-set (X, d) (see A.12). Assign $\rho_0 := \mathscr{F}^\sim(p)$, $\oplus := \mathscr{F}^\sim(+)$, and $\odot := \mathscr{F}^\sim(\cdot)$, where $\mathscr{F}^\sim$ is the immersion defined in A.12 (3). The mappings $\oplus$ and $\odot$ define the structure of a vector space over the field $\mathbb{C}^\wedge$ in the set $\mathscr{X}_0$, and the function $\rho_0 : \mathscr{X}_0 \times \mathscr{X}_0 \to \mathscr{R}$ is a norm. In view of the Maximum Principle, there exist elements $\mathscr{X}, \rho \in \mathbf{V}^{(B)}$, for which $[\![(\mathscr{X}, \rho)$ is a complex Banach space that is the completion of the normed space $(\mathscr{X}_0, \rho_0)]\!] = \mathbf{1}$. Moreover, we may assume that $[\![\mathscr{X}_0$ is a dense $\mathbb{C}^\wedge$-subspace of $\mathscr{X}]\!] = \mathbf{1}$. Let $\imath : X \to X_0 := \mathscr{X}_0{\downarrow}$ be the canonical injection (see A.12). Since $+$ is a nonexpanding mapping from $X \times X$ into X, the addition in X_0, i.e. $+ := \oplus{\downarrow}$, is uniquely determined by the relation $\imath \circ + = + \circ (\imath \times \imath)$, where $\imath \times \imath : (x, y) \mapsto (\imath x, \imath y)$ is the canonical injection of the B-set $X \times X$. However, it is equivalent to additivity of $\imath$. Similarly, for the operation $\cdot := \odot{\downarrow}$, we have $\imath \circ \cdot = \cdot \circ (\varkappa \times \imath)$, where $\varkappa \times \imath : (\lambda, x) \mapsto (\lambda^\wedge, \imath x)$ $(\lambda \in \mathbb{C}, x \in X)$. Thus, $\imath$ is a linear operator. By repeating the same reasoning for $p_0 := \rho_0{\downarrow}$, we obtain $\imath_E \circ p_0 = p_0 \circ \imath$, where $\imath_E$ is the canonical injection of E. This means that $\imath$ is an isometry, i.e., $\imath$ preserves the vector norm. Consider a subspace $Y \subset \mathscr{X}{\downarrow}$, $\imath X \subset Y$, that is a universally complete Banach–Kantorovich space with the norm $q(y) := \rho{\downarrow}(y)$ $(y \in Y)$. From decomposability of the norm q and disjoint completeness of Y, it follows that $X_0 \subset Y$. Indeed, $X_0 = \operatorname{mix}(\imath X)$, and, due to (c) of 8.3.1 (2), given an $x \in \mathscr{X}{\downarrow}$, we have $x = \operatorname{mix}(b_\xi \imath x_\xi)$ if and only if $x = bo\text{-}\sum \chi(b_\xi) \imath x_\xi$. On the other hand, Y is decomposable and d-complete; hence, according to 2.1.3 and 2.2.1, Y is invariant under every projection $x \mapsto \chi(b)x$ and contains all sums of the indicated type. Arguing analogously, we have $Y = \operatorname{mix} Y$. If $\mathscr{Y} := Y{\uparrow}$ then $[\![\mathscr{X}_0 \subset \mathscr{Y} \subset \mathscr{X}]\!] = \mathbf{1}$; moreover, $\mathscr{Y}{\downarrow} = Y$. Let $\sigma : \omega^\wedge \to \mathscr{Y}$ be a Cauchy sequence and let s be its modified descent. Then s is a bo-fundamental sequence in Y; consequently, $y = \lim s$ exists. From 8.1.4 (3) it is clear that $[\![y = \lim \sigma]\!] = \mathbf{1}$. Completeness of the set $\mathscr{Y}$ is thus established; therefore, $\mathscr{X} = \mathscr{Y}$ and $X = Y$.

Let $\mathscr{Z}$ be a Banach space inside $\mathbf{V}^{(B)}$; moreover, let $\mathscr{Z}{\downarrow}$ be the universal extension of the lattice-normed space X. If $\imath'$ is the corresponding isometric embedding of X into $\mathscr{Z}{\downarrow}$, then $\imath' \circ \imath$ uniquely extends to a linear isometry of X_0 onto a disjointly complete subspace $Z_0 \subset \mathscr{Z}{\downarrow}$. The spaces $\mathscr{X}_0$ and $\mathscr{Z}_0 := Z_0{\uparrow}$ are isometric. Therefore, their completions $\mathscr{X}$ and $\mathscr{Y} \subset \mathscr{Z}$ are isometric too. Since $\mathscr{Y}{\downarrow}$ is a Banach–Kantorovich space and $\imath X \subset \mathscr{Y}{\downarrow} \subset \mathscr{Z}{\downarrow}$, it follows that $\mathscr{Y}{\downarrow} = \mathscr{Z}{\downarrow}$. Therefore, $\mathscr{Y} = \mathscr{Z}$ and, thus, $\mathscr{X}$ and $\mathscr{Z}$ are linearly isometric. $\triangleright$

8.3.3. Theorem. *Let $\mathscr{X}$ and $\mathscr{Y}$ be Boolean-valued realizations for Banach–Kantorovich spaces X and Y normed by some universally complete K-space E. Let $\mathscr{L}^B(\mathscr{X}, \mathscr{Y})$ be the space of bounded linear operators from $\mathscr{X}$ into $\mathscr{Y}$ inside $\mathbf{V}^{(B)}$, where $B := \mathfrak{B}(E)$. The immersion mapping $T \mapsto T^\sim$ of the operators implements a linear isometry between the lattice-normed spaces $\mathscr{L}_E(X, Y)$ and $\mathscr{L}^B(\mathscr{X}, \mathscr{Y}){\downarrow}$.*

$\triangleleft$ By Theorem 8.3.2, without loss of generality we may assume that $E = \mathscr{R}{\downarrow}$,

$X = \mathscr{X}{\downarrow}$, and $\mathscr{Y}{\downarrow} = Y$ (see 4.1.3 (5) for definition of $\mathscr{L}_E(X, Y)$). Take a mapping $\mathscr{T} : \mathscr{X} \to \mathscr{Y}$ inside $\mathbf{V}^{(B)}$ and put $T := \mathscr{T}{\downarrow}$. Let ρ and θ be the norms of the Banach spaces $\mathscr{X}$ and $\mathscr{Y}$, let $p := \rho{\downarrow}$ and $q := \theta{\downarrow}$, and let $+$ stand for all addition in $\mathscr{X}$, $\mathscr{Y}$, X, and Y. The linearity and boundedness of $\mathscr{T}$ imply validity for the relations

$$\mathscr{T} \circ + = + \circ (\mathscr{T} \times \mathscr{T}), \quad \theta \circ \mathscr{T} \leq k\rho,$$

where $0 \leq k \in \mathscr{R}{\downarrow}$. The descent and ascent rules for composition allow us to write down the relations in the following equivalent form:

$$T \circ + = + \circ (T \times T), \quad q \circ T \leq kp.$$

But this means that T is linear and bounded. Let K be the set constituted of $0 \leq k \in \mathscr{R}{\downarrow}$ such that $q(Tx) \leq kp(x)$ $(x \in X)$. Then $K{\uparrow} = \{k \in \mathscr{R}_+ : \theta \circ \mathscr{T} \leq k\rho\}$ inside $\mathbf{V}^{(B)}$.

Appealing to 8.1.4 (2), we derive

$$\mathbf{V}^{(B)} \models |T| = \inf K = \inf(K{\uparrow}) = \|\mathscr{T}\|.$$

Hence, the mapping $\mathscr{T} \mapsto \mathscr{T}{\downarrow}$ preserves the vector norm. To justify the linearity of the mapping, it suffices to check its additivity. Given $\mathscr{T}_1, \mathscr{T}_2 \in \mathscr{L}^B(\mathscr{X}, \mathscr{Y}){\downarrow}$, we have

$$\begin{aligned}(\mathscr{T}_1 + \mathscr{T}_2){\downarrow}(x) &= (\mathscr{T}_1 + \mathscr{T}_2)(x) = \mathscr{T}_1 x + \mathscr{T}_2 x \\ &= \mathscr{T}_1{\downarrow}x + \mathscr{T}_2{\downarrow}x = (\mathscr{T}_1{\downarrow} + \mathscr{T}_2{\downarrow})x\end{aligned}$$

inside $\mathbf{V}^{(B)}$ for every $x \in X$. Consequently, $(\mathscr{T}_1 + \mathscr{T}_2){\downarrow} = \mathscr{T}_1{\downarrow} + \mathscr{T}_2{\downarrow}$. So, the descent implement a linear isometry of $\mathscr{L}^B(\mathscr{X}, \mathscr{Y}){\downarrow}$ onto the space of all extensional bounded linear operators from X into Y. It remains to observe that every bounded linear operator from X into Y is nonexpanding, or which is the same, satisfies the inequality $[\![x = 0]\!] \leq [\![Tx = 0]\!]$. Indeed, if $b := [\![x = 0]\!]$ then $\chi(b)x = 0$ by 8.3.1 (2); therefore,

$$\chi(b)q(Tx) \leq \chi(b)kp(x) = kp(\chi(b)x) = 0.$$

Hence, $q(\chi(b)Tx) = 0$ or $\chi(b)Tx = 0$ and, employing 8.3.1 (2) again, we conclude that $b \leq [\![Tx = 0]\!]$. $\triangleright$

8.3.4. Assume that X is a normed space and $\widetilde{X}$ is the completion of X. Let $\mathscr{X}$ be the completion of the $\mathbb{R}^\wedge$-normed space $X^\wedge$ inside $\mathbf{V}^{(B)}$.

(1) *Theorem.* *The universally complete Banach–Kantorovich space $\mathscr{X}{\downarrow}$ is linearly isometric to the space $C_\infty(Q, \widetilde{X})$, with Q the Stone space of B.*

◁ Identify the K-spaces $\mathscr{R}{\downarrow}$ and $C_\infty(Q)$, and apply Theorem 8.3.2 to the lattice-normed space $(X, p, \mathscr{R}{\downarrow})$, with $p(x) = \|x\| \cdot \mathbf{1}$. Using the notation of the proof of 8.3.2, note that $\mathscr{X}_0 = X^{\wedge}$. Hence, $\mathscr{X}{\downarrow} := (\mathscr{X}{\downarrow},\ q,\ \mathscr{R}{\downarrow})$ is the universal completion of $(X, p, \mathscr{R}{\downarrow})$. For simplicity, assume that $X \subset \mathscr{X}{\downarrow}$. Since $[\![X^{\wedge}$ is norm-dense in $\mathscr{X}]\!] = \mathbf{1}$, we deduce that to $u \in C_\infty(Q, \widetilde{X})$ and $\varepsilon > \mathbf{0}$ there are a family $(x_\xi) \subset X$ and a partition of unity $(Q_\xi) \subset \operatorname{Clop}(Q)$ for which the step-function u_ε, equal to x_ξ on the set Q_ξ, obeys the estimate $|u - u_\varepsilon| \leq \varepsilon \mathbf{1}$. Put $\mathscr{T}(u_\varepsilon) = \operatorname{mix}(b_\xi x_\xi)$ where b_ξ stands for the member of B corresponding to the clopen set Q_ξ. Now, $|\mathscr{T}(u_\varepsilon)| = |u_\varepsilon|$. Hence, $\mathscr{T}$ is a linear isometric embedding of the subspace of all vector-functions of the shape u_ε. If $\varepsilon \to \mathbf{0}$ then $|u_\varepsilon - u| \xrightarrow{(r)} \mathbf{0}$, and so $\big(\mathscr{T}(u_{1/n})\big)$ is an r-fundamental sequence. Since $\mathscr{X}{\downarrow}$ is complete, $\mathscr{X}{\downarrow}$ contains the limit $v := r\text{-}\lim \mathscr{T}(u_{1/n})$. Assigning $\mathscr{T}(u) := v$, obtain a linear isometric embedding $\mathscr{T} : C_\infty(Q, \widetilde{X}) \to \mathscr{X}{\downarrow}$. If $Z := \operatorname{im}(\mathscr{T})$ then Z is a decomposable *bo*-complete subspace of $\mathscr{X}{\downarrow}$ and $X \subset Z$. By Theorem 8.3.2 and the definition of 2.2.6, $Z = \mathscr{X}{\downarrow}$. ▷

(2) *Assume that $\mathscr{X}'$ is the dual of X inside $\mathbf{V}^{(B)}$. Then the spaces $\mathscr{X}'{\downarrow}$ and $E_s(X')$, with $E = C_\infty(Q)$, are linearly isometric.*

◁ Apply Theorem 8.3.3 to $Y := E$ and $X := (X, p, E)$, with $p(x) = \|x\|\mathbf{1}$. Deduce so that the spaces $\mathscr{X}'{\downarrow} := \mathscr{L}^{(B)}(\mathscr{X}, \mathscr{R}){\downarrow}$ and $L_A(X, E)$ are linearly isometric. To complete the proof, refer to 5.5.1 (1). ▷

8.3.5. An isometry $\imath$ between normed B-spaces is *B-isometry* if $\imath$ is linear and commutes with every projection in B. Say that Y is a *B-cyclic completion* of a B-space X if Y is B-cyclic and there is a B-isometry $\imath : X \to Y$ such that every B-cyclic subspace of Y containing $\imath(X)$ coincides with Y.

(1) *Each Banach B-space possesses a B-cyclic completion unique up to B-isometry.*

◁ The claim follows from 2.2.8, 7.1.3, and 7.3.3 (1). ▷

Let Λ be the bounded part of the universally complete K-space $\mathscr{C}{\downarrow}$, i.e. Λ is the order-dense ideal in $\mathscr{C}{\downarrow}$ generated by the order-unity $\mathbf{1} := 1^{\wedge} \in \mathscr{C}{\downarrow}$. Take a Banach space $\mathscr{X}$ inside $\mathbf{V}^{(B)}$. Denote

$$\mathscr{X}{\downarrow}^{\infty} := \{x \in \mathscr{X}{\downarrow} : |x| \in \Lambda\}.$$

Then $\mathscr{X}{\downarrow}^{\infty}$ is a Banach–Kantorovich space called the *bounded descent* of $\mathscr{X}$. Since Λ is an order complete AM-space with unity, $\mathscr{X}{\downarrow}^{\infty}$ is a Banach space with mixed norm over Λ, and hence, B-cyclic Banach space, see 7.3.4. If $\mathscr{Y}$ is another Banach space and $\mathscr{T} : \mathscr{X} \to \mathscr{Y}$ is a bounded linear operator inside $\mathbf{V}^{(B)}$ with $|\mathscr{T}{\downarrow}| \in \Lambda$ then the bounded descent of $\mathscr{T}$ is the restriction of $\mathscr{T}{\downarrow}$ to $\mathscr{X}{\downarrow}^{\infty}$. Clearly, the bounded descent of $\mathscr{T}$ is a bounded linear operator from $\mathscr{X}{\downarrow}^{\infty}$ to $\mathscr{Y}{\downarrow}^{\infty}$.

(2) Theorem. *A Banach space X is linearly isometric to the bounded descent of some Banach space inside $\mathbf{V}^{(B)}$ if and only if X is B-cyclic.*

$\lhd$ Cf. 8.3.1, 8.3.2, 7.1.3, and 7.3.3 (1). $\rhd$

(3) Take a normed B-space X. Denote by $\widetilde{X}$ the norm completion of X. Then $\widetilde{X}$ is a Banach B-space for every projection $b \in B$ admits a unique extension to the whole of $\widetilde{X}$ which preserve the norm of b. By (1), $\widetilde{X}$ possesses a cyclic B-completion which we denoted by $\overline{X}$. Applying Theorem (2), we now take a Banach space $\mathscr{X}$ inside $\mathbf{V}^{(B)}$ whose bounded descent is B-isometric with $\overline{X}$. The element $\mathscr{X} \in \mathbf{V}^{(B)}$ is called the *Boolean-valued representation* of X.

(4) Let X and Y be normed space such that $B \subset \mathscr{L}(X)$ and $B \subset \mathscr{L}(Y)$. An operator $T : X \to Y$ is *B-linear*, if T commutes with every projection in B; i.e. $b \circ T = T \circ b$ for all $b \in B$.

Denote by $\mathscr{L}_B(X, Y)$ the set of all bounded B-linear operators from X to Y. In this event $W := \mathscr{L}_B(X, Y)$ is a Banach space and $B \subset W$. If Y is B-cyclic then so is W. A projection $b \in B$ acts in W by the rule $T \mapsto b \circ T$ $(T \in W)$.

We call $X^{\#} := \mathscr{L}_B(X, \Lambda)$ the *B-dual* of X. If $X^{\#}$ and Y are B-isometric to each other then we say that Y is a *B-dual space* and X is a *B-predual* of Y. In symbols, $X = Y_{\#}$.

8.3.6. Theorem. *Assume that X is a normed B-space and Y is a B-cyclic Banach space. Let $\mathscr{X}$ and $\mathscr{Y}$ stand for the Boolean-valued representation of X and Y. The space $\mathscr{L}_B(X, Y)$ is B-isometric to the bounded descent of the space $\mathscr{L}^B(\mathscr{X}, \mathscr{Y})$ of all bounded linear operators from $\mathscr{X}$ to $\mathscr{Y}$ inside $\mathbf{V}^{(B)}$. Moreover, to $T \in \mathscr{L}_B(X, Y)$ there corresponds the member $\mathscr{T} := T\uparrow$ of $\mathbf{V}^{(B)}$ determined from the formulas $[\![\mathscr{T} : \mathscr{X} \to \mathscr{Y}]\!] = \mathbf{1}$, $[\![\mathscr{T}\imath x = \imath T x]\!] = \mathbf{1}$ $(x \in X)$, where $\imath$ stands for a mapping that embeds X into $\mathscr{X}\downarrow$ and Y into $\mathscr{Y}\downarrow$.*

$\lhd$ Without loss of generality, assume that X and Y the bounded descents of some Banach spaces $\mathscr{X}$ and $\mathscr{Y}$ (cf. 8.3.5 (1) and 8.3.5 (2)). Put $X_0 := \mathscr{X}\downarrow$ and $Y_0 := \mathscr{Y}\downarrow$. By 8.3.3, the spaces $\mathscr{L}^B(\mathscr{X}, \mathscr{Y})\downarrow$ and $\mathscr{L}_{\mathscr{R}}(X_0, Y_0)$ are linearly isometric. Moreover, the restriction of $\mathscr{L}_{\mathscr{R}}(X_0, Y_0)$ relative to $\mathscr{S}(B)$ coincides with the bounded descent of $\mathscr{L}(\mathscr{X}, \mathscr{Y})$. It suffices to note that each member T of $\mathscr{L}_{\mathscr{R}}(X, Y)$ admits a unique extension which preserves the norm of T. $\rhd$

8.3.7. Let $\mathscr{X}^*$ be the dual of $\mathscr{X}$ inside $\mathbf{V}^{(B)}$. Denote by $\simeq$ and $\simeq_B$ the relations of isometric isomorphy and isometric B-isomorphy between Banach spaces. Suppose also that X, Y, $\mathscr{X}$, and $\mathscr{Y}$ are the same as in 8.3.6.

(1) *The following equivalence holds:* $X^{\#} \simeq_B Y \Leftrightarrow [\![\mathscr{X}^* \simeq \mathscr{Y}]\!] = \mathbf{1}$.

(2) *If $\overline{X}$ is the B-cyclic completion of X then $X^{\#} = \overline{X}^{\#}$.*

8.3.8. Theorem. *The bounded descent of an arbitrary Hilbert space in $\mathbf{V}^{(B)}$ is a Kaplansky–Hilbert module over the Stone algebra $\mathscr{S}(B)$. Conversely, if X is* *a Kaplansky–Hilbert module over $\mathscr{S}(B)$, then there is a Hilbert space $\mathscr{X}$ in $\mathbf{V}^{(B)}$ whose bounded descent is unitarily equivalent with X. This space is unique to within unitary equivalence inside $\mathbf{V}^{(B)}$.*

$\triangleleft$ Without loss of generality, we may assume that $\mathscr{S}(B) \subset \mathscr{C}\downarrow$. Suppose that $\mathscr{X}$ is a Hilbert space inside $\mathbf{V}^{(B)}$ and X is the bounded descent of $\mathscr{X}$. Then the pair $(X, |\cdot|)$, with $|\cdot|$ the descent of the norm of $\mathscr{X}$ is a Banach–Kantorovich space and the pair $(X, |||\cdot|||)$, with $|||x||| = \||x|\|$ $(x \in X)$, is a B-cyclic Banach space (cf. 8.3.5 (2)). In particular, X is a unitary module over $\mathscr{S}(B)$. Suppose that $(\cdot\,|\,\cdot) \in \mathbf{V}^{(B)}$ is the inner product in $\mathscr{X}$ and $\langle\cdot\,|\,\cdot\rangle$ is the descent of $(\cdot\,|\,\cdot)$. It is easy to check that $\langle\cdot\,|\,\cdot\rangle$ satisfies 7.4.3 (1–4) for all $x,\ y,\ z \in \mathscr{X}\downarrow$ and $a \in \mathscr{C}\downarrow$. If $x,\ y \in X$ then $[\![\, |(x\,|\,y)| \le \|x\| \cdot \|y\| \,]\!] = \mathbf{1}$. Hence, $|\langle x\,|\,y\rangle| \le |x| \cdot |y|$. Since $|x|, |y| \in \mathscr{S}(B)$; therefore, $\langle x\,|\,y\rangle \in \mathscr{S}(B)$. Thus, the restriction of $\langle\cdot\,|\,\cdot\rangle$ to $X \times X$, denoted by the same symbol, is a $\mathscr{S}(B)$-valued inner product on X. It suffices to note that $|x| = \sqrt{\langle x\,|\,x\rangle}$, since $[\![\, \|x\| = \sqrt{(x\,|\,x)} \,]\!] = \mathbf{1}$ and the descent of the function $\surd : \mathscr{R}^+ \to \mathscr{R}^+$ depicts the square root in $\mathscr{S}(B)$.

Now, consider a Kaplansky–Hilbert module X over $\mathscr{S}(B)$. By Theorem 8.3.2, the Boolean-valued representation $\mathscr{X} \in \mathbf{V}^{(B)}$ of the Banach–Kantorovich space $(X, |\cdot|, \mathscr{S}(B))$ is a Banach space inside $\mathbf{V}^{(B)}$. We may thus assume that $X \subset \mathscr{X}\downarrow$. Let $(\cdot\,|\,\cdot)$ stand for the ascent of the $\mathscr{S}(B)$-valued inner product $\langle\cdot\,|\,\cdot\rangle$ in X. Then $(\cdot\,|\,\cdot)$ is an inner product on $\mathscr{X}$ inside $\mathbf{V}^{(B)}$. Arguing as above, we see that $[\![\|x\| = \sqrt{(x\,|\,x)} \quad (x \in \mathscr{X})]\!] = \mathbf{1}$, since $|x| = \sqrt{\langle x\,|\,x\rangle} \quad (x \in X)$.

Suppose that $\mathscr{Y}$ is another Hilbert space inside $\mathbf{V}^{(B)}$ and the bounded descent Y of $\mathscr{Y}$ is unitarily equivalent with X. If $U : X \to Y$ is a unitary isomorphism then $u := U\uparrow$ is a linear bijection from $\mathscr{X}$ to $\mathscr{Y}$. Since U enjoys the property $\langle\cdot\,|\,\cdot\rangle \circ (U \times U) = \langle\cdot\,|\,\cdot\rangle$, we note inside $\mathbf{V}^{(B)}$ that

$$(\cdot\,|\,\cdot) \circ (u \times u) = \langle\cdot\,|\,\cdot\rangle \uparrow \circ (U\uparrow \times U\uparrow) = (\langle\cdot\,|\,\cdot\rangle \circ (U \times U))\uparrow = \langle\cdot\,|\,\cdot\rangle\uparrow = (\cdot\,|\,\cdot).$$

Hence, u is a unitary equivalence between $\mathscr{X}$ and $\mathscr{Y}$. This end the proof. $\triangleright$

As usual, we call $\mathscr{X}$ the *Boolean-valued representation* of a Kaplansky–Hilbert module X.

Suppose that $\mathscr{L}^B(\mathscr{X}, \mathscr{Y})$ the space of bounded linear operators from $\mathscr{X}$ to $\mathscr{Y}$ inside $\mathbf{V}^{(B)}$. Let $\mathrm{Hom}(X, Y)$ stand for the space of all bounded Λ-linear operators from X to Y where X and Y are Kaplansky–Hilbert modules over the commutative AW^*-algebra $\mathscr{S}(B) = \Lambda$. As before, we let $\mathscr{S}(B)$ stand for the bounded descent of the field $\mathbb{C}$. It is easy to see that $\mathrm{Hom}(X, Y) = \mathscr{L}_\Lambda(X, Y)$ (cf. 4.1.3 (5), 8.3.3).

8.3.9. *Theorem.* *Suppose that $\mathscr{X}$ and $\mathscr{Y}$ are Hilbert spaces inside $\mathbf{V}^{(B)}$. Let X and Y stand for the bounded descents of $\mathscr{X}$ and $\mathscr{Y}$. For every bounded Λ-linear operator $\Phi : X \to Y$ the element $\varphi := \Phi\uparrow$ is a bounded linear operator from $\mathscr{X}$ to $\mathscr{Y}$ inside $\mathbf{V}^{(B)}$.*

Moreover, $[\![\,\|\varphi\| \le c^{\wedge}\,]\!] = \mathbf{1}$ *for some* $c \in \mathbb{R}$. *The mapping* $\Phi \mapsto \varphi$ *is a* B*-linear isometry between the* B*-cyclic Banach spaces* $\operatorname{Hom}(X, Y)$ *and* $\mathscr{L}^B(\mathscr{X}, \mathscr{Y}){\downarrow}^{\infty}$.

$\lhd$ Appealing to 8.3.5 (2) and 8.3.6, we complete the proof. $\rhd$

8.3.10. *Theorem.* *For a Kaplansky–Hilbert module* X *to be* λ*-homogeneous it is necessary and sufficient that* $[\![\dim(\mathscr{X}) = |\lambda^{\wedge}|\,]\!] = \mathbf{1}$.

$\lhd$ By Theorem 8.3.2 we may assume, that $X \subset \mathscr{X}{\downarrow}$. The mapping $\langle\cdot\,\cdot\rangle$ and the descent of the form $(\cdot\,|\,\cdot)$ agree on $X \times X$. Therefore, for all $x, y \in X$ and $a \in \Lambda$ the following are equivalent $\langle x|y\rangle = a$ and $[\![\,(x|y) = a\,]\!] = \mathbf{1}$. We thus see that the orthogonality relation on X is the restriction to X of the descent of the orthogonality relation on $\mathscr{X}$. From these observations it follows that a subset $\mathscr{E}$ of X is orthonormal if and only if $[\![\,\mathscr{E}{\uparrow}$ is an orthonormal set in $\mathscr{X}\,]\!] = \mathbf{1}$. Applying the descent rules for polars to orthogonal complements in X and $\mathscr{X}$, we obtain $(\mathscr{E}{\uparrow})^{\perp}{\downarrow} = (\mathscr{E}{\uparrow\downarrow})^{\perp}$. Observe also that $\mathscr{E}^{\perp} = (\mathscr{E}{\uparrow\downarrow})^{\perp}$. Hence, $\mathscr{E}^{\perp}{\uparrow} = (\mathscr{E}{\uparrow})^{\perp}$. In particular, $\mathscr{E}^{\perp} = \mathbf{0}$ if and only if $[\![\,(\mathscr{E}{\uparrow})^{\perp} = \{\mathbf{0}\}\,]\!] = \mathbf{1}$. Thus, $\mathscr{E}$ is a basis for X only on condition that $[\![\,\mathscr{E}$ is a basis for $\mathscr{X}\,]\!] = \mathbf{1}$. If $|\mathscr{E}| = \lambda$ and $\varphi : \lambda \to \mathscr{E}$ are bijections then the modified ascent $\varphi{\Uparrow}$ is a bijection of $\lambda^{\wedge}$ to $\mathscr{E}{\uparrow}$. Conversely, suppose that $\mathscr{D}$ is a basis for $\mathscr{X}$ and $[\![\,\psi : \lambda^{\wedge} \to \mathscr{D}$ is a bijection$\,]\!] = \mathbf{1}$ for some cardinal λ. In this case the modified descent $\varphi := \psi{\Downarrow} : \lambda \to \mathscr{D}{\downarrow}$ is injective. Consequently, the set $\mathscr{E} := \operatorname{im}(\varphi)$ has cardinality λ. Moreover, as shown above, it is orthonormal. We are left with observing that $\mathscr{D}{\downarrow} = \operatorname{mix}(\mathscr{E}) = \mathscr{E}{\uparrow\downarrow}$, i.e., $[\![\,\mathscr{E}{\uparrow} = \mathscr{D}\,]\!] = \mathbf{1}$. Finally, $\mathscr{E}$ is a basis for X, which completes the proof. $\rhd$

8.3.11. *Theorem.* *For a Kaplansky–Hilbert module* X *to be strictly* λ*-homogeneous it is necessary and sufficient that* $[\![\dim(\mathscr{X}) = \lambda^{\wedge}\,]\!] = \mathbf{1}$.

$\lhd$ Suppose that X is strictly λ-homogeneous module. By Theorem 8.3.10 $[\![\dim(\mathscr{X}) = |\lambda^{\wedge}|\,]\!] = \mathbf{1}$. On the other hand, there is a partition of unity $(b_{\alpha})_{\alpha<\beta}$ in the Boolean algebra B such that $|\lambda^{\wedge}| = \operatorname{mix}_{\alpha<\beta}(b_{\alpha}\alpha^{\wedge})$. Since $b_{\alpha} \le [\![\,\mathscr{X} = b_{\alpha}\mathscr{X}\,]\!]$; therefore, $b_{\alpha} \le [\![\dim(b_{\alpha}\mathscr{X}) = \alpha^{\wedge}\,]\!]$. Consider the set $B_{\alpha} := [\mathbf{0}, b_{\alpha}] := \{b' \in B : b' \le b_{\alpha}\}$. If $b_{\alpha} \ne \mathbf{0}$ then B_{α} is a complete Boolean algebra. In the respective universe $\mathbf{V}^{(B_{\alpha})}$ we observe that $\mathbf{V}^{(B_{\alpha})} \models$ "$b_{\alpha}\mathscr{X}$ is a Hilbert space and $\alpha^{\wedge} = \dim(b_{\alpha}\mathscr{X})$." The space $b_{\alpha}X$ is the bounded descent of the Hilbert space $b_{\alpha}\mathscr{X}$ inside $\mathbf{V}^{(B_{\alpha})}$ is $b_{\alpha}X$. Consequently, $b_{\alpha}X$ is an α-homogeneous Kaplansky–Hilbert module. Furthermore, $\mathbf{V}^{(B_{\alpha})} \models$ "$\alpha^{\wedge}$ is a cardinal" and so α is a cardinal too. By the definition of strict homogeneity, $\lambda \le \alpha$. Thus, $b_{\alpha} = \mathbf{0}$ for $\alpha < \lambda$. Therefore, $[\![\,\lambda^{\wedge} \le |\lambda^{\wedge}|\,]\!] = \mathbf{1}$. We so see that $[\![\,\lambda^{\wedge} = |\lambda^{\wedge}|\,]\!] = \mathbf{1}$, since the formula $[\![\,|\lambda^{\wedge}| \le \lambda^{\wedge}\,]\!] = \mathbf{1}$ is valid by the definition of cardinality. We may now conclude that $[\![\dim(\mathscr{X}) = \lambda^{\wedge}\,]\!] = \mathbf{1}$.

Assume the last equality valid. Then λ is a cardinal, since $\lambda^{\wedge}$ is a cardinal inside $\mathbf{V}^{(B)}$. By 8.3.10 X is λ-homogeneous. If X is γ-homogeneous for some γ then, appealing to 8.3.10 again, we obtain $[\![\dim(\mathscr{X}) = |\gamma^{\wedge}|\,]\!] = \mathbf{1}$. Hence, $[\![\,\lambda^{\wedge} =$

$|\gamma^\wedge| \leq \gamma^\wedge]\!] = \mathbf{1}$ and so $\lambda \leq \gamma$. The same arguments will apply to every AW^*-algebra bX with $\mathbf{0} \neq b \in B$ provided that we substitute $\mathbf{V}^{([\mathbf{0},b])}$ for $\mathbf{V}^{(B)}$. Thus, X is a strictly λ-homogeneous Kaplansky–Hilbert module. $\triangleright$

8.4. Involutive Banach Algebras

In this section we develop the Boolean-valued transfer principle for involutive Banach algebras. By way of application, we give a functional representation for type I AW^*-algebras. This also implies that every type I AW^*-algebra has a unique decomposition into the direct sum of strictly homogeneous subalgebras.

8.4.1. *Theorem.* *The bounded descent of a Banach algebra inside $\mathbf{V}^{(B)}$ is a B-cyclic Banach algebra. Conversely, to each B-cyclic Banach algebra A there is a Banach algebra $\mathscr{A}$ inside $\mathbf{V}^{(B)}$ such that A is isometrically B-isomorphic to the bounded descent of $\mathscr{A}$. Moreover, this algebra $\mathscr{A}$ is unique up to isomorphism.*

$\triangleleft$ Take a B-cyclic Banach algebra A. By Theorem 8.3.5 (2) there is a Banach space $\mathscr{A}$ in $\mathbf{V}^{(B)}$ whose bounded descent A_0 is a B-cyclic Banach space admitting an isometrical B-isomorphism with A. Without loss of generality, we may assume that $A_0 = A$. Multiplication on A is extensional. Indeed, if $b \leq [\![x = u]\!] \wedge [\![y = v]\!]$ with x, y, u, $v \in A$ then, by item (b) of 8.3.1 (2),

$$\mathbf{0} = x\chi(b)(y - v) + \chi(b)(x - u)v$$
$$\Rightarrow \chi(b)(xy - uv) = \mathbf{0} \Rightarrow \chi(b)(xy) = \chi(b)uv \Rightarrow b \leq [\![xy = uv]\!].$$

Let $\odot$ stand for the ascent of the multiplication $\cdot$ on A. It is easy that $\odot$ is a binary operation on $\mathscr{A}$ and the space $\mathscr{A}$ with the operation $\odot$ is an algebra. If p is the vector norm of the space A then $\|a\| = \|p(a)\|_\infty$ and $[\![p(a) = \rho(a)]\!] = \mathbf{1}$ $\quad (a \in \mathscr{A})$ where ρ is the norm of $\mathscr{A}$ (cf. 7.3.2). Show that p is a submultiplicative norm, i.e. $p(xy) \leq p(x)p(y)$. To this end, recall (cf. 8.3.1 (2) and 7.3.2), that A is a Banach module over the ring $\Lambda := \mathscr{S}(B)$ (the bounded part of $\mathscr{C}\!\downarrow$). Furthermore, p maintains the equality

$$p(x) = \inf\{\alpha \in \Lambda_+ : x \in \alpha U_A\} \quad (x \in A).$$

Hence, submultiplicativity of p follows from the fact that the unit ball U_A is closed under multiplication, i.e. the containments x, $y \in U_A$ imply $xy \in U_A$. Therefore, $p \circ (\cdot) \leq (\cdot) \circ (p \times p)$. Using the rules for ascending mappings (cf. A.10), we see that $[\![\rho \circ \odot \leq \odot \circ (\rho \times \rho)]\!] = \mathbf{1}$, i.e., $[\![\rho$ is a submultiplicative norm $]\!] = \mathbf{1}$. We finally infer that $\mathscr{A}$ is a Banach algebra inside $\mathbf{V}^{(B)}$. To show uniqueness of A argue as follows: Assume that $\mathscr{A}_1$ and $\mathscr{A}_2$ are Banach algebras inside $\mathbf{V}^{(B)}$. Let g be an isometric isomorphism between the bounded descents of $\mathscr{A}_1$ and $\mathscr{A}_2$. Then g

is an extensional mapping and $\psi := g\!\uparrow$, the ascent of g, is a linear isometry between the Banach spaces $\mathscr{A}_1$ and $\mathscr{A}_2$. Multiplicativity of ψ follows from the formulas

$$\psi \circ \odot = g\!\uparrow \circ (\cdot)\!\uparrow = (g \circ (\cdot))\!\uparrow = ((\cdot) \circ (g \times g))\!\uparrow = (\cdot)\!\uparrow \circ (g\!\uparrow \times g\!\uparrow) = \odot \circ (\psi \times \psi)$$

with $\odot$ standing for the operations of multiplication on $\mathscr{A}_1$ and $\mathscr{A}_2$ and $(\cdot)$ symbolizing the operations of multiplication in the bounded descents of $\mathscr{A}_1$ and $\mathscr{A}_2$.

Assume now that $\mathscr{A}$ is a Banach algebra inside $\mathbf{V}^{(B)}$ and A is the bounded descent of $\mathscr{A}$. We know that A is a B-cyclic Banach space (cf. 8.3.5 (2)). If χ is the canonical isomorphism of B onto the base $\mathfrak{E}(\Lambda)$ then $b \leq [\![x = \mathbf{0}]\!] \Leftrightarrow \chi(b)x = \mathbf{0}$ for all $x \in A$ (cf. 8.3.1 (2)). Considering the definition of χ and the obvious implication

$$\chi(b) = \mathbf{0} \vee \chi(b) = \mathbf{1} \Rightarrow \chi(b)xy = (\chi(b)x)y = x\,(\chi(b)y) \quad (x, y \in A),$$

and taking arbitrary x, $y \in A$, we may write

$$[\![\chi(b)xy = x\chi(b)y = (\chi(b)x)y]\!] \geq [\![\chi(b) = \mathbf{1}]\!] \vee [\![\chi(b) = \mathbf{0}]\!] = b \vee b^* = \mathbf{1}.$$

This shows that the projection $\pi_b : x \mapsto \chi(b)x$ $(x \in A)$ enjoys the sought property: $\pi_b xy = (\pi_b x)y = x\,(\pi_b y)$ $(x, y \in A)$. Thus, A is a B-cyclic algebra. $\triangleright$

8.4.2. (1) Theorem. *The bounded descent of a C^*-algebra inside $\mathbf{V}^{(B)}$ is a B-cyclic C^*-algebra. Conversely, to each B-cyclic C^*-algebra A there is a C^*-algebra $\mathscr{A}$ inside $\mathbf{V}^{(B)}$ such that the bounded descent of $\mathscr{A}$ is $*$-B-isomorphic with A. Moreover, this C^*-algebra $\mathscr{A}$ is unique up to $*$-isomorphism.*

$\triangleleft$ If A is a B-cyclic C^*-algebra then the structure of a Banach $\mathscr{S}(B)$-module on A possesses the additional property that $(\alpha x)^* = \alpha x^*$ $(\alpha \in Z,\ x \in A)$, where Z stands for the real part of the complex Banach algebra $\mathscr{S}(B)$. Indeed, if $\alpha := \sum_{k=1}^{n} \lambda_k \pi_k$ with $\lambda_1, \dots, \lambda_n \in \mathbb{R}$ and $\pi_1, \dots, \pi_n \in \mathfrak{E}(\mathscr{S}(B))$, then

$$(\alpha x)^* = \sum_{k=1}^{n} \lambda_k\,(\pi_k x)^* = \sum_{k=1}^{n} \lambda_k \pi_k x^* = \alpha x^*.$$

Involution is an isometry on every C^*-algebra and so $U_A^* = U_A$. We may conclude that

$$x \in \alpha U_A \Leftrightarrow xx^* \in \alpha^2 U_A \quad (x \in A,\ \alpha \in Z).$$

Hence, $p(xx^*) = p(x)^2$. In particular, the involution on A is an isometry with respect to the vector norm p, i.e. $p(x^*) = p(x)$ $(x \in A)$. Note also that if $(\mathscr{A}, \rho)$ is a Banach algebra inside $\mathbf{V}^{(B)}$, A is the bounded descent of $\mathscr{A}$, and p is the restriction of $\rho\!\downarrow$ to A; then the descent of the involution on A obeys the conditions $[\![(\forall x \in \mathscr{A}) \rho(xx^*) = \rho(x)^2]\!] = \mathbf{1}$ if and only if $p(xx^*) = p(x)^2$ $(x \in A)$. It suffice to appeal to Theorem 8.4.1 and add a few elementary arguments. $\triangleright$

(2) *Assume that $\mathscr{A}$ is an involutive Banach algebra inside $\mathbf{V}^{(B)}$ and A is the bounded descent of $\mathscr{A}$. Then $x \in A$ is a hermitian element or a positive (central) projection if and only if $[\![x$ is a hermitian element or a positive (central) projection $]\!] = \mathbf{1}$.*

$\lhd$ This is obvious. $\rhd$

8.4.3. We agree to let B-Hom(A_1, A_2) stand for the set comprising B-homomorphisms from A_1 to A_2. We also agree that $\mathrm{Hom}^B(\mathscr{A}_1, \mathscr{A}_2)$ symbolizes the element of $\mathbf{V}^{(B)}$ which depicts the set of all homomorphisms from $\mathscr{A}_1$ to $\mathscr{A}_2$.

(1) *Assume that $\mathscr{A}_1$ and $\mathscr{A}_2$ are Banach algebras inside $\mathbf{V}^{(B)}$ and A_1 and A_2 are their respective bounded descents. If $\Phi \in B\text{-}\mathrm{Hom}^B(A_1, A_2)$ and $\varphi := \Phi\!\uparrow$ then $[\![\varphi \in \mathrm{Hom}^B(\mathscr{A}_1, \mathscr{A}_2)]\!] = \mathbf{1}$ and $[\![\|\varphi\| \leq C^{\wedge}]\!] = \mathbf{1}$ for some $C \in \mathbb{R}$. The mapping $\Phi \mapsto \varphi$ is an isometrical bijection between B-Hom(A_1, A_2) and $\mathrm{Hom}^B(\mathscr{A}_1, \mathscr{A}_2){\downarrow}^{\infty}$.*

$\lhd$ All claims with the exception of multiplicativity ensue from 8.3.3. The fact that φ and Φ are multiplicative may be justified in much the same way as in the proof of uniqueness in 8.4.2. $\rhd$

(2) *Assume that $\mathscr{A}_1$ and $\mathscr{A}_2$ are involutive Banach algebras inside $\mathbf{V}^{(B)}$, while $\Phi \in B\text{-}\mathrm{Hom}(A_1, A_2)$ and $\varphi \in \mathrm{Hom}^B(\mathscr{A}_1, \mathscr{A}_2)$ correspond to one another under the bijection of* (1)*. Then the equality $[\![\varphi$ is $*$-preserving$]\!] = \mathbf{1}$ holds if and only if Φ is $*$-preserving.*

$\lhd$ Appealing to A.9 (7) and 8.4.2 completes the proof. $\rhd$

8.4.4. (1) Theorem. *Assume that $\mathscr{A}$ is an AW^*-algebra inside $\mathbf{V}^{(B)}$ and A is the bounded descent of $\mathscr{A}$. Then A is also an AW^*-algebra and, moreover, $\mathfrak{P}_c(A)$ has an order-closed subalgebra isomorphic with B. Conversely, let A be an AW^*-algebra such that B is an order-closed subalgebra of the Boolean algebra $\mathfrak{P}_c(A)$. Then there is an AW^*-algebra $\mathscr{A}$ in $\mathbf{V}^{(B)}$ whose bounded descent is $*$-B-isomorphic with A. This algebra $\mathscr{A}$ is unique up to isomorphism inside $\mathbf{V}^{(B)}$.*

$\lhd$ By Theorems 7.5.5 and 8.4.2 (1) we only need to prove that the C^*-algebras A and $\mathscr{A}$ are Baer. The last claim is immediate on using 8.4.2 (2) and the rules for ascending and descending polars which are annihilators in our case (cf. A.18). $\rhd$

(2) Theorem. *If $\mathscr{A}$ is an AW^*-factor inside $\mathbf{V}^{(B)}$ then the bounded descent A of $\mathscr{A}$ is an AW^*-algebra whose Boolean algebra of central projections is isomorphic with B. Conversely, if A is an AW^*-algebra and $B := \mathfrak{P}_c(A)$ then there is an AW^*-factor $\mathscr{A}$ in $\mathbf{V}^{(B)}$ whose bounded descent is isomorphic with A. This factor is unique up to $*$-isomorphism inside $\mathbf{V}^{(B)}$.*

$\lhd$ Apply (1) and recall that the descent of the two-element Boolean algebra $\{\mathbf{0}, \mathbf{1}\}$ is isomorphic with B (cf. A.16 (1)). $\rhd$

We will now demonstrate that immersion in a Boolean-valued model preserves the type of an AW^*-algebra.

8.4.5. Theorem. *Suppose that $\mathscr{A}$ is an AW^*-algebra inside $\mathbf{V}^{(B)}$ and A is the bounded descent of $\mathscr{A}$. For every projection $\pi \in \mathfrak{P}(A)$ the following hold:*

(1) *π is abelian $\Leftrightarrow [\![\pi$ is abelian$]\!] = \mathbf{1}$;*

(2) *π is finite $\Leftrightarrow [\![\pi$ is finite$]\!] = \mathbf{1}$;*

(3) *π is purely infinite $\Leftrightarrow [\![\pi$ is purely infinite$]\!] = \mathbf{1}$.*

$\lhd$ The claim of (1) is obvious. Furthermore, note that for $\pi, \rho \in \mathfrak{P}(A)$ the formulas $\pi \sim \rho$, $\pi \leq \rho$ and $\pi \lesssim \rho$ may be rewritten as algebraic identities (cf. 7.5.6):

$$\pi \sim \rho \Leftrightarrow xx^* = \pi \wedge x^*x = \rho,$$
$$\pi \leq \rho \Leftrightarrow \pi\rho = \rho\pi = \pi,$$
$$\pi \lesssim \rho \Leftrightarrow \pi \sim \pi_0 \wedge \pi_0 \leq \rho.$$

Multiplication, involution and equality in A appear as descents of the corresponding objects in $\mathscr{A}$. Therefore,

$$\pi \sim \rho \Leftrightarrow [\![\pi \sim \rho]\!] = \mathbf{1},$$
$$\pi \leq \rho \Leftrightarrow [\![\pi \leq \rho]\!] = \mathbf{1},$$
$$\pi \lesssim \rho \Leftrightarrow [\![\pi \lesssim \rho]\!] = \mathbf{1}.$$

To prove (2), take into account the formula

$$[\![(\forall x \in \mathscr{A})\varphi(x) \to \psi(x)]\!] = \bigwedge\{[\![\psi(x)]\!] : x \in \mathscr{A}{\downarrow},\ [\![\varphi(x)]\!] = \mathbf{1}\},$$

and the equality $\mathfrak{P}(\mathscr{A}){\downarrow} = \mathfrak{P}(A)$. We may write down the chain of equivalences:

$$[\![\pi \text{ is finite}]\!] = \mathbf{1}$$
$$\Leftrightarrow [\![(\forall \rho \in \mathfrak{P}(\mathscr{A}))\, \pi \sim \rho \leq \pi \to \pi = \rho]\!] = \mathbf{1}$$
$$\Leftrightarrow (\forall \rho \in \mathfrak{P}(A))\, [\![\pi \sim \rho \leq \pi]\!] = \mathbf{1} \Rightarrow [\![\pi = \rho]\!] = \mathbf{1}$$
$$\Leftrightarrow (\forall \rho \in \mathfrak{P}(A))\, \pi \sim \rho \leq \pi \Rightarrow \pi = \rho.$$

We arrive at (3) similarly, thus completing the proof. $\rhd$

8.4.6. Theorem. *Suppose that algebras A and $\mathscr{A}$ are the same as in 8.4.5. Then the following hold:*

(1) *A is finite $\Leftrightarrow [\![\mathscr{A}$ is finite$]\!] = \mathbf{1}$;*

(2) *A has type I $\Leftrightarrow [\![\mathscr{A}$ has type I$]\!] = \mathbf{1}$;*

(3) *A has type II $\Leftrightarrow [\![\mathscr{A}$ has type II$]\!] = \mathbf{1}$;*

(4) *A has type III $\Leftrightarrow [\![\mathscr{A}$ has type III$]\!] = \mathbf{1}$.*

$\lhd$ All claims are immediate from 8.4.5 and definitions. $\rhd$

8.4.7. We call a partition of unity $(b_\gamma)_{\gamma\in\Gamma}$ in B the *B-dimension* of a Kaplansky–Hilbert module X provided that Γ is a nonempty set of cardinals, $b_\gamma \neq 0$ for all $\gamma \in \Gamma$, and $b_\gamma X$ is a strictly γ-homogeneous Kaplansky–Hilbert module for every $\gamma \in \Gamma$. In this event we write $B\text{-dim}(X) = (b_\gamma)_{\gamma\in\Gamma}$. Note that the members of the B-dimension of a Kaplansky–Hilbert module are pairwise distinct by the definition of strict homogeneity. We say that the B-dimension of X equals γ (in symbols, $B\text{-dim}(X) = \gamma$) if $\Gamma = \{\gamma\}$ and $b_\gamma = \mathbf{1}$. The equality $B\text{-dim}(X) = \gamma$ evidently means that X is strictly γ-homogeneous. We may define the multiplicity function $\varkappa$ of 7.4.6 in the case of an arbitrary Kaplansky–Hilbert module X by the formula $\varkappa(b) = \sup\{\varkappa(b') : b' \leq b,\ b' \in hb\}$, where the set hb comprises $b' \leq b$ such that $b'X$ is homogeneous. Clearly, if $B\text{-dim}(X) = (b_\gamma)_{\gamma\in\Gamma}$ then $\varkappa(b) = \sup\{\gamma \in \Gamma : b \wedge b_\gamma \neq \mathbf{0}\}$.

8.4.8. Theorem. *Suppose that $(b_\gamma)_{\gamma\in\Gamma}$ is a partition of unity in B, with $b_\gamma \neq \mathbf{0}$ $(\gamma \in \Gamma)$ and Γ a set of cardinals. Then $B\text{-dim}\, X = (b_\gamma)_{\gamma\in\Gamma}$ if and only if $[\![\dim(\mathscr{X}) = \text{mix}_{\gamma\in\Gamma}(b_\gamma\gamma^\wedge)]\!] = \mathbf{1}$.*

$\lhd$ As was noted above, we may identify $b_\gamma X$ with the bounded descent of the Hilbert space $b_\gamma\mathscr{X}$ inside $\mathbf{V}^{(B_\gamma)}$ where $B_\gamma := [\mathbf{0}, b_\gamma]$. By virtue of 8.3.11 strictly γ-homogeneity for $b_\gamma X$ amounts to the formula $b_\gamma = [\![\dim(b_\gamma\mathscr{X}) = \gamma^\wedge]\!]^{B_\gamma} \leq [\![\dim(\mathscr{X}) = \gamma^\wedge]\!]^B$. But then the equality $B\text{-dim}(X) = (b_\gamma)_{\gamma\in\Gamma}$ holds if and only if $b_\gamma \leq [\![\dim(\mathscr{X}) = \gamma^\wedge]\!]$ $(\gamma \in \Gamma)$, since $b_\gamma \leq [\![\mathscr{X} = b_\gamma\mathscr{X}]\!] = [\![\dim \mathscr{X} = \dim b_\gamma\mathscr{X}]\!]$. In turn, the last formulas imply that $[\![\dim(X) = \text{mix}_{\gamma\in\Gamma}(b_\gamma\gamma^\wedge)]\!] = \mathbf{1}$. This ends the proof. $\rhd$

8.4.9. Theorem. *A partition of unity $(b_\gamma)_{\gamma\in\Gamma}$ in a complete Boolean algebra B which consists of pairwise distinct elements serves as the B-dimension of some Kaplansky–Hilbert module if and only if Γ consists of cardinals and b_γ is γ-stable element for every $\gamma \in \Gamma$.*

$\lhd$ Put $\lambda := \text{mix}_{\gamma\in\Gamma}(b_\gamma\gamma^\wedge)$. Inside $\mathbf{V}^{(B)}$ we may find a Hilbert space $\mathscr{X}$, satisfying $[\![\dim(\mathscr{X}) = |\lambda|]\!] = \mathbf{1}$. By 8.4.8 $B\text{-dim}(X) = (b_\gamma)_{\gamma\in\Gamma}$ if and only if $[\![\,|\lambda| = \lambda\,]\!] = \mathbf{1}$. The last relation amounts to the simultaneous inequalities

$$b_\gamma \leq [\![\,|\gamma^\wedge| = \gamma^\wedge\,]\!] \quad (\gamma \in \Gamma).$$

The inequality $b_\gamma \leq [\![\,|\gamma^\wedge| = \gamma^\wedge\,]\!]$ for a nonzero b_γ means that $\mathbf{V}^{([\mathbf{0},b_\gamma])} \models \gamma^\wedge = |\gamma^\wedge|$. Consequently, it remains to demonstrate that the γ-stability of the Boolean algebra $B_0 = [\mathbf{0}, b]$ and the relation $\mathbf{V}^{(B_0)} \models \gamma^\wedge = |\gamma^\wedge|$ are valid or are not valid simultaneously. Note that

$$\begin{aligned} [\![\,\gamma^\wedge = |\gamma^\wedge|\,]\!] &= [\![\,(\forall \alpha \in \mathrm{On})\,(\gamma^\wedge \sim \alpha \to \gamma^\wedge \leq \alpha)\,]\!] \\ &= \bigwedge\{[\![\,\gamma^\wedge \sim \alpha^\wedge\,]\!] \Rightarrow [\![\,\gamma^\wedge \leq \alpha\,]\!] : \alpha \in \mathrm{On}\}. \end{aligned}$$

Clearly, $[\![\gamma^\wedge = |\gamma^\wedge|]\!] = \mathbb{1}$ if and only if $c := [\![\gamma^\wedge \sim \alpha^\wedge]\!] \le [\![\gamma^\wedge \le \alpha^\wedge]\!]$ for every ordinal α. If $c \ne \mathbb{0}$ then $\gamma \le \alpha$. Furthermore, the inequality $c \le [\![\gamma^\wedge \sim \alpha^\wedge]\!]$ means that $c(\gamma) \simeq c(\alpha)$. Thus, the equality $[\![\gamma^\wedge = |\gamma^\wedge|]\!] = \mathbb{1}$ amounts to the γ-stability of the Boolean algebra B_0. $\rhd$

8.4.10. We need an extra auxiliary fact. Denote by $\mathbb{P}$-lin(A) the set of all linear combinations of the members of A with coefficients in $\mathbb{P}$.

Suppose that X is a vector space over $\mathbb{F}$ and $\mathbb{P}$ is a subfield of $\mathbb{F}$. Then $X^\wedge$ is a vector space over the field $\mathbb{F}^\wedge$ and $(\mathbb{P}\text{-lin}(A))^\wedge = \mathbb{P}^\wedge\text{-lin}(A^\wedge)$ for every $A \subset X$.

$\lhd$ The first claim is evident, since the proposition "X is a vector space over $\mathbb{F}$" presents a bounded formula. By the same reason, $(\mathbb{P}\text{-lin}(A))^\wedge$ is a $\mathbb{P}^\wedge$-linear subspace in $X^\wedge$ which contains $A^\wedge$. Therefore, $\mathbb{P}^\wedge\text{-lin}(A^\wedge) \subset (\mathbb{P}\text{-lin}(A))^\wedge$. Conversely, suppose that an element x in X has the form $\sum_{k\in n} \alpha(k)\, u(k)$, where $n \in \mathbb{N}$, $\alpha : n \to \mathbb{P}$, and $u : n \to A$. Then $\alpha^\wedge : n^\wedge \to \mathbb{P}^\wedge$, $u^\wedge : n^\wedge \to A^\wedge$ and $x^\wedge = \sum_{k\in n^\wedge} \alpha^\wedge(k) u^\wedge(k)$. Consequently, $x^\wedge \in \mathbb{P}^\wedge\text{-lin}(A^\wedge)$, which proves the inclusion $(\mathbb{P}\text{-lin}(A))^\wedge \subset \mathbb{P}^\wedge\text{-lin}(A^\wedge)$, see A.8 (1). $\rhd$

8.4.11. *Theorem.* *Suppose that H is a Hilbert space and $\lambda = \dim(H)$. Suppose further that $\mathscr{H}$ is the completion of the metric space $H^\wedge$ inside $\mathbf{V}^{(B)}$. Then $[\![\mathscr{H}$ is a Hilbert space and $\dim(\mathscr{H}) = |\lambda^\wedge|]\!] = \mathbb{1}$.*

$\lhd$ By definition, $\mathscr{H}$ is a Banach space. If $b(\cdot\,,\,\cdot)$ is the inner product on H then $b^\wedge : H^\wedge \times H^\wedge \to \mathbb{C}^\wedge$ is a uniformly continuous function admitting a unique continuation on the whole of $\mathscr{H} \times \mathscr{H}$. We let $(\cdot\,|\,\cdot)$ stand for this continuation. Clearly, $(\cdot\,|\,\cdot)$ is an inner product on $\mathscr{H}$ and

$$\mathbf{V}^{(B)} \models \|x\| = \sqrt{(x\,|\,x)} \quad (x \in \mathscr{H}).$$

Hence, $[\![\mathscr{H}$ is a Hilbert space$]\!] = \mathbb{1}$. Suppose that $\mathscr{E}$ is a Hilbert basis for H. Show that $[\![\mathscr{E}^\wedge$ is a basis for $\mathscr{H}]\!] = \mathbb{1}$. Orthonormality for $\mathscr{E}^\wedge$ ensues from the definition of inner product on $\mathscr{H}$. Indeed, this is seen from the following calculations (see A.8 (1, 5)):

$$[\![(\forall x \in \mathscr{E}^\wedge)\ (x\,|\,x) = 1]\!] = \bigwedge_{x\in\mathscr{E}} [\![(x^\wedge|x^\wedge) = 1]\!] = \bigwedge_{x\in\mathscr{E}} [\![b(x,x)^\wedge = 1^\wedge]\!] = \mathbb{1};$$

$$[\![(\forall x, y \in \mathscr{E}^\wedge)\ (x \ne y \to (x\,|\,y) = 0)]\!] = \bigwedge_{x,y\in\mathscr{E}} [\![x^\wedge \ne y^\wedge]\!]$$

$$\Rightarrow [\![(x^\wedge|y^\wedge) = 0]\!] = \bigwedge_{\substack{x,y\in\mathscr{E}\\ x\ne y}} [\![b^\wedge(x^\wedge, y^\wedge) = 0]\!] = \bigwedge_{\substack{x,y\in\mathscr{E}\\ x\ne y}} [\![b(x,y)^\wedge = 0^\wedge]\!] = \mathbb{1}.$$

Since $H^\wedge$ is dense in $\mathscr{H}$ and $\mathbb{C}^\wedge\text{-lin}(\mathscr{E}^\wedge) \subset \mathscr{C}\text{-lin}(\mathscr{E}^\wedge)$; therefore, we are left with showing only that $\mathbb{C}^\wedge\text{-lin}(\mathscr{E}^\wedge)$ is dense in $H^\wedge$. Take $x \in H$ and $\varepsilon > 0$. Since $\mathscr{E}$ is a basis for H, there is $x_\varepsilon \in \mathbb{C}\text{-lin}(\mathscr{E})$ satisfying $\|x - x_\varepsilon\| < \varepsilon$. Hence, $[\![\, \|x^\wedge - x_\varepsilon{}^\wedge\| < \varepsilon^\wedge \,]\!] = \mathbf{1}$ and $[\![\, x_\varepsilon^\wedge \in (\mathscr{C}\text{-lin}(\mathscr{E}))^\wedge \,]\!] = \mathbf{1}$. Recalling 8.4.10, we conclude that the formula

$$(\forall x \in H)\ (\forall 0 < \varepsilon \in \mathbb{R}^\wedge)\ (\exists x_\varepsilon \in \mathbb{C}^\wedge\text{-}\operatorname{lin}(\mathscr{E}^\wedge))\ (\|x - x_\varepsilon\| < \varepsilon)$$

is valid inside $\mathbf{V}^{(B)}$; i.e., $[\![\, \mathbb{C}^\wedge\text{-lin}(\mathscr{E}^\wedge)$ is dense in $H^\wedge \,]\!] = \mathbf{1}$. It remains to note that if φ is a bijection between the set $\mathscr{E}$ and the cardinal λ then $\varphi^\wedge$ is a bijection between $\mathscr{E}^\wedge$ and $\lambda^\wedge$ inside $\mathbf{V}^{(B)}$. This ends the proof. $\rhd$

We list a few corollaries.

8.4.12. (1) *In the hypotheses of Theorem* 8.4.11 *the bounded descent of a Hilbert space $\mathscr{H}$ inside $\mathbf{V}^{(B)}$ is unitarily equivalent to the Kaplansky–Hilbert module $C_\#(Q, H)$, where Q is the Stone space of B.*

$\lhd$ This ensues for 7.4.8 (1) and 8.3.3 (1). $\rhd$

(2) *Let M be a nonempty set. The bounded descent of the Hilbert space $l_2(M^\wedge)$ inside $\mathbf{V}^{(B)}$ is unitarily equivalent to the Kaplansky–Hilbert module $C_\#(Q, l_2(M))$, where Q is the Stone space of B.*

$\lhd$ Assign $H = l_2(M)$ in Theorem 8.4.11 and recall the formula $[\![\, \dim(\mathscr{H}) = |M^\wedge| \,]\!] = \mathbf{1}$. We now see that $[\![\, \mathscr{H}$ and $l_2(M^\wedge)$ are unitarily equivalent $]\!] = \mathbf{1}$. This completes the proof. $\rhd$

(3) *Suppose that $\lambda = \dim(H)$ is an infinite cardinal. The Kaplansky–Hilbert module $C_\#(Q, H)$ is strictly λ-homogeneous if and only if Q is a λ-stable compact space.*

$\lhd$ Apply 8.3.11, 8.4.9, and 8.4.11 to complete the proof. $\rhd$

(4) *Granted an arbitrary infinitely dimensional Hilbert spaces H_1 and H_2, we may find an extremal compact space Q so that the Kaplansky–Hilbert modules $C_\#(Q,\ H_1)$ and $C_\#(Q, H_2)$ are unitarily equivalent.*

$\lhd$ Put $\lambda_k := \dim(H_k)$ $(k = 1, 2)$. There exists a complete Boolean algebra B such that the ordinals $\lambda_1^\wedge$ and $\lambda_2^\wedge$ have the same cardinality inside $\mathbf{V}^{(B)}$ (see A.19). The claim follows from 8.4.11 and (1). $\rhd$

(5) *Let $k = 1, 2$. Suppose that H_k is a Hilbert space and $\lambda_k := \dim(H_k) \geq \omega$ Suppose further that the Kaplansky–Hilbert module $C_\#(Q,\ H_k)$ is strictly λ_k-homogeneous. If the modules $C_\#(Q, H_1)$ and $C_\#(Q, H_2)$ are unitarily equivalent then the Hilbert spaces H_1 and H_2 are unitarily equivalent too.*

$\lhd$ From 8.3.11, 8.4.11 and (1) we see that $[\![\, \lambda_1^\wedge = |\lambda_1^\wedge| = |\lambda_2^\wedge| = \lambda_2^\wedge \,]\!] = \mathbf{1}$. Therefore $\lambda_1 = \lambda_2$. $\rhd$

A Kaplansky–Hilbert module X is *B-separable,* if there is a sequence $(x_n) \subset X$ such that the Kaplansky–Hilbert submodule, generated by the set $\{bx_n : n \in \mathbb{N},\ b \in B\}$, coincides with X. Obviously, if H is a B-separable Hilbert space then the Kaplansky–Hilbert module $C_\#(Q, H)$ is B-separable.

(6) *To every infinitely dimensional Hilbert space H, there exists an extremal compact space Q such that the Kaplansky–Hilbert module $C_{\#}(Q, H)$ is B-separable, with B standing for the Boolean algebra of the characteristic functions of clopen subsets of Q.*

$\lhd$ Put $H_1 := l_2(\omega)$ and $H_2 := H$ in (4) and use the separability of $l_2(\omega)$ to complete the proof. $\rhd$

8.4.13. Theorem. *To each Kaplansky–Hilbert module X there is a family of nonempty extremal compact spaces $(Q_\gamma)_{\gamma\in\Gamma}$, with Γ a set of cardinals, such that Q_γ is γ-stable for all $\gamma \in \Gamma$ and the following unitary equivalence holds:*

$$X \simeq \sum_{\gamma\in\Gamma}^{\oplus} C_{\#}\big(Q_\gamma, l_2(\gamma)\big).$$

If some family $(P_\delta)_{\delta\in\Delta}$ of extremal compact spaces satisfies the above conditions then $\Gamma = \Delta$, and P_γ is homeomorphic with Q_γ for all $\gamma \in \Gamma$.

$\lhd$ By Theorem 8.3.8 we may assume that X is the bounded descent of a Hilbert space $\mathscr{X}$ inside $\mathbf{V}^{(B)}$. Suppose further that B-$\dim(X) = (b_\gamma)_{\gamma\in\Gamma}$ and Q_γ is the clopen subset of the Stone space of B which corresponds to $b_\gamma \in B$; i.e., the support of b. We make use of the fact that X is the direct sum of the spaces of the form $b_\gamma X$, with $b_\gamma X$ unitarily equivalent to the bounded descent of the space $b_\gamma \mathscr{X}$ inside $\mathbf{V}^{(B_\gamma)}$, where $B_\gamma = [\mathbf{0}, b_\gamma]$. By 8.4.9, note that $b_\gamma \le [\![\dim(b_\gamma \mathscr{X}) = \gamma^\wedge]\!]$. Consequently, given a nonzero b_γ, conclude that $\mathbf{V}^{(B_\gamma)} \models$ "$b_\gamma \mathscr{X}$ is a Hilbert space of dimension $\gamma^\wedge$." Appealing to the Transfer Principle, we infer that $\mathbf{V}^{(B_\gamma)} \models$ "$b_\gamma \mathscr{X}$ is unitarily equivalent to $l_2(\gamma^\wedge)$." By virtue of 8.4.12 (2), the bounded descent of $l_2(\gamma^\wedge)$ in the model $\mathbf{V}^{(B_\gamma)}$ is unitarily equivalent to the Kaplansky–Hilbert module $C_{\#}(Q_\gamma, l_2(\gamma))$. Suppose that $u_\gamma \in \mathbf{V}^{(B_\gamma)}$ is a unitary isomorphism from $b_\gamma \mathscr{X}$ onto $l_2(\gamma^\wedge)$ inside $\mathbf{V}^{(B_\gamma)}$, and U_γ is the bounded descent of u_γ. Then U_γ establishes unitary equivalence between the Kaplansky–Hilbert modules $b_\gamma X$ and $C_{\#}(Q_\gamma, l_2(\gamma))$. By definition, the element $b_\gamma \in B$, together with the compact space Q_γ, is γ-stable.

Assume now that some family of extremal compact spaces $(P_\delta)_{\delta\in\Delta}$ obeys the same conditions as $(Q_\gamma)_{\gamma\in\Gamma}$. Then P_δ is homeomorphic with some clopen subset P'_δ of the Stone space of B. Moreover, P'_δ is δ-stable. If $P_{\delta\gamma} := P'_\delta \cap Q_\gamma$ and $b_{\gamma\delta}$ is the corresponding element of B then the Kaplansky–Hilbert modules $C_{\#}(P_{\delta\gamma}, l_2(\delta))$ and $C_{\#}(P_{\delta\gamma}, l_2(\gamma))$ are unitarily equivalent to the same member $b_{\delta\gamma} X$. Furthermore, the compact space $P_{\delta\gamma}$ must be δ- and γ-stable simultaneously.

According to 8.4.12 (3) and 8.4.12 (5), $P_{\delta\gamma} = \varnothing$ or $l_2(\delta) \sim l_2(\gamma)$, implying $\delta = \gamma$. Therefore, $P'_\gamma = Q_\gamma$ $(\gamma \in \Gamma)$. $\rhd$

8.4.14. We now are able to give a Boolean-valued proof of Theorems 7.5.11 and 7.5.12 including the uniqueness.

◁ By Theorem 8.4.4 (2) we may assume, that A is the bounded descent of an AW^*-factor $\mathscr{A}$ in $\mathbf{V}^{(B)}$. In this event $\mathscr{A}$ has type I (see 8.4.6), and so $\mathscr{A} \simeq B(\mathscr{X})$, where $\mathscr{X}$ is a Hilbert space inside $\mathbf{V}^{(B)}$. Hence, we see that A is $*$-isomorphic with $\mathrm{End}(X)$, where X stands for the bounded descent of $\mathscr{X}$. This proves 7.5.11.

Suppose that B-dim$(X) = (b_\gamma)_{\gamma\in\Gamma}$, and Q_γ is the clopen subset of the Stone space Q of the Boolean algebra B which corresponds to $b_\gamma \in B$. By virtue of 8.4.9, Q_γ is a γ-stable compact space. So (1) holds. By Theorem 8.4.13 we note the following unitary equivalence $X \simeq \sum^{\oplus}_{\gamma\in\Gamma} C_\#(Q, l_2(\gamma))$. But then we see the next $*$-isomorphism of AW^*-algebras:

$$\mathrm{End}(X) \simeq \sum_{\gamma\in\Gamma}^{\oplus} \mathrm{End}(C_\#(Q_\gamma, l_2(\gamma))).$$

Appealing to Theorem 7.5.10, we arrive at the sought condition (2).

Uniqueness ensues from 8.4.13. ▷

8.4.15. *The following are valid:*

(1) *Every type I AW^*-algebra splits into the direct sum of strictly homogeneous bands. This decomposition is unique up to $*$-isomorphism;*

(2) *Two type I AW^*-algebras are $*$-isomorphic to one another if and only if they have the congruent multiplicity functions or, which is the same, the congruent strict decomposition series.*

◁ This assertion ensues from (1) on observing that in representation 7.5.11 the dimension of A is congruent to the partition of unity $(\chi_\gamma)_{\gamma\in\Gamma}$, with χ_γ the characteristic function of the set Q_γ in the disjoint sum Q of the family (Q_γ). ▷

(3) *Suppose that Γ is a set of cardinals and (b_γ) is a partition of unity in B which consists of nonzero pairwise distinct elements. Then $(b_\gamma)_{\gamma\in\Gamma}$ is a strict decomposition series of some AW^*-algebra if and only if b_γ is γ-stable for all $\gamma \in \Gamma$.*

◁ This ensues from 8.4.9 and 7.5.10. ▷

8.5. Cyclically Compact Operators

The Boolean-valued interpretation of compactness gives rise to the new notions of *cyclically compact sets* and *operators* which deserves an independent study. A part of the corresponding theory is presented in this section.

8.5.1. Let B be a complete Boolean algebra and let A be a nonempty set. Recall that $B(\mathrm{A})$ denotes the set of all partitions of unity in B with the fixed index set A, see 2.2.2. The sets $B(\mathrm{A})$ and $\mathrm{A}^{\wedge}{\downarrow}$ are bijective, so that they are frequently identified.

Consider a normed B-space X and a net $(x_\alpha)_{\alpha\in\mathrm{A}}$ in it. For every $\nu \in B(\mathrm{A})$ put $x_\nu := \mathrm{mix}_{\alpha\in\mathrm{A}}(\nu(\alpha)x_\alpha)$. If all mixings exist then we come to a new net $(x_\nu)_{\nu\in B(\mathrm{A})}$ in X. Every subnet of the net $(x_\nu)_{\nu\in B(\mathrm{A})}$ is called a *cyclical subnet* of the original net $(x_\alpha)_{\alpha\in\mathrm{A}}$. If $s : \mathrm{A} \to X$ and $\varkappa : \mathrm{A}' \to B(\mathrm{A})$ then the mapping $s \bullet \varkappa : \mathrm{A}' \to X$ is defined by $s \bullet \varkappa(\alpha) := x_\nu$ where $\nu = \varkappa(\alpha)$. A *cyclical subsequence* of a sequence $(x_k)_{k\in\mathbb{N}} \subset X$ is a sequence of the form $(x_{\nu_k})_{k\in\mathbb{N}}$ where $(\nu_k)_{k\in\mathbb{N}}$ is a sequence in $B(\mathbb{N})$ with $\nu_k \ll \nu_{k+1}$ for all $k \in \mathbb{N}$.

According to 8.3.5 there is no loss of generality in assuming that X is a decomposable subspace of the Banach–Kantorovich space $\mathscr{X}{\downarrow}$, where $\mathscr{X}$ is a Banach space inside $\mathbf{V}^{(B)}$ and every projection $b \in B$ coincides with the restriction of $\chi(b)$ onto X, see 8.3.1. More precisely, we will assume that X is the bounded descent of $\mathscr{X}$, i.e., $X = \{x \in \mathscr{X}{\downarrow} : |x| \in \Lambda\}$, where Λ is the Stone algebra $\mathscr{S}(B)$ identified with the bounded part of the complex algebra $\mathscr{C}{\downarrow}$. In this event a subset $C \subset X$ is mix-complete if and only if $C = C{\uparrow}{\downarrow}$.

(1) *Given a sequence $\sigma : \mathbb{N}^\wedge \to C{\uparrow}$ and $\varkappa : \mathbb{N}^\wedge \to \mathbb{N}^\wedge$, the composite $\sigma{\downarrow} \circ \varkappa{\downarrow}$ is a cyclical subsequence of the sequence $\sigma{\downarrow} : \mathbb{N} \to C$ if and only if $[\![\sigma \circ \varkappa$ is a subsequence of $\sigma]\!] = \mathbb{1}$. Given a sequence $s : \mathbb{N} \to C$ and $\varkappa : \mathbb{N} \to B(\mathbb{N})$, the composite $s{\uparrow} \circ \varkappa^\wedge$ is a subsequence of the sequence $\sigma{\uparrow} : \mathbb{N}^\wedge \to C{\uparrow}$ inside $\mathbf{V}^{(B)}$ if and only if $s \bullet \varkappa$ is a cyclical subsequence of the sequence s.*

A subset $C \in X$ is said to be *cyclically compact* if C is mix-complete (see 7.3.1) and every sequence in C has a cyclic subsequence that converges (in norm) to some element of C. A subset in X is called *relatively cyclically compact* if it is contained in a cyclically compact set.

(2) *A set $C \subset X$ is cyclically compact (relatively cyclically compact) if and only if $C{\uparrow}$ is compact (relatively compact) in $\mathscr{X}$.*

$\lhd$ It suffices to prove the claim about cyclical compactness. In view of Theorem 8.3.2 we may assume that $X = \mathscr{X}{\downarrow}$. Suppose that $[\![C{\uparrow}$ is compact $]\!] = \mathbb{1}$. Take an arbitrary sequence $s : \mathbb{N} \to C$. Then $[\![s{\uparrow} : \mathbb{N}^\wedge \to C{\uparrow}$ is a sequence in $C{\uparrow}]\!] = \mathbb{1}$. By assumption $C{\uparrow}$ is compact inside $\mathbf{V}^{(B)}$, so that there exist $\rho, x \in \mathbf{V}^{(B)}$ with $[\![\rho$ is a subsequence of $s{\uparrow}]\!] = [\![x \in C{\uparrow}]\!] = [\![\lim(\rho) = x]\!] = \mathbb{1}$. Since C is mix-complete, we obtain that $\rho{\downarrow}$ is a cyclical subsequence of s and lim $(\rho{\downarrow}) = x \in C$. Conversely, suppose that C is a cyclically compact set. Take a sequence $\sigma : \mathbb{N}^\wedge \to C{\uparrow}$ in C. By assumption the sequence $\sigma{\downarrow} : \mathbb{N} \to C$ has a cyclic subsequence $\rho : B(\mathbb{N}) \to C$ converging to some $x \in C$. It remains to observe that $[\![\rho{\uparrow}$ is a subsequence of the sequence $\sigma]\!] = \mathbb{1}$ and $[\![\lim(\rho{\uparrow}) = x]\!] = \mathbb{1}$. $\rhd$

8.5.2. Theorem. *A mix-complete set C in a Banach B-space X is relatively cyclically compact if and only if for every $\varepsilon > 0$ there exist a countable partition of unity (π_n) in the Boolean algebra $\mathfrak{B}(X)$ and a sequence (θ_n) of finite subsets $\theta_n \subset C$ such that the set $\pi_n(\mathrm{mix}(\theta_n))$ is an ε-net for $\pi_n(C)$ for all $n \in \mathbb{N}$. The* *last means that if* $\theta_n := \{x_{n,1}, \dots, x_{n,l(n)}\}$ *then for every* $x \in \pi_n(C)$ *there exists a partition of unity* $\{\rho_{n,1}, \dots, \rho_{n,l(n)}\}$ *in* $\mathfrak{B}(X)$ *with*

$$\left\| x - \sum_{k=1}^{l(n)} \pi_n \rho_{n,k} x_{n,k} \right\| \leq \varepsilon.$$

$\triangleleft$ According to 8.3.5 we may assume that $X := \mathscr{X}{\downarrow}$ for some Banach space $\mathscr{X}$ inside $\mathbf{V}^{(B)}$. By 8.5.1 (2) a set $C \subset X$ is relatively cyclically compact if and only if $[\![C{\uparrow}$ is relatively compact $]\!] = \mathbf{1}$. By applying the Hausdorff Criterion to $C{\uparrow}$ inside $\mathbf{V}^{(B)}$, we obtain that relative cyclical compactness of $C{\uparrow}$ is equivalent to $[\![C{\uparrow}$ is totally bounded $]\!] = \mathbf{1}$ or, what amounts to the same, the following formula is valid inside $\mathbf{V}^{(B)}$:

$$(\forall 0 < \varepsilon \in \mathbb{R}^{\wedge})\,(\exists n \in \mathbb{N}^{\wedge})\,(\exists f : n \to \mathscr{X})\,(\forall x \in C{\uparrow})\,(\exists k \in n)(\|x - f(k)\| \leq \varepsilon).$$

Writing out Boolean truth values for the quantifiers, we see that the last claim can be stated in the following equivalent form: for every $0 < \varepsilon \in \mathbb{R}$ there exist a countable partition of unity (b_n) in B and a sequence (f_n) of elements of $\mathbf{V}^{(B)}$ such that $[\![f_n : n^{\wedge} \to \mathscr{X}]\!] \geq b_n$ and

$$[\![(\forall x \in C{\uparrow})\,(\exists k \in n^{\wedge})(\|x - f_n(k)\| \leq \varepsilon^{\wedge})]\!] \geq b_n.$$

Substitute f_n for $\operatorname{mix}(b_n f_n, b_n^* g_n)$, where g_n is an element of $\mathbf{V}^{(B)}$ with $[\![g_n : n^{\wedge} \to \mathscr{X}]\!] = \mathbf{1}$. Then f_n meets the above properties and obeys the additional requirement $[\![f_n : n^{\wedge} \to \mathscr{X}]\!] = \mathbf{1}$. Denote $h_n := f_n{\downarrow}$. So, the above implies that for every $x \in C$ holds

$$\bigvee \{ [\![\|x - h_n(k)\| \leq \varepsilon^{\wedge}]\!] : k \in n \} \geq b_n.$$

Let $\chi : B \to \mathfrak{B}(X)$ be the isomorphism from 8.3.1 and put $\pi_k := \chi(b_k)$. If $b_{n,k} := [\![\|x - h_n(k)\| \leq \varepsilon^{\wedge}]\!]$ and $x' := \sum_{k=0}^{n-1} \chi(b_{n,k}) h_n(k)$ then $[\![\|x' - x\| \leq \varepsilon^{\wedge}]\!] = \mathbf{1}$, or equivalently $|\pi_n(x - x')| \leq \varepsilon \mathbf{1}$. Thus, putting $\theta_n := \{h_n(0), \dots, h_n(n-1)\}$, we obtain the desired sequence θ_n of finite subsets of C. $\triangleright$

8.5.3. Let X be a B-normed space and let $X^{\#}$ be its B-dual, see 7.3.1. For every $f \in X^{\#}$ define a seminorm p_f on X by $p_f : x \mapsto \|f(x)\|_{\infty}$ $(x \in X)$. Denote by $\sigma_{\infty}(X, X^{\#})$ the topology in X generated by the family of seminorms $\{p_f : f \in X^{\#}\}$.

A mix-complete convex set $C \subset X$ *is cyclically* $\sigma_{\infty}(X, X^{\#})$*-compact if and only if* $C{\uparrow}$ *is* $\sigma(\mathscr{X}, \mathscr{X}^*)$*-compact inside* $\mathbf{V}^{(B)}$.

$\triangleleft$ The algebraic part of the claim is easy. Let the formula $\psi(\mathscr{A}, u)$ formalize the sentence: u belongs to the weak closure of $\mathscr{A}$. Then the formula can be written as

$$(\forall n \in \mathbb{N})\,(\forall \theta \in \mathscr{P}_{\mathrm{fin}}(\mathscr{X}))\,(\exists v \in \mathscr{A})\,(\forall y \in \theta)\,|(x \,|\, y)| \leq n^{-1},$$

where ω is the set of naturals, $(\cdot\,|\,\cdot)$ is the inner product in $\mathscr{X}$, and $\mathscr{P}_{\text{fin}}(\mathscr{X})$ is the set of all finite subsets of X. Suppose that $[\![\psi(\mathscr{A},u)]\!] = \mathbf{1}$. Using the Maximum Principle and the relation A.9 (6), we may calculate Boolean truth values and arrive at the following assertion: For any $n \in \omega$ and any finite collections $\theta := \{y_1, \dots, y_m\}$ in $X^\#$, there exists $v \in \mathscr{A}{\downarrow}$ such that

$$[\![(\forall y \in \theta^\wedge)\,|(u - v\,|\,y)| \leq 1/n^\wedge]\!] = \mathbf{1}.$$

Moreover, we may choose v so that the extra condition $[\![\|v\| \leq |u|]\!] = \mathbf{1}$ holds. Therefore,

$$|v| \leq |u|, \quad |\langle (u - v)\,|\,y_l\rangle| < n^{-1}\mathbf{1} \quad (k := 1, \dots, n;\ l := 1, \dots, m).$$

There exists a fixed partition of unity $(e_\xi)_{\xi \in \Xi} \subset B$ which depends only on u and is such that $e_\xi|u| \in \Lambda$ for all ξ. From here it is seen that $e_\xi u \in A$ and $e_\xi v \in A$. Moreover,

$$\|\langle e_\xi(u - v)\,|\,y_l\rangle\|_\infty < n^{-1} \quad (k := 1, \dots, n;\ l := 1, \dots, m).$$

Repeating the above argument in the opposite direction, we come to the following conclusion: The formula $\psi(\mathscr{A},u)$ is true inside $\mathbf{V}^{(B)}$ if and only if there exist a partition of unity $(e_\xi)_{\xi\in\Xi}$ in B and a family $(u_\xi)_{\xi\in\Xi}$ such that u_ξ belongs to the σ_∞-closure of A and $u = \operatorname{mix}(e_\xi u_\xi)$.

Now, assume that A is σ_∞-closed and the formula $\psi(\mathscr{A},u)$ is true inside $\mathbf{V}^{(B)}$. Then u_ξ is contained in A by assumption and $[\![u_\xi \in \mathscr{A}]\!] = \mathbf{1}$. Hence $e_\xi \leq [\![u \in \mathscr{A}]\!]$ for all ξ, i.e., $[\![u \in \mathscr{A}]\!] = \mathbf{1}$. Therefore,

$$\mathbf{V}^{(B)} \models (\forall u \in \mathscr{L}(\mathscr{X}))\psi(\mathscr{A},u) \to u \in \mathscr{A}.$$

Conversely, assume $\mathscr{A}$ to be weakly closed. If u belongs to the σ_∞-closure of A, then u is contained in the weak closure of $\mathscr{A}$. ▷

8.5.4. Consider $X^{\#\#} := (X^\#)^\# := \mathscr{L}_B(X^\#, \Lambda)$, the *second B-dual* of X. Given $x \in X$ and $f \in X^\#$, put $x^{\#\#} := \iota(x)$ where $\iota(x) : f \mapsto f(x)$. Undoubtedly, $\iota(x) \in L(X^\#, \Lambda)$. In addition,

$$\begin{aligned} |x^{\#\#}| = |\iota(x)| &= \sup\{|\iota(x)(f)| : |f| \leq \mathbf{1}\} \\ = \sup\{|f(x)| : (\forall x \in X)\,|f(x)| \leq |x|\} &= \sup\{|f(x)| : f \in \partial(|\cdot|)\} = |x|. \end{aligned}$$

The last equality follows from 3.1.9 (1). Thus, $\iota(x) \in X^{\#\#}$ for every $x \in X$. It is evident that the operator $\iota : X \to X^{\#\#}$, defined as $\iota : x \mapsto \iota(x)$, is linear and isometric. The operator ι is referred to as the *canonical embedding* of X into the second B-dual. As in the case of Banach spaces, it is convenient to treat x and $x^{\#\#} := \iota x$ as the same element and consider X as a subspace of $X^{\#\#}$. A B-normed space X is said to be *B-reflexive* if X and $X^{\#\#}$ coincide under the indicated embedding ι.

Theorem. *A normed B-space is B-reflexive if and only if its unit ball is cyclically $\sigma_\infty(X, X^\#)$-compact.*

$\lhd$ The Kakutani Criterion claims that a normed space is reflexive if and only if its unit ball is weakly compact. Hence, the result follows from 8.5.3. $\rhd$

8.5.5. Let X and Y be normed B-spaces. An operator $T \in \mathscr{L}_B(X, Y)$ is called *cyclically compact* (in symbols, $T \in \mathscr{K}_B(X, Y)$) if the image $T(C)$ of any bounded subset $C \subset X$ is relatively cyclically compact in Y. It is easy to see that $\mathscr{K}_B(X, Y)$ is a decomposable subspace of the Banach–Kantorovich space $\mathscr{L}_B(X, Y)$.

Let $\mathscr{X}$ and $\mathscr{Y}$ be Boolean-valued representations of X and Y. Recall that the immersion mapping $T \mapsto T^\sim$ of the operators is a linear isometric embedding of the lattice-normed spaces $\mathscr{L}_B(X, Y)$ into $\mathscr{L}^B(\mathscr{X}, \mathscr{Y}){\downarrow}$, see 8.3.6. Assume that Y is a B-cyclic space.

(1) *A bounded operator T from X into Y is cyclically compact if and only if $[\![T^\sim$ is a compact operator from $\mathscr{X}$ into $\mathscr{Y}]\!] = \mathbf{1}$.*

$\lhd$ Observe that C is bounded in X if and only if $[\![C^\sim$ is bounded in $\mathscr{X}]\!] = \mathbf{1}$. Moreover, according to A.12 (4),

$$\mathbf{V}^{(B)} \models T(C)^\sim = T^\sim(C^\sim).$$

It remains to apply 8.5.1 (2). $\rhd$

(2) *$\mathscr{K}_B(X, Y)$ is a bo-closed decomposable subspace in $\mathscr{L}_B(X, Y)$.*

$\lhd$ Let $\mathscr{X}$ and $\mathscr{Y}$ in $\mathbf{V}^{(B)}$ be the same as above and let $\mathscr{K}^{(B)}(\mathscr{X}, \mathscr{Y})$ be the space of compact operators from $\mathscr{X}$ into $\mathscr{Y}$ inside $\mathbf{V}^{(B)}$. As was shown in 8.3.3 the mapping $T \to T^\sim$ is an isometric embedding of $\mathscr{L}_B(X, Y)$ into $\mathscr{L}^{(B)}(\mathscr{X}, \mathscr{Y}){\downarrow}$. It follows from (1) that this embedding maps the subspace $\mathscr{K}_B(X, Y)$ onto the bounded part of the space $\mathscr{K}^{(B)}(\mathscr{X}, \mathscr{Y}){\downarrow}$. By the ZFC-theorem claiming the closure of the subspace of compact operators, we have $[\![\mathscr{K}^{(B)}(\mathscr{X}, \mathscr{Y})$ is a closed subspace in $\mathscr{L}^{(B)}(\mathscr{X}, \mathscr{Y})]\!] = \mathbf{1}$. From this we deduce that $\mathscr{K}^{(B)}(\mathscr{X}, \mathscr{Y}){\downarrow}$ is *bo*-closed and decomposable in $\mathscr{L}^{(B)}(\mathscr{X}, \mathscr{Y}){\downarrow}$. Thus, the bounded part of $\mathscr{K}^{(B)}(\mathscr{X}, \mathscr{Y}){\downarrow}$ is also *bo*-closed and decomposable. $\rhd$

(3) *Let $T \in \mathscr{L}_B(X, Y)$ and $S \in \mathscr{L}_B(Y, Z)$. If either T or S is cyclically compact then $S \circ T$ is also cyclically compact.*

$\lhd$ We need only to immerse the composite $S \circ T$ inside $\mathbf{V}^{(B)}$ and, taking into account (1) and A.12 (4), apply therein the ZFC-theorem about compactness of the composite of a bounded operator and a compact operator. The subsequent descent leads immediately to the desired result. $\rhd$

(4) *A bounded operator T is cyclically compact if and only if its adjoint T^* is cyclically compact.*

◁ Apply the above procedure, immersion into a Boolean-valued model and the subsequent descent. Observe that the operator $(T^*)^\sim$ is the adjoint of $T^\sim$ inside $\mathbf{V}^{(B)}$ and use the corresponding ZFC-theorem on compactness of the adjoint of a compact operator. ▷

Now we consider general form of a cyclically compact operators in Kaplansky–Hilbert modules.

8.5.6. Theorem. *Let T in $\mathscr{K}_B(X,Y)$ be a cyclically compact operator from a Kaplansky–Hilbert module X to a Kaplansky–Hilbert module Y. There are orthonormal families $(e_k)_{k\in\mathbb{N}}$ in X, $(f_k)_{k\in\mathbb{N}}$ in Y, and a family $(\mu_k)_{k\in\mathbb{N}}$ in Λ such that the following hold:*

(1) $\mu_{k+1} \le \mu_k$ $(k \in \mathbb{N})$ *and* $o\text{-}\lim_{k\to\infty} \mu_k = 0$;

(2) *there exists a projection π_∞ in Λ such that $\pi_\infty\mu_k$ is a weak order-unity in $\pi_\infty\Lambda$ for all $k \in \mathbb{N}$;*

(3) *there exists a partition $(\pi_k)_{k=0}^\infty$ of the projection $\pi_\infty^\perp$ such that $\pi_0\mu_1 = 0$, $\pi_k \le [\mu_k]$, and $\pi_k\mu_{k+1} = 0$, $k \in \mathbb{N}$;*

(4) *the representation is valid*

$$T = \pi_\infty bo\text{-}\sum_{k=1}^\infty \mu_k e_k^\# \otimes f_k + bo\text{-}\sum_{n=1}^\infty \pi_n \sum_{k=1}^n \mu_k e_k^\# \otimes f_k.$$

◁ By virtue of 8.3.8 we may assume that X and Y coincide with the bounded descents of Hilbert spaces $\mathscr{X}$ and $\mathscr{Y}$, respectively. The operator $\mathscr{T} := T\!\uparrow : \mathscr{X} \to \mathscr{Y}$ is compact and we may apply inside $\mathbf{V}^{(B)}$ the ZFC-theorem on the general form of a compact operator in Hilbert space. Working inside $\mathbf{V}^{(B)}$ we may choose orthonormal sequences $(e_k)_{k\in\mathbb{N}}$ in $\mathscr{X}$, $(f_k)_{k\in\mathbb{N}}$ in $\mathscr{Y}$, and a decreasing numeric sequence $(\mu_k)_{k\in\mathbb{N}}$ in $\mathscr{R}_+ \setminus 0$ such that $\lim \mu_k = 0$ and the presentation holds:

$$\mathscr{T} = \sum_{k=1}^\infty \mu_k e_k^* \otimes f_k.$$

Moreover, either $(\forall k \in \mathbb{N})\, \mu_k > 0$ or $(\exists k \in \mathbb{N})\, \mu_k = 0$. Since $[\![\mu_1 \le \|\mathscr{T}\|]\!] = \mathbf{1}$ we have $\mu_1 \le |T| \in \Lambda$, whence $(\mu_k) \subset \Lambda$. Let $\pi_\infty := [\![\mathscr{T}$ be an infinite-rank compact operator from a Hilbert space $\mathscr{X}$ to a Hilbert space $\mathscr{Y}]\!] = \mathbf{1}$. If $\mu'_k := \pi_\infty\mu_k$ then $[\![\mu'_k > 0]\!] = [\![\mu'_k \ge \mu'_{k+1}]\!] = [\![\lim \mu'_k = 0]\!] = \pi_\infty$, so that μ'_k is a weak order-unity in $\pi_\infty\Lambda$, $\mu'_k \ge \mu'_{k+1}$, and $o\text{-}\lim \mu'_k = 0$. From the above-indicated presentation for $\mathscr{T}$ we deduce

$$\pi_\infty T = bo\text{-}\sum_{k=1}^\infty \mu'_k e_k^\# \otimes f_k.$$

Consider the fragment $\pi_\infty^\perp T$. From the definition of π_∞ it follows that $\pi_\infty^\perp = [\![\, \mathscr{T}$ is a finite-rank operator$\,]\!] = \mathbf{1}$. The operator $\mathscr{T}$ has finite rank if and only if $\mu_n = 0$ for some $n \in \mathbb{N}$. Thus,

$$\pi_\infty^\perp = [\![\, (\exists n \in \mathbb{N}^\wedge)\, \mu_n = 0 \,]\!] = \bigvee_{n=1}^{\infty} [\![\, \mu_n = 0 \,]\!].$$

Put $\rho_n := [\![\, \mu_n = 0 \,]\!]$, $\pi_0 := \rho_1$, $\pi_n := \rho_{n+1} - \rho_n$ $(n \in \mathbb{N})$. Since $\pi_n = [\![\, \mu_{n+1} = 0 \,\&\, \mu_n \neq 0 \,]\!]$, we have construct a countable partition $(\pi_n)_{n=0}^{\infty}$ of the projection $\pi_\infty^\perp$ with $\pi_n \mu_{n+1} = 0$. Therefore, $\pi_n T = \sum_{k=1}^{n} \pi_n \mu_k e_k^{\#} \otimes f_k$ for all $n \in \mathbb{N}$. It remains to observe that $T = \pi_\infty T + bo\text{-}\sum_{n=0}^{\infty} \pi_n T$. $\triangleright$

8.5.7. A variant of the Fredholm Alternative holds for cyclically compact operators. We will call it the *Fredholm B-Alternative.* Let X be a Banach space with the dual X^*. Take a bounded operator $T : X \to X$ and consider the equation of the first kind

$$Tx = y \quad (x, y \in X)$$

and the conjugate equation

$$T^* y^* = x^* \quad (x^*, y^* \in X^*).$$

The corresponding homogeneous equations are defined as $Tx = 0$ and $T^*y^* = 0$. Let $\varphi_0(T)$, $\varphi_1(n, T)$, $\varphi_2(n, T)$, and $\varphi_3(n, T)$ be set-theoretic formulas formalizing the following statements.

$\varphi_0(T)$: The homogeneous equation $Tx = 0$ has a sole solution, zero. The homogeneous conjugate equation $T^*y^* = 0$ has a sole solution, zero. The equation $Tx = y$ is solvable and has a unique solution given an arbitrary right side. The conjugate equation $T^*y^* = x^*$ is solvable and has a unique solution given an arbitrary right side.

$\varphi_1(n, T)$: The homogeneous equation $Tx = 0$ has n linearly independent solutions $x_1, \dots, x_n$. The homogeneous conjugate equation $T^*y^* = 0$ has n linearly independent solutions $y_1^*, \dots, y_n^*$.

$\varphi_2(n, T)$: The equation $Tx = y$ is solvable if and only if $y_1^*(y) = \dots = y_n^*(y) = 0$. The conjugate equation $T^*y^* = x^*$ is solvable if and only if $x^*(x_1) = \dots = x^*(x_n) = 0$.

$\varphi_3(n, T)$: The general solution x of the equation $Tx = y$ is the sum of a particular solution x_0 and the general solution of the homogeneous equation; i.e., it has the form

$$x = x_0 + \sum_{k=1}^{n} \lambda_k x_k \quad (\lambda_k \in \mathbb{C}).$$

The general solution y^* of the conjugate equation $T^*y^* = x^*$ is the sum of a particular solution y_0^* and the general solution of the homogeneous equation; i.e., it has the form

$$y^* = y_0^* + \sum_{k=1}^{n} \mu_k y_k^* \quad (\mu_k \in \mathbb{C}).$$

Using this notation, the Fredholm Alternative can be written as follows:

$$\varphi_0(T) \vee (\exists n \in \mathbb{N})\, \varphi_1(n,T) \,\&\, \varphi_2(n,T) \,\&\, \varphi_3(n,T).$$

Thus, the conventional Fredholm Alternative distinguishes the cases $n = 0$ and $n \neq 0$. (If $n = 0$ then the formula $\varphi_1(n,T) \,\&\, \varphi_2(n,T) \,\&\, \varphi_3(n,T)$ is equivalent to $\varphi_0(T)$.)

8.5.8. Consider now a B-cyclic Banach space X and a bounded B-linear operator T in X. In this case X and $X^\#$ are modules over the Stone algebra $\Lambda := \mathscr{S}(B)$ and T is Λ-linear (= module homomorphism). A subset $\mathscr{E} \subset X$ is said to be *locally linearly independent* if whenever $e_1, \dots, e_n \in \mathscr{E}$, $\lambda_1, \dots, \lambda_n \in \mathbb{C}$, and $\pi \in B$ with $\pi(\lambda_1 e_1 + \dots + \lambda_n e_n) = 0$ we have $\pi \lambda_k e_k = 0$ for all $k := 1, \dots, n$. We say that the *Fredholm B-Alternative* is valid for an operator T if there exists a countable partition of unity (b_n) in B such that the following conditions are fulfilled:

(1) The homogeneous equation $b_0 \circ Tx = 0$ has a sole solution, zero. The homogeneous conjugate equation $b_0 \circ T^\# y^\# = 0$ has a sole solution, zero. The equation $b_0 \circ Tx = b_0 y$ is solvable and has a unique solution given an arbitrary $y \in X$. The conjugate equation $b_0 \circ T^\# y^\# = b_0 x^\#$ is solvable and has a unique solution given an arbitrary $x^\# \in X^\#$.

(2) For every $n \in \mathbb{N}$ the homogeneous equation $b_n \circ Tx = 0$ has n locally linearly independent solutions $x_{1,n}, \dots, x_{n,n}$ and the homogeneous conjugate equation $b_n \circ T^\# y^\# = 0$ has n locally linearly independent solutions $y_{1,n}^\#, \dots, y_{n,n}^\#$ (hence have nonzero solutions).

(3) The equation $Tx = y$ is solvable if and only if $b_n \circ y_{k,n}^\#(y) = 0$ $(n \in \mathbb{N},\ k \leq n)$. The conjugate equation $T^\# y^\# = x^\#$ is solvable if and only if $b_n \circ x^\#(x_{k,n}) = 0$ $(n \in \mathbb{N},\ k \leq n)$.

(4) The general solution x of the equation $Tx = y$ has the form

$$x = bo\text{-}\sum_{n=1}^{\infty} b_n \left(x_n + \sum_{k=1}^{n} \lambda_{k,n} x_{k,n} \right),$$

where x_n is a particular solution of the equation $b_n \circ Tx = b_n y$ and $(\lambda_{k,n})_{n \in \mathbb{N}, k \leq n}$ are arbitrary elements in Λ.

The general solution $y^\#$ of the conjugate equation $T^\# y^\# = x^\#$ has the form

$$y^\# = bo\text{-}\sum_{n=1}^{\infty} b_n \left(y_n^\# + \sum_{k=1}^{n} \lambda_{k,n} y_{k,n}^\# \right),$$

where $y_n^\#$ is a particular solution of the equation $b_n \circ T^\# y^\# = b_n x^\#$ and $\lambda_{k,n}$ are arbitrary elements Λ for $n \in \mathbb{N}$ and $k \leq n$.

8.5.9. *Theorem.* *If S is a cyclically compact operator in a B-cyclic space X then the Fredholm B-Alternative is valid for the operator $T := I_X - S$.*

$\triangleleft$ Again we assume, without loss of generality, that X is the bounded part of the descent of a Banach space $\mathscr{X} \in \mathbf{V}^{(B)}$ and T is the restriction onto X of the descent of a bounded linear operator $\mathscr{T} \in \mathbf{V}^{(B)}$. Moreover, $[\![\mathscr{T} = I_{\mathscr{X}} - \mathscr{S}]\!] = \mathbf{1}$ and $[\![\mathscr{S}$ is a compact operator in $\mathscr{X}]\!] = \mathbf{1}$. It follows from 8.3.7 (1) that we may assume that also $X = \mathscr{X}^*{\downarrow}^\infty$ and $T = \mathscr{T}^*{\downarrow}^\infty$. The Fredholm Alternative 8.5.7 is fulfilled for $\mathscr{T}$ inside $\mathbf{V}^{(B)}$ by virtue of the Transfer Principle. In other words, the following relations hold:

$$\begin{aligned} \mathbf{1} &= [\![\varphi_0(\mathscr{T}) \vee (\exists n \in \mathbb{N}^\wedge)\, \varphi_1(n, \mathscr{T}) \,\&\, \varphi_2(n, \mathscr{T}) \,\&\, \varphi_3(n, \mathscr{T})]\!] \\ &= [\![\varphi_0(\mathscr{T})]\!] \vee \bigvee_{n \in \mathbb{N}} [\![\varphi_1(n^\wedge, \mathscr{T})]\!] \wedge [\![\varphi_2(n^\wedge, \mathscr{T})]\!] \wedge [\![\varphi_3(n^\wedge, \mathscr{T})]\!]. \end{aligned}$$

Denote $b_0 := [\![\varphi_0(\mathscr{T})]\!]$ and $b_n := [\![\varphi_1(n^\wedge, \mathscr{T})]\!] \wedge [\![\varphi_2(n^\wedge, \mathscr{T})]\!] \wedge [\![\varphi_3(n^\wedge, \mathscr{T})]\!]$. Since the formulas $\varphi_0(\mathscr{T})$ and $\varphi_1(n, \mathscr{T}) \,\&\, \varphi_2(n, \mathscr{T}) \,\&\, \varphi_3(n, \mathscr{T}))$ for different n are inconsistent, the sequence $(b_n)_{n=0}^\infty$ is a partition of unity in B. We will now prove that 8.5.8 (1–4) are valid.

(1): The claim 8.5.8 (1) is equivalent to the identities $\ker(T) = \{0\}$ and $\operatorname{im}(T) = X$ that are ensured by the following easy relations:

$$\mathbf{V}^{(B)} \models \ker(T){\uparrow} = \ker(\mathscr{T}) = \{0\}, \quad \mathbf{V}^{(B)} \models \operatorname{im}(T){\uparrow} = \operatorname{im}(\mathscr{T}) = \mathscr{X}.$$

(2): The part of the assertion $\varphi_1(n^\wedge, \mathscr{T})$ concerning the solution of the equation $Tx = 0$ is formalized as

$$\begin{aligned} &(\exists x)\, \big((x : \{1, \dots, n\}^\wedge \to \mathscr{X}) \,\&\, (\forall k \in \{1, \dots, n\}^\wedge)\, (\mathscr{T} x(k) = 0) \\ &\qquad \&\ \text{the set } x(\{1, \dots, n\}^\wedge) \text{ is linearly independent})\big). \end{aligned}$$

Moreover, there is no loss of generality in assuming that $\|x(k)\| \leq 1$, $k \in \{1, \dots, n\}^\wedge$. Using the Maximum Principle and the properties of the modified descent we may find a mapping $\mathbf{x}$ from $\{1, \dots, n\}$ to X such that the image of the mapping $b_n \mathbf{x}$:

$k \mapsto b_n \mathbf{x}(k)$ is a locally linearly independent set in X and $[\![\mathscr{T}\mathbf{x}(k) = 0]\!] \geq b_n$ for each $k \in \{1, \dots, n\}$. Put $x_{k,n} := b_n \mathbf{x}(k)$. Further,

$$[\![Tx_{k,n} = 0]\!] = [\![\mathscr{T}\mathbf{x}(k) = 0]\!] \wedge [\![\mathbf{x}(k) = x_{k,n}]\!] \geq b_n,$$

so that $b_n T x_{k,n} = 0$. The conjugate homogeneous equation is handled in the same fashion.

(3): Necessity of the stated conditions can be easily checked; prove sufficiency. We confine exposition to the equation $Tx = y$, since the conjugate equation is considered along similar lines. Suppose that $y^{\#}_{k,n}(y) = 0$ for $k, n \in \mathbb{N}$ and $k \leq n$. Then

$$b_n \leq [\![y^{\#}_{k,n}(y) = 0]\!] = [\![y^{\#}_{k,n}\!\uparrow (y) = 0]\!] \quad (k \in \{1, \dots, n\}).$$

At the same time, in view of (2), $[\![\{y^{\#}_{k,n} : k = 1, \dots, n\}\!\uparrow$ is a maximal linearly independent set of solutions of the equation $\mathscr{T}^* y^* = 0]\!] = \mathbf{1}$. All this implies that $[\![$ the equation $\mathscr{T}x = y$ is solvable $]\!] \geq b_n$, whence the equation $b_n \circ Tx = b_n y$ has at least one solution x_n. It is then easy to check that $\bar{x} := \sum_{n=1}^{\infty} b_n x_n$ is a solution of the equation $Tx = y$.

(4): If x is a solution of the equation $Tx = y$ then $[\![\mathscr{T}x = y]\!] = \mathbf{1}$. Taking into account the inequality $[\![\varphi_3(n^\wedge, \mathscr{T})]\!] \geq b_n$, we arrive at

$$b_n \leq [\![(\exists \lambda)\, (\lambda : \{1, \dots, n\}^\wedge \to \mathscr{R} \,\&\, x = x^* + \sum_{k=1}^{n^\wedge} \lambda(k) u(k))]\!],$$

where u is the ascent of the mapping $k \mapsto x_{k,n}$ $(k = 1, \dots, n)$. The Maximum Principle guarantees the existence of a mapping $\boldsymbol{\lambda}_n$ from $\{1, \dots, n\}$ to Λ such that

$$[\![x = \bar{x} + \sum_{k=1}^{n^\wedge} \boldsymbol{\lambda}_n\!\uparrow (k) u(k)]\!] = \mathbf{1}.$$

Putting $\lambda_{k,n} := b_n \boldsymbol{\lambda}_n(k)$, we obtain

$$b_n x = b_n x_n + \sum_{k=1}^{n} \lambda_{k,n} b_n x_{k,n},$$

whence the desired representation follows. The general form of the solution of the conjugate equation is established by similar arguments. ▷

8.6. Comments

8.6.1. (1) As was mentioned in 1.6.3 (2), the heuristic transfer principle proposed by L. V. Kantorovich in connection with the concept of K-space was substantiated by the inventor as well as by his successors. Essentially, this principle turned out to be one of those profound ideas that, playing an active and leading role in the formation of a new branch of analysis, led eventually to a deep and elegant theory of K-space rich in various applications. At the very beginning of the development of the theory, attempts were made at formalizing the above heuristic argument. In this direction, there appeared the so-called theorems of relation preservation which claimed that if some proposition involving finitely many functional relations is proven for the reals then an analogous fact remains valid automatically for the elements of every K-space (see [163, 388]).

However, the inner mechanism responsible for the phenomenon of relation preservation still remained vague and the applicability range for such assertions are not found nor the general causes of numerous analogies and parallelism with the classical function theory. The depth and universality of Kantorovich's principle were demonstrated within Boolean-valued analysis.

(2) The Boolean-valued status of the concept of K-space is established in Gordon's Theorem 8.1.2 obtained in [105]. This fact can be interpreted as follows: *A universally complete K-space is the interpretation of the reals in an appropriate Boolean-valued model.* Moreover, it turns out that every theorem on reals (in the framework of ZFC) has an analog for the corresponding K-space. Theorems are transferred by means of precisely-defined procedures: *ascent, descent,* and *canonical embedding,* that is, algorithmically as a matter of fact. Thereby Kantorovich's prophesy that "the elements of a K-space are generalized numbers" acquires a precise mathematical meaning in Boolean-valued analysis. On the other hand, Boolean-valued analysis makes rigorous and omnipresent the heuristic transfer principle which played a secondary role in the pre-Boolean-valued history of K-space.

(3) Applications of Boolean-valued models to functional analysis stem from the works by E. I. Gordon [105, 106] and G. Takeuti [367]. If B in 8.1.2 is the algebra of measurable sets modulo sets of zero measure μ then $\mathscr{R}{\downarrow}$ is isomorphic to the universally complete K-space $L^0(\mu)$ of measurable functions. This fact (for the Lebesgue measure on an interval) was already known to Scott and Solovay (see [354]). If B is a complete Boolean algebra of projections in a Hilbert space then $\mathscr{R}{\downarrow}$ is isomorphic to the space of selfadjoint operators $\mathfrak{A}(B)$. The two indicated particular cases of Gordon's Theorem were intensively and fruitfully exploited by G. Takeuti (see [367] and the bibliography in [212]). The object $\mathscr{R}{\downarrow}$ for general Boolean algebras was also studied by T. Jech [141–143] who in fact rediscovered Gordon's Theorem. The difference is that in [140] a (complex) universally complete

K-space with unity is defined by another system of axioms and is referred to as a complete Stone algebra. The interconnections between properties of numeric objects and the corresponding objects in the K-space $\mathscr{R}\downarrow$, indicated in 8.1.3–8.1.5 were actually obtained by E. I. Gordon [105, 106].

(4) The Representation Theorem 8.1.6 was obtained by A. G. Kusraev [198]. A close result (in other terms) is presented in T. Jech's article [143], where Boolean-valued interpretation of the theory of linearly ordered sets is developed. Corollaries 8.1.7 (7, 8) are well known (see [163, 388]). Tests 8.1.8 (2, 4) for o-convergence (in the case of sequences) were obtained by L. V. Kantorovich and B. Z. Vulikh (see [163]). It was shown in 8.1.8 that, in fact, they are merely the interpretation of convergence properties of numeric nets (sequences).

(5) As was mentioned in (1), the first attempts at formalizing the Kantorovich heuristic principle resulted in theorems of relation preservation (see [163, 388]). The contemporary forms of such theorems, based on the method of Boolean-valued models, may be found in [107, 142] (see also [212]).

(6) Subsystems of the field $\mathscr{R}$ appear not only as Boolean-valued representation of Archimedean vector lattices (see 8.1.6 (1)). For instance, the following fact are stated in [198] and proved in [206, 212].

The following assertions are valid:

(i) *the Boolean-valued realization of an Archimedean lattice-ordered group is a subgroup of the additive group of $\mathscr{R}$;*

(ii) *an Archimedean f-ring contains two complementary bands one of which has the zero multiplication and is realized as in* (1) *and the other, as a subring of $\mathscr{R}$;*

(iii) *an Archimedean f-algebra contains two complementary bands one of which is realized as in 8.1.6 and the other, as a sublattice and subalgebra of the field $\mathscr{R}$ considered as a lattice-ordered algebra over the field $\mathbb{R}^{\wedge}$ (see also* [143]*).*

(7) A. E. Gutman's result presented in 5.1.7 (see [122]) settles simultaneously the question of whether $\mathbb{R}^{\wedge}$ and $\mathscr{R}$ coincide inside $\mathbf{V}^{(B)}$: This property amounts to the σ-distributivity of B. Indeed, it can be seen easily that a universally complete K-space is locally one-dimensional if and only if $\mathscr{R}$ is a one-dimensional vector space over $\mathbb{R}^{\wedge}$ inside $\mathbf{V}^{(B)}$, i.e. $\mathbf{V}^{(B)} \models \mathscr{R} = \mathbb{R}^{\wedge}$. Thus, the same result provides an example of an atomless Boolean algebra B with the property $\mathbf{V}^{(B)} \models \mathscr{R} = \mathbb{R}^{\wedge}$.

8.6.2. (1) The results of this section, with minor exception, are well known to the specialists in the theory of vector lattices. The novelty consists in the method of proving: all basic facts are derived by interpreting simple properties of the reals in a Boolean-valued model. Concerning 8.1.2–8.1.5, see comments 1.6.3 and 1.6.4.

(2) It follows from 8.2.11 that every resolution of unity with values in a σ-algebra determines a spectral measure on the Borel σ-algebra of the real axis. This fact was indicated for the first time by V. I. Sobolev in [353]. However, he assumed that such a measure can be obtained by means of the Carathéodory extension method. D. A. Vladimirov [383] showed that the Carathéodory extension of a complete Boolean algebra of countable type is possible if and only if the algebra is regular. Thus, the extension method of 8.2.11, grounded on the Loomis–Sikorski representations for Boolean σ-algebras, differs essentially from the Carathéodory extension. In [404], M. Wright obtained 8.2.11 (for $n = 1$) as a consequence of Theorem 6.2.7 (1). It should be noted in this connection that every F-valued countably additive positive Borel measure on a σ-compact topological space (and, in particular, every countably additive measure $\mu : \mathscr{B}or(\mathbb{R}) \to F$) is a quasi-Radon measure (see [216]).

(4) V. I. Sobolev was apparently the first who considered Borel functions defined on an arbitrary K_σ-space with unity (see [353, 388]). Theorems 8.2.13 and 8.2.14 presented here were obtained in [214, 216], where there was also constructed the Borel functional calculus for (countable or uncountable) collections of elements of an arbitrary K-space. A Boolean-valued proof of Theorem 8.2.14 was given in [141]. More details can be found also in [206, 212].

(5) Observe an interesting corollary of the Borel functional calculus for sequences in a K_σ-space obtained by S. A. Malyugin [216].

Theorem. *A sequentially order complete vector lattice E is weakly σ-distributive if and only if for every $e \in E$ the factor algebra $\mathscr{B}or(\mathbb{R})/\mu_e^{-1}(0)$ is weakly σ-distributive.*

8.6.3. (1) Theorems 8.3.1, 8.3.2 and 8.3.3 on Boolean-valued representation of lattice-normed spaces and bounded operators in such spaces were obtained by A. G. Kusraev in [196, 197].

(2) Theorem 8.3.4 (1) is a particular case of the general construction of the Boolean completion of a uniform space which was suggested by E. I. Gordon and V. A. Lyubetskiĭ [110]. Proposition 8.3.4 (2) is a simple consequence of 8.3.4 (1) and a relevant result by E. I. Gordon on representation of operators with abstract norm [106].

(3) The restricted descent of 8.3.5 appeared in the research by G. Takeuti into von Neumann algebras and C^*-algebras within Boolean-valued models [370, 371] and in the research by M. Ozawa into Boolean-valued interpretation of the theory of Hilbert spaces [303].

(4) Theorems 8.3.8 and 8.3.9 on Boolean-valued representation of Kaplansky–Hilbert modules and bounded module homomorphisms in them were proven by M. Ozawa [303]. Theorems 8.3.10 and 8.3.11 are demonstrated in [305] and [201], respectively.

8.6.4. (1) G. Takeuti started studying C^*-algebras and von Neumann algebras by using Boolean valued models with [370, 371]. Theorem 8.4.2 belongs to him [371]; Theorems 8.4.5 and 8.4.6 must be also attributed to G. Takeuti [370]. Theorems 8.4.4 (1, 2) were obtained by M. Ozawa.

(2) The definition of the Boolean dimension of a Kaplansky–Hilbert module given in 8.4.7 was studied by A. G. Kusraev in [201]. Prior to this research, M. Ozawa had defined the Boolean dimension of a Kaplansky–Hilbert module as the dimension of any Hilbert space serving as a Boolean-valued representation of the module in question, i.e., as an internal object of a Boolean-valued model [305]. So, the definition of B-dimension in 8.4.7 is an external decoding of the definition by M. Ozawa. Theorem 8.4.9 may be found in [201] and its Boolean-valued version, in [305].

(3) The result on functional representation, Theorem 8.4.13, was established by A. G. Kusraev in [201]. Classification of type I AW^*-algebras was given in [305] and [201] in somewhat different forms. The true difference lies in the fact that the invariant characterizing a type I AW^*-algebra type I to within $*$-isomorphism in [307] is a Boolean-valued cardinal, i.e., an internal object of the Boolean-valued universe in question. The definition in [201] does not appeal to the construction of the Boolean-valued universe.

(4) Observe that 8.4.12 (4) implies a negative solution to the I. Kaplansky problem of unique decomposition of a type I AW^*-algebra into the direct sum of homogeneous bands. M. Ozawa gave this solution in [306, 307]. The lack of uniqueness is tied with the effect of the cardinal shift that may happens during immersion into a Boolean-valued model $\mathbf{V}^{(B)}$ A.19 (3). The cardinal shift is impossible in the case when the Boolean algebra of central idempotents B under study satisfies the countable chain condition, (cf. A.19 (4)) and so the decomposition in question is unique. I. Kaplansky established uniqueness of the decomposition on assuming that B satisfies the countable chain condition and conjectured that uniqueness fails in general [166].

(5) There exist various classes of ordered and involutive algebras to which we may apply the technique of Chapter 8 (see [65, 129, 334]). Among most important of them is the class of JB-algebras; see [30, 17, 129]. A considerable contribution to Boolean-valued analysis of von Neumann algebras and AW^*-algebras was made by H. Nishimura [298–301].

8.6.5. (1) Cyclically compact sets and operators in lattice-normed spaces were introduced in [194] and [197], respectively. A standard prof of Theorem 5.3.4 can be extracted from [197] wherein more general approach is developed. Certain variants of Theorems 8.5.6 and 8.5.9 for operators in Banach–Kantorovich spaces can be also found in [197].

(2) The famous result by P. G. Dodds and D. H. Fremlin [81] asserts that if a positive operator acting from a Banach lattice whose dual has order continuous norm to a Banach lattice with order continuous norm is dominated by a compact operator then the initial operator is also compact, see [23] for proof and related results. As regards cyclical compactness, we observe the conjecture of [199] that if a dominated operator T between spaces with mixed norm is cyclically compact and $|T| \leq S$ with S compact then T is also compact on assuming some conditions on the norm lattices like in the Dodds–Fremlin Theorem. This problem remains open.

Appendix. Boolean-Valued Models

In the section we briefly present necessary information on the theory of Boolean-valued models. Details may be found in [37, 140, 206, 212, 275, 332, 372].

The most important feature of the method of Boolean-valued models consists in comparative analysis of standard and nonstandard (Boolean-valued) models which uses a special technique of ascending and descending. Moreover, it is often necessary to carry out some syntax comparison of formal texts. Therefore, before we launch into ascending and descending, it is necessary to grasp a more clear idea of the status of mathematical objects in the framework of a formal set theory.

A.1. At present, the most widespread axiomatic foundation for mathematics is Zermelo–Fraenkel set theory. We will briefly recall some of its concepts, outlining the details we need in the sequel. Observe that, speaking of a formal set theory, we will freely (because it is in fact unavoidable) adhere to the level of rigor accepted in mathematics and introduce abbreviations by means of the *definor*, *assignment operator*, $:=$ without specifying subtleties.

(1) The alphabet of Zermelo–Fraenkel theory (ZF or ZFC if the presence of choice is stressed) comprises the symbols of variables; the parentheses (and); the propositional connectives (= the signs of propositional calculus) $\vee$, $\wedge$, $\rightarrow$, $\leftrightarrow$, and $\neg$; the quantifiers $\forall$ and $\exists$; the equality sign $=$; and the symbol of a special binary predicate of *containment* or *membership* $\in$. In general, the domain of the variables of ZF is thought as the world or *universe* of sets. In other words, the universe of ZF contains nothing but sets. We write $x \in y$ rather than $\in (x, y)$ and say that x is an *element* of y.

(2) The formulas of ZF are defined by means of a routine procedure. In other words, the formulas of ZF are finite texts resulting from the *atomic* formulas $x = y$ and $x \in y$, where x and y are variables of ZF, by reasonably placing parentheses, quantifiers, and propositional connectives. So, if φ_1 and φ_2 are formulas of ZF and x is a variable symbol then the texts $\varphi_1 \rightarrow \varphi_2$ and $(\exists\, x)\,(\varphi_1 \rightarrow (\forall y)\,\varphi_2) \vee \varphi_1$ are formulas of ZF, whereas $\varphi_1\, \exists x$ and $\forall\,(x \exists \varphi_1 \neg \varphi_2$ are not. We attach the natural meaning to the terms *free* and *bound variables* and the term *domain of a quantifier.*

For instance, in the formula $(\forall x)\,(x \in y)$ the variable x is bound and the variable y is free, whereas in the formula $(\exists y)\,(x = y)$ the variable x is free and y is bound (for it is bounded by a quantifier). Henceforth, in order to emphasize that the only free variables in a formula φ are the variables $x_1, \dots, x_n$, we write $\varphi(x_1, \dots, x_n)$. Sometimes such a formula is considered as a "function"; in this event, it is convenient to write $\varphi(\cdot, \dots, \cdot)$ or $\varphi = \varphi(x_1, \dots, x_n)$, implying that $\varphi(y_1, \dots, y_n)$ is a formula of ZF obtained by replacing each free occurrence of x_k by y_k for $k := 1, \dots, n$.

(3) Studying ZF, it is convenient to use some expressive tools absent in its formal language. In particular, in the sequel it is worthwhile employing the concepts of class and definable class and also the corresponding symbols of *classifiers* like $A_\varphi := A_{\varphi(\cdot)} := \{x : \varphi(x)\}$ and $A_\psi := A_{\psi(\cdot,y)} := \{x : \psi(x, y)\}$, where φ and ψ are formulas of ZF and y is a distinguished collection of variables. If it is desirable to clarify or eliminate the appearing records then we may assume that use of classes and classifiers is connected only with the conventional agreement on introducing abbreviations. This agreement, sometimes called the *Church schema*, reads:

$$z \in \{x : \varphi(x)\} \leftrightarrow \varphi(z),$$
$$z \in \{x : \psi(x, y)\} \leftrightarrow \psi(z, y).$$

Working within ZF, we will employ some notations that are widely spread in mathematics. Some of them are as follows:

$$(\exists! z)\,\varphi(z) := (\exists z)\,\varphi(z) \wedge ((\forall x)\,(\forall y)\,(\varphi(x) \wedge \varphi(y) \to x = y));$$
$$x \neq y := \neg x = y, \quad x \notin y := \neg x \in y;$$
$$\varnothing := \{x : x \neq x\};$$
$$\{x, y\} := \{z : z = x \vee z = y\}, \quad \{x\} := \{x, x\},$$
$$(x, y) := \{x, \{x, y\}\};$$
$$(\forall x \in y)\,\varphi(x) := (\forall x)\,(x \in y \to \varphi(x));$$
$$(\exists x \in y)\,\varphi(x) := (\exists x)\,(x \in y \wedge \varphi(x));$$
$$\bigcup x := \{z : (\exists y \in x)\, z \in y\};$$
$$\bigcap x := \{z : (\forall y \in x)\, z \in y\};$$
$$x \subset y := (\forall z)\,(z \in x \to z \in y);$$
$$\mathscr{P}(x) := \text{the class of all subsets of } x := \{z : z \subset x\};$$
$$\mathbf{V} := \text{the class of all sets} := \{x : x = x\}.$$

Note also that in the sequel we accept more complicated descriptions in which much is presumed:

$$\begin{aligned}\mathrm{Funct}(f) &:= f \text{ is a function};\\ \mathrm{dom}(f) &:= \text{the domain of } f;\\ \mathrm{im}(f) &:= \text{the range of } f;\\ \varphi \vdash \psi &:= \varphi \to \psi := \psi \text{ is derivable from } \varphi;\\ \text{a class } A \text{ is a set} &:= A \in \mathbf{V} := (\exists x)\,(\forall y)\,(y \in A \leftrightarrow y \in x).\end{aligned}$$

Such simplifications will be used in rendering more complicated formulas without special stipulations. For instance, instead of some rather involved formulas of ZF we simply write

$$\begin{aligned} f : x \to y &\equiv \text{“}f \text{ is a function from } x \text{ to } y\text{”};\\ &\text{“}E \text{ is a } K\text{-space”};\\ U \in \mathscr{L}(X,Y) &\equiv \text{“}U \text{ is a bounded operator from } X \text{ to } Y.\text{”}\end{aligned}$$

A.2. In ZFC, we accept the usual axioms and derivation rules of a first-order theory with equality which fix the standard means of classical reasoning (syllogisms, the law of the excluded middle, modus ponens, generalization, etc.). Moreover, we accept the following *special* or *proper* axioms:

(1) The axiom of extensionality

$$(\forall x)\,(\forall y)\,(x \subset y \wedge y \subset x \to x = y).$$

(2) The axiom of union

$$(\forall x)\,(\exists y)\,\left(y = \bigcup x\right).$$

(3) The axiom of the powerset

$$(\forall x)\,(\exists y)\,(y = \mathscr{P}(x)).$$

(4) The axiom of replacement

$$(\forall x)\,((\forall y)\,(\forall z)\,(\forall u)\,\varphi(y,z) \wedge \varphi(y,u) \to z = u) \to (\exists v)\,(v = \{z : (\exists y \in x)\,\varphi(y,z)\}).$$

(5) The axiom of foundation

$$(\forall x)\,(x \neq \varnothing \to (\exists y \in x)\,(y \cap x = \varnothing)).$$

(6) The axiom of infinity

$$(\exists \omega)\,(\varnothing \in \omega) \wedge (\forall x \in \omega)\,(x \cup \{x\} \in \omega).$$

(7) The axiom of choice

$$(\forall F)\,(\forall x)\,(\forall y)\,((x \neq \varnothing \wedge F : x \to \mathscr{P}(y))$$
$$\to ((\exists f)\, f : x \to y \wedge (\forall z \in x)\, f(z) \in F(z)).$$

Grounding on the above axiomatics, we acquire a clear idea of the class of all sets, the *von Neumann universe* $\mathbf{V}$. As the initial object of all constructions we take the empty set. The elementary step of introducing new sets consists in taking the union of the powersets of the sets already available. Transfinitely repeating these steps, we exhaust the class of all sets. More precisely, we assign $\mathbf{V} := \bigcup_{\alpha \in \mathrm{On}} \mathbf{V}_\alpha$, where On is the class of all ordinals and

$$\mathbf{V}_0 := \varnothing,$$
$$\mathbf{V}_{\alpha+1} := \mathscr{P}(\mathbf{V}_\alpha),$$
$$\mathbf{V}_\beta := \bigcup_{\alpha<\beta} \mathbf{V}_\alpha \quad (\beta \text{ is a limit ordinal}).$$

The class $\mathbf{V}$ is the standard model of ZFC.

A.3. Now, we describe the construction of a Boolean-valued universe. Let B be a complete Boolean algebra. Given an ordinal α, put

$$\mathbf{V}_\alpha^{(B)} := \{x : \mathrm{Funct}(x) \wedge (\exists \beta)\,(\beta < \alpha \wedge \mathrm{dom}(x) \subset \mathbf{V}_\beta^{(B)} \wedge \mathrm{im}(x) \subset B)\}.$$

Thus, in more detail we have

$$\mathbf{V}_0^{(B)} := \varnothing,$$
$$\mathbf{V}_{\alpha+1}^{(B)} := \{x : x \text{ is a function with domain in } \mathbf{V}_\alpha^{(B)} \text{ and range in } B\};$$
$$\mathbf{V}_\alpha^{(B)} := \bigcup_{\beta<\alpha} \mathbf{V}_\beta^{(B)} \quad (\beta \text{ is a limit ordinal}).$$

The class

$$\mathbf{V}^{(B)} := \bigcup_{\alpha \in \mathrm{On}} \mathbf{V}_\alpha^{(B)}$$

is a *Boolean-valued universe.* An element of the class $\mathbf{V}^{(B)}$ is a *B-valued set.* It is necessary to observe that $\mathbf{V}^{(B)}$ consists only of functions. In particular, $\varnothing$ is the function with domain $\varnothing$ and range $\varnothing$. Hence, the "lower" levels of $\mathbf{V}^{(B)}$ are organized as follows:

$$\mathbf{V}_0^{(B)} = \varnothing, \quad \mathbf{V}_1^{(B)} = \{\varnothing\}, \quad \mathbf{V}_2^{(B)} = \{\varnothing, (\{\varnothing\}, b) : b \in B\}.$$

It is worth stressing that $\alpha \leq \beta \to \mathbf{V}_\alpha^{(B)} \subset \mathbf{V}_\beta^{(B)}$ holds for all ordinals α and β. Moreover, the following *induction principle* is valid in $\mathbf{V}^{(B)}$:

$$(\forall x \in \mathbf{V}^{(B)})\,((\forall y \in \mathrm{dom}(x))\,\varphi(y) \to \varphi(x)) \to (\forall x \in \mathbf{V}^{(B)})\,\varphi(x),$$

where φ is a formula of ZFC.

A.4. Take an arbitrary formula $\varphi = \varphi(u_1, \dots, u_n)$ of the ZFC theory. If we replace the elements $u_1, \dots, u_n$ by elements $x_1, \dots, x_n \in \mathbf{V}^{(B)}$ then we obtain some statement about the objects $x_1, \dots, x_n$. It is to this statement that we intend to assign some *truth-value*. Such a value $[\![\psi]\!]$ must be an element of the algebra B. Moreover, it is naturally desired that the theorems of ZFC be true, i.e., attain the greatest truth-value, unity.

We must obviously define truth-values by double induction, taking into consideration the way in which formulas are built up from atomic formulas and assigning truth-values to the above formulas $x \in y$ and $x = y$, where $x, y \in \mathbf{V}^{(B)}$ in accord with the way in which $\mathbf{V}^{(B)}$ is constructed.

It is clear that if φ and ψ are evaluated formulas of ZFC and $[\![\varphi]\!] \in B$ and $[\![\psi]\!] \in B$ are their truth-values then we should put

$$\begin{aligned}
[\![\varphi \wedge \psi]\!] &:= [\![\varphi]\!] \wedge [\![\psi]\!],\\
[\![\varphi \vee \psi]\!] &:= [\![\varphi]\!] \vee [\![\psi]\!],\\
[\![\varphi \to \psi]\!] &:= [\![\varphi]\!] \Rightarrow [\![\psi]\!],\\
[\![\neg\varphi]\!] &:= [\![\varphi]\!]^*,\\
[\![(\forall x)\,\varphi(x)]\!] &:= \bigwedge_{x \in \mathbf{V}^{(B)}} [\![\varphi(x)]\!],\\
[\![(\exists x)\,\varphi(x)]\!] &:= \bigvee_{x \in \mathbf{V}^{(B)}} [\![\varphi(x)]\!],
\end{aligned}$$

where the right-hand sides involve Boolean operations corresponding to the logical connectives and quantifiers on the left-hand sides: $\wedge$ is the taking of an infimum, $\vee$ is the taking of a supremum, $*$ is the taking of the complement of an element, and the operation $\Rightarrow$ is introduced as follows: $a \Rightarrow b := a^* \vee b$ $(a, b \in B)$. Only such definitions provide the value "unity" for the classical tautologies.

We turn to evaluating the atomic formulas $x \in y$ and $x = y$ for $x, y \in \mathbf{V}^{(B)}$. The intuitive idea consists in the fact that a B-valued set y is a "(lattice) fuzzy set," i.e., a "set that contains an element z in $\mathrm{dom}(y)$ with probability $y(z)$." With this in mind and intending to preserve the logical tautology of $x \in y \leftrightarrow (\exists z \in y)\,(x = z)$ as well as the axiom of extensionality, we arrive at the following definition by recursion:

$$[\![x \in y]\!] := \bigvee_{z \in \mathrm{dom}(y)} y(z) \wedge [\![z = x]\!],$$

$$[\![x = y]\!] := \bigwedge_{z \in \mathrm{dom}(x)} x(z) \Rightarrow [\![z \in y]\!] \wedge \bigwedge_{z \in \mathrm{dom}(y)} y(z) \Rightarrow [\![z \in x]\!].$$

A.5. Now we are able to attach some meaning to formal expressions of the form $\varphi(x_1, \dots, x_n)$, where $x_1, \dots, x_n \in \mathbf{V}^{(B)}$ and φ is a formula of ZFC; i.e., we may define exactly in which sense the set-theoretic proposition $\varphi(u_1, \dots, u_n)$ is valid for elements $x_1, \dots, x_n \in \mathbf{V}^{(B)}$. Namely, we say that the *formula* $\varphi(x_1, \dots, x_n)$ *is valid inside* $\mathbf{V}^{(B)}$ or the *elements* $x_1, \dots, x_n$ *possess the property* φ if $[\![\varphi(x_1, \dots, x_n)]\!] = \mathbf{1}$. In this event, we write $\mathbf{V}^{(B)} \models \varphi(x_1, \dots, x_n)$.

It is easy to convince ourselves that the axioms and theorems of the first-order predicate calculus are valid in $\mathbf{V}^{(B)}$. In particular,

(1) $[\![x = x]\!] = \mathbf{1}$,

(2) $[\![x = y]\!] = [\![y = x]\!]$,

(3) $[\![x = y]\!] \wedge [\![y = z]\!] \le [\![x = z]\!]$,

(4) $[\![x = y]\!] \wedge [\![z \in x]\!] \le [\![z \in y]\!]$,

(5) $[\![x = y]\!] \wedge [\![x \in z]\!] \le [\![y \in z]\!]$.

It is worth observing that for each formula φ we have

$$\mathbf{V}^{(B)} \models x = y \wedge \varphi(x) \to \varphi(y),$$

i.e., in detailed notation

(6) $[\![x = y]\!] \wedge [\![\varphi(x)]\!] \le [\![\varphi(y)]\!]$.

A.6. In a Boolean-valued universe $\mathbf{V}^{(B)}$, the relation $[\![x = y]\!] = \mathbf{1}$ in no way implies that the functions x and y (considered as elements of $\mathbf{V}$) coincide. For example, the function equal to zero on each layer $\mathbf{V}_\alpha^{(B)}$, where $\alpha \ge 1$, plays the role of the empty set in $\mathbf{V}^{(B)}$. This circumstance may complicate some constructions in the sequel. In this connection, we pass from $\mathbf{V}^{(B)}$ to the *separated Boolean-valued universe* $\overline{\mathbf{V}}^{(B)}$ often preserving for it the same symbol $\mathbf{V}^{(B)}$; i.e., we put

$\mathbf{V}^{(B)} := \overline{\mathbf{V}}^{(B)}$. Moreover, to define $\overline{\mathbf{V}}^{(B)}$, we consider the relation $\{(x, y) : [\![x = y]\!] = \mathbf{1}\}$ in the class $\mathbf{V}^{(B)}$ which is obviously an equivalence. Choosing an element (a representative of least rank) in each class of equivalent functions, we arrive at the separated universe $\overline{\mathbf{V}}^{(B)}$. Note that

$$[\![x = y]\!] = \mathbf{1} \to [\![\varphi(x)]\!] = [\![\varphi(y)]\!]$$

is valid for an arbitrary formula φ of ZF and elements x and y in $\mathbf{V}^{(B)}$. Therefore, in the separated universe we can calculate the truth-values of formulas paying no attention to the way of choosing representatives. Furthermore, working with the separated universe, for the sake of convenience, one often considers (exercising due caution) a concrete representative rather than a class of equivalence as it is customary, for example, while dealing with function spaces.

A.7. The most important properties of a Boolean-valued universe $\mathbf{V}^{(B)}$ are stated in the following three principles:

(1) Transfer Principle. All theorems of ZFC are true in $\mathbf{V}^{(B)}$; i.e., the *Transfer Principle*, written in symbols as

$$\mathbf{V}^{(B)} \vDash \text{a theorem of ZFC},$$

is valid.

The Transfer Principle is established by rather laboriously checking that all axioms of ZFC have truth-value $\mathbf{1}$ and the rules of derivation preserve the truth-values of formulas. Sometimes, the Transfer Principle is worded as follows: "$\mathbf{V}^{(B)}$ is the Boolean-valued model of ZFC."

(2) Maximum Principle. For each formula φ of ZFC there exists $x_0 \in \mathbf{V}^{(B)}$ for which

$$[\![(\exists x)\, \varphi(x)]\!] = [\![\varphi(x_0)]\!].$$

In particular, if it is true in $\mathbf{V}^{(B)}$ that there is an x for which $\varphi(x)$ then there is an element x_0 in $\mathbf{V}^{(B)}$ (in the sense of $\mathbf{V}$) for which $[\![\varphi(x_0)]\!] = \mathbf{1}$. In symbols,

$$\mathbf{V}^{(B)} \vDash (\exists x)\, \varphi(x) \to (\exists x_0)\, \mathbf{V}^{(B)} \vDash \varphi(x_0).$$

In other words, the *Maximum Principle*

$$(\exists x_0 \in \mathbf{V}^{(B)})\, [\![\varphi(x_0)]\!] = \bigvee_{x \in \mathbf{V}^{(B)}} [\![\varphi(x)]\!]$$

is valid for each formula φ of the ZFC theory.

The last equality accounts for the origin of the term "maximum principle." The proof of the principle represents a simple application of the following

(3) Mixing Principle. Let $(b_\xi)_{\xi\in\Xi}$ be a *partition of unity* in B, i.e. a family of elements of a Boolean-valued algebra B such that

$$\bigvee_{\xi\in\Xi} b_\xi = \mathbf{1}, \quad (\forall \xi, \eta \in \Xi)\,(\xi \neq \eta \to b_\xi \wedge b_\eta = 0).$$

For every family of elements $(x_\xi)_{\xi\in\Xi}$ of the universe $\mathbf{V}^{(B)}$ and a partition of unity $(b_\xi)_{\xi\in\Xi}$ there exists a (unique) mixing of (x_ξ) by (b_ξ), i.e. an element x of the separated universe $\mathbf{V}^{(B)}$ such that $b_\xi \leq [\![x = x_\xi]\!]$ for all $\xi \in \Xi$.

The *mixing* of x of a family (x_ξ) by (b_ξ) is denoted as follows:

$$x = \operatorname{mix}_{\xi\in\Xi}(b_\xi x_\xi) = \operatorname{mix}\{b_\xi x_\xi : \xi \in \Xi\}.$$

A.8. The comparative analysis mentioned above presumes that there is some close interconnection between the universes $\mathbf{V}$ and $\mathbf{V}^{(B)}$. In other words, we need a rigorous mathematical apparatus that would allow us to find out the interplay between the interpretations of one and the same fact in the two models $\mathbf{V}$ and $\mathbf{V}^{(B)}$. The base for such apparatus is constituted by the operations of canonical embedding, descent, and ascent to be presented below. We start with the canonical embedding of the von Neumann universe. Given $x \in \mathbf{V}$, we denote by the symbol $x^\wedge$ the *standard name* of x in $\mathbf{V}^{(B)}$; i.e., the element defined by the following recursion schema:

$$\varnothing^\wedge := \varnothing, \quad \operatorname{dom}(x^\wedge) := \{y^\wedge : y \in x\}, \quad \operatorname{im}(x^\wedge) := \{\mathbf{1}\}.$$

$$\varnothing^\wedge := \varnothing, \quad \operatorname{dom}(x^\wedge) := \{y^\wedge : y \in x\}, \quad \operatorname{im}(x^\wedge) := \{\mathbf{1}\}.$$

Observe some properties of the mapping $x \mapsto x^\wedge$ we need in the sequel.

(1) For an arbitrary $x \in \mathbf{V}$ and a formula φ of ZF we have

$$[\![(\exists y \in x^\wedge)\,\varphi(y)]\!] = \bigvee\{[\![\varphi(z^\wedge)]\!] : z \in x\},$$
$$[\![(\forall y \in x^\wedge)\,\varphi(y)]\!] = \bigwedge\{[\![\varphi(z^\wedge)]\!] : z \in x\}.$$

(2) If x and y are elements of $\mathbf{V}$ then, by transfinite induction, we establish

$$x \in y \leftrightarrow \mathbf{V}^{(B)} \models x^\wedge \in y^\wedge,$$
$$x = y \leftrightarrow \mathbf{V}^{(B)} \models x^\wedge = y^\wedge.$$

In other words, the standard name can be considered as an embedding of $\mathbf{V}$ into $\mathbf{V}^{(B)}$. Moreover, it is beyond a doubt that the standard name sends $\mathbf{V}$ onto $\mathbf{V}^{(2)}$, which fact is demonstrated by the next proposition:

(3) The following holds:

$$(\forall u \in \mathbf{V}^{(2)})\,(\exists! x \in \mathbf{V})\ \mathbf{V}^{(B)} \models u = x^{\wedge}.$$

(4) A formula is called *bounded* or *restricted* if each bound variables in it is restricted by a bounded quantifier; i.e., a quantifier ranging over a specific set. The latter means that each bound variable x is restricted by a quantifier of the form $(\forall x \in y)$ or $(\exists x \in y)$ for some y.

Restricted Transfer Principle. For each bounded formula φ of ZFC and every collection $x_1, \dots, x_n \in \mathbf{V}$ the following equivalence holds:

$$\varphi(x_1, \dots, x_n) \leftrightarrow \mathbf{V}^{(B)} \models \varphi(x_1^{\wedge}, \dots, x_n^{\wedge}).$$

Henceforth, working in the separated universe $\overline{\mathbf{V}}^{(B)}$, we agree to preserve the symbol $x^{\wedge}$ for the distinguished element of the class corresponding to x.

(5) Observe as an example that the Restricted Transfer Principle yields the following assertions:

$$\text{“}\Phi \text{ is a correspondence from } x \text{ to } y\text{”}$$
$$\leftrightarrow \mathbf{V}^{(B)} \models \text{“}\Phi^{\wedge} \text{ is a correspondence from } x^{\wedge} \text{ to } y^{\wedge}\text{”};$$
$$\text{“}f \text{ is a function from } x \text{ to } y\text{”} \leftrightarrow \mathbf{V}^{(B)} \models \text{“}f^{\wedge} \text{ is a function from } x^{\wedge} \text{ to } y^{\wedge}\text{”}$$
$$(\text{moreover, } f(a)^{\wedge} = f^{\wedge}(a^{\wedge}) \text{ for every } a \in x).$$

Thus, the standard name can be considered as a covariant functor of the category of sets (or correspondences) in $\mathbf{V}$ to the appropriate subcategory of $\mathbf{V}^{(2)}$ in the separated universe $\mathbf{V}^{(B)}$.

(6) A set X is *finite* if X coincides with the image of a function on a finite ordinal. In symbols, this is expressed as $\mathrm{Fin}(X)$; hence,

$$\mathrm{Fin}(X) := (\exists n)(\exists f)(n \in \omega \wedge \mathrm{Fnc}\,(f) \wedge \mathrm{dom}(f) = n \wedge \mathrm{im}(f) = X).$$

Obviously, the above formula is not bounded. Nevertheless there is a simple transformation rule for the class of finite sets under the canonical embedding. Denote by $\mathscr{P}_{\mathrm{fin}}(X)$ the class of all finite subsets of X:

$$\mathscr{P}_{\mathrm{fin}}(X) := \{Y \in \mathscr{P}(X) : \mathrm{Fin}(Y)\}.$$

For an arbitrary set X the following holds:

$$\mathbf{V}^{(B)} \models \mathscr{P}_{\mathrm{fin}}(X)^{\wedge} = \mathscr{P}_{\mathrm{fin}}(X^{\wedge}).$$

A.9. Given an arbitrary element x of the (separated) Boolean-valued universe $\mathbf{V}^{(B)}$, we define the *descent* $x{\downarrow}$ of x as

$$x{\downarrow} := \{y \in \mathbf{V}^{(B)} : [\![y \in x]\!] = \mathbf{1}\}.$$

We list the simplest properties of descending:

(1) The class $x{\downarrow}$ is a set, i.e., $x{\downarrow} \in \mathbf{V}$ for each $x \in \mathbf{V}^{(B)}$. If $[\![x \neq \varnothing]\!] = \mathbf{1}$ then $x{\downarrow}$ is a nonempty set.

(2) Let $z \in \mathbf{V}^{(B)}$ and $[\![z \neq \varnothing]\!] = \mathbf{1}$. Then for every formula φ of ZFC we have

$$[\![(\forall x \in z)\, \varphi(x)]\!] = \bigwedge\{[\![\varphi(x)]\!] : x \in z{\downarrow}\},$$
$$[\![(\exists x \in z)\, \varphi(x)]\!] = \bigvee\{[\![\varphi(x)]\!] : x \in z{\downarrow}\}.$$

Moreover, there exists $x_0 \in z{\downarrow}$ such that $[\![\varphi(x_0)]\!] = [\![(\exists x \in z)\, \varphi(x)]\!]$.

(3) Let Φ be a correspondence from X to Y in $\mathbf{V}^{(B)}$. Thus, Φ, X, and Y are elements of $\mathbf{V}^{(B)}$ and, moreover, $[\![\Phi \subset X \times Y]\!] = 1$. There is a unique correspondence $\Phi{\downarrow}$ from $X{\downarrow}$ to $Y{\downarrow}$ such that

$$\Phi{\downarrow}(A{\downarrow}) = \Phi(A){\downarrow}$$

for every nonempty subset A of the set X inside $\mathbf{V}^{(B)}$. The correspondence $\Phi{\downarrow}$ from $X{\downarrow}$ to $Y{\downarrow}$ involved in the above proposition is called the *descent* of the correspondence Φ from X to Y in $\mathbf{V}^{(B)}$.

(4) The descent of the composite of correspondences inside $\mathbf{V}^{(B)}$ is the composite of their descents:

$$(\Psi \circ \Phi){\downarrow} = \Psi{\downarrow} \circ \Phi{\downarrow}.$$

(5) If Φ is a correspondence inside $\mathbf{V}^{(B)}$ then

$$(\Phi^{-1}){\downarrow} = (\Phi{\downarrow})^{-1}.$$

(6) Let I_X be the identity mapping inside $\mathbf{V}^{(B)}$ of the set $X \in \mathbf{V}^{(B)}$. Then

$$(I_X){\downarrow} = I_{X{\downarrow}}.$$

(7) Suppose that $X, Y, f \in \mathbf{V}^{(B)}$ are such that $[\![f : X \to Y]\!] = \mathbf{1}$, i.e., f is a mapping from X to Y inside $\mathbf{V}^{(B)}$. Then $f{\downarrow}$ is a unique mapping from $X{\downarrow}$ to $Y{\downarrow}$ for which

$$[\![f{\downarrow}(x) = f(x)]\!] = \mathbf{1} \quad (x \in X{\downarrow}).$$

By virtue of assertions (1)–(7), we can consider the descent operation as a functor from the category of B-valued sets and mappings (correspondences) to the category of the usual (i.e., in the sense of $\mathbf{V}$) sets and mappings (correspondences).

(8) Given $x_1, \dots, x_n \in \mathbf{V}^{(B)}$, denote by $(x_1, \dots, x_n)^B$ the corresponding ordered n-tuple inside $\mathbf{V}^{(B)}$. Assume that P is an n-ary relation on X inside $\mathbf{V}^{(B)}$; i.e., $X, P \in \mathbf{V}^{(B)}$ and $[\![P \subset X^{n^\wedge}]\!] = \mathbf{1}$ $(n \in \omega)$. Then there exists an n-ary relation P' on $X{\downarrow}$ such that

$$(x_1, \dots, x_n) \in P' \leftrightarrow [\![(x_1, \dots, x_n)^B \in P]\!] = \mathbf{1}.$$

Slightly abusing notation, we denote the relation P' by the same symbol $P{\downarrow}$ and call it the *descent* of P.

A.10. Let $x \in \mathbf{V}$ and $x \subset \mathbf{V}^{(B)}$; i.e., let x be some set composed of B-valued sets or, in other words, $x \in \mathscr{P}(\mathbf{V}^{(B)})$. Put $\varnothing{\uparrow} := \varnothing$ and

$$\mathrm{dom}(x{\uparrow}) = x, \quad \mathrm{im}(x{\uparrow}) = \{\mathbf{1}\}$$

if $x \neq \varnothing$. The element $x{\uparrow}$ (of the separated universe $\mathbf{V}^{(B)}$, i.e., the distinguished representative of the class $\{y \in \mathbf{V}^{(B)} : [\![y = x{\uparrow}]\!] = \mathbf{1}\}$) is called the *ascent* of x.

(1) The following equalities hold for every $x \in \mathscr{P}(\mathbf{V}^{(B)})$ and every formula φ:

$$[\![(\forall z \in x{\uparrow})\, \varphi(z)]\!] = \bigwedge_{y \in x} [\![\varphi(y)]\!],$$

$$[\![(\exists z \in x{\uparrow})\, \varphi(z)]\!] = \bigvee_{y \in x} [\![\varphi(y)]\!].$$

Introducing the ascent of a correspondence $\Phi \subset X \times Y$, we have to bear in mind a possible difference between the domain of departure X and the domain $\mathrm{dom}(\Phi) := \{x \in X : \Phi(x) \neq \varnothing\}$. This difference is inessential for our further goals; therefore, we assume that, speaking of ascents, we always consider everywhere-defined correspondences; i.e., $\mathrm{dom}(\Phi) = X$.

(2) Let $X, Y, \Phi \in \mathbf{V}^{(B)}$, and let Φ be a correspondence from X to Y. There exists a unique correspondence $\Phi{\uparrow}$ from $X{\uparrow}$ to $Y{\uparrow}$ inside $\mathbf{V}^{(B)}$ such that

$$\Phi{\uparrow}(A{\uparrow}) = \Phi(A){\uparrow}$$

is valid for every subset A of the set $\mathrm{dom}(\Phi)$ if and only if Φ is *extensional*; i.e., satisfies the condition

$$y_1 \in \Phi(x_1) \rightarrow [\![x_1 = x_2]\!] \leq \bigvee_{y_2 \in \Phi(x_2)} [\![y_1 = y_2]\!]$$

for $x_1, x_2 \in \mathrm{dom}(\Phi)$. In this event, $\Phi{\uparrow} = \Phi'{\uparrow}$, where $\Phi' := \{(x, y)^B : (x, y) \in \Phi\}$. The element $\Phi{\uparrow}$ is called the *ascent* of the initial correspondence Φ.

(3) The composite of extensional correspondences is extensional. In addition, the ascent of a composite is equal to the composite of the ascents (inside $\mathbf{V}^{(B)}$): On the condition that $\operatorname{dom}(\Psi) \supset \operatorname{im}(\Phi)$ we have

$$\mathbf{V}^{(B)} \models (\Psi \circ \Phi)\uparrow = \Psi\uparrow \circ \Phi\uparrow.$$

Note that if Φ and Φ^{-1} are extensional then $(\Phi\uparrow)^{-1} = (\Phi^{-1})\uparrow$. However, in general, the extensionality of Φ in no way guarantees the extensionality of Φ^{-1}.

(4) It is worth mentioning that if an extensional correspondence f is a function from X to Y then its ascent $f\uparrow$ is a function from $X\uparrow$ to $Y\uparrow$. Moreover, the extensionality property can be stated as follows:

$$[\![x_1 = x_2]\!] \le [\![f(x_1) = f(x_2)]\!] \quad (x_1, x_2 \in X).$$

Given a set $X \subset \mathbf{V}^{(B)}$, we denote by the symbol $\operatorname{mix} X$ the set of all mixings of the form $\operatorname{mix}(b_\xi x_\xi)$, where $(x_\xi) \subset X$ and (b_ξ) is an arbitrary partition of unity. The following assertions are referred to as the *rules for canceling arrows* or the "*descent-ascent*" and "*ascent-descent*" *rules.*

(5) Let X and X' be subsets of $\mathbf{V}^{(B)}$ and $f : X \to X'$ be an extensional mapping. Suppose that $Y, Y', g \in \mathbf{V}^{(B)}$ are such that $[\![Y \ne \varnothing]\!] = [\![g : Y \to Y']\!] = \mathbf{1}$. Then the following relations are valid:

$$X\uparrow\downarrow = \operatorname{mix} X, \quad Y\downarrow\uparrow = Y;$$
$$f\uparrow\downarrow = f, \quad g\downarrow\uparrow = g.$$

(6) From A.8 (6) follows the useful relation:

$$\mathscr{P}_{\text{fin}}(X\uparrow) = \{\theta\uparrow : \theta \in \mathscr{P}_{\text{fin}}(X)\}\uparrow.$$

A.11. Suppose that $X \in \mathbf{V}$, $X \ne \varnothing$; i.e., X is a nonempty set. Let the letter ι denote the standard name embedding $x \mapsto x^\wedge$ $(x \in X)$. Then $\iota(X)\uparrow = X^\wedge$ and $X = \iota^{-1}(X^\wedge\downarrow)$. Using the above relations, we may extend the descent and ascent operations to the case in which Φ is a correspondence from X to $Y\downarrow$ and $[\![\Psi$ is a correspondence from $X^\wedge$ to $Y]\!] = \mathbf{1}$, where $Y \in \mathbf{V}^{(B)}$. Namely, we put $\Phi\Uparrow := (\Phi \circ \iota)\uparrow$ and $\Psi\Downarrow := \Psi\downarrow \circ \iota$. In this case, $\Phi\Uparrow$ is called the *modified ascent* of the correspondence Φ and $\Psi\Downarrow$ is called the *modified descent* of the correspondence Ψ. (If the context excludes ambiguity then we simply speak of ascents and descents using simple arrows.) It is easy to see that $\Psi\Uparrow$ is a unique correspondence inside $\mathbf{V}^{(B)}$ satisfying the relation

$$[\![\Phi\Uparrow(x^\wedge) = \Phi(x)\uparrow]\!] = \mathbf{1} \quad (x \in X).$$

Similarly, $\Psi\downarrow$ is a unique correspondence from X to $Y\downarrow$ satisfying the equality

$$\Psi\downarrow(x) = \Psi(x^\wedge)\downarrow \quad (x \in X).$$

If $\Phi := f$ and $\Psi := g$ are functions then the indicated relations take the form

$$[\![f\uparrow(x^\wedge) = f(x)]\!] = \mathbf{1}, \quad g\downarrow(x) = g(x^\wedge) \quad (x \in X).$$

A.12. (1) A *Boolean set* or a *set with B-structure* or just a *B-set* is a pair (X, d), where $X \in \mathbf{V}$, $X \neq \varnothing$, and d is a mapping from $X \times X$ to the Boolean algebra B which satisfies the following conditions for arbitrary $x, y, z \in X$:

(a) $d(x, y) = 0 \leftrightarrow x = y$;

(b) $d(x, y) = d(y, x)$;

(c) $d(x, y) \leq d(x, z) \vee d(z, y)$.

An example of a B-set is given by any $\varnothing \neq X \subset \mathbf{V}^{(B)}$ if we put

$$d(x, y) := [\![x \neq y]\!] = [\![x = y]\!]^* \quad (x,\ y \in X).$$

Another example is a nonempty X with the "discrete B-metric" d; i.e., $d(x, y) = \mathbf{1}$ if $x \neq y$ and $d(x, y) = \mathbf{0}$ if $x = y$.

(2) Let (X, d) be some B-set. There exist an element $\mathscr{X} \in \mathbf{V}^{(B)}$ and an injection $\iota : X \to X' := \mathscr{X}\downarrow$ such that $d(x, y) = [\![\iota x \neq \iota y]\!]$ $(x, y \in X)$ and every element $x' \in X'$ admits the representation $x' = \mathrm{mix}_{\xi \in \Xi}(b_\xi \iota x_\xi)$, where $(x_\xi)_{\xi \in \Xi} \subset X$ and $(b_\xi)_{\xi \in \Xi}$ is a partition of unity in B. The element $\mathscr{X} \in \mathbf{V}^{(B)}$ is referred to as the *Boolean-valued realization* of the B-set X. If X is a discrete B-set then $\mathscr{X} = X^\wedge$ and $\iota x = x^\wedge$ $(x \in X)$. If $X \subset \mathbf{V}^{(B)}$ then $\iota\uparrow$ is an injection from $X\uparrow$ to $\mathscr{X}$ (inside $\mathbf{V}^{(B)}$).

A mapping f from a B-set (X, d) to a B-set (X', d') is said to be *nonexpanding* if $d(x, y) \geq d'(f(x), f(y))$ for all $x, y \in X$.

(3) Let X and Y be some B-sets, $\mathscr{X}$ and $\mathscr{Y}$ be their Boolean-valued realizations, and ι and $\varkappa$ be the corresponding injections $X \to \mathscr{X}\downarrow$ and $Y \to \mathscr{Y}\downarrow$. If $f : X \to Y$ is a nonexpanding mapping then there is a unique element $g \in \mathbf{V}^{(B)}$ such that $[\![g : \mathscr{X} \to \mathscr{Y}]\!] = \mathbf{1}$ and $f = \varkappa^{-1} \circ g\downarrow \circ \iota$. We also accept the notations $\mathscr{X} := \mathscr{F}^\sim(X) := X^\sim$ and $g := \mathscr{F}^\sim(f) := f^\sim$.

(4) Moreover, the following are valid:

(1) $\mathbf{V}^{(B)} \models f(A)^\sim = f^\sim(A^\sim)$ for $A \subset X$;

(2) If $g : Y \to Z$ is a contraction then $g \circ f$ is a contraction and $\mathbf{V}^{(B)} \models (g \circ f)^\sim = g^\sim \circ f^\sim$;

(3) $\mathbf{V}^{(B)} \models$ "$f^\sim$ is injective" if and only if f is a B-isometry;

(4) $\mathbf{V}^{(B)} \models$ "$f^\sim$ is surjective" if and only if $\bigvee\{d(f(x), y) : x \in X\} = \mathbf{1}$ for every $y \in Y$.

(5) We present an example of a B-set important for the sequel. Let E be a vector lattice and $B := \mathfrak{B}(E)$. Set

$$d(x, y) := \{|x - y|\}^{\perp\perp} \quad (x, y \in E).$$

It is easy that d meets the conditions (b, c) of A.12. At the same time, (a) of A.12 (1) is valid only for E is Archimedean (see 1.1.3).

Thus, (E, d) is a B-set if and only if the vector lattice E is Archimedean.

A.13. Recall that a *signature* is a 3-tuple $\sigma := (F, P, \mathfrak{a})$, where F and P are some (possibly, empty) sets and $\mathfrak{a}$ is a mapping from $F \cup P$ to ω. If the sets F and P are finite then σ is a *finite signature*. In applications we usually deal with algebraic systems of finite signature.

An *n-ary operation* and an *n-ary predicate* on a B-set A are contractive mappings $f : A^n \to A$ and $p : A^n \to B$ respectively. By definition, f and p *are contractive mappings* provided that

$$d(f(a_0, \dots, a_{n-1}), f(a'_0, \dots, a'_{n-1})) \le \bigvee_{k=0}^{n-1} d(a_k, a'_k),$$

$$d_s(p(a_0, \dots, a_{n-1}), p(a'_0, \dots, a'_{n-1})) \le \bigvee_{k=0}^{n-1} d(a_k, a'_k)$$

for all $a_0, a'_0, \dots, a_{n-1}, a'_{n-1} \in A$, where d is the B-metric of A, and d_s is the *symmetric difference* on B; i.e., $d_s(b_1, b_2) := b_1 \triangle b_2$ (cf. 1.1.4).

Clearly, the above definitions depend on B and it would be cleaner to speak of B-operations, B-predicates, etc. We adhere to a simpler practice whenever it entails no confusion.

An *algebraic B-system* $\mathfrak{A}$ of signature σ is a pair (A, ν), where A is a nonempty B-set, the *underlying set* or *carrier* or *universe* of $\mathfrak{A}$, and ν is a mapping such that (a) $\mathrm{dom}(\nu) = F \cup P$; (b) $\nu(f)$ is an $\mathfrak{a}(f)$-ary operation on A for all $f \in F$; and (c) $\nu(p)$ is an $\mathfrak{a}(p)$-ary predicate on A for every $p \in P$.

It is in common parlance to call ν the *interpretation* of $\mathfrak{A}$, in which case the notation f^ν and p^ν are common substitutes for $\nu(f)$ and $\nu(p)$.

The signature of an algebraic B-system $\mathfrak{A} := (A, \nu)$ is often denoted by $\sigma(\mathfrak{A})$; while the carrier A of $\mathfrak{A}$, by $|\mathfrak{A}|$. Since $A^0 = \{\varnothing\}$, the nullary operations and predicates on A are mappings from $\{\varnothing\}$ to the set A and to the algebra B respectively. We agree to identify a mapping $g : \{\varnothing\} \to A \cup B$ with the element $g(\varnothing)$. Each nullary operation on A thus transforms into a unique member of A. Analogously, the set of all nullary predicates on A turns into the Boolean algebra B. If $F := \{f_1, \dots, f_n\}$ and $P := \{p_1, \dots, p_m\}$ then an algebraic B-system of signature σ is often written down as $(A, \nu(f_1), \dots, \nu(f_n), \nu(p_1), \dots, \nu(p_m))$ or even $(A, f_1, \dots, f_n, p_1, \dots, p_m)$. In this event, the expression $\sigma = (f_1, \dots, f_n, p_1, \dots, p_m)$ is substituted for $\sigma = (F, P, \mathfrak{a})$.

A.14. We now address the B-valued interpretation of a first-order language. Consider an algebraic B-system $\mathfrak{A} := (A, \nu)$ of signature $\sigma := \sigma(\mathfrak{A}) := (F, P, \mathfrak{a})$. Let $\varphi(x_0, \dots, x_{n-1})$ be a formula of signature σ with n free variables. Assume given $a_0, \dots, a_{n-1} \in A$. We may readily define the truth value $|\varphi|^{\mathfrak{A}}(a_0, \dots, a_{n-1}) \in B$ of a formula φ in the system $\mathfrak{A}$ for the given values $a_0, \dots, a_{n-1}$ of the variables $x_0, \dots, x_{n-1}$. The definition proceeds as usual by induction on the complexity of φ: Considering propositional connectives and quantifiers, we put

$$
\begin{aligned}
|\varphi \wedge \psi|^{\mathfrak{A}}(a_0, \dots, a_{n-1}) &:= |\varphi|^{\mathfrak{A}}(a_0, \dots, a_{n-1}) \wedge |\psi|^{\mathfrak{A}}(a_0, \dots, a_{n-1}); \\
|\varphi \vee \psi|^{\mathfrak{A}}(a_0, \dots, a_{n-1}) &:= |\varphi|^{\mathfrak{A}}(a_0, \dots, a_{n-1}) \vee |\psi|^{\mathfrak{A}}(a_0, \dots, a_{n-1}); \\
|\neg\varphi|^{\mathfrak{A}}(a_0, \dots, a_{n-1}) &:= |\varphi|^{\mathfrak{A}}(a_0, \dots, a_{n-1})^*; \\
|(\forall x_0)\varphi|^{\mathfrak{A}}(a_1, \dots, a_{n-1}) &:= \bigwedge_{a_0 \in A} |\varphi|^{\mathfrak{A}}(a_0, \dots, a_{n-1}); \\
|(\exists x_0)\varphi|^{\mathfrak{A}}(a_1, \dots, a_{n-1}) &:= \bigvee_{a_0 \in A} |\varphi|^{\mathfrak{A}}(a_0, \dots, a_{n-1}).
\end{aligned}
$$

Now, the case of atomic formulas is in order. Suppose that $p \in P$ symbolizes an m-ary predicate, $q \in P$ is a nullary predicate, and $t_0, \dots, t_{m-1}$ are terms of signature σ assuming values $b_0, \dots, b_{m-1}$ at the given values $a_0, \dots, a_{n-1}$ of the variables $x_0, \dots, x_{n-1}$. By definition, we let

$$
\begin{aligned}
|\varphi|^{\mathfrak{A}}(a_0, \dots, a_{n-1}) &:= \nu(q), && \text{if } \varphi = q^\nu; \\
|\varphi|^{\mathfrak{A}}(a_0, \dots, a_{n-1}) &:= d(b_0, b_1)^*, && \text{if } \varphi = (t_0 = t_1); \\
|\varphi|^{\mathfrak{A}}(a_0, \dots, a_{n-1}) &:= p^\nu(b_0, \dots, b_{m-1}), && \text{if } \varphi = p^\nu(t_0, \dots, t_{m-1}),
\end{aligned}
$$

where d is a B-metric on A.

Say that $\varphi(x_0, \dots, x_{n-1})$ is *valid* in $\mathfrak{A}$ at the given values $a_0, \dots, a_{n-1} \in A$ of $x_0, \dots, x_{n-1}$ and write $\mathfrak{A} \models \varphi(a_0, \dots, a_{n-1})$ provided that $|\varphi|^{\mathfrak{A}}(a_0, \dots, a_{n-1}) = \mathbf{1}_B$. Alternative expressions are as follows: $a_0, \dots, a_{n-1} \in A$ *satisfies* $\varphi(x_0, \dots, x_{n-1})$; or $\varphi(a_0, \dots, a_{n-1})$ holds true in $\mathfrak{A}$. In case $B := \{\mathbf{0}, \mathbf{1}\}$, we arrive at the conventional definition of the validity of a formula in an algebraic system.

Recall that a closed formula φ of signature σ is *tautology* if φ is valid on every algebraic $\mathbf{2}$-system of signature σ.

A.15. Consider algebraic B-systems $\mathfrak{A} := (A, \nu)$ and $\mathfrak{C} := (C, \mu)$ of the same signature σ. The mapping $h : A \to C$ is a *homomorphism* of $\mathfrak{A}$ to $\mathfrak{C}$ provided that, for all $a_0, \dots, a_{n-1} \in A$, the following are valid:

(1) $d_B(h(a_1), h(a_2)) \leq d_A(a_1, a_2)$;

(2) $h(f^\nu) = f^\nu$, $\mathfrak{a}(f) = 0$;

(3) $h(f^\nu(a_0, \dots, a_{n-1})) = f^\nu(h(a_0), \dots, h(a_{n-1}))$, $0 \neq n := \mathfrak{a}(f)$;

(4) $p^\nu(a_0, \dots, a_{n-1}) \leq p^\mu(h(a_0), \dots, h(a_{n-1}))$, $n := \mathfrak{a}(p)$.

A homomorphism h is called *strong* if

(5) $\mathfrak{a}(p) := n \neq 0$ for all $p \in P$, and, for any $c_0, \dots, c_{n-1} \in C$ the following inequality holds:

$$p^\mu(c_0, \dots, c_{n-1})$$
$$\geq \bigvee_{a_0,\dots,a_{n-1} \in A} \{p^\nu(a_0, \dots, a_{n-1}) \wedge d_C(c_0, h(a_0)) \wedge \dots \wedge d_C(c_{n-1}, h(a_{n-1}))\}.$$

If a homomorphism h is injective and (1) and (4) are fulfilled with equality holding; then h is said to be a *isomorphism from* $\mathfrak{A}$ *to* $\mathfrak{C}$. Undoubtedly, each surjective isomorphism h and, in particular, the identity mapping $I_A : A \to A$ are strong homomorphisms. The composite of (strong) homomorphisms is a (strong) homomorphism. Clearly, if h is a homomorphism and h^{-1} is a homomorphism too, then h is an isomorphism.

Note again that in the case of the two-element Boolean algebra $B := \{\mathbf{0}, \mathbf{1}\}$ we come to the conventional notions of homomorphism, strong homomorphism, and isomorphism.

A.16. Before giving a general definition of the descent of an algebraic system, consider the descent of a very simple but important algebraic system, the two-element Boolean algebra. Choose two arbitrary elements, $0,\ 1 \in \mathbf{V}^{(B)}$, satisfying $[\![0 \neq 1]\!] = \mathbf{1}_B$. We may for instance assume that $0 := \mathbf{0}_B^\wedge$ and $1 := \mathbf{1}_B^\wedge$.

(1) *The descent C of the two-element Boolean algebra $\{0, 1\}^B \in \mathbf{V}^{(B)}$ is a complete Boolean algebra isomorphic to B. The formulas*

$$[\![\chi(b) = 1]\!] = b, \quad [\![\chi(b) = 0]\!] = b^* \quad (b \in B)$$

defines an isomorphism $\chi : B \to C$.

Consider now an algebraic system $\mathfrak{A}$ of signature $\sigma^\wedge$ inside $\mathbf{V}^{(B)}$, and let $[\![\mathfrak{A} = (A, \nu)^B]\!] = \mathbf{1}$ for some $A,\ \nu \in \mathbf{V}^{(B)}$. The *descent* of $\mathfrak{A}$ is the pair $\mathfrak{A}{\downarrow} := (A{\downarrow}, \mu)$, where μ is the function determined from the formulas:

$$\mu : f \mapsto (\nu{\downarrow}(f)){\downarrow} \quad (f \in F),$$
$$\mu : p \mapsto \chi^{-1} \circ (\nu{\downarrow}(p)){\downarrow} \quad (p \in P).$$

Here χ is the above isomorphism of the Boolean algebras B and $\{0,1\}^B{\downarrow}$.

In more detail, the modified descent $\nu{\downarrow\!\!\!\!\!\downarrow}$ is the mapping with domain $\mathrm{dom}(\nu{\downarrow\!\!\!\!\!\downarrow}) = F \cup P$. Given $p \in P$, observe $[\![\mathfrak{a}(p)^\wedge = \mathfrak{a}^\wedge(p^\wedge)]\!] = \mathbf{1}$, $[\![\nu{\downarrow\!\!\!\!\!\downarrow}(p) = \nu(p^\wedge)]\!] = \mathbf{1}$ and so

$$\mathbf{V}^{(B)} \models \nu{\downarrow\!\!\!\!\!\downarrow}(p) : A^{\mathfrak{a}(f)^\wedge} \to \{0,1\}^B.$$

It is now obvious that $(\nu{\downarrow\!\!\!\!\!\downarrow}(p)){\downarrow} : (A{\downarrow})^{\mathfrak{a}(f)} \to C := \{0,1\}^B{\downarrow}$ and we may put $\mu(p) := \chi^{-1} \circ (\nu{\downarrow\!\!\!\!\!\downarrow}(p)){\downarrow}$.

Let $\varphi(x_0, \dots, x_{n-1})$ be a fixed formula of signature σ in n free variables. Write down the formula $\Phi(x_0, \dots, x_{n-1}, \mathfrak{A})$ in the language of set theory which formalizes the proposition $\mathfrak{A} \models \varphi(x_0, \dots, x_{n-1})$. Recall that the formula $\mathfrak{A} \models \varphi(x_0, \dots, x_{n-1})$ determines an n-ary predicate on A or, which is the same, a mapping from A^n to $\{0,1\}$. By the Maximum and Transfer Principles, there is a unique element $|\varphi|^{\mathfrak{A}} \in \mathbf{V}^{(B)}$ such that

$$[\![|\varphi|^{\mathfrak{A}} : A^{n^\wedge} \to \{0,1\}^B]\!] = \mathbf{1},$$
$$[\![|\varphi|^{\mathfrak{A}}(a{\uparrow}) = 1]\!] = [\![\Phi(a(0), \dots, a(n-1), \mathfrak{A})]\!] = \mathbf{1}$$

for every $a : n \to A{\downarrow}$. Instead of $|\varphi|^{\mathfrak{A}}(a{\uparrow})$ we will write $|\varphi|^{\mathfrak{A}}\,(a_0, \dots, a_{n-1})$, where $a_l := a(l)$. Therefore, the formula

$$\mathbf{V}^{(B)} \models \text{“}\varphi(a_0, \dots, a_{n-1}) \text{ is valid in } \mathfrak{A}\text{”}$$

holds true if and only if $[\![\Phi(a_0, \dots, a_{n-1}, \mathfrak{A})]\!] = \mathbf{1}$.

(2) *Let $\mathfrak{A}$ be an algebraic system of signature $\sigma^\wedge$ inside $\mathbf{V}^{(B)}$. Then $\mathfrak{A}{\downarrow}$ is a universally complete algebraic B-system of signature σ. In this event,*

$$\chi \circ |\varphi|^{\mathfrak{A}\downarrow} = |\varphi|^{\mathfrak{A}}{\downarrow}$$

foe each formula φ of signature σ.

(3) *Let $\mathfrak{A}$ and $\mathfrak{B}$ be algebraic systems of the same signature $\sigma^\wedge$ inside $\mathbf{V}^{(B)}$. Put $\mathfrak{A}' := \mathfrak{A}{\downarrow}$ and $\mathfrak{B}' := \mathfrak{B}{\downarrow}$. Then, if h is a homomorphism (strong homomorphism) inside $\mathbf{V}^{(B)}$ from $\mathfrak{A}$ to $\mathfrak{B}$ then $h' := h{\downarrow}$ is a homomorphism (strong homomorphism) of the B-systems $\mathfrak{A}'$ and $\mathfrak{B}'$.*

Conversely, if $h' : \mathfrak{A}' \to \mathfrak{B}'$ is a homomorphism (strong homomorphism) of algebraic B-systems then $h := h'{\uparrow}$ is a homomorphism (strong homomorphism) from $\mathfrak{A}$ to $\mathfrak{B}$ inside $\mathbf{V}^{(B)}$.

A.17. *Let $\mathfrak{A} := (A, \nu)$ be an algebraic B-system of signature σ. Then there are $\mathscr{A}$ and $\mu \in \mathbf{V}^{(B)}$ such that the following are fulfilled:*

(1) $\mathbf{V}^{(B)} \models$ "$(\mathscr{A}, \mu)$ *is an algebraic system of signature* $\sigma^{\wedge}$";

(2) *If* $\mathfrak{A}' := (A', \nu')$ *is the descent of* $(\mathscr{A}, \mu)$ *then* $\mathfrak{A}'$ *is a universally complete algebraic* B*-system of signature* σ;

(3) *There is an isomorphism* $\imath$ *from* $\mathfrak{A}$ *to* $\mathfrak{A}'$ *such that* $A' = \mathrm{mix}(\imath(A))$;

(4) *For every formula* φ *of signature* σ *in* n *free variables, the equalities hold*

$$\begin{aligned}|\varphi|^{\mathfrak{A}}(a_0, \dots, a_{n-1}) &= |\varphi|^{\mathfrak{A}'}(\imath(a_0), \dots, \imath(a_{n-1})) \\ &= \chi^{-1} \circ (|\varphi|^{\mathfrak{A}^{\sim}}){\downarrow}(\imath(a_0), \dots, \imath(a_{n-1}))\end{aligned}$$

for all $a_0, \dots, a_{n-1} \in A$ *and* χ *the same as in* A.16 (1).

A.18. Comments. (1) Boolean-valued analysis is a branch of functional analysis which uses a special model-theoretic technique that is embodied in the Boolean-valued models of set theory. The term was coined by G. Takeuti [368]. It is interesting to observe that the invention of Boolean-valued models was not connected with the theory of ordered vector spaces. The necessary language and technical tools were available within mathematical logic since the early 1960s. However, there was no general idea to breathe life into the already-created mathematical apparatus and promote rapid progress in model theory. Such an idea appeared along with P. J. Cohen's discovery; in 1963 he established that the classical continuum hypothesis is absolutely undecidable (in a rigorous mathematical sense), see [67]. It was the Cohen forcing method whose comprehension gave rise to the Boolean-valued models of set theory. Their appearance is commonly associated with the names of D. Scott, R. Solovay, and P. Vopěnka (see [37, 140, 212, 332, 372]).

(2) The forcing method splits naturally into two parts: general and special. The general part comprises the apparatus of Boolean-valued models of set theory, i.e., construction of a Boolean-valued universe $\mathbf{V}^{(B)}$ and interpretation of the set-theoretic propositions in $\mathbf{V}^{(B)}$. Here, a complete Boolean algebra B is arbitrary. The special part consists in constructing specific Boolean algebras B providing some special (usually, pathological and exotic) properties of the objects (for example, K-spaces) obtained from $\mathbf{V}^{(B)}$. Both parts are of independent interest, but their combination yields the most impressive results. In Chapter 8, like in most investigations in Boolean-valued analysis, we use mostly the general part of the forcing method. The special part is widely employed for proving independence or consistency (see [37, 72, 140, 332, 372]). The further progress in Boolean-valued analysis will almost surely be connected with applying the forcing method at full strength.

References

1. Abramovich Yu. A., "On the maximal normed extension of semiordered normed spaces," Vestnik Leningrad Univ., No. 1, 7–17 (1970).
2. Abramovich Yu. A., "Some theorems on normed structures," Vestnik Leningrad Univ., No. 13, 5–11 (1971).
3. Abramovich Yu. A., "Injective envelopes of normed structures," Dokl. Akad. Nauk SSSR, **197**, No. 4, 743–745 (1971).
4. Abramovich Yu. A., "Multiplicative representation of disjointness preserving operators," Indag. Math. (N.S.), **45**, No. 3, 265–279 (1983).
5. Abramovich Yu. A., "Two results about the multiplicative representation of Riesz homomorphisms," An. Univ. Graiova Ser. Mat. Fiz. Chim., **14**, 17–21 (1986).
6. Abramovich Yu. A., "On isometries of normed lattices," Optimization, **43**, 74–80 (1988).
7. Abramovich Yu. A., "Operators preserving disjointness on rearrangement invariant spaces," Pacific J. Math., **148**, No. 2, 201–206 (1991).
8. Abramovich Yu. A., Arenson E. L., and Kitover A. K., "Operators in Banach $C(K)$-modules and their spectral properties," Dokl. Akad. Nauk SSSR, **301**, No. 3, 525–529 (1988).
9. Abramovich Yu. A., Arenson E. L., and Kitover A. K., Banach $C(K)$-Modules and Operators Preserving Disjointness [Preprint, No. 05808–91], Math. Sci. Res. Inst., Berkeley (1991).
10. Abramovich Yu. A., Arenson E. L., and Kitover A. K., Banach $C(K)$-Modules and Operators Preserving Disjointness, Harlow, Longman Sci. Tech. (1992). (Pitman Res. Notes Math. Ser.; **277**.)
11. Abramovich Yu. A., Veksler A. I., and Koldunov A. V., "Disjointness preserving operators, their continuity, and multiplicative representation" in: Linear Operators and Their Applications [in Russian], Leningrad, Leningrad Ped. Inst., 1981.
12. Abramovich Yu. A., Veksler A. I., and Koldunov A. V., "On disjointness-preserving operators," Dokl. Akad. Nauk SSSR, **289**, No. 5, 1033–1036 (1979).

13. Abramovich Yu. A. and Wickstead A. W., "The regularity of order bounded operators into $C(K)$. II," Quart. J. Math. Oxford Ser. 2, **44**, 257–270 (1993).

14. Akilov G. P., Kolesnikov E. V., and Kusraev A. G., "The Lebesgue extension of a positive operator," Dokl. Akad. Nauk SSSR, **298**, No. 3, 521–524 (1988).

15. Akilov G. P. and Kutateladze S. S., Ordered Vector Spaces, Novosibirsk, Nauka (1978).

16. Alexandrov A. D., "Additive set functions in abstract spaces," Mat. Sb. (N.S), **8**, 307–348 (1940); **9**, 563–628 (1941); **13**, 169–238 (1943).

17. Alfsen E. M., Shultz F. W., and Störmer E., "A Gel′fand-Neumark theorem for Jordan algebras," Adv. in Math. **28**, No. 1, 11–56 (1978).

18. Aliprantis C. D. and Burkinshaw O., "On order properties of order bounded transformations," Canad. J. Math., **27**, 666–678 (1977).

19. Aliprantis Ch. D. and Burkinshaw O., Locally Solid Riesz Spaces, Academic Press, New York (1978).

20. Aliprantis C. D. and Burkinshaw O., "The components of a positive operator," Math. Z., **184**, No. 2, 245–257 (1983).

21. Aliprantis C. D. and Burkinshaw O., "On positive order continuous operators," Indag. Math. (N.S.), **45**, No. 1, 1–6 (1983).

22. Aliprantis C. D. and Burkinshaw O., "Projecting onto the band of kernel operators." Houston J. Math., **11**, No. 1, 7–14 (1985).

23. Aliprantis C.D. and Burkinshaw O., Positive Operators, Academic Press, New York (1985).

24. Amemiya J., "A general spectral theory in semi-ordered linear spaces," J. Fac. Sci. Hokkaido Univ. Math., **12**, No. 3, 111–156 (1953).

25. Ando T., "Banachverbände und Positive Projectionen,"Math. Z., **109**, 121–130 (1969).

26. Arendt W., "Spectral properties of Lamperti operators," Indiana Univ. Math. J., **32**, 193–215 (1983).

27. Arendt W. and Hart D. R., "The spectrum of quasi-invertible disjointness preserving operators," J. Funct. Anal., **68**, 149–167 (1986).

28. Arveson W., "Operator algebras and invariant subspaces," Ann. of Math., **100**, No. 2, 433–532 (1974).

29. Arveson W., An Invitation to C^*-Algebras, Berlin etc., Springer-Verlag (1976).

30. Ayupov Sh. A., Classification and Representation of Ordered Jordan Algebras [in Russian], Fan, Tashkent (1986).

31. Badé W. G.,"On Boolean algebras of projections and algebras of operators," Trans. Amer. Math. Soc., **80**, 343–359 (1955).

32. Badé W. G.,"A multiplicity theory for Boolean algebras of projections in Banach spaces," Trans. Amer. Math. Soc., **92**, 508–530 (1958).

33. Basile A., Bukhvalov A. V., Yakubson M. Ya., "The generalized Yosida–Hewitt theorem," Math. Proc. Cambridge Philos. Soc., **116**, 527–233 (1994).

34. Batt J., "Die Verallgemeinerungen des Darstellungssatzes von F. Riesz und ihre Anwendungen," Jahresber. Deutsch. Math.-Verein., **74**, 147–181 (1973).

35. Batty C. J. K. and Robinson D. W., "Positive one-parameter semigroups on ordered Banach spaces," Acta Appl. Math., **74**, 221–296 (1984).

36. Behrends E., M-Structure and Banach-Stone Theorem, Springer-Verlag, Berlin etc. (1979). (Lecture Notes in Math., **736**.)

37. Bell J. L., Boolean-Valued Models and Independence Proofs in Set Theory, Clarendon Press, New York etc. (1985).

38. Berberian S. K., Baer $*$-Rings, Springer-Verlag, Berlin (1972).

39. Bernau C. B., "Orthomorphisms of Archimedean vector lattices," Math. Proc. Cambridge Philos. Soc., **89**, 119–126 (1981).

40. Bernau C. B., "Extensions of vector lattice homomorphisms," J. London Math. Soc., **33**, No. 2, 33–45 (1986).

41. Bernau C. B., "Sums and extensions of vector lattice homomorphisms," Acta Appl. Math., **27**, 33–45 (1992).

42. Bernau C. B., Huijsmans C. B., and de Pagter B., "Sums of lattice homomorphisms," Proc. Amer. Math. Soc., **115**, No. 1, 151–156 (1992).

43. Berz E., "Verallgemeinierung eines Satzes von F. Riesz," Manuscripta Math., Jahresber. Deutsch. Math.-Verein., **2**, 285–299 (1970).

44. Bigard A., Keimel K., and Wolfenstein S., Groupes et Anneaux Réticulés, Springer-Verlag, Berlin etc. (1977). (Lecture Notes in Math., **608**.)

45. Bigard A. and Keimel K., "Sur les endomorphismes conservant les polaires d'un groupe réticulé archimédien," Bull. Soc. Math. France, **97**, 381–398 (1969).

46. Birkhoff G., Lattice Theory, 3rd ed., Amer. Math. Soc. Colloq. Publ., No. 25, Providence (1967).

47. Bohnenblust H. F., "An axiomatic characterization of L_p-spaces," Duke Math. J., **6**, 627–640 (1940).

48. Bonnice W. and Silvermann R., "The Hahn–Banach extension and the least upper bound properties are equivalent," Proc. Amer. Math. Soc., **18**, No. 5, 843–850 (1967).

49. Boole G., Selected Manuscripts on Logic and Its Philosophy, Birkhauser-Verlag, Basel (1997).

50. Boole G., An Investigation of the Laws of Thought on Which Are Founded the Mathematical Theories of Logic and Probabilities, Dover, New York (1957).

51. Bourbaki N., Integration. Ch. 1–8 [in French], Hermann, Paris (1967).

52. Bratteli O. and Robinson D., Operator Algebras and Quantum Statistical Mechanics. Vol. 1, Springer-Verlag, New York etc. (1982).

53. Buck R. C., "Multiplication operators," Pacific J. Math., **11**, 95–103 (1961).

54. Bukhvalov A. V., "Spaces of vector-functions and tensor products," Sibirsk. Mat. Zh., **13**, No. 6, 1229–1238 (1972).

55. Bukhvalov A. V., “On analytical representation of operators with abstract norm,” Dokl. Akad. Nauk SSSR, **208**, No. 5, 1012–1015 (1973).

56. Bukhvalov A. V., “On integral representation of linear operators,” Zap. Nauchn. Sem. Leningrad. Otdel. Mat. Inst. Steklov. (LOMI), **47**, 5–14 (1974).

57. Bukhvalov A. V., “A criterion for integral representability of linear operators,” Funktsional. Anal. i Prilozhen., **9**, No. 1, 51 (1975).

58. Bukhvalov A. V., “On complex interpolation method in spaces of vector-functions and generalized Besov spaces,” Dokl. Akad. Nauk SSSR, **260**, No. 2, 265–269 (1981).

59. Bukhvalov A. V., “Applications of the methods of the theory of order bounded operators to the theory of operators in L_p spaces,” Uspekhi Mat. Nauk, **38**, No. 6, 37–83 (1983).

60. Bukhvalov A. V., “Order bounded operators in vector lattices and spaces of measurable functions,” in: Mathematical Analysis [in Russian], VINITI, Moscow, 1988, pp. 3–63 (Itogi Nauki i Tekhniki, **26**).

61. Bukhvalov A. V., Veksler A. I., and Geĭler V. A., “Normed lattices,” in: Mathematical Analysis [in Russian], VINITI, Moscow, 1980, pp. 125–184 (Itogi Nauki i Tekhniki, **18**).

62. Bukhvalov A. V., Veksler A. I., and Lozanovskiĭ G. Ya., “Banach lattices; some Banach aspects of the theory,” Uspekhi Mat. Nauk, **34**, No. 2, 137–183 (1980).

63. Buskes G. J. H. M., The Hahn–Banach Theorem Surveyed, Inst. Mat. PAN, Warszawa (1993) (Diss. Mat., CCCXXVII).

64. Buskes G. J. H. M. and van Rooij A. C. M., “Hahn–Banach for Riesz homomorphisms,” Indag. Math. (N.S.), **51**, No. 1, 25–34 (1989).

65. Chilin V. I., “Partially ordered involutive Baer algebras,” in: Contemporary Problems of Mathematics. Newest Advances [in Russian], VINITI, Moscow, **27**, 1985, pp. 99–128.

66. Clément Ph., Heijmans H. J. A. M., Angenent S., van Duijn C. J., and de Pagter B., One-Parametr Semigroups, North-Holland, Amsterdam etc. (1987).

67. Cohen P. J., Set Theory and the Continuum Hypothesis, W. A. Benjamin Inc., New York and Amsterdam (1966).

68. Conrad P. F. and Diem J. E., “The ring of polar preserving endomorphisms of an abelian lattice-ordered group,” Illinois J. Math., **15**, 222–240 (1971).

69. Cooper J. L. B., “Coordinated spaces,” Proc. London Math. Soc., **3**, No. 3, 305–327 (1953).

70. Cooper J. L. B., “On generalization of the Köthe coordinated spaces,” Math. Ann., **162**, No. 3, 351–363 (1966).

71. Crenshaw J. A., “Extreme positive linear operators,” Math. Scand., **25**, No. 2, 195–217 (1969).

72. Dales H. and Woodin W., An Introduction to Independence for Analysts, Cambridge University Press, Cambridge (1987).
73. Diestel J., Geometry of Banach Spaces — Selected Topics, Springer-Verlag, Berlin etc. (1975).
74. Diestel J. and Uhl J. J., Vector Measures, Amer. Math. Soc., Providence (1977) (Series: Math. Surveys; **15**).
75. Diestel J. and Uhl J. J., "Progress in vector measures — 1977–1983," in: Measure Theory and Applications (J. M. Belley, J. Dubois, and P. Morales, eds.), Springer-Verlag, Berlin etc., 1983, pp. 144–192. (LectureNotes in Math., **1033**.)
76. Diestel J., Sequences and Series in Banach Spaces, Springer-Verlag, Berlin etc. (1984).
77. Dikanova Z. T., "On some properties of the semiordered space of continuous functions on a bicompactum," Uspekhi Mat. Nauk, **14**, No. 6, 166–172 (1959).
78. Dinculeanu N., Vector Measures, VEB Deutscher Verlag der Wissenschaften, Berlin (1966).
79. Dixmier J., C^*-Algebras and Their Representations [in French], Gauthier-Villars, Paris (1969).
80. Dixmier J., Les Algebres d'Operateurs dans l'Espace Hilbertien (Algebres de von Neumann) [in French], Gauthier-Villars, Paris (1969).
81. Dodds P. G. and Fremlin D. H., "Compact operators in Banach lattices," Israel J. Math., **34**, 287–320 (1979).
82. Dodds P. G. and B. de Pagter, "Orthomorphisms and Boolean algebras of projections," Math. Z., **187**, 361–381 (1984).
83. Dodds P. G. and de Pagter B., "Algebras of unbounded scalar-type spectral operators," Pacific J. Math., **130**, No. 1, 41–74 (1987).
84. Dodds P. G., de Pagter B., and Ricker W., "Reflexivity and order properties of scalar-type spectral operators in locally convex spaces," Trans. Amer. Math. Soc., **293**, No. 1, 355–380 (1986).
85. Dodds P. G. and Ricker W., "Spectral measures and the Bade reflexivity theorem," J. Funct. Anal., **61**, 136–163 (1985).
86. Duhoux M. and Meyer M., "A new proof of the lattice structure of orthomorphisms," J. London Math. Soc., **25**, No. 2, 375–378 (1982).
87. Dunford N. and Schwartz J. T., Linear Operators. Vol. 1: General Theory, John Wiley & Sons, Inc., New York (1988).
88. Dunford N. and Schwartz J. T., Linear Operators. Vol. 2: Spectral Theory. Selfadjoint Operators in Hilbert Space, John Wiley & Sons, Inc., New York (1988).
89. Edwards R., Functional Analysis. Theory and Applications, Holt, Rinehart and Winston, New York etc. (1965).

90. Ellis A. J., "Extreme positive operators," Quart. J. Math. Oxford Ser. 2, **15**, 342–344 (1964).
91. Ellis H. W., "A note on Banach function spaces," Proc. Amer. Math. Soc., **9**, 75–81 (1958).
92. Ellis H. W. and Halperin I., "Function spaces determined by a levelling length function," Canad. J. Math., **5**, No. 4, 576–592 (1953).
93. Espelie M. S., "Multiplicative and extreme positive operators," Pacific J. Math., **48**, 57–66 (1973).
94. Fakhoury H., "Représentations d'opérateurs à valeurs dans $L^1(X,\Sigma,\mu)$," Math. Ann., **240**, No. 3, 203–212 (1979).
95. Fourman M. P., Mulvey C. J., and Scott D. S. (eds.), Applications of Sheaves (Proc. Res. Sympos. Appl. Sheaf Theory to Logic, Algebra and Anal., Univ. Durham, Durham, 1977), Springer-Verlag, Berlin, 1979.
96. Frank M., "Geometrical aspects of Hilbert C^*-modules," Kobenhavns Universitet, Matematisk. Institut. Prep. Series, No. 22 (1993).
97. Fremlin D. H., "Abstract Köthe spaces. II," Proc. Cambridge Philos. Soc., **63**, No. 4, 951–956 (1967).
98. Fremlin D. H., "Tensor product of Archimedean vector lattices," Amer. J. Math., **94**, 777–798 (1972).
99. Fremlin D. H., "Tensor products of Banach lattices," Math. Ann., **211**, 87–106 (1974).
100. Fremlin D. H., Topological Riesz Spaces and Measure Theory, Cambridge University Press, Cambridge (1974).
101. Fremlin D. H., "A direct proof of the Matthes–Wright integral extension theorem," J. London Math. Soc., **11**, No. 3, 276–284 (1975).
102. Freudenthal H., "Teilwise geordnete moduln," Proc. Konink. Nederl. Akad. Wetensch., **39**, 641–651 (1936).
103. Fuchssteiner B. and Lusky W., Convex Cones, North-Holland, Amsterdam etc. (1981).
104. Gierz G., Bundles of Topological Vector Spaces and Their Duality, Springer-Verlag, Berlin, etc. (1982).
105. Gordon E. I., "Real numbers in Boolean-valued models of set theory and K-spaces," Dokl. Akad. Nauk SSSR, **237**, No. 4, 773–775 (1977).
106. Gordon E. I., "K-spaces in Boolean-valued models of set theory," Dokl. Akad. Nauk SSSR, **258**, No. 4, 777–780 (1981).
107. Gordon E. I., "To the theorems of relation preservation in K-spaces," Sibirsk. Mat. Zh., **23**, No. 5, 55–65 (1982).
108. Gordon E. I., Rationally Complete Semiprime Commutative Rings in Boolean Valued Models of Set Theory [in Russian], Gor′kiĭ, VINITI, No. 3286-83 Dep (1983).
109. Gordon E. I., Elements of Boolean Valued Analysis [in Russian], Gor′kiĭ State University, Gor′kiĭ (1991).

110. Gordon E. I. and Lyubetskiĭ V. A., "Some applications of nonstandard analysis in the theory of Boolean valued measures," Dokl. Akad. Nauk SSSR, **256**, No. 5, 1037–1041 (1981).

111. Gordon E. I. and Morozov S. F., Boolean Valued Models of Set Theory [in Russian], Gor′kiĭ State University, Gor′kiĭ (1982).

112. Gray J. D., "The shaping of the Riesz representation theorem: A chapter in the history of analysis," Arch. Hist. Exact. Sci., **31**, 127–187 (1984).

113. Greim P., "Isometries and L^p-structure of separably valued Bochner L^p-spaces," in: Measure Theory and Its Applications (J. M. Belley, J. Dubois, and P. Morales, eds.), Springer-Verlag, Berlin etc., 1983, pp. 209–218. (Lecture Notes in Math., **1033**.)

114. Greim P. and Jameson J. E., "Hilbert spaces have the strong Banach–Stone property," Math. Z., **196**, No. 4, 511–515 (1987).

115. Gribanov Yu. I., "On the measurability of a certain function," Izv. Vyssh. Uchebn. Zaved. Mat., No. 3, 22–26 (1970).

116. Gribanov Yu. I., "Linear operators in perfect function spaces. III," Izv. Vyssh. Uchebn. Zaved. Mat., No. 9, 37–44 (1970).

117. Gribanov Yu. I., "On the measurability of the kernels of integral operators," Izv. Vyssh. Uchebn. Zaved. Mat., No. 7, 31–34 (1972).

118. Gutman A. E., "Banach bundles in the theory of lattice-normed spaces. I. Continuous Banach bundles," Siberian Adv. Math., **3**, No. 3, 1–55 (1993).

119. Gutman A. E., "Banach bundles in the theory of lattice-normed spaces. II. Measurable Banach bundles," Siberian Adv. Math., **3**, No. 4, 8–40 (1993).

120. Gutman A. E., "Banach bundles in the theory of lattice-normed spaces. III. Approximating sets and bounded operators," Siberian Adv. Math., **4**, No. 2, 54–75 (1994).

121. Gutman A. E., "Banach bundles in the theory of lattice-normed spaces," in: Linear Operators Compatible with Order [in Russian], Sobolev Institute Press, Novosibirsk, 1995, pp. 63–211.

122. Gutman A. E., "Locally one-dimensional K-spaces and σ-distributive Boolean algebras," Siberian Adv. Math., **5**, No. 2, 99–121 (1995).

123. Gutman A. E., "Disjointness preserving operators," in: Vector Lattices and Integral Operators (Ed. S. S. Kutateladze), Kluwer Academic Publishers, Dordrecht etc., 1996, pp. 361–454.

124. Gutman A. E. and Koptev A. V., "On the notion of a dual Banach bundle," Siberian Adv. Math., **9**, No. 2, 99–121 (1999).

125. Gutman A. E. and Losenkov G. A., "Function representation of the Boolean-valued universe," Siberian Adv. Math., **4**, No. 1, 99–120 (1998).

126. Hackenbroch W., "Measures admitting extremal extension," Arch. Math., **49**, No. 3, 257–266 (1987).

127. Halmos P. R., Lectures on Boolean Algebras, Van Nostrand, Toronto, New York, and London (1963).

128. Halmos P. and Sunder V., Bounded Integral Operators on L^2 Spaces, Springer-Verlag, New York (1978).

129. Hanshe-Olsen H. and Störmer E., Jordan Operator Algebras, Pitman Publ. Inc., Boston etc. (1984).

130. Hart D. R., "Some properties of disjointness preserving operators," Indag. Math. (N.S.), **47**, No. 2, 183–197 (1985).

131. Hofman K. H. and Keimel K., "Sheaf theoretical concepts in analysis: bundles and sheaves of Banach spaces, Banach $C(X)$-modules," in: Applications of Sheaves, 1979, Springer-Verlag, Berlin. (Lectures Notes in Math.,**753**.)

132. Holmes R. B., Geometric Functional Analysis and Its Applications, Springer-Verlag, Berlin etc. (1975).

133. Horn A. and Tarski A., "Measures in Boolean algebras," Trans. Amer. Math. Soc., **64**, No. 3, 467–497 (1948).

134. Huijsmans C. B. and Luxemburg W. A. J., "An alternative proof of a Radon–Nikodým theorem for lattice homomorphisms," Acta Appl. Math., **27**, 67–71 (1992).

135. Huijsmans C. B. and de Pagter B., "Ideal theory in f-algebras," Trans. Amer. Math. Soc., **269**, 225–245 (1982).

136. Huijsmans C. B. and de Pagter B., "Disjointness preserving and diffuse operators," Compositio Math., **79**, 351–374 (1991).

137. Huijsmans C. B. and de Pagter B., "Some remarks on disjointness preserving operators," Acta Appl. Math., **27**, 73–78 (1992).

138. Ioffe A. D., "A new proof of the equivalence of the Hahn–Banach extension and the least upper bound properties," Proc. Amer. Math. Soc., **82**, No. 3, 385–389 (1981).

139. Jameson G. J. O., Ordered Linear Spaces, Springer-Verlag, Berlin etc. (1970).

140. Jech T., Lectures in Set Theory with Particular Emphasis on the Method of Forcing, Springer-Verlag, Berlin (1971).

141. Jech T. J., "Abstract theory of abelian operator algebras: an application of forcing," Trans. Amer. Math. Soc., **289**, No. 1, 133–162 (1985).

142. Jech T. J., "First order theory of complete Stonean algebras (Boolean-valued real and complex numbers)," Canad. Math. Bull., **30**, No. 4, 385–392 (1987).

143. Jech T. J., "Boolean-linear spaces," Adv. in Math., **81**, No. 2, 117–197 (1990).

144. Johnstone P. T., Stone Spaces, Cambridge University Press, Cambridge and New York (1986).

145. Jonge De and Van Rooij A. C. M., Introduction to Riesz Spaces, Mathematisch Centrum, Amsterdam (1977).

146. Kadison R. V. and Ringrose J. R., Fundamentals of the Theory of Operator Algebras. Vol. 1 and 2, Academic Press, New York (1983, 1986).

147. Kalton N. J., "The endomorphisms of L_p $(0 \leq p \leq 1)$," Indiana Univ. Math. J., **27**, No. 3, 353–381 (1978).

148. Kalton N. J., "Linear operators on L_p for $0 < p < 1$," Trans. Amer. Math. Soc., **259**, No. 2, 319–355 (1980).

149. Kalton N. J., "Representation of operators between function spaces," Indiana Univ. Math. J., **33**, No. 5, 639–665 (1984).

150. Kalton N. J., "Endomorphisms of symmetric function spaces," Indiana Univ. Math. J., **34**, No. 2, 225–247 (1985).

151. Kalton N. J. and Saab P., "Ideal properties of regular operators between Banach lattices. III," J. Math., **29**, No. 3, 382–400 (1985).

152. Kamowitz H., "Compact weighted endomorphisms of $C(X)$," Proc. Amer. Math Soc., **83**, No. 3, 517–521 (1981).

153. Kantorovich L. V., "On semiordered linear spaces and their applications to the theory of linear operations," Dokl. Akad. Nauk SSSR, **4**, No. 1–2, 11–14 (1935).

154. Kantorovich L. V., "On a class of functional equations," Dokl. Akad. Nauk SSSR, **4**, No. 5, 211–216 (1936).

155. Kantorovich L. V., "On some classes of linear operations," Dokl. Akad. Nauk SSSR, **3**, No. 1, 9–13 (1936).

156. Kantorovich L. V., "General forms of some classes of linear operators," Dokl. Akad. Nauk SSSR, **3**, No. 9, 101–106 (1936).

157. Kantorovich L. V., "To the general theory of operations in semiordered spaces," Dokl. Akad. Nauk SSSR, **1**, No. 7, 271–274 (1936).

158. Kantorovich L. V., "On functional equations," Trudy LGU, **3**, No. 7, 17–33 (1937).

159. Kantorovich L. V., "Concerning the problem of moments for a finite interval," Dokl. Akad. Nauk SSSR, **14**, No. 9, 531–536 (1937).

160. Kantorovich L. V., "Sur la continuite et sur la prolongement des operations lineaires," C. R. Acad. Sci. Paris Ser. I Math., **206**, 833–835 (1938).

161. Kantorovich L. V., "The method of successive approximation for functional equations," Acta Math., **71**, 63–97 (1939).

162. Kantorovich L. V. and Akilov G. P., Functional Analysis [in Russian], Nauka, Moscow (1984).

163. Kantorovich L. V., Vulikh B. Z., and Pinsker A. G., Functional Analysis in Semiordered Spaces [in Russian], Gostekhizdat, Moscow and Leningrad (1950).

164. Kaplansky I., "Projections in Banach algebras," Ann. of Math. (2), **53**, 235–249 (1951).

165. Kaplansky I., "Algebras of type I," Ann. of Math. (2), **56**, 460–472 (1952).

166. Kaplansky I., "Modules over operator algebras," Amer. J. Math., **75**, No. 4, 839–858 (1953).

167. Kelley J., "Measures on Boolean algebras," Pacific J. Math., **9**, No. 4, 833–835 (1959).

168. Kevin T. A., "Representation of compact and weakly compact operators on the space of Bochner integrable functions," Pacific J. Math., **92**, No. 2, 257–267 (1981).

169. Khudalov V. T., "On tensor products of ordered Banach space and an arbitrary Banach space," Vestnik Leningrad. Univ., N19, 114-116 (1979).

170. Khudalov V. T., Ordered Banach Spaces and Their Applications, Iriston, Vladikavkaz (1999).

171. Khurana S. S., "Lattice-valued Borel measures," Rocky Mountain J. Math., **6**, No. 2, 377–382 (1976).

172. Khurana S. S., "Lattice-valued Borel measures. II," Trans. Amer. Math. Soc., **235**, No. 2, 205–211 (1978).

173. Kitover A. K., "On disjoint operators in Banach lattices," Dokl. Akad. Nauk SSSR, **250**, No. 4, 800–803 (1980).

174. Kitover A. K., "On mappings of extremally disconnected compacta and disjointnes preserving operators," Optimization, **40**, 130–140 (1987).

175. Kolesnikov E. V., "The fragments of a positive operator," Optimization, **40**, 141–146 (1987).

176. Kolesnikov E. V., "On the fragments of a majorized operator," Optimization, **43**, 81–85 (1988).

177. Kolesnikov E. V., "Decomposition of a positive operator," Sibirsk. Mat. Zh., **30**, No. 5, 70–73 (1989).

178. Kolesnikov E. V., "Some order projections generated by ideals of a vector lattice," Sibirsk. Mat. Zh., **36**, No. 6, 1342–1349 (1995).

179. Kolesnikov E. V., "The shadow of a positive operator," Sibirsk. Mat. Zh., **37**, No. 3, 513–592 (1996).

180. Kolesnikov E. V. and Kusraev A. G., "On functional representation of the Lebesgue extension," Sibirsk. Mat. Zh., **32**, No. 1, 190–194 (1991).

181. Kolesnikov E. V., Kusraev A. G., and Malyugin S. A., On Dominated Operators [in Russian] [Preprint, No. 26], Sobolev Inst. Mat., Novosibirsk (1988).

182. Kollatz L., Funktionalanalysis und Numerische Mathematik, Springer-Verlag, Berlin etc. (1964).

183. Korotkov V. B., Integral Operators [in Russian], Nauka, Novosibirsk (1983).

184. Krasnosel′skiĭ M. A., Positive Solutions to Operator Equations. Chapters of Nonlinear Analysis [in Russian], Fizmatgiz, Moscow (1962).

185. Krasnosel′skiĭ M. A., Lifshits E. A., and Sobolev A. V., Positive Linear Systems. The Method of Positive Operators [in Russian], Nauka, Moscow (1985).

186. Krasnosel′skiĭ M. A. and Rutitskiĭ Ya. B., Convex Functions and Orlicz Spaces, Noordhoff, Groningen (1961).

187. Krasnosel′skiĭ M. A. and Zabreĭko P. P., Geometric Methods of Nonlinear Analysis [in Russian], Nauka, Moscow (1975).

188. Krasnosel'skiĭ M. A., Zabreĭko P. P., Pustyl'nik E. I., and Sobolevskiĭ P. E., Integral Operators in Spaces of Summable Functions [in Russian], Nauka, Moscow (1966).

189. Kreĭn S. G., Petunin Yu. I., and Semënov E. M., Interpolation of Linear Operators [in Russian], Nauka, Moscow (1978).

190. Krengel U., "Uber den absolutbetrag stetiger linearer operatoren und seine anwendung auf ergodische zerlegungen," Math. Scand., **13**, 151–187 (1963).

191. Krivine J. L., "Theoremes de factorisation dans les espaces reticulés," Seminar Maurey-Schwartz, 1973–74, Ecole Politech., Exp 22–23, 22p.

192. Kurepa G., "Tableaux ramifiés d'ensembles, Espaces pseudodistanciés," C. R., **198**, 1563–1565 (1934).

193. Kusraev A. G., On a Property of the Base of the K-space of Regular Operators and Some of Its Applications [in Russian], Sobolev Inst. Mat., Novosibirsk (1977).

194. Kusraev A. G., "Boolean valued analysis of duality between universally complete modules," Dokl. Akad. Nauk SSSR, **267**, No. 5, 1049–1052 (1982).

195. Kusraev A. G., "General desintegration formulas," Dokl. Akad. Nauk SSSR, **265**, No. 6, 1312–1316 (1982).

196. Kusraev A. G., "On Banach–Kantorovich spaces," Sibirsk. Mat. Zh., **26**, No. 2, 119–126 (1985).

197. Kusraev A. G., Vector Duality and Its Applications [in Russian], Nauka, Novosibirsk (1985).

198. Kusraev A. G., "Numeric systems in Boolean-valued models of set theory," in: Proceedings of the VIII All-Union Conference on Mathematical Logic (Moscow), Moscow, 1986, p. 99.

199. Kusraev A. G., "Linear operators in lattice-normed spaces," in: Studies on Geometry in the Large and Mathematical Analysis. Vol. 9 [in Russian], Trudy Inst. Mat. (Novosibirsk), Novosibirsk, 1987, pp. 84–123.

200. Kusraev A. G., "On integral representation of majorized operators," Dokl. Akad. Nauk SSSR, **293**, No. 4, 788–792 (1987).

201. Kusraev A. G., "On functional representation of type I AW^*-algebras," Sibirsk. Mat. Zh., **32**, No. 3, 78–88 (1991).

202. Kusraev A. G., "Dominated operators. I," Siberian Adv. Math., **4**, No. 3, 51–82 (1994).

203. Kusraev A. G., "Dominated operators. II," Siberian Adv. Math., **4**, No. 4, 24–59 (1994).

204. Kusraev A. G., "Dominated operators. III," Siberian Adv. Math., **5**, No. 1, 49–76 (1995).

205. Kusraev A. G., "Dominated operators. IV," Siberian Adv. Math., **5**, No. 2, 99–121 (1995).

206. Kusraev A. G. and Kutateladze S. S., Nonstandard Methods of Analysis, Nauka, Novosibirsk (1990); Kluwer Academic Publishers, Dordrecht (1994).

207. Kusraev A. G. and Kutateladze S. S., "Nonstandard methods for Kantorovich spaces," Siberian Adv. Math., **2**, No. 2, 114–152 (1992).
208. Kusraev A. G. and Kutateladze S. S., "Nonstandard methods in geometric functional analysis," Amer. Math. Soc. Transl. Ser. 2, **151**, 91–105 (1992).
209. Kusraev A. G. and Kutateladze S. S., Subdifferentials: Theory and Applications, Nauka, Novosibirsk (1992); Kluwer Academic Publishers, Dordrecht (1995).
210. Kusraev A. G. and Kutateladze S. S., "The Kreĭn-Mil′man theorem and Kantorovich spaces," Optimization, **51(68)**, 5–18 (1992).
211. Kusraev A. G. and Kutateladze S. S., "Nonstandard methods in functional analysis," in: Interaction Between Functional Analysis, Harmonic Analysis, and Probability Theory, Marcel Deccer Inc., New York, 1995, pp. 301–306.
212. Kusraev A. G. and Kutateladze S. S., Boolean Valued Analysis, Nauka, Novosibirsk (1999); Kluwer Academic Publishers, Dordrecht (1999).
213. Kusraev A. G. and Malyugin S. A., "On order continuous component of a majorizable operator," Sibirsk. Mat. Zh., **28**, No. 4, 127–139 (1987).
214. Kusraev A. G. and Malyugin S. A., Some Questions of the Theory of Vector Measures [in Russian], Inst. Mat. (Novosibirsk), Novosibirsk (1988).
215. Kusraev A. G. and Malyugin S. A., "On atomic decomposition of vector measures," Sibirsk. Mat. Zh., **30**, No. 5, 101–110 (1989).
216. Kusraev A. G. and Malyugin S. A., "The product and projective limit of vector measures," in: Contemporary Problems of Geometry and Analysis. [in Russian], Nauka, Novosibirsk, 1989, pp. 132–152.
217. Kusraev A, G. and Malyugin S. A., "On extension of finitely additive measures," Mat. Zametki, **48**, No. 1, 56–60 (1990).
218. Kusraev A. G. and Strizhevskiĭ V. Z., "Lattice-normed spaces and dominated operators," in: Studies on Geometry and Functional Analysis. Vol. 7 [in Russian], Trudy Inst. Mat. (Novosibirsk), Novosibirsk, 1987, pp. 132–158.
219. Kutateladze S. S., "Supporting sets of sublinear operators," Dokl. Akad. Nauk SSSR, **230**, No. 5, 1029–1032 (1976).
220. Kutateladze S. S., "Subdifferentials of convex operators," Sibirsk. Mat. Zh., **18**, No. 5, 1057–1064 (1977).
221. Kutateladze S. S., "Extreme points of subdifferentials," Dokl. Akad. Nauk SSSR, **242**, No. 5, 1001–1003 (1978).
222. Kutateladze S. S., "Convex operators," Uspekhi Mat. Nauk, **34**, No. 1, 167–196 (1979).
223. Kutateladze S. S., "The Kreĭn-Mil′man theorem and its converse," Sibirsk. Mat. Zh., **21**, No. 1, 130–138 (1980).
224. Kutateladze S. S., "On convex analysis in modules," Sibirsk. Mat. Zh., **22**, No. 4, 118–128 (1981).
225. Kutateladze S. S., Fundamentals of Functional Analysis, Nauka, Novosibirsk (1983).

226. Kutateladze S. S., "Caps and faces of operator sets," Dokl. Akad. Nauk SSSR, **280**, No. 2, 285–288 (1985).
227. Kutateladze S. S., "Fragments of a positive operators," Sibirsk. Mat. Zh., **30**, No. 5, 111–119 (1989).
228. Kutateladze S. S. (ed.), Vector Lattices and Integral Operators, Kluwer Academic Publishers, Dordrecht (1996).
229. Kutateladze S. S., "Nonstandard tools for convex analysis," Math. Japon., **43**, No. 2, 391–410 (1996).
230. Kutateladze S. S. and Rubinov A. M., The Minkowski Duality and Its Applications [in Russian], Nauka, Novosibirsk (1976).
231. Lacey H. E., The Isometric Theory of Classical Banach Spaces, Springer-Verlag, Berlin etc. (1974).
232. Lamperti J., "On the isometries of some function-spaces," Pacific J. Math., **8**, No. 3, 459-466 (1958).
233. Lance E. C., Hilbert C^*-Modules — A Toolkit for Operator Algebraists, Lecture Notes Univ. of Leeds, School of Math., Leeds, England (1993).
234. Lavrič B., "On Freudenthal's spectral theorem," Indag. Math. (N.S.), **48**, 411–421 (1986).
235. Lessner L., "A lattice theoretic characterization of an integral operator," Proc. Amer. Math. Soc., **53**, No. 2, 391–395 (1975).
236. Lessner L., "Lattice properties of integral operators," Proc. Amer. Math. Soc., **72**, No. 3, 497–500 (1978).
237. Levin V. L., "Tensor products and functors in categories of Banach spaces determined by *KB*-lineals," Dokl. Akad. Nauk SSSR, **163**, No. 5, 1058–1060 (1965).
238. Levin V. L., "Tensor products and functors in categories of Banach spaces determined by *KB*-lineals," Trudy Moskov. Mat. Obshch., **20**, 43–82 (1969).
239. Levin V. L., Convex Analysis in Spaces of Measurable Functions and Its Application in Mathematics and Economics [in Russian], Nauka, Moscow (1985).
240. Lindenstrauss J. and Tzafriri L., Classical Banach Spaces. Vol. 1. Sequence Spaces, Springer-Verlag, Berlin etc. (1977).
241. Lindenstrauss J. and Tzafriri L., Classical Banach Spaces. Vol. 2. Function Spaces, Springer-Verlag, Berlin etc. (1979).
242. Lindenstrauss J. and Tzafriri L., Classical Banach Spaces, Springer-Verlag, Berlin etc. (1973).
243. Lipecki Z., "Extensions of positive operators and extreme points. II," Colloq. Math., **42**, 285–289 (1979).
244. Lipecki Z., "Extensions of vector lattice homomorphisms," Proc. Amer. Math. Soc., **79**, 247–248 (1980).
245. Lipecki Z., "Extensions of positive operators and extreme points. III," Colloq. Math., **46**, 263–268 (1982).

246. Lipecki Z., "Extension of vector-lattice homomorphisms revisited," Indag. Math. (N.S.), **47**, 229–233 (1985).

247. Lipecki Z., "Riesz type representation theorems for positive operators," Math. Nachr., **131**, 351–356 (1987).

248. Lipecki Z., Plachky D., and Thomsen W., "Extensions of positive operators and extreme points. I," Colloq. Math., **42**, 279–284 (1979).

249. Lipecki Z. and Thomsen W., "Extensions of positive operators and extreme points. IV," Colloq. Math., **46**, 267–273 (1982).

250. Loomis L. H., An Introduction to Abstract Harmonic Analysis, D. Van Nostrand Company Inc., Princeton (1953).

251. Loś J. and Marczewski E., "Extensions of measure," Fund. Math. Soc., **36**, 267–276 (1949).

252. Lotz H. P., "Extensions and liftings of positive linear mappings in Banach lattices," Trans. Amer. Math. Soc., **211**, 85–100 (1975).

253. Lozanovskiĭ G. Ya., "On almost integral operators in KB-spaces," Vestnik Leningrad Univ. Mat. Mekh. Astronom., No. 7, 35–44 (1966).

254. Lozanovskiĭ G. Ya., "On Banach lattices of Calderon," Dokl. Akad. Nauk SSSR, **172**, No. 5, 1018–1020 (1967).

255. Lozanovskiĭ G. Ya., "On a realization of the space of regular functionals and some of its applications," Dokl. Akad. Nauk SSSR, **188**, No. 3, 522–524 (1969).

256. Lozanovskiĭ G. Ya., "On localized functionals in vector structures," in: Function Theory, Functional Analysis, and Their Applications [in Russian], Khar′kov, 1974, No. 19, pp. 66–80.

257. Lozanovskiĭ G. Ya., "On discrete functionals in Marcinkevicz and Orlicz spaces," in: Study of the Theory of Function in Several Real Variables [in Russian], Yaroslavsk. Univ., Yaroslavl′, No. 2, 132–147 (1978).

258. Luxemburg W. A. J., "Notes on functional Banach spaces. XV, XVI," Indag. Math. (N.S.), **27**, No. 3–4, 415–467 (1965).

259. Luxemburg W. A. J., Some Aspects of the Theory of Riesz Spaces (Univ. Arkansas Lecture Notes Math., **4**), Fayetteville, Arkansas, 1979.

260. Luxemburg W. A. J. and Schep A., "An extension theorem for Riesz homomorphisms," Indag. Math. (N.S.), **41**, 145–154 (1979).

261. Luxemburg W. A. J. and Schep A., "A Radon–Nikodým type theorem for positive operators and a dual," Indag. Math. (N.S.), **40**, 357–375 (1978).

262. Luxemburg W. A. J. and Zaanen A. C., Riesz Spaces. Vol. 1, North-Holland, Amsterdam and London (1971).

263. Luxemburg W. A. J. and Zaanen A. C., "The linear modulus of an order bounded linear transformation," Proc. Ned. Acad. Wetensch., **A74**, No. 5, 422–447 (1971).

264. Maharam D., "On homogeneous measure algebras," Proc. Nat. Acad. Sci., **28**, 108–111 (1942).

265. Maharam D., "An algebraic characterization of measure algebras," Amer. Math., **48**, No. 1, 154–167 (1948).

266. Maharam D., "The representation of abstract measure functions," Trans. Amer. Math. Soc., **65**, No. 2, 279–330 (1949).

267. Maharam D., "Decompositions of measure algebras and spaces," Trans. Amer. Math. Soc., **69**, No. 1, 142–160 (1950).

268. Maharam D., "The representation of abstract integrals," Trans. Amer. Math. Soc., **75**, No. 1, 154–184 (1953).

269. Maharam D., "On kernel representation of linear operators," Trans. Amer. Math. Soc., **79**, No. 1, 229–255 (1955).

270. Maharam D., "On a theorem of von Neumann," Proc. Amer. Math. Soc., **9**, 987–994 (1958).

271. Maharam D., "On positive operators," Contemp. Math., **26**, 263–277 (1984).

272. Malyugin S. A., "On the Hamburger vector moments problem," Optimization, No. 48(65), 124–141 (1990).

273. Malyugin S. A., "Quasi-Radon measures," Sibirsk. Mat. Zh., **32**, No. 5, 101–111 (1991).

274. Malyugin S. A., "The problem of moments in K_σ-space," Sibirsk. Mat. Zh., **34**, No. 2, 110–120 (1993).

275. Manin Yu. I., Provable and Unprovable [in Russian], Sov. Radio, Moscow (1979).

276. Markov A. A., "On mean values and exterior densities," Mat. Sb. (N.S), **4**, 165–191 (1938); **9**, 563–628 (1941); **13**, 169–238 (1943).

277. Matthes K., "Uber die Ausdehnung positiver linear Abbildungen," Math. Nachr., **23**, 223–257 (1961).

278. Maurey B., "Type et cotype dans les espaces munis de structures locales incondicionnelles," Seminar Maurey–Schwartz, 1973–74, Ecole Politech., Exp 24–25.

279. McPolin P. T. N. and Wickstead A. W., "The order boundedness of band preserving operators on uniformly complete vector lattices," Math. Proc. Cambridge Philos. Soc., **97**, No. 3, 481–487 (1985).

280. McShane E. J., "Order-preserving maps and integration process," Ann. of Math. Stud., **31**, (1953).

281. Meyer M., "Le stabilisateur d'un espace vectoriel réticulé," C. R. Acad. Sci. Paris Sér. A, **283**, 249–250 (1976).

282. Meyer-Nieberg P., Banach Lattices, Springer-Verlag, Berlin etc. (1991).

283. Monk J. D. and Bonnet R. (eds.), Handbook of Boolean Algebras. Vol. 1–3, North-Holland, Amsterdam etc. (1989).

284. Murphy G., C^*-Algebras and Operator Theory, Academic Press Inc., Boston (1990).

285. Nagel R. (ed.), One-parameter Semigroups of Positive Operators, Lecture Notes in Math., **1184** (1986).

286. Naĭmark M. A., Normed Rings, Noordhoff, Groningen (1959).

287. Nakano H., "Teilweise geordnete algebra," Japan J. Math., **17**, 425–511 (1950).

288. Nakano H., Modulated Semi-Ordered Linear Spaces, Maruzen Co., Tokyo (1950).

289. Nakano H., Semi-ordered Linear Spaces, Japan Society for the Promotion of Science, Tokyo (1955).

290. Navodnov V. G., "On integral representation of operators acting from a Banach space of measurable vector-functions into a Banach space," Izv. Vyssh. Uchebn. Zaved. Mat., No. 3, 82–84 (1983).

291. Navodnov V. G., "On the vector-valued analog of the Dunford–Pettice theorem," Izv. Vyssh. Uchebn. Zaved. Mat., No. 3, 55–60 (1987).

292. Neumann J., von, "Charakterisierung des Spectrums eines Integraloperators," Actualités Sci. Indust. (Paris), No. 229 (1935).

293. Neumann J., von, Collected Works. Vol. 3: Rings of Operators, Pergamon Press, New York etc. (1961).

294. Neumann J., von, Collected Works. Vol. 4: Continuous Geometry and Other Topics, Pergamon Press, Oxford etc. (1962).

295. Neumann J., von, Selected Works on Functional Analysis [in Russian], Nauka, Moscow (1987).

296. Nielsen N. J., "On Banach ideals determined by Banach lattices and their applications," Dissertationes Math. (Rozprawy Mat.), **109**, 1–66 (1973).

297. Nielsen N. J. and Szulga J., "p-lattice summing operators," Math. Nachr., **119**, 119–130 (1984).

298. Nishimura H., "Some connections between Boolean valued analysis and topological reduction theory for C^*-algebras," Z. Math. Logik Grundlag. Math., **36**, No. 5, 471–479 (1990).

299. Nishimura H., "On a duality between Boolean valued analysis and topological reduction theory," Math. Logic Quart., **39**, No. 1, 23–32 (1993).

300. Nishimura H., "On the duality between Boolean-valued analysis and reduction theory under the assumption of separability," Internat. J. Theoret. Phys., **32**, No. 3, 443–488 (1993).

301. Nishimura H., "Boolean valued and Stone algebra valued measure theories," Math. Logic Quart., **40**, No. 1, 69–75 (1994).

302. Ogasawara T., "Vector lattices I and II [in Japanese]," Sci. Hirishima Univ. Ser. A., **12**, 37–100 (1942); **13**, 41–161 (1944).

303. Ozawa M., "Boolean valued interpretation of Hilbert space theory," J. Math. Soc. Japan, **35**, No. 4, 609–627 (1983).

304. Ozawa M., "Boolean valued analysis and type I AW^*-algebras," Proc. Japan Acad. Ser. A Math. Sci., **59A**, No. 8, 368–371 (1983).

305. Ozawa M., "A classification of type I AW^*-algebras and Boolean valued analysis," J. Math. Soc. Japan, **36**, No. 4, 589–608 (1984).

306. Ozawa M., "A transfer principle from von Neumann algebras to AW^*-algebras," J. London Math. Soc., **32**, No. 2, 141–148 (1985).
307. Ozawa M., "Nonuniqueness of the cardinality attached to homogeneous AW^*-algebras," Proc. Amer. Math. Soc., **93**, 681–684 (1985).
308. Ozawa M., "Boolean-valued interpretation of Banach space theory and module structures of von Neumann algebras," Nagoya Math. J., **117**, 1–36 (1990).
309. de Pagter B., "The components of a positive operator," Indag. Math. (N.S.), **48**, No. 2, 229–241 (1983).
310. de Pagter B., "A note on disjointness preserving operators," Proc. Amer. Math. Soc., **90**, No. 4, 543–549 (1984).
311. de Pagter B., "The space of extended orthomorphisms in Riesz space," Pacific J. Math., **112**, No. 1, 193–210 (1984).
312. de Pagter B. and Wnuk W., "Some remarks on Banach lattices with non-atomic duals," Indag. Math. (N.S.), **1**, No. 3, 391–394 (1990).
313. Panchapagesan T. V. and Palled Sh. V., "On vector lattice-valued measures. I," Math. Slovaca, **33**, No. 3, 269–292 (1983).
314. Panchapagesan T. V. and Palled Sh. V., "On vector lattice-valued measures. II," J. Austral. Math. Soc. (Ser. A), **40**, No. 2, 234–252 (1986).
315. Pannenberg M., "Two notions of stable rank for Banach lattices containing quasi-interior positive elements," Indag. Math. (N.S.), **3**, No. 4, 403–418 (1992).
316. Paschke W. L., "Inner product spaces over B^*-algebras," Trans. Amer. Math. Soc., **182**, 443—468 (1973).
317. Peressini A. I., Ordered Topological Vector Spaces, Harper&Row, New York (1967).
318. Phelps R. R., "Extreme positive operators and homomorphisms," Trans. Amer. Math. Soc., **108**, 265–274 (1963).
319. Phuong-Cac, "Generalized Köthe function spaces. I," Math. Proc. Cambridge Philos. Soc., **65**, No. 3, 601–611 (1969).
320. Plachky D., "Extremal and monogenic set function," Proc. Amer. Math. Soc., **54**, 193–196 (1976).
321. Putten B., van, Disjunctive Linear Operators and Partial Multiplication in Riesz Spaces, Meppel, Krips Repro, 1980.
322. Radnaev V. A., "On latticially nonatomic subdifferentials," Sibirsk. Mat. Zh., **35**, No. 4, 853–859 (1994).
323. Radnaev V. A., "On n-disjoint operators," Siberian Adv. Math., **7**, No. 4, 44–78 (1997).
324. Radnaev V. A., On Metric n-Indecomposability in Ordered Lattice Normed Spaces and Its Applications, PhD Thesis, Sobolev Institute Press, Novosibirsk (1997).
325. Radnaev V. A., On Metric n-Indecomposability and n-Disjoint Operators [in Russian] [Preprint, No. 28], Sobolev Institute Press, Novosibirsk (1997).

326. Rao K. P. S. B. and Rao M. B., Theory of Charges. A study of Finitely Additive Measures, Academic Press, London and New York (1983).

327. Rice N. M., "Multiplication in vector lattices," Canad. J. Math., **20**, 1136–1149 (1968).

328. Riečan B., "An extension of the Daniel integration scheme," Mat. Čas, **25**, No. 3, 211–219 (1975).

329. Riečan B., "A simplified proof of the Daniel integral extension theorem in ordered spaces," Math. Slovaca, **32**, No. 1, 75–79 (1982).

330. Riečan B., "On the Kolmogorov consistency theorem for Riesz space valued measures," Math. Slovaca, **48/49**, 173–180 (1986).

331. Riesz F., "Sur la decomposition des operations fonctionnelles," Atti Congresso Internaz. Bologna 1928, **3**, 143–148 (1930).

332. Rosser J. B., Simplified Independence Proofs. Boolean Valued Models of Set Theory, Academic Press, New York and London (1969).

333. Sakai S. C^*-Algebras and W^*-Algebras, Springer-Verlag, Berlin etc. (1971).

334. Sarymsakov T. A. et al., Ordered Algebras [in Russian], Fan, Tashkent (1983).

335. Schaefer H. H., Topological Vector Spaces, Springer-Verlag, New York etc. (1971).

336. Schaefer H. H., Banach Lattices and Positive Operators, Springer-Verlag, Berlin etc. (1974).

337. Schep A. R., "Order continuous components of operators and measures," Indag. Math. (N.S.), **40**, 110–117 (1978).

338. Schep A. R., "Kernel operators," Proc. Konink. Nederl. Akad. Wetensch., **A82**, No. 1, 39–53 (1979).

339. Schochetman I. E., Kernels and Integral Operators for Continuous Sums of Banach Spaces, Amer. Math. Soc., Providence, RI (1978). (Memoirs, vol. 14, No. 202.)

340. Schröder J., "Das Iterationsverfahren bei allgemeinierem Abshtandsbegriff," Math. Z., **66**, 111–116 (1956).

341. Schwarz H.-V. Banach Lattices and Operators, Teubner, Leipzig (1984).

342. Semadeni Zb., Banach Spaces of Continuous Functions, Polish Scientific Publishers, Warsaw (1971).

343. Shamaev I. I., "On countable extension of measures and σ-integrals with values in vector lattices," Mat. Zametki, **39**, No. 5, 757–765 (1986).

344. Shamaev I. I., "On decomposition and representation of regular operators," Sibirsk. Mat. Zh., **30**, No. 2, 192–202 (1989).

345. Shamaev I. I., "An operator disjoint from lattice homomorphisms, Maharam operators, and integral operators," Optimization, **46**, 154–159 (1990).

346. Shamaev I. I., "Operators of diffuse type," Dokl. Akad. Nauk SSSR, **316**, No. 1, 47–50 (1991).

347. Shamaev I. I., Topological Methods for the Theory of Regular Operators in Kantorovich Spaces (Doctoral Thesis) [in Russian], Yakutsk (1994).
348. Shilov G. E. and Gurevich B. L., Integral, Measure, and Derivative [in Russian], Nauka, Moscow (1967).
349. Shotaev G. N., "On bilinear functionals in lattice normed spaces," Optimization, **37**, 38–50 (1986).
350. Schotaev G. N., "On tensor product of lattice normed spaces," Sibirsk Mat. Zh., **29**, No. 4, 195-201 (1988).
351. Schotaev G. N., "On tensor product of Banach spaces with mixed norm," Optimization, **45(62)**, No. 4, 117-129 (1989).
352. Sikorski R., Boolean Algebras [Russian translation], Mir, Moscow (1969).
353. Sobolev V. I., "On a poset valued measure of a set, measurable functions and some abstract integrals," Dokl. Akad. Nauk SSSR, **91**, No. 1, 23–26 (1953).
354. Solovay R. M., "A model of set theory in which every set of reals is Lebesgue measurable," Ann. of Math., **92**, No. 2, 1–56 (1970).
355. Solovay R. and Tennenbaum S., "Iterated Cohen extensions and Souslin's problem," Ann. Math., **94**, No. 2, 201–245 (1972).
356. Solov′ëv Yu. P. and Troitskiĭ E. V., C^*-Algebras and Elliptic Operators in Differential Topology [in Russian], Factorial, Moscow (1996).
357. Sourour A. R., "Operators with absolutely bounded matrices," Math. Z., **162**, No. 2, 183–187 (1978).
358. Sourour A. R., "The isometries of $L^p(\Omega, X)$," J. Funct. Anal., **30**, No. 2, 276–285 (1978).
359. Sourour A. R., "Pseudo-integral operators," Trans. Amer. Math. Soc., **253**, 329–363 (1979).
360. Sourour A. R., "A note on integral operators," Acta Sci. Math., **41**, No. 43, 375–379 (1979).
361. Sourour A. R., "Characterization and order properties of pseudo-integral operators," Pacific J. Math., **99**, No. 1, 145–158 (1982).
362. Strizhevskiĭ V. Z., "Disjointness of some classes of order continuous operators," Dokl. Akad. Nauk SSSR, **280**, No. 3, 556–559 (1985).
363. Sunder V. S., An Invitation to von Neumann Algebras, Springer-Verlag, New York etc. (1987).
364. Synnatzchke J., "On almost integral operators in K-spaces," Vestnik Leningrad. Univ. Mat., No. 13, 81–89 (1971).
365. Synnatzchke J., "On the adjoint to a regular operator and some of its applications to the question of complete continuity and weak complete continuity of regular operators," Vestnik Leningrad. Univ. Mat., No. 1, 60–65 (1972).
366. Takesaki M., Theory of Operator Algebras. Vol. I, Springer-Verlag, New York (1979).
367. Takeuti G., Two Applications of Logic to Mathematics, Princeton Univ. Press, Princeton, (1978).

368. Takeuti G., "Boolean valued analysis," Applications of Sheaves (Proc. Res. Sympos. Appl. Sheaf Theory to Logic, Algebra and Anal., Univ. Durham, Durham, 1977), Springer-Verlag, Berlin etc., 1979, pp. 714–731. (Lecture Notes in Math., **753**.)

369. Takeuti G., "Boolean completion and m-convergence," in: Categorical Aspects of Topology and Analysis (Ottawa, Ont., 1980), Springer-Verlag, Berlin etc., 1982, pp. 333–350. (Lecture Notes in Math., **915**.)

370. Takeuti G., "Von Neumann algebras and Boolean valued analysis," J. Math. Soc. Japan, **35**, No. 1, 1–21 (1983).

371. Takeuti G., "C^*-algebras and Boolean valued analysis," Japan. J. Math. (N.S.), **9**, No. 2, 207–246 (1983).

372. Takeuti G. and Zaring W. M., Axiomatic Set Theory, Springer-Verlag, New York (1973).

373. Tibilov K. T., "On pseudointegral operators in the spaces of measurable vector-functions," Sibirsk. Mat. Zh., **31**, No. 5, 149–156 (1990).

374. To T-O., "The equivalence of the least upper bound property in ordered vector spaces," Proc. Amer. Math. Soc., **30**, No. 2, 287–296 (1970).

375. Ionescu Tulcea A. and Ionescu Tulcea C., Topics in the Theory of Lifting, Springer-Verlag, Berlin etc. (1969).

376. Veksler A. I., "On homomorphisms between classes of regular operators in K-lineals and their completions," Izv. Vyssh. Uchebn. Zaved. Mat., **14**, No. 1, 48–57 (1960).

377. Veksler A. I., "On lattice orderability of algebras and rings," Dokl. Akad. Nauk SSSR, **164**, No. 2, 259–262 (1965).

378. Veksler A. I., "On a new construction of the Dedekind completion of some lattices and l-groups with division," Sibirsk. Mat. Zh., **10**, No. 6, 1206–1213 (1969).

379. Veksler A. I., "Banach cyclic spaces and Banach lattices," Dokl. Akad. Nauk SSSR, **213**, No. 4, 770–773 (1973).

380. Veksler A. I., "Projection properties of vector lattices and the Freudenthal theorem," Math. Nachr., **74**, 7–25 (1976).

381. Veksler A. I. and Geĭler V. A., "On the order and disjoint completeness of semiordered linear spaces," Sibirsk. Mat. Zh., **13**, No. 1, 43–51 (1972).

382. Vladimirov D. A., "On completeness of semiordered space," Uspekhi Mat. Nauk, **15**, No. 2, 165–172 (1960).

383. Vladimirov D. A., "On countable additivity of Boolean measure," Vestnik Leningrad Univ., No. 19, 5–15 (1961).

384. Vladimirov D. A., Boolean Algebras [in Russian], Nauka, Moscow (1969).

385. Vulikh B. Z., "On multiplicative linear operations," Dokl. Akad. Nauk SSSR, **41**, No. 4, 148–151 (1943).

386. Vulikh B. Z., "Analytic representation of multiplicative linear operations," Dokl. Akad. Nauk SSSR, **41**, No. 5, 197–201 (1943).

387. Vulikh B. Z., "Multiplication in semiordered linear spaces and its application to the theory of operations. II," Mat. Sb., **22**, No. 2, 267–317 (1948).

388. Vulikh B. Z., Introduction to the Theory of Partially Ordered Spaces [in Russian], Fizmatgiz, Moscow (1961).

389. Vulich B. Z., Introduction in the Theory of Cones in Normed Spaces [in Russian], Kalinin Univ. Press, Kalinin (1977).

390. Vulich B. Z., Special Problems of Geometry of Cones in Normed Spaces [in Russian], Kalinin Univ. Press, Kalinin (1978).

391. Vulikh B. Z. and Lozanovskiĭ G. Ya., "On representation of completely linear and regular functionsals in semiordered spaces," Mat. Sb., **84**, No. 3, 331–352 (1971).

392. Weis L., "Integral operators and changes of density," Indiana Univ. Math. J., **31**, No. 1, 83–96 (1982).

393. Weis L., "Decompositions of positive operators and some of their applications," in: Functional Analysis: Survey and Recent Results. Vol. 3: Proc. 3rd Conf., Paderborn, 24–29 May, 1983, Amsterdam (1984).

394. Weis L., "On the representation of order continuous operators by random measures," Trans. Amer. Math. Soc., **285**, No. 2, 535–563 (1984).

395. Weis L., "The range of an operator in $C(X)$ and its representing stochastic kernel," Arch. Math., **46**, 171–178 (1986).

396. Wickstead A. W., "Representation and duality of multiplication operators on Archimedean Riesz spaces," Compositio Math., **35**, No. 3, 225-238 (1977).

397. Wickstead A. W., "Extreme structure of cones of operators," Quart. J. Math. Oxford., **32**, 239-253 (1981).

398. Wickstead A. W., "Stone algebra valued measures: integration of vector-valued functions and Radon–Nikodým type theorems," Proc. London Math. Soc., **45**, No. 2, 193-226 (1982).

399. Wong Y.-Ch. and Ng K.-F., Partially Ordered Topological Vector Spaces, Clareandon Press, Oxford (1973).

400. Wright J. D. M., "A Radon–Nikodým theorem for Stone algebra valued measures," Trans. Amer. Math. Soc., **139**, 75–94 (1969).

401. Wright J. D. M., "Stone algebra valued measures and integrals," Proc. London Math. Soc. Proc. London Math. Soc., **19**, No. 3, 107-122 (1969).

402. Wright J. D. M., "The measure extension problem for vector lattices," Ann. Inst. Fourier (Grenoble), **21**, 65-68 (1971).

403. Wright J. D. M., "The solution of a problem of Sikorski and Mattes," Bull. London Math. Soc., **3**, 52–54 (1971).

404. Wright J. D. M., "Vector lattice measure on locally compact spaces," Math. Z. **120**, 1903-203 (1971).

405. Wright J. D. M., "An extension theorem," J. London Math. Soc. **7**, No. 2, 531-539 (1973).

406. Wright J. D. M., "An algebraic characterization of vector lattices with the Borel regularity property,"**7**, No. 2, 277-285 (1973).

407. Yanovskiĭ L. P., "Summing, order summing, and characterization of *AL*-spaces," Sibirsk. Mat. Zh., **20**, No. 2, 402–408 (1979).

408. Zaanen A. C., "Examples of orthomorphisms," J. Approx. Theory, **13**, No. 2, 192–204 (1975).

409. Zaanen A. C. Riesz Spaces. Vol. 2, North-Holland, Amsterdam etc. (1983).

410. Zaanen A. C., "Measurable functions and integral operators, Nieuw arch. wisk., **3**, No. 2, 167–205 (1985).

411. Zaanen A. C., Introduction to Operator Theory in Riesz Spaces, Springer-Verlag, Berlin etc. (1997).

412. Zhdanov S. I., Integral Representations of Operators in Locally Convex and Partially Ordered Spaces [in Russian], Dis. Kand. Fiz.-Mat. Nauk, Novosibirsk Univ., Novosibirsk (1975).

Symbol Index

$x \triangle y$, 6
B/J, 7
$\prod_{\xi \in \Xi} B_\xi$, 8
$\mathscr{P}(X)$, 9
2^X, 9
$\mathrm{Clop}(X)$, 9
$\mathrm{RC}\,(X)$, 9
$\mathrm{RO}\,(X)$, 9
$\mathscr{B}or(X)$, 9
$\mathbb{R}$, 10
$(\Omega, \mathscr{B}, \mu)$, 10
$B(\Omega)$, 10
$B(\Omega, \mathscr{B}, \mu)$, 10
$\mathrm{Clop}_\sigma(Q)$, 14
$x_1 \vee \cdots \vee x_n$, 18
$x_1 \wedge \cdots \wedge x_n$, 18
x^+, 18
x^-, 18
$|x|$, 18
$[a, b]$, 19
$x \perp y$, 19
$\mathfrak{E}(u)$, 19
$M^\perp$, 19
$\mathfrak{B}(E)$, 19
$[K]$, 20
π_K, 20
$\mathfrak{P}(E)$, 20
$[u]$, 20
$x = o\text{-}\lim x_\alpha$, 20
$x_\alpha \xrightarrow{(o)} x$, 20
$x = r\text{-}\lim_{\alpha \in \mathrm{A}} x_\alpha$, 20
$x_\alpha \xrightarrow{(r)} x$, 20
$\mathscr{I}(E)$, 21
$\mathscr{I}_p(E)$, 21
$E(u)$, 21
e_λ^x, 23
$\{f < \lambda\}$, 25
$\{f \leq \lambda\}$, 25
$C_\infty(Q)$, 26
$C(Q)$, 27
$\mathfrak{K}(B)$, 27
$\mathbb{C}$, 31
$M(\Omega, \mathscr{A}, \mathscr{N})$, 31
$L^0(\Omega, \mathscr{A}, \mu)$, 31
$M(\Omega, \mathscr{A}, \mu)$, 31
$L^\infty(\Omega, \mathscr{A}, \mu)$, 31
$\mathscr{L}^\infty(\Omega, \mathscr{A}, \mu)$, 31
$E^\sim$, 34
$E_+^\sim$, 34
$E_n^\sim$, 34
E', 34
$\mathscr{H}(\mathbb{R}^l)$, 37
$\mathrm{ba}(\mathscr{A})$, 40
$\mathrm{ba}_+(\mathscr{A})$, 40
$x \perp y$, 46
$M^\perp$, 46

$\mathscr{B}(X)$, 46
$|M|$, 46
$\mathscr{P}(X)$, 47
$\mathbb{B}(A)$, 54
X^*, 56
$r(U)$, 57
$o(U)$, 57
$d(U)$, 57
$r_0(U)$, 57
$d_0(U)$, 57
mX, 57
$\varkappa : X \to X^{**}$, 57
h-lim, 60
$\hat{X}$, 61
$C_b(Q, Y)$, 62
$C_\infty(Q, X)$, 63
$C_\#(Q, X)$, 63
$E \otimes X$, 63
$C_\infty(Q, X|Z)$, 65
$E_w(X, Z)$, 65
$E_w(X')$, 65
$\mathscr{L}(X, Y)$, 65
$\mathfrak{M}_Q(X, Y)$, 65
$E_w(\mathscr{L}(X, Y), Z)$, 66
$E_w(\mathscr{L}(X, Y'))$, 66
$X \hat{\otimes} Y$, 66
$L^0(\mu, X)$, 66
$L^0(\Omega, \Sigma, \mu, X)$, 66
$E(X)$, 66
$E[F]$, 67
$L^0(\mu, X|Z)$, 68
$L^0_Z(\Omega, \Sigma, \mu, X)$, 69
$E_w(X, Z)$, 69
$E_w(X')$, 69
$E_w(\mathscr{L}(X, Y), Z)$, 70
$E_{ws}(\mathscr{L}(X, Y))$, 70
$S(Q, \mathscr{X})$, 72
$\|\cdot\|$, 72
$E(\mathscr{X})$, 73
$C_\#(Q, \mathscr{X})$, 74
$\mathrm{Hom}_Q(\mathscr{X}, \mathscr{Y})$, 75
$L^0(\Omega, \Sigma, \mu, \mathscr{X})$, 80
$\mathscr{L}^\infty(\Omega, \mathscr{X})$, 81
$L^\infty(\Omega, \mathscr{X})$, 81
$L^r(E, F)$, 90
$L^\sim(E, F)$, 90
$L_+(E, F)$, 90
$L^{\mathrm{ext}}(G, F)$, 93
$\mathscr{N}(U)$, 93
$\mathscr{C}_T$, 93
π_G, 93
$\pi_{[e]}$, 94
π_e, 94
∂p, 97
$L(X, E)$, 98
$\mathrm{epi}(p)$, 98
$\ker(T)$, 99
$L^\sim_n(E, F)$, 99
$L^\sim_{n\sigma}(E, F)$, 99
$L^\sim_s(E, F)$, 100
$L^\sim_{s\sigma}(E, F)$, 100
$L_{\mathscr{A}}(E, F)$, 102
$\mathscr{I}_d(E)$, 102
$\mathscr{I}_{\sigma d}(E)$, 102
T_n, 102
$T_{n\sigma}$, 102
$\mathrm{Hom}(E, F)$, 109
$L^\sim_a(E, F)$, 109
$L^\sim_d(E, F)$, 109
$\mathrm{Orth}(D, D')$, 109
$\mathrm{Orth}(D, E)$, 109
$\mathrm{Orth}^\infty(E)$, 109
$\mathrm{Orth}(E)$, 110
$\mathscr{Z}(E)$, 110
$\mathrm{Ch}(p)$, 112
$\partial_h p$, 115
$\mathscr{C}_T$, 118
$S \ll T$, 118
$l_1(\mathfrak{A}, F)$, 118
$M^\downarrow$, 129

$M^{\uparrow}$, 129
$M^{\downharpoonleft}$, 129
$M^{\upharpoonleft}$, 129
$M^{\downarrow\uparrow}$, 129
$M^{\downharpoonleft\upharpoonleft}$, 129
$\mathscr{S}(\overline{X})$, 131
$\mathscr{S}(\Phi)$, 131
$\langle K \rangle$, 131
$\langle x \rangle$, 131
$\mathscr{A}(\Phi)$, 132
$\mathscr{E}_0(P)$, 136
$\mathrm{maj}(T)$, 142
$M(X, Y)$, 142
$\mathscr{L}(X, Y)$, 143
$\mathscr{L}_\Lambda(X, Y)$, 144
$\mathrm{ba}(\mathscr{A}, F)$, 154
$\mathrm{ba}_+(\mathscr{A}, F)$, 154
$\mathrm{da}(\mathscr{A}, Y)$, 154
$\mathrm{maj}\,(\mu)$, 154
$S(\mathscr{A})$, 154
$C(\mathscr{A})$, 154
$M_n(X, Y)$, 156
$M_{\sigma n}(X, Y)$, 156
$M_G(X, Y)$, 161
$L_G^{\sim}(E, F)$, 161
$J(E, F)$, 162
$M_J(X, Y)$, 162
$M_{*n}(X, Y)$, 162
$L_{*n}(E, F)$, 162
$M_{os}(X, Y)$, 164
$L_{os}^{\sim}(E, F)$, 164
$M_s(X, Y)$, 166
$L_s^{\sim}(E, F)$, 166
$\mathscr{N}_T$, 166
$\mathbf{1}_{\mathscr{E}}/f$, 206
e/f, 206
$C_0(Q, P)$, 214
$e \bullet s$, 215
$\mathscr{L}^1(\mu)$, 240
$\mathscr{N}(\mu)$, 244
$\mathrm{dqa}(\mathscr{B}or(Q), F)$, 256
$C_0(P)$, 257
χ_C, 257
$\mathscr{I}^{\sim}(E, F)$, 267
$\mathscr{M}(X, Y)$, 297
$\mu(T)$, 297
$\mathscr{R}(E, F)$, 297
$\mu^p(T)$, 297
$\mathscr{M}^p(X, Y)$, 297
$\sigma_{pq}(T)$, 302
$\mathfrak{S}_{p,q}(X, Y)$, 302
$\mathfrak{P}(A)$, 316
$\mathfrak{P}_c(A)$, 316
$\mathrm{Sp}(x)$, 316
$\mathscr{R}_x$, 317
$\mathscr{S}(B)$, 318
$M^{\perp}$, 320
$|M|$, 320
$\mathscr{Z}(A)$, 326
$*$, 327
$SC_\infty(Q, B(H))$, 331
$SC_\#(Q, B(H))$, 332
$\mathscr{R}$, 340
$\mathscr{R}{\downarrow}$, 340
$\mathbb{Q}$, 359
$\mathscr{L}^B(\mathscr{X}, \mathscr{Y})$, 365
$\mathscr{X}{\downarrow}^{\infty}$, 367
$\mathscr{L}_B(X, Y)$, 368
$X^{\#}$, 368
B-$\mathrm{dim}(X)$, 375
$\sigma_\infty(X, X^{\#})$, 381
$X^{\#\#}$, 382
$x^{\#\#}$, 382
$\mathscr{K}_B(X, Y)$, 383
$\mathscr{P}_{\mathrm{fin}}(X)$, 402

Subject Index

$*$-convergent net, 162
$*$-isomorphism, 317
$*$-representation, 317

A-discrete element, 202
AL-space, 37
AL^p-space, 36
AM-space, 36
AM-space with unity, 36
AW^*-algebra, 326
AW^*-factor, 326
AW^*-module, 319

B-cyclic C^*-algebra, 327
B-cyclic Banach algebra, 327
B-cyclic completion, 367
B-cyclic involutive algebra, 327
B-cyclic normed space, 309
B-dimension, 375
B-dual, 368
B-dual space, 368
B-homomorphism, 327
B-isometry, 367
B-linear operator, 368
B-predual, 368
B-separable module, 377
B-set, 406
B-valued set, 397
$\mathscr{B}$-bounded set, 310
$\mathscr{C}$-step-section, 80
$(\mathscr{B}, p)$-cyclic Banach space, 310

bo-complete space, 48
bo-continuous norm, 295
bo-continuous operator, 156
bo-convergence, 48
bo-fundamental net, 48
bo-summable family, 48
br-complete space, 48
br-convergence, 48
br-fundamental net, 48

$\mathscr{C}$-measurable section, 80
C^*-algebra, 317
C^*-module, 319

d-complete space, 48
d-decomposable hull, 62
d-decomposable norm, 45

e-uniform convergence, 20
E-valued norm, 45

f-algebra, 24

G-continuity, 161
G-convergence, 161

h-o-continuous operator, 198
h-approximating set, 60
h-approximation of a set, 60
h-approximation of an element, 60
h-convergence in a Boolean algebra, 60
h-convergence in a vector lattice, 60
h-convergence in an LNS, 60
h-limit of a net in a vector lattice, 60
h-limit of a net in an LNS, 60
h-modular measure, 244

KB-space, 35
K_σ-space, 22

n-ary operation, 407
n-ary predicate, 407
n-disjoint, 195

o-bounded operator, 90
o-complete K-space, 22
o-completion, 57
o-convergent net, 20
o-extreme point, 136
o-ideal, 21, 48
o-limit, 20
o-semicontinuous, 48
o-sum, 21
o-summable family, 21

p-superadditive operator, 306
(p, q)-concave operator, 306
(p, q)-convex operator, 306
(p, q)-regular operator, 307
(p, q)-summing operator, 302

q-superadditive operator, 306

r-limit, 20

T-extreme point, 136

WSW-representation, 208
WSW-representation of an operator in LNSs, 212

Z-measurable vector-function, 68
Z-weakly measurable function, 69

$*$-B-homomorphism, 327
$*$-B-isomorphism, 327
$*$-algebra, 316
$*$-preserving element, 327

λ-homogeneous algebra, 329
λ-homogeneous Kaplansky–Hilbert module, 320
λ-stable Boolean algebra, 325
λ-stable Stone space, 325
Λ-valued inner product, 318

μ-integrable element, 240

Φ-measurable set, 261
Φ-integrable function, 260
Φ-negligible set, 258
Φ-summable function, 260
Φ-summable set, 261

ρ-indistinguishable points, 82
ρ-invariant measurable Banach bundle, 82
ρ-isometric bundles, 81

σ-algebra, 4
σ-complete Boolean algebra, 4
σ-complete subalgebra, 6
σ-depressed towards zero, 172
σ-distributivity, 190
σ-inductive Boolean algebra, 192
σ-order continuous norm, 35
σ-order-dense ideal, 99
σ-regular subalgebra, 6
σ-regular subalgebra generated by a set, 6
σ-singular operator, 100
$\sigma(X, Z)$-measurable vector-function, 68
(σ, ∞)-depressed towards zero, 179

abelian projection, 328
abnormal operator, 164
absolutely continuous operator, 118, 247
abstract M-space, 36
abstract norm, 143
additive function, 16
adjoint of an operator, 330
algebra, 316
algebra of Borel sets modulo meager sets, 9
algebra of measurable sets by measure zero sets, 10
algebraic B-system, 407
almost global section, 71
almost integral operator, 106, 162
ample Banach bundle, 74
ample hull, 76
ample vector measure, 244
anti-isomorphism, 4
antichain, 8
antitonic mapping, 7
approximating set, 59
Archimedean ordered vector space, 18
ascending and descending, 373
ascent of a correspondence, 404
ascent of an element, 404
ascent-descent rule, 405
descent-ascent rule, 405
associated algebra, 10
associativity, 4
atom, 31
atomic formula, 394
atomic operator, 109
atomic vector lattice, 22

Badé-complete Boolean algebra, 310
Baer $*$-algebra, 325
Baire property, 9
Baire space, 10
Banach algebra, 317
Banach ideal space, 37
Banach lattice, 33
Banach space with mixed norm, 293
Banach–Kantorovich lattice, 54
Banach–Kantorovich space, 54
band, 6, 19, 46
band of essential positivity, 93
band preserving operator, 109, 144
band preserving operator in LNS, 193
band projection, 20, 46
base of a vector lattice, 20
basis of a Kaplansky–Hilbert module, 320
Birkhoff–Ulam Theorem, 15
Bonnice–Silvermann–To Theorem, 98
boolean, 9
Boolean algebra, 4
Boolean algebra of projections, 47
Boolean homomorphism, 7
Boolean operations, 4
Boolean ring, 5
Boolean set, 406
Boolean space, 11
Boolean-valued analysis, 338
Boolean-valued realization, 348, 406
Boolean-valued representation, 368
Boolean-valued representation of
a module, 369
Boolean-valued universe, 338, 397
Borel extendable measure, 185
bound variable, 394
bounded descent, 367
bounded formula, 402
bounded vector measure, 154
Brothers Kreĭn–Kakutani Theorem, 37
Bukhvalov Theorem, 272
bundle associated with a multinormed
space, 72
bundle, 71

canonical embedding, 382
canonical embedding in the second
order dual, 34
canonical immersion, 15
canonical mapping, 7
canonical sublinear operator method, 135
Cantor discontinuum, 11
carrier, 118
carrier of a system, 407
carrier of an operator, 93
Cartesian product, 8
center of an AW^*-algebra, 326
central element, 326
central projection, 316
character of an algebra, 10
character space, 11
characteristic of an element, 23
Church schema, 395
classifier, 395
clopen set, 9
coinitial set, 8
comeager set, 63
commutativity, 4
complement of an element, 4
complemented lattice, 4
complete Banach bundle, 74
complete Boolean algebra, 4
complete homomorphism, 7
complete lattice, 4
complete subalgebra, 6
completely additive function, 16
completely additive operator, 160
completely additive vector measure, 170
completely disjoint elements, 24
completion, 76, 193
complex vector lattice, 25
component, 6, 19, 46
congruent families, 333
connected space, 11
continuity structure, 72
continuous Banach bundle, 71
continuous functional calculus in
C^*-algebras, 317
continuous vector lattice, 22
contractive mapping, 407
coordinatewise operation, 8
coordinatewise order, 7
countable additivity, 10
countable order complete vector lattice, 22
countable sup property, 168
countably additive function, 16
countably additive vector measure, 170
cover of a Boolean algebra, 190
cyclic Banach space, 335
cyclical compactness, 379

cyclical subnet, 380
cyclical subsequence, 380
cyclically compact operator, 383
cyclically compact set, 380

De Morgan formulas, 5
decomposable norm, 45
decomposable operator, 225
decreasing net, 20
dense subalgebra, 192
descent of an algebraic system, 409
descent of an element, 402
descent of a correspondence, 403
descent of a relation, 404
descent of the Banach space, 364
diagonal property, 183
diffuse operator, 109, 137
diffuse vector lattice, 22
dilatator, 135
direct sum property, 10
discrete element, 22
discrete operator, 111
discrete vector lattice, 22
disintegrated kernel, 279
disjoint elements, 4, 19, 46
disjoint functional, 139
disjoint set, 4
disjointly σ-complete vector lattice, 38
disjointly complete vector lattice, 38
disjointly-decomposable norm, 45
distributive lattice, 4
distributive laws, 4
domain, 394
dominant, xi, 142
dominant measure, 154
dominated extension, 98
dominated extension property, 98
dominated operator, xi, 142, 306, 333
dominated vector measure, 154
dual isomorphism, 4
dual space, 268

Egorov property, 168
epigraph, 98
equivalent projections, 328
essentially positive operator, 93, 135
evaluation mapping, 34
exact dominant, 142, 154
exact ordered algebra, 24
Exhaustion Principle, 8
extendable measure, 184
extended K-space, 22
extended orthomorphism, 109
extended section, 73
extensional correspondence, 404
extremal space, 13
extremally disconnected space, 13
extreme extension of a measure, 287
extreme operator, 112

factor algebra, 7
factor homomorphism, 7
faithful f-algebra, 24
faithful Kaplansky–Hilbert module, 320
filter, 11
finite algebra, 329
finite measure, 16
finite projection, 328
finite set, 402
finite signature, 407
first category set, 9
formula valid inside $\mathbf{V}^{(B)}$, 399
Fourier coefficient family, 323
Fourier series, 324
Fourier transform relative to a basis, 323
fragment, 19
fragment of a dominated operator, 154
Fredholm B-Alternative, 385, 386
free variable, 394
Freudenthal Spectral Theorem, 24, 355
full-valued operator, 138
function integrable with respect to
 a spectral measure, 357
functional calculus in vector lattices, 37

Gelfand transform, 12
general form of a cyclically compact
 operator in Kaplansky–Hilbert
 modules, 384
generating set of projections, 95
global section, 71
Gordon Theorem, 341
greatest lower bound, 3

Hahn–Banach formula, 99
Hahn–Banach–Kantorovich Theorem, 98
hermitian element, 316
heuristic transfer principle, 42
homomorphism of B-systems, 408
homomorphism of continuous Banach
 bundles, 75

hyperstonian space, 16

ideal, 21
ideal center, 110
ideal of a Boolean algebra, 6
ideal space, 32
increasing net, 20
induction principle, 398
infimum, 3
infinite distributive laws, 5, 18
infinite operations, 5
infinite projection, 328
inner weight of a WSW-representation in LNSs, 212
inner weight of a WSW-representation in vector lattices, 208
integrable element, 240
integrable element by μ, 240
integrable element of a K-space, 240
integral operator, 266
integral representation, 267
interpretation mapping, 407
interval preserving operator, 225
invariant measurable Banach bundle, 82
involution, 316
involutive algebra, 316
involutive Banach algebra, 317
isometric $*$-isomorphism, 317
isometric $*$-representation, 317
isometry of continuous Banach bundles, 75
isomorphic lattices, 109
isomorphism, 7
isomorphism to a B-system, 409
isotonic mapping, 7

join, 3

Kakutani Criterion, 383
Kantorovich norm, 45
Kantorovich principle, 134
Kantorovich space, 22
Kantorovich Theorem, 97
Kantorovich–Pinsker space, 32
Kaplansky–Hilbert module, 319
Kaplansky–Hilbert submodule generated by a subset, 320
kernel of an integral operator, 267
Kreĭn–Milman Theorem, 136
Krull Theorem, 12
Kutateladze Theorem, 110
Kutateladze–Milman Theorem, 114

Köthe space, 43

Lamperti operators, 235
laterally σ-continuous norm, 35
lattice, 3
lattice homomorphism, 108
lattice isomorphism, 109
lattice monomorphism, 108
lattice norm, 33
lattice rich in σ-order-dense ideals, 100
lattice-normed space, 45
lattice-ordered algebra, 24
least dominant, 142, 154
least upper bound, 3
left annihilator, 325
liftable measurable Banach bundle, 81
lifting, 31, 81
lifting of a factor algebra, 15
local countable chain condition, 288
local Hamel basis, 188
local homomorphism, 232
local section, 71
locally constant element, 188
locally finite function, 16
locally linearly independent set, 188, 386
locally one-dimensional vector lattice, 189
Loomis–Sikorski Theorem, 14
lower bound, 3
Luxemburg–Schep Theorem, 125

Maharam operator, 118
Maharam property, 118
majorant, xi, 142
majorizable operator, 142
majorized operator, xi
majorizing sublattice, 21
massive sublattice, 21
maximal ideal space, 11
Maximum Principle, 400
meager set, 9
measurable section, 80
measurable space, 31
measurable space of countable type, 257
measure, 10, 16
measure extension property, 185, 287
measure kernel, 278
measure space, 10
meet, 3
Meyer Theorem, 109
minimal extension, 93

minimal extension operator, 93
minorant set, 8
minorizing set, 8
minorizing sublattice, 21
mix-complete set, 308
mixed norm, 292
mixed order continuous norm, 295
mixing, 401
mixing of a family, 308
modified ascent of a correspondence, 405
modified descent of a correspondence, 405
modular measure, 244
modular measure with respect to h, 244
module-discrete element, 202
modulus, 18
monomorphism, 7, 12
monotone mapping, 7
monotone norm, 48
monotone set, 178
monotonically complete, 35
multinormed Boolean algebra, 16
multiplication operator, 135
multiplicative operator, 233
multiplicative projection, 327

negative part, 18
negligible set, 258
nonexpanding mapping, 406
nonextending operator, 109
norm disjointness, 51
norm lattice, 45
norm order continuous operator, 156
norm transformation of an LNS, 62
norm-n-decomposable element, 51
norm-bounded set, 48
norm-indecomposable element, 51
normal element, 316
normal subdominant, 149
normed B-space, 307
normed Boolean algebra, 16
normed ideal space, 37
normed vector lattice, 33
norming subspace, 65
null ideal, 93, 244

Ogasawara Theorem, 13
operator extension property, 185
operator of norm transformation of
an LNS, 62
opposite order, 3
order σ-continuous operator, 99
order complete vector lattice, 22
order completion, 57
order continuous functional, 34
order continuous norm, 35
order continuous operator, 99
order continuous vector norm, 48, 295
order disjointness, 51
order dual space, 34
order ideal, 21
order interval, 19
order isomorphism, 109
order projection, 20
order semicontinuous norm, 35, 48
order semicontinuous vector norm, 48
order singular operator, 164
order summable family, 21
order-bounded functional, 34
order-bounded operator, 90
order-bounded set, 21
order-dense ideal, 21
order-unity, 22
ordered algebra, 24
ordered set, 3
ordered vector space, 17
orthogonal complement, 320
orthogonal projection, 316
orthomodular lattice, 326
orthomorphism, 110, 144
orthonormal set in a Kaplansky–Hilbert
module, 320
outer weight of a WSW-representation
in LNSs, 212
outer weight of a WSW-representation
in vector lattices, 208

Parseval identity, 324
part, 19
partial isometry, 328
partially ordered set, 3
partition of a measurable set, 257
partition of an element, 8
partition of unity, 8, 401
polar preserving endomorphisms, 135
poset, 3
positive cone, 18
positive element, 18
positive element of a $*$-algebra, 316
positive measure, 16
positive operator, 90

positive part, 18
positive vector measure, 154
positively decomposable operator, 225
positively homogeneous function, 37
powerset, 9
predominant, 297
predominated operator, 297
preservation of suprema and infima, 7
principal band, 19
principal ideal, 6
principal projection, 20
principal projection property, 20
product algebra, 8
product of bundles, 71
product order, 7
projection band, 46
projection in $*$-algebra, 316
projection property, 20, 46
proper axiom, 396
proper filter, 11
proper ideal, 6
Property (A), 35
Property (B), 35
Property (C), 35
pseudointegral operator, 279
purely atomic measure, 31
purely infinite projection, 329

quasi-Radon measure, 249
quasiextremal space, 13
quasiextremally disconnected space, 13
quasiregular measure, 249
quotient algebra, 7

Radon measure, 249
refined function, 190
refined subset, 190
reflexive B-space, 382
regular K-space, 162
regular closed set, 9
regular integral operator, 267
regular measure, 249
regular open set, 9
regular operator, 90
regular subalgebra, 6
regular subalgebra generated by a set, 6
regulator of convergence, 20
relative uniform convergence, 20
relatively cyclically compact set, 380
relatively uniformly complete vector lattice, 20
representation by means of a section, 218
representation of a ring homomorphism, 215
representation of a shift operator, 216
representation of a weighted shift operator, 217
representation of an orthomorphism, 215
representing measure, 278
reproducing cone, 112
resolution of unity in a Boolean algebra, 27
restricted formula, 402
restriction operator, 93
reverse order, 3
Riesz Decomposition Property, 19
right annihilator, 325
ring homomorphism, 7
rules for canceling arrows, 405

satisfaction of a formula, 408
scalar WSW-factorization, 212
scalar WSW-representation, 212
scalar orthomorphism, 212
second B-dual, 382
section, 71
semibounded operator, 194
semivector WSW-representation, 212
separated Boolean-valued universe, 399
sequential G-continuity, 161
sequentially bo-continuous operator, 156
set with B-structure, 406
shadow of an operator, 135, 197
shearing net, 150
shift by a homomorphism, 208
shift of a WSW-representation in LNSs, 212
shift of a WSW-representation in vector lattices, 208
shift operator in LNSs, 211
shift operator in vector lattices, 208
signature, 407
Sikorski Theorem, 13
simply measurable, 70
simultaneous extension operator, 139
singular operator, 100, 166
Sourour Theorem, 281
space with mixed norm, 293
special axiom, 396
spectral function, 23

spectral integral, 357
spectral measure, 356
spectral theorem for C^*-algebras, 317
spectral value of an element of an algebra, 316
spectrum of an element of an algebra, 316
stabilisateur, 135
stalk, 71
stalkwise dense set, 71
standard Borel space, 279
standard name, 401
step-section, 80
Stone algebra, 318
Stone representation method, 41
Stone Representation Theorem, 12
Stone space, 12
Stone space of a Boolean algebra, 11
Stone transform, 13, 83
Stone transform of a bundle, 84
strictly γ-homogeneous module, 320
strictly positive measure, 16
strong form of the Yosida–Hewitt decomposition, 142, 164
strong Freudenthal property, 24
strong homomorphism, 409
strong measure extension property, 185
strong order-unity, 22
strong unity, 22
strongly disjoint operators, 209
strongly disjoint sum of operators, 209
subalgebra, 6
subalgebra generated by a set, 6
subbundle, 76
subdominant, 149
subdominated operator, 149
sublinear operator, 97
submorphism, 115
submultiplicative norm, 317
summation in order, 21
summing operator, 303, 306
support of a function, 33
support of a subset, 33
support set, 97
supporting operator, 98
supremum, 3
symmetric difference, 6, 407

tautology, 408
tight algebra, 256
topology generated by sublattice, 254
total variation, 40
totally disconnected space, 11
trace of an element, 23
Transfer Principle, 400
true formula, 408
truth-value, 398
type I algebra, 329
type II algebra, 329
type III algebra, 329

ultrafilter, 11
underlying set, 407
unit element, 19
unity, 4
universal algebra, 41
universal completion, 57
universally complete K-space, 22
universe, 394, 407
up-down theorem, 140
upper bound, 3

valid formula, 408
vector WSW-factorization, 212
vector WSW-representation, 212
vector lattice, 18
vector lattice of bounded elements, 22
vector measure, 154
vector order, 18
vector sublattice, 21
vector-valued norm, 45
von Neumann universe, 338, 397

weak form of the Yosida–Hewitt decomposition, 142, 164
weak Freudenthal property, 24
weak order-unity, 22
weakly σ-distributive lattice, 168, 172
weakly (σ, ∞)-distributive lattice, 171
weakly $\sigma\sigma$-distributive vector lattice, 183
weight-shift-weight factorization, 208, 212
wide operator at an element, 206
wide operator on a subset of a vector lattice, 206

zero, 4
zero of a lattice, 4